KU-075-369

FOREWORD

The OECD/NEA Working Party on the Physics of Plutonium Fuels and Innovative Fuel Cycles (WPPR) was established in 1993 and reports to the OECD/NEA Nuclear Science Committee. Its main activity has been to analyse physics code benchmarks for problems related to the physics of plutonium fuels. Past volumes of published work have examined the physics of plutonium-fuelled pressurised water reactors (PWRs), as well as the physics of metal- and oxide-fuelled fast reactors.

Some of the questions that the Working Group has attempted to address in the present and previous volumes include:

- What is the effect on the physics of the degradation in plutonium fissile quality that occurs with increasing burn-up and age of spent fuel?
- What is the impact on the physics of multiple recycling of plutonium if MOX assemblies are themselves reprocessed? What are the limitations associated with multiple recycling? Is there a point beyond which multiple recycling becomes no longer practicable in the present generation of PWRs?
- Are the present nuclear data libraries and neutronics codes capable of predicting the physics performance of MOX made from plutonium derived from high burn-up and/or multiple-recycling scenarios?
- Is thermal MOX recycling ultimately compatible with a fast-reactor fuel cycle? Can fast reactors be used in a flexible manner to address whatever requirements there might be to consume plutonium from a thermal MOX programme? Can plutonium-burning fast reactors be changed to plutonium breeding to maintain self-sufficiency in the longer term?
- How do the nuclear codes compare with real data from experiments?

The "Physics of Plutonium Recycling" series currently comprises the following titles:

- Volume I: Issues and Perspectives (OECD/NEA, 1995);
- Volume II: Plutonium Recycling in Pressurised Water Reactors (OECD/NEA, 1995);
- Volume III: Void Reactivity Effect in Pressurised Water Reactors (OECD/NEA, 1995);
- Volume IV: Fast Plutonium Burner Reactors: Beginning of Life (OECD/NEA, 1995);
- Volume V: Plutonium Recycling in Fast Reactors (OECD/NEA, 1996);
- Volume VI: Multiple Plutonium Recycling in Advanced PWRs (OECD/NEA, 2002).

Volumes VII and VIII, in preparation, will be devoted to a theoretical benchmark of a boiling water reactor (BWR) assembly containing mixed-oxide fuel rods, and to plutonium fuel in high-temperature reactors.

While all of the earlier work consisted of theoretical benchmarks comparing different nuclear codes and nuclear data libraries, comparisons against experimental measurements were made possible by SCK-CEN using data from the VENUS-2 reactor. The VENUS-2 data concerned an experimental mock-up of a PWR core containing UO_2 and mixed-oxide (MOX) assemblies. The results of this benchmark were published in *Benchmark on the VENUS-2 MOX Core Measurements* (OECD/NEA, 2000). The benchmark was carried out under the joint auspices of the WPPR and the Task Force on Reactor-based Plutonium Disposition (TFRPD). Another benchmark was undertaken for three critical core configurations of the KRITZ reactor: two with UO_2 fuel and one with MOX fuel. Measurements were performed for room temperature as well as elevated temperatures (~245°C). The results are due to be published soon.

UNIVERSITY OF STRATHCLYDE
30125 00700661 1
S/O ML
OCM 33379415
BIB 487484.

Nuclear Science

Physics of Plutonium Recycling

ANDERSONIAN LIBRARY
WITHDRAWN
FROM
LIBRARY
STOCK
UNIVERSITY OF STRATHCLYDE

Volume VI

dvanced PWRs

NUCLEAR ENERGY AGENCY
ORGANISATION FOR ECONOMIC CO-OPERATION AND DEVELOPMENT

D
621.48335
PHY

ORGANISATION FOR ECONOMIC CO-OPERATION AND DEVELOPMENT

Pursuant to Article 1 of the Convention signed in Paris on 14th December 1960, and which came into force on 30th September 1961, the Organisation for Economic Co-operation and Development (OECD) shall promote policies designed:

- to achieve the highest sustainable economic growth and employment and a rising standard of living in Member countries, while maintaining financial stability, and thus to contribute to the development of the world economy;
- to contribute to sound economic expansion in Member as well as non-member countries in the process of economic development; and
- to contribute to the expansion of world trade on a multilateral, non-discriminatory basis in accordance with international obligations.

The original Member countries of the OECD are Austria, Belgium, Canada, Denmark, France, Germany, Greece, Iceland, Ireland, Italy, Luxembourg, the Netherlands, Norway, Portugal, Spain, Sweden, Switzerland, Turkey, the United Kingdom and the United States. The following countries became Members subsequently through accession at the dates indicated hereafter: Japan (28th April 1964), Finland (28th January 1969), Australia (7th June 1971), New Zealand (29th May 1973), Mexico (18th May 1994), the Czech Republic (21st December 1995), Hungary (7th May 1996), Poland (22nd November 1996), Korea (12th December 1996) and the Slovak Republic (14 December 2000). The Commission of the European Communities takes part in the work of the OECD (Article 13 of the OECD Convention).

NUCLEAR ENERGY AGENCY

The OECD Nuclear Energy Agency (NEA) was established on 1st February 1958 under the name of the OEEC European Nuclear Energy Agency. It received its present designation on 20th April 1972, when Japan became its first non-European full Member. NEA membership today consists of 28 OECD Member countries: Australia, Austria, Belgium, Canada, Czech Republic, Denmark, Finland, France, Germany, Greece, Hungary, Iceland, Ireland, Italy, Japan, Luxembourg, Mexico, the Netherlands, Norway, Portugal, Republic of Korea, Slovak Republic, Spain, Sweden, Switzerland, Turkey, the United Kingdom and the United States. The Commission of the European Communities also takes part in the work of the Agency.

The mission of the NEA is:

- to assist its Member countries in maintaining and further developing, through international co-operation, the scientific, technological and legal bases required for a safe, environmentally friendly and economical use of nuclear energy for peaceful purposes, as well as
- to provide authoritative assessments and to forge common understandings on key issues, as input to government decisions on nuclear energy policy and to broader OECD policy analyses in areas such as energy and sustainable development.

Specific areas of competence of the NEA include safety and regulation of nuclear activities, radioactive waste management, radiological protection, nuclear science, economic and technical analyses of the nuclear fuel cycle, nuclear law and liability, and public information. The NEA Data Bank provides nuclear data and computer program services for participating countries.

In these and related tasks, the NEA works in close collaboration with the International Atomic Energy Agency in Vienna, with which it has a Co-operation Agreement, as well as with other international organisations in the nuclear field.

© OECD 2002

Permission to reproduce a portion of this work for non-commercial purposes or classroom use should be obtained through the Centre français d'exploitation du droit de copie (CCF), 20, rue des Grands-Augustins, 75006 Paris, France, Tel. (33-1) 44 07 47 70, Fax (33-1) 46 34 67 19, for every country except the United States. In the United States permission should be obtained through the Copyright Clearance Center, Customer Service, (508)750-8400, 222 Rosewood Drive, Danvers, MA 01923, USA, or CCC Online: http://www.copyright.com/. All other applications for permission to reproduce or translate all or part of this book should be made to OECD Publications, 2, rue André-Pascal, 75775 Paris Cedex 16, France.

UNIVERSITY OF STRATHCLYDE
13 NOV 2002
UNIVERSITY LIBRARY

TABLE OF CONTENTS

CONTRIBUTORS

Chair	*K. Hesketh*	BNFL	UK
Secretary	*E. Sartori*	OECD/NEA	
Prinicpal Authors	*M. Delpech*	CEA	France
	K. Hesketh	BNFL	UK
	D. Lutz	IKE	Germany
	E. Sartori	OECD/NEA	
Problem Specification	*M. Delpech*	CEA	France
Special Contribution	*G. Schlosser*	Siemens/KWU	Germany
Data Compilation and Analysis	*M. Delpech*	CEA	France
	I. Champtiaux	CEA	France
	M. Goutefarde	CEA	France
	M. Juanola	CEA	France
Special Study (Chapter 6)	*S. Cathalau*	CEA	France
	D. Lutz	IKE	Germany
	W. Bernnat	IKE	Germany
	M. Mattes	IKE	Germany
	T.T.J. M. Peeters	ECN	The Netherlands
	Y. Peneliau	CEA	France
	A. Puill	CEA	France
	H. Takano	JAERI	Japan
Editor	*K. Hesketh*	BNFL	UK
Text Processing and Layout	*A. McWhorter*	OECD/NEA	
	H. Déry	OECD/NEA	

BENCHMARK PARTICIPANTS

Name	*Establishment*	*Country*
MALDAGUE, Thierry	Belgonucléaire	Belgium
HOEJERUP C. Frank	Risø	Denmark
BARBRAULT Patrick	EDF-Clamart	France
VERGNES Jean	EDF-Clamart	
DELPECH Marc	CEA-Cadarache	
COSTE Mireille	CEA-Saclay	
LEE Yi-Kang	CEA-Saclay	
PENELIAU Yannick	CEA- Saclay	
PUILL André	CEA- Saclay	
ROHART Michelle	CEA- Saclay	
TSILANIZARA Aimé	CEA- Saclay	
POINOT-SALANON Christine	FRAMATOME	
HESSE Ulrich	GRS	Germany
MOSER Eberhard	GRS	
LEHMANN Sven	U-Braunschweig	
BERNNAT Wolfgang	U-Stuttgart-IKE	
LUTZ Dietrich	U- Stuttgart -IKE	
MATTES Margerete	U- Stuttgart -IKE	
SCHWEIZER M.	U- Stuttgart -IKE	
KANEKO K.	ITIRO	Japan
AKIE Hiroshi	JAERI	
TAKANO Hideki	JAERI	
MISAWA Tsuyoshi	U-Nagoya	
TAKEDA Toshikazu	U-Osaka	
YAMAMOTO Toshihisa	U-Osaka	
FUJIWARA Daisuke	U-Tohoku	
IWASAKI Tomohiko	U-Tohoku	
KIM Young-Jin	KAERI	Republic of Korea
KLOOSTERMAN Jan Leen	Petten	The Netherlands
TSIBOULIA Anatoli	IPPE-Obninsk	Russian Federation
PARATTE Jean-Marie	PSI	Switzerland
CROSSLEY Steve	BNFL	United Kingdom
HESKETH Kevin	BNFL	
BLOMQUIST Roger N.	ANL	USA
GRIMM Karl	ANL	
HILL Robert N.	ANL	

EXECUTIVE SUMMARY

The recycling of plutonium in LWRs has for quite some time been seen as an important interim step prior to the large scale introduction of fast reactors; several countries have, in fact, been using MOX fuel in PWRs on a commercial scale for many years. Now that fast reactors are no longer expected to be introduced for some time, the need to manage existing plutonium stocks puts even more emphasis on LWR MOX. Even though there is sufficient experience of MOX in PWRs to be assured of satisfactory performance under existing conditions, the situation is not static and it is very important to address future requirements.

There are two main trends which need to be accounted for. First, discharge burn-ups are still increasing steadily. Second, the fissile quality of the plutonium is generally becoming poorer. The decline in fissile quality is partly a consequence of the higher burn-ups, but another major consideration is the effect of recycling MOX fuel itself. Multiple recycle scenarios cause a significant decline in fissile quality in each recycle generation. One consequence is that the initial plutonium content of MOX fuel is considerably increased in each recycle generation and this poses a considerable challenge to current nuclear design codes and the associated nuclear data libraries.

The OECD Working Group on the Physics of Plutonium Recycling (WPPR) has considered the issue of multiple recycle of MOX in PWRs in Volumes I and II of its report published in 1995. The main effort was directed towards an international benchmark exercise in which participants were invited to submit solutions to a fully defined problem representative of a multiple recycle scenario in a PWR. Such benchmark exercises are a valuable first step towards understanding the problems to be overcome before a future scenario can be realised in practice. They cannot, of course, identify the correct answer; this must wait until there is experimental or operational experience to learn from. However they can highlight the areas in which current calculational methods and nuclear data start to breakdown and thereby provide a rational basis for directing the research effort.

The earlier benchmark work identified major discrepancies between the various nuclear design codes and nuclear data libraries, specifically differences of up to 4%Δk in reactivities. A portion of the range of discrepancy was explained because some of the codes did not apply self-shielding to the resonance treatment of the higher plutonium isotopes; with the large concentration of ^{242}Pu specified in the benchmark, self-shielding in this nuclide was very important. Nevertheless the remaining discrepancy was considered sufficiently large that further work would clearly be necessary before the multiple recycle scenario envisaged in the specification could be implemented.

A shortcoming of the initial benchmark was that the cases considered only included one corresponding to today's situation (with plutonium of good isotopic quality), to one which might arise after many generations of MOX recycle (with extremely poor isotopic quality). Analysis of the intermediate steps was missing, and therefore there was no possibility of determining precisely where the nuclear codes and data libraries would start to lose their applicability. Accordingly, CEA suggested a benchmark in which five consecutive generations of multiple recycle in a PWR would be followed. In the specification of the benchmark, attempts were made to make it as realistic a scenario as possible, taking account of such details as the length of time between recycle generations (accounting for the time delays in pond storage, MOX fabrication, etc.) and the dilution effect when MOX and UO_2 assemblies are co-reprocessed.

As in the previous benchmark exercise, the benchmark was restricted to the level of the lattice codes. This is the logical first step because it does not make sense to progress to the three-dimensional whole core codes until the underlying nuclear data and lattice code calculations show adequate agreement. Two cases were considered, one for a standard 17×17 PWR lattice such as used in many of today's PWRs (designated the STD PWR) and one for a lattice with an increased moderator/fuel ratio (3.5:1 compared with 2:1 for the STD PWR). The latter case, designated the HM PWR (for highly moderated), was intended to cover a proposed PWR design dedicated for use with MOX only. In such a PWR it would be possible to optimise the lattice to give increased moderation, thereby improving the efficiency with which the plutonium can be used.

Parameters examined in the benchmark exercise included end-of-cycle reactivity, which determines cycle length, the variation of reactivity with burn-up, reactivity coefficients, microscopic cross-sections, isotopic evolution and isotopic toxicity evolution with time. The detailed comparison of the results is very extensive and is not given in full in this report. The following are the broad conclusions and observations that have emerged:

Since the earlier benchmarks, considerable progress has been made in nuclear data libraries and methods. The discrepancies between the different data libraries and lattice methods, when applied to multiple recycle scenarios in PWR, are now generally within reasonable bounds. The observed spread of results is now consistent with the uncertainties in the underlying nuclear data and will require further experimental validation prior to practical implementation of multiple recycle. Multiple recycle scenarios therefore appear to be practicable and feasible in conventional PWRs, at least in the near term. Questions arising from a possible positive void coefficient would almost certainly preclude recycle beyond the second generation and even possibly even in the second generation in this particular scenario.

In the longer term, the HM PWR concept shows some merit. However, the principal advantage, that of needing a lower initial plutonium content with the same dilution ratio, is largely eroded in later recycle generations; in spite of the much improved moderation, the HM PWR degrades the plutonium isotopic quality more rapidly than the STD PWR, to a large extent negating the benefit of the softer spectrum. The HM PWR case also seems to pose more difficulties for present nuclear data libraries and codes, as evidenced by the larger number of discrepant results seen in the HM PWR benchmark. Therefore, even the HM PWR is of questionable practicability with respect to the later recycle generations.

In view of these considerations, at its last meeting the WPPR agreed that there is no compelling reason to continue further benchmark studies at the level of the lattice codes. Suggestions for future work include examining existing experimental configurations where the specifications are freely available, and also the re-examination of the problem at the level of whole-core calculations, in order to be able to understand more clearly how the lattice codes results relate to whole-core reactivity coefficients and other important parameters for the design and safety analyses. With such studies it is hoped to be able to make a definitive judgement as to where the limits of multiple recycle in PWRs are.

Chapter 1

INTRODUCTION

Background

The recycling of plutonium as PuO_2/UO_2 mixed oxide (MOX) fuel is already established in pressurised water reactors (PWR) in several countries on a commercial scale. The discharge burn-up of MOX fuel, and indeed its overall performance, is essentially the same as that of UO_2 fuel. Thus the MOX fuel currently being irradiated in PWRs is typically intended to be discharged at burn-ups of 40 to 45 MWd/t. The initial plutonium content needed to achieve such burn-ups varies depending on the precise source of the plutonium, but is typically in the range 7 to 8 w/o total plutonium, expressed as an average over the whole assembly. The experience of MOX utilisation in PWRs has been positive and there are no outstanding operational or safety issues to be resolved. However the situation is not static, and such issues will have to be addressed in the near future as the background conditions change.

The fundamental changes are that discharge burn-ups are continuing to trend upwards, while there will also arise a need to recycle the plutonium from discharged MOX assemblies. Both these changes will manifest themselves as a decrease in isotopic quality of the plutonium that is available for recycle. For thermal reactors the even isotopes of plutonium (238, 240 and 242) do not contribute significantly to fissions. The ratio (^{239}Pu + ^{241}Pu)/(total plutonium) is thus denoted the fissile fraction of the plutonium and is a measure of plutonium quality for thermal reactor MOX. The problem is that plutonium quality decreases as the discharge burn up increases and decreases yet further following recycle of the plutonium recovered from MOX. Combined with the self-evident need to increase plutonium content to reach higher burn-ups, it will be necessary to significantly increase the total plutonium content of the MOX fuel.

Compared with conventional UO_2 fuel, MOX fuel is already significantly different from a neutronic point of view, there being a much smaller thermal flux for a given rating. This is due to the combined effects of the higher fission and absorption cross-sections of ^{239}Pu and ^{241}Pu compared with ^{235}U, exacerbated by the significant absorption of the ^{240}Pu and the ^{242}Pu. The difference in spectrum affects the core performance because the control, reactivity coefficient and transient behaviours are all altered. Increasing the total plutonium content beyond present levels exaggerates all these effects further. Ultimately, the deterioration in parameters such as control rod reactivity worth, boron reactivity worth and moderator void and temperature coefficients may become a barrier to further utilisation of MOX in PWRs, at least in conventional lattices. The question is, therefore, at what precise point will such considerations present a barrier. Also, it is important to ask whether present nuclear data libraries and nuclear codes agree as to where this point occurs.

It was against this background that the OECD/NEA Working Group on the Physics of Plutonium Recycle (WPPR) initiated an earlier benchmark studies for PWR MOX. This was reported in 1995 (refer to Vol. I and Vol. III). The benchmarks in question considered first a PWR MOX infinite lattice cell and a multi-assembly arrangement designed to test the nuclear data libraries and lattice codes and to test core-wide calculational methods respectively.

The infinite lattice benchmark considered the identical geometrical arrangement of the lattice both a good and a poor isotopic quality for the plutonium as part of the specification. The good quality case was considered in order to provide a reference point for comparison, by showing how well the various solutions were in agreement for a situation that was prototypic of the present generation MOX. The poor isotopic quality case was intended to be an extreme test of the nuclear data and codes. While the various solutions were in reasonable (but not perfect agreement) for the former case, they were very discrepant in the case of the latter. Predictions of k-infinity versus burn-up were different by more than 4% Δ-k. This was a very serious discrepancy, and called into doubt the ability of the nuclear data libraries and codes to model the future situation. An explanation for some of the discrepancy, due to some of the codes ignoring resonance self-shielding in ^{242}Pu, was noted. Nevertheless, the remaining discrepancies were considered unacceptable for design and licensing applications.

The multi-assembly Benchmark was intended to explore the void reactivity coefficient, to see whether the various codes were in agreement as to the point where it becomes positive. The void coefficient is very important for safety and should be negative, or at least non-positive to ensure negative feedback. The void coefficient tends to become less negative the higher the total plutonium content and in the conventional PWR lattice changes sign from negative to positive at a total plutonium concentration of between 10 and 12 w/o. The benchmark comparisons showed reasonably satisfactory level of agreement between the various solutions submitted.

The present benchmark was intended to complement the earlier two as explained in the next section.

Current objectives

Multiple recycle scenarios presently form a very important topic for MOX recycle in LWRs. With the prospects for large scale deployment of fast reactors having receded in recent years, MOX is becoming more significant as a means of utilising plutonium stocks that were originally intended to start fast reactor cycles. The existence of surplus ex-weapons plutonium and establishing a means for its consumption or disposal has, at the same time, added further urgency and importance. The question of how many times plutonium recovered from MOX assemblies can be re-used in PWRs is important strategically and logistically. Strategically, it is important because it affects the energy potential that is available from plutonium. Logistically it determines whether there will be a need to store or dispose of MOX assemblies and/or plutonium and some future point if indefinite recycle does not prove practical.

Each recycle generation involves irradiation of MOX fuel (lasting typically 4-5 years), followed by pond cooling (typically 5 years), followed by reprocessing and re-fabrication as MOX (taking a further 2 years). Thus each generation of multiple recycle will last at least 11 years. Multiple recycle scenarios therefore extend over very long periods measured in decades. It is clear that over such extended timescales there will be ample opportunity for major changes in world energy requirements and strategies. It may well be the case that the scenario considered here is overtaken by events well before even the first or second generations of recycle are completed. Nevertheless, it is important to analyse such scenarios, just to be sure that they are practical technically, strategically, logistically and to establish their impact on environmental and safety considerations.

The present benchmark is for a standard lattice PWR and a highly moderated PWR operating with a moderately high burn-up fuel cycle (51 MWd/t discharge burn-up), which is expected to be representative of PWR operation in the next decade. The specification calls for the reprocessing of MOX fuel along with a certain fraction of UO_2 assemblies, as will be required in the current generation of reprocessing plants for technical reasons. The emphasis was on specifying a benchmark problem that was as realistic as possible, but keeping within the bounds of what is known from current technology and not relying on extrapolation to an uncertain future. This latter constraint may mean that the benchmark may turn out to have been pessimistic by not accounting for technological developments, but this is unavoidable. A fuller description of the benchmark is provided in Chapter 2, while Appendix A gives the complete specification as issued to the individual contributors.

The primary objectives of the benchmark were:

- to compare reactivities, reactivity coefficients and isotopic evolution calculations obtained with different lattice codes and their associated nuclear data libraries;
- to determine at what point, if any, the calculations diverge to such an extent that the physics predictions must be considered unreliable;
- to determine at what point, if any, further generations of recycle are excluded on technical grounds such as unacceptable reactivity coefficient characteristics;
- to evaluate the environmental impact of multiple recycle.

The rationale behind including a highly moderated PWR lattice is that theoretically, such a lattice may show technical advantages in multiple recycle scenarios. The idea would be to dedicate a small number of new PWRs to MOX usage only. With no need to accommodate UO_2 fuel, the reactor designer could then choose to optimise the moderator/fuel volume ratio for plutonium. It turns out that the optimum occurs at moderator/fuel volume ratio of about 3.5, compared with 2.0 for uranium fuel in a standard lattice. This could be achieved by preserving the fuel rod design and dimensions and simply increasing the rod to rod pitch. The reactor core would be marginally bigger in its radial dimensions, but otherwise the reactor design and the associated equipment would be much as for a conventional PWR.

The present benchmark was naturally specified for PWRs, as all commercial experience of MOX usage has so far been obtained in them. However, the lessons learned are expected to be broadly applicable to BWRs as well, although the details will inevitably be different.

Important physics issues

The key physics issues that this collaboration was intended to address are those of:

- The variation of k-infinity with burn-up for the MOX infinite lattice calculations with burn-up during each recycle generation. It is important for nuclear designers to be able to predict the k-infinity behaviour with a good degree of confidence if cycle lengths predictions are to be accurate enough for a utility's requirements.
- The behaviour of reactivity coefficients with recycle generation. Reactivity coefficients such as boron, fuel temperature (Doppler), moderator temperature and moderator void are key parameters that will largely determine the practicability of multiple recycle. Determining

precisely at what point any of these coefficients would become unacceptable is outside the scope of this study, as it necessarily requires a core-wide spatial analysis. Nevertheless, establishing the underlying trends with the number of recycle generations is a valuable first step.

- The calculation of the isotopic evolution in each recycle generation. Particularly in respect to plutonium, the burn-up calculations will determine the dependence of initial plutonium content with recycle generation. In turn, the initial plutonium content has major impacts for fuel fabrication, reprocessing, fuel thermo-mechanical and physics behaviour.

- The comparison of the highly moderated MOX lattice with a standard MOX lattice. To test, in particular, whether such lattices would be advantageous in a multiple recycle scenario.

Outline of this volume

Chapter 2 provides a summary description of the present benchmark. Chapter 3 lists the participants in the benchmark and describes the lattice codes and nuclear data libraries. Chapter 4 presents a summary of the principal results of the benchmark. Full details of the results are too voluminous for inclusion in this report, but are available on request from the OECD/NEA on computer disks. Chapter 5 discusses the results of the benchmarks and their significance to the practical situation. Chapter 6 describes a supplementary benchmark carried out in parallel to the main benchmark that offers further insights to the problem. Finally, Chapter 7 presents the conclusions and recommendations.

Chapter 2
BENCHMARK SPECIFICATION

Pin cell geometry

This chapter provides an outline description of the benchmark and comments on it. The full specification, as given to participants beforehand, is reproduced here in Appendix A. The purpose was to test the various nuclear data libraries and lattice codes in the simplest possible geometry of an infinite lattice cell or pin cell. Two pin cells were examined, as illustrated in Figure 2.1:

Figure 2.1. Schematic of standard (STD) lattice and highly moderated (HM) lattice pin cells

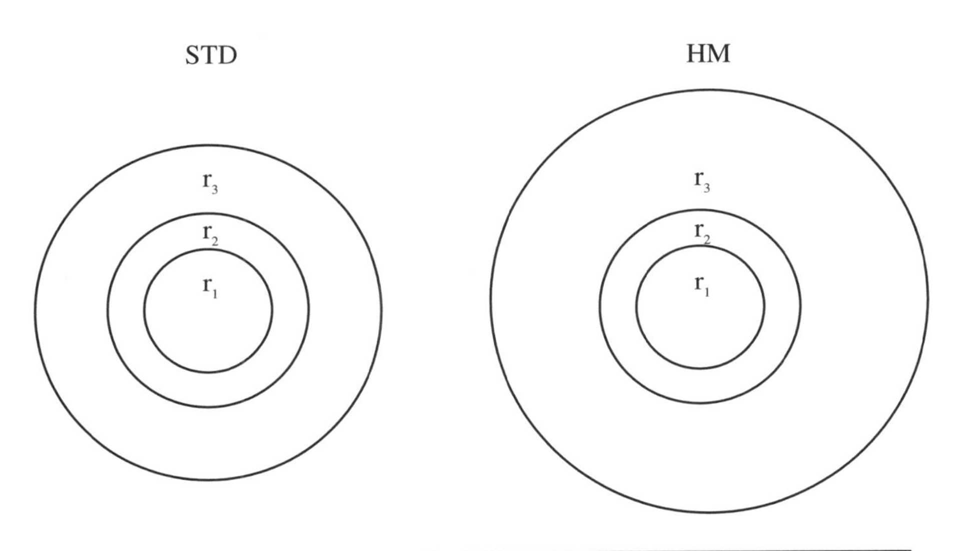

r_1: MOX fuel, 0.4127 cm r_2: Zirconium clad, 0.4744 cm r_3: Water + 500 ppm dissolved boron, 0.7521 cm STD, 0.9062 cm HM

The first was intended to be representative of a modern PWR assembly design and is designated the STD pincell in this report. Specifically, it represents a 17×17 MOX assembly design with 25 guide and instrument tubes and 264 fuel rods, illustrated schematically in Figure 2.2. Table 2.1 lists details of the main geometrical data for the assembly, which is modelled on the design in use in EdF's three and four loop plants. The outer diameter of the moderator region of the pincell was specified so that the overall moderator/fuel ratio (at approximately 2:1) matches that of the full assembly, including the guide and instrument thimbles and outside water gaps. Complications such as the absorption effect of grids, guide thimble tubes burnable poisons, etc. were avoided by ignoring them in the pincell specification. Their neglect does not affect the benchmark's aims, and so is not considered important.

Figure 2.2. Schematic of multiple recycle logistics

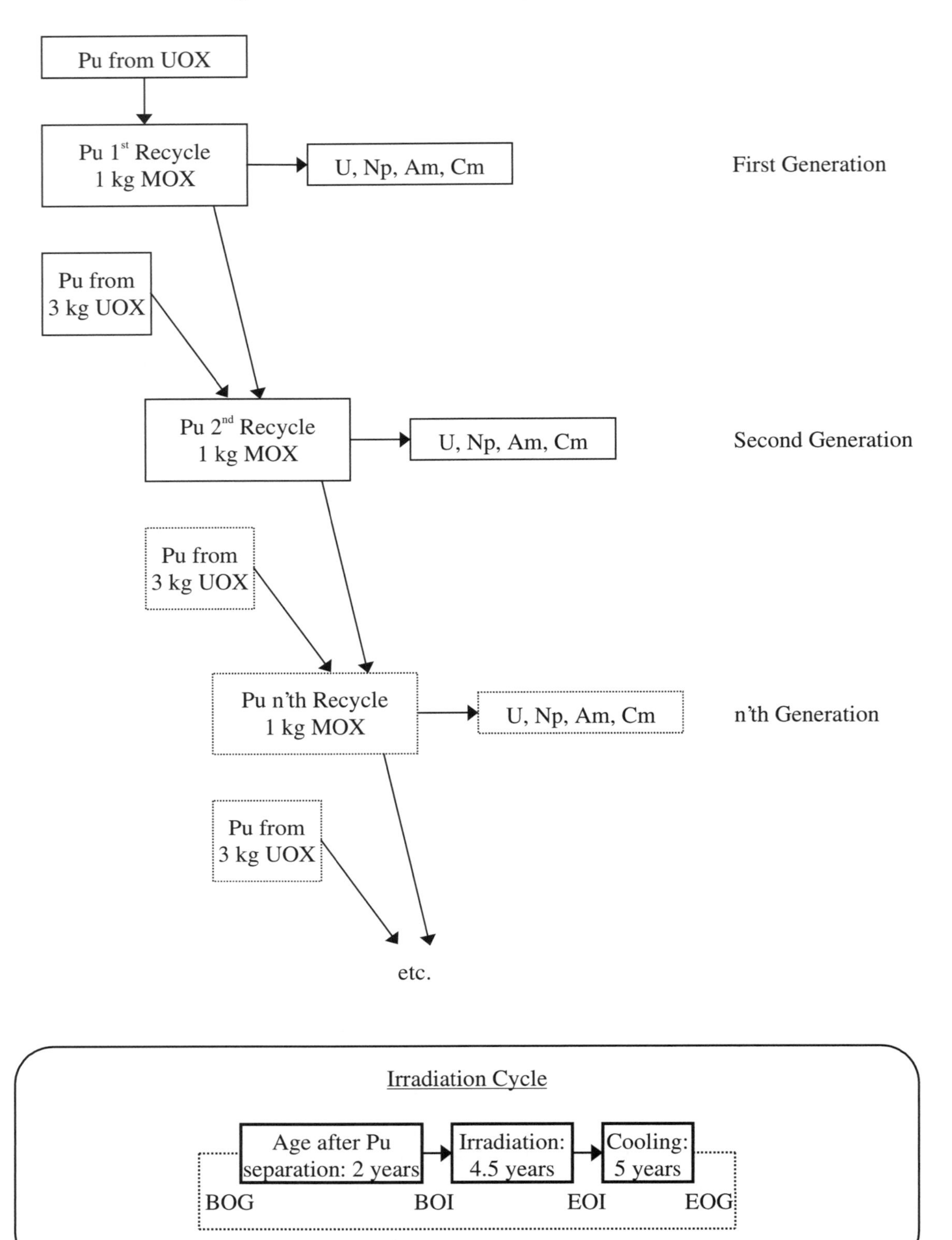

The second pincell was intended to represent a proposed assembly design for a highly moderated MOX fuelled PWR, with the designation HM pincell. This design has been proposed for a small family of future PWRs that would be wholly fuelled with MOX. In the conventional lattice assembly, the moderator/fuel ratio is too low to be optimal for plutonium fuel; a higher value of around 3.5:1 has been shown to be better suited for plutonium, as it is nearer the optimum for reactivity. In theory, a MOX fuelled reactor with a moderator/fuel ratio of 3.5 would need a lower plutonium concentration and would have better reactivity control and shutdown characteristics. The exclusive use of MOX assemblies would eliminate the boundary effects between UO_2 and MOX assemblies that necessitates multiple plutonium enrichments in a conventional MOX assembly.

The highly moderated design is most easily implemented by simply increasing the pin to pin pitch while retaining the pin design unmodified. This implies that the radial dimension of the core needs to be increased and with it the diameter of the pressure vessel. Other than this relatively minor change, the remainder of the plant would be based on existing, well proven components. Table 2.1 identifies where the HM assembly differs geometrically from the STD assembly.

Table 2.1. Main geometrical parameters of standard (STD) and highly moderated (HM) assemblies

	STD Pincell	HM Pincell
Lattice	17×17-25*	17×17-25*
Fuel pellet radius (cm)	0.4127	0.4127
Clad outer radius (cm)	0.4744	0.4744
Clad thickness (cm)	0.0617	0.0617
Pin to pin pitch (cm)		
Outer radius of equivalent pin-cell (cm)	0.7521	0.9062
Moderator/fuel ratio	2.0	3.5

* Guide thimbles (see Appendix A)

Material compositions

Table 2.2 summarises the material composition data which are common to the two pincells. For the three main components of fuel pellet, clad and moderator the following observations apply.

Pellet

The fuel pellet consists of UO_2/PuO_2 mixed oxide at a density of 10.02 g/cm^3. This corresponds to a nominal density of 95% of the theoretical density of 10.96 g/cm^3, with allowance for the loss of volume due to end-dimples and pellet chamfers. The fuel temperature, used to specify the Doppler broadening parameters, is representative of nominal full power operation.

Clad

For the purposes of the benchmark, the clad is assumed to be composed of natural zirconium. In reality, of course, Zircaloy is used, but the neglect of the minor alloying elements has no implications for the benchmarks. The temperature is again representative of full power operation.

Table 2.2. Material compositions common to STD and HM pincells

	STD & HM Pincells
MOX Fuel Pellet	
Density (g/cm^3)	10.02
Temperature (K)	900
Uranium carrier enrichment (w/o)	0.25
Clad	
Material	Zirconium
Density (g/cm^3)	6.55
Temperature (K)	600
Zr number density (b cm)	4.3248E-2
Moderator	
Material	H_2O + dissolved boron
Density (g/cm^3)	0.71395
Temperature (K)	573.16
Dissolved boron concentration (ppm)	500
H_2O Number density (barn cm)	2.386013E-2
B-10 Number density (barn cm)	3.92943E-6
B-11 Number density (barn cm)	1.59162E-5

1 barn cm = 1E-24 cm^2

Moderator

The moderator is light water at a density of 0.71395 g/cm^3, representative of the average density in nominal full power operation. At the nominal working pressure of 155.5 Bars, this density corresponds with a moderator temperature of 306°C (579.16 K). However, the specification sets the temperature at precisely 300°C (573.16 K) so that MCNP data libraries can be used by participants wishing to contribute a Monte Carlo solution, this being one of the standard temperature tabulations. The difference in temperature has a small effect on the neutron spectrum that is of no significance for the benchmarks. The specification calls for the pincell depletions to be carried out with 500 ppm dissolved boron. This is roughly the lifetime average value in a modern PWR. It is important to carry out the depletion in this manner in order to correctly capture the average spectral history effect. However, certain calculations called for the boron to be set to zero, as discussed in the following section.

Fuel depletion

The fuel depletion specifications are the same for each generation of recycle. The underlying assumption is that the fuel management remains the same throughout all five generations of recycle considered. Of course, it is unlikely to be the case in practice, but it is necessary in order to define a manageable problem.

The fuel management is a three batch 18-month fuel cycle, with a discharge burn-up of 51 MWd/kg. Thus at each refuelling outage, the 1/3 of the fuel assemblies resident longest in the core are discharged and replaced with fresh fuel. Such a fuel management scheme is a little outside present experience with both UO_2 and MOX, but is certainly realistic for a few years hence.

Table 2.3 summarises the fuel depletion parameters. The rating is at nominal full power for the reactor. The depletion to 51 MWd/kg, at constant 500 ppm boron, is followed by five years' pond cooling before reprocessing and recycle back into the next generation after a further two years. The pond cooling and recycle delays are important because of the decay of ^{241}Pu to ^{241}Am, and the consequent impact on the physics. The effect of refuelling shutdowns are neglected. Since each refuelling cycle lasts for 18 months, the complete fuel depletion lasts for 4.5 years and therefore the complete cycle from fresh fuel of one generation to fresh fuel of the next generation occupies 11.5 years.

Table 2.3. Fuel depletion parameters

	STD Pincell	HM Pincell
Ageing after fabrication and before irradiation (years)	2	2
Linear rating during irradiation (W/cm)	178	280
Mass rating (W/g of heavy metal)	37.7	59.3
Constant concentration of dissolved boron in moderator (ppm)	500	500
Cooling time after irradiation before reprocessing (years)	5	5

The linear and mass ratings for the STD pincell are the nominal full power values for the current 17×17 fuel assembly. Those for the HM pincell are a factor 1.57 higher than for the STD pincell. The design intent for the HM PWR would be to maintain the enthalpy rise in the moderator roughly in line with that of the STD PWR, otherwise the thermal hydraulic conditions of the primary system would be affected and thermal efficiency reduced. The enthalpy rise depends on the energy deposited by the fuel rods and the coolant flow rate. Optimisation between the coolant flow speed and the fuel linear heating allows to reach a moderation ratio of 4 in standard PWR (EPR) with a 19×19 lattice.

Multiple recycle logistics

This is the essence of these benchmarks and requires careful explanation. The benchmark calls for the plutonium to be followed through five recycle generations, as illustrated in Figure 2.2. Each recycle generation consists of a two-year ageing period from recovery of the plutonium from reprocessing to loading of MOX fuel in reactor. There then follows the irradiation of the fuel, which lasts 4.5 years and the subsequent five-year pond cooling period. The bottom part of Figure 2.2 is a schematic showing the three phases. The points denoted BOG, BOI, EOI and EOG stand for Beginning of Generation, Beginning of Irradiation, End of Irradiation and End of Generation.

The first generation simply involves taking the plutonium obtained from reprocessing a number of UOX assemblies and recycling it as MOX. The UOX assemblies, assumed to be discharged at 51 MWd/kg, contain 12.4 kg/tU total plutonium at discharge (where it is understood that the mass of uranium refers to the initial value). With the isotopic composition of this plutonium, the initial plutonium content of the first generation MOX needed to attain the 51 MWd/kg fuel cycle was determined by CEA to be 10.15 w/o or 101.5 kg/t HM). This value was taken to be the specification for the first generation MOX assembly, with all participants expected to use the same value. The reasoning here was that since there was no expectation of major discrepancies in the first generation (which is already proven in commercial reactors), it would be better to specify the starting point in order to be able to have a reference point for comparison in the later generations.

In the second, third, fourth and fifth generations, the plutonium from the MOX assemblies was assumed to be recovered by co-reprocessing with conventional UOX assemblies in a ratio of 3 UOX: 1 MOX. There are two main reasons why this is appropriate:

1. It is not envisaged that reprocessing plants specifically dedicated to MOX will be built, so that the MOX assemblies would necessarily have to be reprocessed in plants primarily intended for reprocessing UOX. This being the case, there are technical reasons why it would not be practicable to recycle MOX assemblies on their own, or why it would be uneconomic to do so. These have to do with factors such as heat generation in the reprocessing liquors (which is higher in MOX) and activity level from minor actinides (also higher). Therefore, the only practical scenario is where UOX and MOX assemblies are reprocessed together, with a 3:1 ratio advised as being a technically sound value.

2. For the STD pincell at least, it is not intended to irradiate MOX assemblies alone in the core; there will also be UOX assemblies. Therefore a utility's spent fuel arisings will consists of a mix of the two types and it would make sense to reprocess them together. For the HM pincell, although the HM core will only consist of MOX assemblies, it is envisaged that the utility operating the HM PWR would have other conventional PWRs running on UOX only, or on UOX and mixed UOX/MOX. Indeed, this is practically a **requirement**, since otherwise the utility would have no source of plutonium to fuel its HM PWRs. Therefore, the utility's spent fuel arisings would still consist of a mix of UOX and MOX and co-reprocessing is the only sensible scenario.

Therefore in the second generation MOX and all later generations, the plutonium from each MOX assembly derives from a number of MOX assemblies and **three** times that number of UOX assemblies. Before discussing how many MOX and UOX assemblies are needed in the (n-1)'th generation to fuel the MOX in the n'th generation, two important points should be made:

1. The plutonium isotopic quality degrades during each MOX irradiation, as the non-fissile even isotopes accumulate. Therefore the role of the UOX assemblies can be seen as providing a measure of dilution of the plutonium, thereby helping to improve the isotopic quality.

2. The MOX assemblies contain at discharge upwards of 80 kg/t HM of plutonium, compared with just 12.4 kg/tU for the UOX assemblies. Therefore in terms of plutonium (rather than assembly masses or volumes), the dilution factor is much smaller than 3:1 and this means that the degradation of plutonium isotopic quality proceeds quickly in spite of the dilution.

The mass logistics from the (n-1)'th to the n'th generation depends on:

1. The initial plutonium content needed in the n'th generation MOX to provide the required 51 MWd/kg lifetime with the isotopic composition of the (n-1)'th generation plutonium.

2. The plutonium mass at discharge of the (n-1)'th generation MOX.

Suppose that the initial plutonium concentration of the n'th generation MOX is required to be p_i^n kg/t HM (how this is determined is explained in the following section). Suppose also that the plutonium concentration at discharge of the (n-1)'th generation MOX is $p_d^{(n-1)}$ kg/t HM. Then assuming that the UOX plutonium content remains at 12.4 kg/tU, for each tonne of MOX discharged in the (n-1)'th generation there will be available ($p_d^{n-1} + 3 \times 12.4$) kg of plutonium. To fabricate one tonne of n'th generation MOX therefore requires $p_i^n / (p_d^{n-1} + 3 \times 12.4)$ tonnes of (n-1)'th generation MOX. In practice, this ratio is always greater than one, so that the number of tonnes of MOX in the n'th generation is less than that in the (n-1)'th. This is a result of the fact that $p_i^n > p_i^{n-1}$ always applies, because of the isotopic degradation. Also, $p_i^n > p_d^n$, because PWR MOX assemblies are incapable of breeding plutonium and the plutonium content at discharge must always be less than that loaded. These inequalities are such that in practice, the infusion of the 3 × 12.4 kg from the UOX assemblies is insufficient to keep the MOX mass from decreasing in each generation.

End of cycle reactivity

The concept of equivalent end of cycle reactivity is used to determine the initial plutonium content of a MOX assembly. A utility would usually wish a MOX assembly to substitute directly for a UOX assembly. Although there are occasionally exceptions, this normally means that the MOX assembly will contribute the same reactivity as a UOX assembly, when **averaged over its lifetime** in the core. MOX and UOX assemblies, having different reactivity characteristics (specifically, very different gradients of k-infinity with burn-up), cannot be matched for reactivity at more than one time in life. Equivalence in terms of end of cycle reactivity defines what time this should be:

Consider the three batch 18 month fuel cycle used in this benchmark. In the equilibrium fuel cycle, there will be 1/3-core of assemblies having been irradiated for one cycle, 1/3-core of two cycle assemblies and 1/3-core three cycle assemblies. Making the gross simplifications that the reactivity versus burn-up is linear and that the burn-up accumulated in each cycle is the same, the reactivity/burn-up relations of a UOX and a MOX assembly are as illustrated in Figure 2.3. Thus the first cycle regions will have accumulated a burn-up of 17 MWd/kg at the end of the cycle, the second cycle regions will be at 34 MWd/kg and the third cycle fuel at 51 MWd/kg.

The reactor will need to be refuelled when the average k-infinity of the three regions (i.e. the average of points A, B and C for the UOX assembly and a, b and c for the MOX assembly) is just sufficient to cover neutron leakage from the core. In a modern PWR, the leakage amounts to approximately 3.7 %Δk, so this implies the end of cycle occurring when the average k-infinity for the points (A, B, C) or (a,b,c) is 1.037.

Figure 2.3 has been contrived so that Points B and b are precisely at 1.037. In this case, the UOX and MOX assemblies would contribute equally to the core's lifetime reactivity and there would be no loss or gain in equilibrium cycle length if a MOX assembly was substituted for a UOX assembly. Although Figure 2.3 is idealised, it captures the concept perfectly. Note that the reactivity of the fuel at zero burn-up does not figure in the calculation, so that the effect of burnable poisons does not affect the lifetime reactivity provided the poison material is substantially used up by Point A or a.

For the present benchmark, the initial plutonium content needed at the start of each MOX generation was defined by requiring k-infinity to be 1.037 at 34 MWd/kg exposure, corresponding to Point b in Figure 2.3. Since the end of the cycle is defined as the point at which the dissolved boron concentration reduces to near zero, it is clear that k-infinity at Point b should be calculated with zero

Figure 2.3. Schematic of reactivity vs. burn-up for UOX and MOX

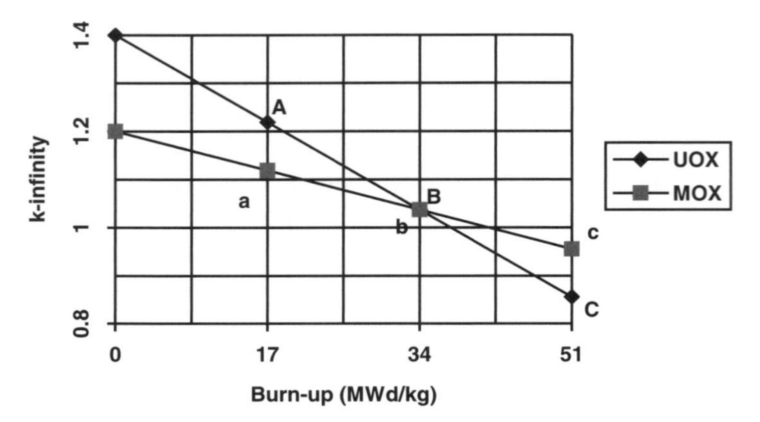

boron. However, it would be incorrect to assume zero boron applies throughout the depletion, since this would introduce a significant error from the incorrect spectral history this implies. A more realistic spectral history can be obtained with the lifetime average boron concentration in the depletion calculations and in this case a nominal value of 500 ppm was applied. The need to calculate Point b at zero ppm still applies, however, and this implies a branch calculation to zero ppm at 34 MWd/kg.

Thus lifetime reactivity equivalence amounts in this case to adjusting the initial plutonium content so that k-infinity for the MOX assembly, at zero ppm, 34 MWd/kg is precisely 1.037.

Plutonium isotopics and concentrations

The benchmark specification called for each participant to start off the first MOX generation with a given total plutonium content (10.15 w/o) and isotopic composition of the plutonium, as shown in Table 2.1. These data were calculated by CEA using the APOLLO code. As first formulated, each participant was then to deplete the MOX assembly to 51 MWd/kg, age for seven years and then recycle the as-calculated plutonium isotopics as detailed in the section entitled *Multiple recycle logistics*. However, it was felt that this might complicate comparisons between different solutions if carried out in isolation, since there is scope for both the reactivity determination and the plutonium isotopic calculation to deviate. Accordingly, it was agreed early on that the multiple recycle strategy would be carried out in two separate stages In the first stage, the isotopic number densities are imposed and in the second they are individually calculated by participants.

- ***Stage I***
  In Stage I the APOLLO code was used as a reference for all other participants, in that the plutonium concentrations and plutonium isotopics calculated by APOLLO were to be treated as given and used as inputs at the start of all five generations of MOX. Table 2.4 lists all the given data. APOLLO was chosen as the reference only because the APOLLO calculations were available prior to other submissions and does not necessarily imply that it is any better than any other code. Indeed, for the purposes of this benchmark, any of the solutions submitted would have served. Carrying out this intermediate step provides a reference point for comparing the reactivity calculations without the complications of having different plutonium contents at the start of each irradiation.

Table 2.4. Initial plutonium (total) contents and plutonium isotopics as calculated with APOLLO

STD PWR

Generation	1	2	3	4	5
Total Pu content (w/o)	10.15	13.625	15.981	17.779	19.228
^{238}Pu (%)	4.0000	4.8260	5.2880	5.5389	5.6633
^{239}Pu (%)	50.4000	42.6180	39.1890	37.1235	35.6755
^{240}Pu (%)	23.0000	26.9270	28.2750	28.9289	29.2940
^{241}Pu(%)	13.5000	13.4140	12.8570	12.3085	11.8480
^{242}Pu (%)	9.1000	12.2150	14.3910	16.1003	17.5181
^{235}U (%)	0.2500	0.2500	0.2500	0.2500	0.2500
^{238}U (%)	99.7500	99.7500	99.7500	99.7500	99.7500

HM PWR

Generation	1	2	3	4	5
Total Pu content (w/o)	6.703	9.636	12.107	14.374	16.583
^{238}Pu (%)	4.0000	4.6484	4.9674	5.1382	5.2151
^{239}Pu (%)	50.4000	36.9325	31.5575	28.5209	26.4344
^{240}Pu (%)	23.0000	29.0647	29.6867	29.3443	28.7947
^{241}Pu (%)	13.5000	13.0907	12.7012	12.1591	11.6091
^{242}Pu (%)	9.1000	16.2637	21.0871	24.8376	27.9467
^{235}U (%)	0.2500	0.2500	0.2500	0.2500	0.2500
^{238}U (%)	99.7500	99.7500	99.7500	99.7500	99.7500

- ***Stage II***
  Stage II called for each participant to submit a set of solutions in which both the initial plutonium content and the plutonium isotopic vector was calculated by each participant independently. In this case, differences in calculating the end of cycle reactivity and the evolution of plutonium isotopics would have the opportunity to accumulate between each generation, providing a test of the relative performances of the complete code systems.

The Stage I Benchmark called for the participants to report:

1. k-infinity versus burn-up at 500 ppm boron, for each of the five MOX generations.

2. k-infinity at 34 MWd/kg, zero boron, for the five generations.

3. Number densities of specified uranium, plutonium, minor actinides and selected fission products at the BOG, BOI, EOI and EOG conditions previously discussed in the section *Multiple recycle logistics*.

4. Net changes in the masses of uranium, plutonium, minor actinides and selected fission products at BOG, BOI, EOI and EOG, in kg/tHM. Also isotopic compositions of same in per cent.

5. Microscopic cross-sections of uranium, plutonium, and minor actinides at 0, 8.5, 17, 25.5, 34, 42.5 and 51 MWd/kg during the irradiation of each generation.

6. Selected macroscopic cross-section at 0 and 51 MWd/kg burn-up (BOI and EOI).

7. Fission energy release for the fissioning nuclides.

8. Delayed neutron fraction (β-eff) data at 51 MWd/kg (EOI).

9. Reactivity balance with constant 500 ppm dissolved boron, being the reactivity difference between the 0.5 and 51 MWd/kg irradiation steps [k(0.5)-k(51)]/50.5, in units of 10^{-5}.

These data were requested largely in order to assist with identifying the reasons for any discrepancies that might have arisen between different participants. In Phase II, the participants were asked to report all of the above and the following additional results:

10. The mass of fuel at BOG, BOI, EOI, and EOG.

11. The isotopic composition of the fuel at BOG, BOI, EOI and EOG.

12. The global activity of the fuel at BOG and EOG for decay timescales from five years to one million years.

13. Radiotoxicities at BOG and EOG for decay timescales between five years and one million years using given half-lives and radiotoxity factors.

These additional data in Stage II largely relate to the question of the long term activity and radiotoxicity, of interest in order to see whether there is significant divergence in predictions of the environmental implications.

Reactivity coefficients

In addition to the calculations just described, the benchmark specification called for reactivity coefficients to be calculated in each generation at irradiations of 0.5, 17, 34 and 51 MWd/kg. Being able to calculate reactivity coefficients with confidence is essential if multiple recycle is to be realised. It is likely that if any limitation to multiple recycle are identified, it will be the reactivity coefficients which will be the determining factors.

The reactivity coefficients examined were the following:

1. Boron efficiency, calculated by perturbing the dissolved boron concentration from 500 to 600 ppm and reporting the boron reactivity coefficient [1/k(500)-1/k(600)]/100, in units of 10^{-5}/ppm. The boron coefficient is important for reactivity balance during normal operation and for emergency shutdown in a PWR.

2. Moderator temperature coefficient calculated by perturbing the moderator temperature and densities between hot and cold states, accordance with the given steam tables (cold state 569.16 K, water density 2.456146E-2 molecules/barn cm, hot state 589.16 K, water density 2.307541E-2 molecules/barn cm). The moderator temperature coefficient is defined as [1/k(cold)-1/k(hot)]/20, in units of 10^{-5}/K.

3. Fuel temperature coefficient calculated by perturbing fuel temperature between 890 and 910 K. The definition of fuel temperature coefficient is [1/k(cold)-1/k(hot)]/20, in units of 10^{-5}/K .

4. Global void effect calculated by changing the moderator density from 100% to 1% of nominal density and defined as [1/k(unvoided)-1/k(voided]/100, in units of 10^{-5}, with given values of moderator density in the unvoided and voided conditions. The calculation is carried out both with a no leakage spectrum and with a critical leakage spectrum in which the buckling parameter has been adjusted to ensure k-eff = 1.

Chapter 3

PARTICIPANTS AND METHODS

Twenty solutions from 10 countries were sent for the benchmark. The main characteristics of method and data used are described in the following table:

		Code	Flux calculation	Self-shielding	Library	HN chain FP chain
1.*	ANL	WIMS-D4M	Monte Carlo		ENDF/B.V, 32 gr. (0-10 MeV)	$^{233}U \rightarrow ^{241}Am$
2.*		CASMO 3		σ_{equi} from 4 eV to 9 kev	ENDF/B.IV, 23 gr. (0-10 MeV)	$^{234}U \rightarrow ^{244}Cm$
3.	Belgonucléaire	WIMS 6	Pij		JEF 2, 172 gr. (0-20 MeV)	
4.	BNFL	WIMS	Pij	sub-group applied to ^{235}U, ^{238}U, ^{239}P and ^{240}P	JEF 2 (0-20 MeV)	$^{233}U \rightarrow ^{245}Cm$
5.	CEA	APOLLO 2	Pij	σ_{equi} + sub-group for E > 0.1 eV	JEF 2, 172 gr. (0-20 MeV)	$^{234}U \rightarrow ^{247}Cm$
6.	ECN	Code system: OCTOPUS	Transport P_3S_g and Monte Carlo	Bondarenko: unresolved region + Nordheim: resolved energy region	JEF 2	Pseudo F.P. for flux ORIGEN-S updated library for depletion
7.	EDF	APOLLO 1	Pij	σ_{equi} for E > 2.6 eV	JEF 1 (0-10 MeV) + ENDF/B.IV, 99 gr.	$^{234}U \rightarrow ^{245}Cm$ 89 F.P.
8.	GRS	Code system: OREST-96			ENDF/B.V	
9.**	IFRR	SPEKTRA			JEF 2	
10.	IKE	RESMOD, RSYST 3, ORIGEN-2	Pij	Self and mutual shielding, 13000 gr. from 3 eV to 1 keV	JEF 2	$^{234}U \rightarrow ^{245}Cm$ 41 HN, 222 F.P.
11.**	IPPE	WIMS (ABBN)			POND-2, 66 gr.	
12.	JAERI	SRAC-95	Pij	Self and mutual shielding below 1 keV	JENDL 3.2	$^{230}Th \rightarrow ^{246}Cm$ 65 F.P.
13.	KAERI	HELIOS	Pij	Subgroups applied to 31 isotopes	ENDF/B.VI, 89 gr.	$^{230}Th \rightarrow ^{246}Cm$ 114 F.P.
14.*		CASMO 3	Pij	σ_{equi} from 4 eV to 9 keV	ENDF/B.VI, 70 gr.	$^{234}U \rightarrow ^{244}Cm$
15.	NAGOYA Univ.	SRAC	Pij	Self and mutual shielding below 130 eV	JENDL 3.1, 107 gr.	$^{233}U \rightarrow ^{245}Cm$ 65 F.P.
16.	PSI	BOXER	Pij	8000 points for 1.3 eV to 907 eV equi, th; for E > 907 eV	JEF 1, 70 gr.	$^{230}Th \rightarrow ^{248}Cm$ 55 F.P., 2 P.F.P
17.	OSAKA Univ.	RESPLA/CP	Pij	Self and mutual shielding below 130 eV	JENDL 3.1	$^{233}U \rightarrow ^{245}Cm$ 65 F.P.
18.		RESPLA/DC				
19.	TOHOKU Univ.	SRAC, ORIGEN 2	Pij	Self and mutural shielding below 1 keV	JENDL 3.2	ORIGEN-2
20.*	RISØ	CCMO-U			ENDF/B.VI, 76 gr.	

* Withdrawn ** Incomplete

Not all participants strictly complied with the benchmark specifications. The two stages of this benchmark imply a large effort, thus, some participants only presented solutions for Stage I. Nevertheless, the majority provided solutions for both stages. Solutions 1, 2, 14 and 20 were withdrawn because the methods used were not considered appropriate for this study.

A large number of solutions is based on flux calculations using the collision probability method, connected to an evolution module or code including more or less isotopes (heavy nuclides and fission products). These codes or systems provide a means of calculating reactivity evolution during depletion.

The main differences stem from the different data and methods used, particularly:

- the data libraries: JENDL, ENDF, JEF;
- the depletion chains: heavy nuclides and fission products;
- the self-shielding methods:
  - equivalent method;
  - sub-group method;
  - ultra-fine group method;
  - Monte Carlo;
- the fission yield and the energy per fission or per reaction.

Based on experience gained from the previous Phase I benchmark, the options used are relatively homogenous:

- the resonance self-shielding is applied to the main plutonium isotopes;
- the chains range from at least ^{234}U to ^{244}Cm.

This shows that there is a common understanding of the physics of plutonium in PWRs and thermal spectrum.

Most of the schemes are similar to those described in Volume II of the *Physics of Plutonium Recycling* series, entitled *Plutonium Recycling in Pressurised Water Reactors*.

Chapter 4

PRESENTATION OF RESULTS

Introduction

This chapter presents a summary of the principal results of the benchmark exercises. This is necessarily incomplete due to the enormous volume of data involved. For the sake of clarity, the Stage I and Stage II exercises are presented in turn and separate discussion provided for each. For Stage I, the results are compared with the APOLLO 2 results regarded as a reference. As discussed earlier, this is purely for convenience and does not imply any special status for APOLLO 2 and serves to highlight the relative spread of results. The Stage II exercise is considered the more fundamental, as the solutions represent what each individual contributor would predict for the evolution of the multi-recycle scenario.

The results presented for Stage I mainly comprise microscopic cross-sections and reactivity balance. Since the initial isotopic content is prescribed at the start of each recycle generation, the reactivity differences can be assumed to be primarily due to:

- data differences (in the nuclear data libraries);
- flux calculations (spectrum);
- self-shielding effects affecting the calculation of resonance absorptions;

Thus, the analysis of the results for Stage I is mainly concentrated on the spreads in microscopic cross-sections and reactivities. For the reactivities, three parameters are linked:

- the initial reactivity value;
- the reactivity balance under irradiation, meaning the change in reactivity with burn-up;
- the end of cycle reactivity value, as defined in Chapter 2.

To analyse all the amount of results, we divide the problem into two distinct parts:

- *Initial reactivity values ⇔ Microscopic cross-sections value*, keeping in mind that ^{239}Pu is the main contributor for fission (75%) and for capture (50%) in the neutronic balance. JAERI performed a comparison on ^{239}Pu source, JEF2, JENDL3 or ENDF/B.VI which shows the impact of data file evaluation and data treatments.
- *Reactivity balance under irradiation ⇔ Microscopic cross-sections value and fission rate.*

The reactivity adjustment value is a combination of the two effects, initial reactivity level and reactivity balance under irradiation.

The analysis of Stage II is mainly focused on the major physical parameters for plutonium recycling in the multiple recycle scenario such as initial plutonium content, reactivity coefficient values and minor actinide production.

Some helpful reference points are given by a Monte Carlo calculation using TRIPOLI 4 and JEF2 (CEA) for initial reactivity values and some reactivity values obtained for the reactivity coefficients conditions.

Standard PWR – Stage I Benchmark

This section presents the results of the benchmark for the standard (STD) PWR, with a moderation ratio of 2 (Vm/Vf) or 4.26 ($n_H/n_{H.N.}$), where n_H is the number density of hydrogen and $n_{H.N.}$ that of the heavy nuclides, with the initial fuel composition at the start of each recycle generation prescribed.

Microscopic cross-sections spreads: ^{238}U, ^{239}Pu, ^{240}Pu, ^{241}Pu, ^{242}Pu

The comparison of microscopic fission and capture cross-sections is shown in Figures 4.1 to 4.7 for all the plutonium isotopes, up to ^{242}Pu. ^{242}Pu becomes increasingly important in the later recycle generations as its concentration increases. The comparison is shown for only the first recycle generation; subsequent recycle generations shows qualitatively similar results.

The spread of values on the microscopic fission cross-sections is about ±2% for ^{239}Pu and -5% to +4% for ^{241}Pu.

The spreads are about ±2% for the microscopic capture cross-section of the main isotopes: ^{239}Pu and ^{238}U. For the latter, the trends with irradiation are rather different, one group which shows a decreasing cross-section trend with burn-up and second one (EDF, ECN, JAERI, PSI, TOHOKU, OSAKA) which has an increasing cross-section tendency with burn-up compared with the CEA.

For higher isotopes, the spreads are larger:

- from -0% to +6% for ^{240}Pu;
- from -4% to +4% for ^{241}Pu;
- from -6% to +2% for ^{242}Pu.

Figure 4.1. STD PWR – Stage I – First recycle:
^{239}Pu fission cross-section – Comparison with CEA values

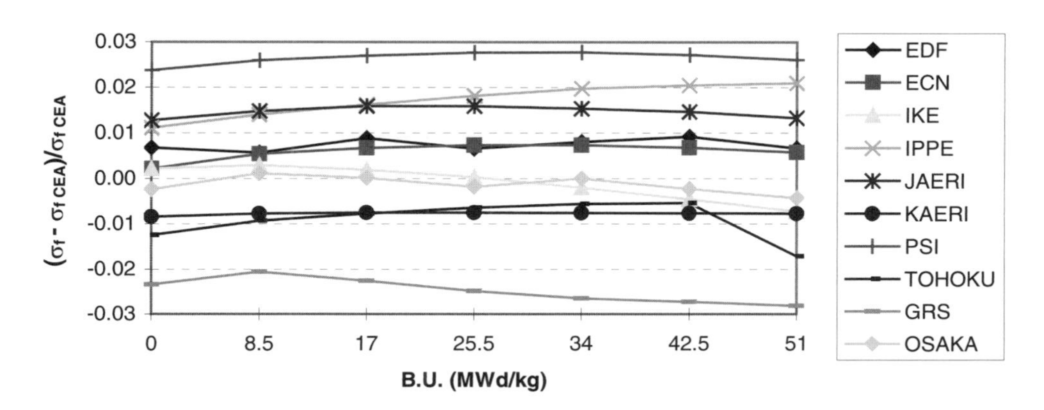

CEA values							
B.U (MWd/kg)	0	8.5	17	25.5	34	42.5	51
σ_f ^{239}Pu (barn)	21.85	22.08	22.90	23.84	24.90	26.06	27.32

Figure 4.2. STD PWR – Stage I – First recycle:
^{241}Pu fission cross-section – Comparison with CEA values

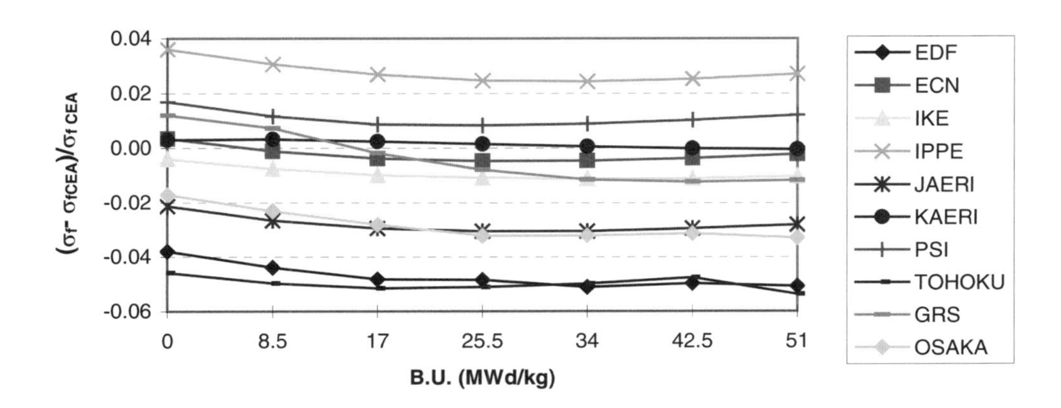

CEA values							
B.U (MWd/kg)	0	8.5	17	25.5	34	42.5	51
σ_f ^{241}Pu (barn)	28.89	28.97	29.73	30.58	31.51	32.52	33.60

Figure 4.3. STD PWR – Stage I – First recycle:
^{238}U capture cross-section – Comparison with CEA values

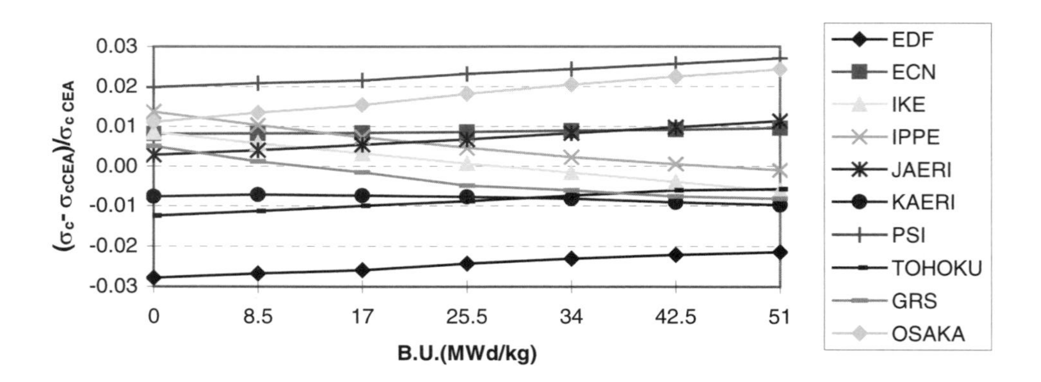

CEA values							
B.U MWd/kg	0	8.5	17	25.5	34	42.5	51
σ_c ^{238}U (barn)	0.82	0.82	0.82	0.83	0.83	0.83	0.83

Figure 4.4. STD PWR – Stage I – First recycle:
^{239}Pu capture cross-section – Comparison with CEA values

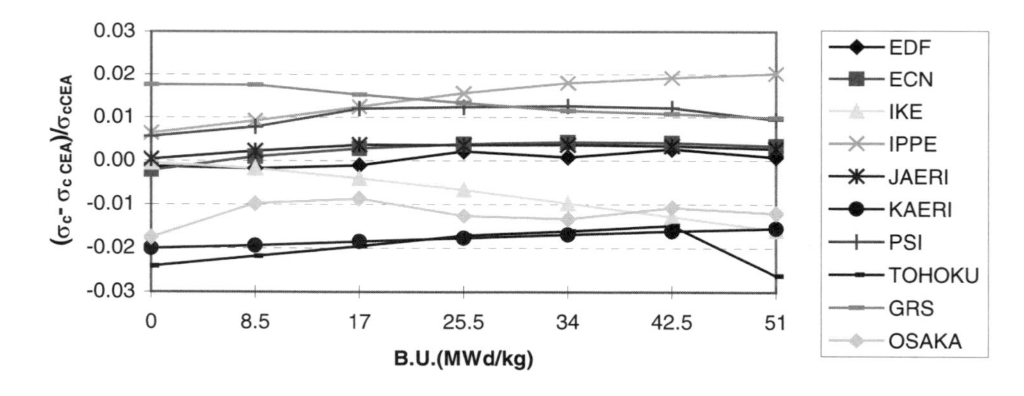

CEA values							
B.U (MWd/kg)	0	8.5	17	25.5	34	42.5	51
σ_c ^{239}Pu (barn)	12.22	12.42	12.91	13.47	14.09	14.76	15.49

Figure 4.5. STD PWR – Stage I – First recycle:
^{240}Pu capture cross-section – Comparison with CEA values

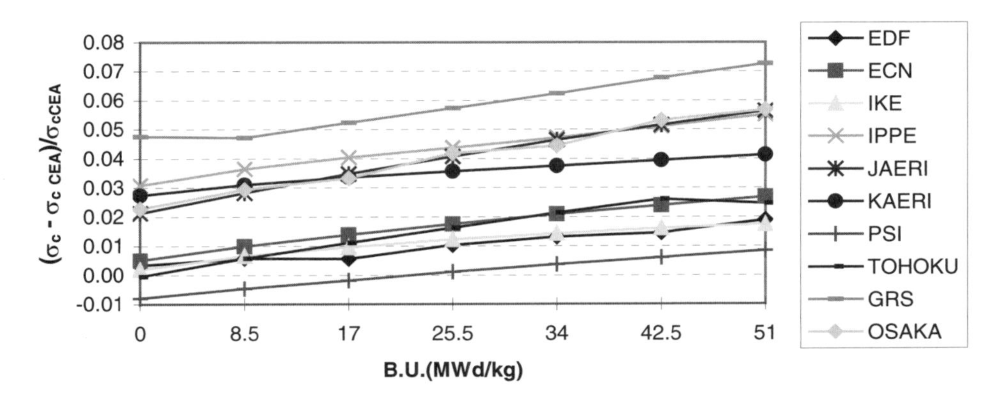

CEA values							
B.U MWd/kg	0	8.5	17	25.5	34	42.5	51
σ_c ^{240}Pu (barn)	25.81	25.35	25.26	25.34	25.57	25.92	26.40

Figure 4.6. STD PWR – Stage I – First recycle:
^{241}Pu capture cross-section – Comparison with CEA values

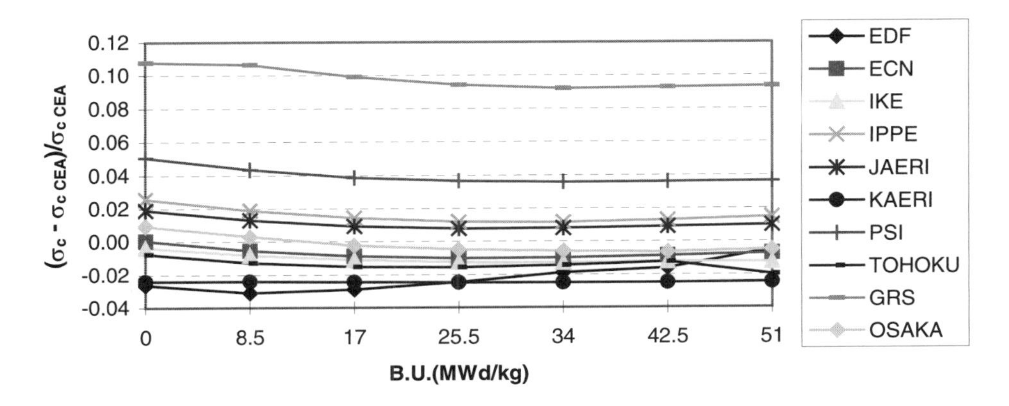

CEA values							
B.U (MWd/kg)	0	8.5	17	25.5	34	42.5	51
σ_c ^{241}Pu (barn)	9.03	9.09	9.37	9.68	10.01	10.37	10.75

Figure 4.7. STD PWR – Stage I – First recycle:
^{242}Pu capture cross-section – Comparison with CEA values

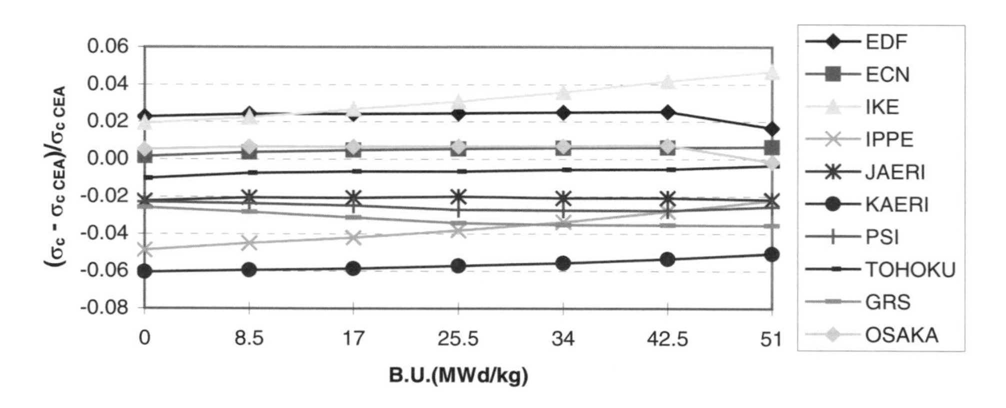

CEA values							
B.U MWd/kg	0	8.5	17	25.5	34	42.5	51
σ_c ^{242}Pu (barn)	11.54	11.42	11.33	11.22	11.12	11.02	10.92

On the microscopic (n,2n) cross-sections

Figures 4.8 and 4.9 show the microscopic (n,2n) cross-sections for ^{238}U and ^{244}Cm. The order of magnitude of the cross-sections for this reaction type in the neutronic balance is small, but it is of interest to show these two for illustration. The differences are large, even between for data based on the same evaluated data file and are largely due to the nuclear data, but there are also contributions from the different treatments of this reaction in each code, at high energy.

Figure 4.8. STD PWR – Stage I – Recycle:
^{238}U (n,2n) cross-section – Comparison with CEA values

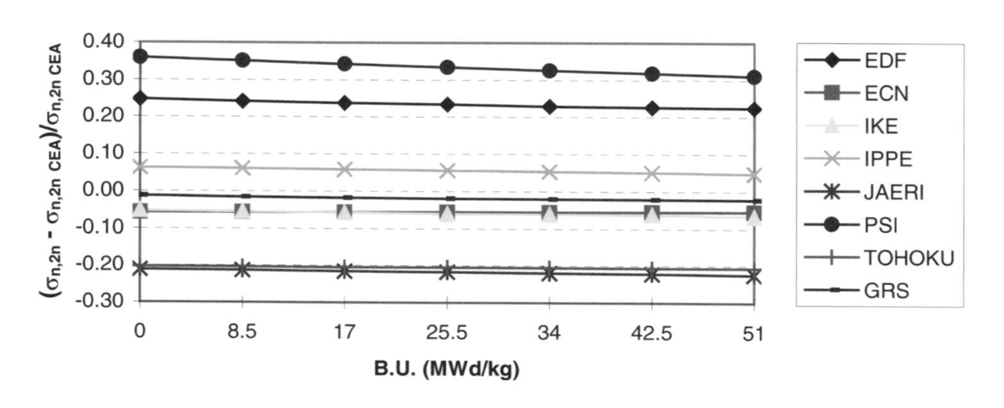

CEA values							
B.U (MWd/kg)	0	8.5	17	25.5	34	42.5	51
$\sigma_{n,2n}$ ^{238}U (barn)	5.72E-03	5.74E-03	5.75E-03	5.76E-03	5.77E-03	5.77E-03	5.78E-03

Figure 4.9. STD PWR – Stage I – First recycle:
^{244}Cm (n,2n) cross-section – Comparison with CEA values

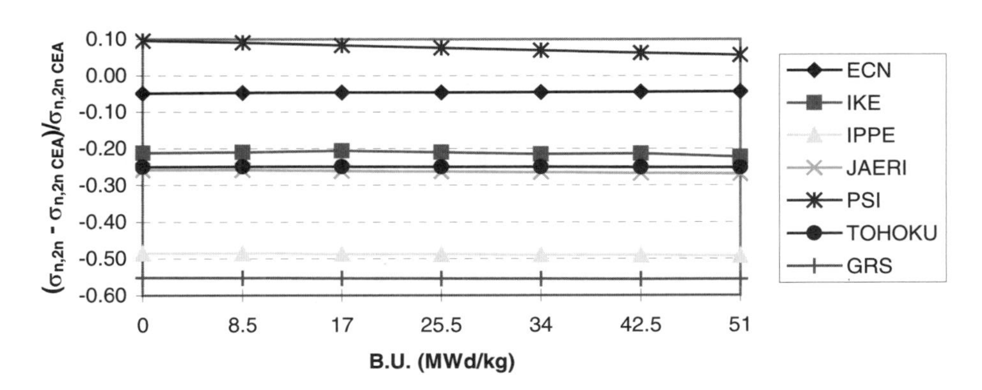

CEA values							
B.U (MWd/kg)	0	8.5	17	25.5	34	42.5	51
$\sigma_{n,2n}$ ^{244}Cm (barn)	2.19E-03	2.19E-03	2.20E-03	2.20E-03	2.20E-03	2.20E-03	2.21E-03

On the ν values

The differences are small for this parameter (the mean number of neutrons per fission). Figure 4.10 shows the comparison of ν values for the main contributing isotope, ^{239}Pu:

Figure 4.10. STD PWR – Stage I – First recycle:
^{239}Pu ν values

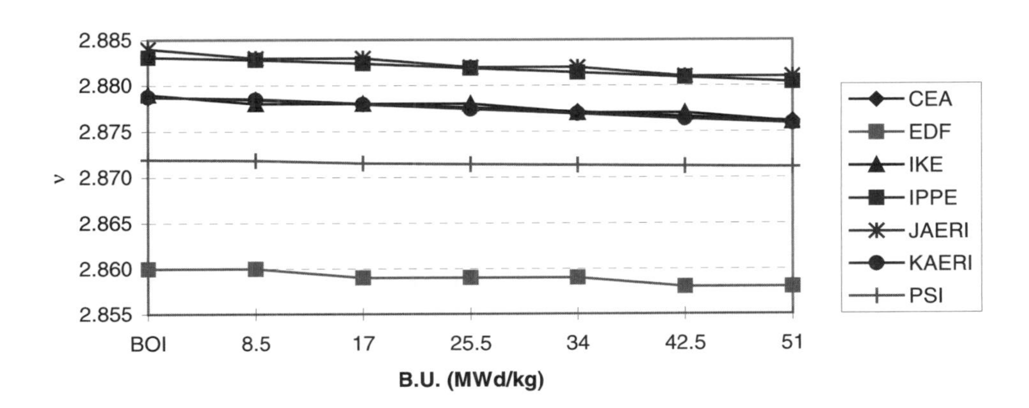

On the fission capture ratio cross-sections

Apart from the GRS values, all of the participants show positive differences between 0 to 2% (Figure 4.11). This implies that the main isotope for the reactivity balance is more fissile and therefore the reactivity value must be larger than CEA for the majority of participants.

For ^{241}Pu, the differences are larger in a range of +2% to -4%.

Figure 4.11. STD PWR – Stage I – First recycle:
^{239}Pu ratio fission/capture – Comparison with CEA values

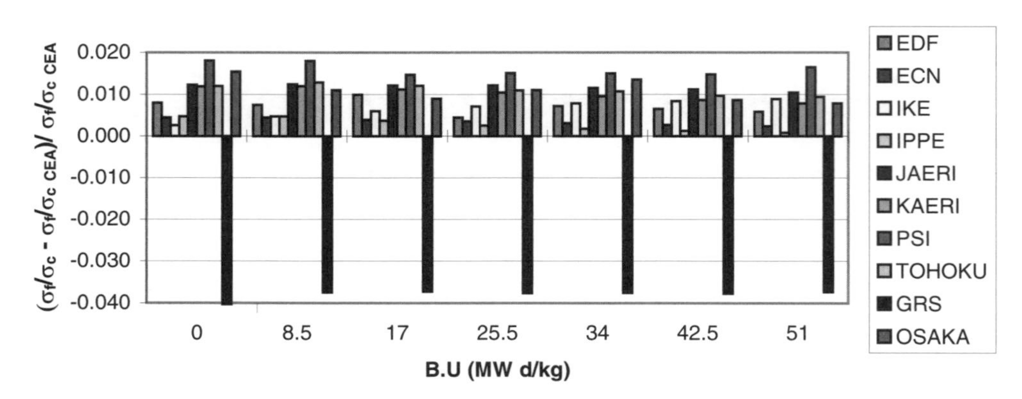

CEA values							
B.U (MWd/kg)	0	8.5	17	25.5	34	42.5	51
σ_f/σ_c ^{239}Pu	1.789	1.777	1.773	1.770	1.767	1.766	1.764

Microscopic differences as per evaluated data file

When the microscopic cross-sections are grouped together by nuclear data library, as in Figures 4.12 and 4.13, more consistent results are observed, as would be expected. The comparison is again with the CEA results, based on JEF:

Figure 4.12. STD PWR – Stage I – First recycle:
^{239}Pu ratio fission/capture – Comparison with CEA values – JENDL library

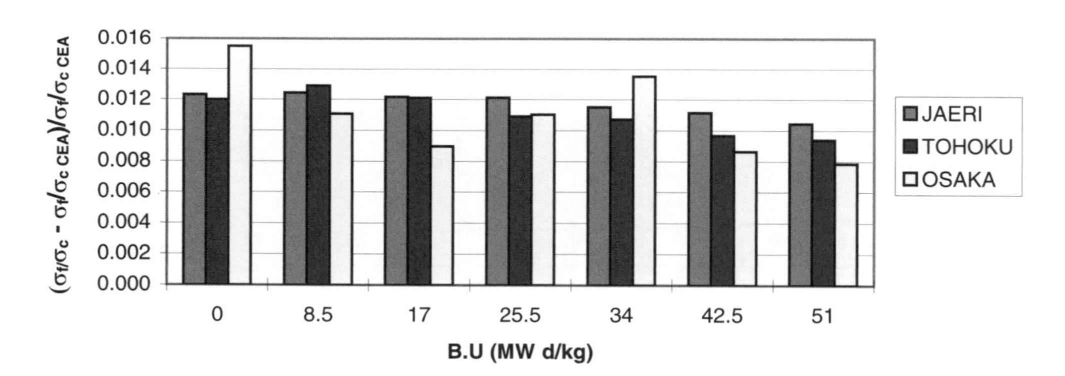

Figure 4.13. STD PWR – Stage I – First recycle:
^{239}Pu ratio fission/capture – Comparison with CEA values – JEF library

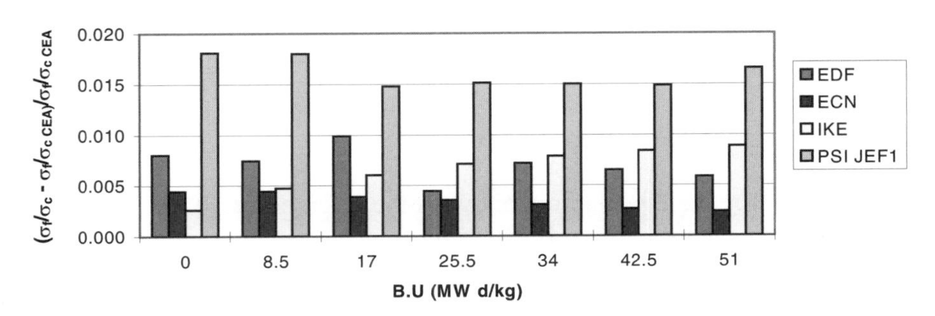

Spread in reactivities

For reactivity and the other major nuclear design parameters, representative results are presented. Although in some cases the results are presented for all five recycle generations, in most cases only the first and fifth generations are shown where there is no exceptional behaviour in the intermediate ones. Thus Figures 4.14 and 4.15 show the comparisons of reactivities versus cell irradiation for the first and fifth recycle generations, compared with CEA values as a reference:

Figure 4.14. STD PWR – Stage I – First recycle:
Reactivity swing – Comparison with CEA values

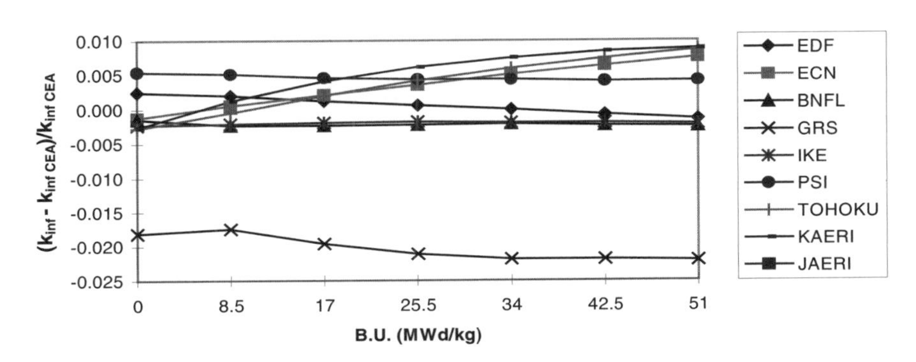

Figure 4.15. STD PWR – Stage I – Fifth recycle:
Reactivity swing – Comparison with CEA values

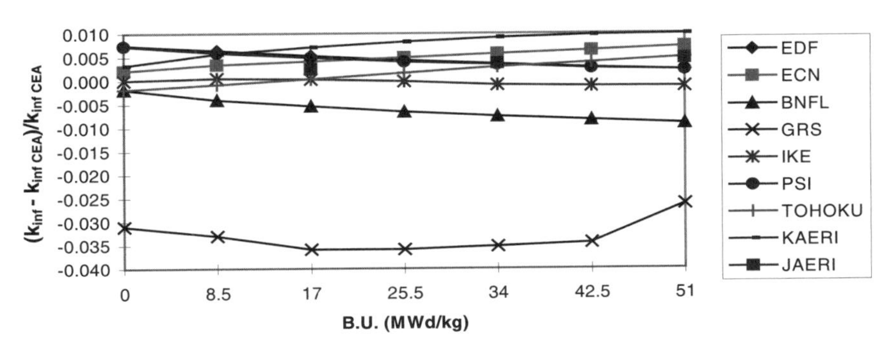

Disregarding the anomalous GRS results, two distinct trends are seen in respect of the reactivity variation with cell irradiation:

1. one in which reactivity decreases less rapidly with burn-up than CEA (KAERI, TOHOKU, ECN);

2. one for which reactivity decreases as per the CEA case.

The figure for the first recycle generation above is the reference case for setting the end-of-cycle reactivity for which we observe two groups of solutions:

1. BNFL, CEA, EDF, IKE, IPPE (smaller than 1.04);

2. ECN, JAERI, OSAKA, PSI, TOHOKU (larger than 1.04).

The differences observed do not increase during later recycle generations; all the results fall within a range of 1% in reactivity. This is small compared with the reduction in reactivity with irradiation, which is about 20% in the first recycle generation and 13% for the fifth generation.

Reactivity balance under irradiation

The value of the reactivity balance, being the change in reactivity with burn-up, is about 0.4% per MWd/kg for the first recycle generation and 0.25% per MWd/kg for the fifth generation as shown in Figure 4.16:

Figure 4.16. STD PWR – Stage I:
Reactivity balance – Comparison with CEA values

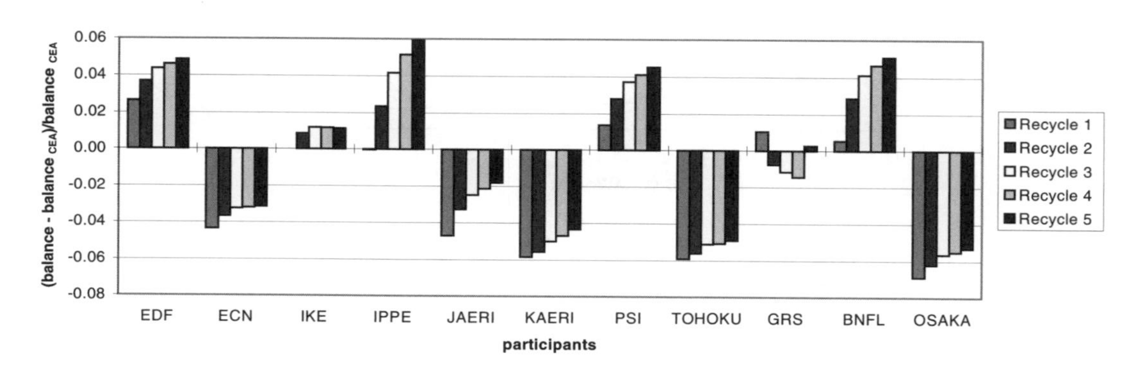

CEA values					
	Recycle 1	Recycle 2	Recycle 3	Recycle 4	Recycle 5
Reactivity balance	363	306	284	272	264

The reactivity balance under irradiation is linked to the fission rate, the cross-section data and the isotopic evolution chain. For most of the participants, the trend compared with CEA is for the reactivity balance to increase with increasing recycle generation, so that those cases starting out more positive than CEA initially become more positive still, while those starting out more negative than CEA become less negative.

Figure 4.17 shows the spread in the reactivity adjustment values, these being the end-of-cycle k-infinities calculated by each participant with the given initial plutonium isotopics and concentrations. Again, the CEA values are regarded as the reference. If there was complete agreement, all participants would return the value 1.037. Table 4.1 lists the k-infinities by participant and recycle generation.

Disregarding the anomalous GRS results, the largest change between first and fifth recycle generation is about 0.55% (BNFL). The spread between participants increases with recycle generation from 0.9% in reactivity up to 1.4% at the fifth recycle. Table 4.2 lists the minimum and maximum values of end-of-cycle k-infinity and the largest differences obtained.

Figure 4.17. STD PWR – Stage I:
Reactivity value adjustment – Comparison with CEA values

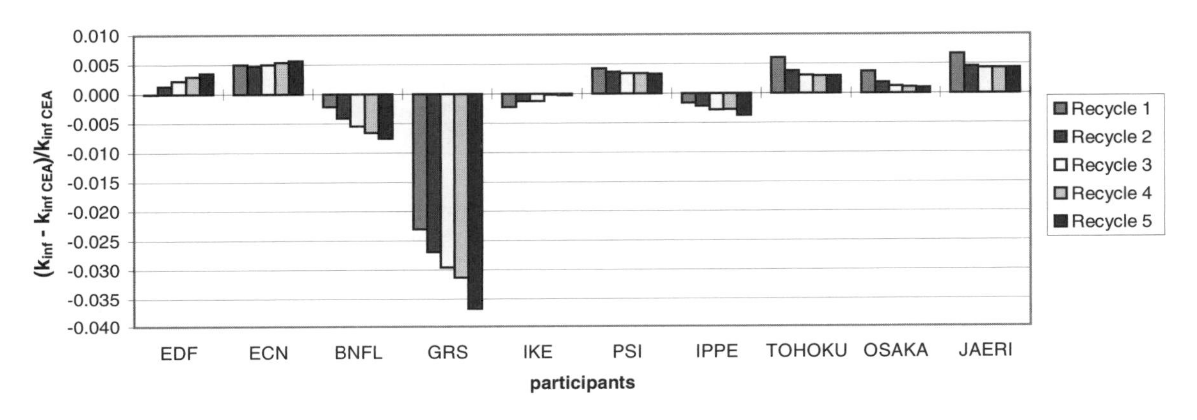

Table 4.1. End-of-cycle k-infinities by participant and recycle generation

	Recycle 1	Recycle 2	Recycle 3	Recycle 4	Recycle 5
CEA	1.03726	1.03726	1.03727	1.03723	1.03728
EDF	1.03720	1.03856	1.03955	1.04031	1.04089
ECN	1.04247	1.04219	1.04247	1.04279	1.04310
BNFL	1.03509	1.03303	1.03162	1.03045	1.02945
GRS	1.01338	1.00934	1.00666	1.00480	0.99919
IKE	1.03500	1.03600	1.03600	1.03700	1.03700
PSI	1.04169	1.04106	1.04086	1.04076	1.04073
IPPE	1.03566	1.03506	1.03435	1.03443	1.03348
TOHOKU	1.04354	1.04124	1.04045	1.04033	1.04031
OSAKA	1.04112	1.03922	1.03856	1.03832	1.03821
JAERI	1.04417	1.04199	1.04168	1.04163	1.04168

Table 4.2. End-of-cycle k-infinities – Extreme values

	Recycle 1	Recycle 2	Recycle 3	Recycle 4	Recycle 5
Min.	1.035	1.033028	1.031616	1.030452	1.02945
Max.	1.044166	1.04219	1.04247	1.04279	1.0431
Largest difference	0.009166	0.009162	0.010854	0.012338	0.01365

Spread in fission rates

Figure 4.18 shows the spread in fission rates calculated by the various participants.

Figure 4.18. STD PWR – Stage I – Fission rate differences (in %)

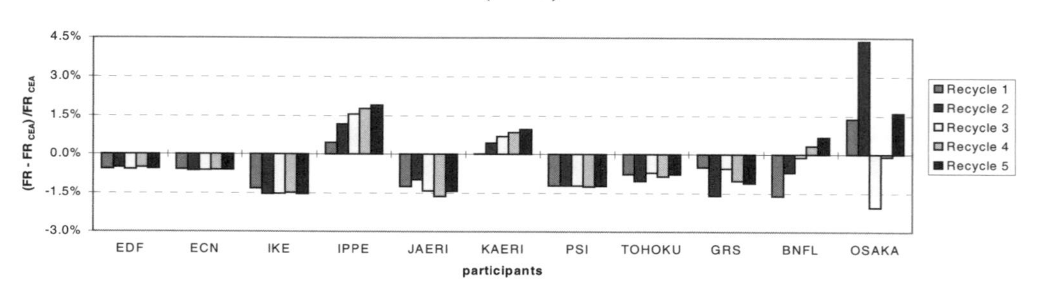

CEA values					
	Recycle 1	Recycle 2	Recycle 3	Recycle 4	Recycle 5
Fission rate (fraction)	0.05215	0.05215	0.05215	0.05217	0.05218

The differences are almost constant for the majority of the participants and fall within a range of 1.5% (EDF, ECN, IKE, PSI), although three participants show an increasing trend with recycle generation. 1% in fission ratio is equivalent to 0.5 MWd/kg, and thus, to 0.2% in reactivity for the first recycle generation and 0.15% for the fifth recycle generation. These differences are therefore insufficient to explain the discrepancies seen in the reactivity balance.

Spread in reaction rates

From the relatively small number of participants who provided reaction rates for comparison, the spread of results (illustrated in Figure 4.19 for the first recycle generation) is quite small and it is difficult to spot any underlying trends. Both absorption and production terms increase from the first recycle generation through to the fifth generation, due to the increasing plutonium content.

These results give us the order of magnitude between the different reactions in the different materials at the end of the life of the fuel (51 MWd/kgHM) for cycles one and five. The values are in a similar range.

Figure 4.19. STD PWR – Stage I – Reaction rates

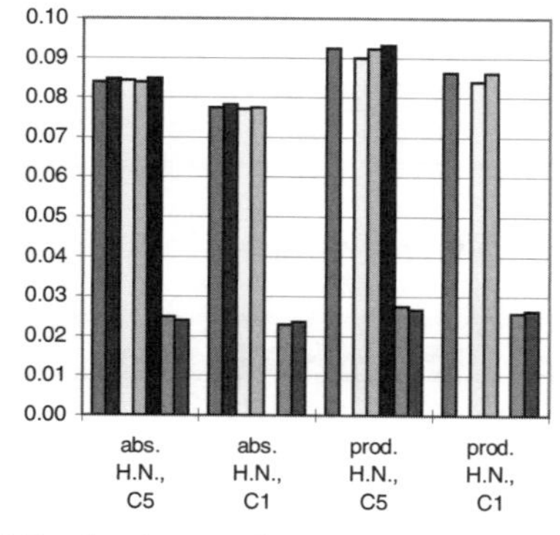

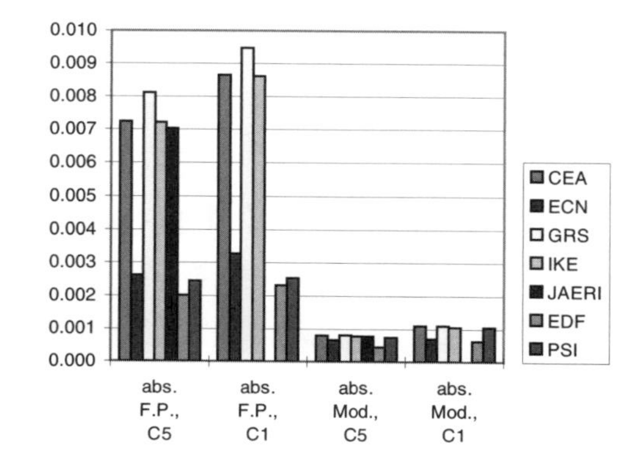

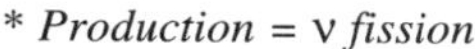
* *Production = ν fission*

Reactivity coefficients

The boron, moderator temperature and fuel temperature coefficients are important feedback parameters which normally have negative values. For convenience, the absolute values of these coefficients are plotted in this section. All three coefficients are actually negative.

The boron reactivity coefficients calculated by the various participants are generally in very good agreement, and show very similar trends as a function of cell burn-up. They are illustrated below in Figures 4.20 and 4.21 for the first and fifth recycle generation cases, plotted versus cell irradiation. Given that in Stage I the initial compositions are given, the 0.5 MWd/kg burn-up steps should be virtually identical in each case, as the reactivity and spectral effects of boron are well defined. Yet the JAERI and GRS solutions stand out as having a lower absolute value, particularly in the fifth recycle generation, implying differences arising from the solution method.

Figure 4.20. STD PWR – Stage I – First recycle: Boron efficiency

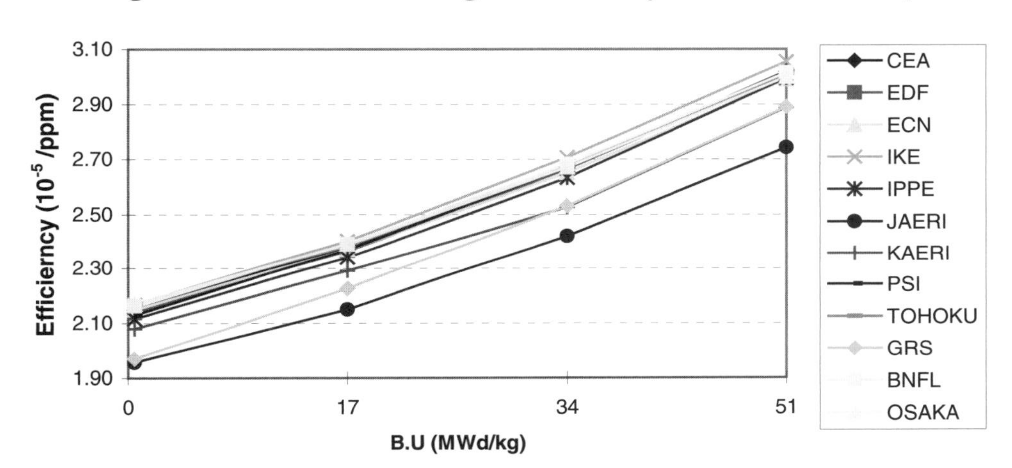

Figure 4.21. STD PWR – Stage I – Fifth recycle: Boron efficiency

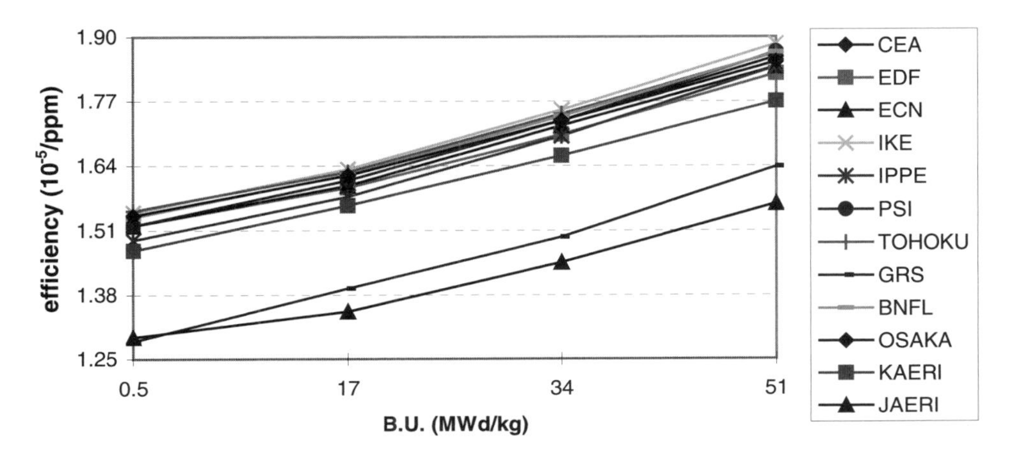

For the majority of participants, the boron coefficients are close for the beginning-of-irradiation step (0 MWd/kg). For the end-of-irradiation step (51 MWd/kg), the differences are about 0.2×10^{-5}/ppm of soluble boron. This implies a difference of 0.4% in reactivity effect for 2 000 ppm of soluble boron. This is an acceptable variation arising from a combination of depletion calculation, flux calculation and nuclear data.

It is notable that there is a large decrease in the magnitude of the boron efficiency between the first and fifth recycle generations, due to the rapid increase in plutonium content and the consequent effect of hardening of the neutron spectrum.

Figures 4.22 and 4.23 show the moderator temperature coefficients for the first and fifth recycle generations as a function of cell irradiation:

Figure 4.22. STD PWR – Stage I – First recycle: Moderator temperature coefficient

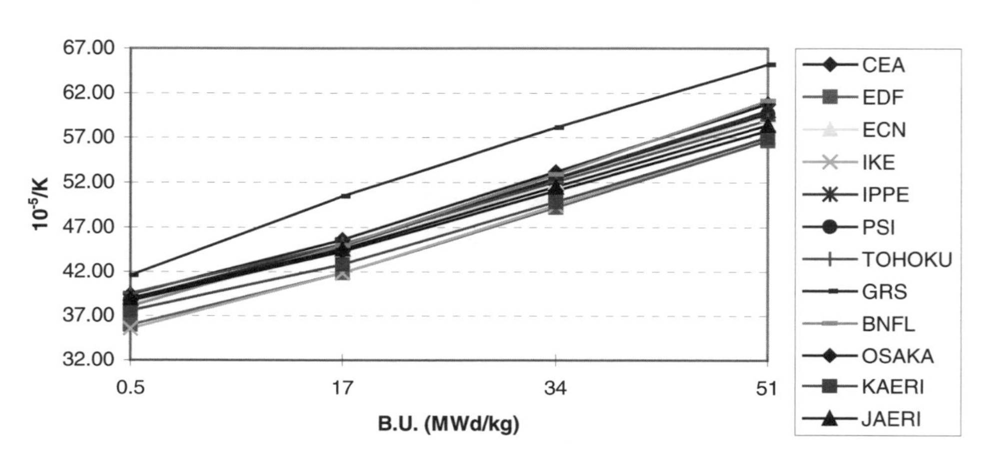

Figure 4.23. STD PWR – Stage I – Fifth recycle: Moderator temperature coefficient

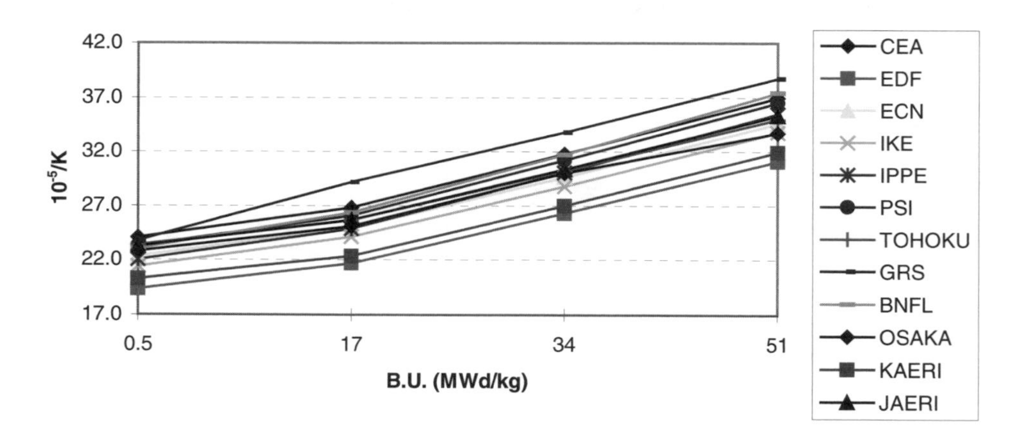

The moderator temperature coefficient values are spread very closely and the general trends are very similar for all participants. The absolute value decreases between the first and the fifth recycle generation, meaning that the moderator temperature coefficient becomes less negative, mainly due to the positive density effect with the increasing plutonium content.

Figures 4.24 and 4.25 show the fuel temperature (Doppler) coefficient as a function of cell irradiation and participant for the first and fifth recycle generations:

Figure 4.24. STD PWR – Stage I – First recycle: Fuel temperature coefficient

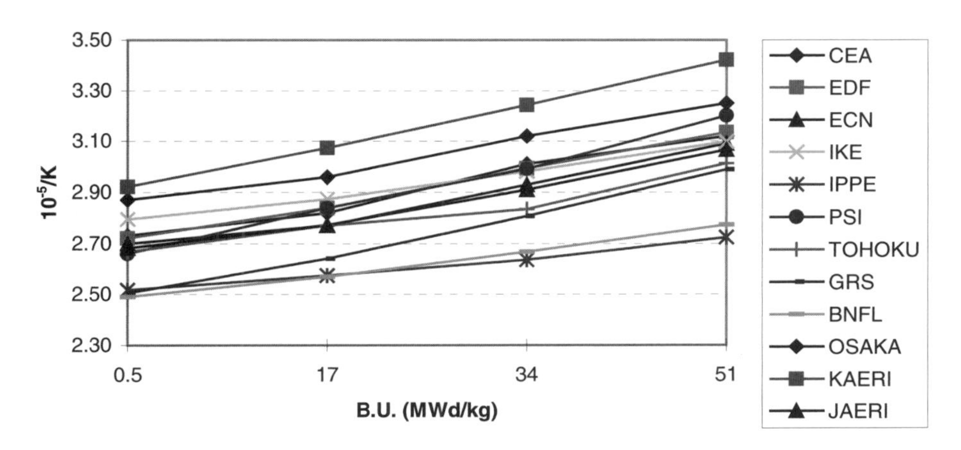

Figure 4.25. STD PWR – Stage I – Fifth recycle: Fuel temperature coefficient

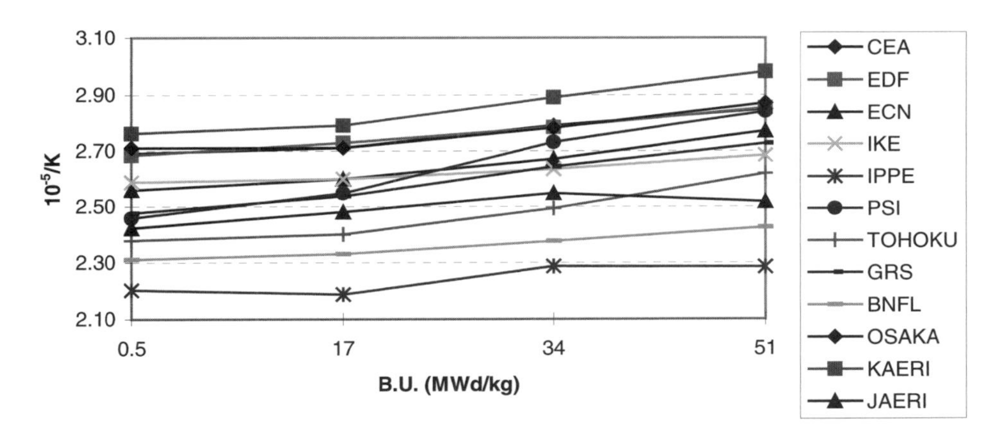

The fuel temperature coefficient values show a larger spread than either the boron coefficient or the moderator coefficient. The underlying trends with cell burn-up are not consistent and the discrepancies tend to increase in the later recycle generations. Isotopic degradation of the plutonium is probably at least partially responsible for these differences. The methods used to treat the temperature dependence of the plutonium resonances are different between the various participants and this may be a contributing factor. Overall it is clear the fuel temperature coefficient is not satisfactory and should be improved. More detailed investigation is required.

Figures 4.26 and 4.27 show the global void effect as a function of cell irradiation. This is the reactivity change caused by completely replacing the moderator by a low density void. A negative value is desirable, as it indicates negative feedback will apply in the event of accidental steam voiding in the core.

Figure 4.26. STD PWR – Stage I – Recycle: Global void effect with no leakage

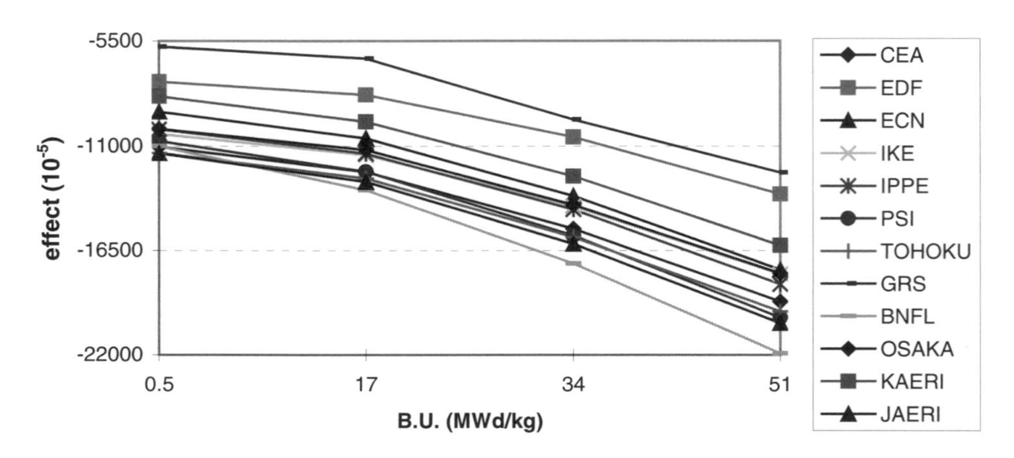

Figure 4.27. STD PWR – Stage I – Fifth recycle: Global void effect with no leakage

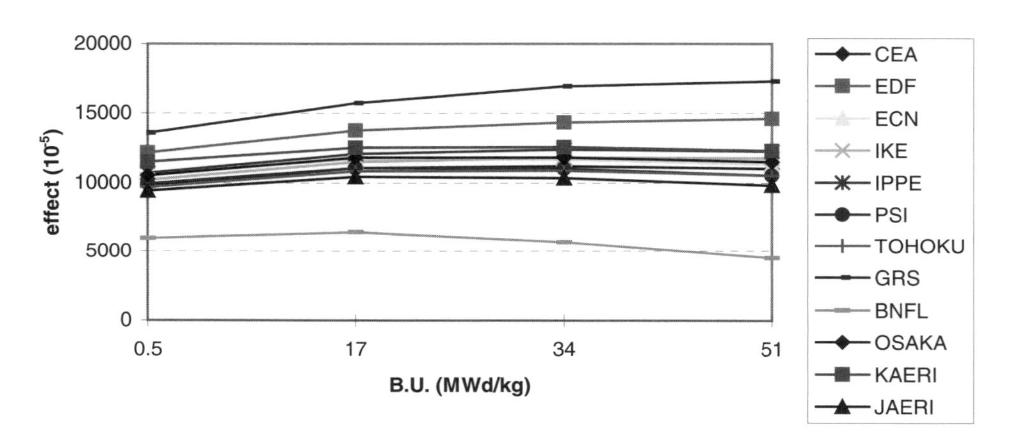

The various participants agree on the magnitude of the void reactivity effect to within a range of 10% in reactivity. The trend with irradiation is similar for all participants, with similar behaviour in all recycle generations.

β_{eff} values

Figure 4.28 shows a comparison of the delayed neutron fraction β_{eff}, measured in 10^{-5} reactivity (pcm). This an important indicator of the transient response of the core. Generally, it should be large in magnitude to slow down the onset of prompt criticality in the event of a reactivity insertion fault.

Figure 4.28. STD PWR – Stage I:
β_{eff} value at end of irradiation

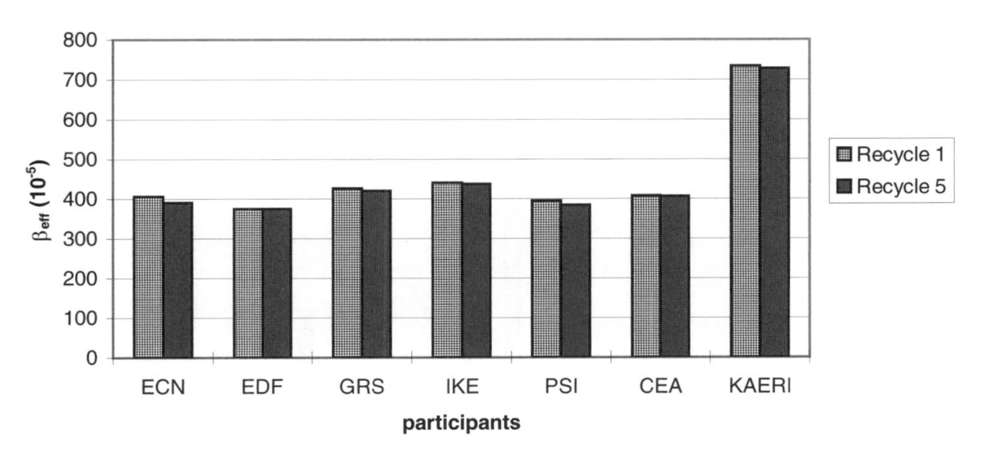

There is a large spread in delayed neutron fractions (β_{eff}), which must arise from underlying discrepancies in the nuclear data.

Standard PWR – Stage II Benchmark

In the Stage II Benchmark, each participant was asked to calculate its own plutonium isotopic assay at the start of irradiation and to fix the total plutonium content in accordance with the method specified in Chapter 2. IPPE and PSI did not keep exactly the agreed specification for the dilution between the plutonium from MOX and UO_2. Their results are not comparable to the others and where appropriate, they are not shown.

Initial plutonium content

Figure 4.29 shows a comparison of the initial plutonium content by recycle generation and participant, with CEA as the reference.

The spread of values increases in later recycle generations, up to a maximum of approximately 1 w/o in the absolute plutonium content. This represents, for a standard PWR of 1 300 MWe, around 1 metric tonne for the plutonium inventory of the whole core.

End-of-cycle reactivity value

Figure 4.30 shows the end-of-cycle k-infinity calculated by each participant in the first recycle generation. The first generation values are subsequently used as targets to be attained in the subsequent recycle generations by adjusting the initial plutonium content. As previously seen in Stage I, the results divide into two populations, one giving a smaller k-infinity than 1.04 and one larger than 1.04. Overall the calculations are fairly consistent, apart from GRS as a clear outlier.

Figure 4.29. STD PWR – Stage II:
Pu content at BOC – Comparison with CEA values

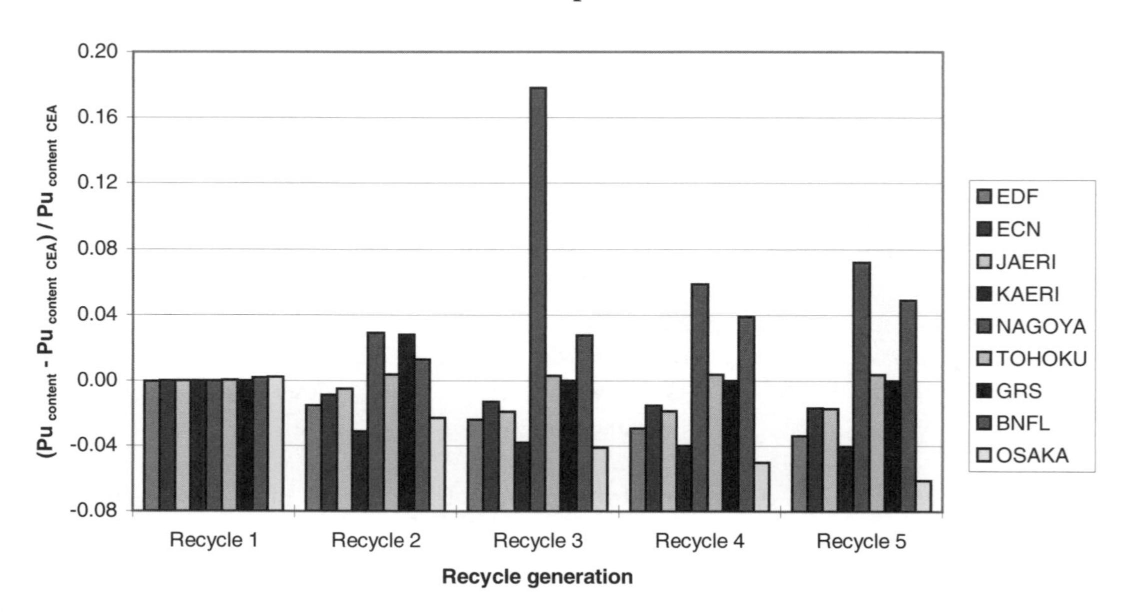

CEA values					
	Recycle 1	Recycle 2	Recycle 3	Recycle 4	Recycle 5
Pu content	10.09%	13.53%	15.87%	17.66%	19.10%

Figure 4.30. STD PWR – Stage II – First recycle: End-of-cycle-reactivity

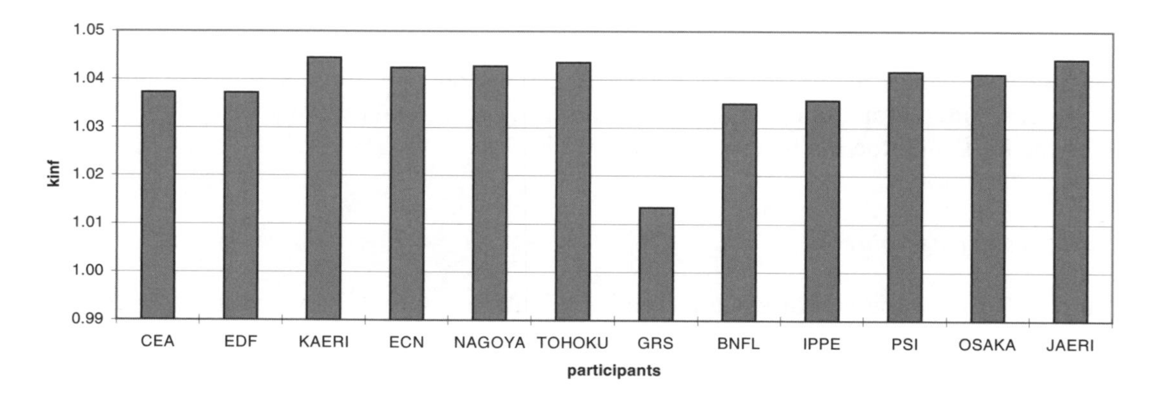

Reactivity parameters values

Figures 4.31 and 4.32 show the boron reactivity coefficients in the first and fifth recycle generations respectively, as a function of cell irradiation:

Figure 4.31. STD PWR – Stage II – First recycle: Boron efficiency

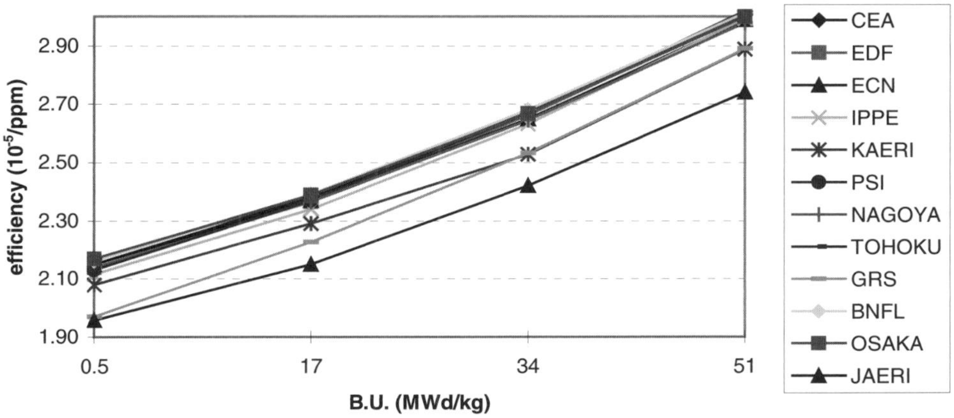

Figure 4.32. STD PWR – Stage II – Fifth recycle: Boron efficiency

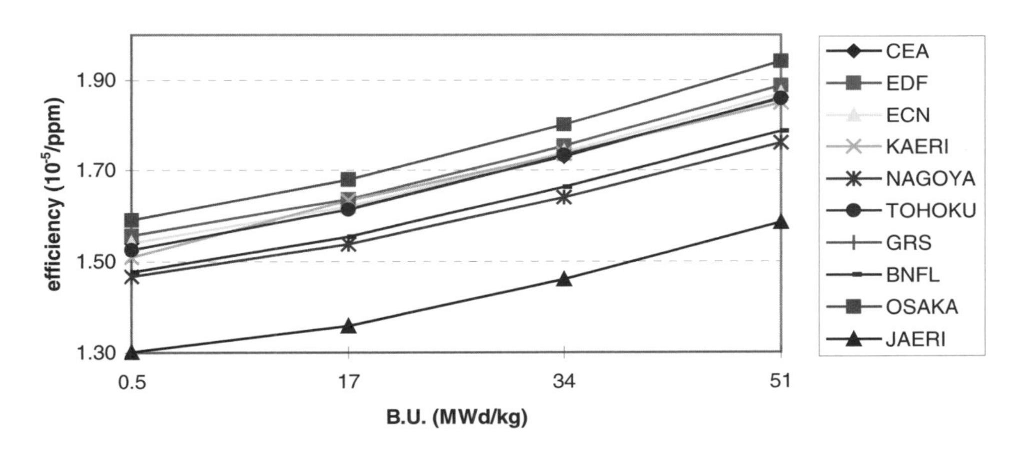

As in the Stage I Benchmark, the boron coefficients are generally in good agreement and show similar trends with cell irradiation. The absolute magnitude of the boron coefficients is very small in all recycle generations (*cf.* typical figure for UO_2: -8 pcm/ppm, 1 pcm = 10^{-5}). This will make satisfactory control and operation of the core more difficult to achieve, particularly in the later recycle generations.

Figures 4.33 and 4.34 show the moderator temperature coefficient for recycle generations one and five respectively.

Figure 4.33. STD PWR – Stage II – First recycle: Moderator temperature coefficient

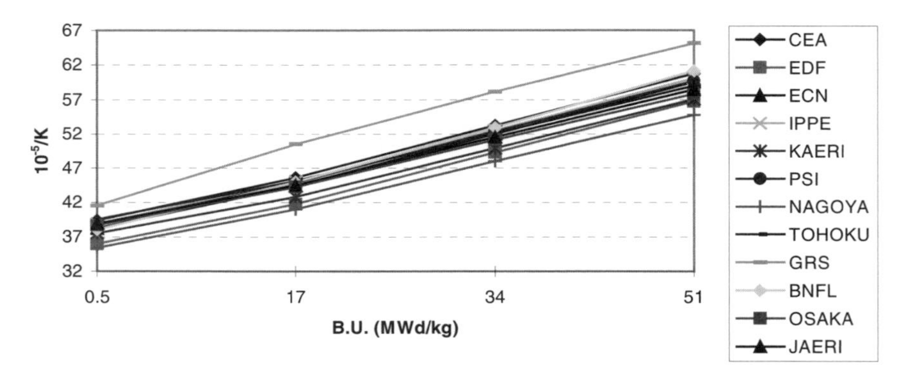

Figure 4.34. STD PWR – Stage II – Fifth recycle: Moderator temperature coefficient

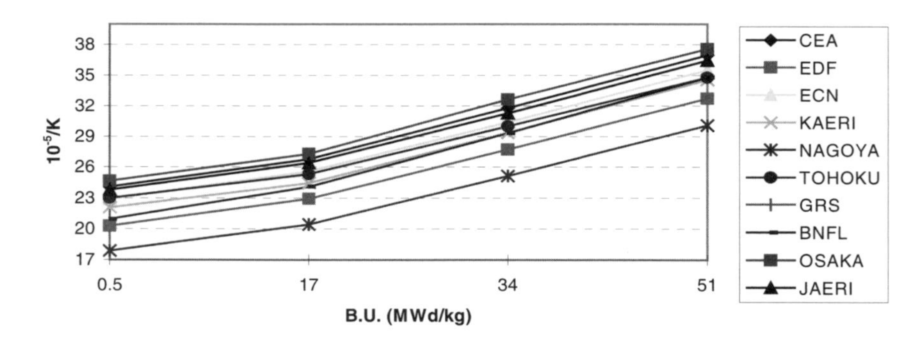

The trends are similar to those seen in the Stage I Benchmark. During later recycle generations, the absolute value moderator temperature coefficient becomes larger, as does the spread between the participants.

Figures 4.35 and 4.36 show the fuel temperature coefficients for the first and fifth recycle generations.

Figure 4.35. STD PWR – Stage II – First recycle: Fuel temperature coefficient

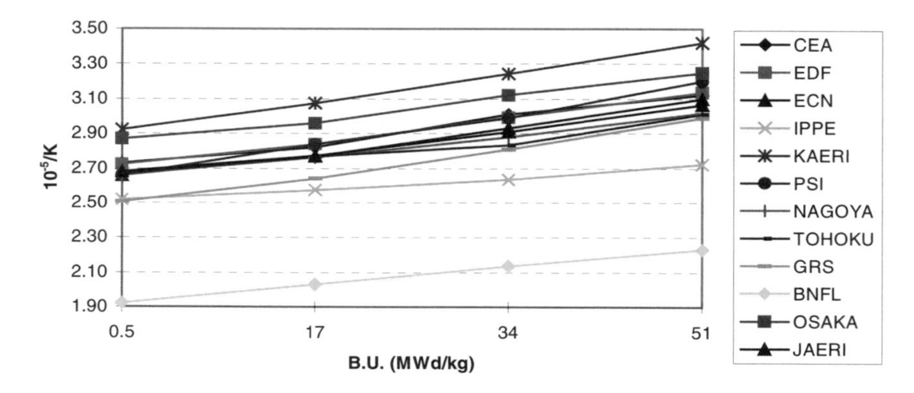

Figure 4.36. STD PWR – Stage II – Fifth recycle: Fuel temperature coefficient

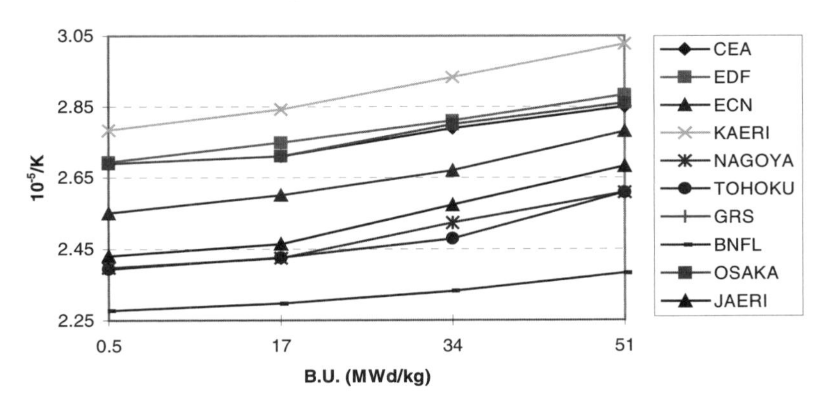

Compared with the Stage I Benchmark, similar trends are obtained, but with the spread of results increased, due to the larger initial differences combined with the plutonium content, which is also different for each participant. The BNFL results are discrepant, probably due to an inconsistent definition of the fuel temperature coefficient.

Figures 4.37 and 4.38 show the global void effect. This parameter is negative in the first recycle generation, but becomes more positive in later recycle generations such that all participants show positive values for the fifth recycle generation.

Figure 4.37. STD PWR – Stage II – First recycle: Global void effect with no leakage

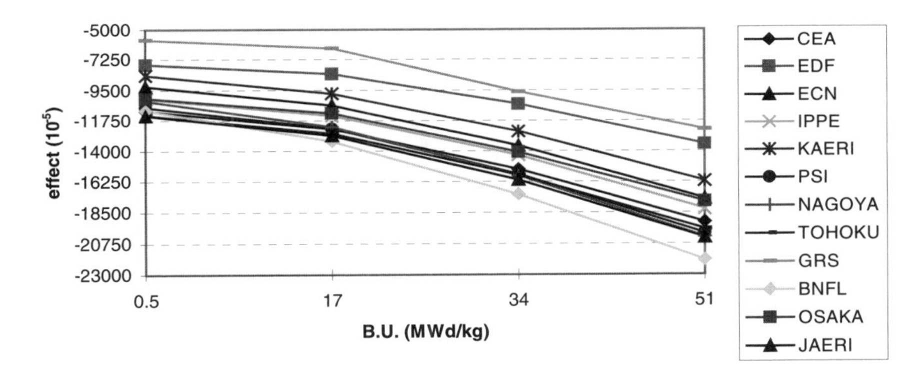

Figure 4.38. STD PWR – Stage II – Fifth recycle: Global void effect with no leakage

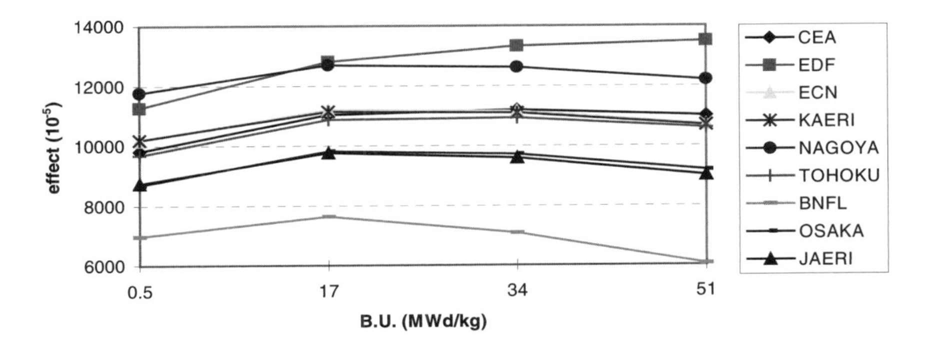

Mass balance

Figures 4.39 and 4.40 show the plutonium mass balance by participant. This is the net **reduction** in plutonium mass during a complete cycle of irradiation of the fuel from initial loading to recycle after pond storage and reprocessing. It therefore accounts for the decay of ^{241}Pu in the period between discharge from the reactor and the subsequent core loading.

Figure 4.39. STD PWR – Stage II – First recycle: Plutonium mass balance (complete cycle)

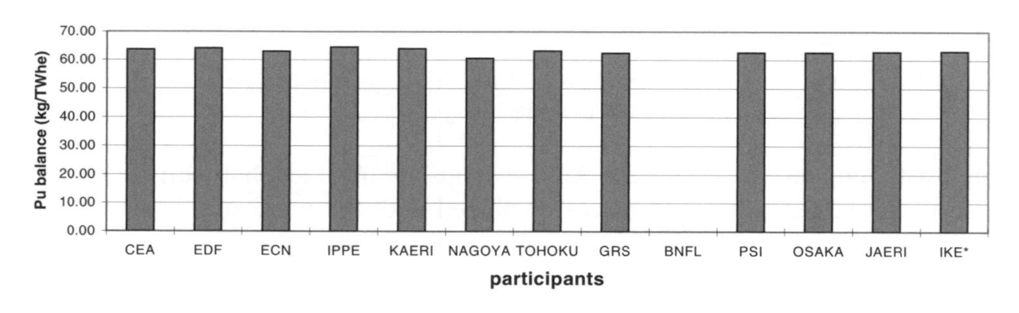

Figure 4.40. STD PWR – Stage II – Fifth recycle: Plutonium mass balance (complete cycle)

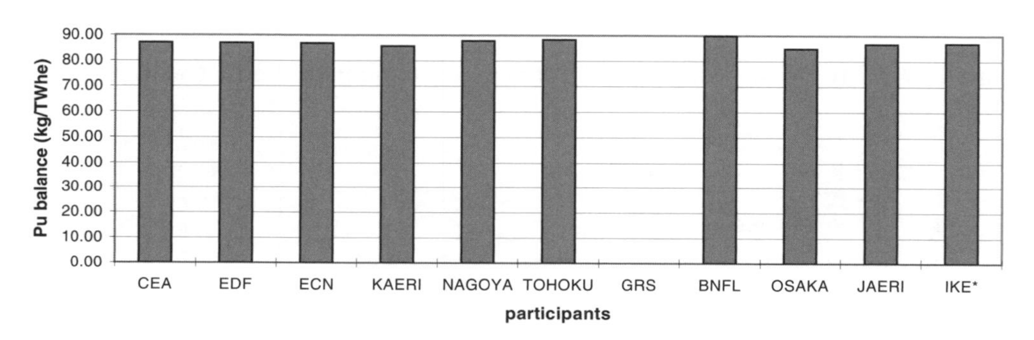

The plutonium mass balance is in good agreement for all participants and is not very sensitive to the initial plutonium content in each recycle generation. The same applies to the americium mass balance, shown in Figures 4.41 and 4.42. For curium, shown in Figures 4.43 and 4.44, the spread is larger in relative terms, though still small in terms of absolute values.

Figure 4.41. STD PWR – Stage II – First recycle: Americium mass balance (complete cycle)

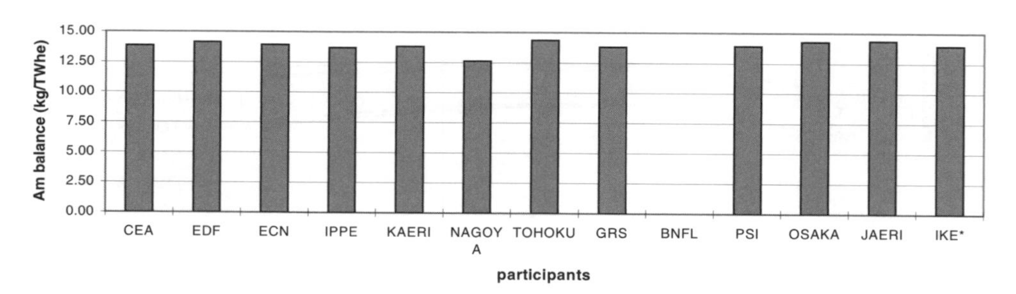

Figure 4.42. STD PWR – Stage II – Fifth recycle: Americium mass balance (complete cycle)

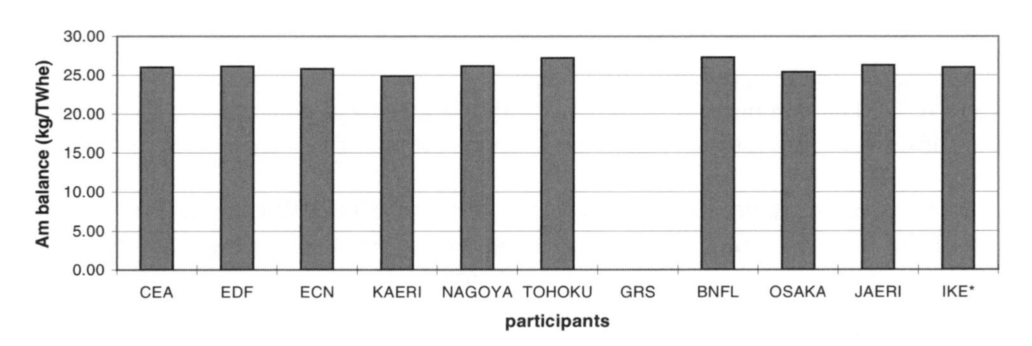

Figure 4.43. STD PWR – Stage II – First recycle: Curium mass balance (complete cycle)

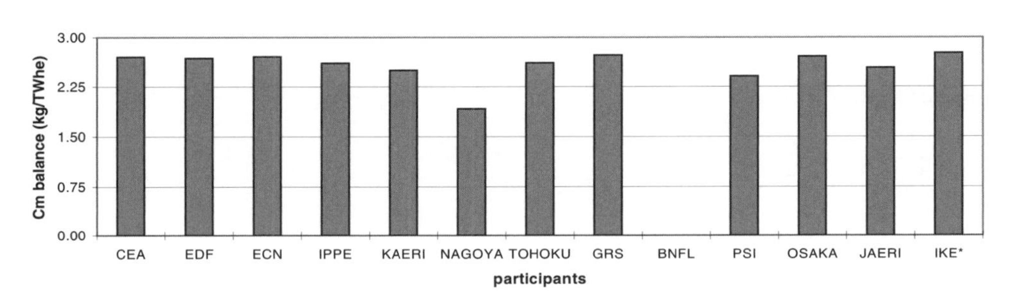

Figure 4.44. STD PWR – Stage II – Fifth recycle: Curium mass balance (complete cycle)

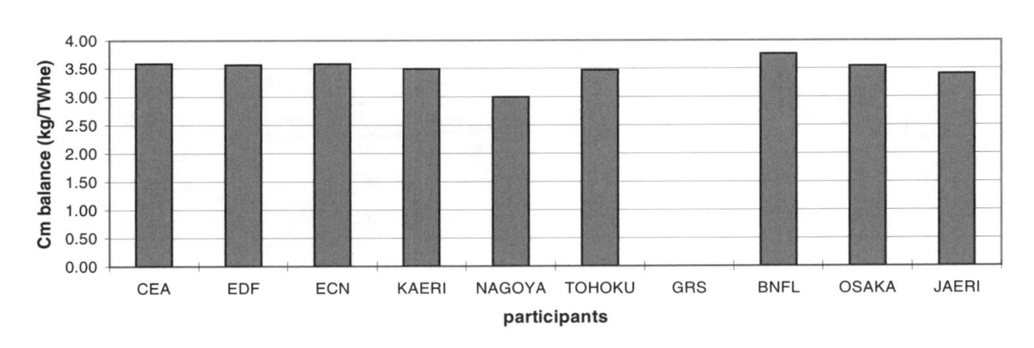

For the long-lived fission products, the production terms are very consistent between all the participants.

Highly moderated PWR – Stage I Benchmark

This section presents the results for the Highly Moderated (HM) PWR, with a moderation ratio of 3.5 (Vm/Vf) or 7.45 ($n_H/n_{H.N.}$), as described in Chapter 2. Only selected results are presented, usually for just the first and fifth recycle generation cases.

Microscopic cross-section spreads: ^{238}U, ^{239}Pu, ^{240}Pu, ^{241}Pu, ^{242}Pu

On the microscopic fission and absorption cross-sections

The comparison of microscopic cross-sections for fission and absorption is shown in Figures 4.45 to 4.51. The comparison is shown for the first recycle generation only, as the results are similar in the subsequent generations. The cross-sections are plotted as a function of cell irradiation.

The spread of values tend to be larger than for the STD PWR case. For the microscopic cross-section of the main isotopes, ^{239}Pu and ^{238}U, spreads over a total range of typically of 6% are seen. For ^{238}U, there are two groups of trends with irradiation, one group which show a tendency for the cross-section to decrease with irradiation and a second one (EDF, TOHOKU, OSAKA) which shows an increasing trend with irradiation compared with CEA.

For higher isotopes, similar spreads are seen:

- from -1% to +4% for ^{240}Pu;
- from ±4% for ^{241}Pu;
- from ±4% for ^{242}Pu.

Figure 4.45. HM PWR – Stage I – First recycle:
^{239}Pu fission cross-section – Comparison with CEA values

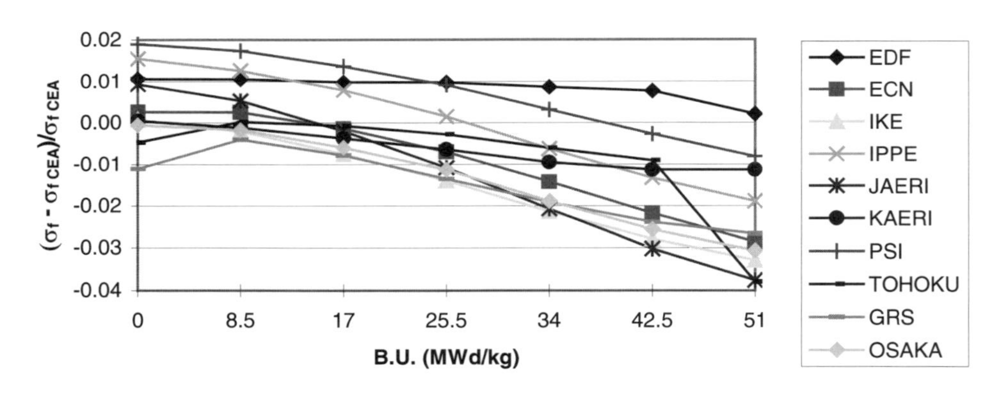

CEA values							
B.U (MWd/kg)	0	8.5	17	25.5	34	42.5	51
σ_f ^{239}Pu (barn)	53.53	57.10	63.48	71.11	80.01	90.02	100.79

Figure 4.46. HM PWR – Stage I – First recycle:
^{241}Pu fission cross-section – Comparison with CEA values

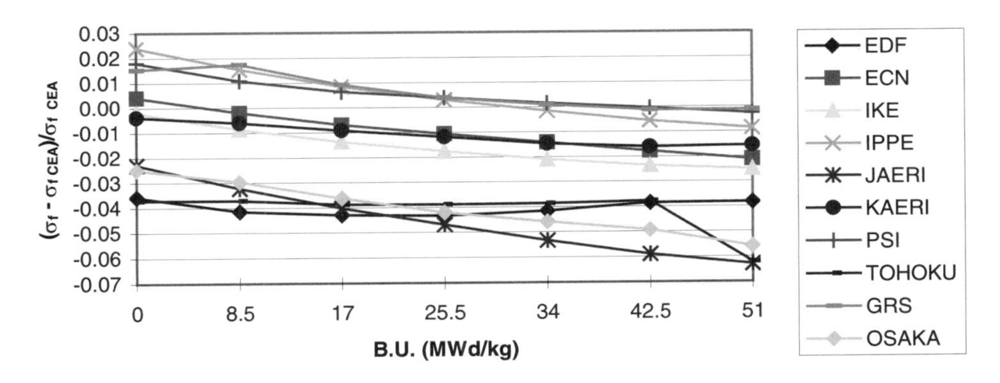

CEA values							
B.U (MWd/kg)	0	8.5	17	25.5	34	42.5	51
σ_f ^{241}Pu (barn)	64.51	68.02	74.71	82.58	91.72	102.04	113.34

Figure 4.47. HM PWR – Stage I – First recycle:
^{238}U capture cross-section – Comparison with CEA values

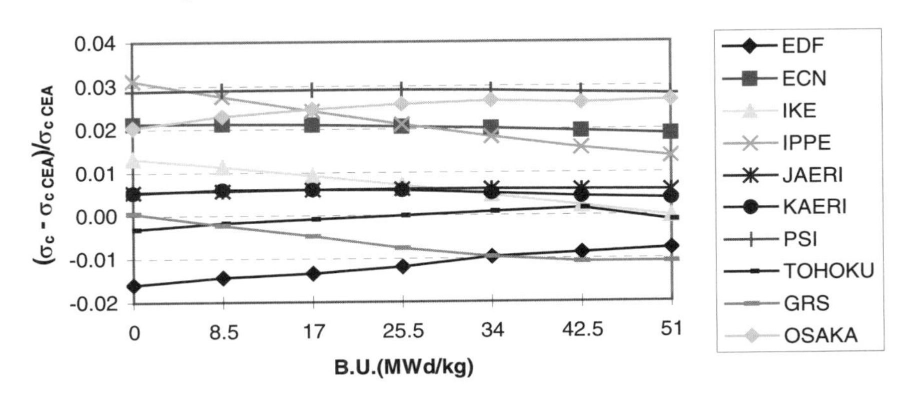

CEA values							
B.U (MWd/kg)	0	8.5	17	25.5	34	42.5	51
σ_c ^{238}U (barn)	0.91	0.91	0.92	0.93	0.94	0.95	0.96

Figure 4.48. HM PWR – Stage I – First recycle:
^{239}Pu capture cross-section – Comparison with CEA value

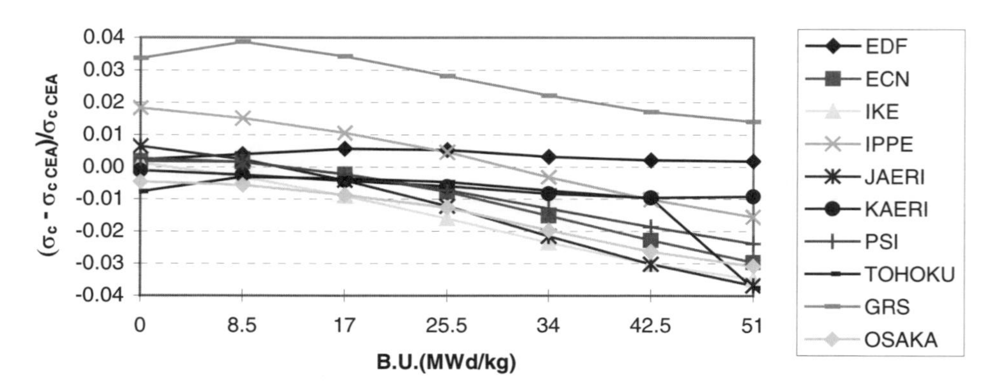

CEA values							
B.U (MWd/kg)	0	8.5	17	25.5	34	42.5	51
σ_c ^{239}Pu (barn)	29.03	31.18	34.70	38.89	43.76	49.19	55.00

Figure 4.49. HM PWR – Stage I – First recycle:
^{240}Pu capture cross-section – Comparison with CEA value

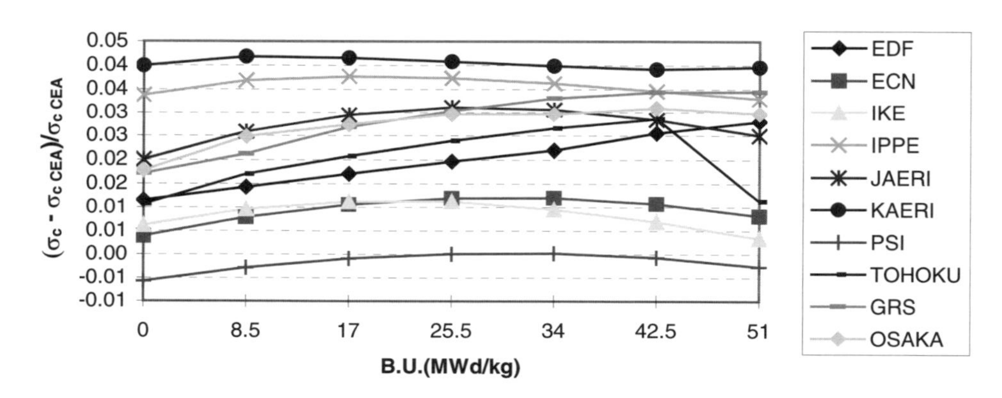

CEA values							
B.U (MWd/kg)	0	8.5	17	25.5	34	42.5	51
σ_c ^{240}Pu (barn)	47.95	47.42	48.18	49.72	52.06	55.19	59.14

Figure 4.50. HM PWR – Stage I – First recycle:
^{241}Pu capture cross-section – Comparison with CEA value

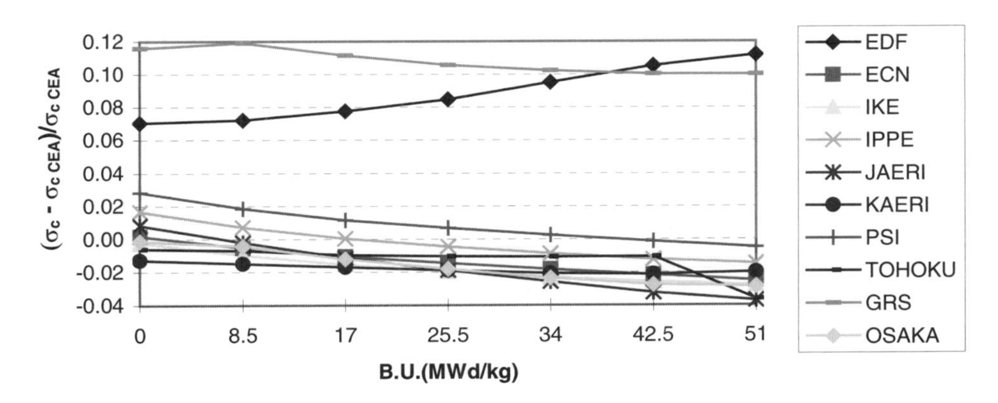

CEA values							
B.U (MWd/kg)	0	8.5	17	25.5	34	42.5	51
σ_c ^{241}Pu (barn)	20.93	22.20	24.50	27.20	30.32	33.84	37.68

Figure 4.51. HM PWR – Stage I – First recycle:
^{242}Pu capture cross-section – Comparison with CEA value

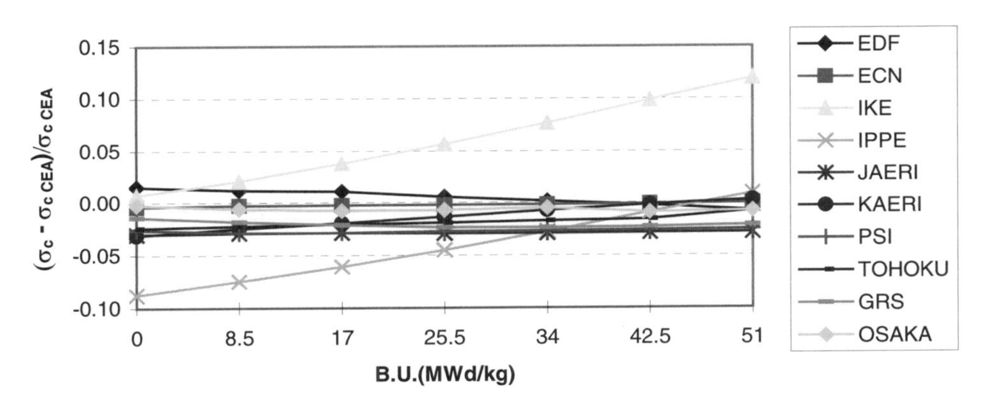

CEA values							
B.U (MWd/kg)	0	8.5	17	25.5	34	42.5	51
σ_c ^{242}Pu (barn)	16.94	16.60	16.23	15.82	15.38	14.94	14.52

On the microscopic (n,2n) cross-sections

Although the higher order reactions are much less important in the overall neutron balance, they nevertheless make a significant contribution. Only ^{238}U and ^{244}Cm are shown in Figures 4.52 and 4.53:

Figure 4.52. HM PWR – Stage I – First recycle:
^{238}U (n,2n) cross-section – Comparison with CEA values

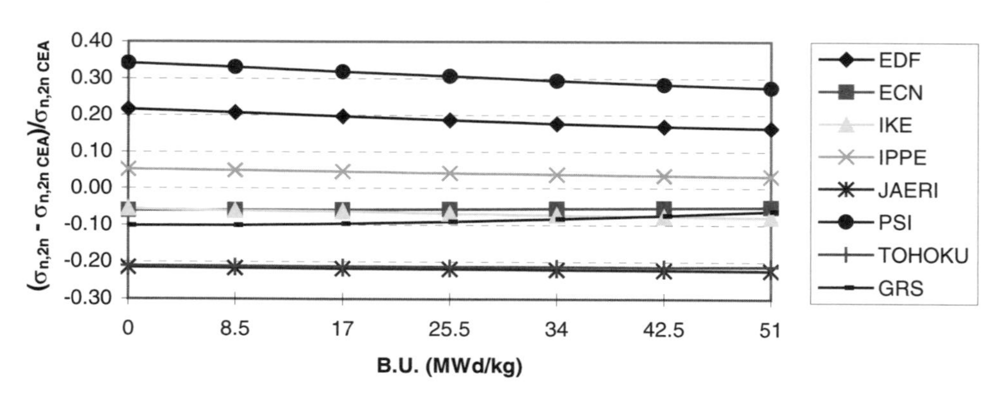

CEA values							
B.U (MWd/kg)	0	8.5	17	25.5	34	42.5	51
$\sigma_{n,2n}$ ^{238}U (barn)	6.20E-03	6.20E-03	6.17E-03	6.13E-03	6.08E-03	6.01E-03	5.94E-03

Figure 4.53. HM PWR – Stage I – First recycle:
^{244}Cm (n,2n) cross-section – Comparison with CEA values

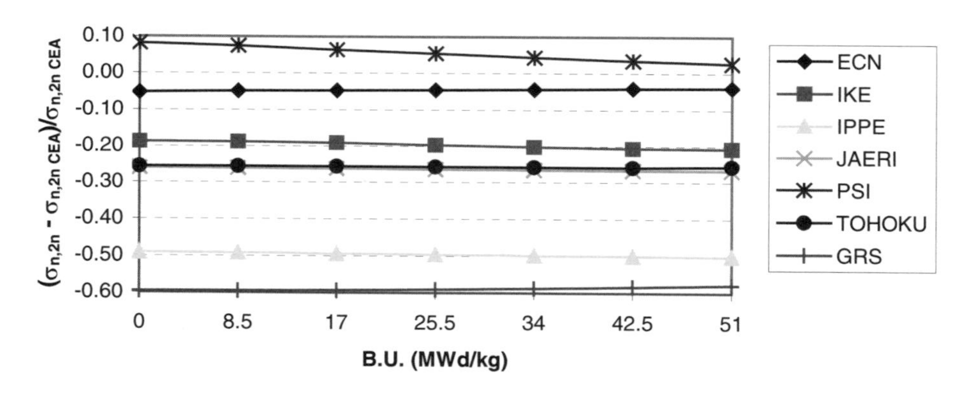

CEA values							
B.U (MWd/kg)	0	8.5	17	25.5	34	42.5	51
$\sigma_{n,2n}$ ^{244}Cm (barn)	2.37E-03	2.37E-03	2.36E-03	2.34E-03	2.33E-03	2.30E-03	2.27E-03

As for the STD PWR, the spread of results is quite large, with the same general trends and similar ranges. The isotopes which show the largest range of values are ^{243}Cm, ^{245}Cm (up to a factor 3) and ^{239}Pu. The other isotopes typically fall within a range of 50%.

On the ν values

Figure 4.54 shows a comparison of ^{239}Pu ν values, the average number of neutrons per fission. These are in reasonably good agreement.

Figure 4.54. HM PWR – Stage I – First recycle: ^{239}Pu ν values

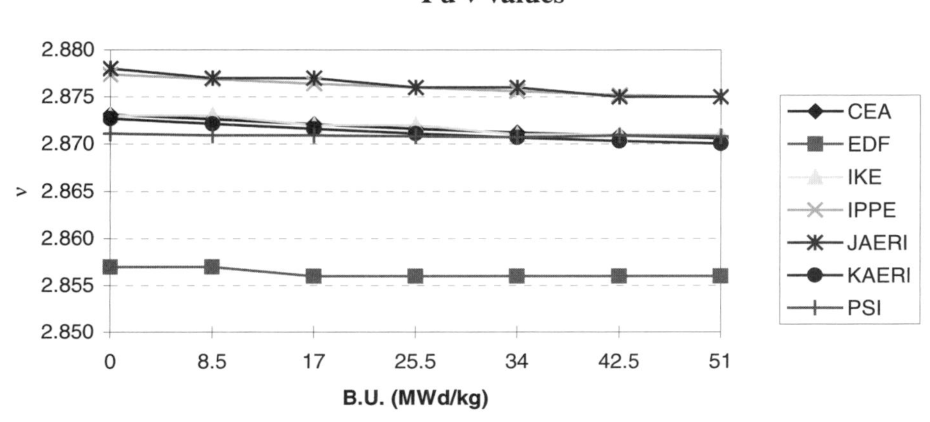

On fission capture cross-sections ratio

The fission/capture ratios are shown for ^{239}Pu in Figure 4.55. Disregarding the clearly discrepant GRS values, all of the participants show positive differentials from 0 to 1% compared with CEA. It implies that the main isotope contributing to reactivity is relatively more fissile and thus the majority of participants should obtain higher reactivities than CEA.

For ^{241}Pu (not plotted), the differences are larger, especially for EDF and GRS (-10%). For the other participants, the differences are in the range +1% to -4%.

Figure 4.55. HM PWR – Stage I – First recycle: ^{239}Pu ratio fission/capture – Comparison with CEA values

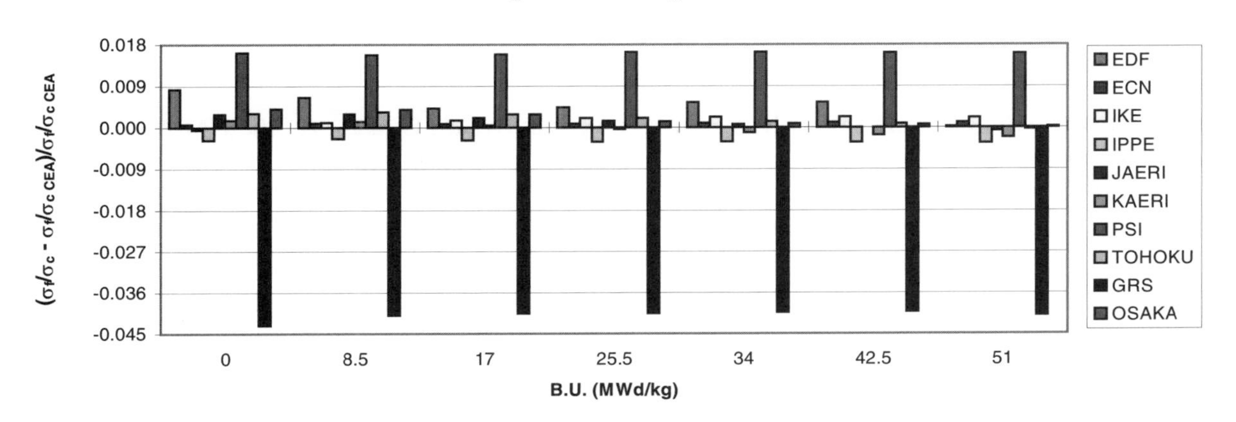

CEA values							
B.U (MWd/kg)	0	8.5	17	25.5	34	42.5	51
σ_f/σ_c ^{239}Pu	1.844	1.831	1.829	1.828	1.828	1.830	1.833

Spread in reactivities

Figures 4.56 and 4.57 show comparisons of reactivities versus cell irradiation, for the first and fifth recycle generations respectively, plotted relative to the CEA values as a reference.

Figure 4.56. HM PWR – Stage I – First recycle: Reactivity swing – Comparison with CEA values

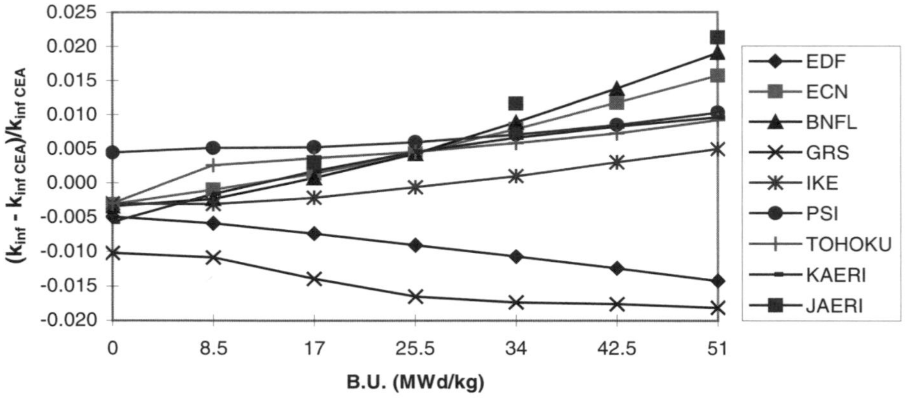

Figure 4.57. HM PWR – Stage I – Fifth recycle: Reactivity swing – Comparison with CEA values

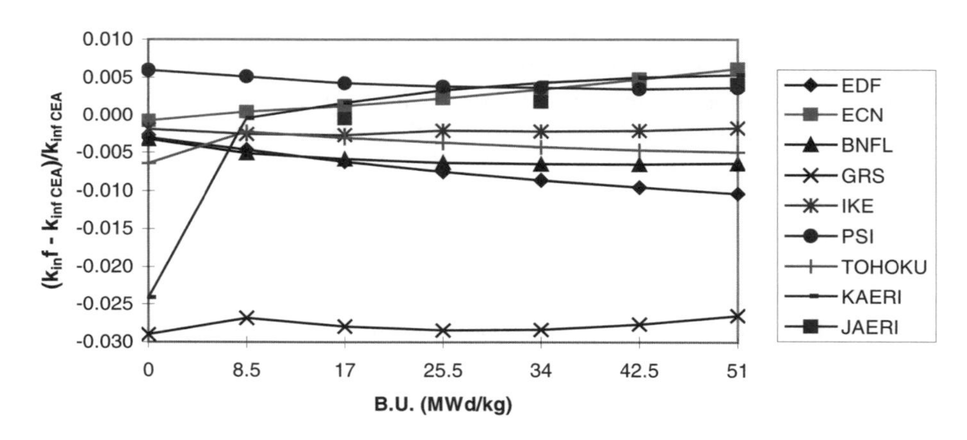

Compared with the STD PWR, the spread of reactivity values is larger for the HM PWR. This is especially evident in the first generation recycle, where there are also significantly discrepant trends with cell irradiation, resulting in a larger spread at high irradiations. There is a tendency for the range of values to converge in subsequent recycle generations.

Reactivity balance under irradiation

The reactivity balance under irradiation is 35% in the first recycle generation and 18% in the fifth generation. These very contrasting behaviours will have an impact on the initial plutonium content and on the reactivity adjustment values. Figure 4.58 shows the comparison of end-of-cycle k-infinities, with the CEA value as the reference. The same data are also listed in Table 4.3, with the minimum and maximum values highlighted in Table 4.4.

Figure 4.58. HM PWR – Stage I:
End-of-cycle reactivity – Comparison with CEA value

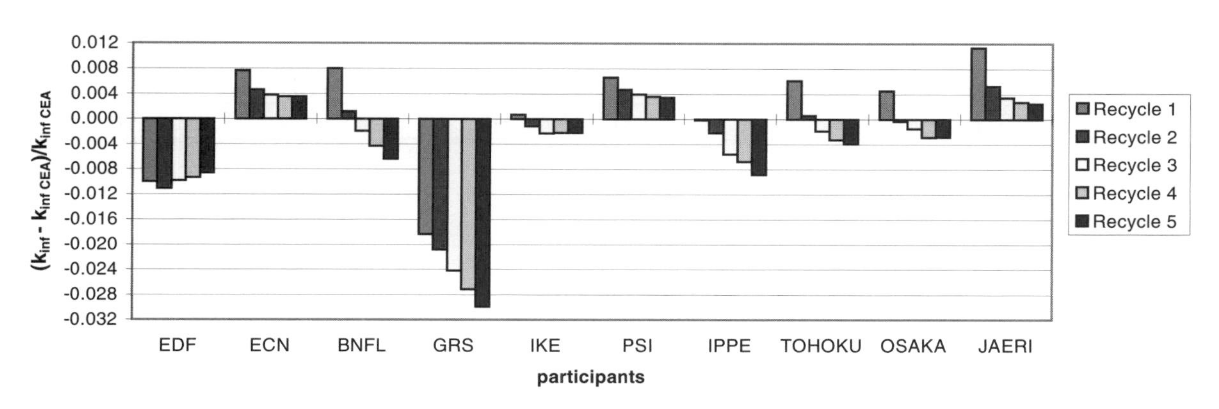

The differences between participants are larger than for the STD PWR due to a larger difference in reactivity values at zero irradiation and the subsequent trend during irradiation.

Table 4.3. End-of-cycle k-infinities by participant and recycle generation

	Recycle 1	Recycle 2	Recycle 3	Recycle 4	Recycle 5
CEA	1.03727	1.03720	1.03732	1.03726	1.03727
EDF	1.02688	1.02573	1.02712	1.02757	1.02835
ECN	1.04522	1.04203	1.04127	1.04095	1.04098
BNFL	1.04560	1.03847	1.03535	1.03280	1.03067
GRS	1.01825	1.01561	1.01224	1.00915	1.00627
IKE	1.03800	1.03600	1.03500	1.03500	1.03500
PSI	1.04411	1.04212	1.04144	1.04101	1.04088
IPPE	1.03713	1.03497	1.03152	1.03027	1.02808
TOHOKU	1.04364	1.03782	1.03537	1.03388	1.03324
OSAKA	1.04198	1.03690	1.03578	1.03431	1.03434
JAERI	1.04904	1.04269	1.04093	1.04014	1.03989

The largest change through the five recycle generations is BNFL (1.5%). The differences between largest and smallest values for each participant are given in the following table.

Table 4.4. End-of-cycle k-infinities – minimum and maximum values by participant

	CEA	EDF	ECN	BNFL	IKE	PSI	IPPE	TOHOKU	OSAKA	JAERI
Min	1.0372	1.02573	1.04095	1.030666	1.035	1.04088	1.02808	1.03324	1.03431	1.039889
Max	1.03732	1.02835	1.04522	1.045604	1.038	1.04411	1.03713	1.04364	1.04198	1.049041
Diff	-0.00012	-0.00262	-0.00427	-0.014938	-0.003	-0.00323	-0.00905	-0.0104	-0.00767	-0.009152

The range covered by the maximum and minimum values, summarised in Table 4.5, tends to decrease during in later recycle generations from 2.2% down to 1.3%. This is opposite to the trend seen with the STD PWR and moreover the spread of values is larger for the HM PWR.

Table 4.5. End-of-cycle k-infinities – extreme values

	Recycle 1	Recycle 2	Recycle 3	Recycle 4	Recycle 5
Min.	1.02688	1.02573	1.02712	1.02757	1.02808
Max.	1.04904	1.04269	1.04144	1.04101	1.04098
Largest difference	0.02216	0.01696	0.01432	0.01344	0.01290

Fission rate discrepancies

Figure 4.59 shows the differences in fission rates relative to CEA as the reference.

Figure 4.59. HM PWR – Stage I:
Differences in fission rates compared with CEA values (in %)

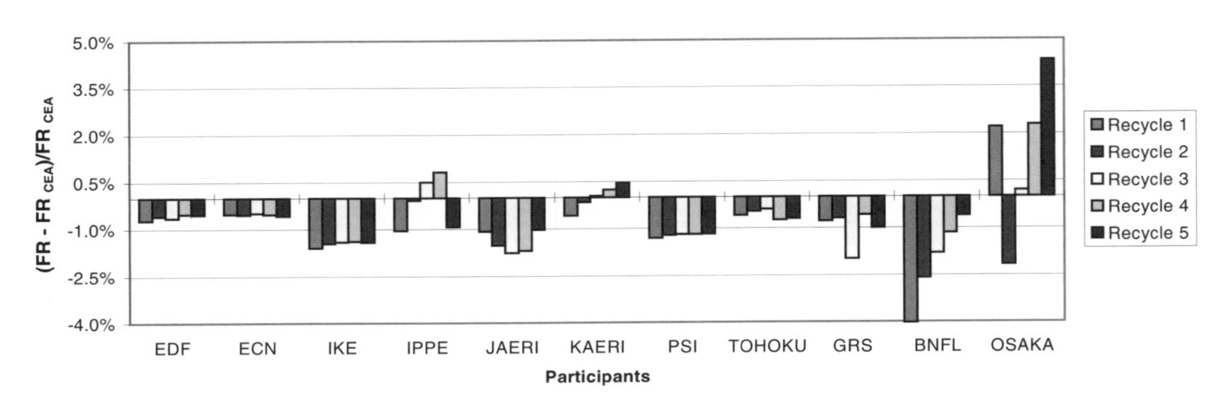

CEA values					
	Recycle 1	Recycle 2	Recycle 3	Recycle 4	Recycle 5
Fission rate (fraction)	0.05216	0.05209	0.05209	0.05209	0.05209

The range of values is larger for the HM PWR than for the STD PWR. 1% in relative difference on the fission rate is equivalent to 0.3% in reactivity in the first generation recycle and to 0.2% for in the fifth generation. The majority of participants are below 2%, thus under 0.6% reactivity difference on reactivity balance under irradiation. Thus the majority of participants obtain smaller fission rates than CEA, which would tend to cause the reactivity balance to be smaller than CEA for the majority of the participants. This actually is not the case, the majority of participants showing a positive increase of reactivity under irradiation.

Spread in reaction rates

Only five participants provided data on the breakdown of reactions rates. These are shown in Figure 4.60. The number is insufficient to reach any definite conclusions, but usefully illustrates the relative contribution of the various reactions and shows order of magnitude of the spread of results. The illustration is for the end of life condition (51 MWd/kgHM) for the first (C1) and fifth (C5) recycle generations. The spread is larger than for the STD PWR.

Absorption and production increase between the first and fifth recycle generations, due to the increasing plutonium content. Also, the reaction rates are increased by the higher moderation ratio and the more thermalised neutron spectrum, compared with the STD PWR.

Figure 4.60. HM PWR – Stage 1 – Reaction rates

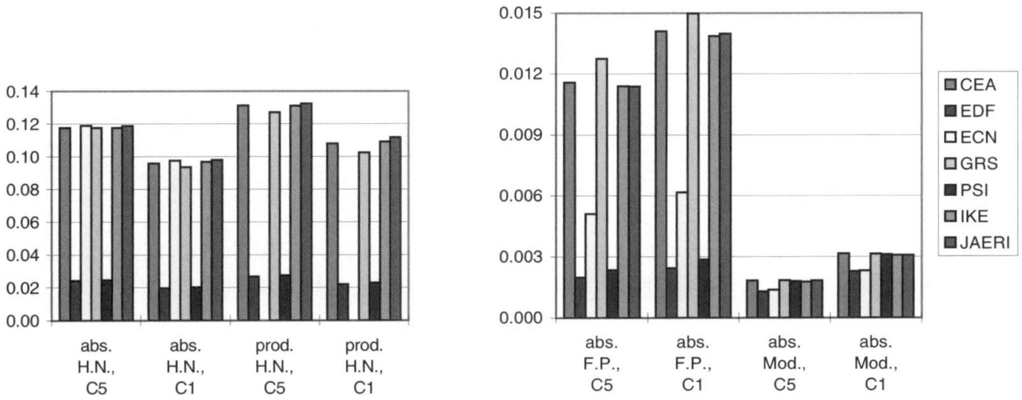

Reactivity coefficients

Figures 4.61 and 4.62 show the boron efficiencies versus cell irradiation for the first and fifth recycle generations. The boron efficiency is in moderately good agreement between the various participants, with acceptable agreement on trends with cell irradiation. In the fifth generation recycle the boron efficiency is reduced compared with the first generation, but is still a factor 2 higher than for the STD PWR. Thus the increase in moderation ratio improves strongly this important control parameter, partly due to the larger mass of boron at a given concentration and partly due to the softer neutron spectrum.

Figure 4.61. HM PWR – Stage I – First recycle: Boron efficiency

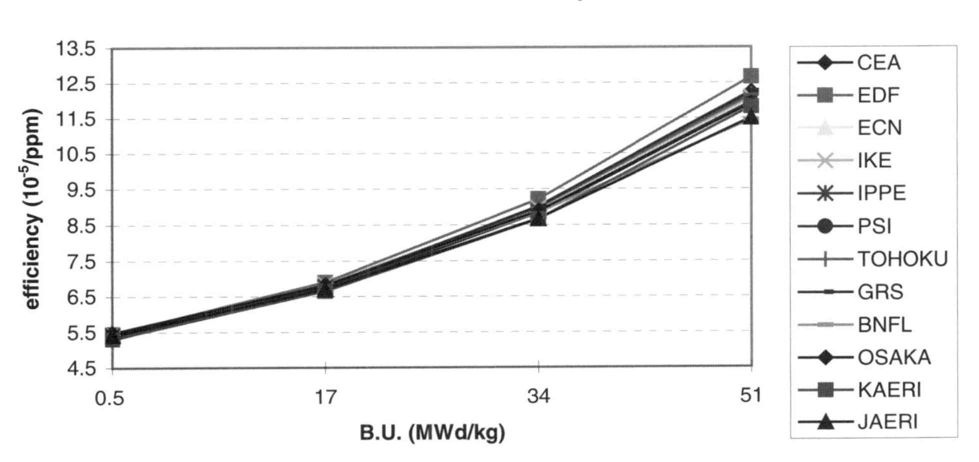

Figure 4.62. HM PWR – Stage I – Fifth recycle: Boron efficiency

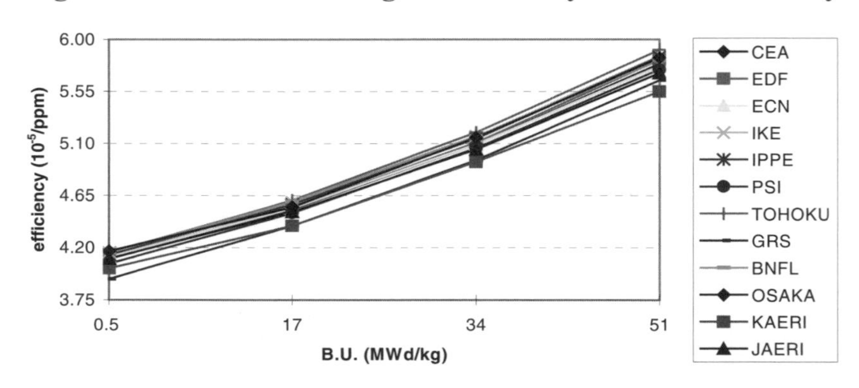

For the first recycle generation at the end of irradiation, the differences are about 10^{-5}/ppm of soluble boron. This corresponds to 2% absolute reactivity difference for 2 000 ppm of soluble boron. This must arise from a combination of discrepancies in the depletion and flux calculations and the nuclear data. The plutonium content is very low and the neutron spectrum is thermal. Nevertheless, the boron efficiency is very high and it will not cause a large impact in terms of safety, only in terms of boron content for the core concept calculations. For the fifth recycle generation, with a harder neutron spectrum, the differences are very small and acceptable.

Figures 4.63 and 4.64 show the moderator temperature coefficients for the first and fifth recycle generations.

Figure 4.63. HM PWR – Stage I – First recycle: Moderator temperature coefficient

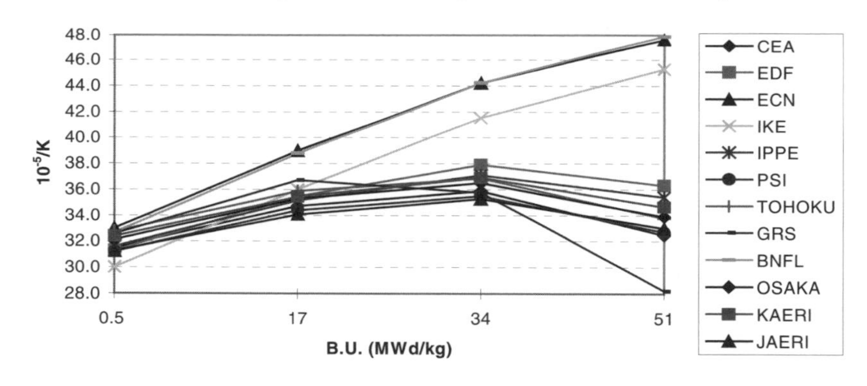

Figure 4.64. HM PWR – Stage I – Fifth recycle: Moderator temperature coefficient

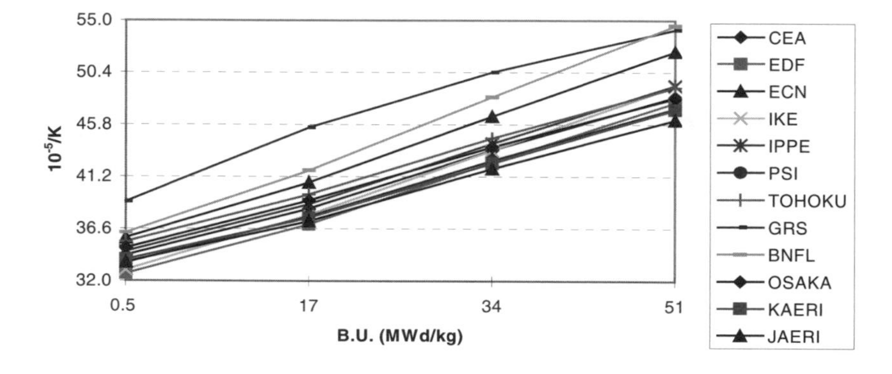

In the first generation recycle, discrepancies are evident in the trend of the moderator temperature coefficient with cell irradiation between ECN, IKE, BNFL and the other participants. This tendency reduces in later recycle generations, disappearing completely by the fourth and fifth generations, although there still remains a large spread in the absolute values three sets of solutions as clear outliers. In terms of absolute values, they are spread over a smaller range than in the STD PWR. The highest absolute values are in the region of -50×10^{-5} /°C. This parameter has an important impact on control and kinetic behaviour for cooling transients; if it is too negative, a faster and a larger increase in reactivity during the transient is implied.

Figures 4.65 and 4.66 show the fuel temperature (Doppler) coefficient for the first and fifth recycle generations. The fuel temperature coefficients are spread over a larger range of values and behavioural trends with cell irradiation than the STD PWR case. Generally, the absolute values are smaller in magnitude, due to the more thermalised spectrum and the increased resonance escape probability. As for the STD PWR, the treatments of temperature dependence of the resonance absorptions and self-shielding differ between the participants.

Figure 4.65. HM PWR – Stage I – First recycle
Fuel temperature coefficient

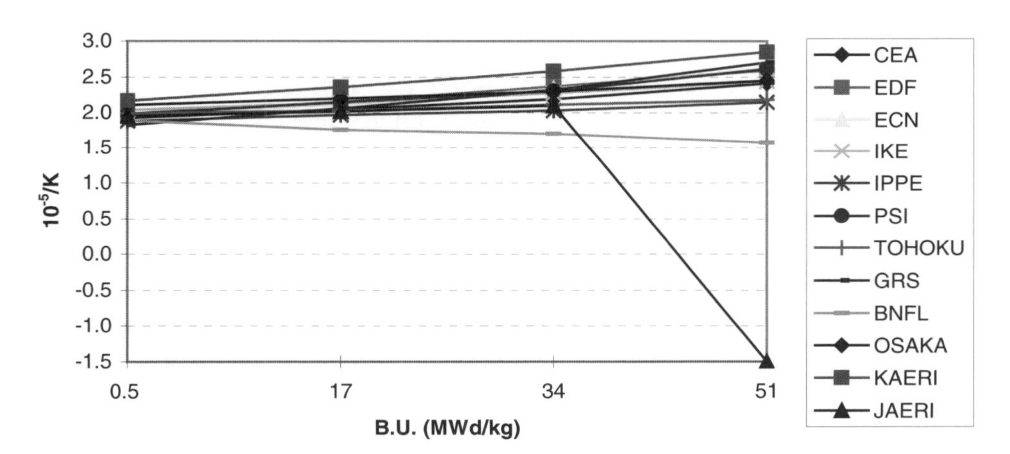

Figure 4.66. HM PWR – Stage I – Fifth recycle:
Fuel temperature coefficient

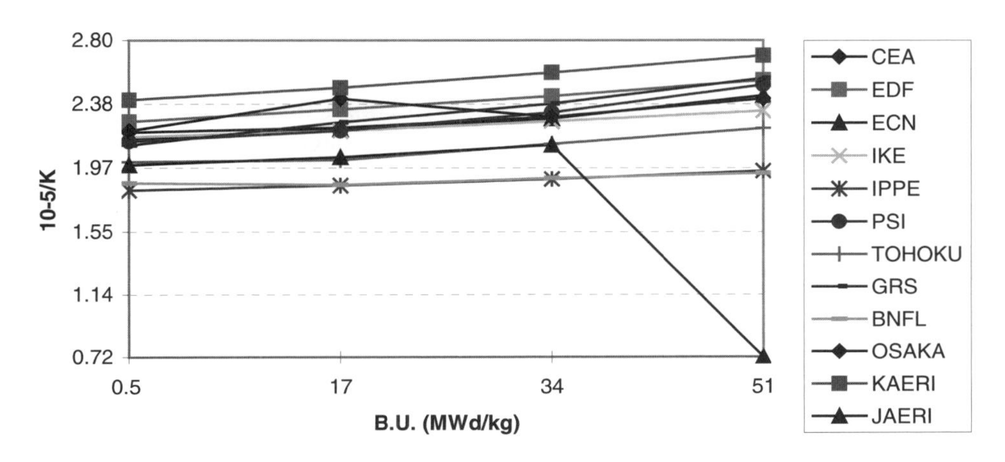

Figures 4.67 and 4.68 show the void effects without leakage. The void effect is the change in reactivity upon complete voiding of the moderator and should be negative to prevent rapid reactivity insertion following void formation in the moderator.

Figure 4.67. HM PWR – Stage I – First recycle: Global void effect with no leakage

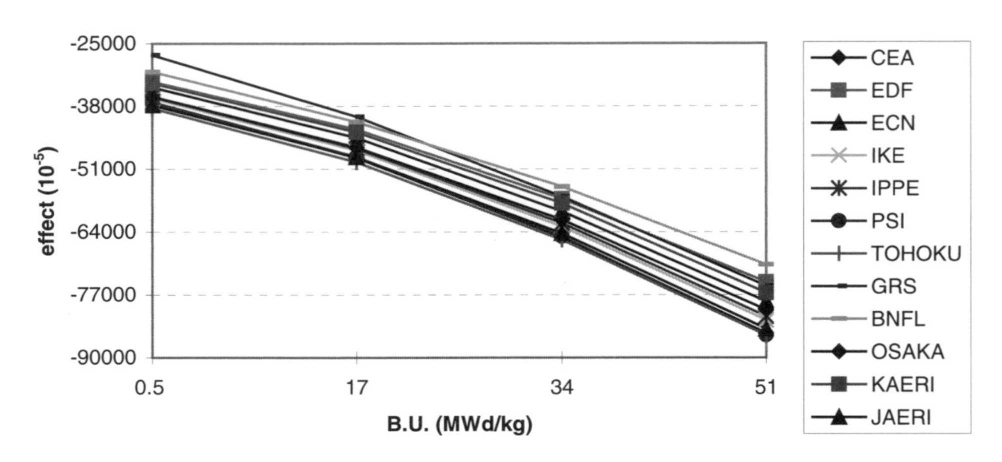

Figure 4.68. HM PWR – Stage I – Fifth recycle: Global void effect with no leakage

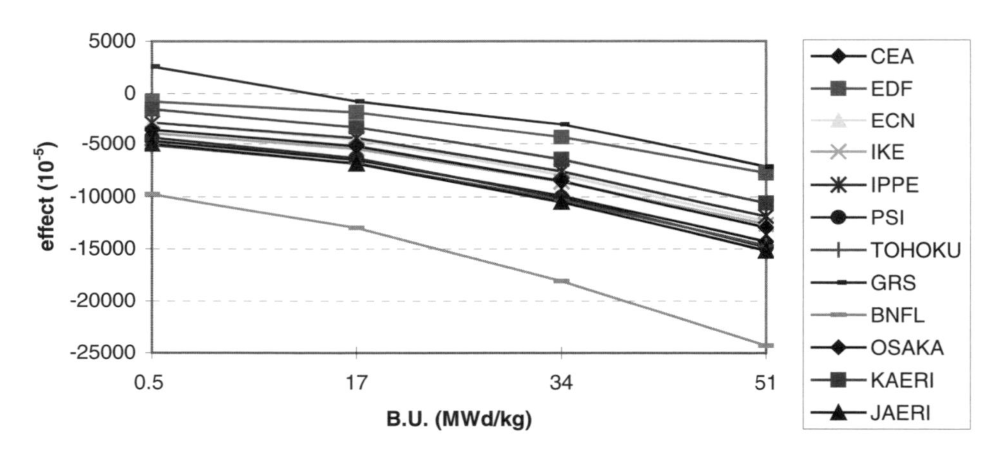

The global void effect falls generally within a range of 10% in reactivity. The trend with cell irradiation is very similar for the various participants, although the BNFL results form an outlier in the later recycle generations. The values remain negative throughout all the recycle generation for almost all the participants. This is a positive advantage of the HM PWR compared to the STD PWR.

Highly moderated PWR – Stage II Benchmark

Since IPPE and PSI did not follow the precise specification with respect to the dilution plutonium from MOX with plutonium from UO_2, their results are not comparable to the others and have been omitted from this section.

Initial plutonium content

Figure 4.69 show the initial plutonium calculated by the various participants relative to CEA as the reference.

Figure 4.69. HM PWR – Stage II: Pu content – Comparison with CEA values

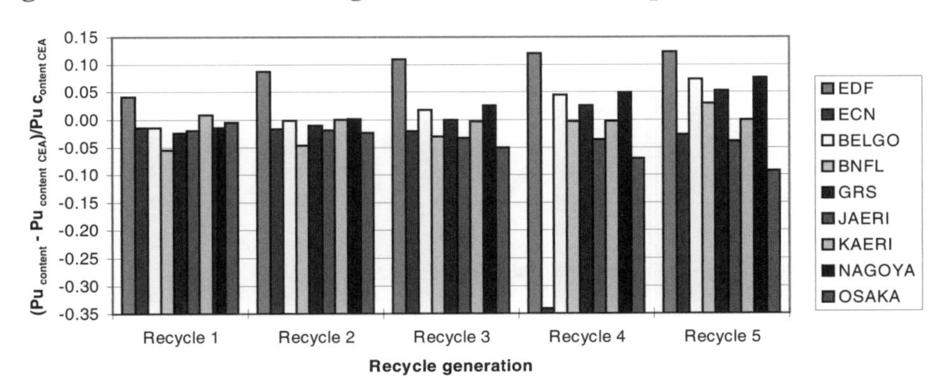

CEA values					
	Recycle 1	Recycle 2	Recycle 3	Recycle 4	Recycle 5
Pu content	6.70%	9.64%	12.11%	14.37%	16.58%

The initial plutonium contents are spread over a range up to ±10% in relative terms and ±1.5% in absolute plutonium content. The differences grow in later recycle generations, as a cumulative effect exists from one recycle generation to the next. The differences represent about one metric tonne for a 1 300 MWe PWR (about 80 metric tonnes of heavy metal), as for the standard PWR.

Adjustment reactivity value

This parameter is a constant for all the cases of plutonium recycling, by definition.

Reactivity coefficients

Figures 4.70 and 4.71 show the boron efficiencies obtained in Stage II as a function of cell irradiation.

Figure 4.70. HM PWR – Stage II – First recycle: Boron efficiency

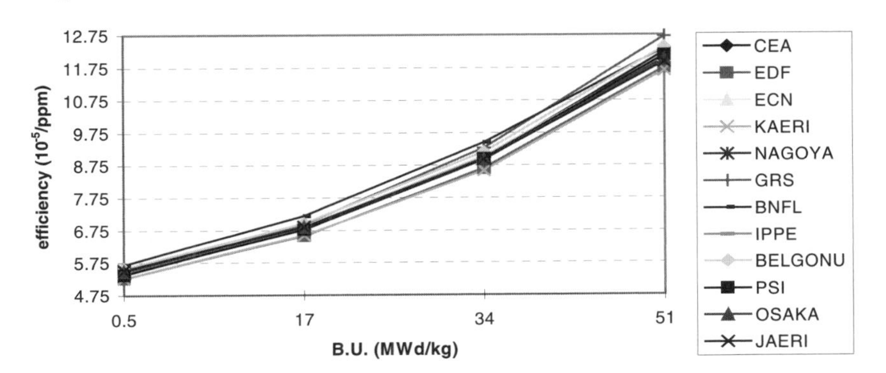

Figure 4.71. HM PWR – Stage II – Fifth recycle: Boron efficiency

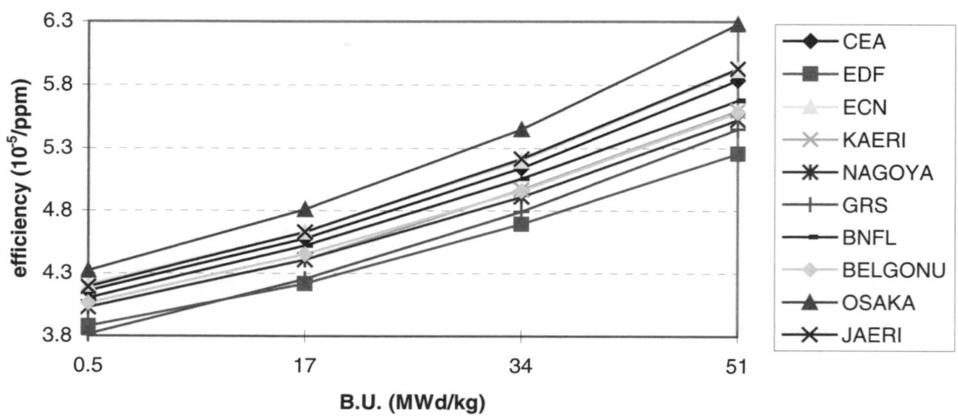

The various participants are in good agreement as to the trend of the boron coefficient with cell irradiation and through the various recycle generations. However, the spread is larger in absolute terms than for the STD PWR, largely because the magnitude of the coefficient is higher. At the end of the irradiation for the fifth generation recycle, the spread is about 10^{-5}/ppm of soluble boron. This difference may be considered as large, and must arise from a combination of plutonium content differences and discrepancies in the depletion calculations. This parameter is sensitive to the dilution factor and initial plutonium content.

Figures 4.72 and 4.73 show the moderator temperature coefficients for the first and fifth recycle generations, as a function of cell irradiation.

Figure 4.72. HM PWR – Stage II – First recycle: Moderator temperature coefficient

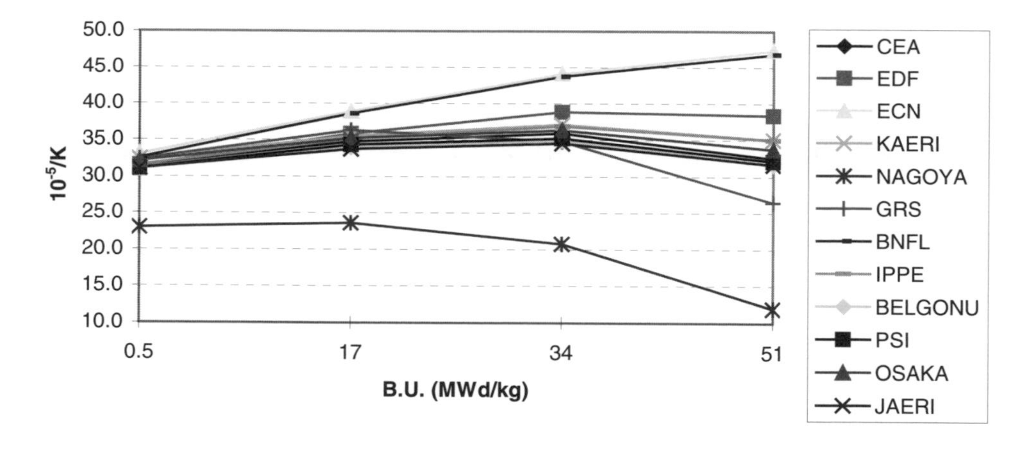

Figure 4.73. HM PWR – Stage II – Fifth recycle: Moderator temperature coefficient

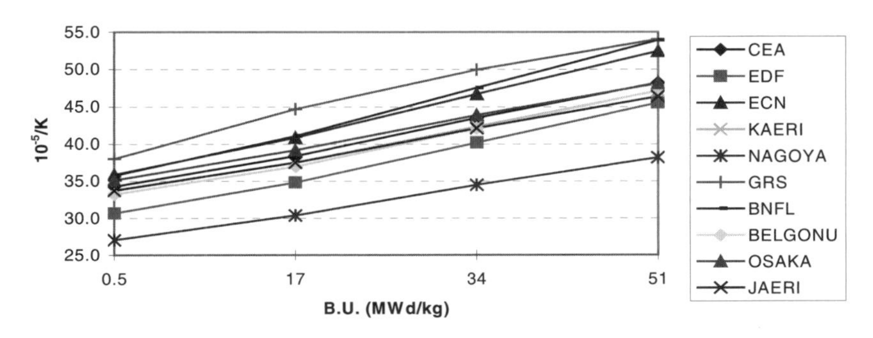

The moderator temperature coefficient has a fairly large spread of results, which tends to increase in the later recycle generations. There is reasonably good agreement regarding the general trend with cell irradiation.

The dilution effect is small for the fifth cycle. As a result, this parameter is more sensitive to the small content of plutonium than for the higher content (more than 12% of initial plutonium content).

Figures 4.74 and 4.75 show the fuel temperature (Doppler) coefficients for the first and fifth recycle generations, as a function of cell irradiation.

Figure 4.74. HM PWR – Stage II – First recycle: Fuel temperature coefficient

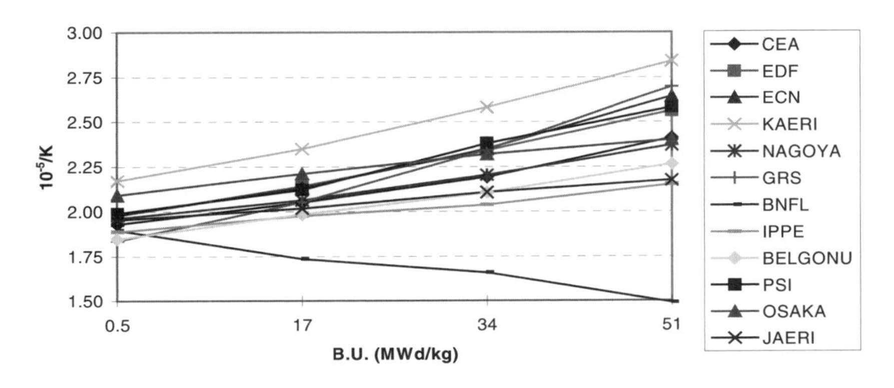

Figure 4.75. HM PWR – Stage II – Fifth recycle: Fuel temperature coefficient

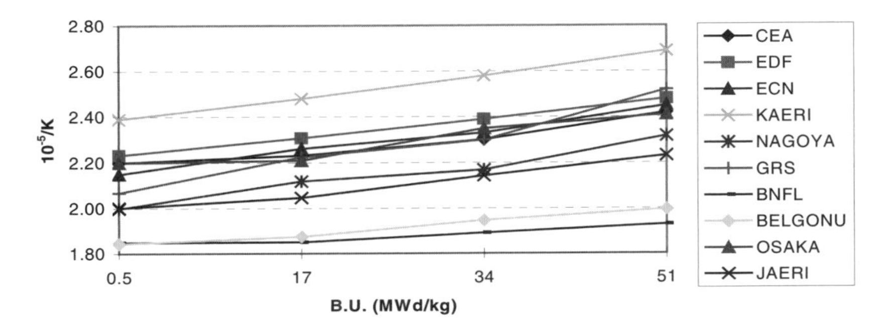

As for the Stage I Benchmark, the fuel temperature coefficient shows large discrepancies. As for the standard case, the BNFL results are clearly anomalous. Even after disregarding the BNFL results, the spread of values is unsatisfactory, approaching 25% of the absolute value of the parameter.

Figures 4.76 and 4.77 show the global void effect (with no leakage) versus cell irradiation.

Figure 4.76. HM PWR – Stage II – First recycle: Global void effect with no leakage

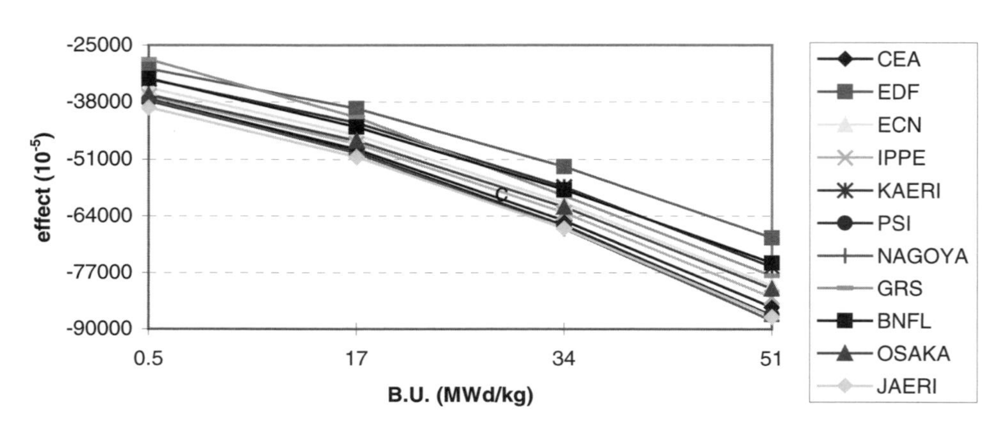

Figure 4.77. HM PWR – Stage II – Fifth recycle: Global void effect with no leakage

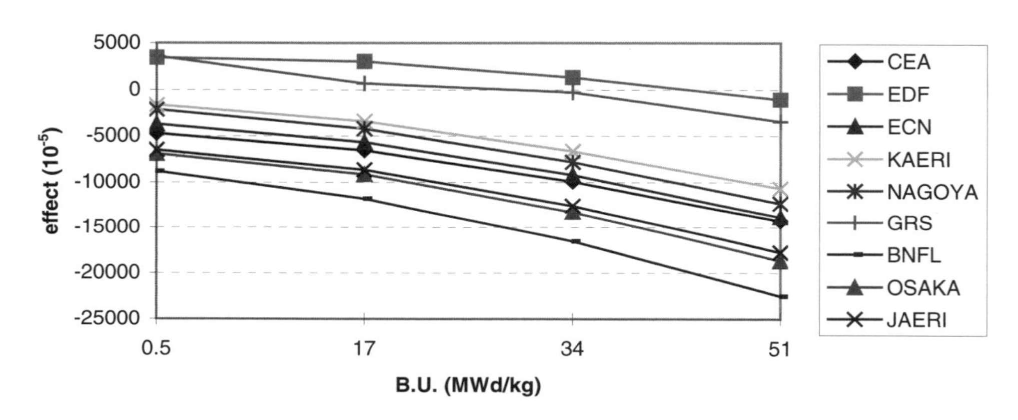

For the global void effect there is good agreement as to the trend with cell irradiation, although there remains a spread in the results which is large in absolute terms. In the fifth recycle generation, EDF, TOHOKU and GRS obtain positive values for the void effect at low irradiations, whereas all the other participants obtain small negative values. The void effect including leakage shows larger discrepancies (about 20%), which must arise from the flux and leakage calculations.

Mass balance

Figures 4.78 and 4.79 show the plutonium mass balances for the first and fifth recycle generations.

Figure 4.78. HM PWR – Stage II – First recycle:
Plutonium mass balance (complete cycle)

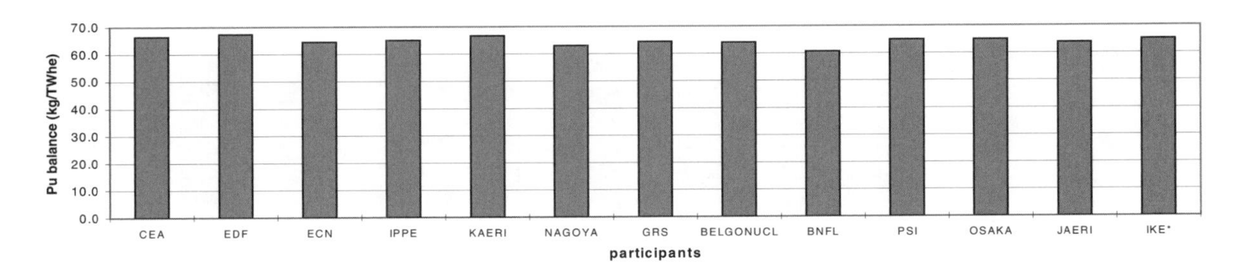

Figure 4.79. HM PWR – Stage II – Fifth recycle:
Plutonium mass balance (complete cycle)

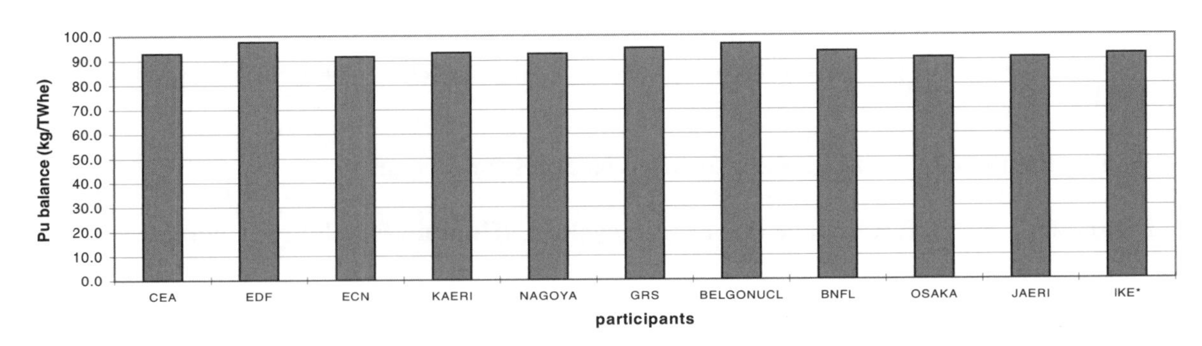

Plutonium consumption (-67 kg/TWhe) is slightly larger in the first recycle generation than for the STD PWR (-64 kg/eTWh) for an initial plutonium content which is some 30% smaller than for the STD PWR. In later recycle generations there is an increase in plutonium consumption to 90 kg/TWhe, higher than attained in the STD PWR. All the participants are in reasonably good agreement.

Figures 4.80 and 4.81 show the mass balances for americium.

Figure 4.80. HM PWR – Stage II – First recycle:
Americium mass balance (complete cycle)

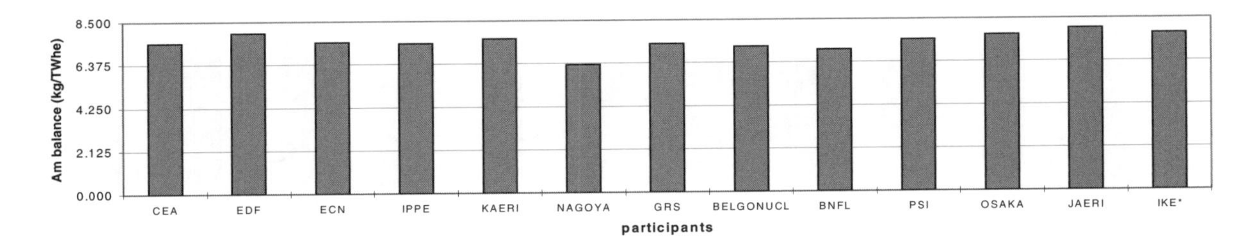

Figure 4.81. HM PWR – Stage II – Fifth recycle: Americium mass balance (complete cycle)

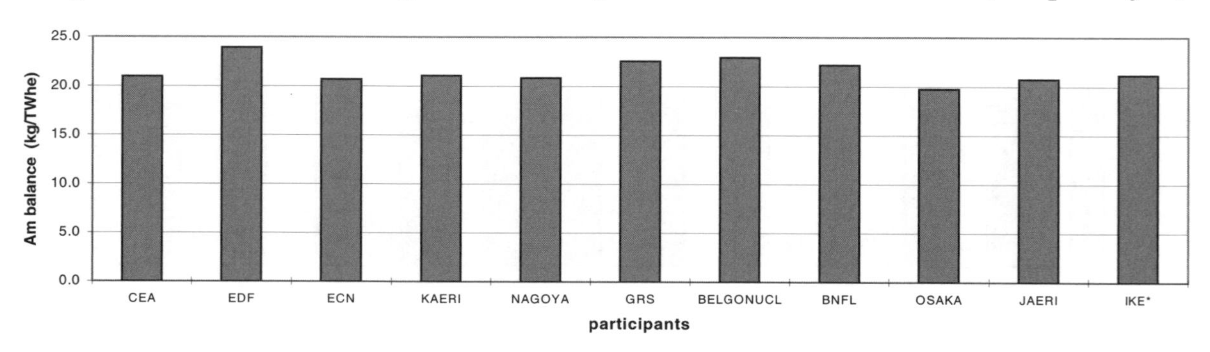

Americium production is a factor 2 lower for the HM PWR, compared with the STD PWR, primarily because of the lower initial plutonium content. The results are in good agreement. For the fifth recycle generation, the americium inventory is some three times that in the first generation, due to the higher initial plutonium content. Generally, the spread of values is larger than for the STD PWR.

Finally, Figures 4.82 and 4.83 show the mass balances for curium.

Figure 4.82. HM PWR – Stage II – First recycle: Curium mass balance (complete cycle)

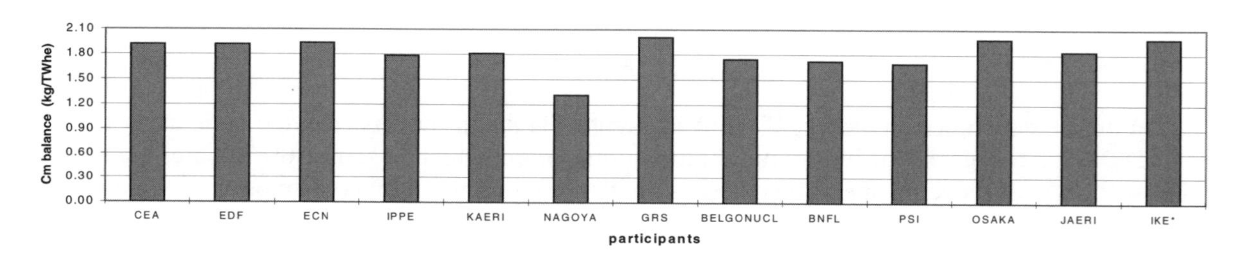

Figure 4.83. HM PWR – Stage II – Fifth recycle: Curium mass balance (complete cycle)

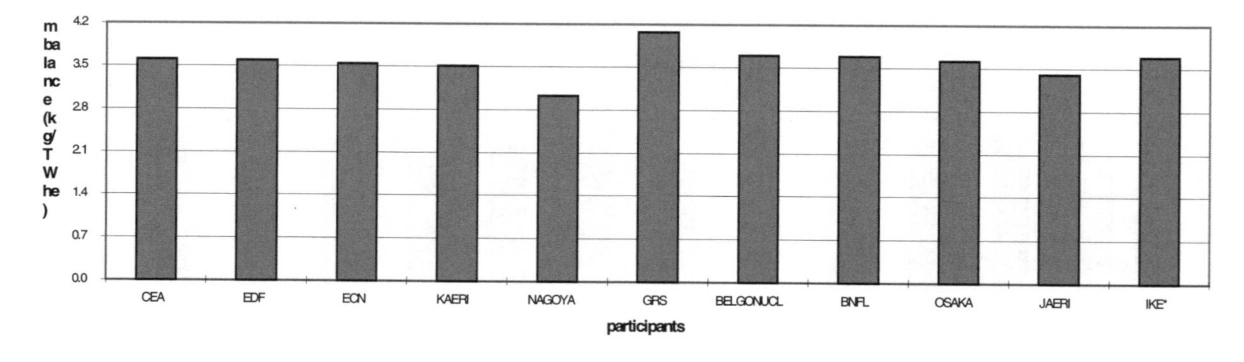

Curium mass balance is smaller for the HM ratio, especially for the first recycling. For the fifth recycling; curium production is similar for the two concepts. The spread between the participants is larger than for the standard PWR and than for the other elements. For the long-lived fission products, the production terms are very close for all the participants.

Activity and radiotoxicity

Comparisons are presented for the fresh fuel (for one metric tonne) and for irradiated fuel after five recycle generations in the STD PWR and in the HM PWR. The comparison is presented for the Stage II Benchmark which accumulates the effects of the different calculations of initial plutonium content and plutonium isotopic evolution.

Activities (for one metric tonne of fresh fuel)

Figure 4.84 shows the global activity levels per metric tonne of initial fuel mass for the first recycle generation fuel of the STD PWR, and Figure 4.85 shows the differences relative to CEA as the reference.

Figure 4.84. STD PWR – Stage II – First recycle: Global activity of one heavy metal metric tonne of initial fuel

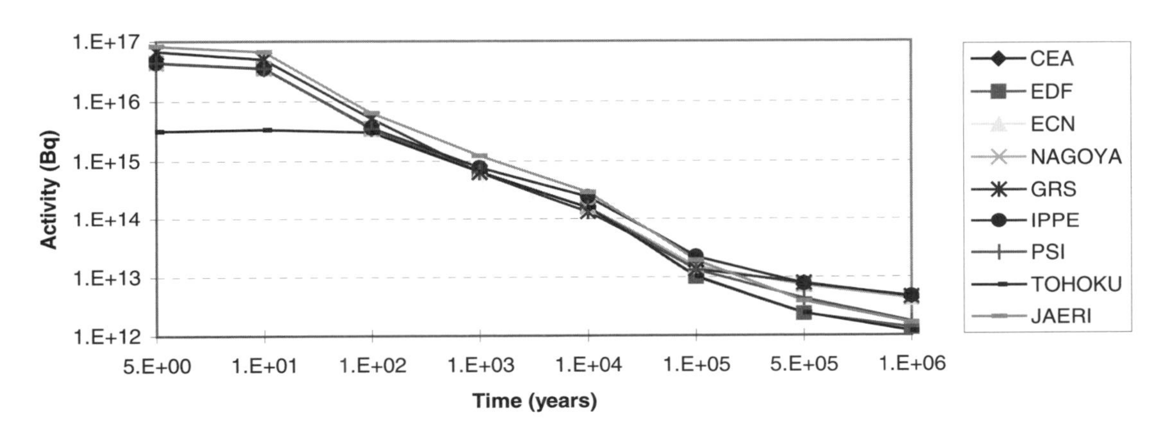

Figure 4.85. STD PWR – Stage II – First recycle: Global activity of tonne of initial fuel – Comparison with CEA values

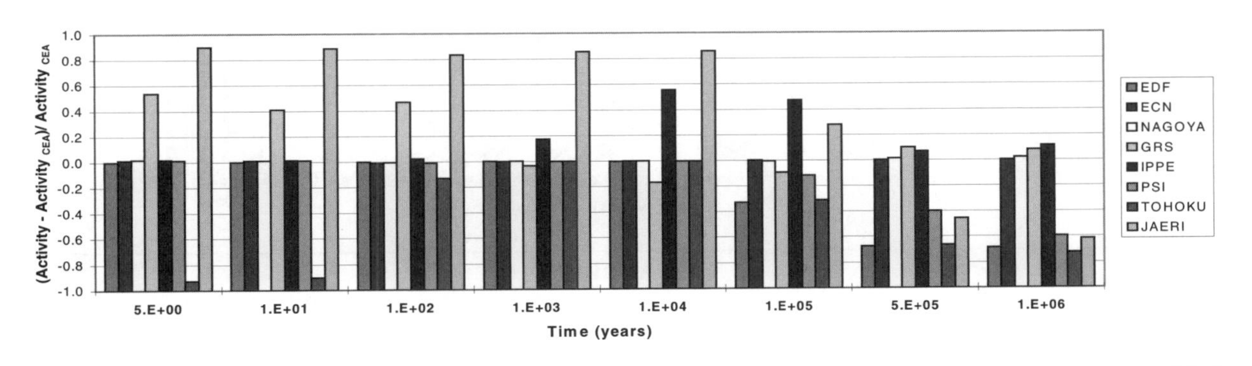

CEA values								
Time (years)	5	10	100	1000	10000	100000	500000	1000000
Activity (Bq)	4.36E+16	3.53E+16	3.46E+15	6.57E+14	1.60E+14	1.50E+13	7.21E+12	4.25E+12

For the shorter term periods, EDF, CEA, ECN, IPPE, PSI are in good agreement. For the long-term period, however, larger discrepancies appear, presumably due mainly to differences the decay chains used. Overall, the agreement is reasonably satisfactory.

Figures 4.86 and 4.87 show the corresponding plots for the STD PWR in the fifth recycle generation. For the STD PWR in the fifth recycle generation, the discrepancies are larger than for the initial step, covering a range +10% to -40%, which may be considered reasonably satisfactory.

Figure 4.86. STD PWR – Stage II – Fifth recycle: Global activity of one heavy metal metric tonne of irradiated fuel

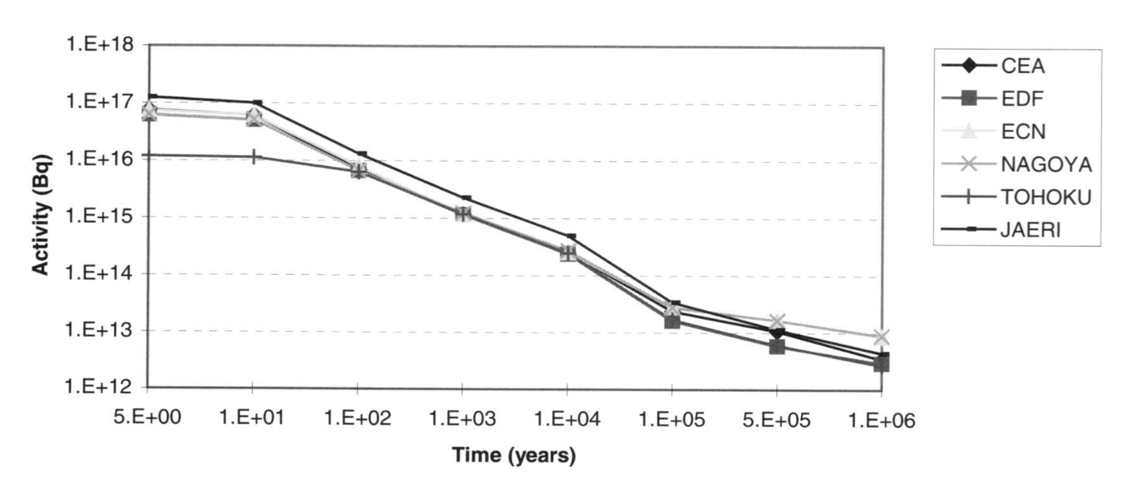

Figure 4.87. STD PWR – Stage II – Fifth recycle: Global activity of tonne of irradiated fuel – Comparison with CEA values

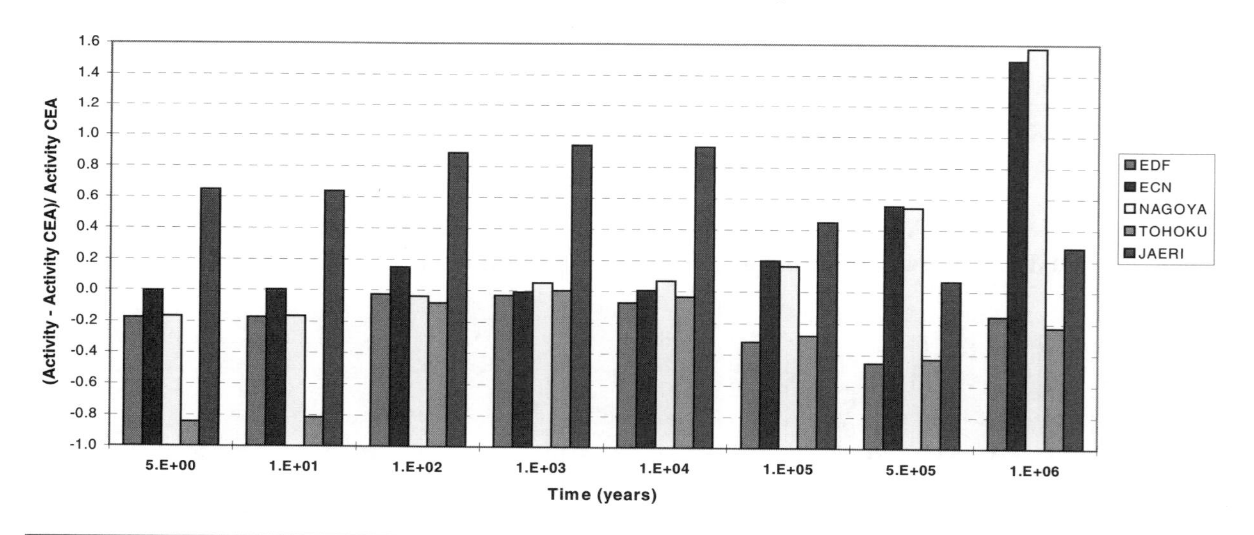

CEA values								
Time (years)	5	10	100	1000	10000	100000	500000	1000000
Activity (Bq)	7.55E+16	6.14E+16	6.93E+15	1.18E+15	2.56E+14	2.36E+13	1.06E+13	3.47E+12

Figures 4.88 and 4.89 present the same plots for the HM PWR in the fifth recycle generation. For the HM PWR, differences are larger, extending over a range of +10% to -60%. Activity is smaller for the irradiated fuel in the HM PWR, about 50% lower, due to the softer neutron spectrum and the lower initial plutonium content.

Figure 4.88. HM PWR – Stage II – Fifth recycle:
Global activity of one heavy metal metric tonne of irradiated fuel

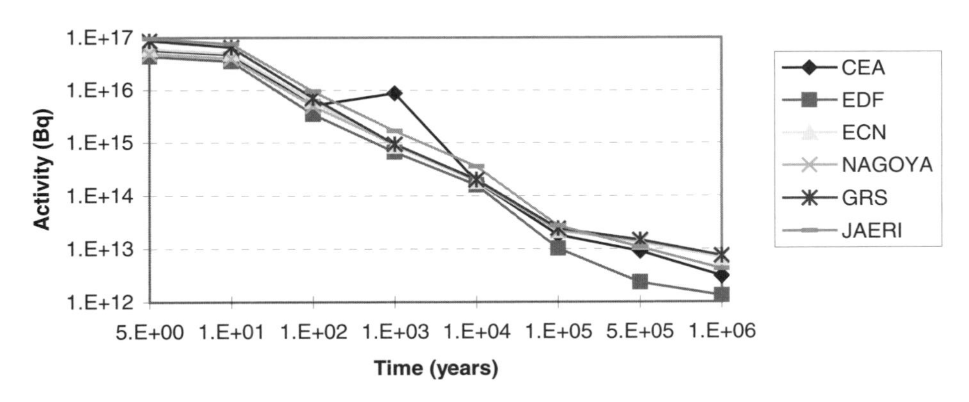

Figure 4.89. HM PWR – Stage II – Fifth recycle:
Global activity of tonne of irradiated fuel – Comparison with CEA values

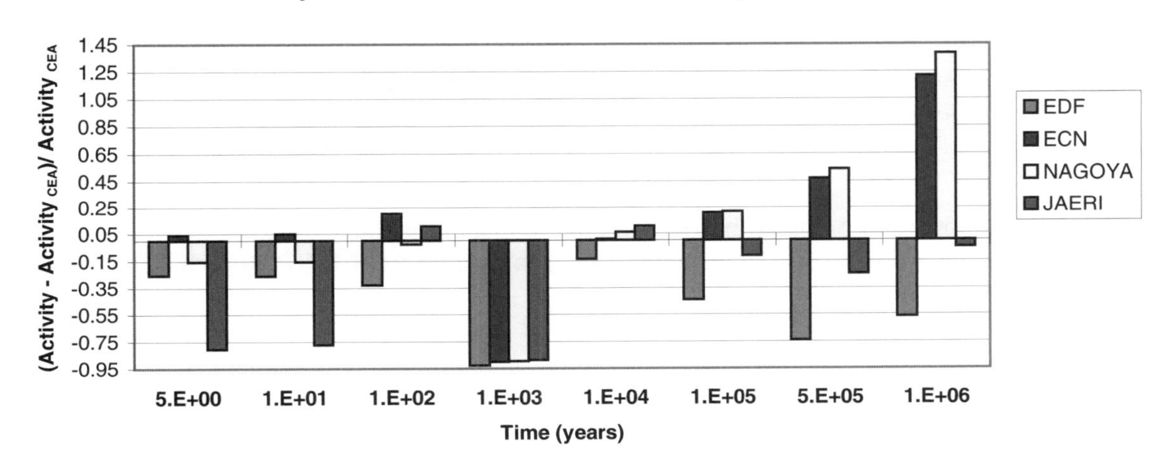

Radiotoxicities

Radiotoxicities depend on the data used for the risk coefficients and on the activities. Only a small number of participants provided results on toxicities and these were presented differently, some of them in Sv, some in CD (Cancer Dose). The following table lists unit each participant used.

Data used	
CEA	Sv
ECN	CD
EDF	Sv
IPPE	Sv
JAERI	CD
PSI	CD

Figures 4.90 and 4.91 plot the radiotoxicities associated with heavy nuclides versus time for the STD PWR. These calculations assume that the multiple recycle is terminated after the fifth recycle generation.

For one metric tonne of fresh fuel

Figure 4.90. STD PWR – Stage II: Radiotoxicity of heavy nuclides of initial fuel

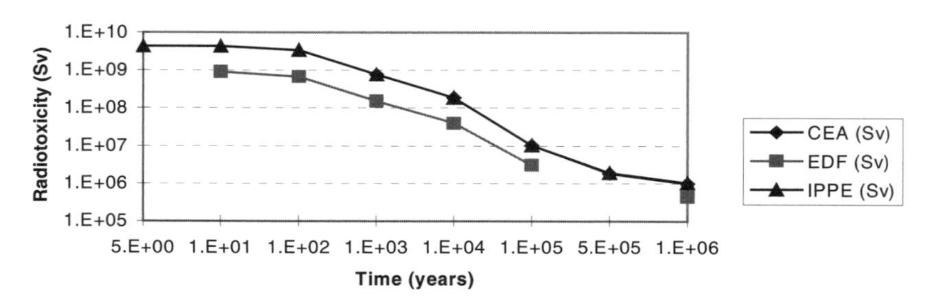

Figure 4.91. STD PWR – Stage II: Radiotoxicity of heavy nuclides of initial fuel

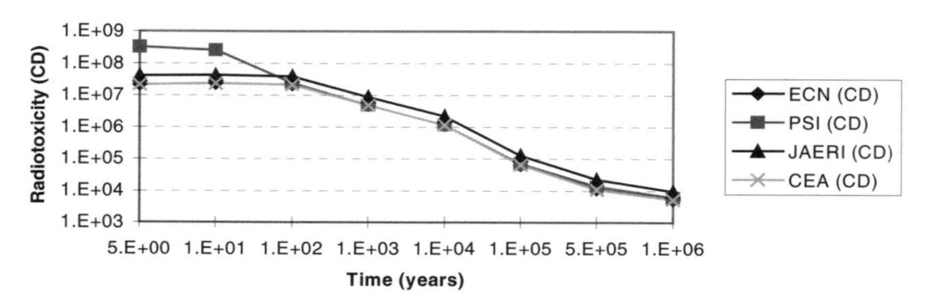

There is a factor up to 5 spread between the participants for the short-term period, reducing to a factor 2 for the long-term period. For comparison, the change from ICRP 30 to ICRP 68 provides a reduction of the radiotoxicity by a factor up to 5.

Annex to Chapter 4

Energy per reaction

This annex provides plots of the energies releases associated with the various reactions obtained by each participant.

Per fission

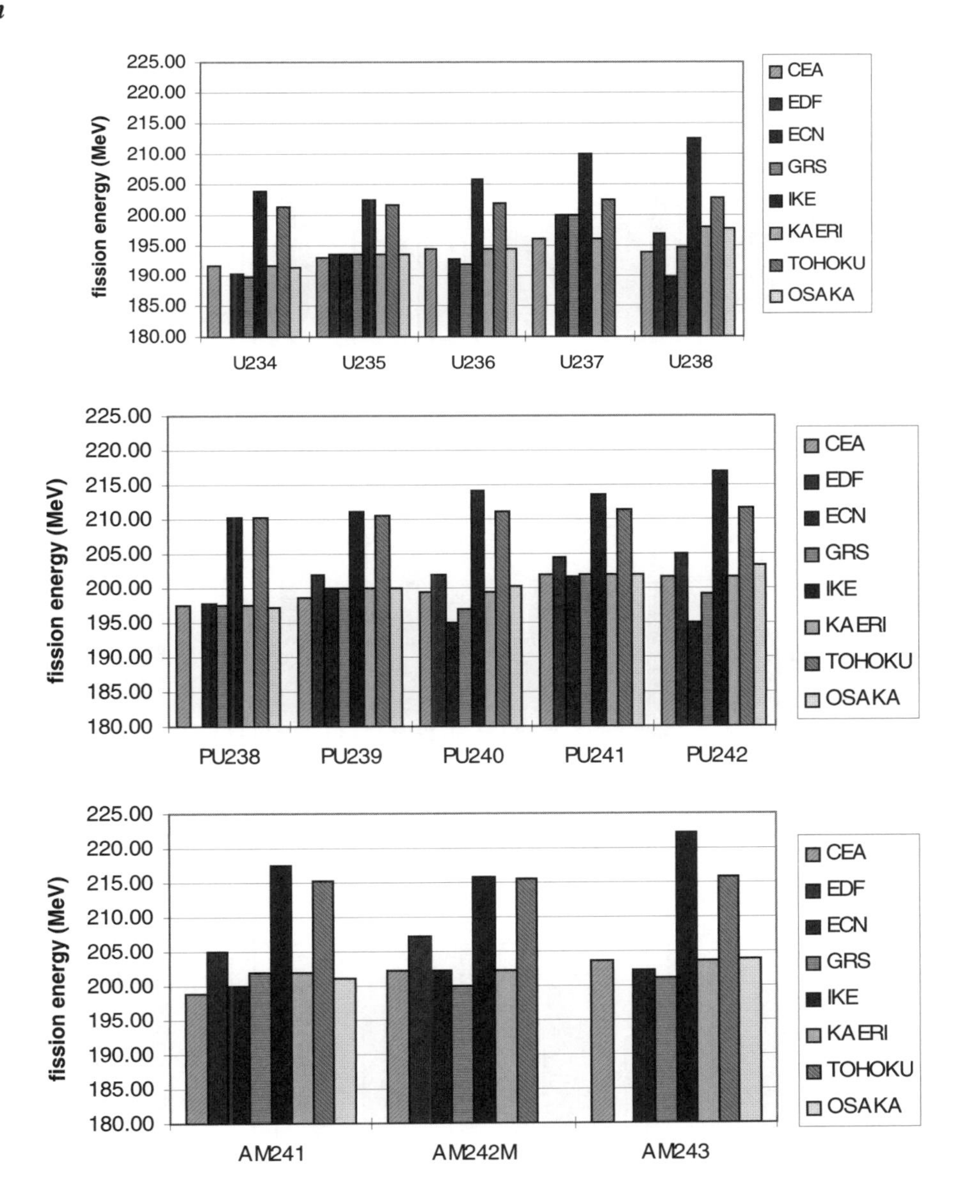

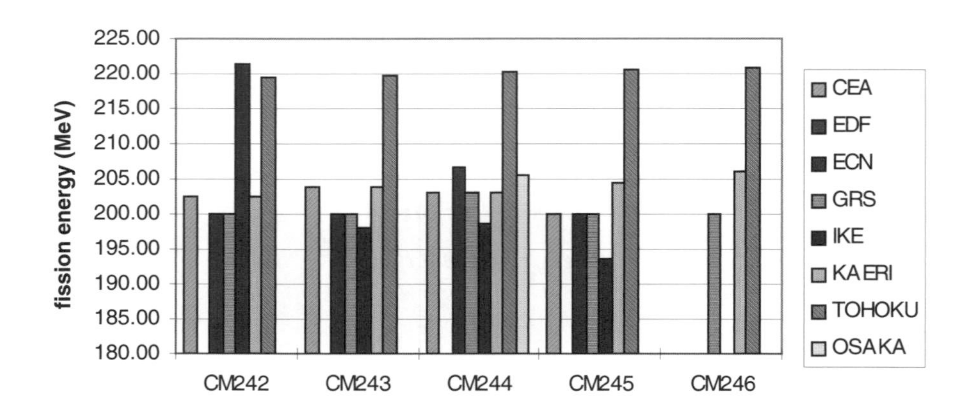
fission energy (MeV)
225.00
220.00
215.00
210.00
205.00
200.00
195.00
190.00
185.00
180.00
CM242
CM243
CM244
CM245
CM246
CEA
EDF
ECN
GRS
IKE
KAERI
TOHOKU
OSAKA

Per capture

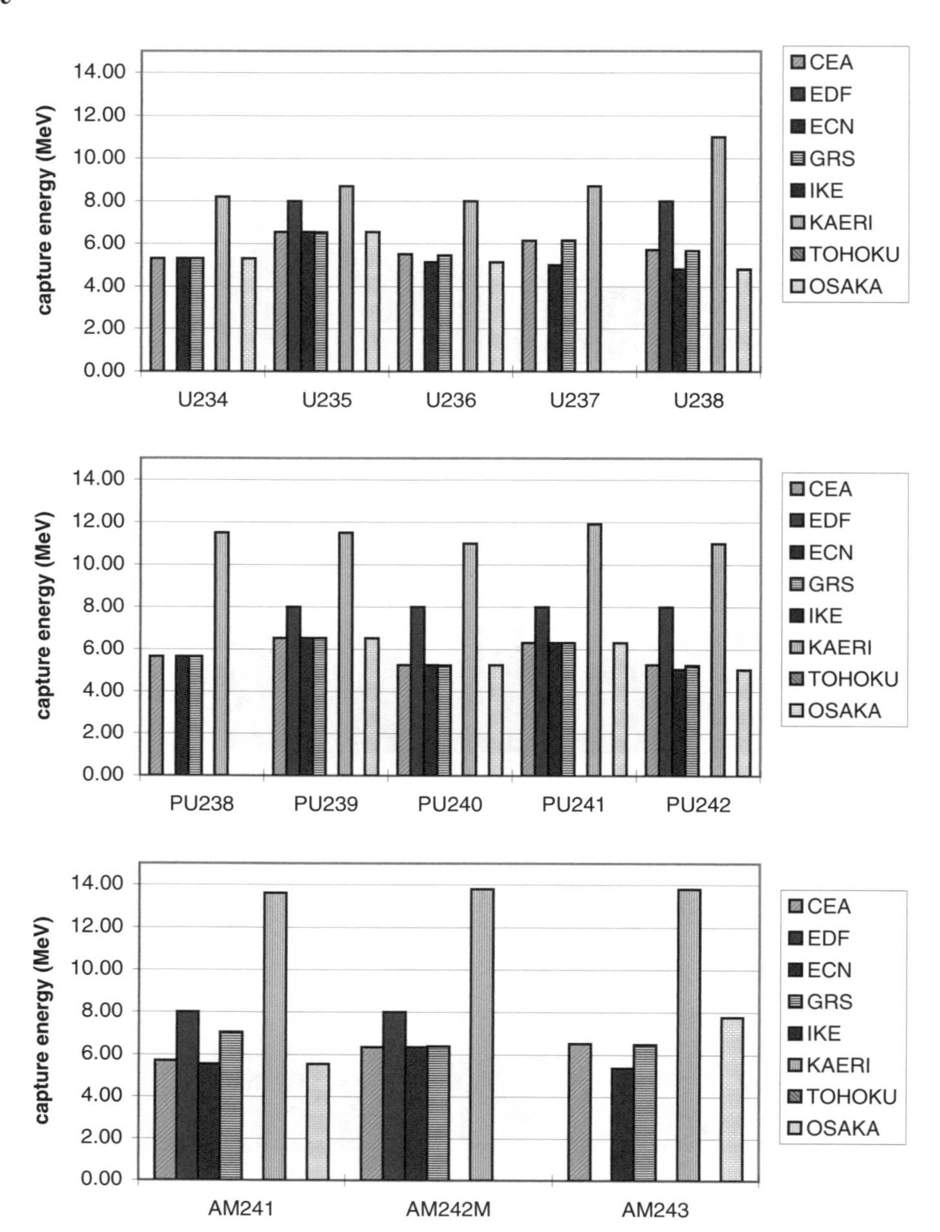
capture energy (MeV)
14.00
12.00
10.00
8.00
6.00
4.00
2.00
0.00
U234
U235
U236
U237
U238
PU238
PU239
PU240
PU241
PU242
AM241
AM242M
AM243
CEA
EDF
ECN
GRS
IKE
KAERI
TOHOKU
OSAKA

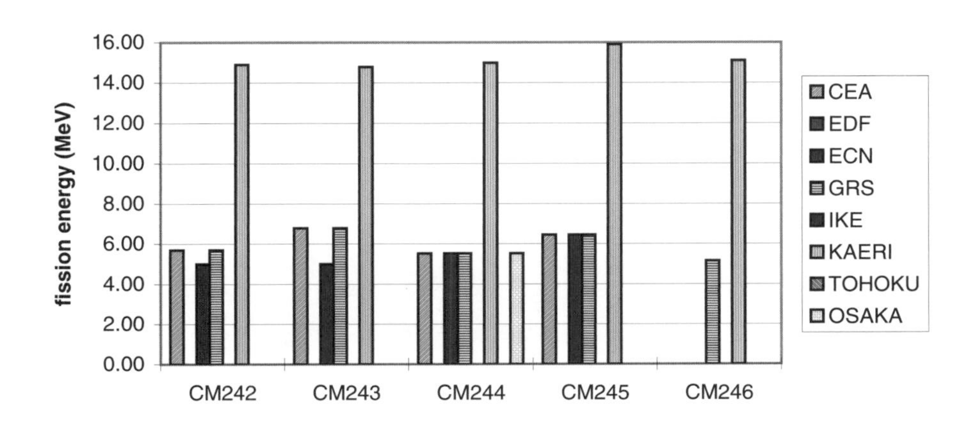

fission energy (MeV)
16.00
14.00
12.00
10.00
8.00
6.00
4.00
2.00
0.00
CM242
CM243
CM244
CM245
CM246
CEA
EDF
ECN
GRS
IKE
KAERI
TOHOKU
OSAKA

Chapter 5

DISCUSSION AND INTERPRETATION OF RESULTS

Objective

The purpose of this chapter is to comment on the results of the STD and HM Benchmarks described in Chapter 4 from the viewpoint of reactor operation and to summarise the lessons that have been learned. This chapter is organised into two parts, the first being a review of all aspects pertaining to the Stage I Benchmark, and the second an equivalent review for the Stage II Benchmark.

Stage I Benchmark

Lifetime average reactivity

This is the determining factor for the fuel cycle length and equivalently, the initial plutonium content needed to achieve a given cycle length. For the STD PWR Benchmark, the spread of reactivities summarised in Table 5.1 (ignoring one outlying solution), varies from 0.9%Δk in the first generation to 1.4%Δk in the fifth generation. For the HM PWR Benchmark, the spreads are somewhat higher, but this time improving from 2.3%Δk in the first generation to 1.3%Δk in the fifth.

Table 5.1. Summary of spread in lifetime average reactivity values

Reactivity	STD PWR		HM PWR	
(% variation)	Cycle 1	Cycle 5	Cycle 1	Cycle 5
Spread (%Δk_{EOI})	0.9	1.4	2.3	1.3

These spreads are considerably better than achieved in the 1994 Benchmark (refer to Vol. I), but are they adequate for operational purposes? This question is not a straightforward one and requires careful consideration: the desired calculational uncertainty on cycle length in conventional PWRs using UOX fuel is usually set at ±20 ppm critical boron, though occasionally errors as high as ±50 ppm are considered acceptable in practice. With a typical boron reactivity coefficient of up to -10 pcm/ppm for UOX, these correspond to reactivity uncertainties in the region of 200 pcm and 500 pcm respectively. Taking the 20 ppm uncertainty as typical, an excess of reactivity in the actual cycle compared with calculation implies that the cycle will end prematurely when the critical boron is still 20 ppm higher than the lowest attainable value (usually about 10 ppm). In the event that the actual reactivity falls short, the 200 pcm reactivity would be made up by continuing operation with the reactor power coasting down gradually, compensating for the reactivity shortfall with the total power coefficient.

Considering, for the sake of argument, an all MOX core, the practicalities are somewhat different. First the boron reactivity coefficient is considerably smaller, as low as -2 pcm/ppm, so that the same reactivity shortfall of 200 pcm would allow the cycle to continue much longer than in the case of UOX because of the smaller reactivity gradient of MOX. Thus the rate at which the reactivity would fall, and therefore the rate at which the reactor power would need to reduce to compensate, would be much lower. Thus, for example in first generation MOX, the reactivity loss between 0 and 51 MWd/kg is approximately 20%Δk, compared with 50%Δk with UOX. Therefore for first generation MOX, a shortfall in reactivity of 500 pcm would have the same impact in terms of power generation loss as 200 pcm in UOX. For later generations the latitude in reactivity is even higher, as the reactivity gradient is smaller. If -0.2 MWd/kg (5 EFPD) is kept as a target, then MOX uncertainty of 100 pcm is required.

This line of argument would suggest that the reactivity target uncertainty for MOX could possibly be relaxed by a factor of at least 2.5 compared with UOX, with a desirable value of 500 pcm and an upper limit of 1250 pcm. With UOX and MOX assemblies co-resident in a mixed core, the same argument would apply. The observed spread in reactivities in this benchmark is actually not far from the desirable value, suggesting that maybe the current codes are already in acceptable agreement.

In practice, of course, without a set of measurement data we are unable to say what the correct answer should be and it is possible that all the results obtained show systematic errors from the "true" values. This underlines the point that benchmark studies such as this are inadequate without operational data to pin-point the correct values. Moreover, the required level of accuracy will most certainly be beyond any of the codes without empirical guidance. This must be obtained from operational experience, starting with demonstration assemblies with appropriate plutonium isotopics and concentrations.

This conclusion should not be surprising, as it is always necessary to gain operational experience when the design parameters progress sufficiently far from existing experience. What is most relevant now is the question of how reliably the existing codes could be used to predict the behaviour of demonstration assemblies. In this respect the requirements on the codes are not so demanding. First of all, since it is usual to load just a few demonstration assemblies in the core to begin with, the impact of errors on end-of-life reactivity are correspondingly diminished. Moreover, it is usual to ensure that such assemblies are kept well away from being the lead assemblies in terms of rating that define the operational and safety limits. For both these reasons, the same degree of confidence is not required for demonstration purposes and could therefore be argued that the level of agreement already achieved would allow a demonstration assembly programme to go ahead with a good degree of confidence. Of course, as the demonstration programme proceeds, the experimental data obtained allows the code systems to be adjusted empirically to ensure a lower calculational uncertainty for later phases.

All the above comments apply to the standard PWR lattice. For the highly moderated lattice, the same arguments will apply, except that the boron reactivity coefficient will be closer to the values seen in a UOX core and the reactivity gradient with burn-up is slightly steeper than for STD MOX. The desirable uncertainty will also depend on the total power coefficient, which may be different from the standard lattice.

In conclusion, therefore, the observed spread of results from the different participants may possibly be adequate for operational purposes and would certainly be sufficient for supporting a demonstration programme. This would be the next logical step required if multiple recycle MOX was to be pursued in practice. With the relatively small spread in end-of-cycle reactivities obtained in these benchmarks, it will be difficult to identify the precise causes in terms of underlying nuclear data and/or calculational methods, as they are likely to be too small to show up obviously even on examining the microscopic cross-sections. Moreover, there are likely to be several factors combining together to give the spreads seen here, which would further complicate at attempt to isolate any of them. There seems little point, therefore, in a follow-up benchmark to the present ones.

Reactivity versus burn-up

Reactivity versus burn-up is the determining factor for the variation of critical boron versus cycle burn-up. It is also inter-related with the lifetime average reactivity. For both the STD and HM PWRs, the reactivity versus burn-up curves plotted in Chapter 4 show essentially the same spreads as do the end-of-cycle reactivities. As in the 1994 Benchmark (reference to Vol. I), the gradients of reactivity versus burn-up are fairly consistent between the various participants, at least at first examination, indicating that the solution of the burn-up equations are reasonably consistent and are not greatly affected by the reactivity spread. Table 5.2 summarises the spread in beginning-of-irradiation (0 MWd/kg) and end-of-irradiation (EOI) cycle reactivity spreads seen for Cycles 1 and 5, with the distant outliers excluded.

Table 5.2. Summary of beginning-of-irradiation and end-of-irradiation reactivity spreads

Reactivity (% variation)	STD PWR		HM PWR	
	Cycle 1	Cycle 5	Cycle 1	Cycle 5
Spread (%Δk_{BOI})	0.8	1.0	1.5	1.4
Spread (%Δk_{EOI})	1.2	1.9	4.0	1.7

For the standard PWR the spreads are always no more than 1.0%Δk at BOI and 1.9%Δk at EOI, which are encouragingly tight. More detailed examination of the reactivity gradients for the STD PWR shows that there is a sub-population represented by the three Japanese solutions with systematically different trends. Thus in all generations the Japanese solutions start off at zero burn-up with a reactivity very close to the CEA solution treated as a reference, but a difference accumulates of around +1%Δk by 51 MWd/kg. This is most probably an effect resulting from the nuclear data libraries, with the Japanese libraries giving a slightly different trend to the others.

The same trends are seen for the HM PWR data, except than in this case the spread of values is somewhat larger in the first generation, at 1.5 to 4.0%Δk. The large spread at EOI is due to two solutions showing the opposite trend with burn-up of the other solutions. If these two data sets are omitted, the spread is more than halved. In later cycles the spread decreases (ignoring one outlying solution).

For both STD and HM cases the spreads in EOI reactivities are larger than those for the lifetime average reactivities of the following section, largely because the latter are determined by the 34 MWd/kg burn-up step; the EOI reactivities are determined by the 51 MWd/kg step, with the spread tending to increase with burn-up.

Taking the spreads in Table 5.2 to be representative, they are higher than would be desirable in an operating PWR. The spread of BOI reactivities up 1.0%Δk in the STD case would adversely affect the prediction of the critical boron. For a UOX core, the ideal target would be 0.2%Δk and the maximum permissible 0.5%Δk. The MOX value is considerably higher and the situation is made worse by the very small boron coefficient, such that the 1.0%Δk would translate into a very large uncertainty on initial critical boron. There are limits to which the initial boron could be allowed to deviate, due to requirements to keep within various limits related to guaranteeing sub-criticality in the shutdown condition and the boration requirements following an emergency shutdown. As previously demonstrated, this further emphasises the need for operational data to tie down the uncertainties.

Boron reactivity coefficient

The boron reactivity coefficient (or boron efficiency) is a simple parameter for the codes to calculate and provided there are no gross spectral differences between the various codes, the spread of values obtained could be expected to be small. Table 5.3 confirms this, giving the spread of values for the first and fifth generation cases and beginning and end of irradiation (BOI & EOI). The boron coefficient is expressed in terms of the reactivity effect in pcm (1E-5 Δk) per ppm increase in boron concentration. Table 5.3 plots the absolute difference (in pcm/ppm) between the highest and the lowest values.

For the STD PWR, the observed spread in values is of the order of 0.3 pcm/ppm or less and shows only a slight trend to worsen in later generations. There is evidence of clustering about a tight spread (ignoring two outlying solutions). The variation of boron coefficient with burn-up is very consistent between the different solutions, and it is quite significant, increasing by around 50% between BOI and EOI. For the STD case the boron coefficients are very small in absolute magnitude, in the region of 20% of that for a normal UOX core. This is the principal reason why current PWRs would not be able to accept a 100% MOX loading, as the boration requirements for normal and emergency shutdown would be excessive.

Table 5.3. Summary of spread in boron reactivity coefficient values

Reactivity Coefficient (pcm/ppm)	STD PWR		HM PWR	
	Cycle 1	Cycle 5	Cycle 1	Cycle 5
BOI	0.21	0.26	0.18	0.24
EOI	0.16	0.32	1.20	0.26

For the HM PWR, the boron efficiencies are much larger in absolute terms, reflecting the more thermalised spectrum and the large volume of moderator in the assembly. The spread of values is generally very small, with the first generation only showing quite a large spread at EOI. Hence the 1.20 pcm/ppm value in Table 5.3 overstates the spread which applies more generally and it is fair to say that with this one exception, there is excellent agreement between the different codes.

An interesting observation is that the boron coefficient varies very dramatically from BOI to EOI in the early generations, but is less sensitive in the later ones. Thus in the first generation, the boron coefficient varies by more than a factor 2 between BOI and EOI. This is a result of the efficient burn-out of the fissile plutonium, with correspondingly lower production of fresh ^{239}Pu from ^{238}U captures, which will affect the level of thermal absorption and the spectrum. In later generations, the non-fissile absorption starts to dominate and the burn-out effect becomes much less evident in the boron coefficient.

In terms of operational requirements, the observed spread in boron coefficient falls within the 10% uncertainty allowance that typically applies to current PWRs and indicates that the boron coefficient calculations are likely to be satisfactory for both the STD and HM cases.

Moderator temperature coefficient

The moderator temperature coefficient (MTC) is important for determining the shutdown margin, the amount of negative feedback in heat-up faults and conversely the amount of reactivity insertion in cool-down faults. The benchmark specifications called for the MTC to be calculated at various cell burn-ups ranging from 0 MWd/kg (BOI) to 51 MWd/kg (EOI). Table 5.4 summarises the absolute spread of values obtained for the STD and HM cases, for the first and fifth generations, in units of pcm/K. The figures quoted exclude one outlying solution.

Table 5.4. Summary of spread in moderator temperature coefficient values

Reactivity Coefficient (pcm/K)	STD PWR		HM PWR	
	Cycle 1	Cycle 5	Cycle 1	Cycle 5
BOI	4.0	4.0	N/A	6.4
EOI	4.5	6.3	N/A	8.4

N/A: Not Applicable

The benchmark specification called for the MTC calculations to be carried out at zero boron, so they are representative of the condition at the end of a refuelling cycle with close to zero boron concentration. For the STD case, the spread in values is quite small, with no evidence of any deterioration with cell burn-up or with recycle generation. There is excellent agreement between the various solutions as to the change in MTC with cell burn-up, all the solutions having virtually the same gradient. This applies to all the recycle generations. Moreover, there is good agreement as to the trend in the absolute value of the MTC in the different generations. In terms of relative error, the spread represents a plus or minus of no more than 5% in the first generation and 10% in the fifth generation. The relative error is higher in the later generations largely because the absolute magnitude of the MTC falls with increasing recycle generation.

For current PWRs, the acceptable uncertainty on MTC in the zero boron end-of-refuelling-cycle condition would typically be set at around 10%. The spread in values obtained here, if they are taken to represent the inherent uncertainty, suggest that the MTC calculations would be acceptable for operational purposes.

In contrast, the situation is less satisfactory for the HM case, especially in the first generation. The problem is that in the first case there is not even agreement between the various solutions as to the trend with increasing cell burn-up. Thus one group of solutions shows a monotonic variation with burn-up, whereas another (larger) group of solutions shows the MTC increasing in magnitude at intermediate burn-ups. This is why no values have been tabulated in Table 5.4 for the first generation case; the spread is meaningless in this case. In the second generation, this behaviour is no longer seen and the dependence on cell burn-up again becomes monotonic for all except one outlying solution. The spread in values tends to improve beyond the second generation, though it remains larger than in the STD case. There is evidence in all generations of two distinct sub-sets of solutions with different gradients of MTC with cell burn-up. Within each of the two sets, the spread is much smaller.

The reason for this less than satisfactory outcome for the HM PWR is not clear, but it must presumably be connected with the different spectral conditions in the HM case. At zero boron, the STD PWR is well undermoderated, so that the competition between moderation and absorption in the water is biased very heavily towards the former. In the HM PWR the balance between these opposing effects is more even, and this would tend to make the calculation more difficult. The suggestion of a maximum in the MTC at intermediate cell burn-ups in the first generation may well be real. In the first generation the plutonium content is very much smaller than even the second generation and it is conceivable that this could affect the burn-up dependence, given that absorption in the fuel would be relatively low.

Fuel temperature coefficient

The fuel temperature (or Doppler) coefficient is important for ensuring rapid negative feedback in transients involving an increase in reactor power. It is also very important in determining the amount of shutdown margin available, since there is positive reactivity insertion from the decrease of fuel temperature between the operating and zero power conditions. Based on the results presented in Chapter 4, Table 5.5 summarises the spread of the calculated values for the first and fifth recycle generations at BOI and EOI conditions. The observed spreads represent between 20 and 30% of the absolute value of the reactivity coefficient and are quite significant.

Table 5.5. Summary of spread in fuel temperature coefficient values

Reactivity Coefficient (pcm/K)	STD PWR		HM PWR	
	Cycle 1	Cycle 5	Cycle 1	Cycle 5
BOI	0.4	0.6	0.3	0.6
EOI	0.7	0.7	0.6	0.8

For both the STD case the various results are fairly consistent in that the trend with burn-up is in close agreement for all the recycle generations and relative rankings remain consistent throughout. The spread of results is most likely due to differences in the treatment of ^{242}Pu resonance absorptions between the various codes. Some of the codes have a temperature dependent tabulation for all resonance isotopes, whereas others may only have a tabulation for a limited range of isotopes. The resonance absorptions in ^{240}Pu and ^{242}Pu are not important contributors to the fuel temperature coefficient in UOX fuel or in MOX fuel with low levels of these isotopes. However, they are very important contributors for these benchmarks. Therefore a code in which, for example, only the ^{238}U and ^{240}Pu resonance parameters are tabulated versus temperature, the fuel temperature coefficient would be expected to be smaller in magnitude. This is precisely the case for the WIMS code, so that the BNFL solution does not incorporate the contribution of ^{242}Pu to the fuel temperature coefficient. We are therefore led to the conclusion that a full temperature dependent resonance tabulation is essential for the situation examined here. Note that if there is only one resonance temperature in the tabulation, it is usual for a fuel material for it to correspond to full power operation. Therefore predictions of reactivity and power in the full power condition will be correct and it is only with deviations from full power where an error will be incurred.

There is a consistent trend for the fuel temperature coefficient to increase in magnitude with burn-up. Perhaps this is due partly to the increase in concentration of the even isotopes with burn-up. There are only small changes between recycle generations, suggesting that changes in the neutron spectrum in the resonance range are modest. There is only a modest deterioration in the spread of results in later recycle generations.

The HM case shows much the same trends, although in this case there are some anomalous results (showing the opposite trend with burn-up) in the first two generations. The magnitude of the fuel temperature coefficient is a little smaller in this case, presumably due to the softer spectrum and the reduction in resonance capture probability that would be expected with the wet lattice.

Reactivity balance

The reactivity balance is the gradient in reactivity with burn-up during an equilibrium fuel cycle (measured in 10^{-5} Δk/MWd/kg or pcm/MWd/kg). It is important for several reasons, one being that it determines the boron letdown rate during the cycle, and another being that it determines the rate of reduction of power during a power stretchout under which the reactor might operate beyond the natural end of cycle. The various results are reasonably consistent and all participants showed the same trend with increasing recycle generation. Thus the reactivity balance falls dramatically between the first and second generations, but thereafter falls much more slowly, with a gradual levelling off to around 250 pcm/MWd/kg. This is considerably lower than is characteristic of UOX fuel, where a value approaching 1 000 pcm/MWd/kg is more typical. This implies a lower rate at which the critical boron needs to be let down during the cycle. It also implies that reactor power will diminish less rapidly during power stretchout operation, which would be beneficial in terms of total generation capacity.

The HM PWR results are similarly fairly consistent between the various participants. However, in this case the let down rate is much higher in the first recycle generation (~600 pcm/MWd/kg). This is due to the more effective burn-out of ^{239}Pu with the soft spectrum. However, in subsequent generations there is a rapid change towards an asymptotic value of just under 300 pcm/MWd/kg, not far removed from the STD PWR value. The two cases converge because the more rapid deterioration of the plutonium isotopic vector in the HM PWR largely cancels outs the effect of the softer spectrum.

Global void defect

The void defect is the change in reactivity associated with voiding the reactor core, as might happen in the event of steam or gas bubbles displacing the moderator. The design objective is to ensure that the overall void defect of the core is negative, in order to ensure a negative feedback mechanism. In the STD PWR, the MOX assemblies will only constitute a fraction of the core loading and so the positive values obtained in the later recycle generations do not necessarily mean that the core would be unsuitable to licence. Nevertheless, it is worrying that the void defect is positive for all except the first recycle generation, due to the high plutonium concentrations needed.

Two sets of calculations were requested, one for an infinite lattice (no leakage) and one which takes account of the finite diffusion length of fast neutrons and the consequent increase in leakage in the voided condition. The spread of results is already large for the no leakage case and larger still for the case where specified leakages have been applied to the unvoided and unvoided conditions. This probably reflects the nature of the problem where the fully moderated condition corresponds to a normal PWR spectrum and where the voided condition corresponds more closely to a fast reactor spectrum.

The spread may be partly a result of difficulties in some of the codes in calculating the resonance absorption of ^{238}U, ^{240}Pu and ^{242}Pu in the voided condition (where the resonance integrals will be considerably higher than in any normal application of the codes). The larger spread in the presence of leakages suggests that the various participants' codes may not necessarily have been consistent in applying the leakage corrections and therefore the void defects with leakages do not contribute usefully to this comparison.

For the HM PWR the void defects with no leakage are generally in closer agreement. The values are more negative in early recycle generations and remain negative even in the fifth generation. The spread of values is also in much closer agreement. As with the STD PWR case, introducing leakage greatly increases the spread of results, though all the values remain negative. This implies that the HM PWR does offer benefits over the STD PWR in that a negative void defect is virtually assured in all recycle generations. This is an essential requirement, since in this case there are no UOX assemblies in the core which might tilt the balance from positive to negative.

Stage II Benchmark

Lifetime average reactivity

For the Stage II Benchmarks, the specification calls for each participant to adjust the initial plutonium content in order that the same lifetime average reactivity matches that obtained in the first generation of Stage I. Thus, taking CEA as an example, the k-infinity at 34 MWd/kg, zero boron in the first generation was 1.037. In subsequent generations, the initial plutonium was adjusted to preserve this lifetime average. Table 5.6 summarises the spread of results obtained. For the STD PWR, the initial plutonium in the first generation (Cycle 1) was defined as a common starting point in the specification of the benchmark and the spread is therefore zero. This is why Table 5.6 starts with the second generation. For the HM PWR, the participants were asked to calculate the initial plutonium in all generations, including the first.

Table 5.6. Summary of spread in initial plutonium values

Initial Plutonium	STD PWR		HM PWR	
(atom % variation)	Cycle 2	Cycle 5	Cycle 1	Cycle 5
Spread (atom %)	0.8	2.5	0.6	2.9

The values quoted in Table 5.6 are selective in that two of the solutions have been disregarded. For these two cases, the initial plutonium content was not changed beyond the second generation, presumably due to a mis-interpretation of the Benchmark specification and it is therefore not appropriate to include them. For both the STD and HM cases, the spread starts off small (< 1 w/o absolute variation in total plutonium). As the number of generations increases, the spread naturally increases, as in this case both the underlying reactivity calculation differs between the various solutions (as was the case in Stage I) and the plutonium isotopic composition varies as well. The latter is a cumulative effect, since a systematic difference in plutonium evolution carries over from one generation to the next and leads to a spread of approaching 3 w/o in the fifth generation. Although quite large in absolute terms, since the initial plutonium content in the fifth generation is around 20 w/o, the spread represents a relative error of plus or minus 7.5%, which is encouragingly small.

An important point to note is that the rate of increase in initial plutonium between generations is much higher for the HM PWR than for the STD PWR. For the HM case, the first generation requires just over 6 w/o plutonium, rising to around 17 w/o in the fifth generation. The corresponding figures for the STD PWR are 10 w/o and 19 w/o respectively. The reason for this is that the HM PWR is better at fissioning the ^{239}Pu and ^{241}Pu in the soft spectrum, so that the isotopic quality degrades more quickly between generations. Thus, although the HM PWR starts off with a low initial plutonium concentration in the first generation, it has a tendency to catch up with the STD case. This is a significant observation, in that the initial advantage of the HM PWR (*i.e.* lower initial plutonium concentration) is eroded in later generations.

Mass balances

The mass balance of Pu, Am and Cm are important for the Stage II Benchmark, as they are a determining factor for the environmental impact. It is therefore important to be confident that they are calculated consistently between the various participants. The mass balances are defined in terms of the build-up or destruction rate in kg per TWh(electrical). The various solutions are generally in very good agreement with one another, except for two and sometimes three solutions which are systematically lower than the remainder by a few percent. This observation applies equally to both the STD and the HM Benchmarks. Some general remarks are useful here.

- *Plutonium*
  The plutonium mass balance is negative for both the STD and HM cases, due to the destruction of the odd fissile isotopes. For both the STD and HM cases, the plutonium destruction rate increases with recycle generation (being some 40% higher in the fifth generation than in the first generation). In terms of kg/TWh(e), the destruction rate is only marginally higher in the HM case. However, this does not imply that the STD and HM cases are equivalent in terms of plutonium destruction. The point to remember is that the fuel mass is considerably smaller for the HM PWR and, at least in the early recycle generations, the concentration of plutonium is much smaller. Therefore, with reference to the initial plutonium content, the HM PWR is more effective at destroying plutonium.

- *Americium & Curium*
  The mass balances for Am and Cm are positive, meaning that they accumulate with burn-up and the build-up rate is higher in the later recycle generations. This result applies both to the STD and HM cases and is a result of the higher initial plutonium concentrations needed in later recycle generations; the higher initial plutonium mass leads to accelerated production of the trans-plutonium nuclides. In terms of kg/TWhe, the production rate is lower for the HM PWR than for the STD PWR in the early recycle generations, though the difference becomes marginal in the later generations.

Global activities

For the STD PWR global activities in the fuel at loading and at discharge are in reasonable agreement for more than half of the participants, but there some outlying results which deviate significantly. The spread of results is about the same in the first and fifth multiple recycle generations, so there is, encouragingly, no evidence of a worsening of the agreement with multiple recycle generation. The subsequent decay of activity is in reasonable agreement right up to the 1 million year cut-off, after allowing for the range of variation of values at discharge.

For the HM PWR, a similar situation is seen in the first recycle generation, but the spread of results increases noticeably by the fifth generation. This is indicative of strong disagreement between the various participants with respect to isotopic evolution for the HM PWR case. In the first generation recycle, the HM PWR case gives almost a factor 2 reduction in activity levels at discharge, and this persists through to the 1 million year calculation. The calculations suggest that for the fifth recycle generation this advantage of the HM PWR is somewhat eroded, but still persists to a lesser extent.

Radiotoxicities

For the radiotoxicity comparison, only a small fraction of the participants submitted calculations and those that were submitted are very discrepant. This is presumably due to lack of consistency in toxicity factor used, since the underlying activity levels are in much better agreement. No useful conclusion can be drawn from the information as submitted.

Chapter 6

RESULTS OF THE SPECIAL BENCHMARK ON PWR MOX PIN CELLS

Introduction

In its first phase of the programme of work the Working Party on Physics of Plutonium Recycling (WPPR) had commissioned several benchmarks. Two in particular concerned PWR pin cells with MOX of isotopic Pu vectors of different quality: from a first recycle (B) and a fifth recycle (A). Although the results from participants were found to agree considerably better than in a similar study carried out a decade earlier, they were still not completely satisfactory. For these benchmarks the spread in k-infinity after removing outlying solutions is in excess of 1%; this value would be unacceptable if it were representative of the uncertainty on lattice design calculations.

It was for this reason that a special investigation was proposed with the aim of better clarifying the sources of discrepancies among participants' results. Special emphasis was directed toward applied cross-section processing methods for users of JEF-2 and JENDL-3 evaluated data libraries (see Annex 1 to Chapter 6). The results of this specific benchmark study are the subject of this article.

The calculations were restricted to fresh fuel as in the originally defined benchmarks. A well defined geometry was chosen with a quadratic cell and 6 or 20 subdivisions in the fuel and 3 subdivisions in the moderator. The temperatures were slightly modified to enable continuous energy Monte Carlo calculations.

The participants were originally restricted to using JEF-2.2 and JENDL-3.1, but later, calculations applying the newer JENDL-3.2 database were accepted (see Table 6.1).

A request providing a four group rate and cross-section output was added to the specification to allow a more detailed association of the discrepancies to the different energy regions. A slightly different upper energy boundary for the thermal group was used in the SRAC calculations. Participants from JAERI verified that this discrepancy does not significantly influence the results.

General discussion

Tables 6.2a and 6.2b contain the global results of the contributions, the k-infinity values and the reaction rates of the three zones, for the degraded MOX pin cell (Case A) and the classical MOX fuel (Case B), respectively. The rates are normalised to the total absorption rate in the cell equal to 1.

In Tables 6.3a and 6.3b the relative differences of the results with respect to the MCNP4 solution of IKE1 are listed to facilitate an easier overview for comparisons. The production, fission and absorption rates are related to the corresponding total rates and converted to pcm by multiplying by a factor of 10^5.

Table 6.1. Characterisation of the contributions

ESTABL.	IDENTIF.	DATA-SOURCE	CODE	FUEL ZONES	REMARKS
IKE	IKE1	JEF-2.2	MCNP4	6	continuous energy
ECN	ECN1	JEF-2.2	WIMSD6	1	
ECN	ECN2	JEF-2.2	WIMSD6	6	
ECN	ECN3	JEF-2.2	WIMSD6	20	
CEA	CEA1	JEF-2.2	APOLLO-2	6	
CEA	CEA2	JEF-2.2	TRIPOLI-4	6	continuous energy
JAERI	JAE1	JENDL-3.1	SRAC	1	
JAERI	JAE2	JENDL-3.1	SRAC	6	
JAERI	JAE3	JENDL-3.2	SRAC	1	
JAERI	JAE4	JENDL-3.2	SRAC	6	
JAERI	JAE5	JENDL-3.2	MVP	1	continuous energy
JAERI	JAE6	JENDL-3.2	MVP	6	continuous energy

The Monte Carlo (MC) results using JEF-2.2 data (CEA2 and IKE1) differ by 200 pcm for Case A and 300 pcm for Case B. The MC results using JENDL-3.2 data are closer to the MCNP4 values in Case A (-80 pcm), but up to 400 pcm higher in the classical Case B. The difference between the 2 MVP results themselves is 137 pcm.

The deterministic codes using JEF-2.2 yield very different eigenvalues: the differences are between -289 and +354 pcm for Case A and between -204 and +606 pcm for Case B. The results calculated with JENDL data differ between 156 and 725 pcm and between 304 and 906 pcm, respectively. The highest differences occur for the SRAC calculations with six subdivisions of the fuel zone (JAE2 and JAE4).

The columns with production rates in Table 6.2 differ from k-infinity when production only by fission is given, but most participants included the (n,2n) and (n,3n) yields into the production rate. Because of these inconsistencies in the contributions, the production rates are not suitable for detailed comparisons. Instead, the fission rates are discussed. The (n,2n) rate itself is between 130 and 180 pcm. The absorption rates in fuel will be discussed in the next section. Surprising are the significant differences in absorption rates in clad and moderator between calculations with JEF-2.2 and JENDL data. The JAERI results show higher absorption rates in the clad of between 100 and 140 pcm and lower ones in the moderator of up to 80 pcm.

The results of a special study addressing the effects due to the use of cross-sections derived from JENDL-3.2, Jef-2.2 and ENDF/B-VI are given in Annex 2 to this chapter.

Discussion of differences in absorption rates

Tables 6.4 to 6.11 give information regarding the absorption rates for the isotopes in the fuel. The first column refers to the total absorption rates, and the following ones contain four group values corresponding to the requested energy subdivision (MeV, unresolved and resolved resonance and thermal region). In the first two lines the MCNP4 values and the corresponding values for the statistical standard deviations (sigma) in pcm are listed (the sigma values of the other MC calculations are of the same order of magnitude). The following ones contain the differences between the other contributions and the MCNP4 results. Horizontal lines divide the sections when different databases are used. This representation shows which isotopes in what energy region are responsible for discrepancies in the results, but it cannot be decided if these are caused by differences in basic cross-sections or by different methods in resonance shielding and spectral calculations, respectively.

Tables 6.4a and 6.4b show the absorption rates of ^{235}U for Cases A and B. All results agree well, except those emerging from JENDL-3.1, which show a difference of about 100 pcm originating mainly from the resonance group.

The MC ^{238}U absorption rates (Table 6.5) of the two calculations with JEF-2.2 agree within the sigma interval, the JAERI results are 120 pcm lower caused by differences in the first and third energy groups. CEA1 shows a value of 65 pcm higher for Case A and a vlaue of 81 pcm lower for Case B, both generated in the first and third group. The ECN solutions differ in the MeV group by -120 pcm and in the resolved resonance group in the range of +250 to +270 pcm. The SRAC results for one fuel zone (JAE1 and JAE3) agree rather well, whereas the solutions with six zones have 300 to 400 pcm lower values in the third group for Cases A and B respectively. In addition the JENDL-3.2 solutions have about 100 pcm lower values in the first group.

For ^{238}Pu Table 6.6a shows a good agreement for all participants. The differences of up to 50 pcm originate from the resolved resonance group.

Tables 6.7a and 6.7b show the results for ^{239}Pu. The absorption rates of CEA2 are 100 and 230 pcm higher than the results of MCNP4; the MC results of JAERI are also higher by up to 326 pcm for Case A and up to 480 pcm for Case B, mainly due to differences in the thermal energy group. CEA1 results agree well for the total value for Case A, but show compensating differences of 92 pcm and -80 pcm in the third and fourth group. For Case B the total difference is 193 pcm originating from 74 and 130 pcm in the third and fourth group. The ECN differences are about -300 pcm for Case A, also caused by deviations in the third and (mainly) fourth group. The agreement for Case B is rather good. The differences of JENDL results are all positive and mostly originate from the thermal group. JENDL-3.1 has 160 and 329 pcm higher values for Cases A and B. JENDL-3.2 shows higher results of up to 376 pcm for Case A and up to 516 pcm higher values for Case B.

Table 6.8 compares the absorption rates of ^{240}Pu. The MC results of CEA2 are 132 pcm and 252 pcm lower for Cases A and B, the MVP differences are 175 pcm generated in the third and fourth group for Case A and are low for Case B, where positive and negative values in the third and fourth group of 70 to 80 pcm cancel. CEA1 has a value of 180 pcm lower, caused by +100 pcm in the resolved resonance group and -281 pcm in the thermal group for Case A. For Case B the differences are similar. The ECN and JAERI results agree rather well with the MC results in both cases, but in Case B the deviations in groups three and four are higher and have a compensating effect.

For ^{241}Pu the agreement of the absorption rates calculated by TRIPOLI-4 and MCNP4 is also good, whereas MVP gives values about 100 pcm lower, originating mainly from the thermal energy region. The CEA1 values are 75 and 91 pcm for A and B summed by about +40 pcm in the third and fourth group. The ECN values are higher for Case A by 230 pcm and 95 pcm for Case B. The discrepancies are mainly produced in the resolved resonance group. The JENDL-3.1 results are up to 226 pcm higher for Case A, which is caused by positive differences in the third and fourth groups. In Case B the differences go up to 138 pcm. The JENDL-3.2 solutions are low and all negative. They add up to -90 pcm.

Tables 6.10a and 6.10b show a rather good agreement for the ^{242}Pu absorption rates. The deviations are all generated in the resolved resonance group. The ECN solutions are 90 pcm higher for Case A and 50 pcm lower for Case B than the other JEF-2.2 appliers. The JENDL results are up to 75 pcm lower than those of IKE1.

Table 6.11 shows the differences among the oxygen absorption rates. All JENDL results are about 100 pcm lower than the results calculated with JEF-2.2; the differences are generated in the first energy group. Comparisons of databases show that the reason for this is a lower threshold for the (n,α) reaction in the JEF-2.2 library than in the JENDL-3 database. The discrepancies observed for the moderator have the same explanation.

Discussion of differences in fission rates

Tables 6.12 to 6.17 give detailed information about the fission rates in fuel. Apart from the (n,2n) and (n,3n) reactions the whole differences in k-infinity are projected onto the fission rates because of the applied normalisation condition. Table 6.12 shows a good agreement of MC results for ^{235}U. The deterministic calculations produced moderate deviations of up to 60 pcm for Case A originating mainly in the third group for JEF-2.2 contributions and in the fourth group for the JENDL-3.2 results. In Case B all differences are very small.

For ^{238}U the discrepancies are of course concentrated in the fast range. All MC results show a rather good agreement. The JEF-2.2 solutions have a different behaviour: APOLLO2 gives 84 pcm higher values and ECN between 78 and 122 pcm lower values than MCNP4. The SRAC calculations with JENDL-3.1 are about 240 pcm higher, but are reduced to between 50 to 85 pcm when applying JENDL-3.2.

Tables 6.8a and 6.8b show a good agreement for the ^{238}Pu fission rates.

The main discrepancies arise for the fission rate of ^{239}Pu in Table 6.15. TRIPOLI-4 calculates 159 and 325 pcm higher fission rates for Cases A and B. The MC calculations applying JENDL-3.2 give 560 to 750 pcm higher results for Case A and between 770 and 900 pcm for Case B, originating in group three (about 250 pcm) and mainly in the thermal energy region. CEA1 shows differences in all groups up to 84 pcm with alternating signs for Case A and a total deviation of 300 pcm for Case B, mainly produced in the thermal region. The ECN results are up to 479 pcm lower for Case A; one-third of the difference is produced in the resolved resonance group, while the rest is produced in the thermal region. In Case B the agreement is rather good with highest deviations of -80 pcm in the third group. JAE1 and JAE2 have rather large differences of 318 and 535 pcm in Case A and B, respectively, which mainly originate from the thermal group. The SRAC calculations with JENDL-3.2 are up to 892 and 997 pcm higher than the MCNP4 solutions, due to different results in the resolved resonance group (200 to 286 pcm) and the thermal group (361 to 789 pcm).

Because of the low contribution of ^{240}Pu to the fission rate effects, low discrepancies are also found for the solutions (Table 6.16). The only remarkable difference emerges from the JAERI calculations, which can is found in the third group and is about -60 pcm.

Tables 6.17a and 6.17b show the fission rate results for ^{241}Pu. The solution of TRIPOLI-4 is 98 and 71 pcm higher than that of MCNP4 for Cases A and B, whereas MVP gives differences of -407 to -478 pcm for Case A and between -342 and -353 pcm for Case B, generated about equally in the third and fourth groups. CEA1 gives 70 pcm higher values for both cases which are caused by differences in the third and fourth groups. The ECN solutions are also higher by 400 pcm in Case A and by 166 pcm in Case B, resulting from about 500 pcm higher fission rates in the resolved resonance region and a value of 100 pcm lower in the thermal group for Case A. In Case B the differences are mainly produced in the third group. SRAC results with JENDL-3.1 data give also higher results between 154 and 259 pcm for Case A and between 90 and 173 pcm for Case B, mainly caused by the thermal group. In contrast to this result, the SRAC solutions applying JENDL-3.2 produce lower fission rates than MCNP4 in the range of 237 and 395 pcm, generated in the third and fourth energy groups.

The contribution of ^{242}Pu to the fission rate is very low and is not a significant contribution to the global discrepancies in both cases.

Local dependency of absorption rates and cross-sections

When defining this benchmark one point of interest has been to obtain information about the influence of the local dependency of spectra and cross-sections in the fuel. In Tables 6.19 to 6.22 the six zone absorption rates are compared for ^{238}U in the second and third groups, for ^{239}Pu in the thermal group and for ^{242}Pu in the third group. Solutions with local and energy dependent result are contributed from IKE1, CEA1, ECN2, JAE2, JAE4 and JAE6. Table 6.19 shows a very good agreement of the solutions for the unresolved resonance absorption of ^{238}U, demonstrating a low influence of the often discussed differences in shielding models in this energy region.

Tables 6.20a and 6.20b show a good agreement for the two MC results in Case A. APOLLO2 has positive and negative differences summed to the values of +40 and -118 pcm reported in Table 6.5.

ECN2 has, in both cases, nearly the same results with +210 pcm in the inner zone decreasing to -130 pcm in the outermost zone. The two SRAC solutions look untypical. The reason for the high deviations in Table 6.5 are deviations of -132 to -354 pcm in the fourth and fifth zones, whereas the agreement with the MC solutions in the outermost zone is rather good.

The absorption rates of ^{239}Pu in the thermal group show a strong local dependency. The surface terms agree well. The deviations increase in the inner zones. Here, APOLLO2 underestimates the absorptions in Case A and overestimates them in Case B. In both cases, ECN2 generates elevated values in the first zone and negative ones in all others. The differences of the JAERI results are all positive for SRAC and MVP calculations.

Table 6.22 shows the absorption rates in the resonance group of ^{242}Pu. The agreement of the MC results is rather good in Case A and very good in Case B. Again the main discrepancies originate from the two innermost zones and the agreement for the surface term is good.

Tables 6.23 to 6.26 show the locally dependent absorption cross-sections as they correspond to the discussed absorption rates. Differences in the databases can be considered responsible for the differences in Tables 6.25 and 6.26. The discrepancies between the SRAC results in Table 6.24 to the other solutions require further explanation by the contributor.

Summary

The main objective of this benchmark – to localise inconsistencies in data and methods applied for MOX pin cell calculations – could be achieved only partially, because a real best-estimate solution is not available. The two MC codes applied with JEF-2.2 data gave differences in k-infinity of 200 to

300 pcm, apparently caused by different conditions in processing the continuous energy libraries with NJOY. (For instance the tolerance for resonance reconstruction uses 0.1% for the results of MCNP4 and 0.5% for TRIPOLI-4. These discrepancies show that the criteria used in the NJOY code to reconstruct resonances to the continuous energy representation are very important and emphasises the necessity to set up a standard procedure to generate data libraries used in calculations.). The results of the two MVP calculations based on JENDL-3.2 data differed by 130 pcm, also well beyond the statistical error bounds. Subdividing the fuel zone in the cell leads to improved results in the SRAC's solutions with six zones seem to show the contrary. However in this case the same Dancoff factor was used for each fuel sub-zone, which is an approximation that cannot be considered adequate.

Detailed comparisons of fission rates of the isotopes in the fuel identified the thermal range of ^{239}Pu as the main source for observed differences between -300 and +270 pcm for the JEF-2.2 users and between +300 and +700 pcm for the JENDL-3.2 calculations. In the resolved resonance group of this isotope the JEF-2.2 results vary from -130 to +84 pcm. The deviations originating from ^{241}Pu are more moderate with about 100 pcm in the resonance and thermal region with the exception of the ECN differences going up to 500 pcm in the resonance group. The results of JENDL-3.2 calculations are about 200 pcm lower in both groups. The fast fission of ^{238}U causes a spread of the JEF-2.2 solutions between -120 and +85 pcm.

Significant contributions to the spread in eigenvalues are caused by the absorptions of the fissile isotopes ^{239}Pu and ^{241}Pu and in addition by ^{240}Pu with differences of -300 pcm for the thermal group in JEF-2.2 solutions and 100 pcm in the resonance group for the users of both databases. The resonance absorption of ^{238}U is still a source of discrepancies at least for the contributions applying JEF-2.2. The resulting differences for group three are between -120 and +280 pcm.

A surprising conclusion resulted from the comparison of the absorption rates of non-actinides in the cell. JENDL-3.2 cell calculations have 110 pcm higher absorption rates in clad and lower ones by 100 pcm for oxygen in fuel and by 80 pcm in the moderator. The reason for the lower oxygen absorption is a different threshold of the (n,α) cross-section in the two databases.

In conclusion, differences emerging from cross-section data used, their processing and differences in the computational schemes still lead to unsatisfactory discrepancies in calculating PWR MOX pin cells by different institutions, although "state-of-the-art" data and methods are used. This means that there is still a need for further analyses of the differences of best estimate solutions with respect to measured data and data processing, and that there is also need for improvement in deterministic methods and basic nuclear data, as well as for better experimental validation.

Table 6.2a. Table of k$_\infty$ and reaction rates, Case A

Participant		Rates				
		Prod.	Fission	Absorption		
	k$_\infty$	Fuel	Fuel	Fuel	Clad	Mod.
IKE1	1.1329	1.1314	0.3936	0.9745	0.0055	0.0200
CEA1	1.1358	1.1341	0.3945	0.9746	0.0052	0.0203
CEA2	1.1352	1.1332	0.3948	0.9754	0.0056	0.0202
ECN1	1.1285	1.1285	0.3927	0.9750	0.0054	0.0196
ECN2	1.1295	1.1295	0.3930	0.9750	0.0054	0.0196
ECN3	1.1295	1.1295	0.3930	0.9750	0.0054	0.0196
JAE1	1.1365	1.1365	0.3951	0.9740	0.0065	0.0195
JAE2	1.1412	1.1412	0.3967	0.9738	0.0065	0.0196
JAE3	1.1347	1.1347	0.3946	0.9735	0.0069	0.0196
JAE4	1.1394	1.1395	0.3962	0.9734	0.0069	0.0197
JAE5	1.1336	1.1352	0.3945	0.9773	0.0066	0.0192
JAE6	1.1320	1.1320	0.3934	0.9750	0.0066	0.0192

Table 6.2b. Table of k$_\infty$ and reaction rates, Case B

Participant		Rates				
		Prod.	Fission	Absorption		
	k$_\infty$	Fuel	Fuel	Fuel	Clad	Mod.
IKE1	1.1839	1.1824	0.4120	0.9634	0.0061	0.0304
CEA1	1.1911	1.1894	0.4144	0.9634	0.0058	0.0309
CEA2	1.1876	1.1874	0.4138	0.9639	0.0062	0.0306
ECN1	1.1815	1.1815	0.4117	0.9639	0.0061	0.0300
ECN2	1.1822	1.1822	0.4119	0.9639	0.0061	0.0301
ECN3	1.1821	1.1821	0.4119	0.9638	0.0061	0.0301
JAE1	1.1892	1.1893	0.4139	0.9630	0.0070	0.0300
JAE2	1.1947	1.1948	0.4159	0.9628	0.0070	0.0302
JAE3	1.1875	1.1875	0.4134	0.9626	0.0073	0.0301
JAE4	1.1930	1.1930	0.4154	0.9624	0.0074	0.0303
JAE5	1.1886	1.1895	0.4140	0.9653	0.0073	0.0297
JAE6	1.1882	1.1875	0.4133	0.9642	0.0074	0.0296

Table 6.3a. Differences of Case A results with respect to the MCNP4 solution (IKE1) in pcm

Participant		Rates				
		Prod.	Fission	Absorption		
	Δk/k	Fuel	Fuel	Fuel	Clad	Mod.
CEA1	253.6	241.4	238.3	10.3	-35.5	24.5
CEA2	196.4	160.4	302.4	93.4	5.6	17.1
ECN1	-389.0	-250.6	-231.4	56.0	-9.6	-47.0
ECN2	-307.2	-168.9	-149.4	49.1	-9.1	-40.7
ECN3	-307.2	-169.7	-150.1	47.9	-9.2	-39.5
JAE1	311.5	450.1	374.0	-51.3	101.1	-50.3
JAE2	724.7	863.1	787.6	-65.7	103.0	-37.9
JAE3	156.0	295.4	253.4	-92.5	135.8	-43.8
JAE4	571.3	710.3	667.4	-107.2	137.9	-31.3
JAE5	56.5	335.0	237.7	283.9	112.7	-80.9
JAE6	-80.4	53.8	-47.5	49.4	112.3	-80.8

Table 6.3b. Differences in Case B results with respect to the MCNP4 solution (IKE1) in pcm

Participant		Rates				
		Prod.	Fission	Absorption		
	Δk/k	Fuel	Fuel	Fuel	Clad	Mod.
CEA1	606.2	581.8	577.4	-5.5	-38.3	43.8
CEA2	307.4	420.9	432.9	50.7	0.5	20.9
ECN1	-204.8	-81.3	-64.2	49.0	-8.2	-40.9
ECN2	-147.2	-24.2	-7.3	42.2	-7.7	-34.5
ECN3	-149.7	-26.4	-9.5	40.3	-7.7	-32.6
JAE1	449.0	572.4	477.9	-43.8	82.4	-38.7
JAE2	906.5	1031.0	934.2	-64.4	84.7	-20.4
JAE3	304.0	428.2	353.7	-85.9	119.9	-34.1
JAE4	764.4	887.4	815.0	-106.7	122.4	-15.9
JAE5	392.1	594.2	494.6	182.5	110.8	-73.3
JAE6	364.4	423.8	327.1	80.2	121.3	-81.0

Table 6.4a. ^{235}U one-and four-group absorption rates of MCNP4 and corresponding differences of the other results in pcm, Case A

Participant	Energy group numbers				
	1/1	1/4	2/4	3/4	4/4
IKE1	0.0260	0.0007	0.0013	0.0137	0.0103
σ	± 3.2	± 0.1	± 0.1	± 2.2	± 2.4
CEA1	-4.2	0.5	-0.3	-16.4	11.9
CEA2	6.5				
ECN1	-18.1	-1.0	0.4	-17.4	-0.1
ECN2	-13.8	-1.1	0.3	-14.6	1.5
ECN3	-14.1	-1.1	0.4	-15.1	1.7
JAE1	87.2	1.5	1.5	73.0	11.4
JAE2	95.1	1.6	1.7	74.1	17.8
JAE3	15.0	0.3	0.7	-4.3	18.6
JAE4	23.1	0.4	0.8	-3.1	25.1
JAE5	9.7	0.1	0.6	0.3	8.8
JAE6	1.9	0.1	0.6	-2.2	3.4

Table 6.4b. ^{235}U one- and four-group absorption rates of MCNP4 and corresponding differences of the other results in pcm, Case B

Participant	Energy group numbers				
	1/1	1/4	2/4	3/4	4/4
IKE1	0.0144	0.0002	0.0005	0.0056	0.0081
σ	± 1.8	± 0.0	± 0.0	± 0.9	± 1.5
CEA1	9.7	0.2	-0.1	-2.9	12.5
CEA2	3.1				
ECN1	1.9	-0.5	0.3	-4.1	6.1
ECN2	3.9	-0.5	0.3	-2.8	7.0
ECN3	3.8	-0.5	0.3	-3.0	7.1
JAE1	37.4	0.9	0.5	28.6	7.4
JAE2	43.1	0.9	0.6	29.4	12.2
JAE3	7.9	0.4	0.2	-3.2	10.4
JAE4	13.6	0.4	0.3	-2.3	15.2
JAE5	4.5	0.0	0.2	-1.9	6.2
JAE6	4.1	-0.1	0.3	-1.5	5.4

Table 6.5a. ^{238}U one- and four-group absorption rates of MCNP4 and corresponding differences of the other results in pcm, Case A

Participant	Energy group numbers				
	1/1	1/4	2/4	3/4	4/4
IKE1	0.2071	0.0312	0.0222	0.1466	0.0071
σ	± 47.0	± 4.6	± 2.3	± 47.2	± 1.5
CEA1	64.7	34.5	-13.6	40.4	3.3
CEA2	46.0				
ECN1	119.4	-137.6	1.1	257.6	-1.7
ECN2	110.6	-140.6	0.6	251.4	-0.7
ECN3	114.0	-140.9	0.9	254.8	-0.7
JAE1	-32.1	-29.9	-26.7	32.4	-7.6
JAE2	-369.3	-22.8	-24.8	-318.1	-3.3
JAE3	-61.1	-98.8	-22.9	56.4	4.2
JAE4	-400.3	-91.9	-21.1	-295.6	8.6
JAE5	-53.4	-36.8	-17.5	2.0	-1.1
JAE6	-41.0	-37.7	-21.1	21.5	-3.7

Table 6.5b. ^{235}U one- and four-group absorption rates of MCNP4 and corresponding differences of the other results in pcm, Case B

Participant	Energy group numbers				
	1/1	1/4	2/4	3/4	4/4
IKE1	0.2348	0.0326	0.0230	0.1638	0.0155
σ	± 50.4	± 4.8	± 2.4	± 50.7	± 2.8
CEA1	-81.0	35.7	-12.8	-118.5	14.8
CEA2	19.4				
ECN1	128.2	-163.9	3.7	280.8	7.6
ECN2	119.6	-166.7	3.2	273.9	9.2
ECN3	124.1	-166.3	3.6	277.6	9.2
JAE1	-65.0	-34.5	-27.2	7.3	-11.1
JAE2	-424.3	-26.7	-25.1	-370.1	-2.3
JAE3	-109.0	-112.1	-23.7	17.3	10.0
JAE4	-469.6	-104.5	-21.7	-362.2	19.0
JAE5	-117.7	-42.6	-20.2	-57.1	2.1
JAE6	-118.8	-55.1	-17.2	-45.5	-1.0

Table 6.6a. ^{238}Pu one- and four-group absorption rates of MCNP4 and corresponding differences of the other results in pcm, Case A

Participant	Energy group numbers				
	1/1	1/4	2/4	3/4	4/4
IKE1	0.0118	0.0009	0.0008	0.0051	0.0050
σ	± 2.5	± 0.1	± 0.1	± 2.0	± 1.4
CEA1	38.7	0.7	-0.3	31.1	7.2
CEA2	2.5				
ECN1	13.6	-0.4	-0.1	12.4	1.7
ECN2	14.8	-0.5	-0.1	12.7	2.7
ECN3	14.7	-0.5	-0.1	12.5	2.8
JAE1	-43.8	4.3	4.4	-53.5	1.0
JAE2	-8.2	4.4	4.5	-21.3	4.2
JAE3	-41.1	2.5	4.4	-52.5	4.5
JAE4	-5.6	2.7	4.6	-20.6	7.7
JAE5	45.0	0.9	4.8	42.3	-2.9
JAE6	39.2	0.9	4.8	39.4	-5.9

Table 6.6b. ^{238}Pu one- and four-group absorption rates of MCNP4 and corresponding differences of the other results in pcm, Case B

Participant	Energy group numbers				
	1/1	1/4	2/4	3/4	4/4
IKE1	0.0037	0.0002	0.0001	0.0012	0.0022
σ	± 0.7	± 0.0	± 0.0	± 0.5	± 0.5
CEA1	6.8	0.1	-0.1	3.0	3.7
CEA2	0.5				
ECN1	0.3	-0.2	0.0	-1.9	2.5
ECN2	0.7	-0.2	0.0	-1.9	2.8
ECN3	0.7	-0.2	0.0	-1.9	2.8
JAE1	-4.6	0.8	0.8	-5.6	-0.6
JAE2	-1.5	0.8	0.9	-3.8	0.7
JAE3	-4.1	0.4	0.9	-5.5	0.1
JAE4	-1.0	0.5	0.9	-3.8	1.4
JAE5	7.8	0.1	0.9	8.4	-1.7
JAE6	7.5	0.1	0.9	8.1	-1.7

Table 6.7a. ^{239}Pu one- and four-group absorption rates of MCNP4 and corresponding differences of the other results in pcm, Case A

Participant	Energy group numbers				
	1/1	1/4	2/4	3/4	4/4
IKE1	0.3651	0.0070	0.0093	0.0984	0.2504
σ	± 59.3	± 0.8	± 0.7	± 19.8	± 57.5
CEA1	-2.3	5.4	-19.1	91.7	-79.7
CEA2	99.1				
ECN1	-325.9	-3.7	6.4	-104.1	-224.0
ECN2	-302.8	-4.5	6.1	-94.9	-208.9
ECN3	-301.3	-4.5	6.3	-98.5	-204.1
JAE1	-3.1	12.6	-14.9	-8.7	9.0
JAE2	155.6	13.8	-13.6	-6.1	162.5
JAE3	216.9	-0.5	-14.4	23.1	209.0
JAE4	376.1	0.7	-13.1	25.3	364.3
JAE5	326.1	-10.7	-12.5	9.0	340.9
JAE6	206.4	-10.9	-12.9	24.2	206.5

Table 6.7b. ^{239}Pu one- and four-group absorption rates of MCNP4 and corresponding differences of the other results in pcm, Case B

Participant	Energy group numbers				
	1/1	1/4	2/4	3/4	4/4
IKE1	0.4590	0.0050	0.0065	0.0803	0.3672
σ	± 69.4	± 0.6	± 0.5	± 16.2	± 67.3
CEA1	193.9	3.5	-13.6	74.1	129.4
CEA2	229.3				
ECN1	-62.0	-5.8	5.1	-30.0	-31.9
ECN2	-41.9	-6.2	4.9	-21.7	-19.5
ECN3	-43.6	-6.1	5.1	-24.4	-18.8
JAE1	121.5	8.1	-10.8	5.8	117.7
JAE2	341.7	9.0	-9.8	14.2	328.9
JAE3	293.5	-1.8	-10.5	20.3	285.7
JAE4	515.8	-0.9	-9.5	28.5	497.4
JAE5	479.2	-8.7	-9.4	35.1	461.4
JAE6	410.7	-9.6	-8.3	47.6	380.4

Table 6.8a. ^{240}Pu one- and four-group absorption rates of MCNP4 and corresponding differences of the other results in pcm, Case A

Participant	Energy group numbers				
	1/1	1/4	2/4	3/4	4/4
IKE1	0.1838	0.0048	0.0025	0.0318	0.1447
σ	± 53.3	± 0.6	± 0.2	± 10.9	± 52.4
CEA1	-180.1	3.9	-1.7	98.9	-281.3
CEA2	-132.0				
ECN1	-28.7	-3.1	6.9	-14.9	-17.6
ECN2	-65.3	-3.6	6.8	-11.4	-57.1
ECN3	-68.8	-3.7	6.8	-12.2	-59.9
JAE1	-51.2	11.3	2.8	87.5	-152.9
JAE2	9.3	12.1	3.2	81.0	-87.1
JAE3	34.8	2.1	3.0	100.4	-70.9
JAE4	95.7	2.9	3.5	94.0	-4.7
JAE5	175.5	-5.5	4.8	92.3	83.9
JAE6	105.8	-5.6	5.0	126.1	-19.8

Table 6.8b. ^{240}Pu one- and four-group absorption rates of MCNP4 and corresponding differences of the other results in pcm, Case B

Participant	Energy group numbers				
	1/1	1/4	2/4	3/4	4/4
IKE1	0.1386	0.0017	0.0009	0.0145	0.1215
σ	± 45.1	± 0.2	± 0.1	± 5.2	± 44.7
CEA1	-228.7	1.3	-0.6	69.5	-298.9
CEA2	-251.8				
ECN1	-36.3	-2.2	0.7	44.3	-79.1
ECN2	-69.0	-2.3	0.7	46.6	-114.0
ECN3	-72.3	-2.3	0.7	46.3	-117.0
JAE1	-88.8	3.7	1.0	62.8	-155.7
JAE2	-32.6	4.1	1.1	60.9	-98.7
JAE3	-50.8	0.3	1.1	65.7	-117.7
JAE4	5.1	0.6	1.2	63.9	-60.6
JAE5	-15.7	-2.3	1.7	73.8	-88.8
JAE6	-18.9	-2.6	1.8	68.1	-86.1

Table 6.9a. ^{241}Pu one- and four-group absorption rates of MCNP4 and corresponding differences of the other results in pcm, Case A

Participant	Energy group numbers				
	1/1	1/4	2/4	3/4	4/4
IKE1	0.1330	0.0022	0.0040	0.0588	0.0681
σ	± 18.9	± 0.3	± 0.3	± 10.3	± 16.1
CEA1	75.3	1.6	-0.9	49.9	24.8
CEA2	42.9				
ECN1	216.9	-3.9	0.0	288.3	-67.6
ECN2	237.6	-4.2	-0.1	300.0	-58.2
ECN3	236.4	-4.2	0.0	297.1	-56.6
JAE1	170.2	-6.7	-10.1	133.7	52.9
JAE2	226.0	-6.3	-9.6	146.1	95.8
JAE3	-90.8	-10.5	-10.5	-37.7	-31.9
JAE4	-35.0	-10.2	-10.0	-25.4	10.7
JAE5	-92.9	-10.6	-9.6	-10.8	-62.0
JAE6	-130.8	-10.6	-9.6	-6.9	-103.7

Table 6.9b. ^{241}Pu one- and four-group absorption rates of MCNP4 and corresponding differences of the other results in pcm, Case B

Participant	Energy group numbers				
	1/1	1/4	2/4	3/4	4/4
IKE1	0.0978	0.0010	0.0018	0.0300	0.0650
σ	± 13.3	± 0.1	± 0.1	± 5.2	± 12.2
CEA1	90.7	0.6	-0.4	41.6	49.0
CEA2	36.6				
ECN1	83.0	-2.3	0.1	100.2	-14.8
ECN2	95.7	-2.4	0.0	107.1	-9.0
ECN3	95.1	-2.4	0.1	105.9	-8.4
JAE1	91.6	-3.1	-4.4	65.5	33.8
JAE2	137.9	-3.0	-4.2	73.4	71.8
JAE3	-93.2	-4.9	-4.6	-22.5	-61.1
JAE4	-47.2	-4.8	-4.4	-14.6	-23.5
JAE5	-102.5	-4.9	-4.3	-16.6	-76.7
JAE6	-107.2	-5.0	-4.0	-12.7	-85.4

Table 6.10a. ^{242}Pu one- and four-group absorption rates of MCNP4 and corresponding differences of the other results in pcm, Case A

Participant	Energy group numbers				
	1/1	1/4	2/4	3/4	4/4
IKE1	0.0451	0.0029	0.0013	0.0391	0.0017
σ	± 24.7	± 0.4	± 0.1	± 24.8	± 0.3
CEA1	16.0	2.3	-1.0	13.9	0.6
CEA2	28.8				
ECN1	94.1	-2.6	-0.5	97.7	-0.5
ECN2	83.5	-2.9	-0.6	87.3	-0.3
ECN3	82.5	-3.0	-0.6	86.3	-0.3
JAE1	-74.3	4.9	0.6	-75.8	-4.0
JAE2	-71.1	5.4	0.8	-74.2	-3.0
JAE3	-64.0	-0.6	0.7	-61.3	-2.9
JAE4	-60.8	-0.1	0.9	-59.8	-1.9
JAE5	-29.3	-4.4	1.7	-24.4	-2.3
JAE6	-35.5	-4.4	1.8	-30.3	-2.6

Table 6.10b. ^{242}Pu one- and four-group absorption rates of MCNP4 and corresponding differences of the other results in pcm, Case B

Participant	Energy group numbers				
	1/1	1/4	2/4	3/4	4/4
IKE1	0.0126	0.0003	0.0001	0.0119	0.0003
σ	± 8.9	± 0.0	± 0.0	± 8.8	± 0.0
CEA1	-1.5	0.2	-0.1	-1.9	0.3
CEA2	10.1				
ECN1	-52.5	-0.4	0.0	-52.2	0.1
ECN2	-53.0	-0.4	0.0	-52.7	0.1
ECN3	-53.6	-0.4	0.0	-53.3	0.1
JAE1	-37.2	0.4	0.1	-37.4	-0.3
JAE2	-31.8	0.5	0.1	-32.3	-0.1
JAE3	-34.9	-0.1	0.1	-34.8	-0.2
JAE4	-29.5	0.0	0.1	-29.5	0.0
JAE5	17.8	-0.4	0.1	18.1	0.0
JAE6	-4.2	-0.5	0.2	-3.7	-0.1

Table 6.11a. Oxygen one- and four-group absorption rates of MCNP4 and corresponding differences of the other results in pcm, Case A

Participant	Energy group numbers				
	1/1	1/4	2/4	3/4	4/4
IKE1	0.0026	0.0026	0.0000	0.0000	0.0000
σ	± 1.2	± 1.1	± 0.0	± 0.0	± 0.0
CEA1	2.3	2.3	0.0	0.0	0.0
CEA2	-0.4				
ECN1	-15.2	-15.2	0.0	0.0	0.0
ECN2	-15.4	-15.4	0.0	0.0	0.0
ECN3	-15.5	-15.5	0.0	0.0	0.0
JAE1	-99.6	-99.6	0.0	0.0	0.0
JAE2	-98.9	-98.9	0.0	0.0	0.0
JAE3	-103.2	-104.4	1.1	0.0	0.0
JAE4	-102.5	-103.7	1.1	0.0	0.0
JAE5	-96.8	-98.0	1.2	0.0	0.0
JAE6	-96.5	-97.7	1.2	0.0	0.0

Table 6.11b. Oxygen one- and four-group absorption rates of MCNP4 and corresponding differences of the other results in pcm, Case B

Participant	Energy group numbers				
	1/1	1/4	2/4	3/4	4/4
IKE1	0.0025	0.0025	0.0000	0.0000	0.0000
σ	± 1.1	± 1.1	± 0.0	± 0.0	± 0.0
CEA1	4.7	4.7	0.0	0.0	0.0
CEA2	1.8				
ECN1	-13.5	-13.5	0.0	0.0	0.0
ECN2	-13.7	-13.7	0.0	0.0	0.0
ECN3	-13.7	-13.7	0.0	0.0	0.0
JAE1	-93.4	-93.4	0.0	0.0	0.0
JAE2	-92.7	-92.7	0.0	0.0	0.0
JAE3	-97.6	-98.7	1.1	0.0	0.0
JAE4	-96.8	-98.0	1.1	0.0	0.0
JAE5	-90.8	-91.9	1.1	0.0	0.0
JAE6	-92.7	-93.9	1.1	0.0	0.0

Table 6.12a. ^{235}U one- and four-group fission rates of MCNP 4 and corresponding differences of the other results in pcm, Case A

Participant	Energy group numbers				
	1/1	1/4	2/4	3/4	4/4
IKE1	0.0194	0.0006	0.0011	0.0089	0.0088
σ	± 2.5	± 0.1	± 0.1	± 1.4	± 2.1
CEA1	-15.0	1.3	-0.6	-41.1	25.5
CEA2	15.6				
ECN1	-42.6	-0.6	0.5	-42.1	-0.4
ECN2	-35.5	-0.7	0.4	-38.3	3.2
ECN3	-35.9	-0.7	0.5	-39.1	3.5
JAE1	-18.9	7.3	-2.4	-36.8	13.0
JAE2	-3.1	7.6	-2.0	-35.4	26.8
JAE3	36.6	4.3	-1.1	-11.9	45.4
JAE4	52.8	4.5	-0.7	-10.3	59.3
JAE5	19.1	1.5	-1.0	-6.0	24.7
JAE6	4.8	1.5	-1.0	-8.9	13.3

Table 6.12b. ^{235}U one- and four-group fission rates of MCNP 4 and corresponding differences of the other results in pcm, Case B

Participant	Energy group numbers				
	1/1	1/4	2/4	3/4	4/4
IKE1	0.0111	0.0002	0.0004	0.0036	0.0069
σ	± 1.4	± 0.0	± 0.0	± 0.6	± 1.3
CEA1	15.4	0.4	-0.2	-10.4	25.6
CEA2	8.5				
ECN1	0.5	-0.5	0.4	-12.0	12.6
ECN2	3.8	-0.6	0.4	-10.4	14.5
ECN3	3.7	-0.6	0.4	-10.7	14.6
JAE1	-8.1	2.5	-0.8	-16.2	6.5
JAE2	3.0	2.6	-0.7	-15.0	16.2
JAE3	19.2	1.3	-0.4	-6.6	24.8
JAE4	30.3	1.4	-0.3	-5.4	34.6
JAE5	11.8	0.4	-0.4	-4.6	16.4
JAE6	9.9	0.3	-0.3	-4.8	14.6

Table 6.13a. ^{238}U one- and four-group fission rates of MCNP4 and corresponding differences of the other results in pcm, Case A

Participant	Energy group numbers				
	1/1	1/4	2/4	3/4	4/4
IKE1	0.0275	0.0274	0.0000	0.0000	0.0000
σ	± 4.4	± 4.4	± 0.0	± 0.0	± 0.0
CEA1	84.4	85.0	-0.4	-0.1	0.0
CEA2	24.9				
ECN1	-70.9	-70.5	-0.4	0.0	0.0
ECN2	-77.9	-77.4	-0.4	0.0	0.0
ECN3	-78.8	-78.3	-0.4	0.0	0.0
JAE1	230.7	232.3	-1.5	0.0	0.0
JAE2	247.3	249.0	-1.5	0.0	0.0
JAE3	67.6	69.2	-1.5	0.0	0.0
JAE4	84.5	85.9	-1.4	0.0	0.0
JAE5	-27.9	-26.9	-1.0	0.0	0.0
JAE6	-31.0	-29.9	-1.0	-0.1	0.0

Table 6.13b. ^{238}U one- and four-group fission rates of MCNP4 and corresponding differences of the other results in pcm, Case B

Participant	Energy group numbers				
	1/1	1/4	2/4	3/4	4/4
IKE1	0.0287	0.0286	0.0000	0.0000	0.0000
σ	± 4.6	± 4.6	± 0.0	± 0.0	± 0.0
CEA1	84.2	84.8	-0.5	-0.1	0.0
CEA2	26.1				
ECN1	-116.6	-116.2	-0.4	0.0	0.0
ECN2	-122.9	-122.4	-0.4	0.0	0.0
ECN3	-122.1	-121.7	-0.4	0.0	0.0
JAE1	226.0	227.6	-1.5	0.0	0.0
JAE2	243.0	244.6	-1.5	0.0	0.0
JAE3	50.2	51.8	-1.5	0.0	0.0
JAE4	67.5	69.2	-1.4	0.0	0.0
JAE5	-36.0	-34.9	-1.0	0.0	0.0
JAE6	-66.6	-65.5	-1.0	-0.1	0.0

Table 6.14a. ^{238}Pu one- and four-group fission rates of MCNP4 and corresponding differences of the other results in pcm, Case A

Participant	Energy group numbers				
	1/1	1/4	2/4	3/4	4/4
IKE1	0.0024	0.0009	0.0005	0.0008	0.0002
σ	± 0.3	± 0.1	± 0.1	± 0.3	± 0.0
CEA1	10.3	1.8	-0.8	8.8	0.5
CEA2	-0.8				
ECN1	1.3	-0.9	-0.3	2.4	0.1
ECN2	1.2	-1.1	-0.3	2.5	0.2
ECN3	1.2	-1.2	-0.3	2.4	0.2
JAE1	-3.5	8.1	0.8	-12.3	-0.2
JAE2	5.1	8.4	1.2	-4.6	0.1
JAE3	-7.2	3.7	0.9	-12.0	0.1
JAE4	1.4	4.1	1.3	-4.3	0.4
JAE5	5.3	-0.2	2.1	3.9	-0.5
JAE6	4.6	-0.3	2.1	3.4	-0.7

Table 6.14b. ^{238}Pu one- and four-group fission rates of MCNP4 and corresponding differences of the other results in pcm, Case B

Participant	Energy group numbers				
	1/1	1/4	2/4	3/4	4/4
IKE1	0.0005	0.0002	0.0001	0.0002	0.0001
σ	± 0.1	± 0.0	± 0.0	± 0.1	± 0.0
CEA1	1.7	0.3	-0.2	1.3	0.3
CEA2	-0.3				
ECN1	-0.6	-0.4	-0.1	-0.3	0.2
ECN2	-0.6	-0.5	-0.1	-0.2	0.2
ECN3	-0.6	-0.5	-0.1	-0.3	0.2
JAE1	0.3	1.4	0.1	-1.3	0.0
JAE2	0.9	1.5	0.2	-0.9	0.1
JAE3	-0.4	0.6	0.2	-1.2	0.1
JAE4	0.1	0.6	0.2	-0.9	0.2
JAE5	0.8	-0.1	0.3	0.7	-0.1
JAE6	0.6	-0.2	0.4	0.5	-0.1

Table 6.15a. ^{239}Pu one- and four-group fission rates of MCNP4 and corresponding differences of other results in pcm, Case A

Participant	Energy group numbers				
	1/1	1/4	2/4	3/4	4/4
IKE1	0.2335	0.0069	0.0076	0.0571	0.1619
σ	± 37.8	± 0.8	± 0.6	± 11.5	± 36.2
CEA1	-21.5	13.6	-44.7	83.9	-74.5
CEA2	159.3				
ECN1	-479.0	-5.8	2.9	-143.3	-333.4
ECN2	-437.9	-7.6	2.1	-128.2	-304.5
ECN3	-435.0	-7.7	2.6	-133.5	-296.7
JAE1	55.2	45.9	-28.6	8.9	29.1
JAE2	318.1	48.8	-25.3	15.5	278.8
JAE3	630.5	13.1	-27.4	281.9	361.1
JAE4	892.2	16.0	-24.1	286.9	613.5
JAE5	755.6	-17.3	-21.5	255.3	538.9
JAE6	559.1	-17.6	-22.5	277.9	321.1

Table 6.15b. ^{239}Pu one- and four-group fission rates of MCNP4 and corresponding differences of other results in pcm, Case B

Participant	Energy group numbers				
	1/1	1/4	2/4	3/4	4/4
IKE1	0.2952	0.0049	0.0053	0.0467	0.2382
σ	± 43.0	± 0.6	± 0.4	± 9.4	± 42.4
CEA1	300.0	8.3	-30.6	52.9	271.0
CEA2	326.8				
ECN1	-86.0	-11.3	2.8	-79.6	2.7
ECN2	-50.9	-12.4	2.3	-66.8	26.6
ECN3	-52.7	-12.2	2.7	-70.5	27.9
JAE1	191.1	29.0	-20.1	-4.4	187.5
JAE2	534.8	31.0	-17.9	7.9	515.3
JAE3	636.4	5.3	-19.3	192.9	458.6
JAE4	984.0	7.3	-17.0	204.3	789.0
JAE5	897.3	-14.2	-15.9	229.8	698.0
JAE6	773.8	-16.5	-14.0	225.0	579.8

Table 6.16a. ^{240}Pu one- and four-group fission rates of MCNP4 and corresponding differences of the other results in pcm, Case A

Participant	Energy group numbers				
	1/1	1/4	2/4	3/4	4/4
IKE1	0.0061	0.0046	0.0010	0.0005	0.0000
σ	± 0.6	± 0.6	± 0.1	± 0.1	± 0.0
CEA1	9.0	9.8	-4.0	3.4	-0.1
CEA2	3.0				
ECN1	3.7	-4.5	-2.3	10.5	0.0
ECN2	2.4	-5.7	-2.5	10.6	0.0
ECN3	2.4	-5.8	-2.4	10.6	0.0
JAE1	-46.7	30.4	-12.1	-64.4	-0.5
JAE2	-43.7	32.3	-11.2	-64.3	-0.5
JAE3	-68.9	7.7	-11.8	-64.3	-0.5
JAE4	-66.0	9.7	-11.0	-64.2	-0.4
JAE5	-86.8	-13.7	-6.7	-66.0	-0.4
JAE6	-86.6	-14.0	-6.2	-66.0	-0.5

Table 6.16b. ^{240}Pu one- and four-group fission rates of MCNP4 and corresponding differences of the other results in pcm, Case B

Participant	Energy group numbers				
	1/1	1/4	2/4	3/4	4/4
IKE1	0.0022	0.0017	0.0004	0.0002	0.0000
σ	± 0.2	± 0.2	± 0.1	± 0.1	± 0.0
CEA1	3.2	3.0	-1.4	1.7	-0.1
CEA2	0.6				
ECN1	-1.9	-4.0	-0.9	3.0	0.0
ECN2	-2.3	-4.4	-1.0	3.1	-0.1
ECN3	-2.2	-4.3	-0.9	3.1	-0.1
JAE1	-20.2	9.6	-4.2	-25.2	-0.4
JAE2	-19.1	10.3	-3.9	-25.2	-0.4
JAE3	-28.3	1.4	-4.1	-25.2	-0.4
JAE4	-27.2	2.1	-3.8	-25.1	-0.4
JAE5	-33.9	-5.5	-2.4	-25.7	-0.4
JAE6	-34.6	-6.3	-2.2	-25.8	-0.4

Table 6.17a. ^{241}Pu one- and four-group fission rates of MCNP4 and corresponding differences of the other results in pcm, Case A

Participant	Energy group numbers				
	1/1	1/4	2/4	3/4	4/4
IKE1	0.1014	0.0020	0.0033	0.0454	0.0508
σ	± 14.3	± 0.2	± 0.3	± 7.3	± 12.1
CEA1	167.5	3.8	-2.1	119.2	46.7
CEA2	97.7				
ECN1	359.2	-1.5	0.0	487.3	-126.5
ECN2	402.2	-2.0	-0.3	512.6	-107.9
ECN3	399.9	-2.1	-0.1	507.0	-104.8
JAE1	154.0	23.4	6.0	16.6	108.3
JAE2	259.3	24.2	7.1	39.3	188.5
JAE3	-395.0	14.0	5.5	-269.5	-144.7
JAE4	-288.8	14.8	6.6	-245.6	-64.7
JAE5	-406.9	5.3	7.3	-212.6	-206.9
JAE6	-478.1	5.3	7.2	-206.6	-283.9

Table 6.17b. ^{241}Pu one- and four-group fission rates of MCNP4 and corresponding differences of the other results in pcm, Case B

Participant	Energy group numbers				
	1/1	1/4	2/4	3/4	4/4
IKE1	0.0739	0.0009	0.0014	0.0231	0.0486
σ	± 10.0	± 0.1	± 0.1	± 3.7	± 9.1
CEA1	172.7	1.4	-0.9	86.5	85.4
CEA2	71.0				
ECN1	141.2	-1.9	0.1	170.0	-26.9
ECN2	166.4	-2.1	0.0	184.5	-15.9
ECN3	165.2	-2.1	0.1	182.1	-14.8
JAE1	90.0	9.5	2.4	2.9	75.3
JAE2	172.8	9.8	2.9	16.5	143.6
JAE3	-321.0	5.3	2.2	-142.7	-185.6
JAE4	-237.5	5.6	2.7	-128.3	-117.3
JAE5	-342.3	1.8	2.9	-130.8	-216.1
JAE6	-352.7	1.4	3.3	-124.8	-232.6

Table 6.18a. ^{242}Pu one- and four-group fission rates of MCNP4 and corresponding differences of the other results in pcm, Case A

Participant	Energy group numbers				
	1/1	1/4	2/4	3/4	4/4
IKE1	0.0033	0.0028	0.0004	0.0001	0.0000
σ	± 0.4	± 0.3	± 0.1	± 0.0	± 0.0
CEA1	3.6	5.8	-2.4	0.1	0.0
CEA2	2.7				
ECN1	-3.0	-2.4	-1.8	1.2	0.0
ECN2	-3.9	-3.2	-1.9	1.1	0.0
ECN3	-3.9	-3.2	-1.8	1.1	0.0
JAE1	3.1	15.5	-5.6	-6.5	-0.3
JAE2	4.7	16.7	-5.2	-6.5	-0.3
JAE3	-10.2	2.1	-5.5	-6.5	-0.3
JAE4	-8.6	3.3	-5.2	-6.5	-0.3
JAE5	-20.5	-11.1	-2.7	-6.5	-0.3
JAE6	-20.3	-11.2	-2.4	-6.5	-0.3

Table 6.18b. ^{242}Pu one- and four-group fission rates of MCNP4 and corresponding differences of the other results in pcm, Case B

Participant	Energy group numbers				
	1/1	1/4	2/4	3/4	4/4
IKE1	0.0003	0.0002	0.0000	0.0000	0.0000
σ	± 0.0	± 0.0	± 0.0	± 0.0	± 0.0
CEA1	0.2	0.4	-0.2	0.0	0.0
CEA2	0.2				
ECN1	-0.8	-0.6	-0.2	-0.1	0.0
ECN2	-0.9	-0.6	-0.2	-0.1	0.0
ECN3	-0.8	-0.6	-0.2	-0.1	0.0
JAE1	-1.2	1.2	-0.5	-1.8	0.0
JAE2	-1.0	1.3	-0.4	-1.8	0.0
JAE3	-2.3	0.0	-0.5	-1.8	0.0
JAE4	-2.2	0.1	-0.4	-1.8	0.0
JAE5	-3.1	-1.0	-0.2	-1.8	0.0
JAE6	-3.2	-1.1	-0.2	-1.8	0.0

Table 6.19a. Six zone absorption rates of ^{238}U in Group 2 calculated with MCNP4 and corresponding differences of other results in pcm, Case A

Participant	Zone numbers					
	1	2	3	4	5	6
IKE1	0.0089	0.0067	0.0022	0.0022	0.0011	0.0011
σ	± 1.0	± 0.7	± 0.2	± 0.2	± 0.1	± 0.1
CEA1	-5.0	-4.4	-1.4	-1.6	-0.5	-0.7
ECN2	-1.8	2.5	0.5	-0.2	-0.2	-0.3
JAE2	-5.6	-8.8	-2.3	-2.9	-2.3	-3.0
JAE4	-4.1	-7.7	-2.0	-2.5	-2.1	-2.8
JAE6	-9.0	-5.9	-2.0	-2.2	-1.0	-1.0

Table 6.19b. Six zone absorption rates of ^{238}U in Group 2 calculated with MCNP4 and corresponding differences of other results in pcm, Case B

Participant	Zone numbers					
	1	2	3	4	5	6
IKE1	0.0092	0.0069	0.0023	0.0023	0.0011	0.0011
σ	± 1.0	± 0.7	± 0.2	± 0.2	± 0.1	± 0.1
CEA1	-5.3	-4.2	-1.1	-1.3	-0.5	-0.4
ECN2	-1.4	3.4	1.1	0.3	-0.1	0.0
JAE2	-6.1	-9.0	-2.1	-2.7	-2.3	-2.9
JAE4	-4.7	-8.0	-1.8	-2.3	-2.1	-2.7
JAE6	-7.4	-5.2	-1.6	-1.6	-0.7	-0.6

Table 6.20a. Six zone absorption rates of ^{238}U in Group 3 calculated with MCNP4 and corresponding differences of other results in pcm, Case A

Participant	Zone numbers					
	1	2	3	4	5	6
IKE1	0.0425	0.0367	0.0146	0.0179	0.0126	0.0224
σ	± 11.5	± 9.2	± 4.8	± 6.3	± 5.4	± 10.1
CEA1	40.9	12.9	-5.7	7.5	17.7	-32.9
ECN2	207.1	128.6	39.4	33.1	-19.0	-137.8
JAE2	5.3	78.7	8.1	-133.5	-322.5	45.8
JAE4	17.8	86.6	10.3	-131.8	-321.8	43.3
JAE6	-13.1	8.8	1.8	5.6	13.8	4.6

Table 6.20b. Six zone absorption rates of ^{238}U in Group 3 calculated with MCNP4 and corresponding differences of other results in pcm, Case B

Participant	Zone numbers					
	1	2	3	4	5	6
IKE1	0.0479	0.0413	0.0162	0.0198	0.0139	0.0248
σ	± 12.0	± 10.3	± 5.2	± 6.7	± 5.8	± 10.7
CEA1	-42.8	-64.5	-22.7	1.8	17.7	-8.0
ECN2	211.5	115.5	43.1	44.4	-9.9	-130.7
JAE2	6.8	65.6	6.6	-145.9	-354.1	50.9
JAE4	15.4	69.8	7.2	-145.9	-354.5	45.8
JAE6	-33.3	-18.2	-9.5	11.3	7.2	-3.0

Table 6.21a. Six zone absorption rates of ^{239}Pu in Group 4 calculated with MCNP4 and corresponding differences of other results in pcm, Case A

Participant	Zone numbers					
	1	2	3	4	5	6
IKE1	0.0766	0.0742	0.0294	0.0328	0.0180	0.0195
σ	± 19.9	± 15.6	± 6.5	± 6.9	± 4.1	± 4.5
CEA1	-29.6	-38.4	-4.8	-12.1	0.4	4.8
ECN2	94.0	-130.2	-56.7	-53.9	-33.5	-28.5
JAE2	7.0	34.5	38.4	43.7	18.4	20.5
JAE4	72.8	94.2	61.4	68.8	32.0	35.1
JAE6	59.2	63.5	26.5	16.5	16.2	24.6

Table 6.21b. Six zone absorption rates of ^{239}Pu in Group 4 calculated with MCNP4 and corresponding differences of other results in pcm, Case B

Participant	Zone numbers					
	1	2	3	4	5	6
IKE1	0.1257	0.1100	0.0407	0.0435	0.0230	0.0242
σ	± 25.1	± 18.7	± 7.3	± 7.4	± 4.1	± 4.6
CEA1	29.0	33.0	22.4	18.5	20.2	6.3
ECN2	205.6	-113.2	-46.6	-28.8	-19.0	-17.5
JAE2	72.0	82.0	58.6	67.8	27.6	20.9
JAE4	130.0	132.0	77.4	87.9	38.2	31.9
JAE6	128.6	103.6	46.5	48.3	27.2	26.2

Table 6.22a. Six zone absorption rates of ^{242}Pu in Group 3 calculated with MCNP4 and corresponding differences of other results in pcm, Case A

Participant	Zone numbers					
	1	2	3	4	5	6
IKE1	0.0113	0.0110	0.0046	0.0054	0.0032	0.0037
σ	± 7.1	± 6.5	± 3.0	± 3.5	± 2.2	± 2.6
CEA1	24.5	-5.5	-3.2	-3.2	-0.2	1.5
ECN2	85.7	2.6	-4.3	-2.5	0.4	5.4
JAE2	-21.9	-28.1	-7.9	-8.4	-5.7	-2.2
JAE4	-17.9	-24.2	-6.2	-6.3	-4.5	-0.6
JAE6	-12.6	-20.6	-2.7	2.3	1.1	2.1

Table 6.22b. Six zone absorption rates of ^{242}Pu in Group 3 calculated with MCNP4 and corresponding differences of other results in pcm, Case B

Participant	Zone numbers					
	1	2	3	4	5	6
IKE1	0.0045	0.0036	0.0012	0.0013	0.0006	0.0006
σ	± 3.5	± 2.5	± 0.9	± 0.9	± 0.5	± 0.5
CEA1	-0.2	-2.2	-0.4	0.1	0.3	0.6
ECN2	-18.9	-19.4	-5.8	-4.7	-2.3	-1.7
JAE2	-12.2	-11.1	-3.3	-2.9	-1.6	-1.2
JAE4	-11.1	-10.3	-3.0	-2.7	-1.4	-1.0
JAE6	0.5	-1.8	-1.1	-0.5	-0.7	-0.1

Table 6.23a. ^{238}U microscopic absorption cross-sections (barn) of Group 2 in six fuel zones, Case A

Participant	Zone numbers					
	1	2	3	4	5	6
IKE1	0.246	0.248	0.248	0.249	0.249	0.250
CEA1	0.246	0.247	0.248	0.248	0.249	0.249
JAE2	0.243	0.244	0.244	0.245	0.245	0.245
JAE4	0.243	0.244	0.245	0.245	0.245	0.246
JAE6	0.243	0.244	0.245	0.245	0.246	0.246

Table 6.23b. ^{238}U microscopic absorption cross-sections (barn) of Group 2 in six fuel zones, Case B

Participant	Zone numbers					
	1	2	3	4	5	6
IKE1	0.247	0.248	0.248	0.249	0.249	0.250
CEA1	0.246	0.247	0.248	0.249	0.249	0.250
JAE2	0.243	0.244	0.245	0.245	0.245	0.245
JAE4	0.243	0.244	0.245	0.245	0.245	0.246
JAE6	0.243	0.244	0.245	0.246	0.246	0.247

Table 6.24a. ^{238}U microscopic absorption cross-sections (barn) of Group 3 in six fuel zones, Case A

Participant	Zone numbers					
	1	2	3	4	5	6
IKE1	1.927	2.183	2.571	3.139	4.391	7.760
CEA1	1.933	2.187	2.561	3.159	4.468	7.694
JAE2	1.922	2.230	2.582	2.903	3.275	7.948
JAE4	1.923	2.230	2.580	2.899	3.270	7.922
JAE6	1.915	2.184	2.570	3.140	4.427	7.751

Table 6.24b. ^{238}U microscopic absorption cross-sections (barn) of Group 3 in six fuel zones, Case B

Participant	Zone numbers					
	1	2	3	4	5	6
IKE1	1.923	2.181	2.553	3.106	4.341	7.724
CEA1	1.891	2.142	2.514	3.110	4.400	7.733
JAE2	1.919	2.218	1.560	2.874	3.240	7.906
JAE4	1.920	2.217	2.558	2.870	3.235	7.880
JAE6	1.907	2.170	2.536	3.118	4.355	7.693

Table 6.25a. ^{239}Pu microscopic absorption cross-sections (barn) of Group 4 in six fuel zones, Case A

Participant	Zone numbers					
	1	2	3	4	5	6
IKE1	622.1	698.2	752.9	790.2	821.7	844.5
CEA1	621.7	697.5	754.5	790.5	822.4	847.5
JAE2	647.3	723.0	778.0	813.3	843.8	868.3
JAE4	648.4	724.1	779.0	814.3	844.8	869.3
JAE6	638.8	716.2	771.3	807.4	840.3	865.9

Table 6.25b. ^{239}Pu microscopic absorption cross-sections (barn) of Group 4 in six fuel zones, Case B

Participant	Zone numbers					
	1	2	3	4	5	6
IKE1	710.1	762.8	799.8	823.3	844.0	860.5
CEA1	709.4	763.4	801.6	825.3	846.6	862.8
JAE2	728.3	781.4	818.6	842.0	862.6	880.2
JAE4	729.2	782.4	819.6	843.0	863.7	881.3
JAE6	725.8	778.5	817.7	839.6	860.3	877.8

Table 6.26a. ^{242}Pu microscopic absorption cross-sections (barn) of Group 3 in six fuel zones, Case A

Participant	Zone numbers					
	1	2	3	4	5	6
IKE1	60.65	61.56	62.26	62.63	62.97	63.40
CEA1	61.21	61.94	62.43	62.79	63.12	63.47
JAE2	61.98	63.02	63.66	64.16	64.71	65.36
JAE4	60.04	61.08	61.74	62.24	62.78	63.43
JAE6	60.38	61.35	61.91	62.40	62.95	63.41

Table 6.26b. ^{242}Pu microscopic absorption cross-sections (barn) of Group 3 in six fuel zones, Case B

Participant	Zone numbers					
	1	2	3	4	5	6
IKE1	65.71	66.31	66.74	66.99	67.37	67.70
CEA1	66.43	66.99	67.43	67.71	67.78	68.04
JAE2	67.14	67.87	68.30	68.66	69.10	69.66
JAE4	65.12	65.85	66.27	66.63	67.06	67.61
JAE6	65.31	65.93	66.36	66.77	67.18	67.41

Annex 1 to Chapter 6

Specifications of the new benchmark to compare MCNP, WIMS, APOLLO2, CASMO4 and SRAC

S. Cathalau
CEA Cadarache, France

The new benchmark devoted to the comparison of WIMS6, APOLLO2, CASMO4, MCNP4.2 and SRAC codes without any burn-up calculations is described in the following:

Available Libraries: JEF2.2 for APOLLO2, WIMS6, CASMO4 and MCNP4.2 JENDL3.1 and JENDL3.2 for SRAC.

The benchmark specification is as follows:
For the two fuel types (A: poor quality Pu and B: classical Pu) the general geometry and temperatures are described below.

Geometry

- FUEL: External Radius = 0.4095 cm
- CLAD: External Radius = 0.4750 cm
- MODERATOR: External Square pitch = 1.3133 cm

Three types of mesh point discretisations for the fuel are proposed:

- FUEL: a) 1 point as: 0 cm ⇒ 0.40950 cm

  b) 6 points as:
  0 cm ⇒ 0.25889 cm
  0.25889 cm ⇒ 0.34261 cm
  0.34261 cm ⇒ 0.36627 cm
  0.36627 cm ⇒ 0.38849 cm
  0.38849 cm ⇒ 0.39913 cm
  0.39913 cm ⇒ 0.40950 cm

  c) 20 equivolumetric rings.

For each calculations, the moderator will be discretised in three zones:

- MODERATOR: 0.475 cm ⇒ 0.580 cm
  0.580 cm ⇒ 0.650 cm
  0.650 cm ⇒ edge of the square cell

Temperatures

- FUEL: = 900.0 K = 626.85 °C
- CLAD: = 600.0 K = 326.85 °C
- MODERATOR: = 573.6 K = 300.45 °C

Isotopic compositions (in atom per barn·cm)

Fuel	*Poor Quality(A)*	*Classical (B)*
^{234}U	0.0000000	2.4626E-7
^{235}U	1.4456E-4	5.1515E-5
^{238}U	1.9939E-2	2.0295E-2
^{238}Pu	1.1467E-4	2.1800E-5
^{239}Pu	1.0285E-3	7.1155E-4
^{240}Pu	7.9657E-4	2.7623E-4
^{241}Pu	3.3997E-4	1.4591E-4
^{242}Pu	5.6388E-4	4.7643E-5
Oxygen	4.5854E-2	4.3100E-2
Clad		
Natural Zr	4.3248E-2	4.3248E-2
Moderator		
H_2O	2.3858E-2	2.3858E-2
^{10}B	3.6346E-6	3.6346E-6
^{11}B	1.6226E-5	1.6226E-5

Results

- Multiplication Factor K infinite;

FOR EACH MEDIUM in the CELL (mesh point in fuel, clad and moderator).

- One and four energy group cross-sections (absorption, fission and production cross-section information);
- One and four energy group reaction rates (absorption, fission, production);

Energy group boundaries:

20.0 MeV,	820.85 keV,	5.5308 keV,	1.5 eV,	0.00001 eV
1	2	3	4	

- Optionally averaged cross-sections and reaction rates in the cell.

Annex 2 to Chapter 6

Effect of different state-of-the-art nuclear data libraries on the PWR MOX pin cells benchmark of Chapter 6

(special study reported in [1])

The two PWR pin cells at operational conditions (T_{fuel}=660 C, T_{mod}=306 C) fuelled with recycled MOX (A) and standard MOX (B), respectively, have been performed applying the code system RSYST3 [2] coupled with ORIGEN-2 in a multigroup approach. The group cross-sections of 20 actinides, of the fission products ^{135}Xe and ^{149}Sm and of moderator and structural materials have been prepared by separate RESMOD calculations for each of the three databases. Cross-sections of the other fission products and some minor actinides have been taken from the JEF-2.2 library and the decay data from ENDF/B-VI.

In Tables 1 and 2 k-infinity and the most important isotopic reaction rates are shown. The reaction rates are normalised in such a way that they directly represent the influence on k-infinity. The k-values at BOL are in both cases rather close together; differences after irradiation are slightly higher in Case A, but about twice as high in Case B. However the production rates of ^{239}Pu and ^{241}Pu show significantly higher discrepancies. Due to different signs there are compensation effects at BOL and also (in Case A) for 50 MWd/kgHM. The spread in the isotopic absorption rates is lower, but exceeds 100 pcm for the main Pu isotopes. Due to compensation effects the total absorption rates are also low. In both benchmarks JENDL-3.2 cross-sections give nearly 100 pcm higher absorption rates in the clad and 90 pcm lower ones for oxygen in the fuel and in the moderator than the other libraries. The multiplication constants at BOL have also been calculated by MCNP-4A (stat. error 0.0007). The results are included in the Tables 1 and 2.

REFERENCES

[1] M. Mattes, D. Lutz, W. Bernnat: *Application of Nuclear Data Libraries Based on JEF-2.2, ENDF/B VI and JENDL-3.2 to LWR Criticality and Burn-up Problems*, Proc. of Intl. Conference on Nuclear Data for Science and Technology, Trieste, 19-24 May 1997.

[2] R.Rühle: *RSYST, an Integrated Modular System for Reactor and Shielding Calculations*, Conf-730 414-12, 1973.

Table 1. k-infinity, absorption and production rates of MOX benchmark A at BOL and at 50 MWd/kgHM, calculated with JEF-2.2 data and differences for ENDF/B-VI and JENDL-3.2

	BOL			50 MWd/kgHM		
	JEF-2.2	ENDF/B-VI	JENDL-3.2	JEF-2.2	ENDF/B-VI	JENDL-3.2
	k-infinity			***k-infinity***		
	1.1253	1.1241	1.1234	0.9553	0.9541	0.9536
MCNP-4A	1.1261	1.1252	1.1251			
	Prod. rate	Differences, pcm		***Prod. rate***	Differences, pcm	
^{239}Pu	0.665	385	528	0.482	295	293
^{241}Pu	0.295	-432	-492	0.321	-394	-261
	Abs. rate	Differences, pcm		***Abs. rate***	Differences, pcm	
fuel	0.976	-38	-7	0.975	-40	-10
^{235}U	0.026	50	5	0.015	-119	-171
^{239}Pu	0.362	77	123	0.263	51	18
^{240}Pu	0.183	-100	111	0.150	-80	24
^{241}Pu	0.131	-166	-127	0.144	-112	21
Ox-fuel	0.003	-6	-96	0.003	-5	-95
clad	0.005	38	91	0.005	38	92
mod	0.030	-3	-84	0.036	-13	-78

Table 2. k-infinity, absorption and production rates of MOX benchmark B at BOL and at 50 MWd/kgHM, calculated with JEF-2.2 data and differences for ENDF/B-VI and JENDL-3.2

	BOL			50 MWd/kgHM		
	JEF-2.2	ENDF/B-VI	JENDL-3.2	JEF-2.2	ENDF/B-VI	JENDL-3.2
	k-infinity			***k-infinity***		
	1.1806	1.1792	1.1813	0.9213	0.9183	0.9188
MCNP-4A	1.1807	1.1787	1.1815			
	Prod.rate	Differences, pcm		***Prod. rate***	Differences, pcm	
^{239}Pu	0.845	269	584	0.508	-21	120
^{241}Pu	0.215	-351	-371	0.300	-382	-266
	Abs.rate	Differences, pcm		***Abs.rate***	Differences, pcm	
fuel	0.965	-37	-14	0.958	-32	-28
^{239}Pu	0.458	39	180	0.276	-45	-35
^{241}Pu	0.097	-150	-112	0.136	-111	6
^{241}Pu	0.012	94	20	0.022	111	55
Ox-fuel	0.002	-6	-93	0.002	-5	-92
clad	0.006	38	96	0.006	38	99
mod	0.030	-1	-82	0.036	-6	-72

Chapter 7
CONCLUSIONS

While acknowledging that MOX recycle in PWRs is already demonstrated practically at the commercial scale in PWRs, there will evidently be important new issues to address in the future arising from the combined impact of increasing discharge burn-ups and multiple recycle. Plutonium recycle in BWRs has yet to be demonstrated on the commercial scale, and while there is no doubt that it will prove as practicable as in PWRs, it is clear that the issue of degrading plutonium isotopic quality will also have to be addressed.

The Working Group considers that considerable progress has been made in the nuclear data libraries and lattice codes since the earlier benchmarks. The data libraries have converged somewhat and more exact calculational methods, notably a more accurate treatment of resonance self-shielding in the higher plutonium isotopes, has considerably reduced the spread of results. The range of results obtained is now generally within acceptable bounds, although some important parameters, such as end-of-cycle reactivity have not converged sufficiently for practical application. Consequently there remains a need to obtain experimental and operational data to confirm the predicted values in practice. The uncertainties in the basic nuclear data are sufficient to account for the range of results seen in the STD PWR, but this is not the case for the HM PWR where the ranges of results are generally larger. This is surprising, as the sensitivity of the HM PWR to nuclear data differences might be expected to be smaller than for the STD PWR.

There was generally good agreement as to the evolution of plutonium inventories through the recycle generations. The HM PWR is clearly more tolerant of the degradation in plutonium isotopic quality between recycle generations, as would be expected with the softer neutron spectrum. However, the reduced generation of fresh plutonium from ^{238}U captures speeds up degradation of plutonium compared with the STD PWR. The HM PWR is therefore arguably more proliferation resistant than the STD PWR.

The range of disagreement was higher for the minor actinides (around 20% for americium and curium), but still acceptable, being within the range of uncertainties of the underlying data. It was noted that the uncertainties in threshold reactions are relatively large, and this may affect the actinide number inventories. As might be expected, relatively large uncertainties were noted for higher order effects such as elastic scattering in the moderator and the variation of fission spectrum with burn-up. Although not a dominant component, some variation was noted in fission product absorption.

Since the boundaries of current knowledge will inevitably be expanded gradually as more expedience of MOX recycle is gained, the Working Group considers that multiple recycle scenarios will prove to be practicable in PWRs at least in the near term. Questions as to the reliability and applicability of the nuclear data libraries and lattice codes are unlikely to present an insurmountable barrier to the implementation of multiple recycle. A more important restriction is likely to be the need to maintain an acceptable moderator void coefficient for the complete core. In the particular scenarios examined here, questions regarding the acceptability of the void coefficient arise as early as the second recycle generation, in the case of the STD PWR, due to the high initial plutonium content (<13 w/o) needed. The HM PWR is less restrictive in this sense, in that the initial plutonium content does not exceed this value until the fourth recycle generation.

The precise point at which the void coefficient becomes unacceptable is not clear because both the present and the earlier benchmarks considered only the nuclear behaviour at the level of lattice cells. To make a definitive evaluation core-wide calculations are necessary, but these are beyond the present scope of work agreed by the Working Group. The Working Group still considers, however, that multiple recycle in the STD PWR case will in practice prove practical up to at least the second generation of recycle, because the scenario examined here is more limiting than can be expected in the immediate future in two senses. First, the burn-up target of 51 MWd/kg in a 1/3-core 18 month cycle exceeds current practice. Second, the plutonium isotopic qualities available from MOX assemblies presently in the discharge ponds or scheduled for discharge are not as poor as those assumed here.

Regarding the need for further benchmark exercises of this kind, the Working Group considers that they would be only of very limited value. This is because the benchmarks presented have converged to the point where it is not possible to clearly identify the underlying causes in the nuclear data and lattice code methods without substantial effort. Even with such effort, the differences are likely to be comparable with the underlying uncertainties in the nuclear data libraries. Whilst there may be some merit in making a comparison on a single nuclear data set, to isolate the contribution to the spread from the lattice code methods, future work on the STD and HM scenarios would be better directed towards whole-core type calculations in an effort to identify the practicable limits such as may be defined by the core-wide reactivity coefficients, transient response and other such behavioural characteristics. Particularly important in this respect would be consideration of the boron coefficient and moderator temperature coefficient and their relevance to such transients as cooldown faults.

Overall the Working Party considers that the latest benchmarks have proved to be very valuable in that they have clearly defined the range of parameters that can be expected in realistic multiple recycle scenarios. There is now a clear understanding of the range of discrepancies that are likely to be expected from the underlying nuclear data and the lattice code methods, which will be immensely useful guidance for the various groups working in the area of plutonium recycle in PWRs. The lessons learned here are likely to be useful as well for BWRs, for which experience of MOX usage lags behind that of PWRs. The HM PWR concept shows clear promise, but generally the range of discrepancies tends to be somewhat larger than for the STD PWR. There is a clear need for experimental support for the HM PWR if it is to be further developed.

REFERENCES

Reports and articles issued from the work of the NSC/WPPR

Physics of Plutonium Recycling – Series of six reports published by the OECD/NEA, 1995/1997.

Issues and Perspectives, ISBN 92-64-14538-9.

Plutonium Recycling in Pressurised-Water Reactors, ISBN 92-64-14560-7.

Void Reactivity Effect in Pressurised-Water Reactors, ISBN 92-64-14591-5.

Fast Plutonium-Burner Reactors: Beginning of Life, ISBN 92-64-14703-9.

Plutonium Recycling in Fast Reactors, ISBN 92-64-14704-7.

Multiple Recycling in Advanced Pressurised-Water Reactors, (this Volume).

A. Tudora: *Evaluation of ^{242}Pu Data for the Incident Neutron Energy Range 5-20 MeV,* NEA/SEN/NSC/WPPR(98)1, December 1998.

Benchmark on the Venus-2 MOX Core Measurements NEA/NSC/DOC(2000)7, OECD/NEA 2000.

Benchmark Calculations of Power Distribution Within Fuel Assemblies, Phase II: Comparison of Data Reduction and Power Reconstruction Methods in Production Codes, NEA/NSC/DOC(2000)3, November 2000.

W. Bernnat, D. Lutz, K. Hesketh, E. Sartori, G. Schlosser, S. Cathalau, M. Soldevila: *PWR Benchmarks from OECD Working Party on Physics of Plutonium Recycling,* Proc. GLOBAL'95, pp. 627-635, 1995.

R.N. Hill, G. Palmiotti, D.C. Wade: *Fast Burner Reactor Benchmark Results from the NEA Working Party on Physics of Plutonium Recycle,* GLOBAL'95, pp. 1367-1373, 1995.

W. Bernnat, S. Cathalau, M. Delpech, K. Hesketh, D. Lutz, M. Mattes, M. Salvatores, E. Sartori, H. Takano: *OECD/NEA Physics Benchmarks on Plutonium Recycling in PWRs,* Proc. PHYSOR'96, Mito, Japan, pp. H61-H74, 1996.

K. Hesketh, M. Delpech, E. Sartori: *Multiple Recycle of Plutonium in PWR: A Physics Code Benchmark Study by the OECD/NEA,* Proc. GLOBAL'97, Yokohama, Japan, pp. 287-294, 1997.

K. Hesketh, M. Delpech, E. Sartori: *The Physics of Plutonium Fuels – A Review of OECD/NEA Activities*, Nuclear Technology, Vol. 131, pg. 385-394, September 2000.

Other reports by the OECD/NEA

Plutonium Fuel: An Assessment, 1989.

The Safety of the Nuclear Fuel Cycle, 1993.

The Economics of the Nuclear Fuel Cycle, 1994.

Management of Separated Plutonium – The Technical Options, A Report by OECD/NEA, 1997, ISBN 92-64-15410-8.

Physics and Fuel Performance of Reactor-Based Plutonium Disposition: Workshop Proceedings, Paris, France, 28-30 September 1998, ISBN: 926417338-1.

Appendix A

PHYSICS OF PLUTONIUM RECYCLING MULTIPLE RECYCLING IN ADVANCED PRESSURISED-WATER REACTORS

Phase II – Benchmark specification

(Revision 2, 15 April 1996)

Table of contents

NEA/SEN/NSC/WPPR(95)1
Revision 2, 15/04/1996

MULTIPLE RECYCLING IN ADVANCED PRESSURISED WATER REACTORS: PHASE 2 BENCHMARK SPECIFICATION

Introduction

The aims of this benchmark are to analyse the physics of the plutonium consumption during its recycling in PWRs, the build-up of minor actinides due to this consumption, the effect on the core kinetic parameters, the activity and radiotoxicity consequences, the mass in the cycle and in the waste disposal in order to evaluate the PWR's potential for plutonium stockpile stabilisation.

Thus, this benchmark proposes plutonium recycling with dilution using the first generation plutonium at each recycling inside two kinds of PWRs with a different moderation ration.

The choice of reactor is a standard PWR with a moderation ratio close to 2 (standard PWR) and a highly moderated PWR (HM-PWR) with a moderation ratio of about 3.5. Recent studies [1] have shown that this last concept has a higher ability to consume the plutonium while minimising the minor actinides production. These two reactors are loaded with 100% MOX fuel assemblies.

The number of plutonium recycles for this benchmark is taken equal to 5 to illustrate the composition evolution for many recyclings.

The benchmark is divided into two parts:

- PART 1 aims at the physics analysis of methods, codes and data used for the recycling studies. For that purpose, the plutonium content is provided with this specification together with the corresponding plutonium isotopic compositions. The outputs are all the microscopic and macroscopic results.

- PART 2 aims at the comparison in a global approach of the results obtained by participants themselves on their own during the plutonium recycling following the prescribed specification for dilution. The outputs are the global parameters such as mass balance, plutonium content.

The analysis of these two parts will be carried out, first separately and then jointly for better understanding. For the second part, an extrapolation to an asymptotic case is proposed.

In order to compare the so-extrapolated asymptotic cases as a function of the results obtained for the calculated cycles, an equilibrium state will be extrapolated by the co-ordinator to evaluate the impact of the differences in the first plutonium recycles on the asymptotic case. Also, the impact on the stock mass and cycle mass will be evaluated for each calculated set for this equilibrium state.

The results must be provided in terms of isotopic vectors, isotopic balance, activity and radiotoxicity. These parameters will be given for each step in the cycle and for each plutonium recycle. Also, the results will be analysed in terms of reactivity parameters (boron efficiency, temperature coefficient, reactivity balance, void effect). This might allow preliminary and global understanding of plutonium recycling in PWRs and evaluation of its feasibility.

1. Data

1.1 Geometry

The dimensions are given for hot conditions.

The calculation of the cylindrised cell is carried out based on the following geometry description:

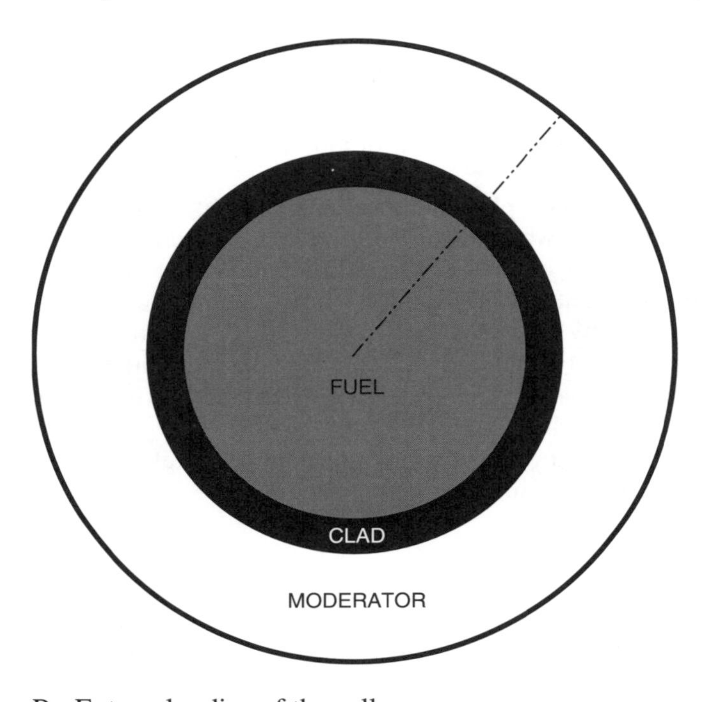

R_3: External radius of the cell.

R_2: External radius of the clad.

R_1: External radius of the fuel pin.

$R_1 = 0.4127$ cm

$R_2 = 0.4744$ cm

For the **standard PWR**........................... $R_3 = 0.7521$ cm

For the **HM-PWR**................................... $R_3 = 0.9062$ cm

In the different zones, the materials are:

- $0 < R < R_1 \Rightarrow$ The fuel material
- $R_1 < R < R_2 \Rightarrow$ The clad
- $R_2 < R < R_3 \Rightarrow$ The moderator (*The water at nominal conditions, pressure 155 bars and temperature 313°C*)

1.2 *Materials*

There are three materials in the model:

- The MOX fuel.
- The clad material.
- The moderator.

Note that the materials temperatures specified in the following are equal to the temperatures of the isotopes in the MCNP libraries used in the previous benchmark of WPPR. They do not correspond exactly to the nominal conditions in a PWR. These temperatures are used only to describe the temperature of the isotopes.

1.2.a MOX fuel

Density: 10.02 g/cm^3

Temperature of isotopes: 900 K (626.84°C)

The initial uranium composition: ^{235}U 0.25%
^{238}U 99.75%

This initial uranium composition is taken unchanged at each beginning of each recycling.

For the first cycle of plutonium recycling, the initial concentration of MOX fuel is (in 10^{24} atoms/cm^3) [after fuel fabrication]:

Initial mass content of plutonium: 10.15%

Isotope	Concentration		Pu mass fraction
^{235}U	$5.0386\ 10^{-5}$		
^{238}U	$2.0028\ 10^{-2}$		
^{238}Pu	$9.0777\ 10^{-5}$	*Pu mass fraction:*	*4.0%*
^{239}Pu	$1.1390\ 10^{-3}$		50.4%
^{240}Pu	$5.1761\ 10^{-4}$		23.0%
^{241}Pu	$3.0255\ 10^{-4}$		13.5%
^{242}Pu	$2.0310\ 10^{-4}$		9.1%
^{16}O	$4.4663\ 10^{-2}$		

It is necessary to take into account the decay due to the ^{241}Pu and ^{238}Pu during the ageing time (2 years) in order to obtain the correct plutonium composition before irradiation including ^{241}Am and ^{234}U.

The plutonium is generated through the irradiation of enriched uranium (4.5% of ^{235}U) with the following fuel management: exit burn-up equal to 55.5 MWd/kg and six batch loading and three years of cooling.*

---

* MWd/kg = Megawatt days per kilogram of initial heavy metal.

For the next cycles, the composition of plutonium will be changed under irradiation and the initial plutonium content is calculated taking into account:

- The level of reactivity adjustment (see next paragraph).
- The density of the MOX fuel is kept constant (10.02 g/cm^3).
- Thus, the initial concentrations will be changed during the multirecycling. *This aspect is explained later in the paper.*

1.2.b Clad

Density: 0.71395 g/cm^3

Temperature of isotopes: 600 K (326.84°C)

The initial concentration is given in the next table is 10^{24} $atoms/cm^3$:

^{nat}Zr	$4.3248\ 10^{-2}$	*In atom %: ^{90}Zr 51.45%, ^{91}Zr 11.32%, ^{92}Zr 17.19%, ^{94}Zr 17.28%, ^{96}Zr 2.76%*

1.2.c Moderator

Density: 0.71395 g/cm^3

Temperature of isotopes: 573.16 K (300°C)

The initial concentration is given in the next table is 10^{24} $atoms/cm^3$:

^{nat}B	$1.98456\ 10^{-5}$	or ^{10}B	$3.92943\ 10^{-6}$
		or ^{11}B	$1.59162\ 10^{-5}$
^{16}O	$2.386013\ 10^{-2}$	or H_2O	$2.386013\ 10^{-2}$
^{1}H	$4.772026\ 10^{-2}$		

These concentrations correspond to realistic data of the light water at 306°C under a pressure of 155.5 bars. As explained before, the temperature of the moderator isotopes does not correspond to this state of the water in order to be able to use MCNP data libraries if necessary.

The boron concentration is kept under constant evolution. It corresponds to a mass fraction of 500 10^{-6} of soluble boron in the moderator. The natural boron is composed by 18.3% in mass fraction of ^{10}B and by 81.7% of ^{11}B (or respectively 19.8% and 80.2% for atom fraction).

For the initial plutonium content calculations and for the equivalence of the reactivity readjustment calculations, the boron concentration must be zero at the end of the averaged final burn-up of the core (referred to in the section on equivalence calculations).

Calculations are performed with no leakage.

2. Burn-up parameters

One cycle is described as follows:

- *Two years of ageing* after fabrication and before irradiation (build-up of ^{241}Am and ^{234}U).
- *Irradiation with exit burn-up of 51 MWd/kg and three batches of loading management.*

  During irradiation, the power is kept constant. The unit is given for a slice of 1 cm of the cell model. The value of the power depends on the reactor type. It is expressed for all the cells (fuel, clad and moderator) and in Watt per gram of heavy metal:

  – *178 W/cm for the cell with the moderation ratio of 2 or 37.7 W/g of heavy metal. These values correspond to a standard square lattice of 17 × 17.*

  – *280 W/cm for the cell with the moderation ratio of 3.5 or 59.3 W/g of heavy metal. These values correspond to a square lattice 17 × 17. For a 19 × 19 lattice, the linear power decreases to 146 W/cm.*

  During irradiation, the soluble boron concentration is kept constant (mass fraction in moderator: 500 10^{-6}). The previous moderator concentration (see 1.2.c) takes this boron quantity into account. The irradiation is done in one time, with no cooling time for reloading at each cycle step. The used fission energy releases will be reported in the results.

- *Five years of cooling* before reprocessing.

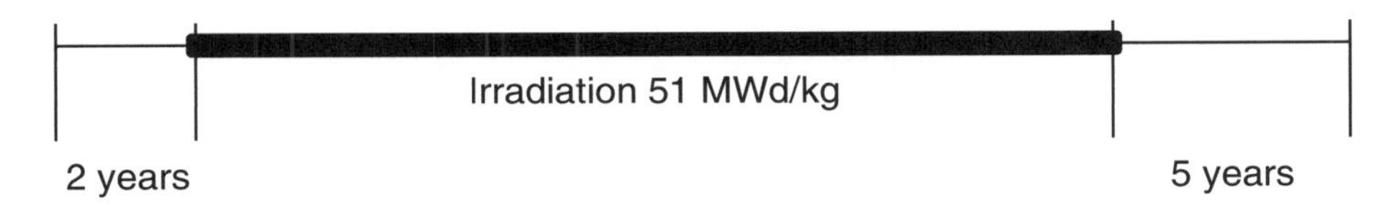

The plutonium recycling with dilution is described as follows:

Only the plutonium is reprocessed at the end of the cycle and put again in reactor after dilution with a first generation plutonium. The rate of the mixture between the first generation plutonium (initial plutonium) and the used plutonium at the end of the cycle is equal to 3. It means that **one assembly** of used plutonium is mixed with **three assemblies** of the first generation plutonium. It means that the electric power share is 25% from the MOX fuel and 75% from the UOX fuel.

The heavy metal mass for one assembly is equal to 540 kg. The plutonium mass per assembly for the UOX irradiated fuel (first-generation plutonium) is equal to 6.7 kg. Thus, we mix 20.1 kg of first generation plutonium with the mass of plutonium contained in one irradiated assembly provided by the MOX fuel.

The other isotopes are extracted, especially the minor actinides (MA) produced by the plutonium under irradiation.

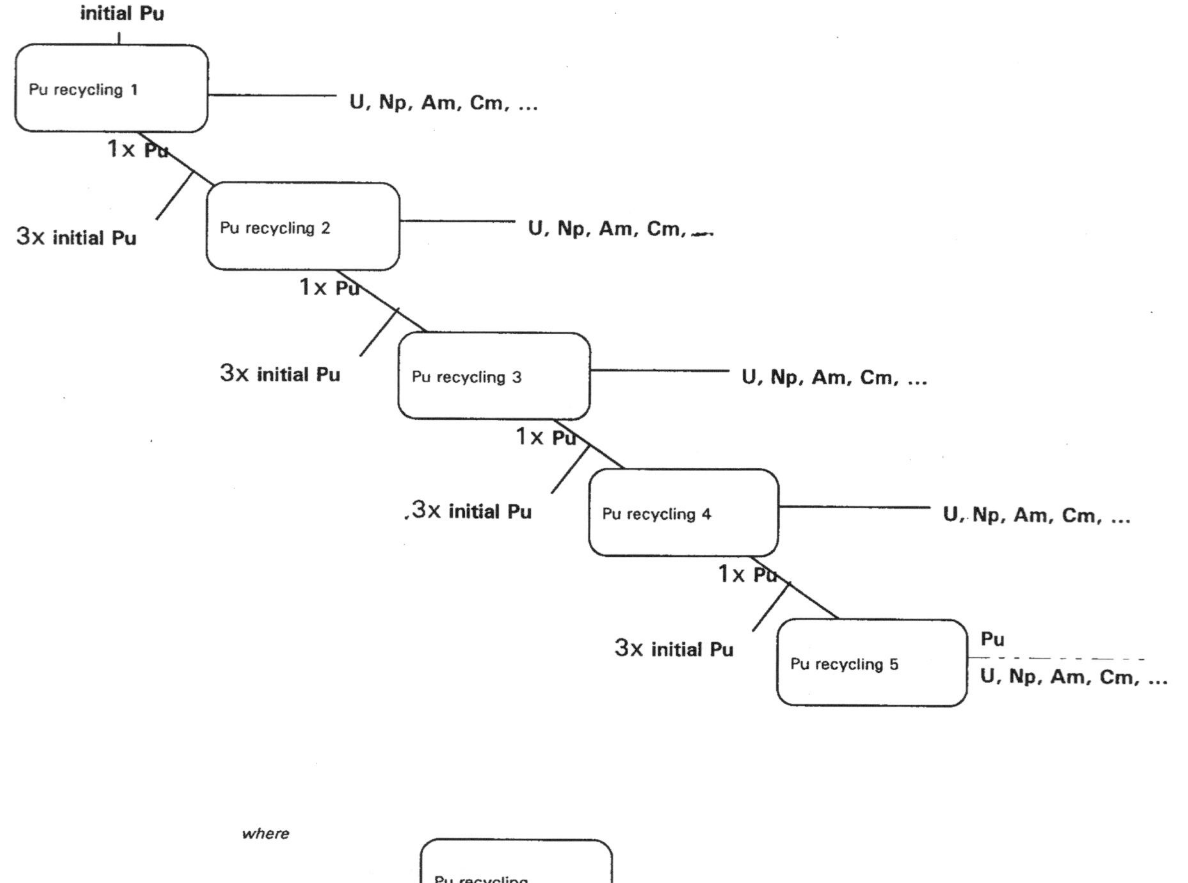

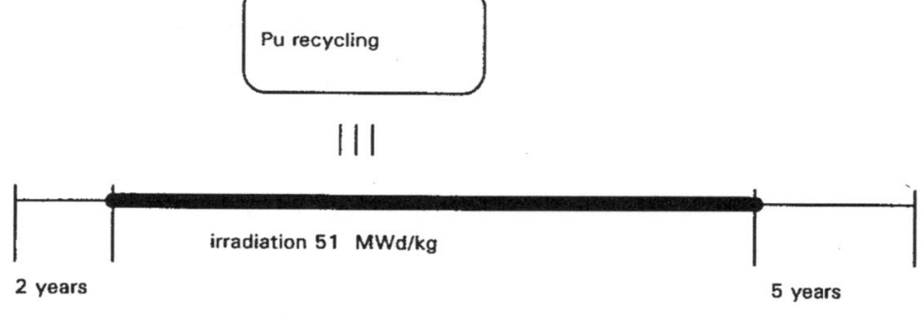

Burn-up model

Usually, the in-core calculations are carried out with boron concentration for all the irradiation time in order to simulate the variation of soluble boron under irradiation. The value of 500 10^{-6} of the mass fraction of soluble boron in the moderator is proper to simulate the effect on the plutonium mass balance.

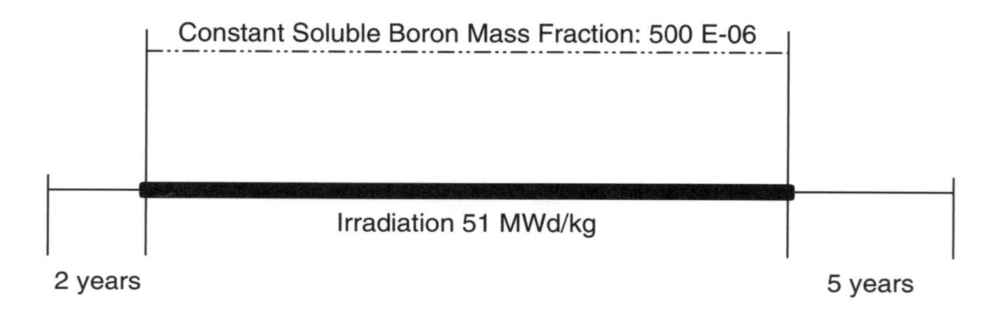

2.1 *Part 1: Physics analysis*

For this part, the initial plutonium content is given for each cycle:

	STANDARD PWR					HM-PWR				
CYCLE	*1*	*2*	*3*	*4*	*5*	*1*	*2*	*3*	*4*	*5*
% Pu	10.1500	13.6250	15.9810	17.7790	19.2280	6.7030	9.6360	12.1070	14.3740	16.5830
^{238}Pu	4.0000	4.8260	5.2880	5.5389	5.6633	4.0000	4.6484	4.9674	5.1382	5.2151
^{239}Pu	50.4000	42.6180	39.1890	37.1235	35.6766	50.4000	36.9325	31.5575	28.5209	26.4344
^{240}Pu	23.0000	26.9270	28.2750	28.9289	29.2940	23.0000	29.0647	29.6867	29.3443	28.7947
^{241}Pu	13.5000	13.4140	12.8570	12.3085	11.8480	13.5000	13.0907	12.7012	12.1591	11.6091
^{242}Pu	9.1000	12.2150	14.3910	16.1003	17.5181	9.1000	16.2637	21.0871	24.8376	27.9467
^{235}U	0.2500	0.2500	0.2500	0.2500	0.2500	0.2500	0.2500	0.2500	0.2500	0.2500
^{238}U	99.7500	99.7500	99.7500	99.7500	99.7500	99.7500	99.7500	99.7500	99.7500	99.7500
at/b•cm ^{238}Pu	9.0777E-5	1.4695E-4	1.8886E-4	2.2008E-4	2.4337E-4	5.9913E-5	1.0010E-4	1.3440E-4	1.6506E-4	1.9328E-4
^{239}Pu	1.1390E-3	1.2922E-3	1.3938E-3	1.4689E-3	1.5267E-3	7.5174E-4	7.9198E-4	8.5028E-4	9.1235E-4	9.7560E-4
^{240}Pu	5.1761E-4	8.1306E-4	1.0014E-3	1.1399E-3	1.2483E-3	3.4162E-4	6.2066E-4	7.9654E-4	9.3477E-4	1.0583E-3
^{241}Pu	3.0255E-4	4.0335E-4	4.5347E-4	4.8297E-4	5.0279E-4	1.9969E-4	2.7838E-4	3.3938E-4	3.8572E-4	4.2489E-4
^{242}Pu	2.0310E-4	3.6578E-4	5.0547E-4	6.291R3-4	7.4033E-4	1.3404E-4	3.4443E-4	5.6111E-4	7.8465E-4	1.0186E-3
^{235}U	5.0386E-5	4.8875E-5	4.7543E-5	4.6527E-5	4.5708E-5	5.2788E-5	5.1131E-5	4.9734E-5	4.8453E-5	4.7204E-5
^{238}U	2.0028E-2	1.9255E-2	1.8730E-2	1.8330E-2	1.8007E-2	2.0797E-2	2.0143E-2	1.9593E-2	1.9089E-2	1.8597E-2

2.2 *Part 2: Global analysis*

Adjustment of the plutonium content at each cycle

This reactivity level adjustment permits a coherent evaluation of initial content of plutonium at each cycle of plutonium recycling using only cell calculations and no core calculations.

The initial content of plutonium must be such that it allows to have criticality at the end of each cycle in the core. Our model is based on one cell model calculation and does not take into account the leakage of the core and other penalties. Thus, we propose a model for the cell calculation:

- To adjust the plutonium content for all the cycles of the plutonium recycling.
- To respect core irradiation conditions (leakage and others).
- To respect the reactivity level at the end of one irradiation cycle in the core.

The core average burn-up

The core average burn-up at the end of an equilibrium cycle represents the mean of all the burn-up values for all the assemblies in the core. One equilibrium irradiation cycle in the core starts at 17 MWd/kg (average burn-up value for all the assemblies in the core) and finishes at 34 MWd/kg (average burn-up value for all the assemblies in the core). The core irradiation length between two loadings is 17 MWd/kg.

The core average burn-up at the end of an equilibrium cycle is calculated as:

$$\overline{B.U.}_{\text{end of cycle}} = 1/n \sum_{i=1}^{n} B.U._{\text{end of cycle}}(i) = 34 \text{ MWd/kg}$$

where: n = 3, number of loading batches

$B.U._{\text{end of cycle}} = i(B.U._{exit}/3)$ for i = 1,2,3, where $B.U._{exit}$ = 51 MWd/kg

$\overline{B.U.}_{\text{end of cycle}}(i)$ are defined as: i = 1: 17 MWd/kg for the first loading
i = 2: 34 MWd/kg for the second loading
i = 3: 51 MWd/kg for the third loading

Comment: $B.U._{exit}/3$ *represents the average burn-up between two loading in the core.*

Reactivity adjustment method

To obtain the initial plutonium content for all cases which allow the fuel management and exit burn-up, we propose a model of in-core irradiation. As described before, the in-core calculations are performed with boron concentration for the full irradiation time in order to take into account a variable quantity of soluble boron under irradiation. The in-core irradiation finishes with no more soluble born in reactor.

The model to calculate the reactivity level at the end of the in-core irradiation is described in two points:

- Settled soluble boron quantity for all the irradiation (500 10^{-6} in mass fraction in the moderator).
- No soluble boron at the end of the averaged burn-up (34 MWd/kg):

$$\overline{B.U.}_{\text{end of cycle}}$$

The following scheme presents the method:

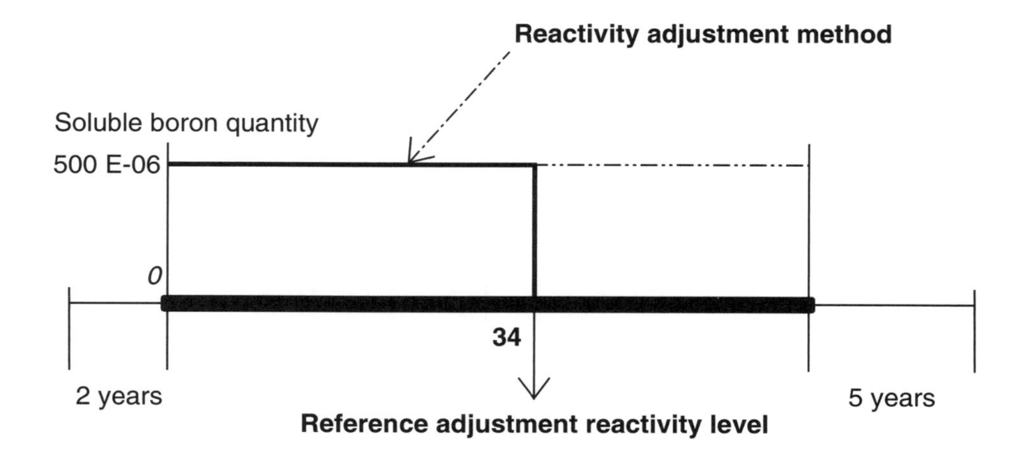

Based on this approach, for each plutonium recycling, we must find the same reactivity level at the averaged burn-up (34 MWd/kg). The first plutonium recycling in standard PWRs (moderation ration equal to 2) gives us this reactivity level. This level takes into account the core leakage and in-core neutron absorption. The plutonium content of the standard PWRs is based on an effective core calculation. In this case, the cell calculation is based on a core calculation with a correct plutonium content, allowing the cycle. This case is named the "Reference Case".

Reactivity readjustment method for the Reference Case

The first plutonium recycling in standard PWR case with the initial plutonium content, previously given in this benchmark, is the Reference Case.

This reference gives the reference adjustment reactivity level at:

$$\overline{B.U.}_{\text{end of cycle}}$$

The reference adjustment reactivity level at $\overline{B.U.}_{\text{end of cycle}}$ is:

$$\rho_{\text{ref}}^{\text{end of cycle}}$$

$\rho_{\text{ref}}^{\text{end of cycle}}$ is calculated as follows:

- An irradiation from 0 MWd/kg to 34 MWd/kg with 500 ppm of soluble boron with no leakage.
- A reactivity calculation at 34 MWd/kg with no soluble boron with no leakage.
- The last step gives the reference adjustment reactivity level: $\rho_{\text{ref}}^{\text{end of cycle}}$.

Reactivity readjustment method for the plutonium recycling cases

All the other cases (others recycling in standard PWR and all the Pu recycling in HM-PWR) are calculated on the base of this Reference Case in a way to find the same reference adjustment reactivity level at:

$$\overline{B.U.}_{\text{end of cycle}} : \rho_{\text{ref}}^{\text{end of cycle}}$$

Conclusions

The adjustment reactivity level is calculated by each of the participants for the Reference Case (given later) and this level will be kept the same for all the other plutonium recycling cases (**standard PWR and HM-PWR**). This method allows specific and coherent calculations for each set of code and data for all plutonium recycling cases.

Main points of the adjustment methods

Reference Case

- Standard PWR (moderation ration of 2).
- First plutonium recycling (see the previous MOX description).
- Evolution with soluble boron (see the previous moderator description).
- No soluble born for adjustment calculations at 34 MWd/kg.

Thus, in the Reference Case, we obtain:

$$\text{Reference Adjustment Reactivity Level} = k_{\infty}^{\text{ref}}\ (34\ \text{MWd/kgk})$$

Other cycles (during plutonium recycling)

One must adjust the initial plutonium content in order to obtain:

$$\overset{\textit{Other cycles}}{k_{\infty}(34\ \text{MWd/kg})} \quad = \quad \overset{\textit{Ref.}}{k_{\infty}(34\ \text{MWd/kg}) \pm 10^{-4}}$$

3. Reactivity coefficients

All of these coefficients are calculated at the beginning and at the end of the irradiation, and also at the beginning of the mean cycle and at the end of the mean cycle:

- At 0.5 MWd/kg.
- At 17 MWd/kg.
- At 34 MWd/kg.
- At 51 MWd/kg.

3.1 Boron efficiency

The boron efficiency is evaluated between two mass boron fractions:

- $500\ 10^{-6}$
- $600\ 10^{-6}$

Moderator composition: boron mass fraction: 500 10^{-6}

B-nat	1.98456 10^{-5}	or B-10	3.92943 10^{-6}
		orB-11	1.59162 10^{-5}
O-16	2.386013 10^{-2}	or H_2O	2.386013 10^{-2}
H-1	4.772026 10^{-2}		

Moderator composition: boron mass fraction: 600 10^{-6}

B-nat	2.381472 10^{-5}	or B-10	4.71531 10^{-6}
		orB-11	1.90994 10^{-5}
O-16	2.386013 10^{-2}	or H_2O	2.386013 10^{-2}
H-1	4.772026 10^{-2}		

The formula for the boron efficiency is:

$$C_{\text{boron}} = \left[1/k_{\infty}(600\ 10^{-6}) - 1/k_{\infty}(500\ 10^{-6})\right]/100$$

The unit is 10^{-5} / ppm.

3.2 Moderator temperature coefficient

The moderator temperature coefficient is calculated between two states. The moderator density variation is calculated when the temperature changes, keeping constant the pressure. The concentrations of the moderator are given as follows:

Unit: 10^{24} atoms/cm^3

	Cold state	**Hot state**
Temperature	569.16 K 296°C	589.16 K 316°C
B-10	0	0
B-11	0	0
H_2O	2.456146 10^{-2}	2.307541 10^{-2}

The formula for the moderator temperature coefficient is:

$$C_{\text{T.Mod.}} = \left[1/k_{\infty}(\text{Hot State}) - 1/k_{\infty}(\text{Cold State})\right]/20$$

The unit is 10^{-5} / K.

3.3 *Fuel temperature coefficient*

The fuel temperature coefficient is evaluated between two states:

- The hot state with a fuel temperature of 910 K (636.84°C).
- The cold state with a fuel temperature of 890 K (616.84°C).

The concentration of all materials are kept equal to those of nominal conditions.

The formula for the fuel temperature coefficient is:

$$C_{T.Comb.} = [1/k_\infty(\text{Hot State}) - 1/k_\infty(\text{Cold State})]/20$$

The unit is 10^{-5} / K.

3.4 *Reactivity balance*

The reactivity balance is defined by:

$$\text{Balance} = \frac{k\infty(0.5\ \text{MWd/kg}) - k\infty(51\ \text{MWd/kg})}{50.5\ \text{MWd/kg}}$$

The unit is 10^{-5} / MWd/kg.

Boron mass fraction in the moderator: 500 10^{-6}.

3.5 *Global void effect*

The global void effect is calculated between an unvoided state and a voided state at 99% of void. The conditions to calculate the void effect are with no leakage or with leakage.

The formula for the void effect is:

$$E_{void} = 1/K\ (\text{unvoided}) - 1/K\ (\text{voided})$$

The unit is 10^{-5}.

For the unvoided state, the moderator concentrations are (unit: 10^{24} atoms/cm^3):

B-nat	0	or B-10 orB-11	0 0
O-16	2.386013 10^{-2}	or H_2O	2.386013 10^{-2}
H-1	4.772026 10^{-2}		

For the voided state, the moderator concentrations are (unit: 10^{24} atoms/cm^3):

B-nat	0	or B-10 orB-11	0 0
O-16	2.386013 10^{-4}	or H_2O	2.386013 10^{-4}
H-1	4.772026 10^{-4}		

The two calculations with and without leakage are carried with the following leakage values:

- With no leakage: $B^2 = 0$ for voided state.
  $B^2 = 0$ for unvoided state.
- With critical leakage ($k_{eff} = 1$).

4. Results – Output data

The results must be sent on computer readable medium in order to process them with Microsoft EXCEL software.

The results necessary for Part 1: Physics analysis, and Part 2: Global analysis of the benchmark are:

4.0 A description of

- The computer program.
- The data libraries (number of groups and energy boundaries).
- The self-shielding models and calculations.

4.1 The reactivity value obtained at the end of the mean batch for the reference case

The reactivity values at 0, 8.5, 17, 25.5, 34, 42.5 and 51 MWd/kg*.

4.2 The reactivity parameters for

- Boron efficiency.
- Moderator temperature coefficient.
- Fuel temperature coefficient.
- Reactivity balance.
- Global void effect.

---

* MWd/kg = Megawatt days per kilogram of initial heavy metal.

4.3 The isotopic composition of the fuel at each cycle step

- Beginning of the cycle.
- Beginning of the irradiation.
- End of the irradiation.
- End of the cooling time (end of the cycle).

4.4 The isotopic composition per elements (uranium, plutonium, neptunium, americium and curium) at each cycle step

- Beginning of the cycle.
- Beginning of the irradiation.
- End of the irradiation.
- End of the cooling time (end of the cycle).

4.5 The one-group microscopic cross-sections and decay values used in the calculations for the first plutonium recycling for all the isotopes. For the microscopic cross-sections, the values are asked at

- Beginning of the irradiation (0 MWd/kg).
- Middle of the first batch (8.5 MWd/kg).
- End of the first batch (17 MWd/kg).
- Middle of the second batch (25.5 MWd/kg).
- End of the second batch (34 MWd/kg).
- Middle of the third batch (42.5 MWd/kg).
- End of the irradiation (51 MWd/kg).

4.6 The macroscopic cross-sections H.N., F.P. and materials (global values for Σ_a and $\nu\Sigma_f$ for the first step of irradiation and the last step of irradiation

4.7 The data of fission energy release of all the isotopes

4.8 The β_{eff} value at the first and fifth cycles for standard PWR and HM-PWR at the end of irradiation

4.9 The mass of the fuel at each cycle step

- Beginning of the cycle.
- Beginning of the irradiation.
- End of the irradiation.
- End of the cooling time (end of the cycle).

4.10 The isotopic composition of the fuel at each cycle step

- Beginning of the cycle.
- Beginning of the irradiation.
- End of the irradiation.
- End of the cooling time (end of the cycle).

4.11 The global activity of one heavy metal metric tonne of

- Initial fuel at the beginning of multirecycling for a time scale between 5 years and 1 million years (time steps*).
- Irradiated fuel at the end of multirecycling for a time scale between 5 years and 1 million years (time steps*).

4.12 The radiotoxicity of fission products and heavy nuclides separately for

- Initial fuel at the beginning of multirecycling for a time scale between 5 years and 1 million of years (time steps*), normalised to one heavy nuclide metric tonne.
- Irradiated fuel at the end of multirecycling for a time scale between 5 years and 1 million of years (time steps*), normalised to one heavy nuclide metric tonne (table with factor is given).

* The time steps are defined as: i 10^j, where i = 1,2,…,9,0 and j = 0,1,2,…,6.

Radiotoxicity data
(CD = Cancer Dose Hazard)

Isotope	Toxicity factor CD/Ci	Half-life Years	Toxicity factor CD/g
Actinides and their daughters			
^{210}Pb	455.0	22.3	3.48E4
^{223}Ra	15.6	0.03	7.99E5
^{226}Ra	36.3	1.60E3	3.59E1
^{227}Ac	1185.0	21.8	8.58E4
^{229}Th	127.3	7.3E3	2.72E1
^{230}Th	19.1	7.54E4	3.94E-1
^{231}Pa	372.0	3.28E4	1.76E-1
^{234}U	7.59	2.46E5	4.71E-2
^{235}U	7.23	7.04E8	1.56E-5
^{236}U	7.50	2.34E7	4.85E-4
^{238}U	6.97	4.47E9	2.34E-6
^{237}Np	197.2	2.14E6	1.39E-1
^{238}Pu	246.1	87.7	4.22E3
^{239}Pu	267.5	2.41E4	1.66E1
^{240}Pu	267.5	6.56E3	6.08E1
^{242}Pu	267.5	3.75E5	1.65E0
^{241}Am	272.9	433	9.36E2
^{242m}Am	267.5	141	2.80E4
^{243}Am	272.9	7.37E3	5.45E1
^{242}Cm	6.90	0.45	2.29E4
^{243}Cm	196.9	29.1	9.96E3
^{244}Cm	163.0	18.1	1.32E4
^{245}Cm	284.0	8.5E3	4.88E1
^{246}Cm	284.0	4.8E3	8.67E1
Short-lived fission products			
^{90}Sr	16.7	29.1	2.28E3
^{90}Y	0.60	7.3E-3	3.26E5
^{137}Cs	5.77	30.2	4.99E2
Long-lived fission products			
^{99}Tc	0.17	2.13E5	2.28E-3
^{129}I	64.8	1.57E7	1.15E-2
^{93}Zr	0.095	1.5E6	2.44E-4
^{135}Cs	0.84	2.3E6	9.68E-4
^{14}C	0.20	5.73E3	8.92E-1
^{59}Ni	0.08	7.6E4	6.38E-3
^{63}Ni	0.03	100	1.70E0
^{126}Sn	1.70	1.0E5	4.83E-2

5. Tables of output data: format and presentation

For greater efficiency, the computed results (1, 2, 3, 4 and 5) must be filled in the following tables for each cycle. For the results concerning activity and radiotoxicity (6 and 7), tables with free format are acceptable.

5.1 *The reactivity value (to be provided for Part 1 and Part 2, Section 4.1)*

At each cycle:

Reactivity value adjustment	

and k_∞ (B.U.)

5.2 *The reactivity parameters for (to be provided for Part 1 and Part 2, Section 4.2)*

At each cycle:

	0.5 MWd/kg	17 MWd/kg	34 MWd/kg	51 MWd/kg
Boron efficiency				
Moderator temperature coefficient				
Fuel temperature coefficient				
Reactivity balance				
Global void effect with no leakage				
Global void effect with leakage (k_{eff} = 1)				

- For each parameter: k_∞.
- For the voided state, and the unvoided state: M^2 values.

5.3 *The isotopic composition of the fuel (to be provided for Part 1 and Part 2, Sections 4.3 & 4.10)*

At each cycle:

- BOC: Beginning of cycle.
- BOI: Beginning of irradiation.
- EOI: End of irradiation.
- EOC: End of cycle.

Unit: 10^{24} atoms/cm^3 at nominal conditions.

	BOC	BOI	EOI	EOC
^{234}U				
^{235}U				
^{236}U				
^{237}U				
^{238}U				
^{237}Np				
^{238}Np				
^{239}Np				
^{238}Pu				
^{239}Pu				
^{240}Pu				
^{241}Pu				
^{242}Pu				
^{241}Am				
^{242}Am				
^{243}Am				
^{242}Cm				
^{243}Cm				
^{244}Cm				
^{245}Cm				
^{246}Cm				
^{79}Se				
^{93}Zr				
^{99}Tc				
^{107}Pd				
^{129}I				
^{135}Cs				
Fission rate				

5.4 ***The mass composition of the fuel*** *(to be provided for Part 2, Section 4.9)*

At each cycle:

- BOC: Beginning of cycle.
- BOI: Beginning of irradiation.
- EOI: End of irradiation.
- EOC: End of cycle.

Unit: kilograms per initial metric tonne of heavy metal.

	BOC	BOI	EOI	EOC
^{234}U				
^{235}U				
^{236}U				
^{237}U				
^{238}U				
^{237}Np				
^{238}Np				
^{239}Np				
^{238}Pu				
^{239}Pu				
^{240}Pu				
^{241}Pu				
^{242}Pu				
^{241}Am				
^{242}Am				
^{243}Am				
^{242}Cm				
^{243}Cm				
^{244}Cm				
^{245}Cm				
^{246}Cm				
^{79}Se				
^{93}Zr				
^{99}Tc				
^{107}Pd				
^{129}I				
^{135}Cs				
Fission rate				

5.5 *The isotopic composition per elements* *(to be provided for Part 1, Section 4.4)*

At each cycle:

- BOC: Beginning of cycle.
- BOI: Beginning of irradiation.
- EOI: End of irradiation.
- EOC: End of cycle.

Unit: atom mass %.

	BOC	BOI	EOI	EOC
^{234}U				
^{235}U				
^{236}U				
^{237}U				
^{238}U				
^{237}Np				
^{238}Np				
^{239}Np				
^{238}Pu				
^{239}Pu				
^{240}Pu				
^{241}Pu				
^{242}Pu				
^{241}Am				
^{242}Am				
^{243}Am				
^{242}Cm				
^{243}Cm				
^{244}Cm				
^{245}Cm				
^{246}Cm				
^{79}Se				
^{93}Zr				
^{99}Tc				
^{107}Pd				
^{129}I				
^{135}Cs				
Fission rate				

5.6 The microscopic cross-section and decay values

For the first plutonium recycling only, the capture, fission, (n,2n), one-group microscopic cross-sections, the nu values and the decay periods used in the calculations. for the one-group microscopic cross-sections, the values are asked at:

- B: Beginning of the irradiation (0 MWd/kg)
- 1: Middle of the first batch (8.5 MWd/kg)
- 2: End of the first batch (17 MWd/kg)
- 3: Middle of the second batch (25.5 MWd/kg)
- 4: End of the second batch (34 MWd/kg)
- 5: Middle of the third batch (42.5 MWd/kg)
- E: End of the irradiation (51 MWd/kg)

	B	1	2	3	4	5	E
^{234}U							
^{235}U							
^{236}U							
^{237}U							
^{238}U							
^{237}Np							
^{238}Np							
^{239}Np							
^{238}Pu							
^{239}Pu							
^{240}Pu							
^{241}Pu							
^{242}Pu							
^{241}Am							
^{242}Am							
^{243}Am							
^{242}Cm							
^{243}Cm							
^{244}Cm							
^{245}Cm							
^{246}Cm							

5.7 The fission energy release, capture energy release (MeV) *(to be provided for Part 1, Section 4.7)*

	By fission	By capture
^{234}U		
^{235}U		
^{236}U		
^{237}U		
^{238}U		
^{237}Np		
^{238}Np		
^{239}Np		
^{238}Pu		
^{239}Pu		
^{240}Pu		
^{241}Pu		
^{242}Pu		
^{241}Am		
^{242}Am		
^{243}Am		
^{242}Cm		
^{243}Cm		
^{244}Cm		
^{245}Cm		
^{246}Cm		

5.8 Macroscopic cross-sections *(to be provided for Part 1, Section 4.6)*

	0 GWd/kg		**51 GWd/kg**	
	Σ_a	$\nu\Sigma_f$	Σ_a	$\nu\Sigma_f$
HN				
FP				
Clad				
Water				

REFERENCES

[1] M. Salvatores, *et al.*, "Nuclear Waste Transmutation: Physics Issues and Potential in Neutron Fields", International Reactor Physics Conference, Tel Aviv, 23-26 January 1994.

Appendix B

LIST OF PARTICIPANTS

(addresses)

BELGIUM

D'HONDT, Pierre
Centre d'Étude de l'Énergie Nucléaire
200 Boeretang
B-2400 MOL

MALDAGUE, Thierry
Belgonucléaire S.A.
Avenue Ariane 4
B-1200 BRUXELLES

CANADA

JONES, Richard T.
Manager, Reactor Physics
AECL Research
Chalk River Nuclear Labs.
CHALK RIVER
Ontario K0J 1J0

DENMARK

HOEJERUP, C. Frank
Senior Scientist
Risoe National Laboratory
P.O. Box 49
DK-4000 ROSKILDE

FRANCE

CATHALAU, Stéphane
CEA – DEN/SPEx/LPE
CEN Cadarache
Bâtiment 230
F-13108 SAINT-PAUL-LEZ-DURANCE CEDEX

COSTE, Mireille
CEA/DEN/DMT
SERMA/LENR
CEA Saclay
F-91191 GIF-SUR-YVETTE CEDEX

DELPECH, Marc
CEA – DER/SERI/LCSI
CEN Cadarache
Bâtiment 230 – B.P. 1
F-13108 SAINT-PAUL-LEZ-DURANCE CEDEX

LEE, Yi-Kang
CEN Saclay
DMT/SERMA/LEPP
F-91191 GIF-SUR-YVETTE CEDEX

PENELIAU, Yannick
CEA/DEN/DMT
SERMA/LEPP
CEA Saclay
F-91191 GIF-SUR-YVETTE CEDEX

POINOT-SALANON, Christine
Société Framatome
EPN
Tour FRAMATOME – Cedex 16
1, Place de la Coupole
F-92084 PARIS LA DÉFENSE

PUILL, André
CEA/DEN/DMT
SERMA/LCA
CEA Saclay
F-91191 GIF-SUR-YVETTE CEDEX

ROHART, Michelle
CEA/DEN/DMT
SERMA/LCA
CEA Saclay
F-91191 GIF-SUR-YVETTE CEDEX

TSILANIZARA, Aimé
CEA/DEN/DMT
SERMA/LEPP
CEA Saclay
F-91191 GIF-SUR-YVETTE CEDEX

GERMANY

BERNNAT, Wolfgang
Universitaet Stuttgart
Institut fuer Kernenergetik und Energiesysteme
Postfach 801140
D-70550 STUTTGART

HESSE, Ulrich
Gesellschaft fuer Anlagen-und Reaktorsicherheit
Forschungsgelaende
Postfach 1328
D-85739 GARCHING

LEHMANN, Sven
Inst. Fuer Raumflugtechnik und Reaktortechnik
Tech. Univ. Braunschweig
Hans-Sommer-Str. 5
D-38106 BRAUNSCHWEIG

LUTZ, Dietrich
Universitaet Stuttgart
Institut fuer Kernenergetik und Energiesysteme
Postfach 801140
D-70550 STUTTGART

MATTES, Margarete
Universitaet Stuttgart
Institut fuer Kernenergetik und Energiesysteme
Postfach 801140
D-70550 STUTTGART

MOSER, Eberhard
Gesellschaft fuer Anlagen-und Reaktorsicherheit
Postfach 1328
D-85739 GARCHING

SCHLOSSER, Gerhard
SIEMENS AG, KWU NBTI
Postfach 3220
Bunsenstr. 43
D-91050 ERLANGEN

JAPAN

AKIE, Hiroshi
Transmutation System Lab.
Dept. of Reactor Engineering
JAERI
TOKAI-MURA, Naka-gun
Ibaraki-ken 319-11

FUJIWARA, Daisuke
Dept. of Quantum Science and
Energy Engineering
Tohoku University
Aoba, Aramaki, Aoba
SENDAI, 980-77

KANEKO, K.
JAERI
Tokai Research Establishment
TOKAI-MURA, Naka-gun
Ibaraki-ken 319-11

MISAWA, Tsuyoshi
Dept. Nuclear Engineering
Nagoya University
Furo-cho, Chikusa-ku
J-NAGOYA-SHI 464-01

TAKANO, Hideki
Pu-Burner Program Team
JAERI
TOKAI-MURA, Naka-gun
Ibaraki-ken 319-11

TAKEDA, Toshikazu
Osaka University
Department of Nuclear Engineering
Graduate School of Engineering
2-1 Yamada-Oka, Suita
OSAKA 565

YAMAMOTO, Toshihisa
Dept.of Nuclear Eng.
Graduate School of Engineering
Osaka University
2-1, Yamada-oka,Suita,
OSAKA 565

KOREA (REPUBLIC OF)

KIM, Young-Jin
Manager, Reactor Physics
Korea Atomic Energy Research
Institute
P.O. Box 105, Yuseong,
TAEJON 305-600

NETHERLANDS

KLOOSTERMAN, Jan Leen
Netherlands Energy Research
Foundation (ECN)
B.U. Nuclear Energy
P.O. Box 1
NL-1755 ZG PETTEN

RUSSIAN FEDERATION

TSIBOULIA, Anatoli
Institute of Physics and Power Engineering (IPPE)
Fiziko-Energiticheskij Inst.
1, Bondarenko Square
249020 OBNINSK

SWITZERLAND

GRIMM, Peter
Paul Scherrer Institute
CH-5232 VILLIGEN PSI

PARATTE, Jean-Marie
Paul Scherrer Institute
CH-5232 VILLIGEN PSI

UNITED KINGDOM

CROSSLEY, Steve
British Nuclear Fuels plc
Springfields
PRESTON
Lancashire PR4 0XJ

HESKETH, Kevin
British Nuclear Fuels plc
Springfields
PRESTON
Lancashire PR4 0XJ

UNITED STATES OF AMERICA

BLOMQUIST, Roger N.
Reactor Analysis Division
Argonne National Laboratory
9700 South Cass Avenue
RA/208
ARGONNE, IL 60439

GRIMM, Karl
Reactor Analysis Division
Argonne National Laboratory
9700 South Cass Avenue
ARGONNE, Il. 60439

HILL, Robert N.
Reactor Analysis Division
Argonne National Laboratory
9700 South Cass Avenue
ARGONNE, Il. 60439

International Organisations

RINEJSKI, Anatoli
Division of Nuclear Power and the Fuel Cycle
I.A.E.A.
P.O. Box 100
A-1400 WIEN

SARTORI, Enrico
OECD/NEA Data Bank
Le Seine-Saint-Germain
12 boulevard des Iles
F-92130 ISSY-LES-MOULINEAUX

List of participants to Annex 1 of Chapter 6

CATHALAU, Stéphane
CEA – DEN/SPEx/LPE
CEN CADARACHE
Bâtiment 230
F-13108 SAINT-PAUL-LEZ-DURANCE CEDEX

LUTZ, Dietrich
Universität Stuttgart
Institut für Kernenergetik une Energiesysteme
Postfach 801140
D-70550 STUTTGART

PEETERS, T.T.J.M.
Netherlands Energy Research Foundation – ECN
P.O. Box 1
NL-1755 ZG PETTEN

PENELIAU, Yannick
CE Saclay
CEA/DRN/DMT
SERMA/LCA
F-91191 GIF-SUR-YVETTE CEDEX

PUILL André
CE Saclay
CEA/DRN/DMT
SERMA/LCA
F-91191 GIF-SUR-YVETTE CEDEX

TAKANO, Hideki
JAERI
Reactor System Laboratory
Tokai Research Establishment
Tokai-Mura, Naka-gun
Ibaraki-ken 319-11

Appendix C

LIST OF SYMBOLS AND ABBREVIATIONS

ANL	Argonne National Laboratory
barn	Nuclear physics' unit for measurement of cross-section (= 10^{-28} m^2)
β_{eff}	Effective delayed neutron fraction
BNFL	British Nuclear Fuels
BWR	Boiling Water Reactor
CEA	Commissariat à l'énergie atomique
ECN	Netherlands Energy Research Foundation
GRS	Gesellschaft fuer Anlagen-und Reaktorsicherheit
HM PWRs	Highly moderated Pressurised Water Reactors
IKE	Institut fuer Kernenergetik und Energiesysteme
IPPE	Institute of Physics and Power Engineering of Obninsk
JAERI	Japan Atomic Energy Research Institute
KAERI	Korea Atomic Energy Research Institute
LWRs	Light Water Reactors
MC	Monte Carlo
MOX	Mixed Oxide (uranium and plutonium)
MWd/t	Mega Watt days per tonne
OECD/NEA	OECD Nuclear Energy Agency

PSI	Paul Scherrer Institute, Switzerland
PWRs	Pressurised Water Reactors
SCK-CEN	Studiecentrum voor Kernenergie-Centre d'études de l'énergie nucléaire
STD PWR	standard 17×17 PWR lattice
TFRPD	Task Force on Reactor-based Plutonium Disposition
UO_2 or UOX	Uranium Oxide
WPPR	Working Party on the Physics of Plutonium Fuels and Innovative Fuel Cycles

Appendix D

CORRIGENDUM FOR PREVIOUS VOLUMES[†]

Corrigendum of corrected specification for Volume II*

Benchmark specification for plutonium recycling in PWRs

Benchmark A: Poor-quality plutonium
J. Vergnes (EdF)

Benchmark B: Better plutonium vector
H. W. Wiese (KfK) and G. Schlosser (Siemens-KWU)

Co-ordinator
H. Küsters, KfK

Benchmark A – poor-isotopic-quality plutonium

The goal of this comparison is to explain the reasons for unexplained differences between results on MOX-PWR cell calculations using degraded plutonium (fifth-stage recycle).

The most important difference is related to the infinite medium multiplication constant k-infinity. We suggest a geometry as simple as possible. We shall describe the proposed options:

- Number of atoms and cell geometry

  Differences could appear for these calculations. So we propose that a number of atoms will be stated for the benchmark.

  For this preliminary calculation, we have taken the geometry of Figure A-1 and the following isotopic balance of plutonium. The plutonium isotopic composition is near the composition at the fifth stage recycle with an average burn-up of 50 MWd/kg.

[†] Corrections are designated by bold characters.

* This section is Appendix A in of *Physics of Plutonium Recycling, Volume II: Plutonium Recycling in Pressurised-Water Reactors*. See pp. 73-78.

Pu-238	4%
Pu-239	36%
Pu-240	28%
Pu-241	12%
Pu-242	20%

The uranium isotopic composition is the following:

U-235	0.711%
U-238	99.289%

The total plutonium concentration proposed is 12.5% (6% of fissile plutonium).

The cladding is only made out of natural zirconium.

In evolution, samarium and xenon concentrations will be self-estimated by each code with a nominal power of 38.3 W/g of initial heavy metal.

- Options of the cell calculation

To ease the comparisons, it is suggested to calculate the cell without any neutron leakage ($B^2 = 0$).

Temperatures will be as follows:

- Fuel 660°C
- Cladding 306.3°C
- Water 306.3°C

Boron concentration is worth **461.4** ppm. Boron composition is as follows:

- B-10 18.3 **w/o**
- B-11 81.7 **w/o**

Table A-1. Number of atoms per cm^3 at irradiation step zero

FUEL	
	ATOMS / cm^3
U-234	0
U-235	$1.4456 \cdot 10^{20}$
U-236	0
U-238	$1.9939 \cdot 10^{22}$
Np-237	0
Pu-238	$1.1467 \cdot 10^{20}$
Pu-239	$1.0285 \cdot 10^{21}$
Pu-240	$7.9657 \cdot 10^{20}$
Pu-241	$3.3997 \cdot 10^{20}$
Pu-242	$5.6388 \cdot 10^{20}$
Am-241	0
Am-242	0
Am-243	0
Cm-242	0
Cm-243	0
CLADDING	
natural Zr	$\mathbf{4.3248} \cdot 10^{22}$
MODERATOR	
H_2O	$2.3858 \cdot 10^{22}$
B-10	$3.6346 \cdot 10^{18}$
B-11	$1.6226 \cdot 10^{19}$

- Options of the evolution calculation

  We propose an evolution calculation from 0 to 50 MWd/kg including the following time steps (0, 0.15, 0.5, 1, 2, 4, 6, 10, 15, 20, 22, 26, 30, 33, 38, 42, 47 and 50 MWd/kg)

  We take into consideration the following fission products:

  Zr-95, Mo-95, Pd-106, Ce-144, Pm-147, Pm-148, Pm-148m, Sm-149, Sm-150, Sm-151, Sm-152, Eu-153, Eu-154, Eu-155, Gd-155, Gd-156, Gd-157, Tc-99, Ag-109, Cd-113, In-115, I-129, Xe-131, Cs-131, Cs-137, Nd-143, Nd-145, Nd-148,

  and four pseudo fission products in which all the other fission products are grouped.

The energy releases from fission are:

NUCLIDE	ENERGY RELEASE (MeV)
U-235	193.7
U-238	197.0
Pu-239	202.0
Pu-241	204.4
Am-242m	207.0

plus 8 MeV for the n-gamma captures of the other non-fissioning (ν-1) neutrons.

- Results

Results should be provided both on paper and computer-processable medium. A short report should be provided describing:

- The computer program(s) used and their precise version,
- The data libraries used and evaluated data file from which they were derived,
- The list of isotopes for which resonance self-shielding was applied and the method used,
- How the build-up of Xenon was treated,
- How the (n,2n)-reaction was taken into account for the k-infinity calculation.

The following data should be provided in tabular form for the following burn-ups:
0, 10, 33, 42 and 50 MWd/kg.

1. Number densities for all nuclides considered:

	burn-up 1	burn-up 2	....	burn-up-n
isotope 1				
isotope 2				
.				
.				
.				
.				
.				
isotope -N				

2. k as a function of burn-up,

3. One energy group cross-section (absorption, fission, nu-bar) as a function of isotope and burn-up (see 1.),

4. Reaction rates (absorption, fission) as a function of isotope and burn-up (see 1.),

5. Applied absolute fluxes used in the evolution calculation (and their normalisation factor),

6. Neutron energy spectrum per unit lethargy as a function of burn-up (and its normalisation factor and group structure).

Benchmark B – better plutonium vector

As a second fuel M2, in agreement both with Dr. G. Schlosser, KWU and Dr. J. Vergnes, EdF, a MOX fuel with first-generation-plutonium as used in [1] with the following specifications is suggested:

- 4.0 wt% **Pu_{fiss}** in uranium tailings (0.25 wt% U-235),

- Composition of plutonium (wt%):

Pu-238	1.8
Pu-239	59.0
Pu-240	23.0
Pu-241	12.2
Pu-242	4.0

- Composition of uranium (wt%):

U-234	0.00119
U-235	0.25
U-238	99.74881

With the heavy material number density normalised to 2.115×10^{22} atoms /cm^3, the following nuclide number densities are determined:

NUCLIDE	ATOMS / cm^3
U-234	$2.4626 \bullet 10^{17}$
U-235	$5.1515 \bullet 10^{19}$
U-238	$2.0295 \bullet 10^{22}$
Pu-238	$2.1800 \bullet 10^{19}$
Pu-239	$7.1155 \bullet 10^{20}$
Pu-240	$2.7623 \bullet 10^{20}$
Pu-241	$1.4591 \bullet 10^{20}$
Pu-242	$4.7643 \bullet 10^{19}$
heavy metal-atoms	$2.155 \bullet 10^{22}$
O	$4.310 \bullet 10^{22}$

All other specifications shall be the same as in the first benchmark – Case A.

REFERENCE

[1] H. W. Wiese, "Investigation of the Nuclear Inventories of High-Exposure PWR Mixed Oxide Fuels with Multiple Recycling of Self-Generating Plutonium", Nuclear Technology, Vol. 102, April 1993, p. 68.

Figure A-1. Cell geometry at 20°C

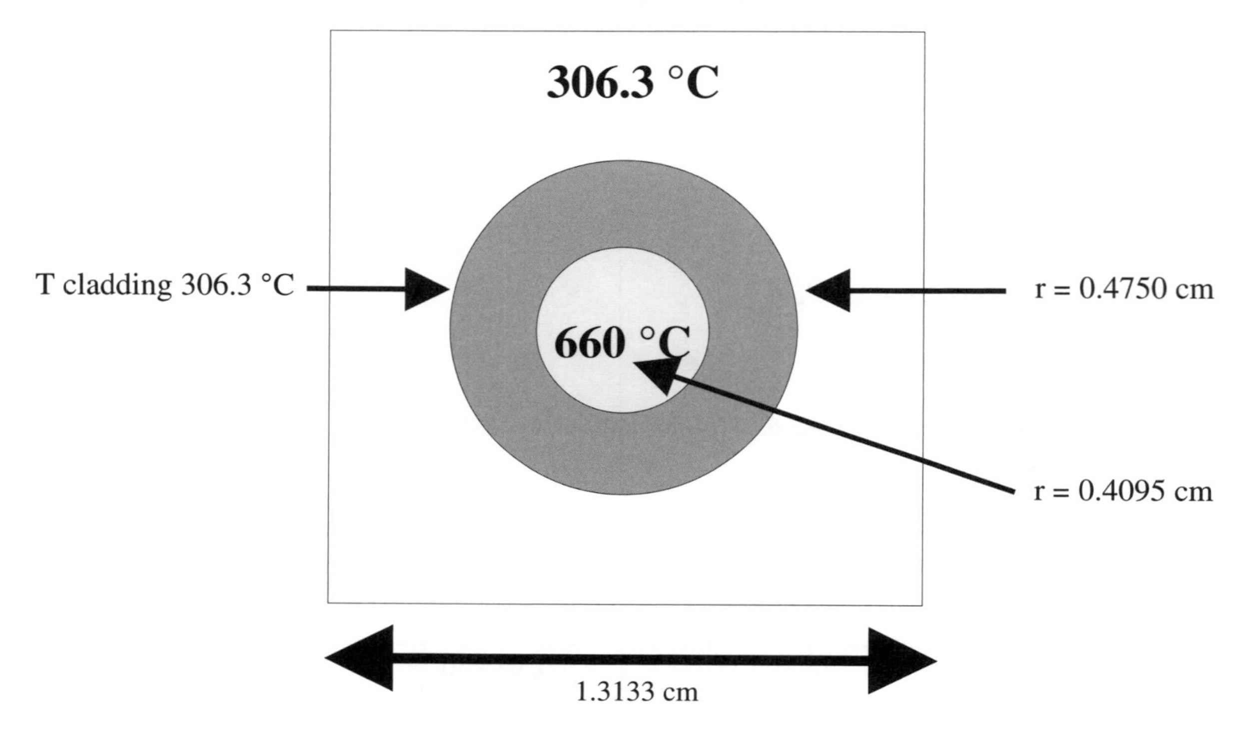

ALSO AVAILABLE

NEA Publications of General Interest

2001 Annual Report (2002) *Free: paper or Web.*

NEA News
ISSN 1605-9581 Yearly subscription: € 40 US$ 45 GBP 26 ¥ 4 800

Geologic Disposal of Radioactive Waste in Perspective (2000)
ISBN 92-64-18425-2 Price: € 20 US$ 20 GBP 12 ¥ 2 050

Nuclear Science

Advanced Reactors with Innovative Fuels (2002)
ISBN 92-64-19847-4 Price: € 130 US$ 113 GBP 79 ¥ 15 000
Basic Studies in the Field of High-temperature Engineering (2002)
ISBN 92-64-19796-6 Price: € 75 US$ 66 GBP 46 ¥ 8 600
Utilisation and Reliability of High Power Proton Accelerators (2002)
ISBN 92-64-18749-9 Price: € 130 US$ 116 GBP 80 ¥ 13 100
Fission Gas Behaviour in Water Reactor Fuels (2002)
ISBN 92-64-19715-X Price: € 120 US$ 107 GBP 74 ¥ 12 100
Shielding Aspects of Accelerators, Targets and Irradiation Facilities – SATIF 5 (2001)
ISBN 92-64-18691-3 Price: € 84 US$ 75 GBP 52 ¥ 8 450
Nuclear Production of Hydrogen (2001)
ISBN 92-64-18696-4 Price: € 55 US$ 49 GBP 34 ¥ 5 550
Pyrochemical Separations (2001)
ISBN 92-64-18443-0 Price: € 77 US$ 66 GBP 46 ¥ 7 230
Evaluation of Speciation Technology (2001)
ISBN 92-64-18667-0 Price: € 80 US$ 70 GBP 49 ¥ 7 600
Comparison Calculations for an Accelerator-driven Minor Actinide Burner (2002)
ISBN 92-64-18478-3 *Free: paper or web.*
Forsmark 1 & 2 Boiling Water Reactor Stability Benchmark (2001)
ISBN 92-64-18669-4 *Free: paper or web.*
Pressurised Water Reactor Main Steam Line Break (MSLB) Benchmark (Volume III)
(2002) *In preparation.*

International Evaluation Co-operation *(Free on request - paper or CD-ROM)*
Volume 1: *Comparison of Evaluated Data for Chromium-58, Iron-56 and Nickel-58* (1996)
Volume 2: *Generation of Covariance Files for Iron-56 and Natural Iron* (1996)
Volume 3: *Actinide Data in the Thermal Energy Range* (1996)
Volume 4: *^{238}U Capture and Inelastic Cross-Sections* (1999)
Volume 5: *Plutonium-239 Fission Cross-Section between 1 and 100 keV* (1996)
Volume 6: *Delayed Neutron Data for the Major Actinides* (2002)
Volume 8: *Present Status of Minor Actinide Data* (1999)
Volume 10: *Evaluation Method of Inelastic Scattering Cross-sections for Weakly Absorbing Fission-product Nuclides* (2001)
Volume 12: *Nuclear Model to 200 MeV for High-Energy Data Evaluations* (1998)
Volume 13: *Intermediate Energy Data* (1998)
Volume 14: *Processing and Validation of Intermediate Energy Evaluated Data Files* (2000)
Volume 15: *Cross-Section Fluctuations and Shelf-Shielding Effects in the Unresolved Resonance Region* (1996)
Volume16: *Effects of Shape Differences in the Level Densities of Three Formalisms on Calculated Cross-Sections* (1998)
Volume 17: *Status of Pseudo-Fission Product Cross-Sections for Fast Reactors* (1998)
Volume 18: *Epithermal Capture Cross-Section of ^{235}U* (1999)

Order form on reverse side.

ORDER FORM

OECD Nuclear Energy Agency, 12 boulevard des Iles, F-92130 Issy-les-Moulineaux, France
Tel. 33 (0)1 45 24 10 15, Fax 33 (0)1 45 24 11 10, E-mail: nea@nea.fr, Internet: www.nea.fr

Qty	Title	ISBN	Price	Amount
			Total*	

* Prices include postage fees.

❑ Payment enclosed (cheque payable to OECD Publications).

Charge my credit card ❑ VISA ❑ Mastercard ❑ Eurocard ❑ American Express

Card No.	Expiration date	Signature
Name		
Address	Country	
Telephone	Fax	
E-mail		

OECD PUBLICATION, 2, rue André-Pascal, 75775 PARIS CEDEX 16
PRINTED IN FRANCE
(66 2002 17 1 P) – No. 52739 2002

Foreword

The long-term safety of spent fuel and high-level radioactive waste (HLW) disposal sites will be assessed according to the quality of safety analyses which are based on both subjective and objective elements. Although hard data and inferences from objective facts are to be preferred whenever possible, in all cases a need exists for validating procedures that demonstrate that the reference models and analyses are not unjustifiably biased.

The topic of validation has been addressed in numerous papers, international projects, workshops, and symposia, and constitutes an important concern within the NEA Radioactive Waste Management Committee and its subcommittees, hence the decision to co-sponsor with the Swedish Nuclear Power Inspectorate (SKI) the GEOVAL '94 Symposium on the subject in Paris in October 1994. An earlier symposium on the same issue, also co-sponsored with SKI, took place in Stockholm in 1990 (GEOVAL '90). That seminar, in turn, followed one held in 1987 (GEOVAL '87) under the sole sponsorship of SKI.

One important aim of GEOVAL '94 has been to bring to a wider audience the lessons learned within the INTRAVAL project which ended officially in 1993. INTRAVAL, the successor to two earlier international projects, INTRACOIN (1981-1984) and HYDROCOIN (1984-1987), was established in 1987 as a forum where experimentalists and modellers could test model concepts quantitatively and, in the process, help develop an approach to validation. The Project ran in two Phases and, at its peak, involved the participation of 46 organisations in 14 countries exploring a wealth of geological data from sites in four continents.

GEOVAL '94 was articulated in six sessions under a common theme: "Validation through model testing with experiments". The papers of Session 1 set the scene by providing a review of the advancement of the state of the art in the past few years. Newer and more detailed information is provided in Sessions 2 to 6: Session 2 is devoted to INTRAVAL; Sessions 3,4, and 5 focus on lessons learned or being learned within integrated experimental studies, including results from underground research laboratories; and Session 6 deals with strategic approaches for including validation in the planning for siting, design, and licensing of deep geologic repositories of long-lived HLW. These proceedings contain also the transcript of the final panel session which discussed the conclusions of the symposium as a whole.

These Proceedings are published on the responsibility of the Secretary-General of the OECD. The views expressed herein do not necessarily reflect those of the OECD or any of its Member governments.

Acknowledgements

The NEA and SKI wish to express their gratitude the Programme Committee, consisting of Mssrs. J. Andersson, T. Äikäs, C. Pescatore, and C.-F. Tsang, for their contribution to the success of the symposium. They also wish to commend the fruitful participation of the panel comprised of Mssrs. C. McCombie (Chairman), J. Andersson, J. Bruno, G. de Marsily, S. Neuman, O. Ollsson, C. Pescatore, and H. Röthemeyer, for there dynamic conduct of discussions.

#33051723

OECD DOCUMENTS

SAFETY ASSESSMENT OF RADIOACTIVE WASTE REPOSITORIES

UNIVERSITY OF STRATHCLYDE
30125 00504050 5

GEOVAL '94
Validation Through Model Testing

Proceedings of an NEA/SKI *Symposium*
Paris, France, 11-14 *October* 1994

organised jointly
by
the OECD *Nuclear Energy Agency*
and
the Swedish Nuclear Power Inspectorate

ANDERSONIAN LIBRARY
WITHDRAWN FROM LIBRARY STOCK
UNIVERSITY OF STRATHCLYDE

UNIVERSITY OF STRATHCLYDE
25 AUG 1995
UNIVERSITY LIBRARY

NUCLEAR ENERGY AGENCY
ORGANISATION FOR ECONOMIC CO-OPERATION AND DEVELOPMENT

ORGANISATION FOR ECONOMIC CO-OPERATION AND DEVELOPMENT

Pursuant to Article 1 of the Convention signed in Paris on 14th December 1960, and which came into force on 30th September 1961, the Organisation for Economic Co-operation and Development (OECD) shall promote policies designed:

— to achieve the highest sustainable economic growth and employment and a rising standard of living in Member countries, while maintaining financial stability, and thus to contribute to the development of the world economy;
— to contribute to sound economic expansion in Member as well as non-member countries in the process of economic development; and
— to contribute to the expansion of world trade on a multilateral, non-discriminatory basis in accordance with international obligations.

The original Member countries of the OECD are Austria, Belgium,Canada, Denmark, France, Germany, Greece, Iceland, Ireland, Italy, Luxembourg, the Netherlands, Norway, Portugal, Spain, Sweden, Switzerland, Turkey, the United Kingdom and the United States. The following countries became Members subsequently through accession at the dates indicated hereafter: Japan (28th April 1964), Finland (28th January 1969), Australia (7th June 1971), New Zealand (29th May 1973) and Mexico (18th May 1994). The Commission of the European Communities takes part in the work of the OECD (Article 13 of the OECD Convention).

NUCLEAR ENERGY AGENCY

The OECD Nuclear Energy Agency (NEA) was established on 1st February 1958 under the name of the OEEC European Nuclear Energy Agency. It received its present designation on 20th April 1972, when Japan became its first non-European full Member. NEA membership today consists of all European Member countries of OECD as well as Australia, Canada, Japan, Republic of Korea, Mexico and the United States. The Commission of the European Communities takes part in the work of the Agency.

The primary objective of NEA is to promote co-operation among the governments of its participating countries in furthering the development of nuclear power as a safe, environmentally acceptable and economic energy source.

This is achieved by:

— *encouraging harmonization of national regulatory policies and practices, with particular reference to the safety of nuclear installations, protection of man against ionising radiation and preservation of the environment, radioactive waste management, and nuclear third party liability and insurance;*
— *assessing the contribution of nuclear power to the overall energy supply by keeping under review the technical and economic aspects of nuclear power growth and forecasting demand and supply for the different phases of the nuclear fuel cycle;*
— *developing exchanges of scientific and technical information particularly through participation in common services;*
— *setting up international research and development programmes and joint undertakings.*

In these and related tasks, NEA works in close collaboration with the International Atomic Energy Agency in Vienna, with which it has concluded a Co-operation Agreement, as well as with other international organisations in the nuclear field.

© OECD 1995
Applications for permission to reproduce or translate all or part
of this publication should be made to:
Head of Publications Service, OECD
2, rue André-Pascal, 75775 PARIS CEDEX 16, France.

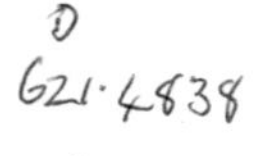

TABLE OF CONTENTS

SESSION 1
(Chairman: T. Äikäs)

SESSION 2
(Chairmen: A. Larsson, L. Dewière)

SESSION 3
(Chairmen: H. Umeki, J.-C. Petit)

SESSION 4
(Chairmen: T. Papp, C. Del Olmo)

SESSION 5

(Chairmen: B. Vignal, A. Van Luik)

SESSION 6

(Chairmen: A.G. Duncan, J. Vigfusson)

PANEL SESSION

OPENING OF THE SYMPOSIUM

S. Norrby, SKI

Ladies and Gentlemen,

This is the third International Geoval Symposium, arranged jointly by the OECD/NEA and the Swedish Nuclear Power Inspectorate. The first Geoval Symposium was in 1987.

The predominant idea for the management and disposal of HLW and spent fuel is still the same, Deep Geological Disposal, but much has happened since 1987.

In many countries the programmes for final disposal have proceeded and are now much more concrete. Procedures for the siting of repositories are in progress. Techniques for manufacturing, welding and control of canisters for spent fuel are being developed. Methods for Safety Assessment of Deep Geological Repositories have developed considerably in the last few years.

The very long time perspectives (hundreds of thousands of years and more for spent fuel) implies effects and burdens on future generations and requires that long-term safety can be evaluated. This is very much a question of Confidence. The need for Confidence is most clearly demonstrated in the process of Site Selection. We, implementors, regulators and the scientific community, must have Confidence that we can evaluate the long term safety. Political decision makers and the general public must have Confidence that there is a careful process for siting, evaluation and licencing of final disposal facilities.

A couple of years ago NEA, IAEA and CEC expressed as a Collective Opinion that methods for the evaluation of Long Term Safety are available but that site specific data is needed and that safety assessment methods can be further developed.

A few weeks ago a workshop organized by the NEA addressed Environmental and Ethical Aspects of Deep Geological Disposal. Very briefly, the conclusions were that Environmental and Ethical Aspects are addressed in ongoing programmes for deep geological disposal. One other observation was that storage of spent fuel and HLW should not be regarded as a final solution. There must be a programme for final disposal in which intermediate storage can be an important part.

My interpretation of this is that we should proceed both as regards the site selection process that will provide the data needed, as regards development of the engineered barriers and as regards the further development of performance assessment methods. To build and preserve Confidence requires an open attitude. A thorough understanding of the system for final disposal and a willingness to see and address problems is a prerequisite for Confidence that remains over the years.

In many countries and in international organizations, great effort is put into final disposal issues and in methods for assessing safety. This symposium can contribute to a better understanding of safety-related matters.

Before I end, I would like to thank the Programme Committee for it's work. The Programme Committee Members are Johan Andersson, Timo Äikes, Claudio Pescatore and Chin-Fu Tsang.

Ladies and Gentlemen! On behalf of the Swedish Nuclear Power Inspectorate, I welcome you all to GEOVAL '94 and wish you a successful Symposium and a nice stay in Paris. Thank you.

OPENING OF THE SYMPOSIUM

K. Uetmatsu, NEA/OECD

Ladies and Gentlemen,

I am glad to welcome you to the 3rd GEOVAL symposium organised jointly by the Swedish Nuclear Power Inspectorate and our Agency.

The NEA's general objective is to be at the forefront of the scientific and technical debate in the field of nuclear energy, and to provide a forum for both technical and policy-oriented issues of major interest to our Member Countries. Validation of geosphere models features prominently on the programme of work of the NEA Radioactive Waste Management Committee, and its two main sub-committees: the Performance Assessment Advisory Group, and the Site Evaluation and Design of Experiments Coordinating Group. Indeed this is an issue which cuts across both the fields of geoscience and mathematical modelling. It is a pleasure to welcome you to a symposium where both types of expertise will be utilised, discussed, and, yes, even confronted against each other. Without doubt much cross-fertilisation across disciplines will occur.

Earlier symposia in this series took place in 1987 and 1990. They were well received, and provided a much welcome opportunity for the waste disposal community to pause and reflect on the meaning of "validation" of geosphere models. Particularly to what extent it can be pursued, and the priorities we must establish before ourselves focusing on our individual goals of developing and licensing a series of national repository sites for high-level, long-lived nuclear waste. The theme of this symposium: "validation through model testing with experiments" is meant to foster a practical view of validation. Indeed, at present, numerous waste management organisations in the NEA member countries are acively developing models and testing them both in the laboratories and in the field. This work in preparation for site evaluations, for planning additional tests, and for generating less complex models is to be used in support of safety assessments through confidence building. All of these considered needs pose different validation requirements.

I would like to take this opportunity to extend a special welcome to the INTRAVAL community. The second session of this symposium is dedicated entirely to the INTRAVAL international project, and it represents the much-awaited culmination of its many years of effort. This session not only provides them with the opportunity to present their work to such a highly specialised audience, but it also challenges them to interact with YOU during the working session, and with the discussion panel. Indeed, I note that this symposium devotes a significant portion to active discussion facilitated by a highly qualified panel.

I also encourage you to exchanges your views further this evening at the reception at the OECD Château. I invite all of you to attend.

I wish you success in your discussions and look forward to learning their results. With the help of the authors, the NEA will take care to publish the proceedings of the symposium so that this new information quickly reaches the technical community.

SESSION I

Chairman:

T. Äikäs

Objectives and scope of GEOVAL-94

Johan Andersson
Swedish Nuclear Power Inspectorate

In many countries the ongoing programmes to site and construct deep geological repositories for high and intermediate level nuclear waste are closer to realization compared to the situation four years ago. A number of studies, including the international INTRAVAL-project demonstrate the potential barrier function of the geosphere, but also that there are many unresolved issues, some of which may never be resolved. It is now high time to identify which issues that indeed are resolved and to develop schemes for handling of the unresolved ones. A key to these problems are the possibilities to gain knowledge by Model Testing with Experiments and this is also the theme of GEOVAL-94. GEOVAL-94 will review the state-of-the-art where model testing with experiments have been specifically aimed at increasing confidence in models used for prediction the performance of the geosphere in a deep repository. The sessions cover conclusions from the INTRAVAL-project, experiences from integrated experimental programs and underground research laboratories as well as the integration between performance assessment and site characterisation. Technical issues ranging from waste and buffer interactions with the rock to radionuclide migration in different geological media will be addressed. The different contributions will all be evaluated with the following questions in mind: How to assess results? How to plan experimental designand analysis of results? Are all sensible alternative conceptual models tested? Is it enough integration of different disciplines? Where are the difficulties? A discussion panel will conclude the symposium.

VALIDATION: AN OVERVIEW OF DEFINITIONS

C. Pescatore

NEA/OECD
(Paris)

Abstract

The term "validation" is featured prominently in the literature on radioactive high-level waste (HLW) disposal. There exists, however, no unique definition of "validation" although it is generally understood to be related to model testing using experiments. The paper reviews the several definitions of "validation" and proposes their categorization into three main classes. The first class links validation to the goal of predicting the physical world as faithfully as possible. This view has been criticized as being unattainable and, in any event, unsuitable for setting goals for the safety analyses. Other definitions (Class 2) are strictly operational, and associate validation to split-sampling or to blind-tests predictions. In this view, the decision to retain a predictive model for use in safety assessments does not belong to the remit of "validation". The latter, however, is the very domain of "validation" according to the third class of definitions, which focuses on the quality of the decision-making process. In this view, one cannot determine when a model or a suite of models are actually "validated". The stress in the decision-making process for retaining a model shifts the debate from validation in the observational sense to that of "reasonable assurance" and "confidence building", which tends to generalize the meaning of "validation". Most prominent in the present review is the observed lack of use of the term "validation" in the field of low-level radioactive waste disposal and, during the first half of this century, in all technical fields. Its first technical appearance dates from the mid-fifties. The term was adopted thereafter in the computer field and elevated to its present status following the computer revolution of the seventies and early eighties. Starting from the late eighties, the term has made its appearance in some HLW safety standards. The continued informal use of the term "validation" in the field of HLW disposal can become cause for misperceptions and endless speculations. The paper proposes either abandoning the use of this term or agreeing to a definition which would be common to all.

INTRODUCTION

In recent years "validation" has become a central subject in the field of radioactive high-level waste (HLW) disposal as it is witnessed by the several international projects, workshops, and symposia, including the present one, which have been organised around it. Recently, the Journal of Advances in Water Resources has dedicated two special issues to this topic [1a, 1b].

Despite the fairly large literature on the subject, there exists no widely accepted definition of what constitutes "validation". It is generally understood to be related to model testing using experiments. However, there seem to exist as many definitions as there are papers on the subject, and the same term may have different meanings in separate sections of the same report[1]. In the field of HLW disposal the concept appears to be still *in fieri*. As one participant to the INTRAVAL project explained: "*The first phase of INTRAVAL has made some progress toward defining what validation means, but participants have not yet reached a consensus on an exact definition. Many participants, through the analysis of test cases have begun to understand what validation is not (i.e., calibration) though what it is remains elusive.*"[3] Adding to the confusion is the fact that the term "validation" is conspicuously absent in the field of low level waste (LLW) disposal.

The present paper reviews some of the definitions of the term "validation" which are reported in the HLW literature and proposes their categorisation within a few classes. At the same time the different meanings of this term, or even its absence, in allied fields are also noted in an effort to provide a perspective on past, present, and future usage.

THE VIEW FROM THE TECHNICAL AND NON-TECHNICAL DICTIONARIES

According to the Oxford English Dictionary (OED) validation is an action: "the action of validating or making valid" [4]. The example quotes from the english literature provided by the OED and spanning the period from 1656 through 1888 suggest a legalistic usage/meaning of the term, i.e., the act of confirmation or corroboration which renders a statement, a document, a deed, etc. legally valid. No literature quotes are provided for the following period up to 1956. The term "validation" is reported thereafter in quotes taken from a *Dictionary of Statistical Terms* (1957), from the *Oxford Computer Explained* (1967), from a *Handbook of Management Technology* (1967), and from the journal *Computer Science* (1980). The quote from 1957 is the earliest one known by this author. It appears very relevant to the HLW disposal debate and reads as follows:

> **Validation**, a procedure that provides, by reference to independent sources, evidence that an inquiry is free from bias or otherwise conforms to its declared purpose. In statistics it is usually applied to a sample investigation with the object of showing that the sample is reasonably representative.

Thus the term "validation" appears to have come in use within the technical community only very recently, and especially in connection with computer program development and implementation. A 1984 bibliography on validation and related key words on behalf of the computer simulation community identifies 308 references mostly from the computer field.[5] The references from the 1960s are only 9, indicating an "explosion" in the use of this and related terminology in the 1970s and early 1980s alongside the computer revolution.

---

1 "*The term validation, which is used throughout this report is ambiguous in the sense that it has different meanings in different contexts.*"[3]

The Chambers Science and Technology Dictionary of 1988 recognises "validation" as a term from computer science. Namely:

> **Validation** (*Comp.*) Input control technique used to detect any data which is inaccurate, incomplete or unreasonable.

A similar attribution is provided by the McGraw-Hill Dictionary of Scientific and Technical Terms of 1978.

The Dictionary of Computing of the Oxford University Press of 1986 does not have a separate entry for "validation" but only for " verification". However, a "verification and validation" entry exists and reads as follows:

> **verification and validation (V&V)** A generic term for the complete range of checks that are performed on a system in order to increase confidence that the system is suitable for its intended purpose. This range might include a rigorous set of functional tests, performance testing, reliability testing, and so on. Although a precise distinction is not always drawn, the verification aspect refers to the completely objective checking of conformity to some well defined specification, while the validation aspect refers to a somewhat subjective assessment of likely suitability in the intended environment.

Other definitions do exist in the computer simulation field but appear to be more nebulous[2].

It is striking, in view of the recent debate within the HLW disposal community, which has extended itself to discuss "*the reality behind models*", that the term "validation" may have been borrowed from the computer field where proving realism through objective methods does not even enter in the definition. Also striking is the fact that the term has not been in use between 1888 and 1956, i.e., at a time when the greatest advances in science have taken place and deep debates were raging on the meaning of the new discoveries and of the scientific method.

MODEL VALIDATION: HOW WIDELY DEMANDED FOR DISPOSAL SAFETY

LLW disposal

The term "validation" is conspicuously absent in the field of radioactive, low-level waste (LLW) disposal. The term has **not** been used during the recent licensing procedures of most LLW waste repositories. Namely: Forsmark (Sweden), El Cabril (Spain), Mosleben (Germany), Rokkasho Mura (Japan), and La Manche and L'Aube (France). The only exception is the VLJ repository at Olkiluoto (Finland).

---

2 In 1979 the Society of Computer Simulation-Technical Committee on Model Credibility defined validation as follows: "*Substantiation that a computer model within its domain of applicability possesses a satisfactory range of accuracy consistent with the intended application of the model*", while verification is the "*substantiation that a computerized model represents a conceptual model within specified limits of accuracy.*"[6]

In the Finnish case, the safety guide reads as follows:

> "The safety analysis shall be based on carefully verified calculational methods and on models that are validated as far as practicable. Except for the optimization analyses referred to below, the safety analyses shall be based on such input data and assumptions that the results of the calculations are, with a high certainty, more unfavourable with respect to safety than the real values."[7]

Unfortunately, this regulatory requirement does not define what "model validation" is, but it can be safely assumed to mean that these models must have an experimental footing. The second part of the requirement is possibly more interesting in that it may be interpreted to suggest an aim for validation. Namely, providing a robust model in the sense explained by McCombie [8], i.e., a model for which "we have confidence that the results are **either** correct (i.e. that they are sufficiently realistic) **or** that they overpredict detriment (i.e. err on the side of conservatism)."

HLW disposal

The term "validation" appears to have entered only very recently in some regulatory documents. For instance, this term does not appear in the German Disposal Safety Criteria (1983) nor it is mentioned explicitly in the NRC (1982) and EPA (1985 and 1993) regulations applicable in the United States for HLW disposal (although the former do emphasize that the safety analyses cannot be only speculative but need to have an experimental basis.[3])

The earliest standard requiring "validation" is the IAEA's of 1989. Namely:

> "Compliance of the overall disposal system with the radiological safety objectives shall be demonstrated by means of safety assessments which are based on models that are validated as far as possible."[9a]

This requirement appears to be based on the view, not necessarily shared by many nowadays, that safety assessments rest almost entirely on the exercise of independent models conceived to predict reality as closely as possible. In fact the IAEA proposes the following definition for validation:

> "Validation is a process carried out by comparison of model predictions with independent field observations and experimental measurements. A model cannot be considered validated until sufficient testing has been performed to ensure an acceptable level of predictive accuracy (note that the acceptable level of accuracy is judgemental and will vary depending on the specific problem or question to be addressed by the model)"[9b]

Thus, according to the IAEA:

1. validation is about model testing with experiments;
2. validation is about achieving predictive accuracy;

---

3 The regulatory requirement is that: "*Analyses and models that will be used to predict future conditions and changes in the geological setting shall be supported by field tests, in-situ tests, laboratory tests which are representative of field conditions, monitoring data, and natural analogue studies.*" [10 CFR 61.21 (c) (1) (ii) (F)]

3. sufficient validation is a judgement call;
4. model validation should be carried out as far as possible.

Propositions 3 and 4 contradict one another and, in the view of this author, proposition 4 should have been spared. The latter, when considered with proposition 2, suggests that, by performing tests "as far as possible", an unlimited degree of accuracy can be achieved. This is not necessarily the case, and the "as far as possible" clause may result in costly delays in radioactive waste disposal programmes.

A very recent standard which appears to have been patterned after the IAEA's is the one formulated by the Safety Authorities in the Nordic Countries (1992). Namely:

> "Compliance of the overall disposal system with the radiation protection criteria shall be convincingly demonstrated through safety assessments which are based on qualitative judgement and quantitative results from models that are validated as far as practicable." [10]

Unfortunately this standard does not define what "model validation" is but, again, it can be safely assumed to mean that these models must have a good experimental basis. The standard implements also the "as far as possible" clause of the IAEA's. However, it recognizes the important role of qualitative judgements in the analysis of safety. Hopefully this recognition may carry on to the evaluation of models.

The Swiss regulations (1993) require "validation" as follows:

> "Each computer code to be used in the safety analysis has to be verified. It also has to be shown that the models used are applicable for the specific repository system (validation), taken both individually and as an overall model chain." [11]

This definition of validation is quite distinct from the previous ones in that it:

1. emphasizes *applicability* for a specific system and for the purpose of safety,

2. does not require, necessarily, to prove a degree of realism or predicting however closely the exact evolution of the system, and

3. brings forward the consideration of the overall model chain. In fact, it can be argued that, from the point of view of safety, single model validation *per se* is uninteresting and that "validation" should be pursued to various degrees depending on the relevance of each model in the overall calculational sequence. It is the latter that will need final validation by the regulator. Unfortunately this view is frequently lost in other definitions, which can become cause for misperceptions and endless speculations.

VALIDATION: THREE CLASSES OF DEFINITIONS

Most of the available definitions of "validation" in the field of HLW disposal focus on single, isolated models and have at their core the point that validation is about model testing with experiments. However, they tend to differ from one another as to the degree of predictive accuracy that is deemed achievable or reasonable. Following are a few such definitions/points of view organised in three classes according to their relation to perceived, achievable accuracy.

Class 1 *Validation: the purist view*

The first class of definitions/views acknowledges that, in principle, validation is linked to the desire to predict the physical world as faithfully as possible. Namely:

> " ... a process whose objective is to ascertain that the code or model indeed reflects the behaviour of the real world." (USDOE, 1986) [12]
>
> "Validation may be defined as the testing of a model in the real world" (SKI/SSI/HSK, 1990) [13]
>
> "Validation is the process of confirming that a model is sufficiently 'close to the truth" (McCombie, 1990) [8]
>
> "Validation is concerned with the genuine understanding of nature, that is, of the reality behind model." (NEA, 1991) [14]
>
> "...models need to be validated to ensure, as much as possible, that they are a good representation of the actual processes occurring in the real system." (Tsang, 1991) [15]
>
> "The term validation refers to the process by which (sub)models or detailed subsystem models, or even the conceptualization of key processes are shown to be realistic representations of the natural system by direct comparison with the observed behaviour within the field or laboratory. Any validation study has a cyclic structure which can be iterated as concepts and models become refined. " (Chapman, 1994) [16]
>
> "...the process of obtaining assurance that a model is a correct representation of the process or system for which it is intended. Ideally, validation is a comparison between predictions derived from a model with empirical observations. However, as this is frequently impractical or impossible owing to the large length and time scales involved in HLW disposal, short term testing supported by other avenues of inquiry such as peer review is used to obtain such assurances" (USNRC, 1985) [17]

In general, the authors of the above quotes do reasonably argue that absolute validation, even of single, isolated models, is not fully possible. Rather, they argue, we are not striving for certainty and perfection but only for "our level best" [18]. Yet, it is precisely this perfectionist view of validation that has motivated others, in the literature, to counterview that:

1. this view of validation is misleading and unsuitable for setting safety analysis goals[4] (Hadermann, 1993)[19];

2. in any fashion, "ground-water models cannot be validated" (Konikow, 1992)[20];

3. when it comes to "Field validation: we have no means to validate geochemical models, we can only invalidate them." (Nordstrom, 1992) [21] ; and

---

[4] *"Der Begriff 'Validierung' erweist sich als missverständlich und ungeeignet für die Zielsetzung"*

4. more generally, "verification and validation of numerical models of natural systems is impossible."(Oreskes, 1994) [22]

In particular, this purist view of validation leads (Nordstrom, 1992) to confine validation only to the realm of laboratory testing. Namely:

> "*Validation*: a test of a model to see that it reproduces experimental laboratory measurements (field measurements are not laboratory measurements)." [21]

Class 2 *Validation: the operational view*

A second class of definitions suggests that validation is accomplished only when the results of "blind" tests *have been* predicted. Thus the use of the term is confined only to situations where the models can be tested again observational data, leaving the long-term analysis outside the remit of "validation".

One comparative approach between model predictions and observations is based on split-sampling in which the model is calibrated with one segment of the observational record. The model is then used to reproduce that part of the observational record that is independent of the data used in calibration. Using this perspective "a model can be considered to have been validated if its prediction uncertainties are sufficiently small. What is meant by the sufficiently small depends on the model objectives and on socio-political issues, which we feel fall beyond scientific or technical arguments." (Usunoff, 1992) [23]

Alternatively, model validation is seen as associated with blind predictions of new observational data following calibration using data from a similar experimental framework. Namely:

> "The classical approach to model validation in geochemistry is calibration of a set of model parameters against a given experiment geometry in a given geological domain for a specific stress of the system. The validation constitutes prediction of model behaviour for another experiment condition (altered stress and/or geometry) in the same geological domain and comparison with field data." (Andersson, 1994) [24]

(Luis, 1991) indicates even more clearly that the decision of accepting a model for the prediction of non-observable data is not "validation" but rather "accuracy assessment". Namely:

> "Model validation addresses the question whether or not a model adequately represents observed phenomena while accuracy assessment addresses the larger question of how well a model will perform under conditions that have not yet been observed" [25]

Class 3 *Validation as a confidence building process*

The realisation that absolute confidence in the ability of models to predict reality over long time scales cannot be guaranteed and that, even on limited time scales, "sufficient validation" entails a subjective judgement, has caused part of the HLW disposal community to shift the emphasis of "validation" onto measures which ensure the reasonableness as to the choice of models for performance assessment. The working definition of "validation" results in further modifications as follows:

"... validation is not just then about comparing predictions of the model with physical observations. It is about establishing whether or not the model is an acceptable representation of physical phenomena. As such it also involves examining the model for consistency with principles that are generally accepted in the scientific community, and presenting the case for (or against) the model for review. Presentation of the case is a very important part of validation ...." (Jackson, 1990) [26]

"... model validation includes not only comparison of modelling results with data from selected experiments, but also evaluation of procedures for the construction of conceptual models and calculational models as well as methodologies for studying data and parameter correlation." (Tsang, 1991) [15]
"The validation of safety assessment models is viewed here as the process of building scientific and public confidence in the methods used to perform such assessments" (Neuman, 1992) [27]

The last definition of validation brings into the discussion "confidence building", which, in turns, opens the question as to what is meant by it. In this regard, the latter definition may be interpreted to mean not only are the models evaluated against each other and against actual observations using various protocols. It also advocates that the process of model selection is open-minded, peer-reviewed, well-documented, and traceable according to the best available technical practices. Intangible aspects are also important for confidence building, such as: openness to discuss differences and alternative views with the scientific community, with interveners, and other interested parties. This perspective closely parallels the debate about "validation" with what is meant by reasonable assurance that a repository will perform over time periods as intended, and how can we demonstrate it.

It must be realised that this approach to validation is possibly open-ended, in the sense that one may not know when a model is validated until, after several iterations among the developers and the regulators, it is judged to be acceptable. However, is this not true of any issue being argued amongst the developers and regulators?

CONCLUSIONS

Validation, as a technical term, appears to have been developed in the computer field and was borrowed when HLW safety assessments were considered large computer-base predictive models. The appearance of this term in HLW regulatory criteria is very recent, and, with one exception, has not been used in the recent licensing of all LLW disposal sites. Nor has the term "validation" entered into discussions related to the major scientific and technical discoveries of this century. Thus, in principle, do we need this term ?

We have shown that even among scientists in the same discipline the term "validation" can be used and understood in different ways. This can lead to serious misunderstanding, and alternative terms, such as "model evaluation", have been proposed and utilised in the field of HLW disposal [20, 21, 28]. If we are to continue using the term "validation" we should give it a limited but clear definition, possibly an operational one. For instance, during its discussions on the subject of model validation in 1993, the NEA Performance Assessment Advisory Group accepted, in principle, the IAEA definition which links validation solely to model testing, but it also indicated that validation is only one aspect of the confidence building process.

A recent trend is to link the term "validation" with the quality of the decision-making process when retaining a model for predictive purposes. Thus, validation becomes synonymous with such terms as "quality assurance", "confidence building", and "reasonable assurance", which blurs its meaning. It must be realised that this view of validation is possibly open-ended, in the sense that one may not know when a model is validated until, after several iterations among the developers and the regulators, it is found acceptable. Is this not true of any issue being argued amongst the developers and regulators? Do we need such an open definition of validation ?

If the term "validation" is to be retained in its broadest meaning, in spite of the above outlined objections, this author's preference is with the earliest definition of validation. Namely:

> **Validation**, a procedure that provides, by reference to independent sources, evidence that an inquiry is free from bias or otherwise conforms to its declared purpose.

The appeal of this definition is that, despite its age, it is very modern. Namely it:

1. does not apply only to "mathematical models" or to single, isolated models;

2. suggests, as it is in fact more and more common practice, that a case is made **not** on the exercise of a single hypothesis or model, but by utilizing several independent sources;

3. acknowledges the intended purpose of the inquiry, although it suggests that an important aim is to reduce bias to the extent needed.

REFERENCES

1a. Special Issue: Validation of Geo-Hydrological Models - Part 1, *Advances in Water Resources*, Vol. 15, No. 1, 1992.

1b. Special Issue: Validation of Geo-Hydrological Models - Part 2, *Advances in Water Resources*, Vol. 15, No.3, 1992.

2. The International Hydrocoin Project, Level 2: Model Validation, NEA/OECD, Paris 1990.

3. L. Lehman., State of Nevada Review of Phase I of the INTRAVAL Project, NWPO-TR-020-93, State of Nevada' s Nuclear Waste Project Office, August 1992.

4. The Compact Edition of the Oxford English Dictionary, Oxford University Press, 1971, and A Supplement to the Oxford English Dictionary, Oxford at the Clarendon Press, 1986.

5. O. Balci and R.G. Sargent, "A bibliography on the credibility assessment and validation of simulation and mathematical models," in *Simuletter*, Vol. 15, No. 3, July 1984.

6. Society for Computer Simulation - Technical Committee on Model Credibility, "Terminology for Model Credibility", in *Simulation*, Vol. 32, No. 3, 1979.

7. Finnish Center for Radiation and Nuclear Safety (STUK), Disposal of reactor waste, YVL-guide 8.1, Helsinki, Finland, 1991.

8. C. McCombie, I.G., McKinley, P. Zuidema, "Sufficient validation: the value of robustness in performance assessment and system design," in Validation of Geosphere Flow and Transport Models (GEOVAL), Proceedings of a NEA/SKI Symposium held in Stockholm 14-17 May, 1990, OECD/NEA, Paris, 1991.

9a. Safety Principles and Technical Criteria for the Underground Disposal of High-Level Radioactive Wastes, IAEA Safety Standards, Safety Series No. 99, IAEA, Vienna, 1989.

9b. Radioactive Waste Management Glossary, IAEA-TECDOC-447, 2nd Edition, Vienna, 1988.

10. The Radiation Protection and Nuclear Safety Authorities in Denmark, Finland, Iceland, Norway, and Sweden, Disposal of High Level Radioactive Waste-Considerations of some Basic Criteria, 1992

11. Swiss Federal Nuclear Safety Inspectorate (HSK), Protection Objectives for the Disposal of Radioactive Waste, HSK-R-21/e, Villigen-HSK, Switzerland, November 1993.

12. US Department of Energy, Environmental Assessment - Yucca Mountain Site, Nevada R&D Area, DOE/RW-0073, Vol. 2, Washington, DC, 1986.

13. SKI/HSK/SSI, Regulatory Guidance for Radioactive Waste Disposal - an Advisory Document, SKI Technical Report 90:15, SKI, Stockholm, Sweden, 1990.

14. Disposal of Radioactive Waste: Review of Safety Assessment Methods, OECD/NEA, Paris, 1991.

15. C.F. Tsang, "The modeling process and model validation," in *Ground Water*, Vol. 29, No. 6, November-December 1991.

16. N. Chapman and P. Grindrod, "A scientific approach to integrating scenarios, simulation and natural systems", in Proceedings of "Geoperspective", Paris, 1994.

17. US NRC Staff, Letter with "List of Defintions", June 24, 1985 (quoted in C. Voss, "Validation Issues...", Proceedings of the HLRWM Conference, Las Vegas, Vol. 2, 1992).

18. G. de Marsily, P. Combes, and P. Goblet, "Comment on 'Ground-water models cannot be validated' by L.F. Konikow & J.D. Brederhoft", in *Advances in Water Resources*, Vol. 15, 367-369, 1992.

19. J. Hadermann, Validierung von Modellen, PSI Report AN-44-93-03, Paul Scherrer Institut, Würenlingen and Villigen, Switzerland, October 1993.

20. L.F. Konikow & J.D. Brederhoft, "Ground-water models cannot be validated", in ref. [1a].

21. D.K. Nordstrom, "On the evaluation and application of geochemical models," in Proceedings of Fifth CEC National Analogue Working Group Meeting, EUR-15176, published by Commission of the European Communities, Bruxelles, 1992.

22. N. Oreskes, K. Shrader-Frechette, and K. Belitz, "Verification, Validation, and Confirmation of Numerical Models in the Earth Sciences," in *Science*, Vol. 263, pp. 641-646, 4 February 1994.

23. E. Usunoff, J. Carrera, and S.F. Mousavi, "An approach to the design of experiments for discriminating among alternative conceptual models," in ref. [1b].

24. P. Andersson and A. Winberg, Eds., INTRAVAL Working Group 2- Summary report on Phase 2 analysis of the Finsjön test case, SKB Tech. Report 94-07, SKB, Stockholm, January 1994.

25. S.J. Luis and D. McLaughlin, "A stochastic approach to model validation," in ref. [1a].

26. C.P. Jackson, D.A. Lever, P.J. Summer, "Validation of transport models for use in repository performance assessments: a view," in Validation of Geosphere Flow and Transport Models (GEOVAL), Proceedings of a NEA/SKI Symposium held in Stockholm 14-17 May, 1990, OECD/NEA, Paris, 1991.

27. S. Neuman, "Validation of safety assessment models as a process of scientific and public confidence building," in Proceedings of HLWM Conference, Vol. 2, Las Vegas, 1992.

28. K.W. Dormuth, "Evaluation of models in performance assessment,", in Proceedings of SAFEWASTE 93, an International Conference on Safe Management and Disposal of Nuclear Waste, June 13-18, Avignon, France.

VALIDATION AND TECHNICAL ISSUES FROM GEOVAL-90 TO GEOVAL-94

Chin-Fu Tsang

Earth Sciences Division
Lawrence Berkeley Laboratory
University of California
Berkeley, California 94720, USA

Abstract

The outstanding issues on model validation and on various scientific and technical problems at the time of GEOVAL-90 are identified and reviewed. Comments are made on the importance of these issues on the safety assessment of nuclear waste geologic disposal. Significant progress has been made on a number of these issues, while much work is still needed on the others. The overview is intended to set the stage for discussions at GEOVAL-94.

INTRODUCTION

The purpose of the present paper is to give an overview of validation and technical issues that were discussed in the previous GEOVAL symposium, GEOVAL-90, which should perhaps be addressed at the present symposium. The material in this paper is based on a detailed review of the papers presented at GEOVAL-90 (NEA/SKI, 1991) and arranged according to my experience gained in the last few years through various international projects such as the INTRAVAL project. It is my hope that this paper may partially provide the stage for the present symposium.

Geosphere is one of the barriers of a multibarrier system for the isolation of nuclear waste in a geologic repository. One reason behind the concept of disposing nuclear waste in a geological repository is that the geosphere is relatively stable and the geological processes mostly go on with geological times, which is much longer than the characteristic times of human constructions. Thus one would be taking advantage of the stability of the geosphere to provide the barrier characteristics for the nuclear waste that will be buried there. The problem we are facing is that in order to evaluate the safety or performance of such a nuclear waste geological repository, we need to attempt to predict into the next 10,000 to 100,000 years. The predictions are necessarily based mainly on small-scale field experiments. The small-scale field experiments are on the order of centimeters to at most a few hundred meters, with a time frame of days to at most ten years. The only exception is the study of natural analog systems, which we will mention a little more later. Furthermore, another major problem is that for the geosphere, we usually do not have complete information and detailed data of every inch of the system. Given the incomplete knowledge and given the need to predict far into the future, the community was talking about the need of some kind of "validation." The main purpose of "validation" is to provide the assurance of a certain confidence level of the performance assessment we are going to make. Also, such validation studies may be required by the regulatory agencies to establish the accuracy or trustworthiness of the models that are used to make these kinds of predictions.

VALIDATION ISSUES AND STATEMENTS FROM GEOVAL-90

During the GEOVAL-90 there were much discussions about how to construct a model to make predictions, and how one should validate the model predictions to test whether they represent the real world. In these discussions, there was already much appreciation on the difficulties associated with the word "validation" as applied to geosphere models. More detailed discussions of the definition and understanding of this word are presented elsewhere in this proceedings (Pescatore, 1995, this volume). I would like to quote a number of statements that were made at the GEOVAL-90 symposium that indicate the state of knowledge or the state of uncertainties at that time. It will be very useful to keep these in mind in the present symposium to see whether these are respected in the work that has since been done; if they are indeed correct; and how far we have progressed since then.

- Validation is a process; there is no such thing as absolute validation.
- Theories are not really proved but they are accepted by a consensus of experts borne out by repeated experiments.
- Validation is the only way to connect a model world to the real world.
- Validation is needed for public acceptance.
- A particular goal of validation is to remove the conceptual uncertainties of the model.

- A nuclear waste geological repository built for high confidence is more acceptable than one built for maximum safety. In other words, even though a repository might be designed for maximum safety, if there is no way you can prove that there is a high level of confidence that is safe, it is not acceptable.

- Model validation to a certain level must be made at certain times. This is related to the availability of relevant data at a given point in time. Based on data availability, one can try to validate the model with the data and the result should state the confidence level to which the model has been validated. This should be done for successive later times.

- At best, validation of predictive models will provide uncertainty bands, within which is the real outcome. In other words, one hopes that the real outcome is within the uncertainty bands calculated by the validated models.

There were also discussions on the validation studies with natural analogs. With natural analog systems such as uranium mines, or the Oklo natural reactor systems, it is recognized that it is hard to get quantified validity level of models by testing them with such data. However, these systems with lifetimes of millions to billions of years represent unique opportunities to study long-term behavior of flow and transport in the geosphere. Performance assessment of nuclear waste repositories is required to be made for time periods up to 10,000 to 100,000 years, but field experimental studies are limited to a few years. Thus natural analogs provide critical information on the long end of the time scale, though not definitely quantitative because of data uncertainties. For example, it is important that natural analog studies show that there is no unexpected phenomena in such a long term natural system, thus ensuring us that we know all the phenomena or processes that may occur in a geologic repository system.

There was a discussion at GEOVAL-90 on the question what could be the factors behind the difference or discrepancy between model predictions and experimental observations. The proposed factors are:

1. The conceptual model is wrong. In other words, there is an error in the features, events, and processes that are being incorporated in the conceptual model.

2. There might be a significantly large error in the process of simplifying the detailed conceptual model to a simpler model in order to do computer calculations.

3. The parameter values used in the modeling is not representative of the larger site which one is trying to predict.

4. The field data used for the validation does not really represent the site.

There is a need to study the residue, or the differences between model predictions and field observations. If there are systematic errors, they may point to missing processes in the models used.

TECHNICAL ISSUES ADDRESSED AT GEOVAL-90

GEOVAL-90 covers a large number of technical issues. From GEOVAL-90 until today, there has not been discovery and identification of significant, new processes, which is a very good sign. That means that we might have captured the major processes that might occur in a geosphere for assessing the performance of a waste repository.

There has been much discussion at GEOVAL-90 on various dispersive mechanisms, such as heterogeneity, channeling, and pathway intersections, and how they cause solute dispersion. There is much progress being made on the study of variable-aperture fractures and fracture networks. The fractures are no longer considered to be representable by parallel plates, but rather the apertures are varying from point to point in the fracture plane. This has a direct bearing on channelization and preferred pathways of flow through such a system. Among the approaches of studying heterogeneous systems, the stochastic models are increasingly being applied. Until recently, stochastic models have been mainly theoretical and to apply these methods attempting to provide answers to a large real system is a major step forward.

A number of retardation mechanisms have been studied. This includes matrix diffusion, anion exclusion, and dissolution and precipitation. On the other hand, one is also concerned with colloidal transport and solute complexation and the effects of density-driven flow and large ionic strengths on transport. This is particularly important for a repository at a salt site, where the salt will create large ionic strength in the fluid. The other technical issue involved is the unsaturated flow systems, with both gas and water transport. This is further complicated by the fact that such multi-phase flow is in complex matrix and fracture systems. Problems in this area include gas transport and liquid infiltration instability, as well as flow fingering, and so on.

The coupled processes, which may occur in the geosphere around a repository because of heat release from the waste, were also discussed. A number of technical issues are addressed, such as coupled hydromechanical processes, coupled thermohydromechanical processes, as well as coupled thermohydrochemical processes. All of these might affect flow properties of the geological system and hence, the transport.

There is also much work done in GEOVAL-90 and in subsequent years on geochemical modeling. It was pointed out that the need is to consider a specific site and disposal design, because it is impossible to study all the chemicals generically. Given a specific site, one can then be specific and study a limited number of chemical species. Currently there is the need to validate databases such as the thermodynamic databases. A lot of work has been done by NEA on this subject. Also, it was pointed out that there seems to be a lack of fundamental theoretical basis for the modeling of sorption and retardation. Another important problem discussed was that the repository site might not be in a steady state right now, rather the in-situ hydrology is slowly transient. How to measure or determine the transient characteristics is a question. Furthermore, how does the repository construction change the geochemical environment of the geosphere? All of these are questions that were raised.

On the whole, different scales have been studied in the different projects. Scales range from centimeter scale, to meters, to hundreds of meters, and timescales in the studies are from days to three or four years. Field experiments are studied based on the porous media system, fracture zones, as well as fracture network conceptual models. There are a number of large projects that use alternative models to study the same set of data in parallel. This was the case of the Finnsjon experiment in the INTRAVAL project, and the Site Characterization and Validation block study in the Stripa project. At the Hard Rock Laboratory at Äspö, the hydrological investigation was also made by using a number of alternative models. The approach of using alternative models in parallel to study the same set of data is a very important one. This will show the uncertainty associated with the construction of conceptual models, which is difficult to quantify.

A VALIDATION STRATEGY

One particular paper from the GEOVAL-90 that I would like to call attention to is on the validation strategy that was developed and practiced in the field of Operations Research and Systems Engineering. Robert G. Sargent presented a paper in which he lists validation and verification techniques from that field (Table 1). It is very useful for those interested in nuclear waste repository performance assessment to carefully consider them. Though direct and full applications of their techniques may not be appropriate, I believe we can learn much from them. For example, there are the so-called historic methods of validation, which have three aspects: rationalism, which means everyone knows the assumptions; empiricism, which means every assumption has been empirically tested; and positive economics, which means so long as a prediction is within reasonable uncertain limits, we really do not need to worry about whether the assumptions are right or wrong. Of course, these three approaches are quite different. One needs to be aware of these alternative approaches and, in one's predictions, one needs to state clearly what one is talking about. Sargent gave the minimum validation steps from his point of view:

1. Agree on the validation processes before the study begins.
2. Identify the assumptions underlying the models where possible.
3. Perform face validity on the conceptual model. That means one takes the conceptual model and thoroughly discusses all the elements and procedures (flowcharts) with knowledgeable people.
4. Explore the model behavior. That means without looking at data, take the model and see how it behaves for typical or extreme cases. One would evaluate the results to see whether they look reasonable and are internally inconsistent.
5. Perform a predictive study and compare with data, for at least two cases. That will help to build confidence in the model.
6. Document the model and results.
7. Plan on periodic review and possible revalidation as one gets more information.

For the field of nuclear waste geological repository, it would be very good to take the experiences from the operations research into account. Even though we might not want to follow every step, we should consider the various techniques that are identified and explicitly relate them to our model validation activities.

CONCLUDING COMMENTS

Some remarks are presented to conclude this paper (Table 2). The first is on different kinds of models, experiments, and predictions. It is very useful to recognize that there are three different kinds of models. One is a single-process model. This is a model where we are only concerned with a single process and where we try to understand whether the calculation can reproduce such a process. For example, when you have a density-driven flow, such as buoyancy, one can write down an equation and provide a numerical solution, and test against laboratory data to see if this is done correctly. The second kind of model is a research model for a given experiment. One would model the behavior of that experiment. Finally, there is a model for long-term predictions. Usually, this means that the model needs a certain completeness in its details in order to make predictions tens of thousands of years into the future. Here we need also to be able to calculate a measure of the associated uncertainty.

Table 1. A Validation Strategy (Sargent, GEOVAL-90, NEA/SKI, 1991)

• Background	Operation Research & Systems Engineering (Authors: Balci, Banks, Davis, Gass, Oren, Sargent, Schlesinger, Shannon, Whitner; 1974–1990) Simulation Conferences; Journal of Simulation
• Verification techniques	Comparison to other models Degenerate test Extreme condition test Fixed value Internal validity (stochastic) Traces
• Validation techniques	Event validity Face validity Historical data validity Multi-stage validation Predictive validation Turing test
• Historic methods	Rationalism: everyone knows assumptions Empiricism: every assumption empirically tested Positive Economics: So long as prediction is OK

Table 2. General Comments

• 3 kinds of models	– Single-process Model – Research Model (of an experiment) – Long-term Predictive Model
• 3 kinds of experiments	– Experiment to measure input parameters – Experiment to study specific processes – Experiment to study system behavior
• 2 kinds of predictions	– Prediction of an experiment – Long-term prediction of system
• Scaling – Do we need to worry about it?	Repository is of large scale – comparable to scale of long-term predictions
• Need to draw all together	– People working in different disciplines – Alternative conceptual models – S.C. and P.A.: coupled and iterative

It is also important to recognize there are three different kinds of experiments. One kind of experiment is for measuring input parameters required by models. The second kind of experiment is that we study specific isolated processes. The third kind of experiment is where we try to understand and check the long term behavior of the system as a whole. An example of this third kind of experiment is that on natural analog systems.

We need to also realize that there are two kinds of predictions. One can predict an experiment. For example, for a five year experiment, one tries to predict the behavior for the five years. However, it is very different from making a prediction thousands of years into the future, where there is no chance of checking on the results. Thus in the latter case you have to be much more careful in being able to make sure that all elements and processes are included and that one has reasonable confidence of the prediction.

Building a nuclear waste geologic repository is typically of a large scale. The repository itself is of kilometer scale and the transport of potential leakage from the repository that one has to predict is also of the kilometer scale. So the information obtained during the construction of the repository has a scale that is comparable to the scale of the predictions. In this sense, the problem of scaling up that is so important in hydrogeology might not be a major problem of concern, because available data is on a scale which is, say, one-half to one-fifth of the system to be predicted. This is not the same as scaling up by two or three orders of magnitude. Thus, the scaling problem might be a less important issue.

Finally, to ensure the proper performance assessment of a nuclear waste repository, it is important to draw people and things together. Not only do people working in site characterization and in performance assessment need to work together, but also alternative conceptual models need to be joined together in the assessment. Indeed, for proper construction of conceptual models, performance assessment and site characterization are not decoupled. They are iterative processes. It is very important to recognize this in any effort of model validation.

REFERENCE

NEA/SKI, Proceedings of GEOVAL-90 Symposium on Validation of Geosphere Flow and Transport Models, Stockholm, Sweden, May 14–17, 1990. Organized by Nuclear Energy Agency, Paris, and Swedish Nuclear Power Inspectorate, Stockholm. Published by the Organization for Economic Cooperation and Development (OECD), Paris, 1991.

ACKNOWLEDGMENT

Discussions with John Andersson and Claudio Pescatore are much appreciated. The paper was prepared with joint support of Swedish Nuclear Power Inspectorate, SKI and the Office of Energy Research, Office of Basic Energy Science, Geoscience and Engineering Division of the U.S. Department of Energy under Contract No. DE-AC03-76SF00098.

Validation: Demonstration of Disposal Safety Requires a Practicable Approach

Piet Zuidema
National Cooperative for the Disposal of Radioactive Waste (Nagra)
(Switzerland)

Abstract

In the first part of this paper the term "validation" is briefly discussed and a very pragmatic definiton is provided, explicitly for application in the field of performance assessment. In the second part of the paper the questions 'what to validate?', 'how to validate?' and 'when to validate?' are briefly discussed and some preliminary answers are presented. Finally, the third part of the paper contains a summary and some provocative conclusions which hopefully will provide some material for discussion during the meeting.

1 Introduction and Structure of Paper

Performance assessment for the final disposal of nuclear waste is based on models which evaluate the evolution of the repository system and estimate impacts that may occur many thousands of years in the future. The modelling results provide key information for the siting, designing and licensing of the disposal facilities. Therefore, models are an important source of information for making decisions regarding the management of radioactive waste and sufficiently reliable models are of key importance for establishing confidence in disposal systems.

In the waste management community, the process of assessing the acceptability of models and building confidence in them is often called "validation". The topic of "validation" is currently a hotly debated issue in performance assessment. According to our understanding, this is due to several factors: Firstly, "validation" is obviously of key importance; however, the process is not easy and represents at least partially a novel task. Fundamental problems result from the fact that the long time-scales involved in waste management are explicitly recognised - in contrast to other areas with similar time-scales. Because the importance of "validation" is recognised, major specific efforts have been devoted to it, e.g. INTRACOIN, HYDROCOIN, INTRAVAL as well as specific conferences such as GEOVAL-87 and -90, FOCUS '93 and the current meeting. Secondly, many misunderstandings exist because the term "validation" is defined differently (especially with respect to the level of rigour required) by different people and in different disciplines. Thirdly, the debate on "validation" is sometimes encouraged by scientists in order to increase funding levels for basic research in areas which are (un)critical for waste disposal projects.

Much has been said about "validation" by many well-respected persons representing a whole variety of disciplines. It is therefore not considered to be worthwhile to add further definitions of "validation"; we will therefore simply select a pragmatic definition. However, much less has been said about "What to validate, how to validate and when to validate in performance assessment". In this paper, all three questions are briefly addressed. To do this, however, it will first be necessary to explain our understanding of the term "validation" and to discuss the nature of performance assessment and the resulting requirements on "validation" of the models used.

2 Different Views on the Meaning of "Validation"

Much of the controversy in the area of "validation" arises from alternative interpretations and perceptions of the term itself. These range from an inherently unachievable "proof of truth" - according to a predominant view of the philosophy of science (e.g. the Popperian view) models can only be falsified - to more pragmatic approaches in waste management, with emphasis on the subjective assessment of whether models and data are "good enough" for the application at hand. Some different views are summarised below:

- From a philosophical point of view, a "validated" model can never be achieved, since " ... to say that a model is verified [or validated] is to say that its truth has been demonstrated ... . However, it is impossible to demonstrate the truth of any proposition, except in a closed system. This conclusion derives directly from the laws of symbolic logic." (Oreskes et al., 1993).

- Therefore, the "... goal of scientific theories is not truth (because that is unobtainable) but empirical adequacy." (Oreskes et al., 1993). A theory or model can thus at best be validated relative to a given body of empirical knowledge (Neuman, 1992).

- Even more extreme, Niederer (1990) states that a model is considered to be validated when a dominant school of relevant experts agrees that it is.

- In the technical jargon of simulation modelling, the term "validation" is more pragmatically defined as "substantiation that a computerised model within its domain of applicability possesses a satisfactory range of accuracy consistent with the intended application of the model" (Schlesinger, 1979).

- In this paper, a model is pragmatically considered to be "validated" when there is confidence that the model is "good enough" for a given purpose. This is also in accordance with the Swiss regulatory guidelines which define "validation" as: "Providing confidence that a computer code used in safety analysis is applicable for the specific repository system." (HSK & KSA, 1993). It should be noted that confidence in a computer code also implies confidence in the underlying model assumptions and data.

According to our definition of "validation", the decision about how much effort must go into the validation process before the model can be considered to be "validated" is necessarily subjective and very dependent on the complexity of the system and on the use to be made of the model and its results. Therefore, the key problem is to define what level of accuracy and what degree of confidence must be achieved in the prediction of specific phenomena. In order to do this, it is first necessary to briefly describe our understanding of the aims, the nature and the limitations of performance assessment. This is done in the next section.

3 The Nature of Performance Assessment and Validation in Geologic Disposal

The main aim of performance assessment in geologic disposal is to determine whether a repository is sufficiently safe. This is done by quantitative modelling. In this context, however, it is of key importance to recognise that, for the demonstration of sufficient safety, a prediction by a model needs not necessarily to be an exact forecast of the value that a given quantity will take; a bounding estimate of the behaviour of a system may well be adequate. This also implies that performance assessment does not aim at one unique answer to how the repository will behave - on the contrary, a whole spectrum of cases (including both alternative scenarios and alternative conceptualisations, see e.g. Nagra, 1994) needs to be analysed.

Furthermore, one has to accept that performance assessment as such will never provide a strict proof of sufficient safety. However, performance assessment can, by its nature, provide a convincing and transparent illustration of the effects of the safety barriers designed or chosen for the repository system. A main attribute of system performance assessment is that it illustrates the integrated behaviour of all safety systems. Therefore, it should be ensured that undue importance is not attached to the numerical outcome of the assessment and that enough importance is attributed to the integrated display of the understanding provided by the assessment (see also Bernero, 1989).

To summarise, performance assessment consists of a careful, well-considered investigation of system behaviour with the aim of analysing if a concept or a facility is sufficiently safe - this should be carried out without any pretence of providing an exact prediction and without making the mistake of trying to provide a rigorous proof of sufficient safety.

To achieve sufficient "validation", from a scientific point of view it is desirable to ensure the empirical adequacy of the models. However, in the case of deep geological disposal, the collection of empirical knowledge can be difficult or even impossible because some of the relevant phenomena cannot be observed for a long enough time to provide direct information on the relevant scales. Therefore, the acceptability of the models has often to be established indirectly on the basis of limited evidence and human subjectivity cannot be eliminated from the process.

However, as stated in the previous section, it is not always necessary to aim at the "truth" (or, more feasible, empirical adequacy); the aim of "validation" in performance assessment is more in the nature of establishing whether there is confidence that models and data are "good enough" to ensure that the results do not underestimate the consequences. For this purpose, supporting arguments require only to demonstrate that models and data give results that err (if at all) on the side of conservatism, i.e. will underestimate system performance and thereby overestimate consequences such as dose to humans.

To conclude, "validation" of predictive models used in performance assessment should provide confidence in the uncertainty band of the results below which we believe the real outcome will fall. The required level of confidence will depend upon the purpose of the performance assessment[1] in hand and will increase as the repository project moves from the initial phase of concept evaluation towards licensing.

4 What to Validate or Where Do We Need to Build Confidence

Obviously, the most important models to "validate" are those that are used to demonstrate sufficient performance in the final safety case. One approach to developing the safety case is to rely on models of those phenomena that are most efficient for providing isolation and/or retention of radionuclides. Another way to develop the safety case is to rely on models of those phenomena for which it is easy to establish confidence in their performance and which will provide sufficient safety of the system. It seems that the modelling of safety-sensitive phenomena is more relevant in the early phases of repository development, whereas provability becomes the dominant issue during the licensing phase.

The latter approach, with emphasis on provability, leads to the concept of robustness (McCombie et al., 1990). In such an approach, the demonstration of safety will primarily rely on those phenomena for which sufficient supporting arguments can be made that give confidence that the quantification is either correct or errs on the conservative side. Thus, for phenomena that contribute to safety, only those are included for which confidence exists that they will operate; this means that even potentially very strong safety factors may have to be neglected in robust analyses. For phenomena detrimental to safety, all those which cannot be confidently ruled out must be included.

This concept implies that the different phenomena acting within and on a repository system can be classified with regard to their treatment in performance assessment. In the Swiss programme, this is done by the way of scenario development procedures (see Sumerling et al., 1993). The resulting classification also gives direct input to the question "what to validate". This process of setting priorities among the different phenomena is equivalent to developing a safety concept. Criteria to be considered for selection of phenomena to be included in robust or other (e.g. "best-guess") analyses, are e.g.:

- level of understanding
- availability of information for a given site/design (or: possibilities and/or plans to obtain information at a later stage)
- availability of independent evidence that supports the effect on the safety case of the phenomenon under consideration

The above-mentioned consideration that the phenomena to be included in the safety case depend upon the data at hand also implies that, for similar geological situations, different safety concepts can be developed depending upon the available information.

[1] The discussion here focuses on the performance assessment models used to determine if a waste disposal system meets regulatory limits. Such models are not, for example, necessarily applicable for optimising the disposal system design, where realistic results are needed.

5 How to Validate

From all that has been said up to now, it is obvious that the level of confidence in performance assessment results will depend upon the care that has been taken in the following steps:

- The development of sufficient understanding to
  - identify all relevant phenomena and not to overlook any critical aspects
  - decide by which basic physical and chemical principles these phenomena can be represented
  - quantify (in a bounding manner) the effects these phenomena have on the performance of the repository by choosing adequate parameter values
  - consider the interactions between these phenomena and quantify the effects resulting from this interaction
  - decide on which of the phenomena are of key importance and on how to integrate them into the safety case according to the desired level of robustness (the so-called "screening process")

  This needs to be done for all the different cases to be considered (both the base-case scenario as well as alternative scenarios) and it may well be possible that, for each of the different cases, a different set of phenomena will be relevant.

- The treatment of the phenomena included in the safety case. Different approaches can be used for the treatment of phenomena which have different requirements concerning their adequate level of confidence (or "validation"), e.g.:
  - one can represent both concept and data in a realistic manner, requiring "validation" on the level of detailed model testing to an extent dependent on how important the phenomenon is for the safety case
  - an alternative is to use a realistic concept but conservative parameter values, requiring model-testing for the concept and supporting arguments that the chosen parameter values lead to conservative results. An example of this type of representation is the conservative choice of depth of rock matrix for the realistic model concept of matrix diffusion, where an incomplete database necessitates this careful approach, see Nagra (1994).
  - sometimes representation in a simplified manner is necessary due to incomplete understanding or incomplete information. This type of representation requires supporting arguments that the representation is either correct (normally not expected) or errs on the conservative side. An example of this type of representation is the selection of solubility limits for many nuclides in Nagra (1994), where solubility limits are estimated using pure solid phases although solid solutions are expected (however, for quantitative modelling of the latter insufficient information is currently available).
  - the conservative approach of ignoring phenomena beneficial to safety requires only supporting arguments that these phenomena will, under no circumstances, be detrimental to safety. An example of this type of phenomenon is the sorption of radionuclides onto iron corrosion products from the HLW canister; see Nagra (1994). The decision not to include this process is due to the current lack of defensible data.

  It is important to recognise that the above-mentioned cases of conservative treatment of phenomena can, at a later stage when more information is available, be replaced by more realistic representations. Those phenomena that are ignored are therefore marked as "reserve phenomena" (see Nagra, 1994), and may be included at a later stage.

- The careful coding and execution of calculations; this is also of key importance but is not discussed further here in the context of "validation".

To summarise, it is the process of modelling, and not the model as such, that is critical to establish the required level of confidence in the modelling results for a specific performance assessment!

6 When to Validate Particular Phenomena

Repository projects are extremely long duration exercises which include activities that differ widely in their nature, e.g. the initial R & D phase with evaluation of the disposal concept, the different repository development phases (site selection, site characterisation, design), the licensing phases and the operational phase before closure of the repository. It is important to recognise that the possibilities for, and requirements on, "validation" will depend upon the phase in which a repository project is for several reasons. For example, some of the models for predicting the long-term performance of a repository are highly site- or system-specific, whereas others are more generic. Generic model "validation" may already be feasible to the required level in the earlier phases of repository development, where fewer binding decisions are needed. For the later site-specific models, the need for local data can delay the possibilities for quantification until well into the repository development process. For these phenomena we therefore have to limit our "validation" efforts for a considerable time to establishing confidence only that predictions are within rather broad bands.

This implies that the phenomena that are considered to make the safety case and which dominate our "validation" activities may well change with progress of the repository programme towards implementation. This is due to the increased availability of site- and system-specific supporting evidence and is especially true for the geosphere transport barrier.

For example, in the HLW repository programme in the crystalline bedrock of Northern Switzerland (Nagra, 1994), the base-case calculations in the current assessment do not give much credit to the geosphere transport barrier; however, with progress of site characterisation work, it is hoped that a modelling approach for the geosphere can be adopted that will demonstrate a more efficient transport barrier. This changing role of the geosphere transport barrier is due to the inherent difficulties involved in characterising the geosphere from the surface with a limited number of boreholes, which are restricted in order not to unnecessarily impair the barrier function of the geosphere. However, once underground investigations begin, it is expected that a more reliable characterisation will be possible. The present approach places more reliance on the near-field (defined here as the system of engineered barriers plus several meters of surrounding geosphere providing adequate boundary conditions for the engineered barriers). In the current safety assessment, this can be done confidently due to massive amounts of relatively well-understood materials used in the current repository design.

7 Summary and Conclusions

a) Don't get lost in semantics!

Several persons consider the use of the term "validation" to be inappropriate for the process of "ensuring sufficient confidence" and suggest using another term. However, the use of other terms may, in the long run, be no less problematic because, semantically, terms such as "validate", "verify", "confirm", "authenticate", "corroborate", and "substantiate" are near-synonyms in ordinary language (eg. Webster, 1987). Furthermore, the chances of widespread acceptance of alternative terminology are expected to be small. Many of the key concepts involved may be summarised with "confidence building".

However, it needs to be recognised that the term "validation" has special disciplinary meanings and modellers themselves should take the lead in clarifying the meaning of "validation". Even more important, they should point out the restrictions and limitations of "validated" models and should remember some important lessons:

- make it clear that "validation" is used in a technical sense;
- don't use the term if it is likely to be misunderstood and create a false sense of truth rather than consensus;
- take care to specify the context of the "validated" model.

b) Be clear and specific about the nature of achievable "confidence"

Absolute truth is not known, so we cannot of course provide a model which provides this. In practice, models will be used to support regulatory and legal decisions, and this will not change no matter how loudly and often it is proclaimed that it cannot be demonstrated that models represent the truth. Thus, our task should be to stop debating the impossibility of model validation in such an absolute sense, and to develop procedures whereby all involved parties can be reasonably assured that models are appropriate and are being used correctly to meet the needs of the problem at hand. Certainty cannot be achieved, we must and should be satisfied with engineering confidence - often it will be sufficient to provide confidence in our ability to bound the outcome of a specific phenomenon!

The level of confidence required in the modelling process is directly related to the importance to public safety of the decisions being made. The decisions become more important as implementation proceeds. However, all uncertainty in model predictions can never be eliminated and absolute knowledge can never be attained; decisions will thus always contain some element of uncertainty.

To summarise, it is important that

- scientists accept that bounding of effects may be sufficient for many phenomena and not all details need to be quantified accurately (no unlimited funding for research on all details!),
- the implementer be ready to investigate his system in sufficient detail to develop an adequate level of understanding,
- the regulator recognises that absolute proof of safety can never be achieved and also conveys this message further to the political decision makers.

Finally, a consensus will be needed between implementer and regulator, the scientific community and ultimately the public, on a reasonable level of needed confidence.

c) Set priorities and don't get lost in details!

In order to advance repository projects, it is important to concentrate efforts on relevant phenomena and to decide on a safety case that is defendable. This requires setting justified priorities and not investigating all details (although these might be very interesting from the scientific point of view). Scientists and engineers thus need to be careful not to miss the opportunity to assist society in making important decisions regarding the management of radioactive waste. As Lewis (1990) puts it "It is painful for a scientist to say this, but we do sometimes get so involved in arguing technical nuances that only a less well-informed person can make any decisions at all".

References

BERNERO, R., 1989: Are You Sure? Performance Assessment Beyond Proof. Proc. of the Symposium "Safety Assessment of Radioactive Waste Repositories", 9-13 October 1989, Paris, France.

HSK & KSA, 1993: HSK-R-21/e, Protection Objectives for the Disposal of Radioactive Waste. Swiss Federal Nuclear Safety Inspectorate (HSK), Villigen, Switzerland.

LEWIS, H.W., 1990: Technological Risk. W.W. Norton & Company, New York, London, 1990.

McCOMBIE, C. et al., 1990: Sufficient Validation: The Value of Robustness in Performance Assessment and System Design. Proc. of the NEA/SKI Symposium "Validation of Geosphere Flow and Transport Model (GEOVAL)", May 1990, Stockholm, Sweden.

NAGRA, 1994: Kristallin-I: Safety Assesment Report. Nagra Techncial Repost NTB 93-22, Nagra, Switzerland, 1994.

NEUMAN, S.P., 1992: Validation of Safety Assessment Models as a Process of Scientific and Public Confidence Building. Proc. 3rd Int. Conf. on High-Level Radioactive Waste Management, Las Vegas, ANS, ASCE, 1404 (1992).

NIEDERER, U., 1990: In Search of Truth: The Regulatory Necessity of Validation. Proc. of the NEA/SKI Symposium "Validation of Geosphere Flow and Transport Model (GEOVAL)", May 1990, Stockholm, Sweden.

ORESKES, N. et al., 1994: Verification, Validation and Confirmation of Numerical Models in Earth Sciences. SCIENCE, Vol. 263, 4 February 1994.

SCHLESINGER, S., 1979: Terminology for Model Credibility. Simulation. V. 32, No. 3, pp 103-104.

SUMERLING, T. et al., 1993: Scenario Development for Safety Demonstration for Deep Geological Disposal in Switzerland. Proc. 4th Int. Conf. on High-Level Radioactive Waste Managment, Las Vegas, USA, 1993.

WEBSTER 1987: Websters's Ninth New Collegiate Dictionary, Springfield MA, USA.

SESSION II

Chairmen:

A. Larsson
L. Dewière

Developing Groundwater Flow and Transport Models for Radioactive Waste Disposal

- Six years of experience from the INTRAVAL Project -

Neil Chapman, Johan Andersson, Peter Bogorinski, Jesús Carrera, Jorg Hadermann, David Hodgkinson, Peter Jackson, Ivars Neretnieks, Shlomo Neuman, Kristina Skagius, Tom Nicholson, Chin-Fu Tsang and Charles Voss

Abstract

The validity of the information and the models used to make predictions is central to the credibility of a performance assessment for a radioactive waste repository. Owing to the large time and space scales involved in natural systems, both the users of models, and the regulatory agencies who must respond to model predictions, have felt the need for some means of demonstrating their 'fitness for purpose' and inspiring confidence that their quantitative output is meaningful. The INTRAVAL project was set up to bring together those working on this issue in many countries; to share experience, to carry out comparison exercises and to help build an internationally accepted approach to developing and applying both the models and the approach to prediction.

This paper first outlines the methodology adopted to evaluating the 18 different test cases of Phases 1 and 2 of INTRAVAL, with summaries of the specific objectives of each. The test cases addressed hard fractured rock, plastic clay, mixed sedimentary and unsaturated geological environments at many scales, with observations and interpretations on a very wide range of space and time scales. The overall, rather than the detailed, findings of INTRAVAL are summarised. Lessons have been learned about analysing experimental data and about building confidence and quantifying uncertainties in models used for making predictions in performance assessment. It is hoped that these lessons can find applications in many waste disposal programmes.

INTRAVAL explored a wealth of geological data from sites in four continents. The great strength of the project was that modelling was tested by multiple groups against real data, at a relatively dense observational scale, with access to all the information available. In addition, the project integrated exercises in both field and laboratory, at a variety of spatial scales. Particularly instructive information came from analyses of field tests which illustrated the real problems of using data from natural systems. Although there were ambitious objectives for some test cases, throughout the project lessons were constantly being learned about what was possible and what was simply unachievable. These lessons should have considerable impact on the incorporation of repository site data into safety assessment.

1. THE INTRAVAL PROJECT

INTRAVAL was set-up to provide a forum where experimentalists and modellers could bring together their respective expertise and disciplines to test model concepts in a quantitative way and, in the process, help to develop an approach to validation. As such, it provided an opportunity for peer review of experimental interpretation and modelling activities. The Project ran for six years, from 1987 to 1993, in two Phases. At its peak it involved 46 organisations in 14 countries. INTRAVAL was the intellectual successor to two earlier international projects, INTRACOIN (1981-84) and HYDROCOIN (1984-7), both of which dealt with the verification of codes and the validation of models for groundwater flow and radionuclide transport (Larsson, 1992). Each project was devised and managed by the Swedish Nuclear Power Inspectorate (SKI). HYDROCOIN and INTRAVAL were carried out under the auspices of the OECD Nuclear Energy Agency.

1.1 Methods and Approach

The structure of INTRAVAL was built upon a series of test cases designed to address specific issues of concern in the development and testing of conceptual models of groundwater flow and chemical transport processes. These test cases ranged from relatively well-controlled, centimetre-scale laboratory experiments to the evaluation of sparse field data at the scale of kilometres, and treated the transport properties of clays, sediments, hard fractured rocks and tuffs (the latter in unsaturated conditions). Phase 1 of INTRAVAL looked at 13 test cases. Six of these were developed and redeployed in Phase 2, which also evaluated 5 entirely new test cases. In retrospect, this was probably too large a number to allow well-focussed discussion. On the other hand, it did allow the treatment of a wide variety of processes in many geological environments.

Apart from field experiments, the first Phase of the Project contained a number of purely laboratory-based test cases, whereas Phase 2 tended to use laboratory experiments only to provide supporting information for field-based test cases. Potentially suitable test cases were assembled and advanced by pilot groups and presented in an identical format to allow their comparison and assessment. They were only incorporated into the Project after they had been thoroughly evaluated, which involved checking for relevance to testing well-defined or alternative conceptual models of processes, the availability and quality of data, and the existence of a support group able to provide additional information, or even run further tests. Ideally, each test case would have used an experiment that had been designed with validation in mind. In reality, most experiments were done to test model concepts, and only one test case was actually tailored specifically to address validation objectives. The remainder of the cases had to make do with what information was available.

In analyzing the different Phase 1 test cases, many of the project teams reported that systematic evaluation of the experimental setup and data was required to detect both unanticipated biases and artefacts introduced by errors in the experimental design. Phase 2 was designed to focus more closely on the development of validation procedures and devote less time to optimizing the design of particular experiments to ensure the generation of data suitable for validating models. For Phase 2, the test cases were divided among four working groups, each of which was expected to develop practical validation strategies appropriate for their set of experiments.

The whole set of test cases selected for both Phases fell naturally into five categories, which effectively implied a set of subsidiary objectives for the project, namely development and testing of the ability to evaluate:

- flow and transport in hard, fractured rocks
- natural radionuclide migration systems ('natural analogues')
- variable density groundwater flow in brine-rich formations
- water, air and solute movement in unsaturated rocks and soils
- advection and diffusion in sediments and sedimentary rocks.

1.2 Validation Strategy

The approach to evaluating these issues was to encourage the different modelling groups to analyze the results of related test cases using different conceptual models. Adoption of these models meant answering the following general questions:

- does the conceptual model address all relevant processes?
- is the geometrical structure of the system well described by the model?
- are the underlying assumptions made in the conceptual model valid for the specific physico-chemical environment being evaluated?
- is the model able to simulate in an adequate manner appropriate experiments and field conditions?
- are the spatial and temporal scales of the experiments or field-tests able to be incorporated into the conceptual model?

The next step involved the more specific tasks of:

- defining clearly the structure of the model in terms of the processes it incorporated, the boundary conditions applied and the initial conditions
- identifying which parameters would be required and a procedure for producing estimated values where data were missing
- constructing sampling strategies to obtain the requisite data (sometimes this also involved devising instrumentation and sampling methods)
- comparing the extracted parameter values with independent information and checking whether a consistent picture emerged.

In addition, for each test case where alternative conceptual models were compared, it was necessary to develop techniques and a strategy for making the comparisons.

The criteria used to judge model validity varied greatly from test case to test case. For Test Case 1b (see section 1.3 for a list of the test cases) on uranium migration through a small crystalline rock core, some modellers suggested that model validity may be assessed simply by evaluating the reasonableness of the parameters in the calibrated model. While close agreement of model and experimental results and the use of physically plausible parameter values do not constitute proof of model validity, the appearance of overall consistency between model and data enhances confidence in the model. This philosophy was adopted in many of the subsequent test cases.

One way to validate a model quantitatively is to split the experimental data into two sets, calibrate the model with one set and compare model predictions[1] to the second data set: if the two sets of data represent quite different conditions then the validation is stronger. The final assessment of model validity using this procedure still depends very much on the quantitative measures used to compare model predictions to experimental results and the criteria used to determine the acceptability of the fit. Some of the project teams in INTRAVAL developed statistical hypothesis-testing procedures to apply quantitative criteria for accepting the predictions of a model.

A validation strategy was suggested for Test Case 1a on radionuclide migration through clay cores wherein the statistical structure of the differences between predicted and observed breakthrough curves was examined. If examination of the differences reveals little or no correlation, the model is presumed adequately to represent the experimental results and at least the model structure is deemed acceptable. If, on the other hand, the differences are strongly correlated, the model structure involves a systematic error. A series of statistical procedures to test the null hypothesis that model error is negligible was applied to Test Case 10, the Las Cruces Trench experiment. The probability of accepting a false model cannot be evaluated by this technique. The desire to minimize the probability of accepting a false model may lead to the adoption of overly strict test criteria which increase the likelihood of rejecting good models. Based on experience modelling Test Case 10, it was suggested that integrated performance measures, such as the first and second moments or total mass flux of a contaminant plume crossing a compliance plane,

---

[1]The word 'prediction' is used frequently in this paper. Although performance assessors attempt to predict the possible behaviour of a repository system, it must be understood that these predictions are intended only as illustrations or approximations of possible behaviour. Some of these approximations may be very conservative. Associated with these predictions are uncertainties which are quantified as part of the performance assessment process. We can thus have little belief in the precise value of any 'prediction'; rather, confidence must be derived from the broad trends that emerge when calculations are made taking uncertainties into account. In this model testing process, the performance assessors must have confidence that they are able to make 'robust' predictions which, whilst scientifically sensible, tend to overpredict consequences and give information on safety margins. Alternatives have been suggested to avoid using the word 'prediction'; for example, 'forecast', 'projection' or 'projection by simulation' might be considered appropriate in some circumstances.

could sometimes be used as acceptance criteria for model validation. In many cases these integrated performance measures are analogous to regulatory standards. Two groups, working on Test Case 1a on radionuclide migration in clay cores and Test Case 1b on uranium migration in crystalline rock cores, applied quantitative model identification methods to distinguish alternative conceptual transport models.

Calibration is often the only viable technique available to determine the physical parameters to be used for long term model predictions, but the resulting parameter values may depend upon the calibration criteria chosen. If an automatic inverse method based on a statistical technique is used, it can be used to rank models and to evaluate confidence intervals of the estimated parameters. The drawback is that this information is only valid under the tested hypothesis. The application of statistical inverse techniques gives no guarantee that the resulting model is a good description of reality, let alone that it is the best description, for the given data and the particular conceptual model analysed.

It was recognised that agreement between experiment and model may, in itself, be deceptive because there may be different mechanisms which may give similar results on the time and space scales tested. If the wrong mechanism is chosen, an extrapolation to longer times and distances might be wrong. The identification of the correct mechanisms is thus crucial to validation.

Apart from the issue of developing validation methodology, other critical scientific issues which were brought forward by the test cases included:

- investigating the possibility of validating stochastic continuum and stochastic fracture network models
- investigating the extent to which the short space and time scales of all types of experiment, whether field or laboratory based, could be extrapolated to the scales required for performance assessment
- the comparison of alternative conceptual models of processes (e.g. continuum, fracture network and channelled models for transport through fractured rocks)
- further testing of the rock matrix diffusion model (the model of diffusion into the rock adjacent to fractures and channels within which groundwater flow occurs).

1.3 The Test Cases

The table below outlines the nature of each of the 18 test cases, the geological formations used, the space and time scales involved and the number of independent organisations which interpreted each case. For full descriptions, see the INTRAVAL Phase 1 and Phase 2 Summary Reports, the individual Phase 1 Test Case reports, and the Phase 2 Working Group Reports (SKI/NEA, 1992-4).

Test Case	Rock Type	Description	Location	Scale (m)	Duration (days)	No. of Groups
1a	Plastic clay (London Clay)	Simulation of diffusion of non-reactive tracers and comparison of alternative models of porosity	Laboratory	0.03	10-40	7
1b	Unfractured granite	Simulation of advection of reactive U tracer and comparison of alternative advection, diffusion and sorption models	Laboratory	0.01-0.02	7-15	7
2	Single fractures in granite	Simulation of advection of reactive and non-reactive tracers and comparison of alternative sorption and diffusive retardation models	Laboratory	0.08-0.3	0.01-0.2	3
4^ Stripa	Fractured granite	Interpretation of advection and dispersion of non-reactive tracers in terms of alternative conceptual models of flow-paths and channels	Underground research site (360m depth)	10-60 (~3 x 10^5 m^3)	750	4
5^ Finnsjön	Major sub-horizontal fracture zone in granite	Interpretation of advection of non-reactive tracers in a complex fracture zone to compare porous medium and fracture transport models	In-situ tracer tests at 200-350m depth	150-190	1-100	7
6	Hypothetical single planar fracture in granite	Numerical experiments to test both predictive capacity of alternative flow and transport models given limited input data and to identify the key processes involved	Synthetic data based on an underground test site	4-10	350 (simulated)	3
7 Poços de Caldas	Extensively fractured, hard volcanic rocks	Development of coupled flow and geochemical models for redox front generation and movement	Borehole observations in an open-pit mine	0.01-100	Simulations to ~10^6 years	1 (5)*
8^ Alligator Rivers	Weathered schists	Testing ability of simple transport codes to reconstruct U transport and retardation and to utilize and account for complex radiochemistry of an ore deposit	Borehole observations to depths of ~50m from surface	100-200	Simulations to ~2 x 10^6 years	3‡
9	Single planar fracture in a granite block	Simulation of reactive and non-reactive tracer advection to explore alternative models of sorption, channelling & diffusion/dispersion	Laboratory	0.9	3-15	4
10^ Las Cruces Trench	Layered calcareous soils (sands to sandy clay loams) in an arid setting.	To test stochastic and deterministic models for water flow and transport in spatially variable unsaturated soils	Field irrigation & tracer tests (instrumented trench & boreholes)	10-30	300	7
11^ Apache Leap	Unsaturated fractured welded tuff	Simulation of the movement of water, solutes and air in an unsaturated rock, including the effects of a temperature gradient	Shallow boreholes at field site & laboratory experiments	30 (field) (~3 x $10^4 m^3$) 0.02-1 (lab)	0.1 -5 0.01-30	2

13	Glass beads simulating a porous medium	Simulation of brine movement through saturated porous medium and evaluation of alternative conceptual flow models	Laboratory	1	0.1-0.2	7
14^ Gorleben	Mixed sediment sequence overlying a salt dome	Simulation of density-driven flow and brine transport at a single borehole and at a regional scale	Regional (300 km^2) site investigation boreholes to 250m depth	~10 km laterally	20 (Simulations to ~10^4 years)	6
Mol	Plastic clay (Boom Clay)	Comparison of alternative models of non-reactive tracer diffusion and diffusion/advection	Underground test site at 220m depth	1	1800	5
WIPP-1	Bedded evaporites at 655m depth	Testing applicability of Darcy's law to model brine flow through evaporites	Underground test site at 650m depth	0.1 - 100	1000	3
WIPP-2	Thin (~8m), continuous fractured dolomite unit at ~220m depth in sedimentary sequence	Study of 2-D stochastic modelling approaches for simulating flow in a heterogeneous porous medium and evaluation of alternative conceptual models (2-D vertical and 3-D models to examine vertical flow).	Extensive areal array of 60 deep boreholes at potential repository site (WIPP)	~30 x 35km (~1000km^2)	~10 (hydraulic tests). Simulations to ~10^5 years.	5
Yucca Mountain	Thick sequence of unsaturated volcanic tuffs	Testing the ability of different models to predict the water content of various lithological horizons in unsaturated tuffs	A series of 4 boreholes to 500m depth	500	n.a.	6
Twin Lake	Aquifer in ~20m thick unconsolidated surficial sediments	Compare models simulating 2-D transport of non-reactive tracers in heterogeneous porous medium and identify model performance measures	Field tracer test site with shallow boreholes	270	271	5

*Four groups outside INTRAVAL also analyzed this test case as part of the Poços de Caldas Project
^These test cases were addressed in both Phase 1 and Phase 2 of the INTRAVAL Project
‡As with Poços de Caldas, a number of groups outside INTRAVAL also analysed this test case.

2. KEY LESSONS LEARNED DURING INTRAVAL

Through its six-year life, INTRAVAL explored a wealth of geological data from sites in four continents. The most instructive exercises were those that involved field rather than laboratory tests, as these approached more closely the real problems of using data from natural systems. Although there were ambitious objectives for some test cases, throughout the project lessons were constantly being learned about what was possible and what was simply unachievable. These lessons will have considerable impact on the incorporation of repository site data into safety assessment.

The overall lessons of INTRAVAL can be summarised as follows:

- By far the greatest issue encountered during the project was how to characterise geological complexity in the field and analyse, theoretically and computationally, its impact on the prediction of flow and transport for the space and time scales of relevance in performance assessment. Other issues paled by comparison. It became even more evident to the modelling community that the geological environment is complex and heterogeneous at all scales of observation. Although it contains both discrete features, and regions which can be treated as a continuum, the scales of both are highly variable. Very few discrete features can be characterised deterministically in sufficient detail to allow satisfactory modelling. Consequently, stochastic approaches are used, in which the models are set up to match the distributions of the properties of the features, rather than to match the properties of all the individual features. It is also crucial to ensure that models incorporate real parameters which are intrinsic and measurable properties of the rock.

- At present, a number of models rely heavily on derived or geometric parameters (such as flowpath length and wetted surface area), understanding of which is currently at an abstract level. Such parameters may never be fully measurable in the field. In some circumstances this may not be a problem. For example, if it is possible to define a parameter and infer its value from separate measurements at one scale, then use it to make predictions at another scale, then this may be useful. If, however, one could only infer the parameter value from the behaviour that one is trying to predict, then the appropriateness of the parameter should be seriously questioned.

- During the course of the project it was realized that a validation strategy must include more than a procedure for comparing model results to experimental data (e.g. Neuman, 1991). Tsang (1991) notes that validation is an iterative **process** which must be carried out at every step of model development, testing and application. It is not a product. The iterative nature of the validation process suggests the concept of 'validation fit for purpose', similar to the ideas discussed by de Marsily et al (1992). In an iterative system one has to exit from the loop as and when demanded by circumstances, with whatever level of validation has been achieved at that time. Such demands might be the requirement to conclude a performance assessment exercise or submit a license application.

- Among INTRAVAL participants, it has generally been agreed that a model cannot be validated generically, in the absence of a context related to its eventual use. However, generic models can be validated in the sense that it can become generally accepted that the models can represent a real system. It is then necessary to demonstrate that the models are applicable to a specific site of interest. This implies strong links to site characterization. By its very nature, INTRAVAL could only take this process so far. It could help to build confidence in generic models and in the approaches adopted to validation, but the work required to demonstrate that a model is applicable to a specific site could only be undertaken to a limited extent using the existing data.

- INTRAVAL recognised the great sensitivity of modelling results to changes in boundary conditions and, for those cases evaluating geological timescales, to the initial conditions of the system. The geological systems studied are dynamic and, over the time periods required by performance assessment, they will respond to an evolving environment such that transport pathways will vary and fluxes will change. In this respect also, it is important to understand that transport cannot be described with any real degree of determinism, and that predictions made by models may be highly sensitive to changes in the underlying assumptions about the stability of boundary conditions.

- In the course of the project, no fundamentally new processes were identified which needed to be incorporated in new conceptual models. In part, this may have been because there was little work on reactive transport systems. Most of the model variation centred on differences in the way in which known processes were treated or coupled, or the degree of significance given to them. Thus, the overall validation problem identified was one of the methodology of an evaluation, rather than the processes it contains. INTRAVAL developed an understanding of the dominant flow and transport properties of various geological formations and an ability to model these in different ways, thus providing a range of possible outcomes which can be compared and discussed.

- There was only limited evaluation of reactive transport mechanisms, as most test cases involved either simple flow, or the use of conservative tracers. Even if we have good confidence in predicting the movement of conservative tracers over long distances, the transport of reactive tracers is, to a large extent, governed by mechanisms other than those which determine the movement of non-reactive tracers. The former are strongly influenced by the reactive capacity of the solid, the latter by pore volume. Access to the reactive capacity of the rock has nor been tested under field conditions and future investigations should aim to address this question.

- With the exception of the clay test cases, transport of conservative tracers under most test conditions was dominated by advection. In analysing this realistically, there is a need to concentrate effort on the heterogeneity and anisotropy of the system. Dispersion, arising from spatial variability of transport properties, is also important. The scale dependence of dispersivity is not fully resolved although, given adequate data on on heterogeneity, it seems possible to predict such dependence quite reliably.

- Demonstrating the effects of matrix diffusion on a repository performance assessment timescale using short-term tracer tests remains problematic. Attempts to demonstrate its impact have been based mainly on the ability to produce a better fit for the tail of tracer breakthrough curves. In most of the fractured crystalline rock test cases its effects appeared inconsequential, although it was a vital component of the WIPP-2 transport model for dolomite.

The short timescale of the tests needs to be borne in mind[2]. In the time frame of safety assessments, matrix diffusion may have a more dominant role, particularly for sorbing species, as illustrated by a number of natural analogue studies.

- It did not prove possible to have a significant number of the test cases purpose-designed for model testing, and the project had to live with the data which were available. Consequently, experimental observations frequently did not fulfil the requirements of the modellers, and some important parameters were not recorded. As a result, even though the 'post-audit' testing of predictions did not feature as largely as we would have wished, we have learned that it is quite possible to design experiments which could be used for such formal validation purposes. A large part of the project thus involved the calibration and fitting of models to test cases, rather than rigorously constructed tests of predictive capacity. Where it was attempted, it was usually found that predictions did not even bound the actual values. Also, models which performed well in predicting one observable parameter, would perform badly at predicting others. In some cases a bounding value would prove adequate for performance assessment purposes, provided it could be shown to be conservative. A significant realization is that we can have little belief in the precise value of any predictions. Confidence must arise from the broad trends that emerge, the ranges of parameter values generated and our understanding of the processes involved.

- The issue of uncertainty is clearly very important. Work within INTRAVAL reinforced this message, with the project identifying and discussing uncertainties in:

  - the choice of conceptual models
  - boundary and initial conditions of a modelled system
  - the scaling of predictions
  - the nature of gross inhomogeneities and discontinuities
  - parameter values.

  At the end of the project, INTRAVAL had developed a better feeling for the actual uncertainties inherent in each test case. Monte Carlo methods can be used to propagate uncertainties through the models used in order to evaluate the effects of uncertainty on predictions. However, within INTRAVAL, such methods were only used to address the uncertainties in parts of the system, not to assess the effect of uncertainty on repository performance measures.

---

[2] We note that both reactive transport and matrix diffusion were addressed directly by Nagra in the MI experiment at Grimsel, although this was not one of the INTRAVAL test cases. The consistency between laboratory and field parameter values and the demonstration of the role of matrix diffusion in the experimental results were impressive.

- It is likely that there will be uncertainty about the conceptual model for a given site; that is, the site can be represented by more than one conceptual model. (These models may have features in common.) This may mean that it is not possible to parameterise the system unambiguously.

- It is important to appreciate that concepts such as 'flowpath' and, by extension, 'transit time' along a flowpath, are largely abstract (similar to stream-tubes in potential flow theory), and great care must be taken when translating predictions made in these terms into performance assessment terminology.

- The desire to develop a methodology for discriminating between alternative conceptual models went unfulfilled within INTRAVAL, largely because the Project was unable to sponsor direct tests of one model against another under well-controlled conditions. In some cases it was found that the data which would be necessary to discriminate between models had not been obtained in the field tests. Consequently, there was considerable impetus within the project to formulate alternatives and develop such a methodology in future. Clearly, insight is gained by analysing an experiment using several different conceptual models. Even where the experimental data do not suffice to discriminate between these models, the spread of different models that could be fitted to an experiment gives information on the degree of uncertainty or non-uniqueness involved. As part of these discussions it was found that a formal and comprehensive approach to conceptual model discrimination would involve making very many measurements of a system at a variety of scales and the use of purpose-designed independent experiments.

- It was found that most tests could be interpreted by models that were relatively simple. A point of diminishing returns of accuracy is reached quite rapidly as a model becomes more complex. This tends to justify the use of simplifications (but not oversimplifications) in performance assessment models. One area where this did not seem to apply was in the interpretation of natural analogue systems involving long-term geochemical transport and deposition. In these cases, the more sophisticated geochemical models were better able to simulate the evolution of the system.

- Highly non-linear problems, such as density dependent flow with salt transport, demand computer resources which, to date, seem to be the limiting factor for the interpretation of site characterisation data. For example, prohibitive requirements for CPU time and memory prevented the fully three dimensional calculation of the flow and transport field for the Gorleben test case.

3. RECOMMENDATIONS FOR SITE CHARACTERIZATION AND PERFORMANCE ASSESSMENT

INTRAVAL brought together many of the leading groups studying radionuclide transport and safety assessment. It formed a valuable focus for the development of ideas both from within the project and from the wider radioactive waste community. As a result of the lessons learned from INTRAVAL, described in the previous section, it has been possible to develop some general recommendations which we hope will help the progressive development of a broadly applicable validation and site characterization methodology which takes account of the intrinsic uncertainties and variability of the natural environment:

- It is important to have some formal mechanism for defining comprehensively all relevant classes of conceptual model known which might be applicable for describing mass transport at a specific site and for defining the dominant processes which are active.

- It must be possible to define clearly what the input parameters are for each conceptual model, and to be sure that these can be measured in the field, or derived transparently from other measurable quantities.

- Site characterization needs to focus on defining the initial conditions, boundary conditions and driving forces for all the conceptual models to be used, and on obtaining the input parameters at as wide a variety of spatial scales as possible.

- More research is required on methodologies for scaling-up from one or many small-scale field tests to make larger scale predictions, and to testing these with larger scale experiments. It is important to consider systems in 3-D as well as 2-D, otherwise potentially important influences may be overlooked.

- Those planning site investigation programmes should consider the length of time required to design and run an iterative series of tests of the type used in INTRAVAL, and such as will inevitably be needed at any potential repository site. Three to five years seems a minimum requirement. Supporting laboratory investigations to investigate specific mechanisms must also be considered.

- Much more effort should be made to develop rigorous testing procedures for conceptual models, *publishing these procedures in advance of the testing process*, then allowing the models to undergo a limited amount of conditioning on site data before being required to make verifiable predictions. Such predictions of flow and transport were made in the Stripa SCV project (Olsson, 1992). As site characterization proceeds, several phases of testing of this type might be envisaged, involving re-calibrating and re-testing initially 'successful' models to make best use of the growing knowledge of the site and to produce a more refined model for eventually making performance assessment predictions. This 'post-audit' approach to making and checking predictions should be used more vigorously than at present. It is also important to check that the parameters of the model are reasonable, to check that the model provides a consistent explanation of all the data and to explore the reasons for discrepancies between predictions of the model and experimental observations, checking whether these have any implications for predictions of repository performance. It is also desirable to consider alternative models. Quantitative approaches for discriminating between alternative conceptual models need to be developed further.

- As part of the validation procedure, conceptual models which are to be used as the basis for making predictions must be expected to include or account for properties of a site related to its past geological evolution. The ability to absorb peripheral 'soft' geological data (for example, explaining the origin and stability of hydrochemical zonation in deep groundwaters) is an essential aspect of model credibility (Chapman & Grindrod, 1994). If a model cannot make a broad interpretation of the past evolution of a site, from which we have copious evidence, then we must have very little faith in its ability to make predictions about future behaviour based only on present conditions.

- Although only a minor part of INTRAVAL, the need to use natural analogues of geochemical systems to carry out rigorous testing of model predictions over long timescales was apparent. The development of coupled chemical kinetic and transport codes for heterogeneous systems will increase this need. The post-audit (or 'blind predictive') approach mentioned above is currently very well developed in testing 'static' equilibrium thermodynamic codes and databases at natural analogue sites and offers a good example of what might be achieved with respect to transport models using this method.

- The performance measures used in safety assessment exercises need to be related transparently to parameters which can be measured, or computed by widely accepted conceptual models. Lack of a clear scientific relationship between a quantity being predicted and a quantity used as a performance target at this stage of a site evaluation would be absurd.

- Site investigators must be urged to make their techniques and their data as transparent and as widely available as possible at the earliest stages of work. This will allow maximum time for the inevitably slow process of peer appraisal which is a vital component of building credibility and ensuring the validity of the final assessment.

- It must be made clear that predictions are not precise, merely indicative. If it is not understood by now that we do not seek certainty, only adequate confidence, and that we only seek to demonstrate acceptable safety of a repository rather than predict actual behaviour, then we should make more effort to drive the points home.

By following these ideas we may move closer to that vital understanding of the evolution and behaviour of a repository site, which underpins making quantitative assessments that are broadly accepted as being pertinent, meaningful and, consequently, valid for making decisions on environmental safety.

ACKNOWLEDGEMENTS

This paper was prepared by the members of the INTRAVAL sub-committee for integration (ISI). It is a condensed version of an overview report prepared by the ISI which is to be issued as an NEA report. The role of the ISI was to provide technical guidance to the managers of the INTRAVAL project and to maintain an overview of the results and the direction of the work. The paper describes the collective effort of a large number of individuals in more than 46 organisations in 14 countries around the world. We would like to thank every one of the more than 120 scientists who have been concerned with making this such a fascinating and stimulating project with which to be involved. Work in the INTRAVAL project was funded by 22 organisations, and the Project was devised and managed by the Swedish Nuclear Power Inspectorate (SKI) under the auspices of the Nuclear Energy Agency of the Organisation for Economic Co-operation and Development (OECD/NEA).

REFERENCES

Chapman, N. A. & Grindrod, P. (1994 in press) A scientific approach to integrating scenarios, simulations and natural systems. In Proc. 'Geoprospective', Paris.

Larsson, A. (1992). The international projects INTRACOIN, HYDROCOIN and INTRAVAL. *Advances in Water Resources*, **15**, 85-87.

de Marsily, G., Combes, P. & Goblet, P. (1992). Comment on 'Ground-water models cannot be validated', by L.F. Konikow & J.D. Bredehoeft. *Advances in Water Resources*, **15**, 367-369.

Neuman, S. P. (1991) Validation of safety assessment models as a process of scientific and public confidence building. Proc. High Level Radioactive Waste Management Conference, Las Vegas.

Olsson, O. (ed) (1992) Stripa Project: Site Characterisation and Validation - Final Report. SKB Stripa Project Technical Report 92-22, SKB, Stockholm, Sweden.

SKI/NEA (1992-4) INTRAVAL Test Case Reports (16 volumes) and Phase 1 and 2 Summary Reports (2 volumes). Swedish Nuclear Power Inspectorate, Stockholm/ OECD Nuclear Energy Agency, Paris.

Tsang, C-F. (1991) The modelling process and model validation. *Ground Water*, **26**, 825-831.

Conclusions from Working Group 1: Partially-Saturated Porous & Fractured Media Test Cases Las Cruces Trench & Apache Leap Tuff Studies

Thomas J. Nicholson
U.S. Nuclear Regulatory Commission

Richard Hills
New Mexico State University

Mark Rockhold
Pacific Northwest Laboratory

Gordon Wittmeyer
Center for Nuclear Waste Regulatory Analyses

Todd C. Rasmussen
University of Georgia

Amado Guzman-Guzman
University of Arizona

Abstract

Two INTRAVAL test case studies were conducted using data from the Las Cruces Trench (LCT), and Apache Leap Tuff Site (ALTS) experiments to evaluate models for water flow and contaminant transport in unsaturated, heterogeneous soils and fractured tuff. The LCT experiments were specifically designed to test various deterministic and stochastic models of water flow and solute transport in heterogeneous, unsaturated soils. Experimental data from the first two LCT experiments, and detailed field characterization studies provided information for developing and calibrating the models. Experimental results from the third experiment were held confidential from the modelers, and were used for model comparison. Comparative analyses included: point comparisons of water content; predicted mean behavior for water flow; point comparisons of solute concentrations; and predicted mean behavior for tritium transport. These analyses indicated that no model, whether uniform or heterogeneous, proved superior. Since the INTRAVAL study, however, a new method has been developed for conditioning the hydraulic properties used for flow and transport modeling based on the initial field-measured water content distributions and a set of scale-mean hydraulic parameters. Very good matches between the observed and simulated flow and transport behavior were obtained using the conditioning procedure, without model calibration. The ALTS experiments were designed to evaluate characterization methods and their associated conceptual models for coupled matrix-fracture continua over a range of scales (i.e., 2.5 centimeter rock samples; 10 centimeter cores; 1 meter block; and 30 meter boreholes). Within these spatial scales, laboratory and field tests were conducted for estimating pneumatic, thermal, hydraulic, and transport property values for different conceptual models. The analyses included testing of current conceptual, mathematical and physical models using the ALTS characterization data. Conclusions drawn were: (1) wetting history has a significant influence on formulating the characteristic curve; (2) thermal conductivity is only poorly related in a linear fashion to water content; and (3) the fracture saturation behind the wetting front initially is very low, perhaps ten percent, but increases to complete saturation during the course of the block wetting experiment contrary to the modeling results which overestimated the fracture imbibition volume by a factor of twenty, and the fracture wetting front advance by a factor of eight. Both the LCT and ALTS studies have demonstrated the value of field-scale experiments and the importance of experimentalists and modelers working together to solve complex flow and transport problems.

Introduction

In 1987, the NRC staff nominated two ongoing field studies for consideration in the INTRAVAL Project. These experiments dealt with the testing and confirmation of water flow and contaminant transport models for unsaturated, heterogeneous soils and fractured tuff. In Phase I of INTRAVAL which focused on laboratory and field studies, the Las Cruces Trench studies were proposed by the NRC staff to examine issues of heterogeneity in unsaturated soils. Specifically, how to characterize it, model it, and assess uncertainty in simulated flow and transport results. The Apache Leap Tuff studies were proposed by the NRC staff to examine issues of non-isothermal, multi-phase flow and transport in unsaturated, fractured tuff. Both projects were continued in Phase II of INTRAVAL which focused on field studies and natural analogues. Both studies have been extensively reported in the literature (see references), particularly in peer reviewed journals, and the NRC's NUREG/CR publication series. This invited review paper addresses a series of questions posed by the GEOVAL-94 Program Committee, which focus on the theme of *"validation through model testing with experiments"*.

Las Cruces Trench Studies

How was the Design Planned and the Experiment Analyzed?

Background

The Las Cruces Trench (LCT) studies originated as a cooperative research effort between field experimentalists at New Mexico State University (NMSU), Pacific Northwest Laboratory (PNL), and modelers at the Massachusetts Institute of Technology (MIT)(Wierenga et al., 1986). The field site was already part of a long-term environmental research study at the NMSU College Ranch (located 64 kilometers northeast of Las Cruces, New Mexico, USA) which was examining soil moisture processes in arid, cattle grazing land. The stated objective of the pre-INTRAVAL LCT study was "to collect a detailed data base for validating stochastic flow and transport models for unsaturated soils" (Nicholson et al., 1989). The original objective went on to state "The theoretical stochastic method; the approach for selecting an adequate data base upon which stochastic models can be tested; and the details of the field program, including the plot design and initial test configuration of the field trench, are all integral parts of the validation process." Central to the LCT study was the need to determine the degree of heterogeneity at the field site, and how it should be represented (e.g., stochastic random fields characterized by means, variances and correlation scales) (Nicholson et al., 1989).

Experimental Objectives

For the INTRAVAL Project, the LCT studies were reformulated using the goal of rigorous model testing (both deterministic and stochastic models) to include multiple field tests. The experimental objectives were that:

- The hydraulic properties for the site should be characterized in sufficient detail so that the site can be modeled using deterministic and stochastic models;
- The boundary conditions on water flow and solute transport should be carefully controlled to minimize ambiguities associated with model testing; and
- The movement of water and solute through the soil profile during infiltration and redistribution should be monitored in sufficient detail so that the effect of spatial variability can be defined (Hills and Wierenga, 1994).

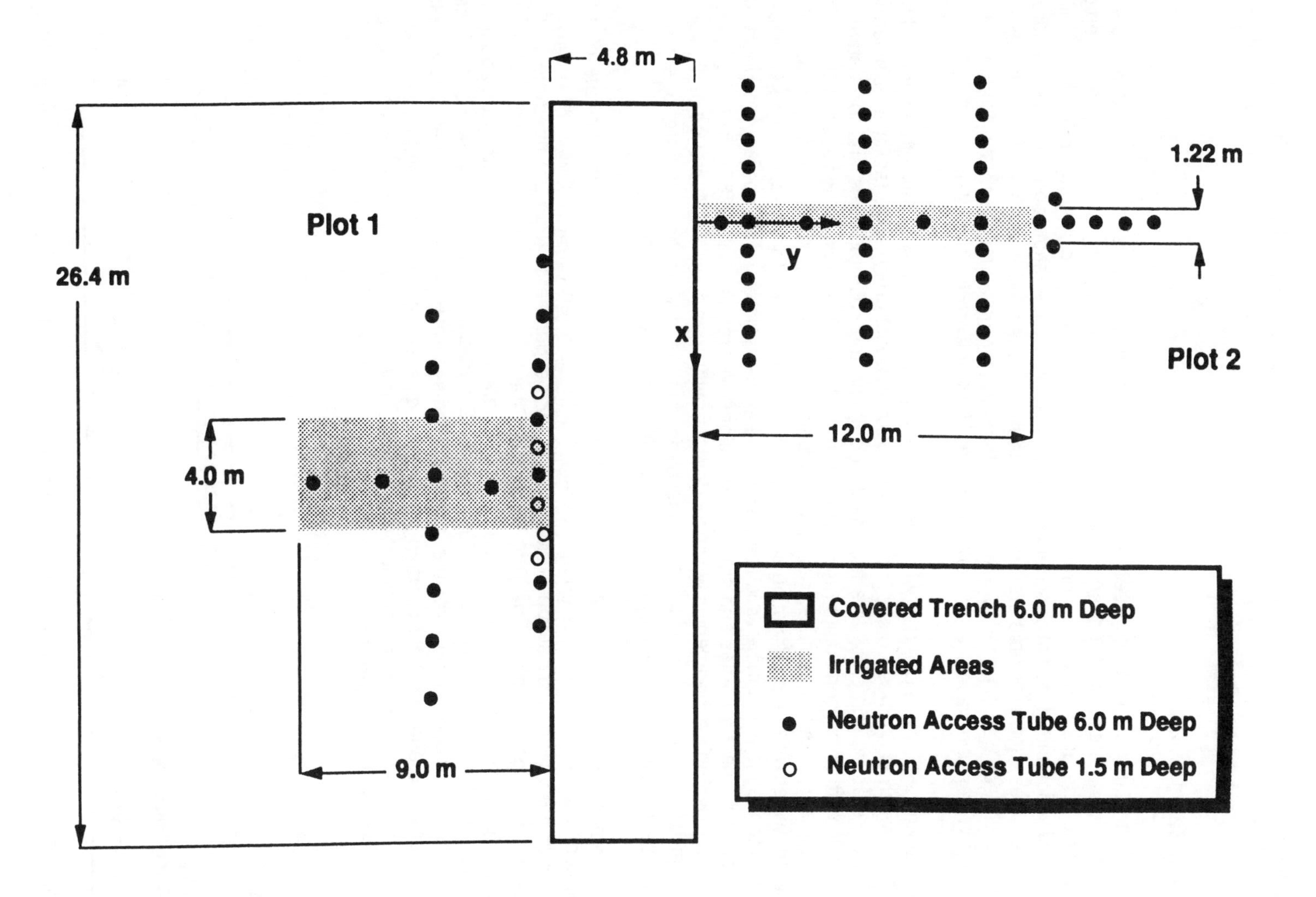

Figure 1. Plan view for the Las Cruces Trench Experiments (from Hills and Wierenga, 1994).

These experimental objectives were successfully met, due in part to the detailed instrumentation and monitoring activities which produced significant experimental datasets (Wierenga et al., 1989 and 1990; and Hills and Wierenga, 1994).

Experimental Setup

Figure 1 illustrates the experimental setup. Water was applied at a rate that would not create saturated conditions, but would enhance lateral spreading to magnify the lateral and vertical heterogeneities of the unsaturated soil profile. The facility was designed for monitoring the applied water and tracer movement at various depths. The field instrumentation consisted of numerous neutron access tubes, and tensiometers and solute samplers installed in the wall of the excavated trench. The trench face was exposed for the first experiment allowing visual observation of the wetting front advance. Water and tracers were applied to two different experimental plots.

A trench measuring 26.5 m long, 4.8 m wide, and 6.0 m deep was excavated through a series of soil horizons. The trench served as an observation, sampling and instrumentation facility. Earlier soil column and lysimeter experiments were important in defining the soil properties and for testing the horizontal versus vertical property differences for various infiltration rates (Wierenga et al., 1986). This information was used to determine the irrigation application rate for the first wetting experiment (Plot 1) using the 4m wide by 9m long irrigation strip shown in Figure 1. The Plot 1 test can be viewed as an initial characterization experiment which along with the earlier lysimeter and laboratory experiments defined the soil horizons, properties and ambient conditions (Wierenga et al., 1989). The Plot 2a test was considered a dynamic experiment in which model calibration was possible (Wierenga et al., 1990). Plot 2b was also a dynamic experiment but was performed as a validation experiment since the experimental results were not given to the modelers (Hills and Wierenga, 1994).

The first experiment used the 4.0 by 9.0 meter plot (Plot 1) with a water application rate of 1.82 cm/day. Tritium was applied with the irrigation water for the first ten days, followed by 76 days of irrigation without the tritium. The second (Plot 2a) and third experiments (Plot 2b) used the 1.22 m wide by 12 m long irrigation strip. The Plot 2a test in Phase I of INTRAVAL involved a reduced application rate (0.43 cm/day) with tritium and bromide tracers for 11.5 days followed by an additional 64 days of water application without tracers. The Plot 2b test, designed by scientists at the University of Arizona and NMSU in consultation with the modeling teams in INTRAVAL, used an application rate identical to the first experiment (1.82 cm/day) but used a different application schedule and tracers [chromium, boron, and pentafluorobenzoic acid (PFBA) initially, and tritium, bromide and 2,6-difluorobenzoic acid (DFBA) later] (see Figure 2).

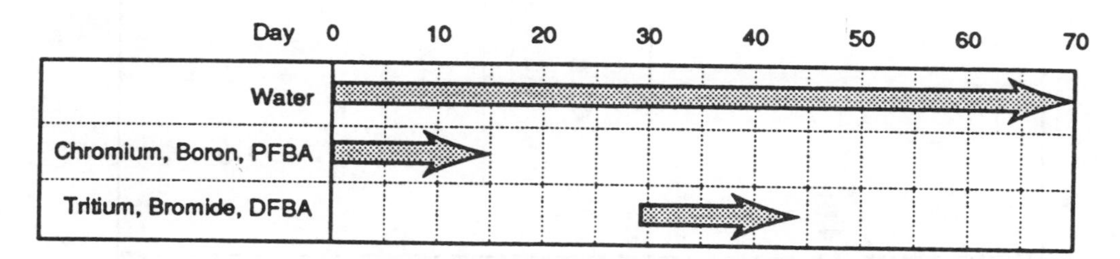

Figure 2. Plot 2b water and chemical tracer application schedule (from Hills and Wierenga, 1994).

Experimental Analysis

The experiments were analyzed by defining: (1) the characterization and dynamic variable properties measured during the experiments (see Table 1); (2) the distributions of water contents using the neutron access holes taking readings at various depths; (3) the tracer plume concentration contours using collected water samples via the soil solution samplers in the face of the trench during the experiments, and destructive core samples collected at the end of the experiments; and (4) the wetting front positions for Plot 1 based upon photographs of the trench face over time. Table 1 also provides the scale and technique used to determine the characterization and dynamic variables (where θ, θ_r and θ_s are the volumetric water content, residual water content, and saturated volumetric water content, α and n are parameters that define the shape of the van Genutchen model soil-water retention curves, and h is pressure head) measured during the experiments (Hills and Wierenga, 1994).

Table 1. Parameters measured during the Las Cruces Trench experiments (from Hills and Wierenga, 1994).

Parameter	Scale	Technique	Reference
		Characterization Variables	
K_{sat}	8 cm	Measured flow through saturated cores	Elrick et al. (1980), Wierenga et al. (1989a)
K_{sat}	10 cm	Borehole permeameter	Reynolds et al. (1984), Wierenga et al. (1989a)
$\theta_r, \theta_s, \alpha, n$	8 cm	Cores and constant pressure apparatus combined with parameter estimation	Wierenga et al. (1989a)
Particle size distribution	8 cm	Soil sieves and modified pipet method	Gee and Bauder (1986). Wierenga et al. (1989a)
		Dynamic Variables	
θ	50 cm	Neutron Probe	Wierenga et al. (1990)
h	2 cm	Tensiometer	Wierenga et al. (1990)
Concentration of tracers	2 cm	Solute samplers and soil sampling	Wierenga et al. (1990)

To characterize the transport parameters of the tracers (tritium, bromide, chromium, boron and chloride) a series of laboratory and field column studies were performed, and are reported in Porro (1989), and Porro and Wierenga (1992). PFBA and DFBA are non-reactive anions and are considered to have the same properties as bromide, and therefore were not tested.

What Models Were Tested?

During Phase I of INTRAVAL, the modeling teams consisted of: (1) the Bureau for Economic Geology, University of Texas (BEG); (2) the Center for Nuclear Waste Regulatory Analysis (CNWRA); (3) HydroGeologic Inc. (HGL): (4) Kemakta Consultant Co. (KKC); (5) Massachusetts Institute of Technology (MIT); (6) New Mexico State University (NMSU); (7) Pacific Northwest Laboratory (PNL); and (8) Sandia National Laboratory (SNL). In Phase II the groups included BEG, CNWRA, MIT, NMSU, and PNL.

Table 2 lists the models used to analyze the Plot 1 and 2a experiments both prior to and during Phase I of INTRAVAL. The models can be grouped into two categories, (1) *deterministic models* that represent heterogeneities as homogenous, uniform property values or as heterogenous with various approaches to portraying distributed property values, and (2) *stochastic models* using Monte Carlo methods or the techniques of Mantoglou and Gelhar (1987) to represent heterogeneities. The process models in Phase I focused on those used to characterize the soil hydraulic properties, and dynamic models used to simulate flow and transport for Plot 1 and 2a experiments as shown in Table 2. Details on the conceptual and numerical aspects of these models is provided in Hills and Wierenga (1994).

Table 2. Modeling of the Plot 1 and 2a experiments (from Hills and Wierenga, 1994).

Group	Model	Comments
CNWRA	BIGFLOW	Finite difference code for high resolution water flow simulations. Modeled a 3-D, randomly heterogeneous and stratified soil with boundary conditions similar to Plot 1 but with wetter initial conditions.
HGL	VAM2D	Finite element code for water flow and transport. Modeled Plot 2a with several levels of heterogeneities (isotropic and anisotropic) in 2-D.
KKC	TRUST + TRUMP	Integrated finite difference code for water flow and transport. Modeled Plot 2a as a homogeneous and a layered soil in 2-D.
MIT		Finite element code for water flow using modified Picard Iteration. Modeled Plot 1 using 3-D effective media stochastic property models in a 2-D simulation.
NMSU		Water Content based finite difference code for water flow & transport. Modeled Plot 2a as a homogenous soil in 2-D.
PNL	UNSAT-H UNSAT2	One and two-dimensional finite difference and finite element codes for water flow. Modeled Plot 1 water flow with several levels of heterogeneities in 2-D.
SNL	VAM2D	Finite element code for water flow and transport. Monte-Carlo simulation of 2-D water flow using VAM2D with multiple realizations of a uniform soil model. Analytic 1-D models are used.

Table 3. Modeling of the Plot 2b experiment (from Hills and Wierenga, 1994).

Group	Models	Comments
BEG	BEG1	**Not Blind.** Finite difference code for two phase flow and multicomponent transport. Modeled water flow in 2-D assuming uniform, isotropic soil.
CNWRA	CNWRA1	**Blind.** Finite volume code for water flow. Modeled water flow in 2-D using a 9 layer,isotropic soil model.
	CNWRA2	**Blind.** Finite volume code for water flow. Modeled water flow using a 2-D, heterogeneous 121 zone (11 x 11 grid), isotropic soil model based on trench face characterization.
	CNWRA3	**Blind.** Finite volume code for water flow. Modeled water flow using a 2-D, heterogeneous 3621 (51 x 71 grid), isotropic soil model based on trench face characterization.
MIT	MIT1	**Blind.** Finite element code for water flow using modified Picard approximation. Modeled water flow using a 3-D effective media stochastic property model (homogeneous, anisotropic) in a 2-D simulation.
NMSU	NMSU1	**Not Blind.** Water content based finite difference code for water flow and tritium transport. Modeled water flow and tritium transport assuming soil is homogeneous and isotropic in 2-D.
	NMSU2-NMSU5	**Not Blind.** Water content based finite difference code for water flow and tritium transport. Modeled hydraulic properties of soil as heterogeneous, isotropic in 2-D using four property realizations sampled from the trench face characterization. Transport properties were modeled as uniform, isotropic.
PNL	PNL1	**Blind.** Finite difference code for two-phase flow and transport. Modeled water flow and tritium transport in 2-D using a composite van Genuchten, uniform, isotropic soil model for the hydraulic properties. Transport properties were modeled as uniform and isotropic.
	PNL2	**Blind.** Finite difference code for two-phase flow and transport. Modeled water flow and tritium transport assuming the hydraulic properties of the soil were uniform, anisotropic, using modified parameters in the standard van Genuchten model. Transport properties were modeled as uniform and isotropic.
	PNL3	**Blind.** Finite difference code for two-phase flow and transport. Modeled water flow and tritium transport assuming the hydraulic properties of the soil were uniform, isotropic, and using van Genuchten parameters estimated from a 1-D inverse analysis of Plot 1 experiment. Transport properties were modeled as uniform and isotropic.
	PNL4	**Blind.** Finite difference code for two-phase flow and transport. Modeled water flow and tritium transport using the 2-D trench face characterization of the water retention parameters and an isotropic, 2-D heterogeneous, conditioned realization of the saturated hydraulic conductivity field. Transport properties were modeled as uniform and isotropic.

For Phase II, the dynamic models simulated flow and transport of the Plot 2b experiment. Table 3 summarizes the models used and the manner in which they represented heterogeneities. (Note: a "blind model" was defined as "one which was formulated before the modeler had seen the experimental results." The NMSU models could not be classified as blind since the modeler was also the party responsible for collecting and analyzing the experimental results.) Detailed discussions of the models listed in both Tables 2 and 3 are provided in Hills and Wierenga (1994).

How were the Models Tested with Experiments?

The following procedure as discussed in Hills and Wierenga (1994) was developed and implemented by the LCT Pilot Team for the Phase II model comparison.

1. The water/tracer application rates, the initial volumetric water contents in the y = 2, 6, and 10 m planes (see Figure 1), and the initial normalized solute concentrations in the y = 0.5 m plane were provided to the modelers.

2. Using the actual water/tracer application histories (see Figure 2), the Plot 2b experiment was simulated by the participating modelers.

3. Several of the modelers provided NMSU with ASCII files of the predicted volumetric water contents and normalized solute concentrations at the measurement locations and times. Since this data was provided to NMSU without the modelers having access to the experimental data, these model predictions were considered blind (see Table 3).

4. Preliminary comparisons between experimental data and model predictions were made and presented at an INTRAVAL workshop. The experimental data was then released to those modelers that had supplied the corresponding model predictions to NMSU.

5. Additional model predictions were provided to NMSU after the above presentations were made. These later modeling predictions were considered non-blind (see Table 3). To be totally unbiased, the model predictions provided by NMSU were also considered non-blind since one of the NMSU personnel had access to the data.

The modelers were periodically updated at Working Group 1 meetings by the Pilot Team who conducted the quantitative comparisons. As shown in Figure 2 several chemicals were applied during the execution of the Plot 2b experiment. However, only tritium was modeled since the data set from tritium proved to be the most complete, and deemed to be the most reliable by the experimentalist.

As discussed in Hills and Wierenga (1994) the initial conditions were provided to the modelers via ASCII files for water content in the y=2, 6, and 10 m planes, and tritium in the y=0.5 m plane. Model predictions were returned to NMSU in the same format for the measurement times and locations. Each record of the initial condition and model prediction files included a time (days since start of Plot 2b experiment), the x, y, z coordinates (m) measured relative to the Plot 2b irrigation centerline, and the value of the predicted variable.

Were Alternative Conceptual Models Tested and Assessed?

As shown in Tables 2 and 3, a great variety of alternative conceptual models were tested and assessed. The alternatives ranged from uniform property to complex distributed property value models. The stochastic models considered spatial variability using spectral analysis approaches which related property distributions to tension dependency.

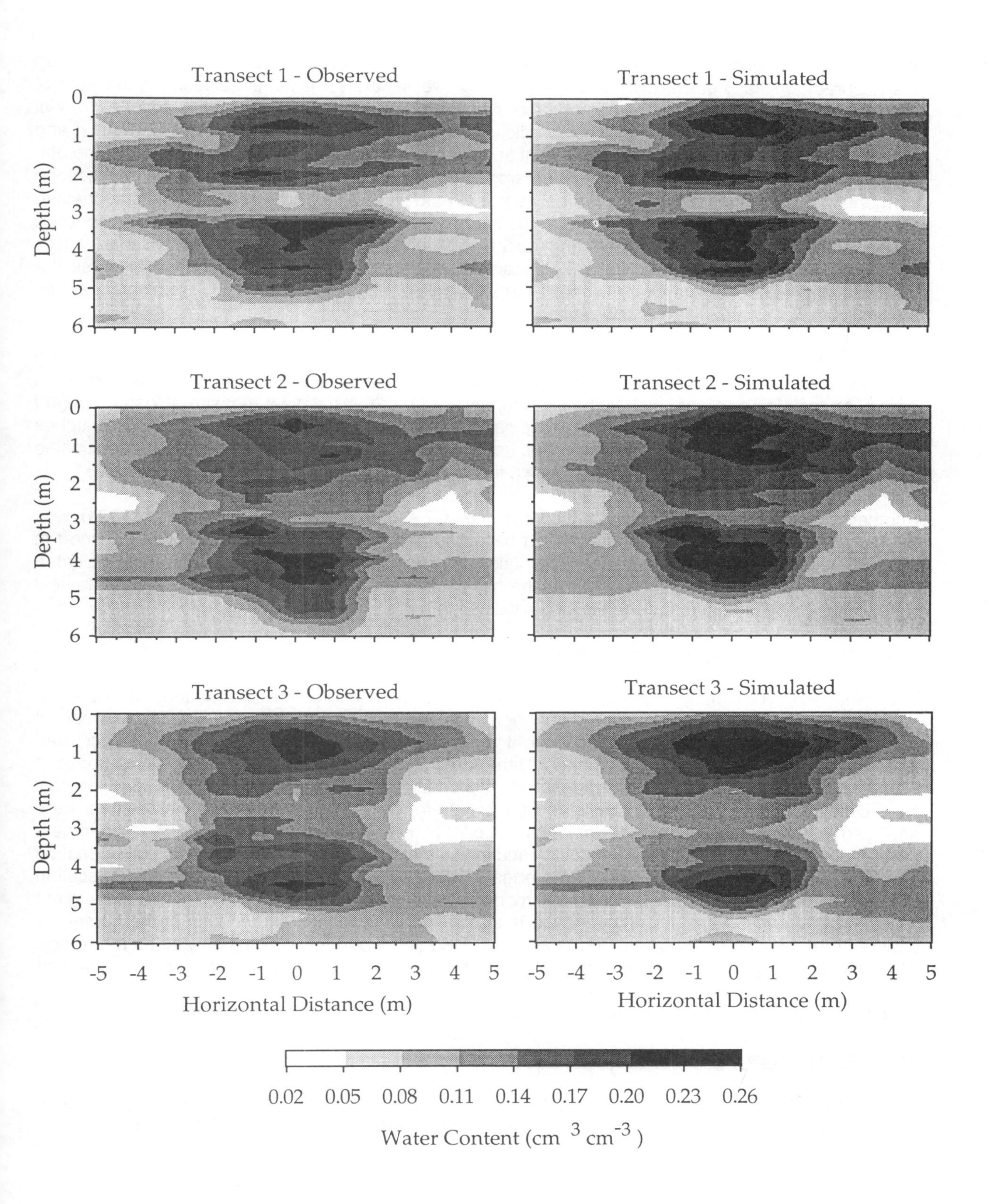

Figure 3. Observed and simulated water content distributions for day 70 (from Rockhold et al., 1994)

Wittmeyer and Sagar (1993) used the LCT data to assess the effect of increased model complexity on the accuracy of the predicted water contents using two separate measures. Based on the sum of squared differences between measured and predicted water contents, the most complex model (CNWRA3) produced the most accurate predictions. However, comparative analysis of the first and second moments of the water content distribution as functions of time lead to equivocal results; no one model was consistently better than the others. That there was less variation among the second moments predicted by the models than between the models and the experimental results suggested that there was a consistent source of bias in the models; an observation confirmed by visual inspection of the propagation of the predicted and observed wetting fronts. Wittmeyer and Sagar (1993) concluded from their study that while increased detail in model structure may indeed increase the accuracy of predictions, site characterization efforts should focus on those specific geologic features whose presence dominates the flow regime.

Following the final modeling comparison studies, a newly developed method focusing on conditioning methods was successfully applied to the LCT datasets for the Plot 2b experiment. This method is based on conditioning the soil hydraulic properties on the initial field-measured water content distributions and a set of scale-mean hydraulic parameters (Rockhold et al., 1994). Very good matches between the observed and simulated flow and transport behavior were obtained using the conditioning procedure, without model calibration as shown in Figure 3.

Another technological advancement derived from the LCT study, as presented in Hills et al. (1994), was the development of a flux-corrected transport based numerical algorithm to model solute transport in heterogeneous variably saturated soils. Application of this new algorithm to the LCT database indicated that the movement of the wetting front may be heavily influenced by the old water, whereas the new water tends to bypass much of the old water indicating preferential flow.

How were Different Disciplines Integrated?

The Pilot Team (University of Arizona and New Mexico State University investigators) consisted of specialists in the areas of soil physics, soil chemistry, soil morphology, hydrogeology and numerical methods. The modeling teams similarly brought varied expertise to the effort. The integration occurred through the four phases of the LCT studies: (1) the site characterization stage in which the soil horizons were mapped in detailed; in situ saturated hydraulic conductivity tests were performed; core samples were collected and tested for hydraulic and transport properties; and particle size distributions were analyzed (Wierenga et al., 1989); (2) the design, construction and conducting of the Plot 1 experiment (Hills et al., 1989a and 1989b); (3) the design and execution of the Plot 2a experiment using information and analysis of the Plot 1 experiment (Wierenga et al. 1990); and (4) the final design and execution of the Plot 2b experiment using the detailed monitored data and analysis from the Plot 2a experiment (Hills and Wierenga, 1994). The integration is also reflected in the joint authorship of the numerous papers produced and the subjects discussed (e.g., field and laboratory studies as well as modeling). Ultimately the integration occurred in the analysis of the site characterization, field experimental, and modeling
results as discussed in Hills and Wierenga (1994).

How was Success Defined and Assessed?

For the INTRAVAL Phase II analysis, the performance measures used to assess the simulation results consisted of both point comparisons, and integrated comparisons with the experimental data.

Point Comparisons

The point comparisons were;

1. Contour plots of observed and predicted water contents and solute concentrations,

2. Scatter plots of observed and predicted water contents and solute concentrations, and

3. First arrival times of the water and solute plumes as a function of depth.

Contour plots as presented in Hills and Wierenga (1994) were very useful in visualizing the behavior of the wetting front, moisture redistribution, and the solute plume. However, caution is advised in interpreting these plots since the contour results often are very dependent on the contouring algorithm.

Therefore, the second point value comparison approach used was scatter plots of observed versus predicted volumetric water contents and normalized solute concentrations. The LCT Pilot Team considered scatter plots to provide a more realistic assessment of point value predictions than the contour plots because contouring methods inherently (sometimes intentionally) average data and because the resulting contours can be very dependent on the analytical techniques used to generate the contours. It was also felt that scatter plots have the added advantage of showing the scatter of the observations about the predictions which give a good indication of bias and uncertainty about the mean (Hills and Wierenga, 1994).

The third point value comparison approach used plots of first arrival times of the water and solute plumes at various depths. The LCT Pilot Team defined the time of arrival of the water plume as that time when the volumetric water content increased by 0.03 from the initial conditions in any of the three measurement planes y = 2, 6, and 10 m (Hills and Wierenga, 1994).

Integrated Comparisons

The integrated quantities used for comparison as discussed in Hills and Wierenga (1994) were:

1. First and second moments of the water and tritium plumes as a function of time;

2. The normalized change in total water volume below each of the z=0, 1,..., 5 m horizontal planes as a function of time; (The LCT Pilot Team felt that the observed changes in water volume below a horizontal plane was a good estimator of the water to pass through that plane while the plume remained fully observable.) and

3. The changes in the sum of relative tritium concentration below each of the z=0,1,..., 5 m horizontal planes as a function of time.

Implementation

The LCT Pilot Team wrote four FORTRAN programs to generate the desired data files, given the point values of the experimental data and model predictions (Hills and Wierenga, 1994). Two programs processed the data for water flow and two processed the data for tritium transport for the point and integrated quantities outlined above.

To What Extent Did the Assessment of Success Take into Account Information Other Than Model Testing?

The assessment of success involved both model testing and the repeated reevaluation of the experimental results. Both field monitoring data and destructive core sampling were used in these reevaluations. The success was also measured by the advance in knowledge and datasets created (Wierenga et al., 1993). The following observations as discussed in the final LCT INTRAVAL report (Hills and Wierenga, 1994) demonstrate this interrelationship between model testing and field analysis.

"The testing of models using data from dry, spatially variable soils is not a trivial task. Not only is it difficult to characterize the site, but it is difficult to obtain sufficient high quality solute samples to obtain good estimates of the movement of solute plumes as a function of time. In contrast, water flow in unsaturated soils is easier to monitor. Neutron probes allow water to be monitored at many locations and the measurements are very consistent day to day over periods of years. For the Plot 2b experiment, the total increase in water observed by the neutron probe just after irrigation was very close to the actual water applied (less than 1% error) suggesting that given the spatial variability of the site, the number and location of the neutron probe measurements were sufficient to resolve the water plume.

There are considerable differences in model predictions even though all the modelers had access to the same very large characterization data set for the Las Cruces Trench experiments. Some of the models presented were fairly simple and assumed uniform soil hydraulic property fields while others conditioned the soil models on the observed two-dimensional spatial heterogeneities observed in the trench face. Even though many models were considered, none of the models stood out as clearly superior. All of the models under predicted first arrival times of the water plume at depths greater than 4.5 m and the models that tended to do well by one measure of mean behavior would often perform poorly by another. Since the probability distributions of the prediction errors do not appear to be well defined and were not clearly distributed as normal or log-normal for all of the models considered, parametric statistical tests were not performed. Less powerful non-parametric tests were used. As a result of these quantitative and graphical comparisons, several observations can be made:

The CNWRA and PNL models consistently provided better predictions of mean or median water contents (i.e., near zero mean or median predictions errors) which suggested a good accounting for the total mass in the system. This was likely due to the extra care CNWRA and PNL exercised in modeling the actual spatial distribution of the initial water contents. However, the improvement in mass balance that resulted did not necessarily lead to a reduction in the spread of the population prediction errors (i.e. RMS error) about zero relative to the other models. NMSU1, for example, showed a lower RMS than the CNWRA models. Thus accurately modeling the initial conditions does not, in itself, always lead to improved predictions for water flow.

Time of first arrival times of the water plume were greater for the experiment than for any of the models once the plume reached 4.5 m. This indicates that none of the models provided conservative estimates for these travel times from a regulatory point of view since they all over-predicted these times. While this is expected for the uniform soil models since they predict mean behavior, this was not expected to be always true for the heterogeneous soil models. Contour plots of experimental water contents show that a dry layer extends throughout the measurement domain. This layer was not well-predicted by any of the models. It is not clear whether this is due to limitations in the experimental characterization procedures, the inability of Mualem's model (1976) to predict unsaturated hydraulic conductivity for this soil layer given the van Genuchten retention model (1980a), or simply due to differences between the soil properties at and away from the trench face.

The observed change in the volume of water below all depths greater than 3 m was significantly greater than that predicted by all of the models except for PNL3. While PNL3 was conservative from a regulatory point of view in the sense that it over predicted the change in water content at depth, its behavior during redistribution was considerably different from that observed in the experiment. The predicted water plume was much more diffuse than the observed plume and more horizontal spreading was predicted. The hydraulic parameters used by PNL3 were obtained with a 1-D inverse procedure (Rockhold, 1993) using the experimental observations obtained during infiltration for the Plot 1 experiment. In contrast, the soil characterization used by the other models was based on outflow data obtained from core and disturbed soil samples. This may explain why the other models (except MIT1) performed better during redistribution.

The two models, PNL2 and MIT1, conceptualized the soil as anisotropic. MIT1 over-predicted the water plume spreading and under-predicted vertical plume movement as did PNL2. PNL2 assumed that the horizontal hydraulic conductivity was twice the vertical whereas MIT1 used a tension dependent anisotropy derived from stochastic theory. For the first 310 days of the experiment, neither model seemed appropriate. The more conventional isotropic models performed as well or better. However, significant heterogeneity induced anisotropy may be present later in redistribution when the plume becomes larger and when the gravitational forces become less important relative to the matric potential forces.

The initial water contents used in the BEG1 model were significantly larger than those observed in the field. While this had the effect of accelerating the downward motion of the water plume which gave good first arrival time estimates, BEG1 performed poorly by many of the other measures.

Results from Phase I of INTRAVAL for the Plot 2a experiment suggested that water movement was easier to model than tritium transport (Voss and Nicholson, 1993). However, most of the model comparisons made during the Plot 2a experiment were side by side comparisons of smoothed contour plots. The more extensive comparisons made here do not support this hypothesis. Tritium transport predictions for the Plot 2b experiment were more acceptable than the corresponding water flow predictions in the sense that the observed tritium behavior (first arrival times, change in tritium concentrations below a horizon) was bounded by the various model predictions whereas the observed water flow was not. It is not clear if this is because tritium transport occurs in the wetter portion of the water plume or simply because the tritium plume did not pass through the anomalous dry layer at 3 m.

The present results indicate that models that appear superior or conservative (from a regulatory point of view) for water flow do not necessarily lead to superior or conservative predictions for tritium transport. For example, the PNL models generally gave better predictions of mean or median water contents while the NMSU models generally gave better predictions of mean or median solute concentrations. Only PNL4 provided conservative estimates of first arrival time down to 4 m for the water plume. In contrast, only the heterogeneous NMSU3 and NMSU4 models provided conservative estimates of first arrival times for the tritium plume introduced during the Plot 2a experiment whereas the heterogeneous models NMSU3 and NMSU5 and the uniform soil model PNL1 provided conservative estimates of arrival time of the plume introduced during the Plot 2b experiment. These results support the idea that using multiple realizations of heterogeneous soil models (i.e., NMSU2-NMSU5) is an appropriate way to bound contaminant plume behavior.

Overall, the results of the present work show that for this particular experiment, traditional uniform soil or heterogeneous soil models conditioned on detailed site characterization data can predict the overall features of water flow and tritium transport. Even though there are considerable differences in how the models conceptualized the soil profile, no model was clearly superior overall. Superior models by one measure were not always superior by another. This suggests that the effect of characterization uncertainty, even when the site is characterized as thoroughly as the Las Cruces Trench Site, may have a greater impact on model predictions than differences in how the models conceptualize the soil."

The LCT studies and the model testing demonstrated one approach to validation through interactive laboratory and field experiments and modeling using a quantitative-based model testing strategy with performance measures tied directly to regulatory significant criteria.

Apache Leap Tuff Site Studies

How was the Design Planned and the Experiment Analyzed?

Background

The Apache Leap Tuff Site (ALTS) studies originated as a research effort by investigators at the University of Arizona to examine site characterization and conceptual model issues associated with high-level radioactive waste repositories. As discussed in Yeh et al., (1987) the objective of this pre-INTRAVAL study was to examine models and strategies for obtaining characterization data. Specifically, the design document stated that "Characterization of fluid flow and solute transport through unsaturated fractured rock requires that site-specific conceptual models be defined, parameters for the models be estimated using field and laboratory data, and validation of the conceptual models be performed" (Yeh et al., 1987). The great difficulty was that an understanding of the fundamental processes, and the development of conceptual models describing flow and transport in unsaturated, fractured rock was in its infancy. As opposed to the Las Cruces Trench studies, the focus needed to be on a much more fundamental and primitive level in which characterization techniques and conceptual model development was very much in question.

Experimental Objectives

For the INTRAVAL Project, the ALTS studies were reformulated to meet the goal of testing conceptual models and their linkage to characterization approaches. The experimental objectives were: (1) to verify the existence of proposed processes, and the parametric form of hypothesized material properties; and (2) to examine conceptual models and characterization approaches over a range of scales and processes using the following approaches;

1. Rock matrix characterization experiments for hydraulic, pneumatic and thermal properties using small (2.5 x 6.0 cm), and large (12 x 10 cm) cores;

2. Non-isothermal core experiments for coupled water, vapor and solute movement using (9.6 x 12 cm) core;

3. Rock fracture characterization studies and analyses of water and gaseous flow properties using a (20 x 21 x 93 cm) block with persistent single fracture;

4. Fracture imbibition experiments for determining imbibition rates and related properties using a (20 x 21 x 93 cm) block with persistent single fracture; and

5. Field air injection experiments for characterizing the heterogeneity of fracture-related permeabilities, and for examining dependency of measured pneumatic properties on measurement support (length of test interval) using inclined 30 m boreholes intersecting fractures.

These experimental objectives were addressed through numerous laboratory and field experiments, and are discussed in detail in Rasmussen et al. (1994), Bassett et al. (1994), and Rasmussen et al. (1990).

Experimental Setup

Figure 4 illustrates the experimental set-up for the core and block studies (numbers 1, and 3 above) used to determine hydraulic, pneumatic, and gaseous transport property values for the matrix and fractures. The core and block specimens used were taken from the Apache Leap Tuff (white unit). The block contained a discrete fracture along the long axis. Details are provided in Rasmussen et al. (1990).

The experimental setup for the core heater experiment (number 2 above) used a large core measuring 9.6 cm in diameter and 12 cm in length that was subjected to a series of coupled heat, water, vapor and solute transport experiments. These experiments involved the use of a one-dimensional thermal gradient (5 to 45°C) applied along the long-axis of the core. The core was hermetically sealed and insulated to provide a closed system for air and water. Dual-gamma attenuation methods were employed to provide water content and solute concentration profiles along the length of the core. Coincident temperatures along the core using thermocouple ports were also measured. Solute concentrations at the end of the experiment were determined (Rassmussen et al., 1994).

Figure 5 provides a three-dimensional portrayal of the 15 inclined and vertical boreholes at the ALTS. Figure 6 provides a schematic representation of the air injection setup (number 5 above) for determining the apparent permeabilities along the borehole intervals. The permeability tests consisted of imposing a sequence of increasing flow rates (a minimum of three), each of which was continued until a steady state pressure response was attained. Air pressure, temperature and relative humidity were measured at the surface and the injection interval. Atmospheric temperature and pressure were also monitored. The flow rate was preset at the surface with the aid of electronic mass flow controllers.

Experimental Analysis

The five distinct groups of studies (see above) produced significant datasets as discussed in Rasmussen et al. (1994). The information produced, and the analysis of these datasets have been summarized in the final INTRAVAL report (Rassmussen et al., 1994) as follows:

1. Laboratory analyses of the first set of experiments on the small core samples provided characterization data related to porosity, characteristic curves, hydraulic conductivity, air permeability and thermal conductivity. The effects of variable water contents, hysteresis and temperature on the physical parameters were examined, as well as, the effects of solute concentrations on ambient matric potential.

2. The core heater experiment demonstrated the presence of an active heat pipe which was observed when the core was brought to an intermediate water content. The resulting latent heat transport was insignificant in comparison to the conductive heat transport in this experiment. When a soluble salt (NaI) was introduced into the experiment, the heat pipe phenomenon was not as active due to the increased osmotic potential near the warm end of the core. The increased osmotic potential lowered the vapor pressure near the warm end and reduced the vapor phase transport of water.

3. The third set of studies conducted on the fractured rock block provided information to characterize the physical properties of the block. Equivalent fracture apertures were obtained using six types of experiments. Three volumetric fracture aperture values were obtained by using a pycnometer, tracer breakthrough volumes, and the ratio of fracture transmissivity to fracture hydraulic conductivity. Two Poiseuille apertures were obtained using a cubic aperture equation applied to gas and water flow rates, and using a quadratic aperture equation gas

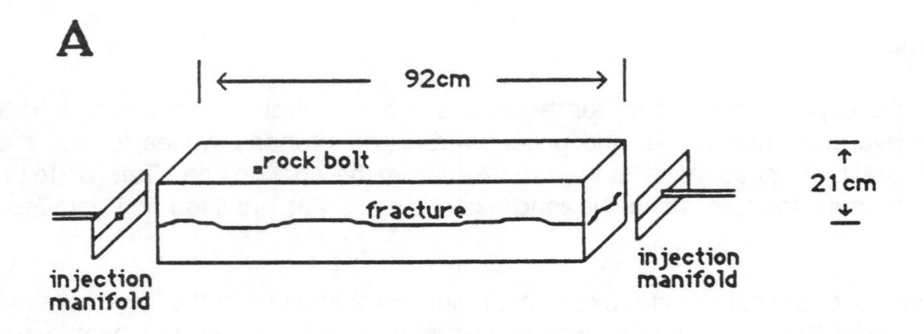

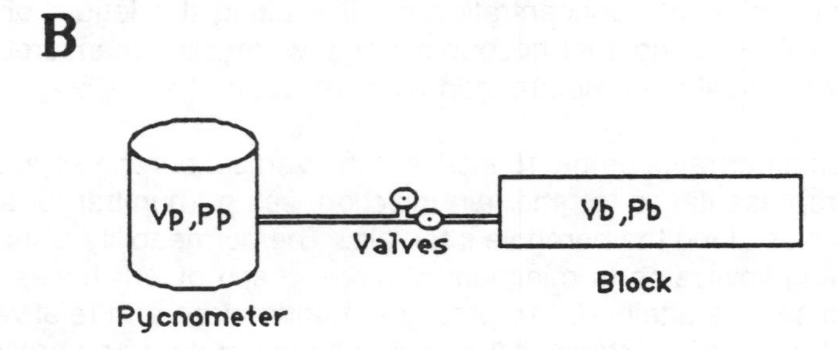

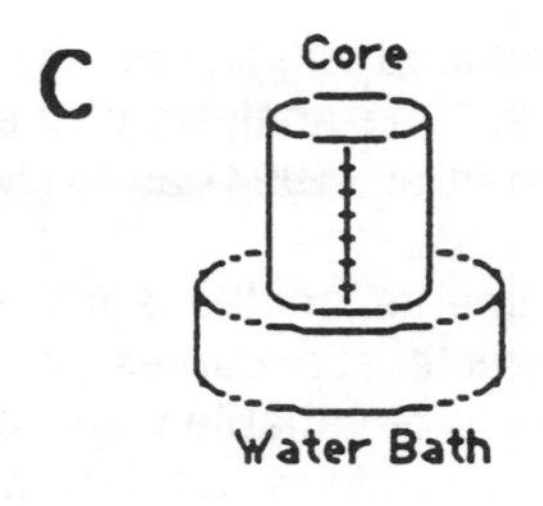

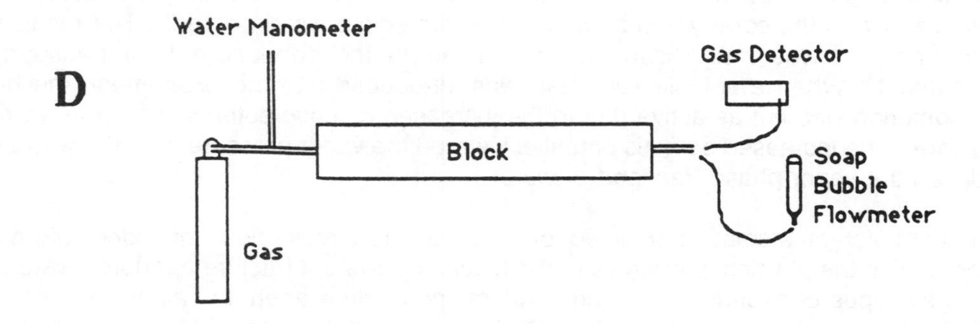

Figure 4. Experimental setups for (A) the wetting front experiment for the fractured block; (B) the rock porosity measurement experiments of the matrix using a pycnometer; (C) hydraulic diffusivity coefficient measurement using a core imbibition experiment; and (D) gas diffusion coefficient measurements and breakthrough curve experiment. (from Bassett et al., 1994).

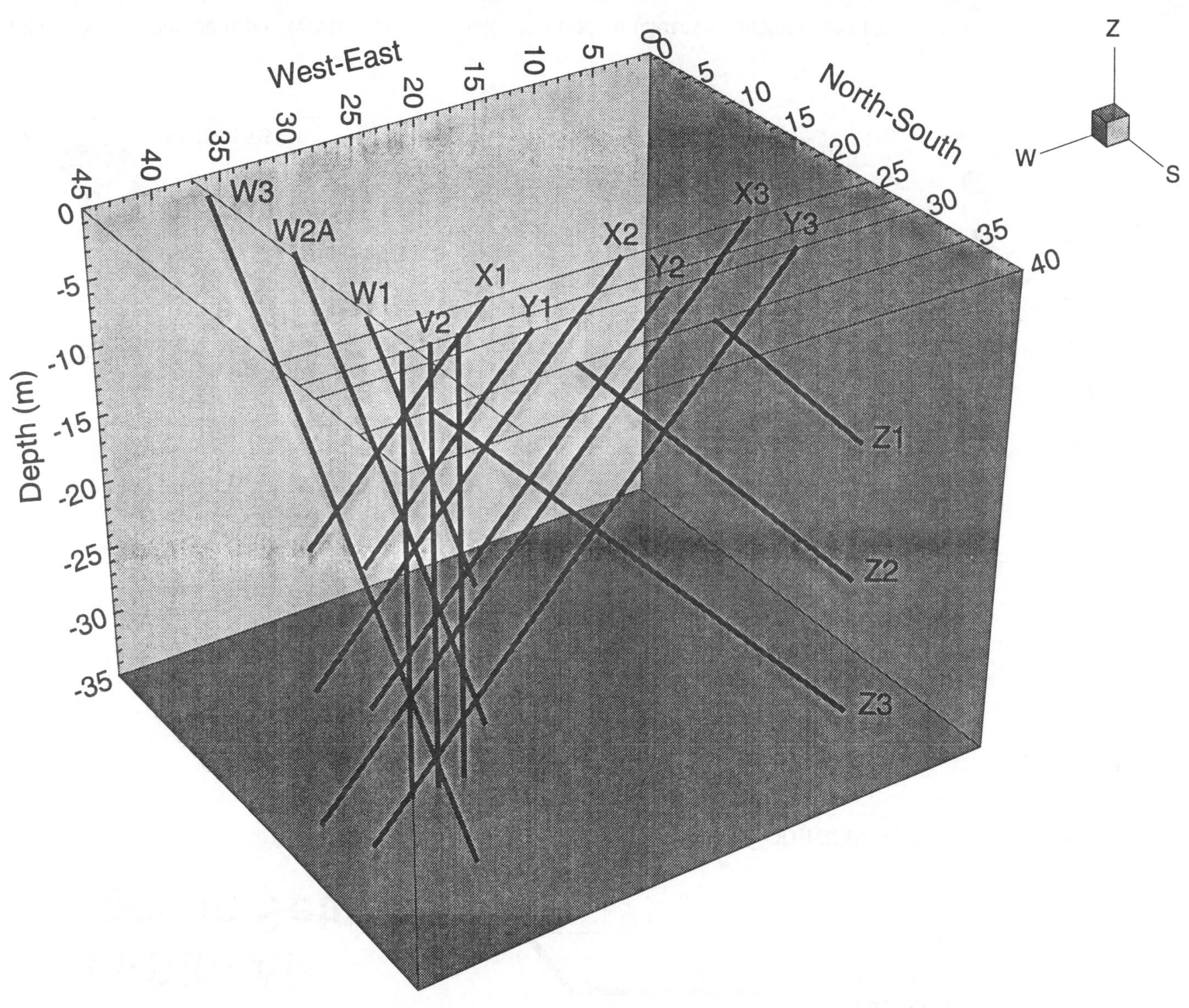

Figure 5. Borehole location at the Apache Leap Tuff Site (from Guzman-Guzman and Neuman, 1994).

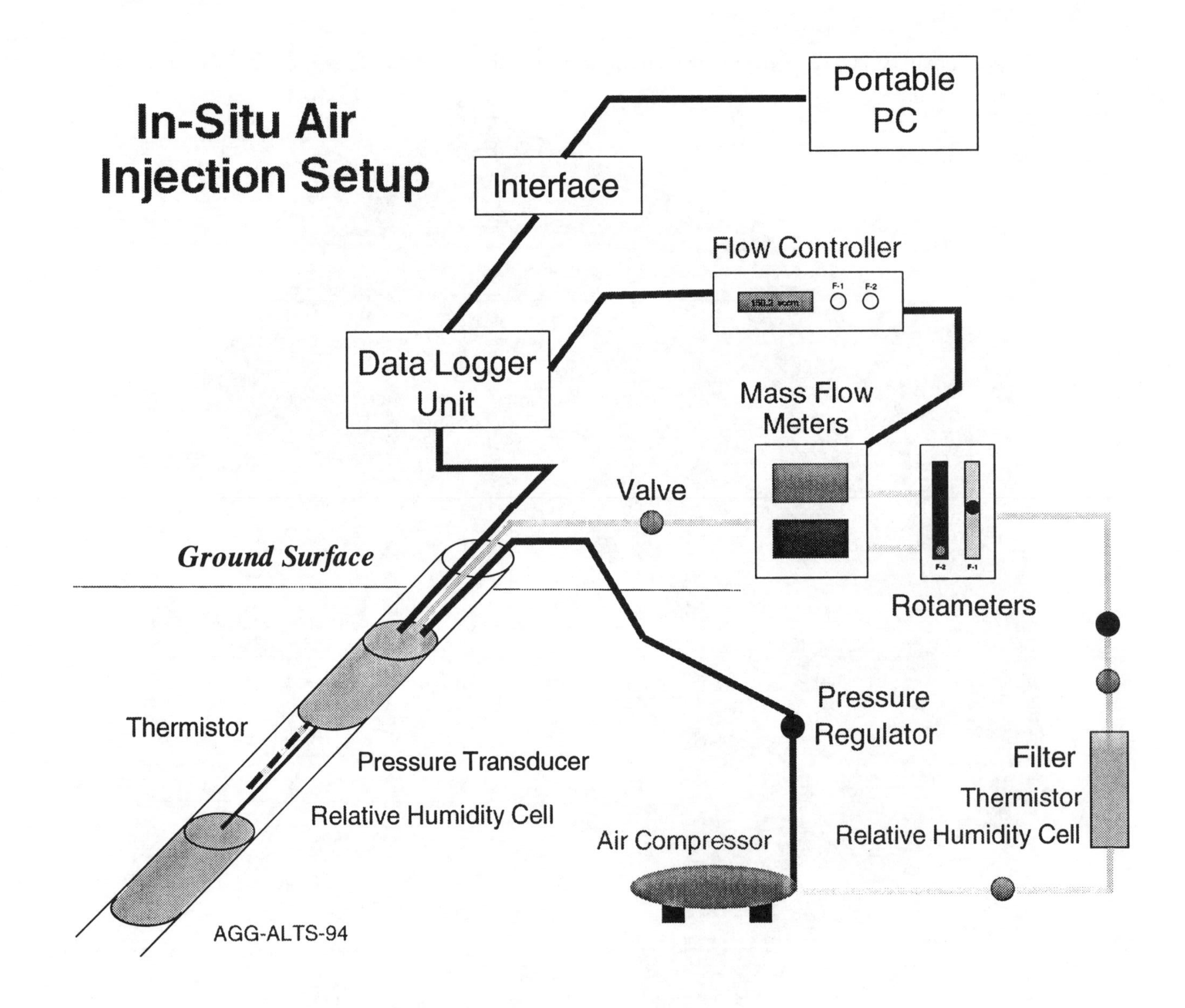

Figure 6. Schematic representation of the air injection system (Guzman-Guzman and Neuman, 1994).

breakthrough velocities. A final estimate of fracture aperture was obtained using the air-entry potential of the saturated fracture.

4. A horizontal fracture imbibition experiment was conducted using water as a fluid imbibed into an initially dry fractured rock to obtain values of cumulative water imbibition volume, and to examine visible wetting front positions.

5. The field air-injection datasets consisted of *in-situ* air-permeability measurements obtained from straddle-packer tests on selected intervals of the boreholes. At the completion of each injection test, there were at least seven different sets of data that could be used to determine the air permeability of the rock surrounding the interval tested; three transient sets during injection; three steady state sets and one recovery set. Datasets consisting of air-permeability measurements at different scales (0.5, 1.0 and 3.0 m) and at multiple-injection rates in six of the boreholes (Guzman-Guzman and Neuman, 1994).

What Models were Tested?

The following models were tested:

1. Characterization models used to define the porosity, characteristic curves, hydraulic conductivity, air permeability and thermal conductivity for the rock matrix using small cores were tested. The models were examined to determine how sensitive they were to a range of water contents, hysteresis and temperatures.

2. Conceptual models describing flow through a single fracture were tested using the large block experiment. Properties values for the matrix and fracture apertures were measured repeatedly and compared to the model estimates.

3. The approximate analytic model of Nitao and Buscheck (1991) was evaluated for its ability to predict the behavior of water imbibition into initially dry fractured rock. The model was evaluated to determine its suitability for use in understanding unsaturated flow.

4. Conceptual models dealing with scale effects and spatial distributions of heterogeneities were tested using datasets from the *in-situ* air-injections studies (Guzman-Guzman and Neuman, 1994).

How were the Models Tested with Experiments?

For the characterization models, the experimental data indicated that variations in temperature affect the shape and position of the moisture-retention curve, and by inference, the shape and position of the relative permeability curve. The wetting history was also shown to have a large influence on the moisture-retention characteristic curve. Thermal conductivity was shown to be only poorly related in a linear fashion to water content. The effects of solute concentrations on ambient matric potential are also demonstrated. It can be concluded that accumulations of saturated salt solutions will increase the osmotic potential which in turn affects the total potential observed under non-isothermal conditions.

For the fracture flow conceptual models, the volumetric apertures estimated using the pycnometer and the tracer breakthrough volumes were closely related. The volumetric aperture determined using the ratio of fracture transmissivity to hydraulic conductivity was less, followed by the apertures determined using the cubic and quadratic equations, respectively. The smallest aperture observed was the capillary aperture. This progression is consistent with the hypothesis that fracture roughness will decrease the effective flow area for the Poiseuille flow, and induce an ink bottle effect at fracture constrictions.

For the approximate analytic model of Nitao and Buscheck (1991), a horizontal fracture imbibition experiment was also conducted using water as a fluid imbibed into an initially dry fractured rock. The imbibition rate was reproduced using a model developed by Nitao and Buscheck [1991]. The form of the model was found to provide a good fit to the shape of the observed data, but the model overestimated the fracture imbibition volume by a factor of twenty and the fracture wetting front advance by a factor of eight. The noted reduction in water inflow may be due to phenomena neglected in the theoretical model, such as fracture surface coatings or enhanced surface weathering, and the inability to accurately determine fracture physical properties *a priori*, such as the fracture water diffusivity. It was shown that fracture saturation behind the wetting front initially is very low, perhaps ten percent, but increases to complete saturation during the course of the experiment. This may indicate that fingers of saturation exist within the fracture during early time, and these fingers expand laterally and dissipate over time.

Data from the imbibition experiment reported here confirm the second imbibition phase as hypothesized by the Nitao and Buscheck model. The experiment was not able to distinguish either the first or third imbibition phase of their model. A new imbibition phase was observed, however, which resulted from the finite length of the fracture within the tuff. The Nitao and Buscheck model should be modified to incorporate the finite extent of discrete fractures. Another concern raised by the experiment was the failure to properly estimate fracture hydraulic properties. It is observed that laboratory estimates of rock fracture hydraulic properties, when used with the Nitao and Buscheck model, substantially overestimated the cumulative imbibition rate, and the rate of advance of the wetting front in the fractured block. Calibrated values of the fracture hydraulic parameter are substantially less than the characterization value. An additional shortcoming of the model is the inability to reproduce the observed fingering of water within the fracture, although the fingering was limited only to the early fracture imbibition period. Fingering may be more important when vertically oriented fractures are present.

For the conceptual models dealing with scale effects and spatial distributions of heterogeneities, in-situ air-injections tests were performed over a range of scales (i.e., 0.5, 1.0 and 3.0 m) and at multiple-injection rates. Field data indicate that the air permeability determinations are strongly affected by two-phase interaction between air and pore water, and in higher permeability zones by inertial flow effects. A 45-degree, 30-meter deep borehole was tested for permeability at three different scales to study the effect of measurement support on permeability estimates and their statistics (Guzman-Guzman and Neuman, 1994). These measurements seem to indicate some dependency of the mean permeability on measurement support (length of test interval), a phenomenon known as "scale effect." Upscaling by weighted arithmetic averaging of the smaller measurement support data produces better estimates than geometric weighted averaging. High permeability values are, however, slightly underpredicted by either upscaling approach. Although the observed variability of air permeabilities at the ALTS is over 3.5 orders of magnitude, the data are amenable to classical geostatistical analysis and yield well-defined semivariograms. The omni-directional semivariogram exhibits a nested structure with two distinct plateaus and correlation scales, and an additional correlation structure whose sill and range are undefined due to the limited extent of the experimental site (Guzman-Guzman and Neuman, 1994). It was also observed that the increase in the variance and correlation scale which grew with the scale, is consistent with the multi-scale continua concept discussed by Burrough (1983), and Neuman (1987, 1990, 1993, 1994). The available fractured rock permeability data can be viewed as a sample from a random (stochastic) field defined over a continuum with multiple scales of heterogeneity (Guzman-Guzman and Neuman, 1994).

Model Comparison and Experimental Results

Results of laboratory experiments conducted to characterize fluid and thermal flow parameters of unsaturated Apache Leap Tuff indicate that hysteresis influences the moisture characteristic curve. Wetting and drying characteristic curves are markedly different, with the wetting curve consistently showing higher matric potentials at equivalent water contents. Efforts to identify the matric potential from water contents of unsaturated rock will require knowledge of the water content history of the site. The successful application of osmotic solutions to maintain constant matric potentials was demonstrated. Saturated salt solutions present in the geologic environment may affect the observed matric potential. Near a repository, accumulations of soluble salts may affect the migration of liquid and vapor due to the osmotic potential induced at high salt concentrations. Coupling of salt concentrations with water activity should be an integral component of simulation models of fluid flow near the waste repository.

Temperature is shown to affect the characteristic curve. Both reduced and increased temperatures cause substantial shifts in the characteristic curve, attributable to the change in the temperature dependence of the fluid surface tension. Coupling of hysteresis effects with temperature changes was not evaluated, nor were changes in the characteristic curves evaluated as a function of dynamic temperature changes. Additional characterization studies will be required to address the effects of temperature fluctuations on characteristic curves.

The relative permeabilities for air and water were determined using rock cores. Estimates of permeabilities were obtained under isothermal conditions. Additional experiments will be required to evaluate the importance of temperature on water and air relative permeability functions.

The influence of water content on the thermal conductivity was examined using a one-dimensional heat cell. A linear relationship between water content and thermal conductivity was not clearly demonstrated. Observed mean thermal conductivities were less than expected for the range of volumetric water contents from 0 to 0.0876. Additional studies will be required to investigate the nature of the unsaturated thermal conductivity relationship, and the influence of hysteresis on the relationship.

Laboratory experiments conducted to observe thermal, liquid, vapor and solute transport through variably saturated, fractured Apache Leap Tuff demonstrate that:

1. Conduction is the dominant heat transport mechanism even when a significant heat pipe effect is present.

2. Water contents increase away from the heat source due to vapor driven advection and condensation.

3. Solutes accumulate near the heat source, but the accumulation of solutes increases the osmotic potential which decreases the heat pipe phenomenon.

4. The heat pipe process may not significantly affect thermal or liquid flow in materials similar to the Apache Leap Tuff samples examined.

5. Solute transport was substantially affected by the heat pipe phenomenon, resulting in the accumulation of significant solutes nearer the heat source than would have occurred if the heat pipe had not been present.

6. Models of heat and liquid flow near high level waste repositories may not need to incorporate heat pipe effects.

7. Models of solute transport should incorporate the heat pipe phenomenon, and should also consider the effects of osmotic potential on liquid and vapor transport.

These observations may only be relevant to the conditions examined. Additional laboratory and computer simulation experiments should be conducted to evaluate the effects of coupled thermal, liquid, vapor and solute transport over a wider range of material properties. Also, the effects of thermomechanical, geochemical, biogeochemical, and radiation-induced changes will also require examination. It is possible that processes not yet considered may significantly affect the migration of radionuclides in the region immediately adjacent to the waste repository. Field and laboratory-scale experiments are necessary to identify these unknown processes.

Table 3 presents estimated characterization properties of the rock matrix and the embedded fracture. Several parameters, including the fracture porosity, liquid saturation changes across the wetting front in the fracture and rock matrix are assumed values. Table 3 also presents characterization parameters with their uncertainties. Uncertainties in the derived parameters were estimated by propagating parameter uncertainties using first-order Taylor series approximations. A first-order approximation of parameter uncertainty propagation was estimated using the Taylor-series expansion of the input errors.

Table 3. Fractured block characterization parameters (from Rasmussen et al., 1994).

		mean ± std. dev.	
Rock Matrix Properties:			
V	rock volume	39,240 ± 0	cm^3
V_t	pore plus fracture volume	4,635 ± 120	cm^3
V_m	matrix pore volume	4,493 ± 127	cm^3
θ_m	porosity	0.115 ± 0.003	
ΔS_m	liquid saturation change	1 ± 0	(1)
D_m	water diffusivity coefficient	3.61 ± 0.28	$cm^2\ hr^{-1}$
D_g	argon gas diffusion coefficient	31.0 ± 0.94	$cm^2\ hr^{-1}$
Rock Fracture Properties:			
V_f	volume	142.3 ± 41.7	cm^3
w	width	20.2 ± 0	cm
a	fracture-boundary distance	10.5 ± 3	cm
b	half-aperture	381 ± 11	µm
L	length	92.5 ± 0	cm
θ_f	porosity	1 ± 0	(1)
ΔS_f	liquid saturation change	1 ± 0	(1)
K_f	hydraulic conductivity	9650 ± 504	$cm\ hr^{-1}$
T_f	transmissivity	490 ± 25.2	$cm^2\ hr^{-1}$
D_f	water diffusivity coefficient		

Note: (1) Assumed value.

Characterization techniques which demonstrate promise for estimating material properties on field scales include the use of a pycnometer to measure fracture and matrix porosities, and gas-phase tracer experiments to estimate the fracture/matrix porosity ratio, the permeability distribution, and the porosity-length distribution. While these indices are only strictly appropriate for gas-phase transport, inferences to liquid phase transport may be derived if relationships between gas and liquid phase transport are known. It is anticipated that gas-phase testing using tracers will become a rapid and effective tool for characterizing macropores on field scales. The interactions between matric storage and advection through fractures have been demonstrated in the laboratory, and field scale experiments are being explored to apply this new technique.

Interpretation of fracture aperture estimates is complicated by the observation that the estimated value is a function of the method employed to provide the estimate. Six measures of fracture aperture were developed and comparisons were made between methods. It was observed that volumetric measures of fracture aperture yield the highest values, with lower estimates provided by measures using Poiseuille's law. The lowest estimate was obtained using capillary theory. It can be concluded that when fracture aperture measurements are reported, the method employed to provide the estimate should also be indicated.

Uncertainty measures of characterization parameters are also presented in Rasmussen et al. (1994). The uncertainty in the measured parameter are required to evaluate the uncertainty in predictions based upon the parameter. Forecasts of flow and transport will require measures of uncertainty in the forecast. Uncertainties in estimated parameters may contribute to a large errors in forecasts.

Based on an extensive data set consisting of steady state apparent air permeability values, the following conclusions are presented in Guzman-Guzman and Neuman (1994). The apparent air permeability from straddle-packer tests is a strong function of the applied pressure. Changes in air permeability with pressure are due to two-phase flow and, in some cases to inertial flow. Computer simulations confirmed the two-phase flow explanation. Upscaling of the apparent permeability is accomplished best via weighted arithmetic averaging. Geostatistical and statistical analyses indicate that the apparent permeability data from ALTS behave as a stochastic multiscale continuum with an echelon and power-law (fractal) structure. The latter is associated with a Hurst coefficient $w = 0.28$ to 0.29 which is remarkably close to the generalized value $w = 0.25$ predicted by Neuman [1990, 1994]. Additional permeability tests spanning larger rock volumes at ALTS would help to determine whether the seemingly fractal behavior extends beyond the scales already tested.

As presented in Guzman-Guzman and Neuman (1994) analysis results strongly suggest that site characterizations must be based on hydrogeologic data collected on a spectrum of scales relevant to performance assessment. They further point out the need to consider two-phase flow and inertia effects in the interpretation of air injection tests. The transient part of these tests may hold the key to site evaluation of functional relationships between rock permeability, fluid pressure and saturation. The ALTS investigators report that the inverse methods hold a promise in this regard, and propose to use them in the context of their ALTS data.

Were Alternative Conceptual Models Tested and Assessed?

The alternative conceptual models considered were an equivalent porous medium model and a model that considered the discrete fracture network embedded in a porous matrix. The models were tested by estimating the physical, hydraulic, pneumatic, and thermal properties for the discrete fracture, and then forming forecasts using the alternate conceptual and mathematical models. The forecasts for each conceptual model includes the 95% forecast confidence interval about the forecast, thus allowing assessment of the accuracy of the forecast when compared against the experimental result. The confidence intervals were generated by first measuring the uncertainties in the input parameters, and then propagating those input parameter uncertainties through the model using a Taylor Series expansion. The model outputs were then compared against experimental results. Because the experimental results also contain uncertainties, the observed confidence intervals were also generated.

How were Different Disciplines Integrated?

The test program utilized expertise from various disciplines as consultants. The core personnel were hydrologists, with support from individuals trained in geology, engineering (civil, chemical, mechanical, nuclear and geological), rock mechanics, and chemistry. The integration occurred through experimental design meetings and reports, discussions of experimental and modeling results, and joint field trips for examining and collecting flow and transport property measurements.

How was Success Defined and Assessed?

Success in experimental design and model performance was evaluated in two ways, heuristically and statistically. The models were tested heuristically by noting whether reasonable relationships between inputs and outputs were observed. Input parameters were varied and outputs were examined to survey the trend in response to combinations of inputs. The experiments were examined heuristically to note whether the measured responses were consistent with *a priori* estimates of system responses. Several defective experiments were found using this method.

The statistical evaluation consisted of comparing model forecast confidence intervals with the experimental results. If the experimental results were found to lie within 95% confidence region for the model forecast, then the model could not be rejected.

Acknowledgement

The authors wish to acknowledge the leadership and organizational work of Dr. Peter J. Wierenga, Chair, Department of Soil and Water Science, University of Arizona who selected the site, designed and conducted the Las Cruces Trench field experiments. Similarly, we wish to credit: (1) the modeling insights and calculations that assisted in the experimental design by Dr. Lynn Gelhar and Dr. Dennis McLaughlin from the Massachusetts Institute of Technology, and Dr. Glendon Gee from Pacific Northwest Laboratory, and (2) the field experimentalists who constructed, monitored and sampled the experiments were Warren Strong, Alex Toorman, David Hudson, Indrek Porro, Mike Kirkland, Maliha Nash, Joe Vinson, Mike Young, Ricardo Carrasco, Jaime Castillo, and David Pickens.

For the Apache Leap Tuff Site studies we wish to acknowledge the leadership of Dr. Daniel D. Evans, Professor Emeritus, University of Arizona who began the work and devoted so much scientific and motivational support. In Phase II, Drs. Randy Bassett, Shlomo P. Neuman and Michael Sully provided scientific direction and support. We also wish to recognize the many laboratory and field experimentalists who assisted in collecting and analyzing the ALTS data; Shirlee Rhodes, Charles Lohrstorfer, James Blanford, Priscilla Sheets, Ingrid Anderson, James Devine, Michael Henrich, Dick Thompson, Michael Getis, Ernie Hardin and Greg Davidson.

Both the LCT and ALTS studies were supported by the U.S. Nuclear Regulatory Commission's Office of Nuclear Regulatory Research involving numerous contractors and NRC staff.

References

Las Cruces Trench

Elrick, D.E., R.W. Sheard, and N. Baumgartner, "A Simple Procedure for Determining the Hydraulic Conductivity and Water Retention of Putting Green Soil Mixtures," in R.W. Sheard (ed), Proceedings of the Fourth International Tuffgrass Research Conference, The Ontario Agricultural College, University of Guelph, Guelph, Ontario, 1980, pp. 189-200.

Gee, G.W. and J.W. Bauder, "Particle-size analysis," In A. Klute (ed.) Methods of Soil Analysis, Part 1, 2nd ed. ASA, SSSA, Madison, WI, Agronomy, Vol. 9, 1986, pp. 383-411.

Hills, R.G., I. Porro, D.B. Hudson and P.J. Wierenga, "Modelling One-Dimensional Infiltration into Very Dry Soils - Part 1: Model Development and Evaluation," Water Resources Research, Vol. 25, No. 6, June 1989a, pp. 1259-1269.

Hills, R.G., D.B. Hudson, I. Porro, and P.J. Wierenga, "Modelling One-Dimensional Infiltration Into Very Dry Soils - Part 2: Estimation of the Soil-Water Parameters and Model Predictions," Water Resources Research, Vol. 25, No. 6, June 1989b, pp. 1271-1282.

Hills, R.G., K.A. Fisher, M.R. Kirkland, and P.J. Wierenga, "Application of flux-corrected transport to the Las Cruces Trench site," Water Resources Research, Vol. 30, No. 8, August 1994, pp. 2377-2386.

Hills, R.G. and P.J. Wierenga, "INTRAVAL Phase II Model Testing at the Las Cruces Trench Site," NUREG/CR-6063, U.S. Nuclear Regulatory Commission, Washington, DC, January 1994.

Mantoglou, A. and L.W. Gelhar, "Stochastic Modeling of Large-Scale Transient Unsaturated Flow Systems," Water Resources Research, Vol. 23, No. 1, January 1987, pp. 37-46.

Nicholson, T.J., P.J. Wierenga, G.W. Gee, E.A. Jacobson, D.J. Polmann, D.B. McLaughlin, and L.W. Gelhar, "Validation of Stochastic Flow and Transport Models for Unsaturated Soils: Field Study and Preliminary Results," in Buxton, B.E. (Editor), Proceedings of the Conference on Geostatistical, Sensitivity, and Uncertainty Methods for Ground-Water Flow and Radionuclide Transport Modeling, September 15-17, 1987, San Francisco, CA, Battelle Press, Columbus, Ohio, 1989.

Voss, C. and T.J. Nicholson, Technical Editors; A. Flint, R. Hills, and T.C. Rasmussen, Primary Contributors, "The International INTRAVAL Project Phase I Test Cases 10, 11 & 12: Flow and Transport Experiments in Unsaturated Tuff and Soil," Nuclear Energy Agency, Organization for Economic Cooperation and Development, Paris, France, 1993.

Porro, I., "Solute Transport through Large Unsaturated Soil Columns," Dissertation, Department of Agronomy and Horticulture, New Mexico State University, Las Cruces, New Mexico, 1989.

Porro, I. and P.J. Wierenga, "Transport of Reactive Tracers by Unsaturated Flow using Field and Column Experiments," Proceedings: Radionuclide Adsorption Workshop, September, Los Alamos, New Mexico, 1992, pp. 111-123.

Reynolds, W.D., D.E. Elrick, N. Baumgartner, and B.E. Clothier, "The Guelph Permeameter for Measuring Field-Saturated Soil Hydraulic Conductivity above the Water Table. II. The Apparatus," Proc. Canadian Hydrology Symposium, Quebec City, Quebec, 1984.

Rockhold, M. L., "Simulation of Unsaturated Flow and Nonreactive Solute Transport in a Heterogeneous Soil at the Field Scale," NUREG/CR-5998, (PNL-8496), U.S. Nuclear Regulatory Commission, Washington, D.C., February 1993.

Rockhold, M.L, R. Rossi, R. Hills and G.W. Gee, "Similar-Media Scaling and Geostatistical Analysis of Soil Hydraulic Properties," in Baalman, R.W. and D. Felton (Editors), Proceedings of the 33rd Hanford Symposium on Health and the Environment, In-Situ Remediation: Scientific Basis for Current and Future Technologies, Richland, WA, November 7-10, 1994 (In Press).

Wierenga, P.J., L.W. Gelhar, C.S. Simmons, G.W. Gee and T.J. Nicholson, "Validation of Stochastic Flow and Transport Models for Unsaturated Soils: A Comprehensive Field Study," NUREG/CR-4622, U.S. Nuclear Regulatory Commission, Washington, DC, August 1986.

Wierenga, P.J., D.B. Hudson, R.G. Hills, I. Porro, J. Vinson, and M.R. Kirkland, "Flow and Transport at the Las Cruces Trench Site: Experiments 1 and 2," NUREG/CR-5607, U.S. Nuclear Regulatory Commission, Washington, DC, 1990.

Wierenga, P.J., A.F. Torman, D.B. Hudson, M. Nash, and R.G. Hills, "Soil Physical Properties at the Las Cruces Trench Site," NUREG/CR-5441, U.S. Nuclear Regulatory Commission, Washington, DC, 1989.

Wierenga, P.J., M.H. Young, G.W. Gee, R.G. Hills, C.T. Kincaid, T.J. Nicholson, and R.E. Cady, "Soil Characterization Methods for Unsaturated Low-Level Waste Sites," NUREG/CR-5988, U.S. Nuclear Regulatory Commission, Washington, DC, February 1993.

Wittmeyer, G.W. and B. Sagar, "Model Complexity and Model Validity: Application to the Las Cruces Trench Experiment, INTRAVAL Test Case 10, " Proceedings Site Characterization and Model Validation, Focus '93, American Nuclear Society, La Grange Park, IL, 1993, pp. 209-216.

Apache Leap Tuff

Bassett, R.L., S.P. Neuman, T.C. Rasmussen, A. Guzman, G.R. Davidson, and C.F. Lohrstorfer, "Validation Studies for Assessing Unsaturated Flow and Transport through Fractured Rock," NUREG/CR-6203, U.S. Nuclear Regulatory Commission, Washington, DC, August 1994.

Burrough, P.A., "Multiscale Sources of Spatial Variation in Soil. I. The Application of Fractal Concepts to Nested Levels of Soil Variation", Journal of Soil Science, Vol. 34, 1983, pp.577-597.

Guzman-Guzman, A. and S.P. Neuman, "Field Air Injection Experiments," Chapter 6 in Rasmussen, T.C., S.C. Rhodes, A. Guzman and S.P. Neuman, "Apache Leap Tuff INTRAVAL Experiments: Results and Lessons Learned," NUREG/CR-6996, U.S. Nuclear Regulatory Commission, Washington, DC, 1994 (In press).

Neuman, S.P., "Stochastic Continuum Representation of Fractured Rock Permeability as an Alternative to the REV and Fracture Network Concepts", in Proceedings, Memoirs of the 28th US Symposium on Rock Mechanics, Tucson, Az, 1987.

Neuman, S.P., "Universal Scaling of Hydraulic Conductivities and Dispersivities in Geologic Media", Water Resources Research, Vol. 26, No. 8, 1990, pp.1749-1758.

Neuman, S.P., Comment on "A Critical Review of Data on Field-Scale Dispersion in Aquifers" by L.W. Gelhar, C. Welty, and K.R. Rehdat, Water Resources Research, Vol. 29, No. 6, 1993, pp.1863-1865.

Neuman, S.P., "Generalized Scaling of Permeabilities: Validation and Effects of Support Scale," Geophysical Research Letters, Vol. 21, No. 5, pp. 349-352, March 1, 1994.

Nitao, J.J. and T.A. Buscheck, "Infiltration of a Liquid Front in an Unsaturated, Fractured Porous Medium", Water Resources Research, Vol. 27, No. 8, 1991, pp. 2099-2122.

Pruess, K. and J.S.Y. Wang, "Numerical Modeling of Isothermal and Non-isothermal Flow in Unsaturated Fractured Rock - A Review," in Evans, D.D. and T.J. Nicholson (Editors), Flow and

Transport Through Unsaturated Fractured Rock, Geophysical Monograph 42, American Geophysical Union, Washington, DC, 1987.

Rasmussen, T.C. and D.D. Evans, "Unsaturated Flow and Transport Through Fractured Rock - Related to High-Level Waste Repositories: Final Report - Phase II," NUREG/CR-4655, U.S. Nuclear Regulatory Commission, Washington, DC, May 1987.

Rasmussen, T.C., D.D. Evans, P.J. Sheets and J.H. Blanford, "Unsaturated Fractured Rock Characterization Methods and Data Sets at the Apache Leap Tuff Site," NUREG/CR-5596, U.S. Nuclear Regulatory Commission, Washington, DC, August 1990.

Rasmussen, T.C., S.C. Rhodes, A. Guzman and S.P. Neuman, "Apache Leap Tuff INTRAVAL Experiments: Results and Lessons Learned," NUREG/CR-6996, U.S. Nuclear Regulatory Commission, Washington, DC, 1994 (In press).

Yeh, T.C., T.C. Rasmussen and D.D. Evans, "Simulation of Liquid and Vapor Movement in Unsaturated Fractured Rock at the Apache Leap Tuff Site: Models and Strategies," NUREG/CR-5097, U.S. Nuclear Regulatory Commission, Washington, DC, March 1988.

Conclusions from WG-2 :
The Analyses of the Finnsjön Experiments

Peter Andersson
GEOSIGMA AB

Anders Winberg
Conterra AB

(Sweden)

Abstract

A comprehensive series of tracer tests on a relatively large scale have been performed by SKB at Finnsjön, Sweden, to increase understanding of transport phenomena which govern migration of radionuclides in major fracture zones. The experimental sequence of tracer tests consisted of; a preliminary tracer test during hydraulic interference tests, a radially converging test and a dipole test. Both sorbing and slightly sorbing tracers were used. The conducted experiments were subsequently selected as a test case for the international INTRAVAL Project, partly because the tests performed at Finnsjön invite to direct address of validation of geosphere models. Work on the Finnsjön test case during Phase 1 of INTRAVAL involved 10 models ranging from advection-dispersion approaches to more complex fracture network approaches. A main conclusion from this phase was that tracer test data from one or two tests at a given site are not sufficient to discriminate between models and active processes. The Phase 2 studies, which involved nine project teams from seven countries, are dominated by porous media approaches in two dimensions, although some project teams utilized one-dimensional and even three-dimensional approaches on a larger scale. The dimensionality employed did not appear to be decisive for the ability to reproduce field responses. It was also demonstrated that stochastic approaches can be used in a validation process. The general conclusion is that flow and transport in the studied fracture zone is governed by advection and that hydrodynamic dispersion is needed to explain the breakthrough curves. Matrix diffusion is assumed to have a small or negligible effect. A variety of validation aspects have been considered. Five teams utilized a model calibrated on one test, to predict another, whereas two teams using the stochastic continuum approach addressed; 1) validity of extrapolating a model calibrated on one transport scale to a larger scale, 2) performance assessment implications of choice of underlying distribution model for hydraulic conductivity, respectively. As a final exercise to assess and illustrate differences in predictive capability between models, a synthetic natural gradient tracer test case was conceived and addressed by some of the modelling teams. The purpose not being to select the best model but rather to quantify the uncertainty associated with alternative calibrated models used to describe the Finnsjön tracer tests when extrapolated to larger transport scales.

1. INTRODUCTION

This INTRAVAL test case is based on three hydraulic interference tests and two tracer experiments performed in a major low angle fracture zone (Zone 2) at the Finnsjön research area, Sweden /Andersson et al., 1989/, /Gustafsson & Andersson, 1991/, /Gustafsson & Nordqvist, 1993/, /Andersson et al., 1993/. The main objectives with the experiments were to determine parameters important for radionuclide transport in major fracture zones and to utilize the results for calibration and verification of radionuclide transport models. These experiments were selected as a test case in the INTRAVAL project since they were designed to study phenomena important in geosphere transport such as advection, dispersion, channelling, dilution, matrix diffusion, heterogeneity on a rather large geometrical scale.

During Phase 1 of the INTRAVAL Project a total of seven project teams studied and analyzed the data with varying amount of detail. The Phase 1 results are reported by Tsang & Neuman (editors) (1992). The major conclusion drawn was that data from one or two tracer tests alone are not sufficient to distinguish between different models and/or processes.

The analysis of the Finnsjön test case was prolonged also during Phase 2 of the INTRAVAL Project, this decision partly augmented by the fact that INTRAVAL Phase 2 put special focus on field experiments. At the end of Phase 2 a comparative study based on a hypothetical natural gradient experiment was initiated to illustrate the predictive uncertainty of the different modelling concepts.

Nine project teams from seven countries, including the Pilot Group (GEOSIGMA) who developed this test case, studied and analyzed the data. Three of the groups also participated in Phase 1, c.f. Table 1. The processes studied and the conceptual approaches used by the eleven teams are summarized in Table 2. The Phase 2 analysis of the Finnsjön test case is more thoroughly described by Andersson & Winberg (1994).

Table 1. **Teams modelling the Finnsjön Test Case in Phase 2 of INTRAVAL.**

Modelling team	Phase 1	Phase 2	Comparative Study
GEOSIGMA / SKB, Sweden	X	X	X
VTT / TVO, Finland	X	X	X
PNC, Japan		X	X
PSI / NAGRA, Switzerland		X	
U. of New Mexico, USA		X	
Hazama Co, Japan	X	X	
Conterra / KTH / SKB, Sweden		X	
BRGM / ANDRA, France		X	
UPV / ENRESA, Spain		X	
JAERI, Japan	X		X
Ecole de Mines de Paris, France	X		X

Table 2. **Conceptual approaches, modelling objectives and processes considered by the teams modelling the Finnsjön Test Case in Phase 2 of INTRAVAL.**

Modelling team	Conceptual approach	Modelling objectives	Processes considered
GEOSIGMA	continuum model	comparison prediction-experimental data (2D) evaluation and comparison of transport parameters (1D)	advection-dispersion, diffusion, sorption, matrix diffusion, radioactive decay
VTT	non-interacting varying aperture channel model	realistic description of flow and transport in rock, relative weight of processes, parameter consistency	advective diffusion, matrix diffusion, generalized Taylor dispersion, multiple flow paths
PNC	dual porosity continuum model with mixing zone, stream tube concept	effect of heterogeneity on transport	advection-dispersion, multiple flow paths
PSI	single and dual porosity continuum model	identification of dominant transport processes, effect of varying flow boundaries	advection-dispersion, multiple flow paths, sorption, matrix diffusion
U. of NM	single and dual porosity continuum model	model discrimination	advection-dispersion, molecular diffusion, matrix diffusion
Hazama	crack tensor theory	flow and transport analysis using REV approach	advection
Conterra/KTH	stochastic continuum multiGaussian	test on a large scale of a stochastic continuum flow model calibrated on a local scale	advection
BRGM	continuum model	evaluation, comparison between analytical and numerical modelling, effects of boundaries and layering	advection-dispersion, kinematic dispersion, radioactive decay
UPV	stochastic continuum, multi-Gaussian, non-multiGaussian	impact of low continuity of extreme values on travel times	advection
JAERI	continuum and variable-aperture channel models	model discrimination	advection-dispersion, local dispersion, multiple flow paths, matrix diffusion
EdM	continuum model	evaluation, comparison of transport parameters, comparison of prediction-experimental data	advection-dispersion

2. DESCRIPTION OF THE FINNSJÖN EXPERIMENTS

At the Finnsjön investigation site, Sweden, two major fracture zones in granitic rock have been identified, the Brändan fracture zone (Zone 1), and a low-angle zone (Zone 2), c.f. Figure 1. The geohydrology of the site is dominated by these two highly conductive zones. Zone 2, which is the zone utilized for the tracer tests, is trending north with a dip of about 16° to the west and consists of sections with high fracture frequency and tectonisation. The zone is well defined in 7 boreholes located within an area of about 500×500 m. In this area the fracture zone is almost planar with the upper surface located between 100 to 240 m below ground surface. The zone is about 100 m thick and consists of three subzones with transmissivities in the order of 10^{-5} to 10^{-3} m^2/s. The parts in between the subzones have transmissivities similar to the country rock. The magnitude of the hydraulic gradient in the zone is about 0.3%, directed towards ENE.

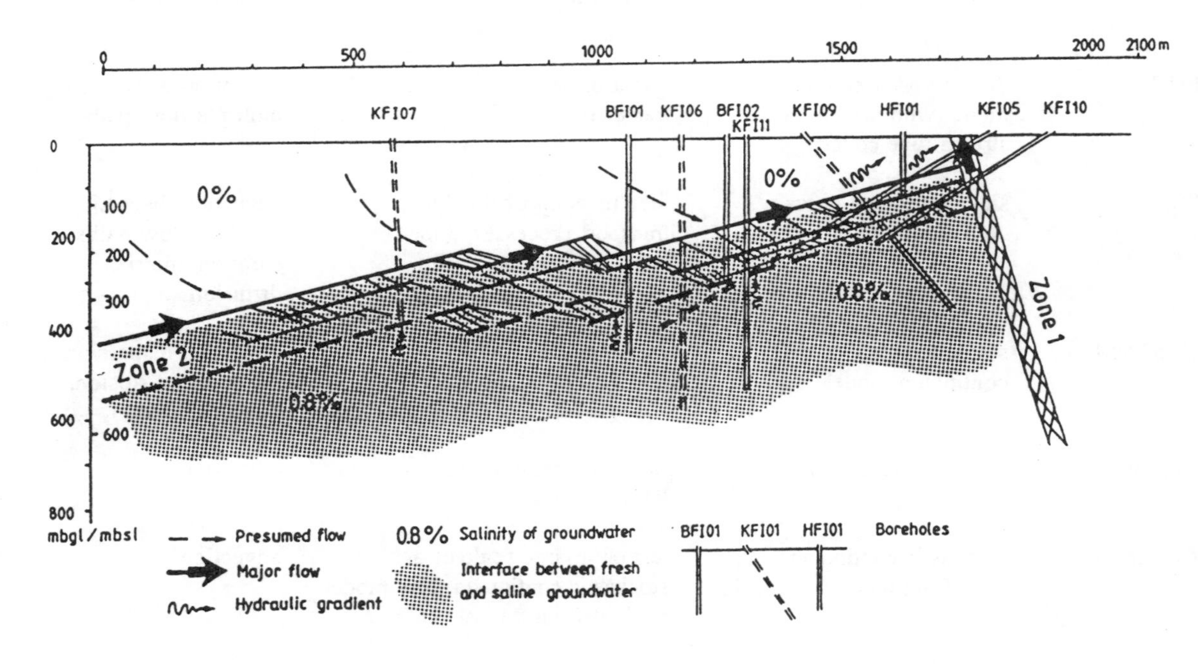

Figure 1. **Cross-section through the Brändan area showing Zone 2.**

Three hydraulic interference tests and two large-scale tracer tests were performed in Zone 2. The interference tests were carried out by withdrawing water from different isolated intervals of Zone 2 in borehole BFI02 /Andersson et al., 1989/. During one of the interference tests a preliminary tracer test was performed. Tracers were injected as pulses in the upper highly conductive subzone of Zone 2 in boreholes BFI01, KFI06 and KFI11. Tracer breakthroughs were monitored in the pumping well BFI02.

A radially converging tracer experiment was performed by pumping in borehole BFI02 and injecting tracers into Zone 2 in three peripheral holes, BFI01, KFI06 and KFI11 /Gustafsson & Nordqvist, 1993/. In each injection hole, three sections were isolated, one in the upper highly conductive part of Zone 2, one at the lower boundary of Zone 2 and one at the most conductive part in between. In the withdrawal hole the isolated section covered the whole thickness of Zone 2. In total 11 different tracers were injected, 8 of them continuously for 5-7 weeks, and three as pulses. A detailed sampling of the withdrawal hole was also made to show the existence of interconnections between highly conductive intervals of Zone 2.

A dipole tracer experiment was performed in a recirculating system between BFI01 (injection) and BFI02 (withdrawal) /Andersson et al., 1993/. Boreholes KFI06 and KFI11 were used as observation holes. Only the upper highly conductive part of Zone 2 was used for tracer injection in borehole BFI01. In total 15 injections of tracers were made in BFI01 including both non-sorbing and sorbing species. Two tracer injections were also made in the upper section of the observation hole KFI11. Tracers were sampled in the upper part of Zone 2 in holes BFI02, KFI06 and KFI11. During the experiment temperature, electrical conductivity and redox potential of the water entering BFI01 was measured showing stable and reducing conditions.

3. MODELLING OBJECTIVES

The objectives stated by the modelling teams carrying out analysis during INTRAVAL Phase 2 are varied in focus and in level of ambition, c.f. Table 2. Keywords in the stated objectives are 1) description of flow and transport, 2) application of simple models, 3) identification of processes and their relative importance, 4) effects of heterogeneity in material properties and hydraulic boundaries, 5) consistency in evaluated transport parameters when analyzing different tracer tests. Surprisingly enough, most of the modelling teams did not explicitly state model validation as an objective although most teams address model validation in some way, c.f. Table 3.

4. CONCEPTUAL APPROACHES

The conceptual models applied by the modelling teams are also in the Phase 2 analysis focused on porous media approaches, c.f. Table 2. In Phase 1, five of the seven teams used porous media approaches while in Phase 2 seven of the nine teams used this concept. The two exceptions, in the latter case, are the VTT team, using a network of channels where transport is assumed to take place in a few non-interacting channels, and the Hazama team using a concept based on a representative elementary volume (REV) obtained by applying the Crack Tensor Theory.

Some of the teams also compare different conceptual approaches. The PNC team uses two different ways to determine the hydraulic conductivity distribution and uses both one-dimensional stream tube and two-dimensional finite difference methods to analyze transport. The UPV team compares two stochastical models, Gaussian/non-Gaussian, and the PSI team applies both fracture and vein flow models. The U. of New Mexico team compares single- versus double porosity models. Some of the teams use the comparison to conclude which conceptual approach that gives the best correspondence to the experimental data and which of the models to be, at least partly, rejected.

The models used are mainly two-dimensional which is an obvious assumption given the two-dimensional character of flow in Zone 2. Some teams use one-dimensional transport concepts and two teams use three-dimensional approaches. If all tracer tests are to be considered, a three-dimensional model may be needed, especially if the large scale head responses are to be incorporated. Experimental evidence of vertical interconnections between different subzones and leakage from the bedrock below the zone also speaks in favour of a three-dimensional approach. However, the relatively simple one-dimensional approaches may also be useful in some cases, e.g. in the analysis of the radially converging test, where variations in source terms and effects of multiple flow paths and matrix diffusion are effectively simulated. The dimensionality of the model does not seem to be decisive for the ability to reproduce the field responses at Finnsjön.

A major difference compared to Phase 1 is that geostatistical approaches have been introduced. In Phase 2 four modelling teams, Conterra/KTH-WRE, UPV, Hazama and PNC have used geostatistical methods to obtain transmissivity or aperture distributions for stochastic travel time analysis. These teams have demonstrated that stochastic approaches may be used within the context of a validation process, although the question remains how to formally validate a stochastic continuum model.

5. PROCESSES STUDIED

In the Phase 1 analysis many of the teams studied several processes, trying to separate between them. However, one of the conclusions from Phase 1 was that the tracer experiments at Finnsjön may not be designed to discriminate between processes /Tsang & Neuman (editors), 1991/. The results were also somewhat ambiguous where e.g. matrix diffusion was considered to be important by one team while others considered it to have none or negligible effect.

In the Phase 2 analysis, only four of the nine teams include more than one process in their analysis, c.f. Table 2. The GEOSIGMA and PSI teams are the only teams considering sorption. The reason for this is probably that there are no independent laboratory or field data for the weakly sorbing tracers used in the dipole test. It should in this context also be noted that the PSI team uses a different data set from Finnsjön, previously used in INTRA-COIN /SKI, 1986/, specially addressing sorption processes. The PSI team conclude that sorption parameters determined from these tests agree well with literature data.

The PSI team also made an analysis of the effect of matrix diffusion and came to the conclusion that matrix diffusion has a small but non-negligible effect. The U. of New Mexico team compares a single and a double porosity model and claims that matrix diffusion is an important process based on the ratio of fracture to matrix porosity. The GEOSIGMA team draw the opposite conclusion based on the same ratio and did not consider matrix diffusion at all in their analysis.

The general conclusion drawn by all teams is that flow and transport in Zone 2 is governed by advection and that hydrodynamic dispersion also is needed to explain the breakthrough curves. Most teams agree that matrix diffusion has no, or very small effect in the experiments, given the high induced velocities and low ratio of fracture to matrix porosity.

6. VALIDATION ASPECTS

The classical approach to model validation in a flow or transport context is calibration of a set of model parameters against a given experiment geometry for a specific stress of the geological system. The validation constitutes prediction of model behavior for another experiment condition (altered stress and/or geometry) in the same modelled system and comparison with experimental data.

During Phase 1, only three out of seven modelling teams (EdM, GEOSIGMA and JAERI) formally addressed validation as defined above by first calibrating their model with the radially converging test and subsequently predicting the dipole test /Tsang & Neuman (editors), 1992/. The EdM team explicitly addressed validation and claimed proper process and parameter identification but acknowledged poor representation of horizontal heterogeneity. The JAERI team did not explicitly address validation but utilized the parameter set obtained from the calibration of the radially converging test to predict the dipole test. GEOSIGMA acting as responsible field organization in succession used the collected field data to enhance their descriptive models to predict the subsequent tracer test, and finally attempted, and succeeded well, to predict the dipole test response.

During Phase 2, various validation aspects have been considered, c.f. Table 3. Five groups address the classical approach to validation, whereas two groups utilizing the stochastic continuum approach address other validation aspects. The modelling teams from GEOSIGMA and VTT address parameter consistency between all three tracer tests performed at Finnsjön. The results show that the average transport behavior of the system can be acceptably described with one single set of transport parameters. However, for a detailed understanding of individual transport paths, transport parameters need to be adjusted for different flow geometries. The PNC team utilizes a porous medium, double porosity model to calibrate the transport parameters using the dipole tracer test. The validity of the model was subsequently checked by simulating the radially converging test. The results show good predictive capability for early times whereas tails in the breakthrough curves are poorly represented. The latter discrepancy was treated by introducing a mixing zone in between the modelled high-conductive layer and an overlying low-conductive layer. The alternate sequence used by the PNC team, in that they start out with the dipole test in their analysis, may have helped to get a better understanding of the heterogeneity between BFI01 and BFI02. The BRGM team uses the classical approach and succeeds in predicting interference test results based on calibrations of early interference tests. The U. of New Mexico team uses a single/double porosity 2D model (in vertical section) to predict the converging test and checks the validity of their model/-s by simulating the dipole test. They conclude that a reasonable agreement between measured and simulated breakthrough is obtained.

The Conterra/KTH-WRE team addresses a performance assessment issue, that of extrapolation of transport models calibrated on a small scale to larger transport scales. Using an exhaustive reference transmissivity field and the stochastic continuum approach they show that a model calibrated on an experimental scale is not validated on a larger far-field scale when being subjected to extrapolation and simulation of a far-field natural gradient test. The reason being that the local scale conditioning data do not suffice also to describe far-field scale heterogeneity.

The UPV team raises an important issue related to the inherent choice of statistical model when generating transmissivity fields in stochastic continuum applications. The UPV work shows that when, and if the multiGaussian model is used this should be preceded by a check whether the data also are bivariate Gaussian and with the understanding that the multi-Gaussian approach intrinsically suppresses connectivity of extreme values. The latter is of paramount interest from a performance assessment perspective since an unwarranted use of the multiGaussian model may transform an "unsafe" site to a "safe" one.

Table 3. **Phase 2 analysis of tracer tests at Finnsjön. Scale of problem and validation aspect of problem studied.**

Modelling team	Scale of problem	Validation aspects of problem
GEOSIGMA / SKB, Sweden	2500x1500m (far-field) 500x500m (local scale) 10x10m discr.	Same conceptual model used for all analyzed tracer tests. Check whether it is possible to simulate tracer tests results in various geometries with the same transport parameters.
VTT / TVO, Finland	< 200m	Same conceptual model possible to explain all tracer test results.
PNC, Japan	< 200m	Calibrated dipole model tested by simulating the radially converging tracer test.
PSI / NAGRA, Switzerland	~ 30m	Consistency in evaluated transport parameters. Comparison between laboratory test results and field test results performed at different sites in granitic rocks and in various geometries.
U. of New Mexico, USA	~ 200x200m	Model(s) calibrated using the radially converging tracer test are tested by simulating the dipole tracer test.
Hazama Co, Japan	1000x1000x100m (far-field) 300x300x100m (local scale) 20x20x20m discr.	Use of REV concept in analysis of flow and transport in fractured rock.
Conterra / KTH / SKB, Sweden	1200x1200m (far-field) 200x200m (local scale) 10x10m discr.	Extrapolation of a model, calibrated on a local scale to a larger far-field scale.
BRGM / ANDRA, France	2500x1200x100m	Consistency in evaluated flow and transport parameters
UPV / ENRESA, Spain	1000x1000m 20x20m discr.	Is use of Gaussian model a conservative choice!?

7. COMPARATIVE STUDY OF A SYNTHETIC CASE

Following the Phase 2 analysis the conclusion still remains that the Finnsjön experiments may not be used to discriminate between different conceptual approaches. However, this fact does not constitute a problem from a performance assessment perspective, if differences between predictions based on these conceptual models are negligible. A need to illustrate the relative predictive uncertainty amongst the different models was identified. To facilitate such a comparison a synthetic natural gradient tracer test was conceived to which all INTRAVAL modelling teams were invited to apply their calibrated models. In order to facilitate analysis by different modelling teams, two transport scales were defined; Case A : 500m and Case B : 1000m. The layout of the test cases and positioning of tracer line sources and line collectors are shown in Figure 2. A mass (M_o = 1000 mg) is introduced as a pulse over a period of 1 hour under a uniform gradient of 0.3%. The mass arrival (breakthrough) as a function of time was monitored along the respective sampling lines. A distribution coefficient (K_d = 1.0 m^3/kg) was defined to address sorption. Representative input data used by the five modelling teams addressing the comparative study are shown in Table 4.

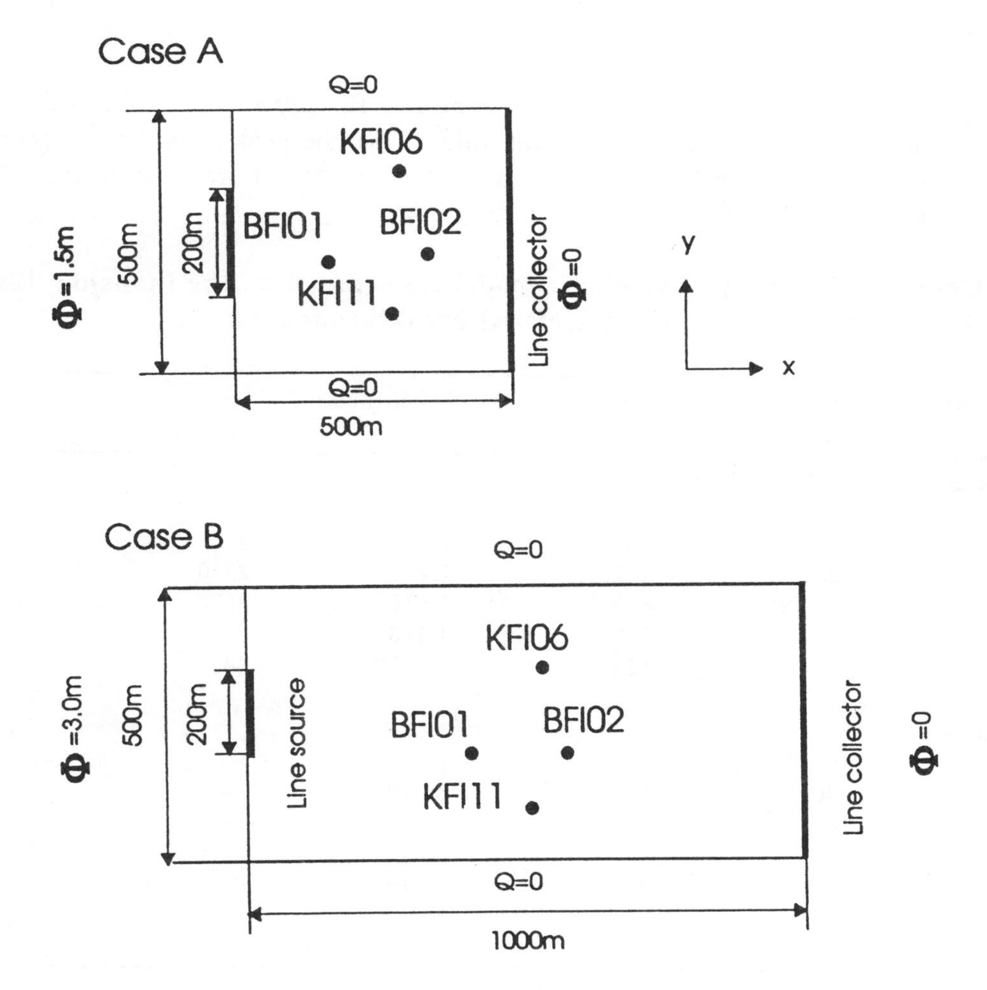

Figure 2. **Layout of the synthetic tracer test used in the comparative study.**

Table 4. **Comparative study of synthetic model cases based on the Finnsjön Test Case. Selected input parameters used by the modelling teams.**

Modelling team	Material property representation	Mean T (m^2/s)	B (m)	ϕ (-)	α_L (m)	α_T (m)	R_d (-)
EdM	K, effective	$1.3 \cdot 10^{-3}$	5	$1.4 \cdot 10^{-3}$	A:50 B:100	-	-
JAERI	2b, stochastic	$6.3 \cdot 10^{-4}$	5	$3 \cdot 10^{-4}$	-	-	-
VTT	channels, # of	$1 \cdot 10^{-3}$	-	-	-	-	-
PNC	T, stochastic	$5.2 \cdot 10^{-4}$	5	$8 \cdot 10^{-3}$	-	-	$3.3 \cdot 10^{5}$
GEOSIGMA	K, effective anisotropic	$T_{max}=1.5 \cdot 10^{-3}$ $T_{min}=1.8 \cdot 10^{-4}$	0.5	0.01	A:5	A:0.2	$2.7 \cdot 10^{5}$

Figure 3a-c shows breakthrough curves for Cases A and B, respectively. The following performance measures were defined; first arrival time t(5%), the peak arrival time t(peak), the maximum relative concentration (M/M_o)max and an index of tracer spreading "α" defined as t(90%)-t(10%), c.f Table 5.

Table 5. **Comparative study of synthetic model cases based on the Finnsjön Test Case. Measures from the calculated breakthrough curves.**

Modelling Team	t(5%) (hours)	t(peak) (hours)	M/M_o (max) (-)	"α" (hours)
CASE A (500m)				
VTT	205	405	$1.4 \cdot 10^{-3}$	1731
PNC	520	1125	$1.4 \cdot 10^{-3}$	2350
JAERI	140	405	$3.8 \cdot 10^{-3}$	863
EdM	75	200	$4.0 \cdot 10^{-3}$	450
GEOSIGMA	295	520	$4.0 \cdot 10^{-3}$	440
CASE B (1000m)				
VTT	470	890	$8.0 \cdot 10^{-4}$	3497
PNC	2050	3600	$4.4 \cdot 10^{-4}$	7710
JAERI	290	810	$2.2 \cdot 10^{-3}$	1420
EdM	150	400	$2.0 \cdot 10^{-3}$	900
GEOSIGMA	-	-	-	-

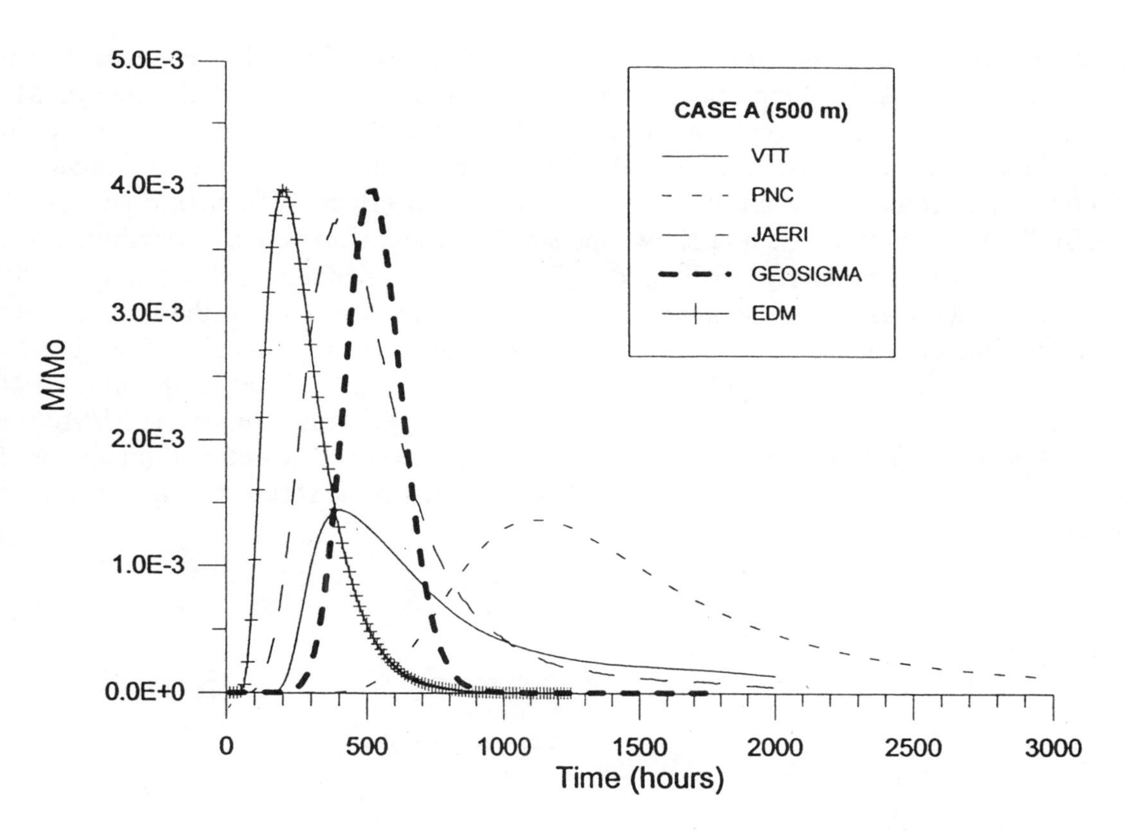

Figure 3a. **Predictions of the natural gradient tracer test, Case A (500 m).**

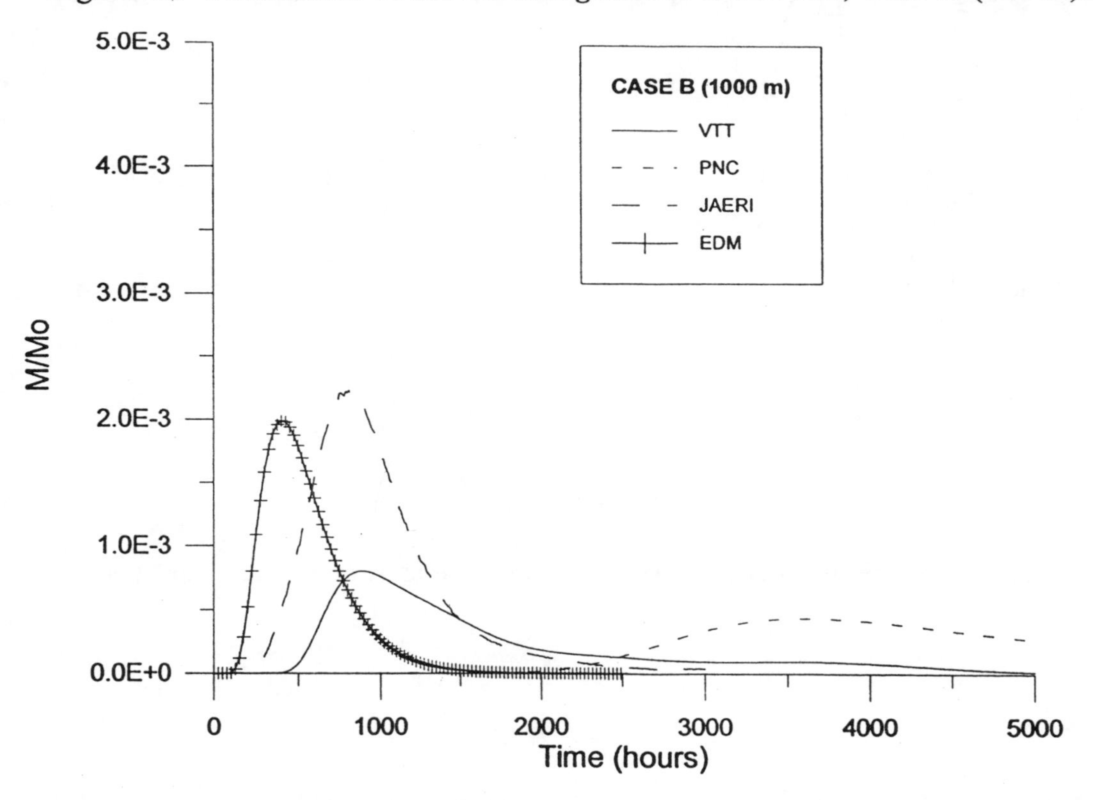

Figure 3b. **Predictions of the natural gradient tracer test, Case B (1000 m).**

The results for Case A shown in Table 5 and Figure 3a exhibit a relatively narrow range of peak travel times, between 200-1125 hours, and an even more limited range of M/M_0 (max), $1.4\text{-}4.0\cdot10^{-3}$. The exception being the PNC breakthrough curve which is more smeared and delayed compared to the other breakthrough curves. This fact is explained by the combined effect of the introduced mixing zone and effects of matrix diffusion. The GEOSIGMA breakthrough curve shows the smallest dispersion which is attributed to the lower values of dispersivity applied, c.f. Table 4. The features and relative qualities identified for the Case A is also noted for Case B with the exception of the PNC curve for which the time dependency of matrix diffusion become more pronounced, c.f. Figure 3a,b. Hence, peak travel time and M/M_0 (max) for the PNC team is 3600 hours and $4.4\cdot10^{-4}$, respectively compared to the other three teams where the peak travel times and M/M_0 (max) range between 400-890 hours and $0.8\text{-}2.2\cdot10^{-3}$, respectively. The retardation factor, R_d, implicit in the assigned K_d value, c.f. Table 4, is reflected in the results taking sorption into account, c.f. Figure 3c.

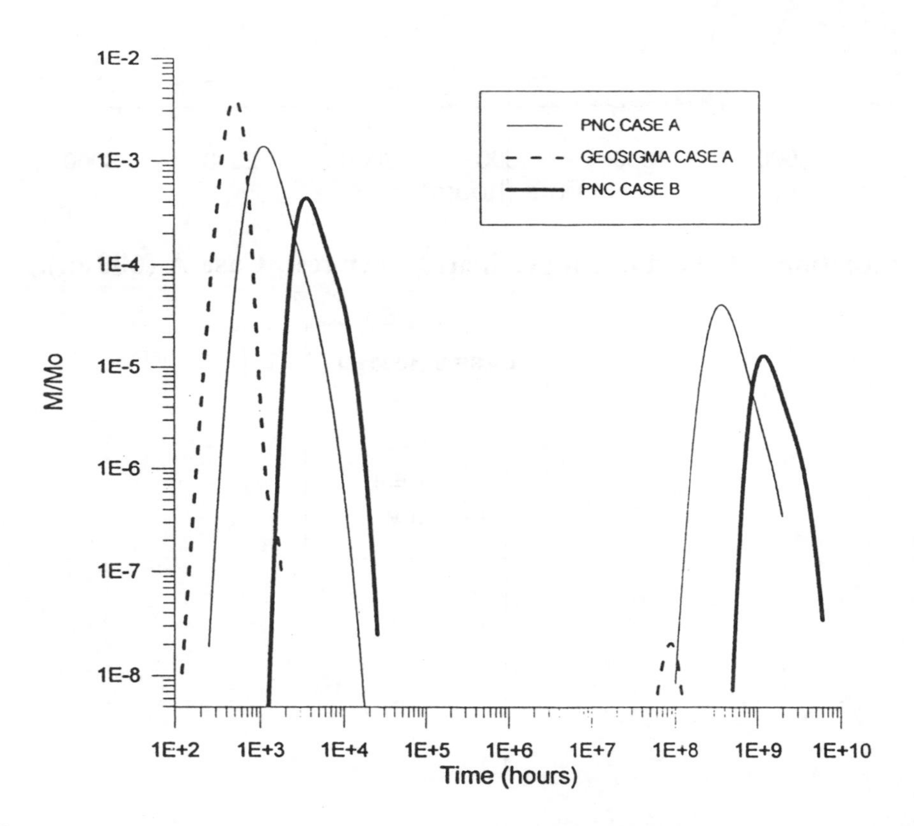

Figure 3c. **Predictions of the natural gradient tracer test, K_d=1, Case A and B.**

8. CONCLUSIONS

A variety of different conceptual approaches have been applied in the analysis of the Finnsjön test case. The results show that the different approaches can be used to successfully model the experiments. The performed comparative study aimed at quantifying the predictive uncertainty when extrapolating the calibrated models to larger transport scales show that the five models, with one exception, provide results which are fairly similar on the two transport scales considered. The exception being a model incorporating a pronounced effect of matrix diffusion. The approaches to model validation are versatile and focus also on other issues than the classical validation issue, which may be equally important for performance assessment apart from improving our predictive capability of solute transport phenomena in crystalline rock. It has been shown that the classical approach to model validation has been successfully applied using a number of conceptual approaches. In addition, one of the stochastic approaches has shown that a heterogeneous model calibrated on a local scale will not necessarily be validated on a larger transport scale. Another stochastic approach has shown that the choice of distribution model for stochastic material property generation has implications for continuity of extreme values and travel time distributions.

9. REFERENCES

Andersson, J-E., Ekman, L., Gustafsson, E., Nordqvist, R., and Tirén, S., 1989: Hydraulic Inter-ference Tests and Tracer Tests within the Brändan Area, Finnsjön Study Site - The Fracture Zone Project Phase 3. SKB Technical Report TR 89-12.

Andersson, P., Nordqvist, R., Persson, T., Eriksson, C-O., Gustafsson, E., and Ittner, T., 1993: Dipole Tracer Experiment in a Low-Angle Fracture Zone at Finnsjön - Results and interpretation. The Fracture Zone Project - Phase 3. SKB Technical Report TR 93-26.

Andersson, P., and Winberg, A., 1994: INTRAVAL Working Group 2 summary report on Phase 2 analysis of the Finnsjön test case. SKB Technical Report TR 94-07.

Gustafsson, E., and Andersson, P., 1991: Groundwater flow conditions in a low-angle fracture zone at Finnsjön, Sweden. Journal of hydrology, Vol 126, pp 79-111.

Gustafsson, E., and Nordqvist, R., 1993: Radially Converging Tracer Experiment in a Low Angle Fracture Zone at the Finnsjön Site, Central Sweden. The Fracture Zone Project Phase 3. SKB Technical Report 93-25.

SKI 1986 : INTRACOIN Final Report Levels 2 and 3. SKI Report 86:2.

Tsang, C-F., and Neuman, S. (editors), 1992: The International Intraval Project. Phase 1, Test Case 5. Studies of Tracer Experiments in a Fracture Zone at the Finnsjön Research area. NEA/SKI, OECD Paris 1992.

Conclusions from Working Group 2 - The Analyses of the WIPP-2 Experiments

C P Jackson
AEA Technology, UK

Abstract

The INTRAVAL WIPP-2 test case is based on data from site investigations carried out at the Waste Isolation Pilot Plant (WIPP) in New Mexico, USA. The site has been chosen as a potential location for a radioactive waste repository. Extensive investigations have been carried out, focused mainly on groundwater flow and transport in the Culebra Dolomite. This is a thin formation extending for many kilometres, which is considered to form the main pathway for transport of radionuclides off the site by groundwater, following an accidental borehole intrusion into the repository itself.

Five INTRAVAL project teams studied the test case. Two teams addressed issues involved in the treatment of heterogeneity. Stochastic models and a Monte Carlo approach were used. Many realizations of the transmissivity of the Culebra were generated with distributions inferred from the data. The groundwater flow and pathlines were calculated for each realization. The uncertainties in predicted quantities of interest were quantified from the range of behaviours of the realizations.

One team quantified the increased uncertainty resulting from fewer data, and explored the issues involved in validation of stochastic models. They recommended a statistical-hypothesis-testing approach, and showed that the differences between some models and the data were statistically significant. The second team developed a new method for conditioning stochastic models on head data, that is generating realizations in which the calculated head matches observations. This team also stressed the need to consider non-Gaussian stochastic models of the heterogeneity.

Two other teams examined issues relating to the choice of conceptual models. Two-dimensional vertical cross-section models were used to explore the importance of vertical flow between the Culebra and overlying formations. A three-dimensional model was also developed. The fifth team advocate the use of a variety of models, including simple scoping and bounding calculations to highlight the most important processes and parameters. On the basis of bounding calculations the team concluded that the distribution of transmissivity was the most important influence on the groundwater flow in the Culebra, and the uncertainties in the salinity were less important.

Overall, the test case provided a valuable opportunity for the INTRAVAL participants to explore the issues involved in validation of field-scale models of groundwater flow and transport using the sort of data that would result from a typical site investigation, and the test case was a strong stimulus to the study and development within INTRAVAL of methods for dealing with heterogeneity.

Introduction

INTRAVAL, which ran from 1987 to 1993, was an international collaboration that addressed the validation of models of radionuclide transport by flowing groundwater for use in performance assessments for underground repositories for radioactive waste. Project teams from many organisations in different countries participated in INTRAVAL, which was structured around a number of test cases based on available experimental data.

The INTRAVAL WIPP-2 test case was based on data obtained from the site investigations carried out at the Waste Isolation Pilot Plant (WIPP) in New Mexico, USA. The site has been chosen as a potential location for an underground repository for the disposal of radioactive waste. Extensive investigations have been carried out in the region of the site. Data from these investigations were made available to INTRAVAL for use as the basis of a test case. This provided a very valuable opportunity for the INTRAVAL participants to explore the issues involved in validation of field-scale models of groundwater flow and transport using the sort of data that would result from investigation of a repository site. In this paper the studies of the WIPP-2 test case within INTRAVAL are summarised.

Data for the test case

Most of the data used in the studies are given in reports by LaVenue et al. [1990], which presents the results of deterministic modelling of flow in the Culebra Dolomite, and Cauffman et al. [1990]. Relevant data are summarized here. The overall stratigraphy of the rocks in the region is shown in Figure 1, and the surface topography is shown in Figure 2. The site investigations mainly focused on data relating to groundwater flow and transport in the Culebra Dolomite. This is a fairly thin formation about 8m thick extending for many kilometres, which is considered to form the main pathway for potential transport of radionuclides off the site, following an accidental borehole intrusion into the repository itself.

Groundwater flow within the Culebra Dolomite is considered to occur mainly in fractures. The average spacing of all fractures is about 10cm, and the average spacing of the fractures carrying flow is considered to be about 50cm. It was considered that a continuum porous-medium model would be a good approximation for modelling flow on the length scales of interest (many kilometres).

The transmissivity of the Culebra Dolomite has been measured at 41 different locations. LaVenue et al. [1990] reviewed the available data and produced a list of recommended values. The data exhibit considerable variability, covering a range of over 7 orders of magnitude. For convenience in visualising the data, Figure 3(a) shows the distribution of transmissivity obtained by interpolating the data using kriging. On the large scale the data appear to exhibit a pronounced trend with high values to the west and low values to the east. However, this trend appears to be reversed on the scale of the site. The matrix porosity of the Culebra Dolomite shows considerably less variability than the transmissivity, and is about 0.16.

LaVenue et al. [1990] also produced a list of recommended values of groundwater head. The general pattern is that the head is high to the north and decreases to the south, corresponding to a general flow from north to south. The groundwater salinity has been measured at various locations. Figure 4 shows an interpolated distribution of salinity. Other data, such as the results of transient cross-hole tests were also available [LaVenue et al., 1990].

INTRAVAL studies

The test case was studied by five project teams within INTRAVAL:

(i) AEA Technology, UK, on behalf of UK Nirex Ltd, (AEA);
(ii) Universidad Politécnica de Valencia, Spain (UPV);
(iii) Bundesanstalt für Geowissenschaften und Rohstoffe, Germany (BGR);
(iv) Sandia National Laboratory, USA (SNL);

(v) Atomic Energy Control Board of Canada (AECB).

The studies undertaken can be divided into three groups:

(i) the AEA and UPV project teams studied stochastic models of heterogeneity;
(ii) the BGR and SNL project teams explored issues relating to the choice of conceptual model;
(iii) the AECB project team used the test case to examine the utility of bounding calculations.

During INTRAVAL the project teams were able to interact with the WIPP project itself. Presentations on the work being undertaken within the WIPP project were made to the INTRAVAL participants, who were given the opportunity to comment on proposals for a new experiment at the site. The proposed experiment was intended to examine transport of both reactive and non-reactive tracers over a range of length scales. It comprised measurements of transport between various combinations of boreholes in a formation with overall dimensions of the order of a hundred metres. It should be stressed that the ultimate responsibility for the design of the proposed experiment, which was not carried out during the course of INTRAVAL, rested with the WIPP project.

The AEA study

The AEA project team used Monte-Carlo stochastic techniques to treat the uncertainty resulting from heterogeneity. The logarithm, Z, of the transmissivity of the Culebra Dolomite was treated as a random spatial process. A statistical model for Z was inferred from the available data, many realizations of Z were generated using a suitable technique, and groundwater flow and transport calculated numerically for each realization. The uncertainty was then quantified from the range of behaviours shown by the realizations.

The project team used Gaussian statistical models, which are characterised by first- and second-order moments: the mean and the variogram. They considered several different statistical models for Z: a model with an exponential covariance, a model with a power-law variogram (which leads to a field with fractal behaviour), a pure-nugget model (that is a model in which Z is completely uncorrelated from point to point), and a model with a linear drift (or trend) and an exponential covariance. They compared different approaches for estimating the parameters for a statistical model of a chosen form (fitting by eye to the experimental variogram, least-squares fitting to the experimental variogram and maximum-likelihood parameter-estimation). They preferred automated methods for parameter estimation, once the form of the variogram had been chosen, which should take geological information into account. Such methods provide a quantitative estimate of the quality of the fit.

For the statistical model with an exponential covariance, the project team also undertook a Monte-Carlo study of the bias in the various automated parameter-estimation methods. They generated many (10000) realizations of Z for known parameters of the statistical model, and then for each realization they used the various methods to estimate the parameters of the statistical model from the values of Z sampled at locations corresponding to the points at which the transmissivity of the Culebra has been measured. In this way the distribution of the estimated parameters could be determined for the given parameter values, and confidence intervals estimated. This was repeated for different selected values to build up a picture of the confidence intervals for the parameters of the underlying statistical model as functions of the estimated parameters. The analysis showed that:

(i) all the methods were biased, that is the expected (or average) values of the estimated parameters (considered over the distribution of these parameters), were not equal to the underlying parameters (which were known in the circumstances of the study);
(ii) maximum-likelihood parameter estimation had tighter confidence intervals (so was preferred);
(iii) if the correlation length was comparable to the size of the domain, then it was not possible to estimate an upper limit to the correlation length. The project team commented that this might not be too significant in uncertainty analyses, provided that the realizations were conditioned on measured values of Z, since they would be strongly controlled by the data in this case.

The project team used the Monte-Carlo approach to quantify the uncertainties in quantities relevant to repository performance, such as the head at a point, the Darcy velocity at a point, the travel time along a pathline, and the position at which a pathline crosses a given boundary (see Figure 5). They used the Turning Bands method [Mantoglou and Wilson, 1982] to generate realizations (see Figure 3(b)). Constant-density groundwater flow for each realization was calculated using the finite-element method, and then pathlines through the flow were calculated (see Figure 6(a)). The realizations were conditioned on the measured values of Z using a technique based on kriging. The project team studied the effect of conditioning the realizations on different numbers of transmissivity measurements. They found that the uncertainties were significantly reduced as the number of measurements used to condition the realizations was increased. For example (see Figure 6) the uncertainties in travel time were five times larger if the realizations were not conditioned, than if they were conditioned on the 39 transmissivity measurements in the domain modelled.

The project team was particularly interested in studying techniques for validating stochastic models. They used the framework of statistical hypothesis testing. A suitable statistic was defined that would be expected to have a small value if the stochastic model was appropriate. The value of the statistic was calculated, and if this value was sufficiently large that it could not reasonably be expected to have arisen by chance, the model was rejected. For example, the project team used the χ^2-test and the Kolmogorov-Smirnov test (KS-test) for normality to assess the pure nugget model, for which the transmissivity data are considered to be uncorrelated. On the basis of the tests the project team concluded that the pure nugget model was unlikely to be an appropriate model for the Z data.

In order to use the tests to assess whether or not the other models were possible explanations of the data it was necessary to take account of the correlations between the data. This was done by computing suitable linear combinations of the data that, given the model, would be independent samples from a unit normal distribution. These combinations were determined by diagonalising the covariance matrix between the measurement points. The χ^2-test and KS-test for normality could then be applied. The project team concluded that the model with an exponential covariance and the model with a power-law (fractal) covariance could be appropriate models for the Z data, whereas the model with a trend and an exponential covariance was unlikely to be an appropriate model. (This does not mean that other models with a trend are not appropriate.)

However, when the head data were considered, they concluded that the models were unlikely to be acceptable. Although the measured heads were generally within one standard deviation of the average over all realizations, when correlations were taken into account, it was considered unlikely that the measured heads would arise as a sample from the ensemble of realizations. The project team noted that although the discrepancies might have arisen because the underlying statistical model was inappropriate, they might also have arisen because the calculations were for constant density flow, or because inappropriate boundary conditions were used. (The prescribed boundary heads were obtained by LaVenue et al. [1990] by extrapolating from measured heads using kriging.)

In most of their work the realizations were conditioned only on the measured values of Z, and the project team used the measured values of head to check the model. They noted that:

(i) although the uncertainties might be less if all the data were used in the construction of the models, no data would be left to check the model against;

(ii) it might be possible to condition an inappropriate statistical model to match all the measurements, and therefore careful statistical tests would have to be carried out to check models that had been conditioned on all the data. It would not be sufficient simply to point out that the model matched all of the data.

However, the project team also explored the use of a technique based on co-kriging for conditioning on measured heads.

The UPV study

The UPV project team used the same basic Monte-Carlo approach to the treatment of heterogeneity as the AEA project team. They began by undertaking an extensive analysis of the data, including an attempt to correlate the variability in transmissivity with geology, and statistical testing of whether a univariate Gaussian model for Z is acceptable. The raw data are correlated so the project team declustered the data using declustering weights proportional to the kriging weights for estimation of the average over the entire domain. First estimates of the experimental variograms were used in this. The KS-test was then used to check for normality. They concluded that the data just passed the test.

The data seem to show anisotropy (see Figure 7) related to the geology so the project team used an anisotropic Gaussian model. However, the project team stressed the need to consider non-Gaussian models as well. In Gaussian models, extreme values are not connected over distances long compared to the correlation length. In a repository performance assessment it is important to consider models in which high transmissivity regions are connected over long distances, as these regions may form fast flow paths, and lead to the main radiological consequences of the repository.

The project team used sequential simulation [Goméz-Hernandez and Journel, 1993] to generate realizations. (This method can also be used to generate realizations of a non-Gaussian model.) Their initial simulations were for constant-density flow, and the realizations were only conditioned on measured values of Z. These simulations did not give a very good match to the measured heads, so the model was improved to take into account the effects of density variation, uncertainties in the boundary heads, and the measured heads. The project team found that, when heterogeneity was taken into account, in some parts of the domain the flow field calculated for variable-density flow differed significantly from the flow field calculated for constant-density flow.

The project team used the following method for conditioning on measured heads. The basic idea of the method is to compute a modification to the transmissivity field that leads to an improved match to the measured heads. The modification is parameterized in terms of the values of the modifications ΔZ_j to the values of Z at selected points j. The modification at other points is obtained by interpolation from these values using kriging. The numerical flow equations are then linearized about the initial transmissivity field, and solved to derive an approximation to the change in head resulting from the modification to the transmissivity as a linear combination of ΔZ_j . Quadratic programming was then used to determine values of ΔZ_j that minimised a weighted sum of the squares of the differences between measured and calculated heads. Figure 8 shows the improved match to the measured heads obtained for one realization. Uncertainties in heads on the boundary can be dealt with using the same approach.

The BGR study

The BGR project team addressed issues relating to the choice of conceptual model. The AEA, UPV and AECB project teams considered flow in two-dimensional areal models of the Culebra Dolomite. Underlying this sort of model is the assumption that the permeabilities of the units above and below the Culebra are sufficiently small that vertical flow can be neglected. The BGR project team considered an alternative model in which all the units above the Salado were represented. They modelled flow in a vertical cross section that runs parallel to the steepest surface gradients and is normal to the margins of the halite beds in the Rustler formation, which represent a potential source of salinity. This model was considered to give a reasonable approximation to the three-dimensional flow.

All available relevant data were incorporated in the model, but nevertheless many parameters had to be estimated, which was done in a manner consistent with the geological setting where possible. Average properties were used for the members of the Rustler formation other than the lower halite beds. (The distribution of permeability used is shown in Figure 9) Density dependent groundwater flow was simulated until an approximate steady-state was reached (see Figure 10).

The plausibility of the solution was checked by comparing the computed density distribution with experimental measurements. The steep gradients in the Culebra west of the site were reproduced (see Figure 11), and the model predicted nearly freshwater in the Magenta, at a position at which this was measured. Pathlines were also tracked back to provide estimates of groundwater age from the travel time along the pathlines. These are only estimates because they do not take account of dispersion. The estimated ages were tens of thousands of years. For comparison, ages of several thousand years were inferred from groundwater chemistry. These results show that the model is reasonable. It should be noted that the model was not calibrated, but based on an initial estimate of the parameters.

The SNL study

The SNL project team also explored the use of different conceptual models for the hydrogeology at the WIPP site. They used a three-dimensional regional model to study the importance of vertical flow between the Culebra Dolomite and the overlying units, and to examine the impact of climate change. The model represented all the rocks above the Salado, and extended laterally to locations that were believed to correspond to groundwater divides for the entire time simulated. Each stratigraphic layer was taken to be homogeneous.

A period of 41000 years was simulated. The initial condition was taken to be the steady-state corresponding to the water table at the surface. This was taken to be representative of the conditions that prevailed in the region at the end of the Pleistocene. For the first 21000 years simulated, infiltration was taken to be zero, representing a dry period. The final 20000 years simulated were taken to correspond to a somewhat wetter period with a maximum potential infiltration of 0.5mm/year. (This is less than 0.2% of current annual rainfall, representing the fact that most rainfall is lost to surface runoff or evapotranspiration.)

The project team noted that there were large uncertainties in many parameters, and various simplifying assumptions had been made, so the results of their modelling should not be taken too literally. Rather the modelling should be considered as helping to understand the regional flow. Their calculations suggest that:

(i) the flow of groundwater in confined units may be strongly influenced by the overlying water table;
(ii) vertical flow through relatively low permeability layers may have an important effect on the flow in relatively permeable layers;
(iii) climate changes may be important.

The AECB study

In order to build confidence in the results of an assessment, the AECB advocates the use of a variety of conceptual models and approaches for demonstrating safety including the use of relatively simple scoping or bounding calculations and sensitivity studies. Validation of detailed models is involved and time consuming. Initial scoping and bounding studies can highlight the most important processes and parameters, and so help to focus more detailed studies, and may even adequately address some aspects of the problem, so avoiding the need for more detailed calculations.

The project team applied this methodology to examine the effects of variable fluid density. They compared the groundwater head and flow for a base case corresponding to the original calibrated model of LaVenue et al. [1990] with results for two variants that were considered to represent bounding distributions for the density:

(i) a distribution obtained by increasing the salinity by 50% everywhere;
(ii) freshwater everywhere.

Their results showed that the groundwater flow and paths from the centre of the WIPP site (which are the important paths for a performance assessment) did not differ greatly for all three cases. They concluded that the flow paths were controlled by the transmissivity distribution, and they suggested that future site investigation should focus on the region to the south of the site that LaVenue at al. [1990] had identified as having high transmissivity.

Overall conclusions

This test case provided a very valuable focus for the development and study of stochastic model for the treatment of heterogeneity in hydrogeological properties. The project teams showed that such approaches can be used in practical cases relevant to repository performance assessments. This should help to build confidence in the application of such methods. The modelling showed how uncertainties resulting from heterogeneity could be quantified, and showed the potential benefits of data in reducing these uncertainties.

During the course of INTRAVAL, the UPV project team developed a new technique for conditioning realizations on measured values of head as well as transmissivity. However, it was noted that, although the uncertainties might be less if all the data were used to construct the model, no data would be left to check the model against.

It seemed that several stochastic models might explain the data, although the resulting uncertainties were similar. It is to be expected that there is no unique stochastic model. Since a repository performance assessment should be robust, it would be desirable to consider a range of stochastic models that are consistent with the data. The UPV project team stressed the need to consider non-Gaussian models. These may may lead to fast flow paths that are unlikely to occur in Gaussian models, and which may be important for the performance of a repository.

The use of a statistical-hypothesis-testing framework was recommended for validation of stochastic models, and it was shown that some models might be rejected using such an approach.

It was shown that estimates of the parameters of stochastic models are biased, and the confidence intervals may be large. Indeed it may not be possible to determine an upper limit on the correlation length if this is comparable to the size of the domain. However this may not significantly affect the uncertainties computed from conditioned realizations, because the transmissivity will be strongly controlled by the data.

All the project teams commented on the need to take account of information about the geology in setting up the models.

It is desirable to consider alternative conceptual models. In particular it was shown that vertical flow through low permeability formations may affect the flow in much higher permeability confined formations, and that it may be necessary to consider the effects of climate change.

Work on the test case, and in particular the discussions of the proposed new experiment provided a valuable opportunity for modellers and experimentalists to interact. The participants stressed the importance of such interactions.

The test case provided a valuable opportunity to explore the issues involved in validation of field-scale models of groundwater flow and transport. Different project teams addressed different aspect of the system, Taken together, the conclusions reached by the project teams were generally consistent, and helped to build up the picture of groundwater flow and transport at the WIPP site. Overall the work undertaken on this test case should help to build confidence in models of the hydrogeology of the site.

Acknowledgment

The financial support of UK Nirex Ltd for the work of the author, which was carried out as part of the Nirex Safety Assessment Research Programme, is gratefully acknowledged.

References

T L Cauffman, A M LaVenue and J P McCord, Groundwater Flow Modelling of the Culebra Dolomite: Volume II: Data Base, Sandia Report SAND89-7068/2, 1990.

J J Goméz-Hernandez and A Journel, Joint Sequential Simulation of MultiGaussian Fields, Proceedings of the Geostatistics Troia 92 conference, edited by A Soares, p85-94, 1993.

A M LaVenue, T L Cauffman and J F Pickens, Groundwater Flow Modelling of the Culebra Dolomite: Volume I: Model Calibration, Sandia Report SAND89-7068/1, 1990.

A Mantoglou and J L Wilson, The Turning Bands Method for Simulation of Random Fields using Line Generation by a Spectral Method, Wat. Resour. Res., 18(5) p1379-1394, 1982.

System	Series	Group	Formation	Member
Recent	Recent		Surficial Deposits	
Quater-nary	Pleisto-cene		Mescalero Caliche	
			Gatuña	
Triassic		Dockum	Undivided	
Permian	Ochoan		Dewey Lake Red Beds	
			Rustler	Forty-niner
				Magenta Dolomite
				Tamarisk
				Culebra Dolomite
				lower
			Salado	
			Castile	
	Guadalupian	Delaware Mountain	Bell Canyon	
			Cherry Canyon	
			Brushy Canyon	

Figure 1 The stratigraphy of the rocks in the WIPP region (from LaVenue et al. [1990])

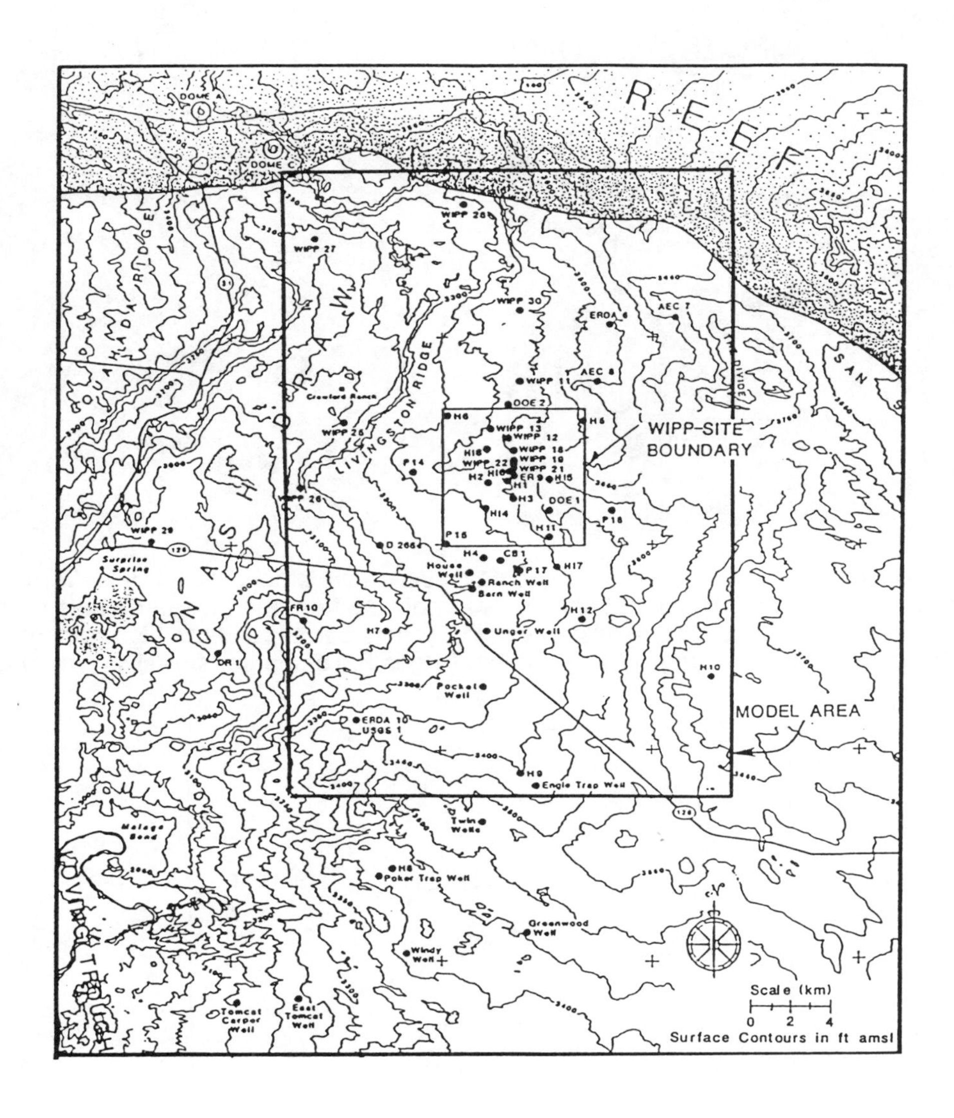

Figure 2. The surface topography in the WIPP region (from Lavenue et al. [1990]). The figure also shows the positions of boreholes, the boundary of the WIPP site, and the domain used in the models of LaVenue et al. [1990] and the AEA and UPV project teams.

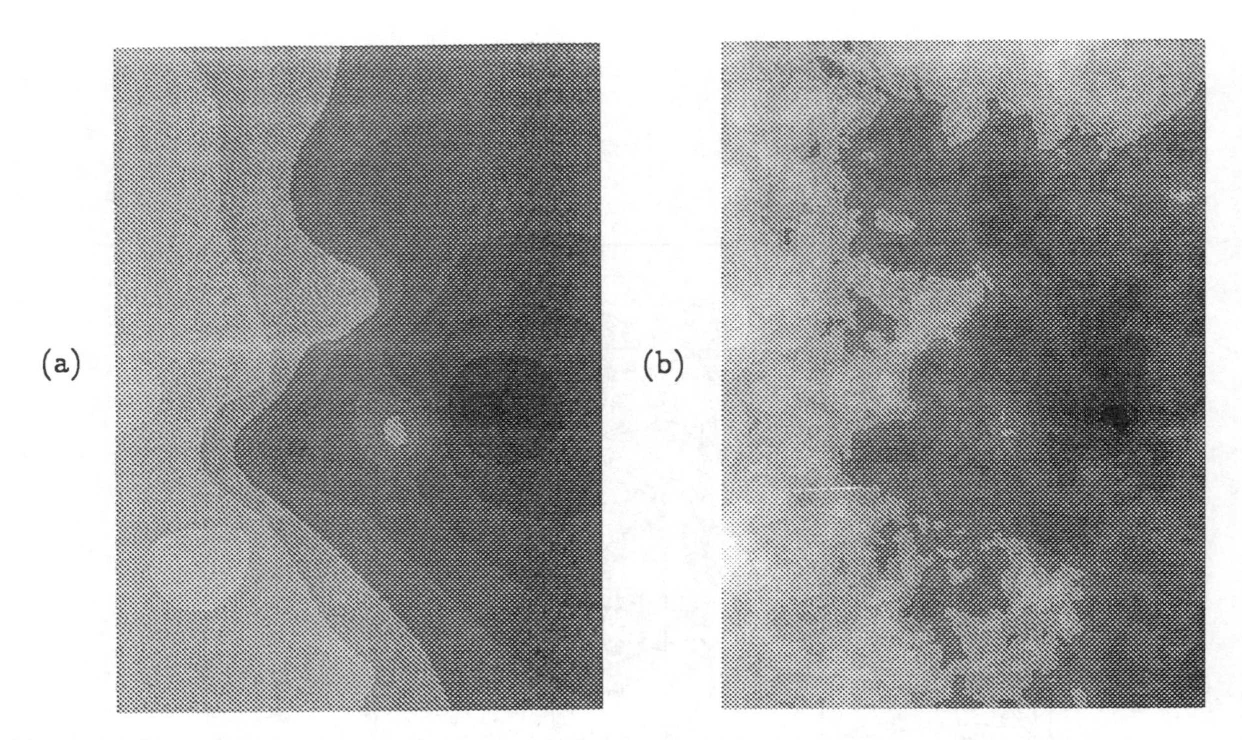

Figure 3. Distributions of transmissivity in the domain used in the stochastic models (from the work of the AEA project team: (a) distribution obtained by interpolating from the measurements using kriging, (b) one realization of a stochastic transmissivity field. (White corresponds to high transmissivity and black to low transmissivity. The range of Z (the natural logarithm of the transmissivity) in (b) is from -28 to 5.2)

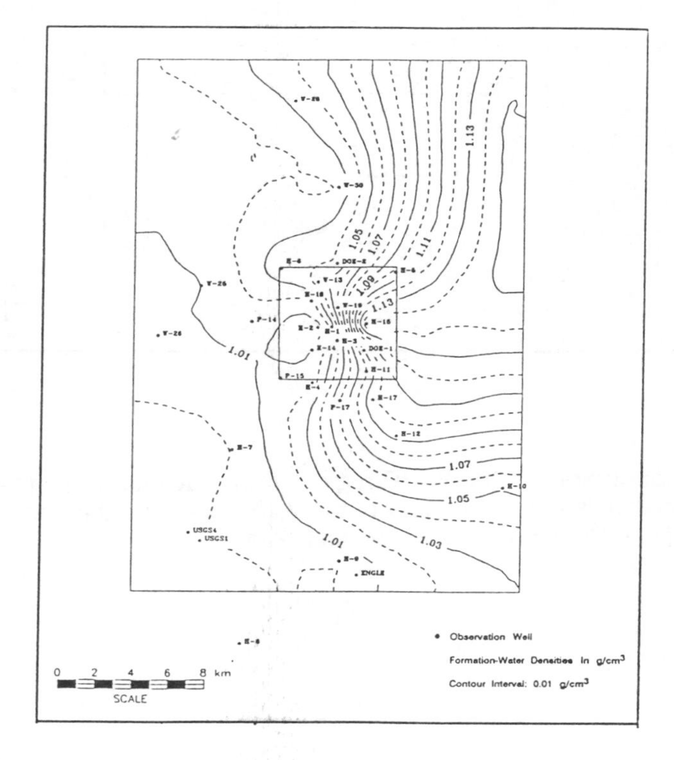

Figure 4. The distribution of density over the domain (from LaVenue et al. [1990]). The distribution was obtained by interpolation between measurements using kriging.

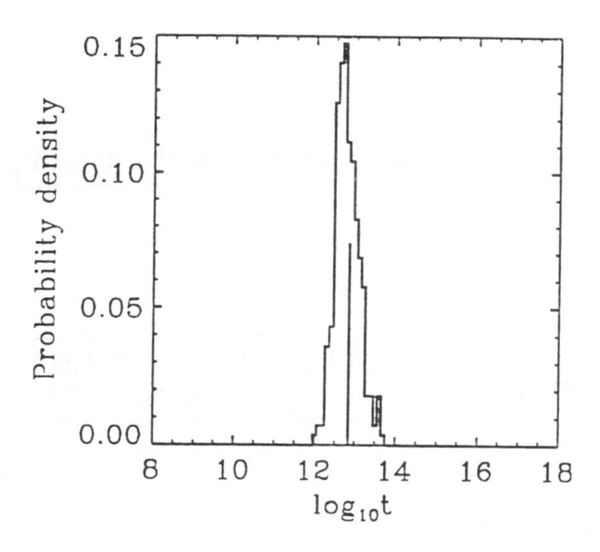

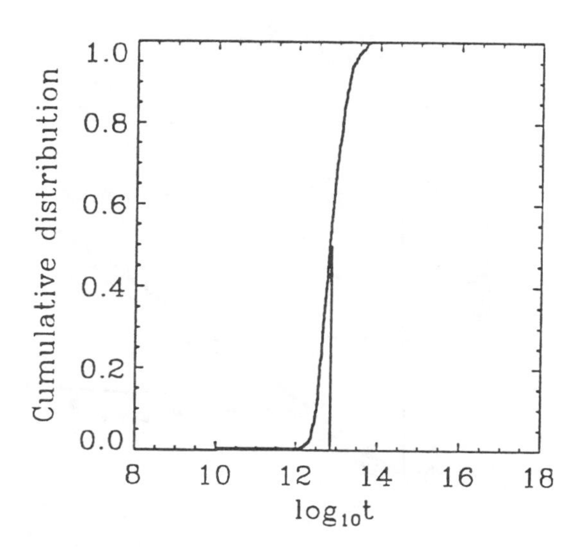

Figure 5. Probability density function and cumulative distribution function for the logarithm of the travel time from the centre of the WIPP site to the boundary of the domain modelled as obtained by the AEA project team from results for 300 realizations. (The vertical line in each plot is at the travel time in the transmissivity field obtained by interpolating the data using kriging (see Figure 3(a)))

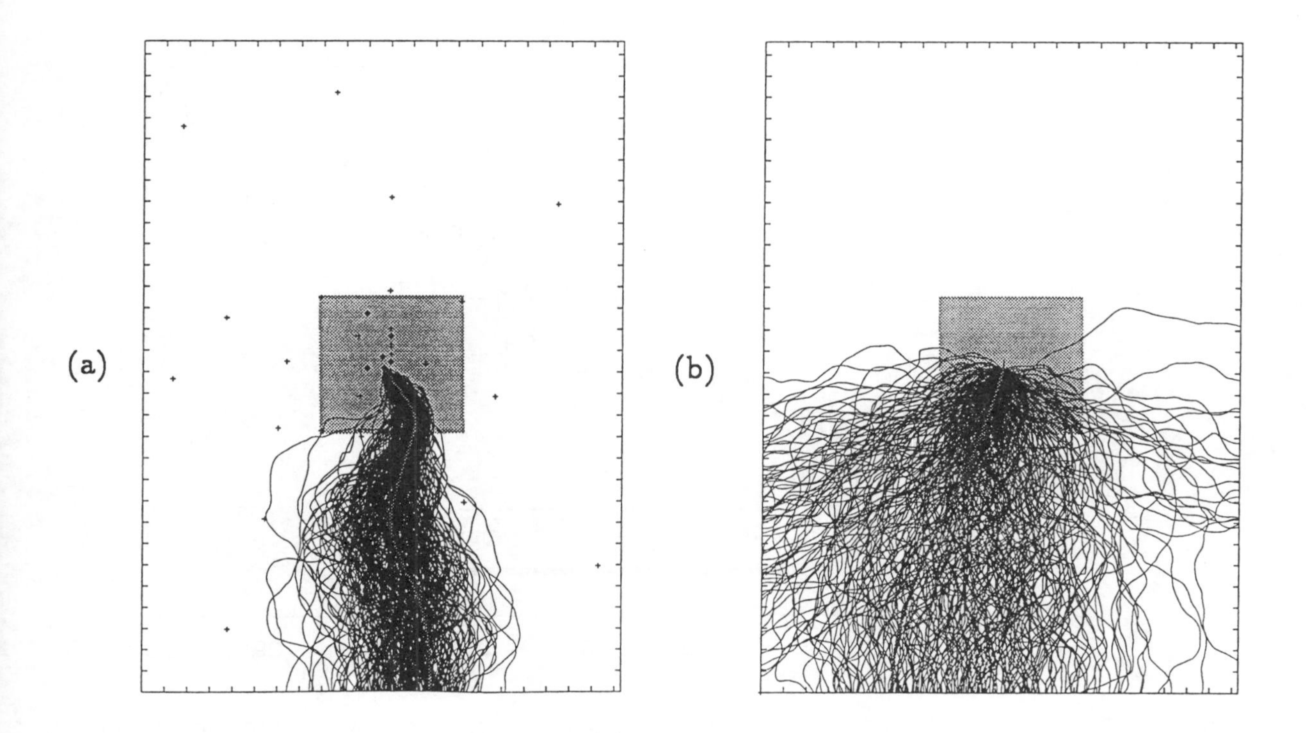

Figure 6. The pathlines obtained by the AEA project team: (a) for realizations conditioned on all the measured values of Z, (b) for unconditioned realizations. (The ticks around the boundary are spaced at 1km intervals, the central square is the WIPP site itself, and + denotes a borehole.)

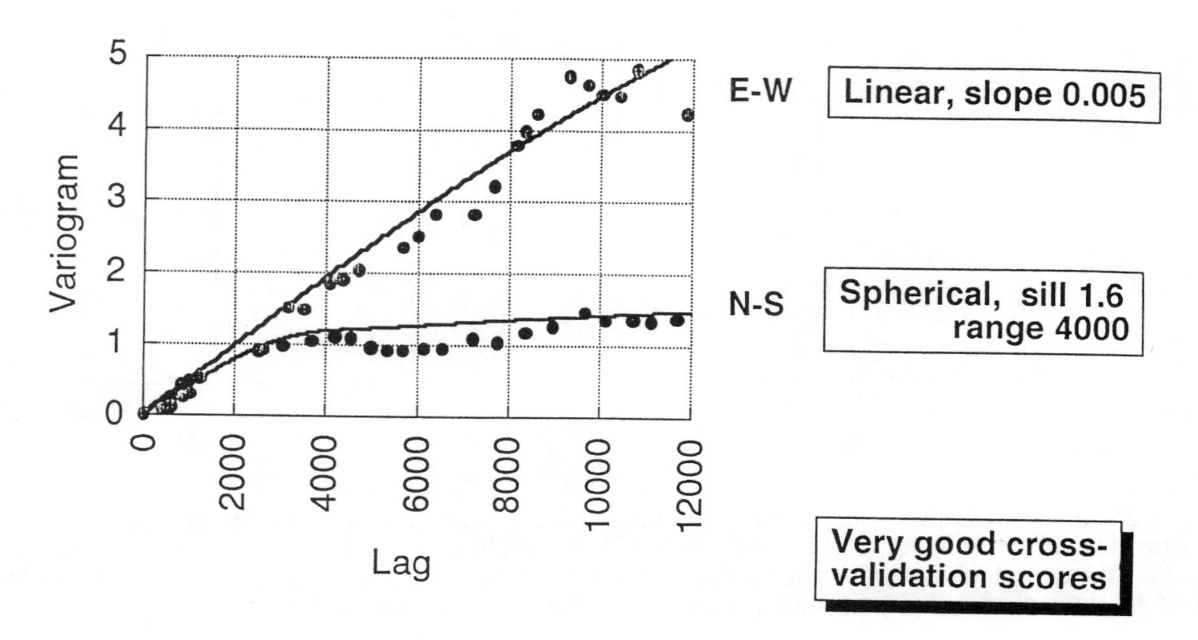

Figure 7. The directional variograms determined by the UPV project team.

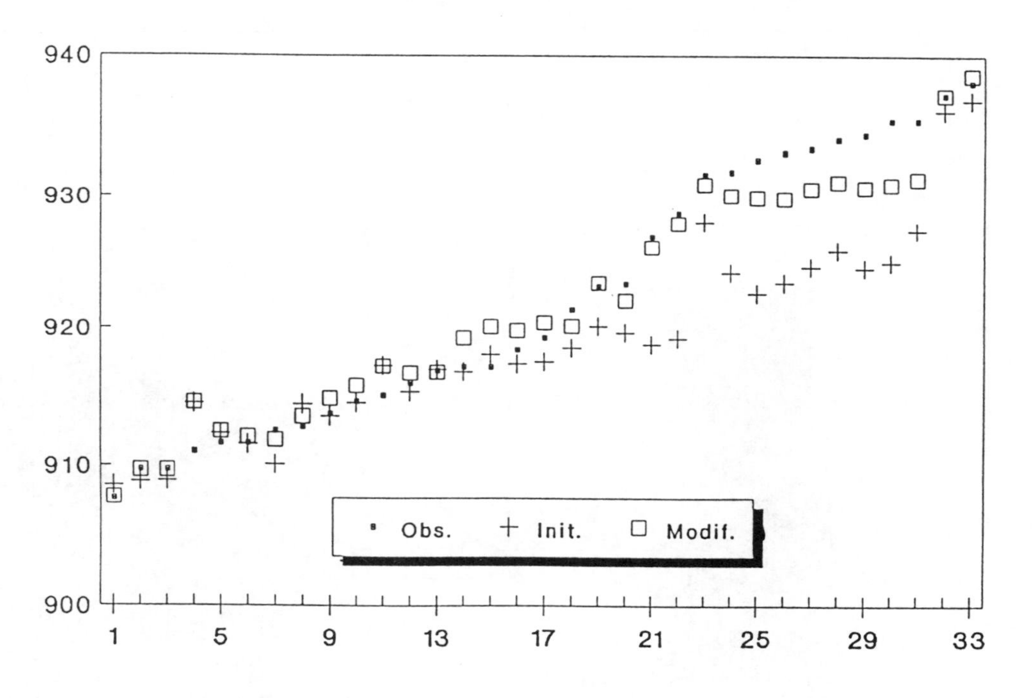

Figure 8. Illustration of the improved match to measured heads resulting from the use of the technique developed by the UPV project team. The figure presents the measured heads in order of size (shown by .), the corresponding heads computed for the realization conditioned on measured values of Z only (shown by +), and the corresponding heads computed after the realization was conditioned on measured heads as well. The improvement is considerable.

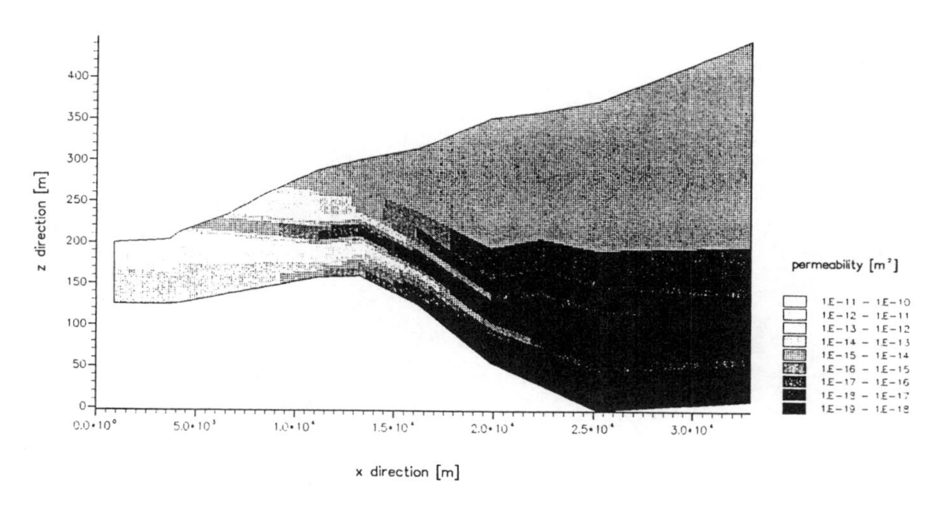

Figure 9. The distribution of permeability in the two-dimensional vertical cross-section model used by the BGR project team.

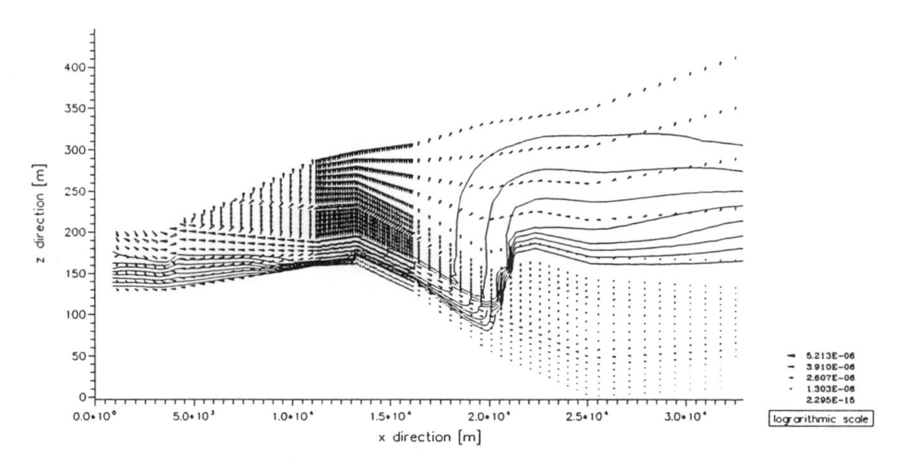

Figure 10. The steady-state flow field and the distribution of salinity for the two-dimensional vertical cross-section model used by the BGR project team.

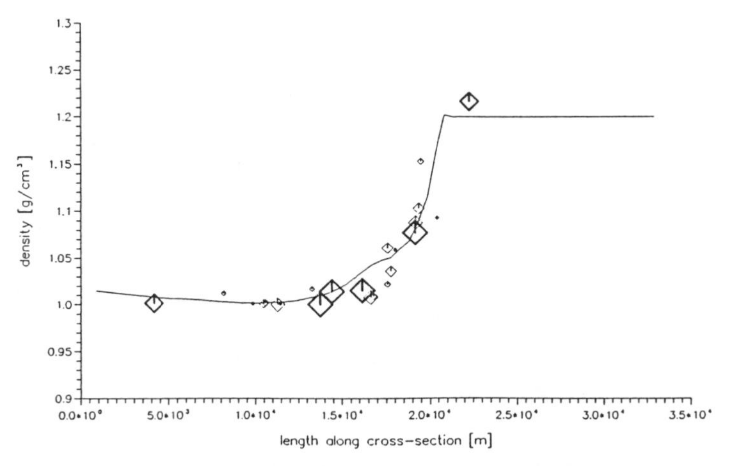

Figure 11. Comparison of computed and measured densities in the Culebra dolomite for the two-dimensional vertical cross-section model used by the BGR project team.

Conclusions from INTRAVAL Working Group 3: Salt- and Clay-Related Cases

Peter Bogorinski
Gesellschaft für Anlagen- und Reaktorsicherheit (GRS) mbH

Germany

Abstract

A number of countries consider sedimentary rocks to host a nuclear waste repository, the isolation potential of which relies mainly on very low permeabilities of those formations. To establish confidence in models used in future safety assessments INTRAVAL working group 3 analysed three test cases addressing the relevant processes which govern the transport of radionuclides in the host formation as well as in the overburden.

The *WIPP 1* test case studied the flow of brine from a bedded salt formation into open excavations under pressure gradients. Experiments carried out at different scales at the Waste Isolation Pilot Plant in New Mexico, USA were carefully analysed and compared to numerical simulations. It could be concluded from the results that brine movement through the formation under test conditions can be described with Darcy's law.

The *Mol* test case studied the main transport mechanisms for solutes through clay. Experiments with radioactive tracers were carried out for several years at the Mol underground laboratory, Belgium. The measurements were compared to numerical simulations. The results revealed that advective transport can be neglected for the time scale considered and molecular diffusion is the relevant process to describe the migration of radionuclides within this formation.

Salt leached from a salt dome influences the groundwater flow field by density variations. These processes were studied in the *Gorleben* test case. Measurements from a pumping test and measurements of salt content with depth collected at boreholes throughout the investigation area at Gorleben, Germany, were compared with the results of analytical and numerical studies. The pumping test could be easily analysed but it failed to show density effects on the flow field. However, it revealed the heterogeneity and anisotropy of the system. Simulations of the regional scale groundwater system in the area concentrated on a two-dimensional representative cross section. The results demonstrated that flow is slowed down in regions with high salt concentrations due to density effects.

None of the test cases involved validation in its true scientific sense. However, despite all limitations the analyses served the purpose to build confidence in the models applied.

1 Introduction

Radioactive waste arising from the operation of nuclear power stations as well as from the use of radioactive material in medicine, industry and research has to be isolated from the biosphere for long time periods to avoid hazards to people and environment. To obtain the long-term isolation required, final disposal of radioactive waste in deep geological formations is being considered in most countries as the safest option. A variety of different formations were chosen as potential host rock for such facilities depending on their availability in the different countries, among them sedimentary rocks like salt and clay. Working group 3 of INTRAVAL dealt with flow and transport processes in these formations.

An integral part of the long-term safety assessment of the nuclear waste disposal system is to study potential releases of radionuclide from the waste into the environment. Groundwater flow and associated transport of radionuclides is regarded as a major mechanism for radionuclides to return to the biosphere. The assessment of these processes requires a detailed understanding of the various physical and chemical phenomena involved. Models serve as a powerful tool for the prediction of releases of radionuclides from the disposal system into the geosphere, movement of water and contaminants through the geosphere, and calculation of radiation doses from radionuclides reaching the biosphere. The question then arises how much one can trust the results of such predictions. Answers to such questions are normally sought by the validation of models. As part of the validation process, models have to be tested on their capability to simulate experiments carried out at various scales.

In the frame of INTRAVAL Working Group 3 three test cases were selected to address those processes which are significant for the release of radionuclides from repositories constructed within salt and clay as host rocks and their transport through the surrounding geologic formations:

- The *WIPP 1* test case studies the movement of fluid within a bedded salt formation under a stress field originating from excavations. A number of in-situ experiments were carried out at the Waste Isolation Pilot Plant, New Mexico. Their scale range from small boreholes to large excavated rooms.

- The *Mol* test case studies the transport of radioactive tracers through a clay formation under pressure gradients at field scale. The experiments with a spatial scale of a few meters and a temporal scale of some years were carried out at the Mol underground laboratory.

- The *Gorleben* test case studies the influence of variable density of groundwater due to its salt content on the flow field. The test case is divided into two parts, namely the pumping test "*Weißes Moor*" and the regional model of the *Gorleben erosion channel.* Both are characterised by a highly heterogeneous geology with large permeability contrasts and a variation of salinity from freshwater near the surface to saturated brine at depth.

The paper summarises the results obtained during phase 2 of INTRAVAL. It is subdivided into four sections, one for each of the test cases *WIPP 1*, *Mol*, and *Gorleben*, describing the test cases, validation issues and results achieved. These are followed bv a summary and conclusions section.

2 *WIPP 1* Test Case

2.1 Description

The WIPP site is located in the northern region of the Delaware Basin in south-eastern New Mexico, USA. The underground facility lies in the lower part of the Salado Formation, which is of Permian age at an approximate depth of 655 m below ground surface. At the WIPP site, the Salado Formation is approximately 600 m thick and is composed largely of halite, with minor amounts of interspersed clay and polyhalite. It also contains interbeds of anhydrite, polyhalite, clay, and siltstone. 45 of the continuous anhydrite and/or polyhalite interbeds are designated as "Marker Beds" numbered downward from 100 to 144. Stratigraphically, the WIPP facility horizon lies between Marker Beds 138 and 139.

The *WIPP 1* experiments provide data on the hydraulic behaviour of the Salado Formation under a variety of testing conditions. The objective was to determine whether or not brine flow through evaporites under different in-situ test conditions could be accurately simulated with a Darcy-type flow model which may therefore be reliably used in safety assessments to predict the rates and volumes of brine inflow to the WIPP repository.

The tests were performed to evaluate formation responses under different conditions. Data from three types of experiments at three different scales, namely spatial from centimetres to meters and temporal from days to years, formed the basis of the test case:

- Small-scale brine-inflow experiments to measure the long-term flow of brine from the formation to boreholes maintained at atmospheric pressure.
- Pore pressure and permeability tests to provide information on pore pressures, permeabilities, and other hydraulic properties at different positions around the repository.
- Large-scale brine-inflow experiments (Room Q) to provide data on inflow to a room-sized opening in the halite, with supporting data provided by pore-pressure and permeability measurements in the surrounding rock.

The focus of the *WIPP 1* test case is to define an appropriate model for the long-term simulation of brine flow. The pore-pressure and permeability tests would provide boundary and initial conditions for the modelling, while the small-scale and large-scale brine-inflow experiments would provide data on the output of the system under the experimental conditions.

2.1.1 Small-Scale Brine-Inflow Experiments

For the small-scale brine-inflow experiments, boreholes with diameters of 10 cm and 1 m were drilled from the repository into the formation oriented either vertically downward or horizontally at lengths from 3 to 6 m. The accumulation of brine seeping into each borehole was measured as a function of time. The boreholes were open over their entire span, but each borehole opening was sealed to prevent moisture loss through evaporation and air circulation. Humidity measurements were made to aid in quantifying the total moisture entering a borehole.

The *WIPP 1* test case data set includes data from four 10 cm boreholes drilled vertically downward (DBT boreholes) intersecting a sequence of pure halite and argillaceous halite (figure 1). Brine-inflow measurements generally showed rapidly declining flow rates for the first few months after drilling, followed by steady or slowly declining flow rates for periods as long as two years. Initial flow rates ranged from about 5 to 25 g/day, while steady flow rates ranged from about 2 to 10 g/day. Inflow rates into some boreholes began to increase two to three years after the holes were drilled.

In addition data from two 10 cm and 36 cm horizontal boreholes drilled into the argillaceous halite in test room L4 were used. One of these boreholes produced as much as 25 g/day during the first few months after it was drilled, but the rate then declined to near zero, the other yielded only about 2 g/day of brine. The humidity in these holes appears to be in equilibrium with free-standing brine.

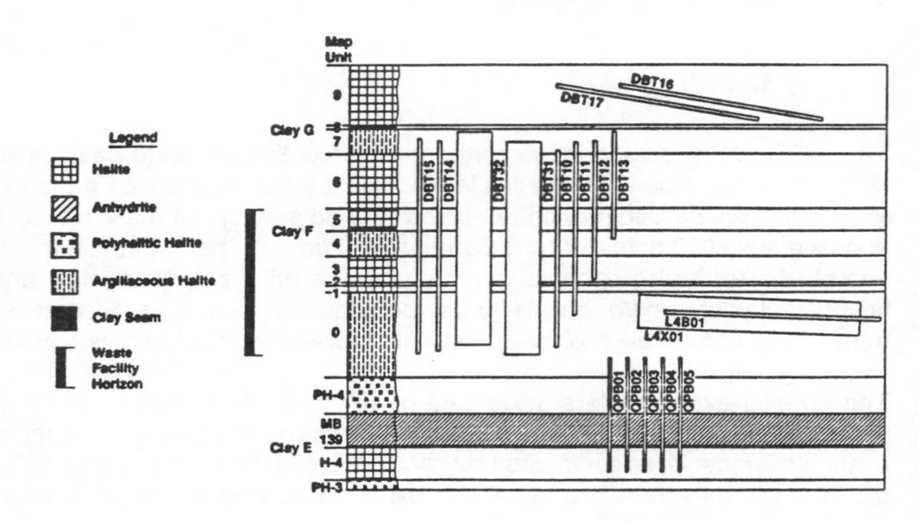

Figure 1. *WIPP 1*: Brine inflow boreholes

2.1.2 Pore-Pressure and Permeability Tests

The *WIPP 1* test case includes sets of permeability and pore-pressure experiments from argillaceous halite layers, anhydrite layers as well as sequences of halite, polyhalitic halite, and clay. Pore-pressure and permeability tests were conducted in 10 cm boreholes drilled in a variety of orientations through different layers at locations of older and younger excavations in the underground facility. The testing sequence in the test zone consisted mostly of an initial build-up period followed by pulse-withdrawal tests.

Pressures are measured in packer-isolated intervals of the boreholes. They range from 0.3 MPa (measured within 2 m of a room) to 12.5 MPa (measured at locations farther than 22 m from a room) whereas the lithostatic pressure at the WIPP repository level is 14.8 MPa, indicating a depressurised zone to exist around the repository. Stratigraphic heterogeneity and stress relief around the excavation may also have significant effects on the pressures observed in the test boreholes.

Permeability experiments in the boreholes involve pressure-pulse, constant-pressure flow, and pressure build-up tests. Pressures, temperatures and flow rates are measured as well as packer pressure, radial borehole deformation, and borehole elongation.

2.1.3 Integrated Large-Scale Experiments in Room Q

The large-scale experiment was conducted in test Room Q, a cylindrical room 110 m long and 2.9 m in diameter bored parallel to bedding in the upper host rock strata of the planned waste disposal stratigraphic interval. Its location represents an isolated, undisturbed area where pre-mining values for pore pressure and fluid flow could be established in a far-field location. Fifteen boreholes were drilled and instrumented to allow for assessment of pre- and post-mining hydrologic conditions. Data sets from four of these boreholes were included in the *WIPP 1* test case. However, no project team other than SNL modelled the room Q. Therefore these experiments will not be discussed further.

2.2 Validation Issues

The goal of the *WIPP 1* test case was to establish the applicability of Darcy's equation or alternative models to describe brine flow through evaporites under a pressure gradient. This test case examined the flow mechanisms at various scales, from boreholes 10 cm diameter to rooms with 3 m diameter. The applicability of a Darcy-flow model to evaporites was investigated, with possible refinements suggested dealing with two-phase flow, rock-creep effects, and coupling of fluid pressures to the stress field in the rock mass. The need for alternative models, such as one in which porosity becomes interconnected and brine is squeezed from the salt as a result of creep, was also considered.

Investigation of the appropriateness of a flow model includes identification of all relevant phenomena and quantification of all necessary parameters that fit into the model. The modelling efforts were expected to identify additional experiments, if any, needed to validate Darcy-flow or alternative models.

2.3 Models Used

The tests were analysed by three international teams who used different subsets of the data as the basis for their individual models, namely

- the Sandia National Laboratories (SNL), USA, as pilot group applied a variety of numerical models and analytical solutions
- the Rijksinstituut voor Volksgezondheid en Milieuhygiene (RIVM), Netherlands, used the METROPOL-2 code, and
- the École Nationale Supérieure des Mines de Paris (EdM) under the auspices of the Commissariat à l'Energie Atomique/Institut de Protection et de Sûreté Nucléaire (CEA/ISPN), France, employed a two-stage mathematical modelling technique using the HYDREF code.

2.4 Results

When comparing the parameters estimated by the three teams interpreting the small-scale brine-inflow experiments, allowances must be made for slight differences in the input parameters, i.e. initial formation pore pressure, brine density, and brine viscosity, used by the three teams and the effects of those differences on the interpreted parameter values. All results discussed here have been compensated appropriately for these differences.

A major difference between the RIVM and EdM interpretations on the one hand and the SNL interpretations on the other hand is the treatment of the strata tested. The RIVM and EdM teams distinguished between pure and impure halite in estimating values for permeability. They assumed that pure halite would have a lower permeability than impure halite and sought single values for each that, when combined with the different thicknesses of pure and impure halite in the different test holes, would allow replication of the entire set of experimental results. The SNL team treated the strata penetrated by each hole as homogeneous and sought values for hydraulic properties, different at each hole, that would best match the observed data. Thus, the RIVM and EdM teams determined values for the permeability of pure and impure halite that provided the best overall match to the brine-inflow data in the aggregate, but not to the brine-inflow data from any individual hole, while the SNL team determined average permeability values for each borehole.

To compare the RIVM and EdM results to the SNL results, the pure and impure halite permeabilities were weighted according to each individual borehole's stratigraphy to determine an average permeability for each borehole. These average permeabilities are shown in Table 1, along with the average permeabilities calculated by the SNL team using different parts of the data.

Table 1. ***WIPP 1***: **Comparison of Permeabilities as Calculated by Different Teams**

Borehole	Calculated Permeabilities, m^2				
	RIVM	EdM	SNL		
			Full Flux	Cum. Vol.	Late-Time
DBT10	$1.8\ 10^{-21}$	$9.9\ 10^{-22}$	$3.7\ 10^{-22}$	$6.6\ 10^{-22}$	$8.2\ 10^{-22}$
DBT11	$2.1\ 10^{-21}$	$1.2\ 10^{-21}$	$1.5\ 10^{-21}$	$1.6\ 10^{-21}$	$2.0\ 10^{-21}$
DBT12	$1.5\ 10^{-21}$	--	$8.3\ 10^{-22}$	$8.4\ 10^{-22}$	$2.6\ 10^{-21}$
DBT13	$8.5\ 10^{-22}$	$3.0\ 10^{-22}$	$2.3\ 10^{-22}$	$3.0\ 10^{-22}$	$4.3\ 10^{-22}$

* Assumes $p_\infty = 1.0 \cdot 10^7$Pa, $\mu = 2.1 \cdot 10^{-3}$Pa·s, $\rho = 1220$kg/m^3

The teams agree about the average permeability at each of the boreholes within a factor of five. The total range of inferred permeability values spans only an order of magnitude, from $2.3 \cdot 10^{-22}$ to $2.6 \cdot 10^{-21}$ m^2. These values are slightly lower than the range of 10^{-21} to 10^{-20} m^2 normally used to predict brine inflows to WIPP disposal rooms. Interpretations by the RIVM and EdM teams were constrained by their determinations of reasonable values for specific capacitance, while the interpretations by the SNL team treated specific capacitance as a free parameter to be fitted.

Figure 2 shows a comparison of the measured data from borehole DTB 10 with the results of the SNL calculation. The curves agree quite well in the first part of the experiment with rapidly decreasing inflow rates and the simulation gives a reasonable average fit for the later part when the inflow rates stay almost constant. None of the teams attempted to model the late-time data from the brine-inflow experiments that showed increasing rates of brine flow. For brine-inflow rates to increase with time, either the permeability around the borehole must increase, or the driving pressure differential must increase, or both. Increasing permeability is the most likely explanation for the increased inflow.

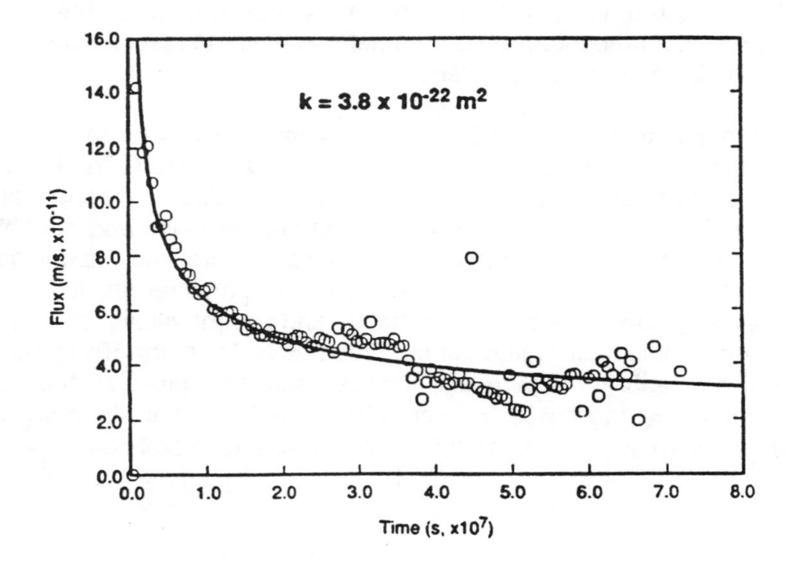

Figure 2. ***WIPP 1***: DBT10: measured and calculated brine inflow rate

Permeability could increase as halite creeps towards the overlying room. Existing fractures may dilate, new fractures may form, and bedding planes may separate as the salt creeps. Increasing permeability in any of these ways would, in effect, allow a borehole to draw brine from a larger volume of rock. However, simulation of creep-induced permeability changes and the resulting changes in brine-inflow rates would require coupling between a geomechanical model and a hydrological flow code.

The goal of the *WIPP 1* test case was to evaluate the use of Darcy's law to describe brine flow through evaporites. In a Darcy-flow regime, pressure and flow behaviour are largely governed by the parameters permeability and specific capacitance. The general approach taken to evaluate the applicability of Darcy's law was, therefore, to try to obtain values of permeability and specific capacitance that would be consistent with other available data and be able to provide reasonable simulations of all of the hydrology experiments performed. Even if it was not to be expected that single values of permeability and specific capacitance would lead to exact replication of all of the experiments, the results could be said to be consistent with a Darcy-flow model if all of them could be matched by values within an order of magnitude.

All of the test teams interpreting the brine-inflow experiments concluded that the average permeability of the halite strata penetrated by the holes was between approximately 10^{-22} and 10^{-21} m^2. The teams that interpreted the permeability experiments, which included a distinct clay seam in addition to halite, found the average permeability to be between 10^{-21} and 10^{-20} m^2. Specific capacitances from the brine-inflow and permeability experiments ranged from 10^{-11} to 10^{-9} Pa^{-1}. However, values in the order of 10^{-10} and 10^{-9}Pa^{-1} are inconsistent with the known constitutive properties of halite, and are attributed to deformation, possibly ongoing, of the halite around the WIPP excavations.

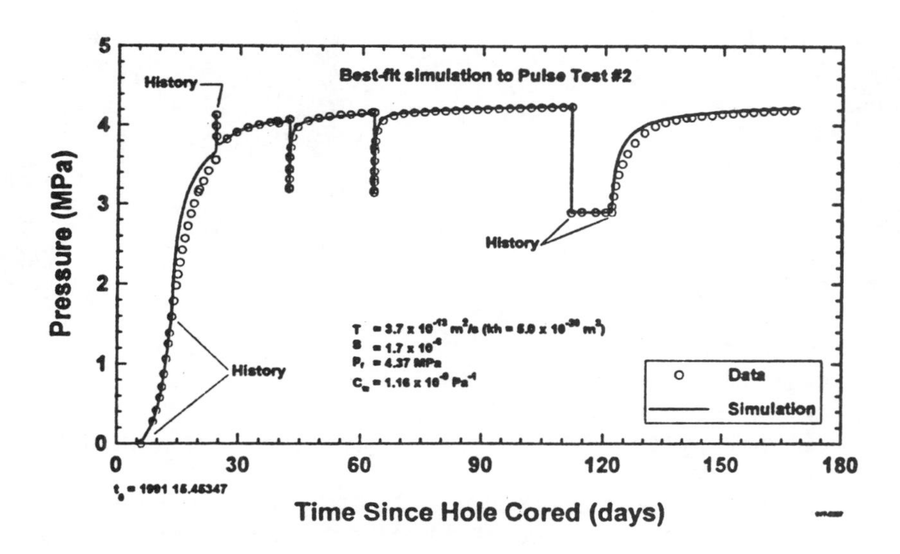

Figure 3. *WIPP 1*: Calculated and measured pressures during S1P73-B testing

Results of one of the permeability tests and the comparison the simulation performed by SNL is shown in figure 3. This test included an anhydrite band (MB-138) and a clay seam (Clay K). A value of $5{\cdot}10^{-20}$ m^2 provided the best fit for the permeability averaged over the total length of the test zone. Assuming the fluid being produced only from the anhydrite and the clay, the permeability of anhydrite was found to be 2.9 x 10^{-19} m^2, roughly two orders of magnitude higher than the permeability of halite. This difference was attributed to the presence of fractures in the anhydrite. The SNL team concluded from their simulations that both numerical and analytical radial-flow models based on Darcy's law provided good fits to both pressure-transient and flow-transient tests using consistent parameter values.

While Darcy's law cannot be said to have been validated by the test case, the interpretations performed by the different teams have demonstrated that models based on Darcy's law are able to replicate the experimental results within reasonable and acceptable bounds. The interpretations presented by the INTRAVAL teams cannot be considered unique in terms of representing the only possible model that could fit the data, but neither do they provide any motivation to look for an alternative model that might fit the data better. To the extent that the simulations differ from the experimental results, refinements of the basic Darcy-flow model are suggested, rather than replacement with a fundamentally different model.

3 *Mol* Test Case

3.1 Description

The *Mol* test case is based on experiments carried out at the Mol Underground Research Facility (URF). The site is located in the northern part of Belgium in a rather level area with maximum height differences of a few meters. The facility has been constructed at a depth of 224 meters in the Boom clay formation underlying the nuclear research centre. The clay layer has a thickness of 110 m. It is the uppermost of an alternating sequence of clay and sand deposits of 30 million years of age. In the facility a multiple screen piezometer has been installed at the end of a horizontal drift, consisting of a pipe with a diameter of 4.6 cm which penetrates 10 m into the clay formation and contains 9 equally spaced filters of 8.5 cm length (figure 4). This installation is used to perform the large scale in-situ tritiated water (HTO) injection experiment. After the clay had settled HTO tracer was injected into the fifth filter with a flow rate of 5.61 ml/day for fifty days. After the end of this injection period the tracer concentration arriving at the other filters was monitored by sampling at regular time intervals.

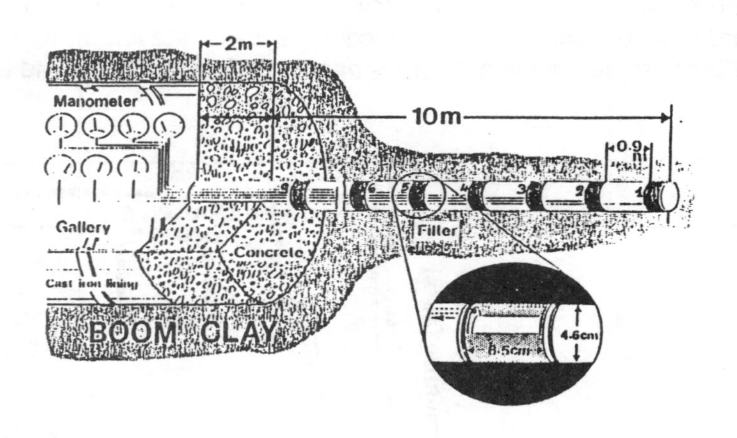

Figure 4. *Mol*: Experimental setup

A vertical experimental shaft lined with concrete bricks that had been built near the end of the URF imposes an enhanced hydraulic pressure gradient in the vicinity of the piezometer nest.

3.2 Validation Issues

The parameters required in safety assessment studies are determined in the laboratory on clay samples taken from the underground. Different types of migration experiments are performed on the samples. Reshaped and reconsolidated clay plugs are used for Flow Through type diffusion experiments (1,2). Clay cores drilled parallel with and perpendicular to the stratification of the formation are used for percolation experiments (3). The values of the migration parameters obtained from the small scale laboratory experiments are used for long term safety assessment calculations (PAGIS and PACOMA studies).

Despite the precautions taken during sample collection and the preparation of the diffusion experiments, parameters determined in the laboratory are subject to uncertainty. To improve the confidence in the laboratory data and to validate the safety assessment model, a large scale in-situ experiment has been set up.

The goal of the in-situ experiment was to obtain a good estimation of the hydraulic parameters of the Boom clay formation and to carry out a large scale in-situ migration test. The purpose of the large scale migration test is to validate the safety assessment model and the migration parameters used for the long time predictive calculations.

3.3 Models Used

Six teams analysed the test case:

- SCK/CEN as the pilot group employed the MICOF code which incorporates a three-dimensional analytical solution of the transport equation;
- the RIVM team from the Netherlands used their finite element code METROPOL;
- the French EdM team used their finite element code METIS,
- the second French team CEA/DMT used TRIO-EF, whereas
- the third team from ANDRA/BRGM studied the test case by means of an analytical solution of the dispersion-advection equation;
- the American SNL team employed the SWIFT code which solves the transport equation by means of a finite difference technique.

3.4 Results

The main effort in simulating the test case stayed with the pilot group of SCK/CEN. Figure 5 shows the measured concentration in the filters CP1/4, CP1/5 and CPI/6 as a function of the time since injection of the tracer and the simulation results as calculated by SCK/CEN with the MICOF code.

The simulations were carried out prior to the experiment. The two different curves (solid and dashed lines) for each monitoring distance upstream and downstream from the injection point reflect the influence of advective transport due to the pressure gradient towards the experimental shaft. However, the fact that the curves are very close indicates that tracer transport is mainly diffusive, a view that is confirmed by the respective measured values (triangles). The analyses of the other teams support this. The SNL team concluded that a purely diffusive model is appropriate for the purpose of long-

Parameters used:

hydraulic conductivity, m/s	3.2×10^{-12}
hydraulic gradient, m/m	-18.9
diffusion accessible porosity	0.35
dispersion length, m	0.002
retardation factor	1
half live HTO, years	12.28
Darcy velocity, m/s	$6.0\ 10^{-11}$
horizontal diffusivity, m^2/s	$4.1\ 10^{-10}$
vertical diffusivity, m^2/s	$2.0\ 10^{-10}$

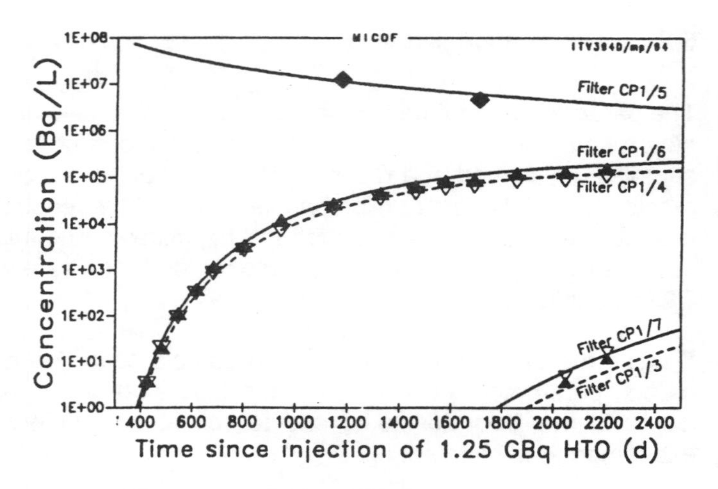

Figure 5. *Mol*: Measurements and modelling results

term safety analyses for a nuclear waste repository in a clay formation, i.e. the diffusive model can be assumed to be validated in this respect.

4 Gorleben Test Case

4.1 Description

The *Gorleben* test case proposed by the Bundesanstalt für Geowissenschaften und Rohstoffe (BGR) is based on the hydrogeological investigation program carried out on behalf of the Bundesamt für Strahlenschutz (BfS) for characterisation of the potential German site for a HLW repository at Gorleben. The site is located in the north-eastern part of Lower Saxony, Germany. The salt dome which is proposed to host the repository is approximately 14 km long, 4 km wide and its base is more than 3000 m below land surface. As result of erosion during the Elsterian glaciation channels were cut into the land surface by melting waters. One of these channels is the Gorleben erosion channel which crosses the salt dome in approximately south-north direction. It is roughly 10 km long, 2 km wide and 300 m deep. Tertiary and Quaternary sandy and gravely sediments form a system of two aquifers which are separated by the Lauenburg clay complex. In some places the lower aquifer is in direct contact with the salt dome. Thus the deep groundwater is highly saline in this region due to salt dissolution. The transition zone from near-surface freshwater to saltwater ranges between 50 and 150 m depth.

The ground surface is quite flat with elevation differences of 5 - 15 m. In the regions of higher elevations, the groundwater table is about 8 m higher than the level of the discharge areas. The recharge is about 180 mm/yr south of the salt dome. The regional groundwater flow is predominantly from the south-east to the north-west with a hydraulic gradient in the uppermost aquifer of 1.5·10-3m/m. Darcy velocities range from 5 to 250 m/yr. The groundwater infiltrates almost to the top of the salt dome at 200 - 250 m depth, where saline water is found. The interface between fresh and saline water is at a depth between 150 m and 50 m. During the site investigation programme 145 boreholes were drilled to carry out field experiments like core sampling and geophysical logging, water sampling, velocity measurements, and pumping tests thus acquiring data on the hydrogeological structure, the hydraulic parameters and the groundwater movement and the salinity distribution.

Two datasets had been selected for validation purposes within the INTRAVAL project, the first concerning the pumping test "*Weißes Moor*" and the second the groundwater flow field and the salinity distribution in the *Gorleben erosion channel.*

4.2 Validation Issues

Usually groundwater near salt formations contains dissolved salt with concentrations varying with depth from saturated brine in the immediate vicinity of the salt formation to freshwater near the surface. The dependency of the fluid density on the salt concentration influences the movement of the groundwater due to buoyancy effects. If this results in slowing down the groundwater velocity in the vertical direction as is commonly assumed, it will have a large impact on the migration of radionuclides. Therefore this phenomenon has to be properly accounted for in safety assessments since simulations must be based on a good knowledge of the groundwater flow field.

To address these issues the *Gorleben* test case studies the groundwater movement in an erosion channel crossing the Gorleben salt dome. The aim is to compare simulation results with the measured salinity distribution on a spatial scale ranging from a few hundred meters in the case of the pumping test "Weißes Moor" to ten kilometres in the regional flow study.

4.3 Models Used

Six project teams carried out simulations on this test case, namely

- Bundesamt für Strahlenschutz (BfS), Germany, by means of an analytical Theis solution and of the finite element codes SUTRA and ROCKFLOW
- Sandia National Laboratories (SNL), USA, with the computer code INTERPRET/2 employing a Theis solution,
- Rijksinstituut voor Volksgezondheid en Milieuhygiene (RIVM), The Netherlands, using the analytical computer code AQ-AT and the finite element code METROPOL,

on the pumping test "*Weißes Moor*" whereas

- Bundesanstalt für Geowissenschaften und Rohstoffe (BGR), Germany, as the pilot group using the finite element code SUTRA
- Gesellschaft für Anlagen- und Reaktorsicherheit (GRS), Germany, by means of the finite element code NAMMU

studied the regional flow system of the *Gorleben erosion channel.* A geostatistical study of the hydrogeologic system was carried out by

- AEA Decommissioning and Waste Management (AEA D&W), United Kingdom.

4.4 Results

4.4.1 Geostatistical Analysis

The geostatistical approach used is based on an indicator methodology which is applied in oil or mining exploration to obtain qualitative information. Similar information is needed at a site like Gorleben where an important issue for safety studies is whether the Lauenburg Clay complex covers the lower aquifer completely or whether connectivities exist between the lower and the upper aquifers near the discharge region.

The same geological cross section along the channel axis which was selected to study the regional flow system has been used to test the methodology. The borehole logs as they were given in the Gorleben data set were analysed in terms of a binary indicator function which depends on whether at a particular depth of the borehole the log consisted of clay, for which a value of 1 was assigned to the indicator function, or of higher permeable material, e.g. sands or silts, for which a value of 0 was assigned. With this information a variogram analysis of the indicator function was carried out, for which the best defined results were obtained when the co-ordinate were transformed in a way that certain stratigraphic horizons were mapped to a horizontal line. These variograms were used to perform indicator kriging which provides an estimate of the probability that a particular type of material is present at a certain location. Assuming the .5 probability contour to be the interface between clay and non-clay the results were then compared to the cross section provided by the hydrogeologists. Some discrepancies were observed which pointed to particular problems in the interpretation of the borehole logs. In one place it was obvious that the original log as it was supplied in the test case data set didn't contain any clay. However, for the preparation of the hydrogeological cross section the data had been re-assessed. The re-interpretation of this particular borehole log identified a certain section as clay rather than silt but this information had not been supplied with the data set. This accounts for the discrepancy between the hydrogeologists interpretation and the cross section obtained by kriging.

Of course the ultimate aim of the work is not simply to reproduce manually generated geological cross sections but to use indicator simulation methods to scope the uncertainty in the results of groundwater flow calculations arising from the uncertainty about the geological model. This was not possible within the frame of the INTRAVAL project, however, an analysis like the one carried out is still a valuable test of the methodology.

4.4.2 Pumping Test "*Weißes Moor*"

The pumping test "*Weißes Moor*" has been evaluated using both analytical solutions and numerical simulations. The analyses carried out by the RIVM and SNL teams employing a Theis approach demonstrated that this analytical solution is only appropriate if modified to account for lateral anisotropy in the aquifer system and for boundary effects. Another problem in the interpretation of the test arises form partially penetrating wells. However, this effects mostly the monitoring boreholes in the immediate vicinity of the pumped well, i.e. at distances of a few tens of meters. Those boreholes should not be included in a Theis type analysis. Figure 6 shows the result of the SNL team for one observation borehole. It was not possible to fit the complete time history of the test. If the drawdown part is fitted the recovery is slower than measured and if the recovery is fitted the initial pressure is calculated to be lower than measured. Both observations indicate a rising water level during the experiment.

The lateral anisotropy and heterogeneity of the system was further studied by RIVM by means of a horizontally two-dimensional numerical model. The hydrogeological data used were based on the available geological information including estimates of the aquifer thickness. They assumed that the main axis of the permeability tensor coincides with the channel axis. The hydrology of the system was characterised by six parameters namely the storativity, the anisotropic permeabilities and multiplication factors to account for the heterogeneous permeability in three different regions. With that is was possible to obtain a reasonable fit for the pumping period of the test.

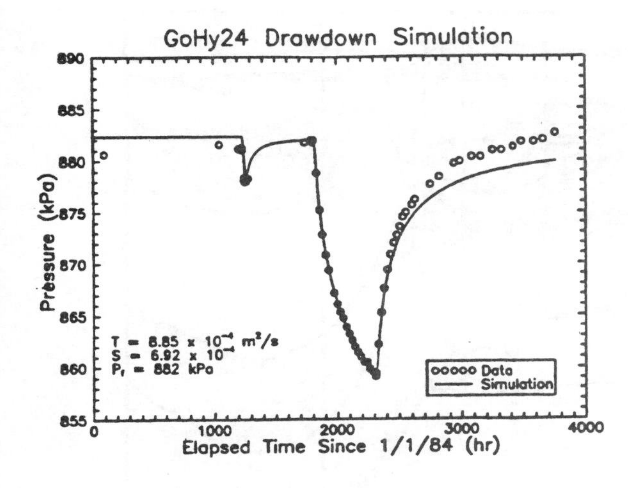

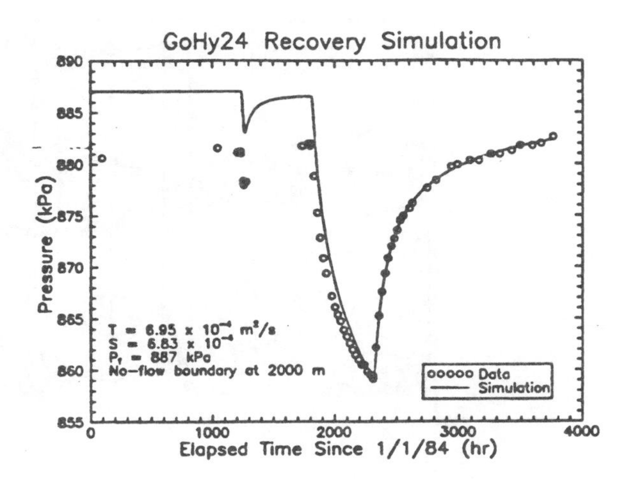

Figure 6. "*Weißes Moor*": Drawdown at observation borehole GoHy24

Despite the limitations mentioned above the Theis solution provided a reasonable estimate for the average values of aquifer transmissivity and storativity for use in numerical simulations even if anisotropy and heterogeneity were not taken into account. This was demonstrated by the BfS team. They used those values to construct a vertically two-dimensional model to study the impact of density variations on the flow field during the pumping test. They used two different codes which gave nearly identical results in the drawdown curves if a constant density approach was used. These drawdown curves in turn agree very well with the Theis solution except for the immediate vicinity of the pumped well. This can be attributed to two-dimensional effects. However, no significant differences were observed between results obtained from constant and from varying density models.

4.4.3 Regional Flow Model of the Gorleben Erosion Channel

The first intention on the study of the regional flow model of the *Gorleben erosion channel* was to study the system with a three-dimensional model. Freshwater studies were carried out by GRS within this frame. But facing the enormous amount of resources needed in both computer memory and computing time which became obvious in these studies as well as in pre-studies carried out by BGR this idea was abandoned. A two-dimensional cross section was selected along the channel axis which was modified to account for connectivities between the lower and upper aquifers not present at the exact location of this particular cross section (figure 7).

Transient salt transport simulations were performed by the BGR and the GRS teams since it is not possible to obtain the current measured salinity distribution with a steady state approach as pre-studies showed. Both teams used different discretisations for their models and different approaches for the initial state of the system.

The BGR team defined a salinity distribution with a transition zone of 20 m from saturated brine to nearly freshwater at a depth of approximately 180 m as initial condition, which was based on assumptions about the salinity distribution under permafrost conditions during the last ice age. A sensitivity study was carried out on the depth of this transition zone. Only one of these variations resulted in a salinity distribution after 10000 years simulation time which has similarity to the measured one (figure 8a). A second sensitivity study was performed on potential connectivities between the upper and the lower aquifers in the discharge region. Although this analysis showed significant influence on

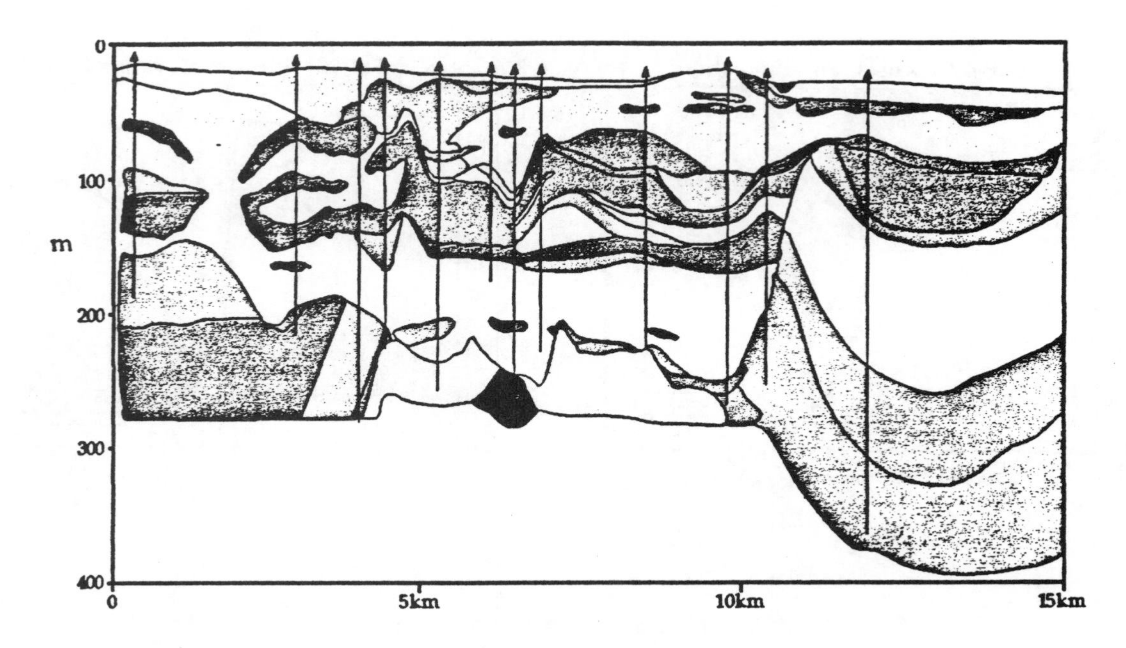

Figure 7. ***Gorleben erosion channel*:** Geological cross section

the calculated salinity distribution. Finally the importance of the transverse dispersivity has been demonstrated. The use of a high value resulted in a bigger spreading of the transition zone than observed in the field investigation.

The GRS team used for their model a very fine discretisation to overcome limitations on the size of the dispersion length due to numerical stability criteria. It was assumed that no salt was present in the system at the beginning of the simulation. Saturated brine was allowed to enter the system at the contact area of the lower aquifer with the salt dome. During the simulation the lower parts of the aquifer system filled with brine slowing down the vertical groundwater velocities. Figure 8b shows the

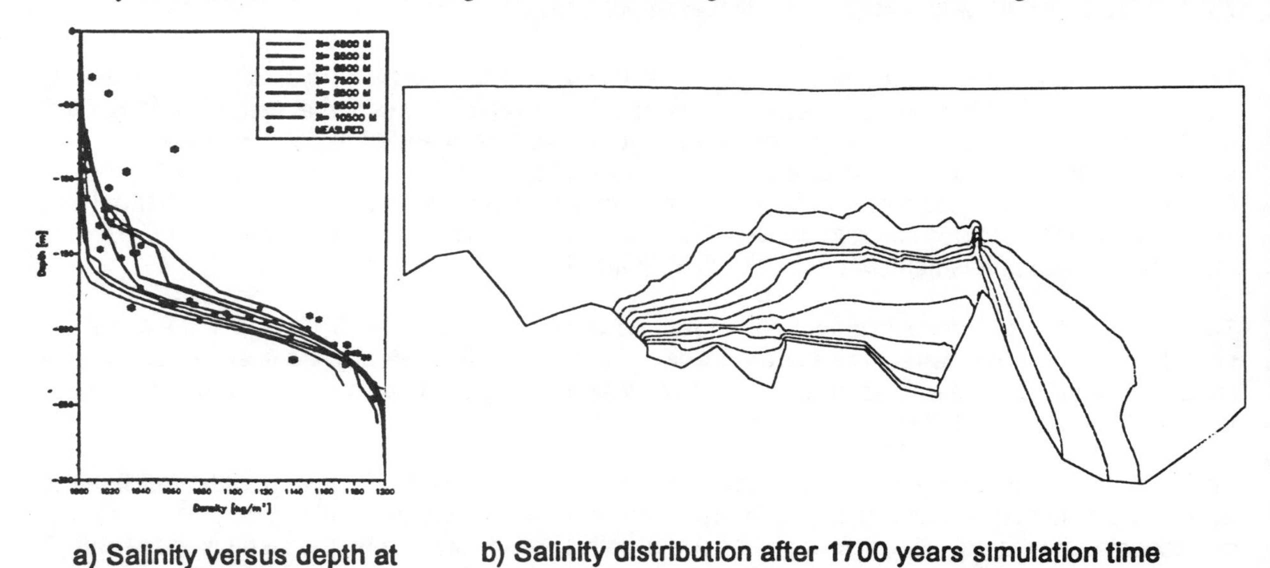

a) Salinity versus depth at 1000 years

b) Salinity distribution after 1700 years simulation time

Figure 8. ***Gorleben erosion channel***

state of the system after 1700 years of simulation time when salt has already entered the right hand side of the model.

Better funded assumptions on the initial salinity distributions using information from different scientific disciplines like paleohydrology will be useful for a better understanding of the current state of the system. To validate variable density flow models specially designed experiments will also be needed.

5 Summary and Conclusions

Three different issues relevant to waste disposal in evaporites and clay were addressed in the analyses performed within INTRAVAL Working Group 3.

WIPP 1 tested the applicability of Darcy's law to the movement of brine under pressure gradients in salt formations. In principle fluid flow in anhydrite and clay layers may be described by a Darcian flow model. However, it cannot be stated that this holds also for pure halite. Coupled hydrological and geomechanical models may provide a better understanding of the system. No evidence had been found, that brine movement alone my lead to significant releases of radionuclides from a repository in bedded salt. Combination with other processes such as gas generation may lead to different conclusions.

The *Mol* test case studied migration of tracers through a clay formation. Blind predictions of the experiments confirmed that advective flow may be neglected in such a system and that pure diffusion models are acceptable on the temporal and spatial scales involved. However, extrapolation to larger distances and longer times remains an open issue. Further experiments to study the importance of the dimensionality are currently under way.

Data from site characterisation were used in the *Gorleben* test case to test models on variable density flow. The analysis of the pumping test "*Weißes Moor*" didn't reveal any effects of density variations on the flow field. However, taking into account heterogeneities and anisotropies of the aquifer system proved to be important for obtaining a realistic interpretation of the flow field. The study of the regional flow system of the *Gorleben erosion channel* demonstrated that fully three-dimensional simulations of salt transport are currently not feasible due to limitations in computer resources. Improvements are needed in the understanding of the past evolution of the system to cope with the long time periods involved in the development of the salinity distribution observed today. Specially designed field experiments will be useful for future efforts in validating variable density flow and salt transport models.

6 Acknowledgements

The author thanks the pilot group leaders Rick Beauheim (SNL), Martin Put (SCK/CEN) and Klaus Schelkes (BGR) for providing the necessary information on the respective test cases and the results. GRS' participation in INTRAVAL was funded by the Federal Minister for the Environment, Nature Conservation and Reactor Safety.

Conclusions from WG-4: The Analyses of the Alligator Rivers Natural Analogue

K. Skagius, L. Birgersson
Kemakta Konsult AB

(Sweden)

Abstract

The Alligator Rivers study is based on work conducted at the Koongarra uranium deposit in the Alligator Rivers Region about 200 km east of Darwin, Australia. The Alligator Rivers Analogue Project (ARAP) was set up in 1987 and was later included as a test case in both phase 1 and 2 of INTRAVAL. The objective was to develop a consistent picture of the processes that have controlled the transport in the weathered zone of the Koongarra ore deposit and the time scale over which they have operated.

Uranium mineralisation occurs at Koongarra in two distinct but related ore bodies. Primary mineralisation in the main ore body is largely confined to quartz-chlorite schists and secondary uranium minerals are present from the surface down to the base of weathering at about 25 m depth and forms a tongue-like body of ore dispersing downslope for about 80 m. The primary ore body at Koongarra is estimated to be 1000 million years old and geomorphological information indicates that weathering started a few million years ago.

The work included in INTRAVAL phase 1 was mainly concentrated to hydrogeological and geochemical modelling which produced results that in INTRAVAL phase 2 were used in modelling simulations of the uranium migration. The model concepts applied in the migration modelling were based on rather simple performance assessment models accounting for advection, dispersion and linear sorption in one or two dimensions. One 1-D model was extended to include α-recoil and transfer between solid phases. The vertical movement of the weathering front was included in the 2-D model.

Studies of the Alligator Rivers Natural Analogue has demonstrated that the system is very complex. The interaction of many geochemical and geohydrological processes occurring over long times makes it difficult to create a quantitative model of the history of groundwater flow and nuclide transport. The study has shown the importance of a joint interpretation of different types of data and an iterative procedure for data collection, data interpretation and modelling in order to get a consistent picture of the evolution of the site. Furthermore, it was shown that sorption is a major retardation mechanism, that uranium fixation in crystalline phases is a potentially important retardation mechanism in geologic media where significant alteration of the rock is expected, and that α-recoil may have an impact on the distribution of uranium isotopes in the water. Modelling simulations indicated migration times in fair agreement with independent geomorphological information. A general conclusion from the INTRAVAL study is that rather simple and robust concepts and models seem able to adequately describe the long range migration processes that have occurred.

Introduction

Mathematical models applied for radionuclide migration calculations from a repository must be evaluated in terms of the validity and reliability of the predictions. One procedure is the comparison of model predictions with experimental observations. Here, information from laboratory experiments, field experiments and natural analogue studies is needed. Laboratory and field experiments yield information on relatively short time scales. Natural analogues may be viewed as naturally occurring experiments that have continued over long time periods. Therefore, they provide important information regarding the understanding of the scientific basis for the long term prediction of radionuclide migration within geological environments that are relevant to radioactive waste repositories.

One of the best studied natural analogues is the uranium deposit at Koongarra in the Alligator Rivers region of the Northern Territory in Australia, where extensive field studies for characterisation have been carried out since 1981. The site was initially investigated with the intention to develop a uranium mine, but was later used as a research facility with the intent to get an understanding of processes that have acted upon the original uranium deposit and have resulted in the present uranium ore.

The modelling work of the Alligator Rivers test case in INTRAVAL phase 1 was a joint venture between The Alligator Rivers Analogue Project (ARAP) and INTRAVAL. ARAP is an international study set up under the auspices of OECD Nuclear Energy Agency. The ARAP was completed in 1992.

The modelling work of this test case during INTRAVAL phase 1 was mainly focused on hydrology, geochemistry and geochemical processes. The main aim of the phase 1 modelling was to obtain a conceptual platform for the migration modelling in phase 2. Participants in the modelling exercises were University of Arizona (USA), RIVM (the Netherlands), AEA Technology Harwell Laboratory (United Kingdom), Johns Hopkins University (USA), CSIRO (Australia), ANSTO (Australia) and Kemakta/SKI (Sweden).

The objective of the Alligator Rivers test case in INTRAVAL phase 2 was to develop a consistent picture of the processes that have controlled the transport in the weathered zone and the time scale over which they have operated. Another objective was to test and evaluate models and concepts used in performance assessment of radioactive waste repositories. The main contributors to the test case in INTRAVAL phase 2 were the National Institute of Public Health and Environmental Protection (RIVM) and Kemakta Konsult. The results of their work are summarised in this report based on Skagius et al. [*Skagius ϵ al. 1993*] and van de Weerd et al. [*van de Weerd et al. 1994*]. It should however be noted that additional migration modelling has been carried out by ANSTO and other organisations in the ARAP. This work is only briefly presented in this paper which has mainly been limited to describe the work that was carried out within INTRAVAL.

Test Case Description

The natural uranium mineralisation occurs in two distinct but related ore bodies, separated by 100 m of barren schists, which strike and dip broadly parallel to a fault, the Koongarra Reverse Fault. The main ore body, which was the subject of the study, persists to 100 m depth. It has a strike length of 450 m and a width that averages 30 m at the top of the unweathered schists (Figure 1).

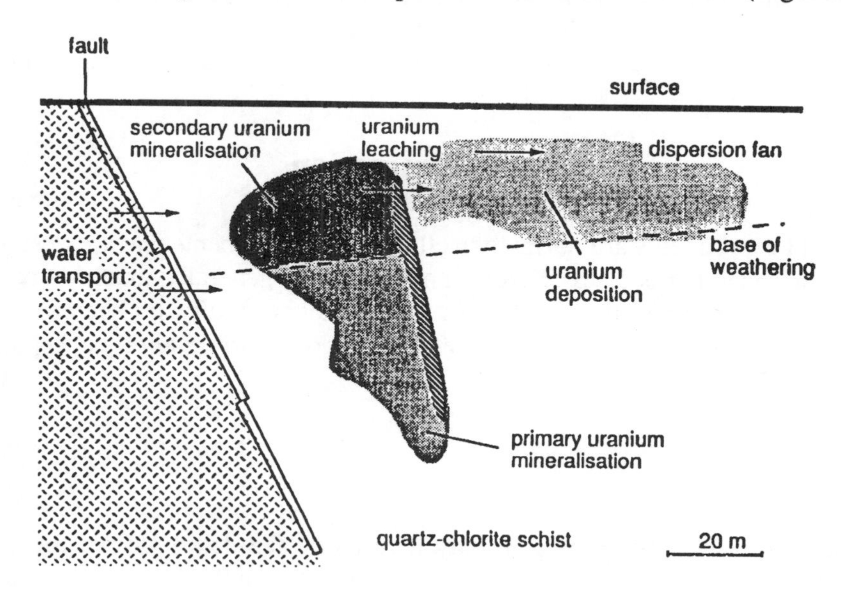

Figure 1. **Cross-section of the main ore body at Koongarra.**

The primary ore consists of pitchblende or uraninite–uranium dioxide veins and veinlets which either follow or cross–cut the layering in the schist. Minor amounts of scattered sulphide minerals, primarily galena (PbS), chalcopyrite ($CuFeS_2$) and pyrite (FeS_2), are associated with the high–grade ore. Uranium in this primary ore zone has been mobilised through oxidation and subsequently immobilised through the formation of a uranyl silicate zone. The schist is weathered from a level near the surface down to a depth of 25–30 m. In this weathered zone, another secondary uranium mineralisation, consisting of uranyl phosphates, forms a tongue-like fan. Away from the tail of the fan, uranium is dispersed in the weathered schists and adsorbed onto clays and iron oxides. The dispersion fan of ore-grade materials extends down-slope for about 80 m. The transformation of the original primary mineralisation zones into the present secondary minerals is perceived to have taken place during a period of well over one million years.

Interpretation of the present hydrology at the site suggests that recharge of groundwater to the weathered Cahill schists occurs via downflow parallel to and in close proximity to the reverse fault in the underlying Kombolgie sandstone and schists [*Duerden, 1992*]. Although the fault zone breccia was found to be practically impermeable, some water has been found to flow from the Kombolgie sandstone into the schists via cross fractures that offset the fault. Once in the schist, the groundwater flow is towards the south and south-east, away from the sandstone cliffs behind the deposit, see Figure 1. Like much of northern Australia, the studied region has a monsoonal climate with almost all rainfall occurring in the wet season between November and March. This causes fluctuations in the water table in the order of 5 to 10 m between the wet and dry seasons.

An extensive experimental programme including both field and laboratory investigations has resulted in a large amount of data characterising the site. Hydrogeological data were taken from drawdown and recovery tests and water pressure tests. Geological data are based on mineralogic and uranium assay logs from 140 percussion holes and 107 drill cores in the immediate vicinity of the uranium deposit. Groundwater chemical data are compiled from more than 70 boreholes and give a general understanding of the groundwater chemistry. Distributions of uranium, thorium and radium isotopes have been determined in the different mineralisation zones. The distribution of uranium and thorium between different mineral phases in the weathered zone has also been studied. Laboratory sorption experiments have been performed, using samples from drill cores retrieved at the site. In addition, distribution coefficients have been measured on natural particles in Koongarra groundwaters.

Data Examination

Laboratory and field experiments are normally well defined experiments that are designed and carried out to study one or several processes. Natural analogues are generally much more complicated to evaluate since the conditions and time scales are usually not well known. Therefore, both Kemakta and RIVM devoted the data examination a lot of effort in order to obtain a conceptual platform for the subsequent migration modelling.

Kemakta

The data examination by Kemakta was primarily focused on the data required for the modelling of the selected system, such as the distribution of uranium in the bulk rock and its individual mineral phases. Radiochemical data were used to study alternative processes of radionuclide migration.

Uranium concentration in solid materials

Uranium concentrations in solids at the Koongarra ore deposit has been collected and reported within ARAP. Most data originate from the transect 6110 mN, see Figure 2. In addition to the ARAP data, a large data set of average uranium concentrations over 5 m drillcore sections is available from exploratory drillings made by a mining company. The observed uranium concentration in solid phase in the transect 6110 mN are shown in Figure 3.

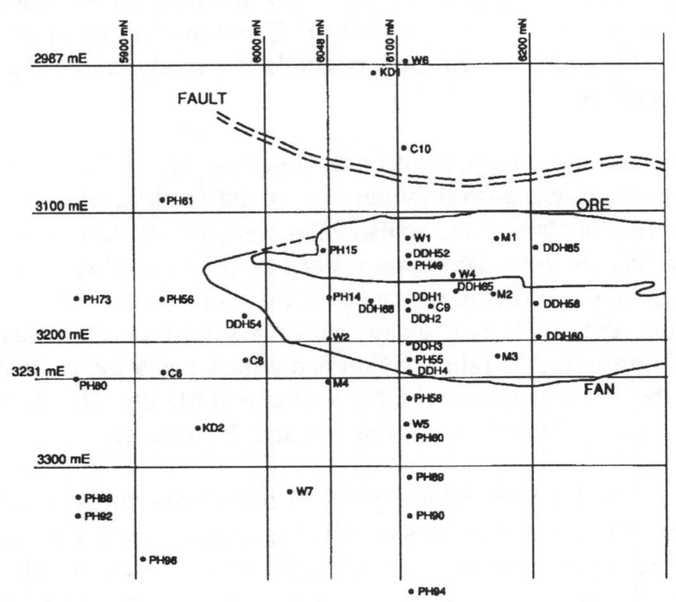

Figure 2. **Area of transect 6110 mN.**

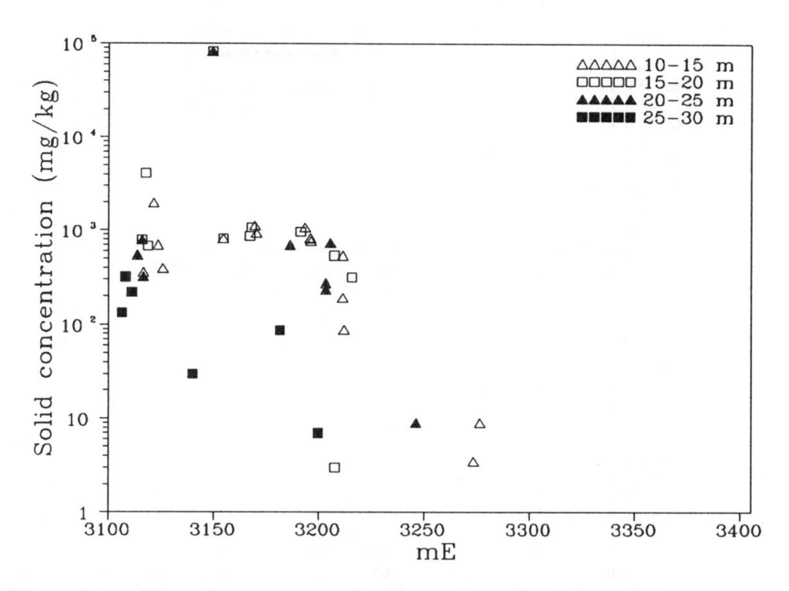

Figure 3. **Kemakta. Uranium concentrations in solid phase in transect 6110 mN.**

Radiochemical data

Uranium and thorium concentrations and activity ratios have been determined in 'amorphous' and 'crystalline' phases of drillcore samples using selective extraction procedures [*Duerden, 1992*]. The amorphous phase was defined as a poorly ordered material, such as ferrihydrite, accessible to and in approximate equilibrium with groundwater. This phase is in the following referred to as the *accessible* phase. The crystalline phases are highly ordered materials such as hematite, goethite, clays and quartz with contained species inaccessible to the groundwater. This phase is in the following referred to as the *inaccessible* phase. Species adsorbed to the surfaces of crystalline phases, mainly iron oxides and clays, are accessible to groundwater and are therefore considered belonging to the accessible phase.

Examples of observed uranium concentrations and $^{234}U/^{238}U$ and $^{230}Th/^{234}U$ activity ratios in the accessible and inaccessible phases along the transect 6110 mN are shown in Figure 4. Most of the uranium seems to be contained in the inaccessible phase. The observed activity ratios, especially in the accessible phase, decrease with distance from the upstream boundary of the ore zone within the dispersion fan, but increase at distances beyond the concentration front, as the uranium concentration reaches background levels. The activity ratios in the accessible phase are generally lower than in the inaccessible phase, indicating a more recent deposition of uranium in the accessible phase and/or a preferential transfer of ^{234}U and ^{230}Th from the accessible to the inaccessible phase, for example as a result of α-recoil.

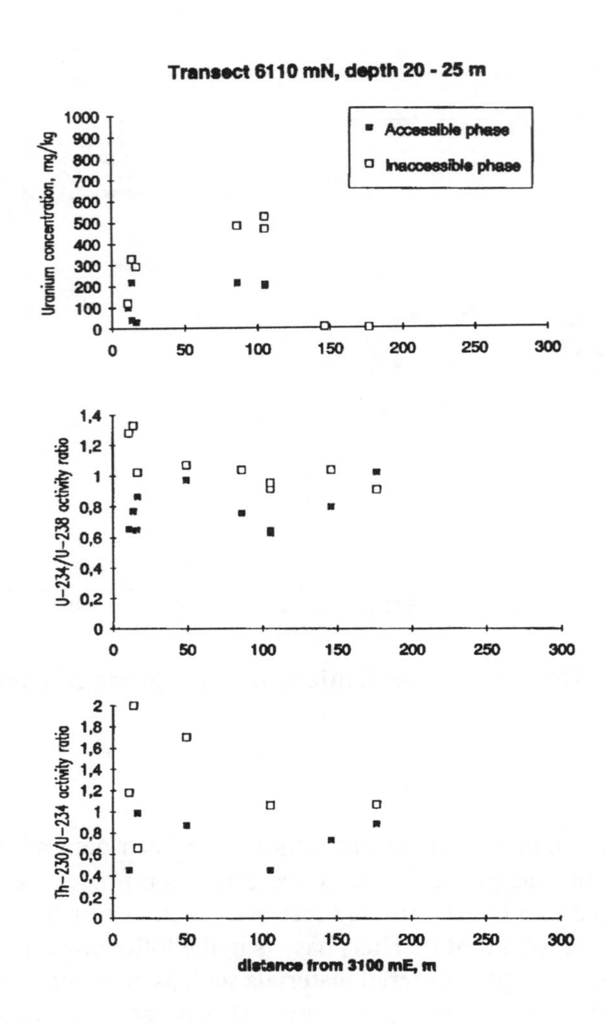

Figure 4. **Kemakta. Uranium concentrations and activity ratios.**

RIVM

One of the most important features of the uranium mineralisation at Koongarra is the occurrence of abundant secondary uranium minerals, principally within the dispersion fan above the ore body. At present, it is believed that the dispersion of uranium started when the lower boundary of the weathered zone reached the top of the ore body and that it has virtually stopped in those parts of the formation that are completely weathered and turned into a clayey material. The dispersive transport is therefore assumed to take place in the transition zone of the weathering, where the flow velocity is the highest and thus the dispersion rate is the largest. Flow and transport are assumed to be mainly horizontal in the transition zone. Thus, one may identify three different layers:

- a top, fully weathered layer, in which no significant dispersion takes place
- an intermediate, partially weathered layer (the transition zone) moving downward as the weathering proceeds, in which the groundwater velocity is the highest and the dispersion rate is the largest
- a lower, unweathered zone, in which no dispersion of uranium occurs.

Changes in dispersion distance and direction with depth

The migration distance and the main direction of dispersion were estimated based on solid phase uranium concentrations that have been analysed and reported within ARAP, see Figures 5a-c.

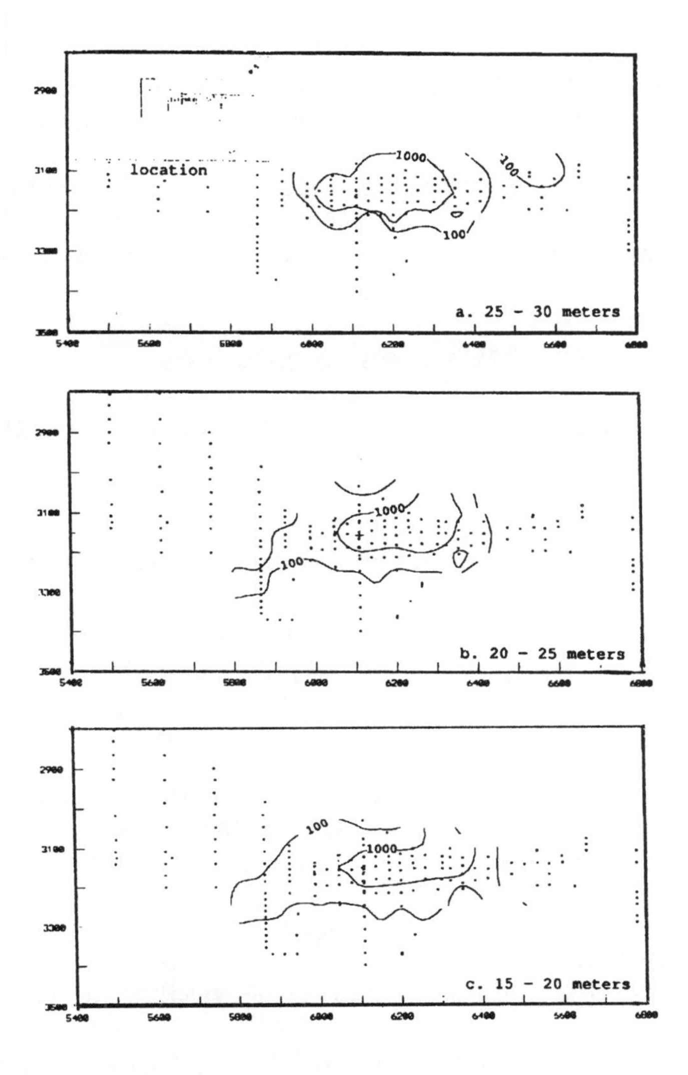

Figure 5a-c. **RIVM. Contours of solid phase uranium concentrations (100 and 1000 mg/kg) at different depths. a) 25-30 m, b) 20-25 m, c) 15-20 m.**

The 100 and 1000 mg/kg contour lines of the solid phase uranium concentrations are given in plan view in Figures 5a-c. The contour lines in each figure represents measurements within a depth interval of 5 m starting at a depth of 30 m, which is where the top of the unweathered schist is found. Figure 5a shows the distribution of uranium in the primary ore body in the unweathered zone as well as in the top of ore body 2 which is visible to the right of the main ore body. Figure 5b shows the uranium concentration contour lines just above the base of weathering. It can be seen that the nuclide transport has occurred in an approximately south-westerly direction. By comparing Figures 5a and 5b, it may be concluded that the base of weathering is the lower boundary for the region in which dispersion of uranium has occurred. The direction and distance of dispersion of uranium is roughly the same at the depths represented in Figures 5b and 5c. The same direction and distance of dispersion of uranium is also found in depth intervals 10-15 m and 5-10 m. No primary ore is however found within the first 5 m. The results of this data analysis show that the direction and distance of dispersion is almost constant with depth.

From this data examination it can be concluded that the shape of the concentration distribution patterns supports the hypothesis that groundwater movement and dispersion of uranium has taken place mainly in the transition zone. As there are no observations at Koongarra of a decrease of the dispersion distance with depth, there are no signs of a significant transport of uranium in the fully weathered top layer. Furthermore, the concentration distribution pattern indicates that the direction of groundwater flow does not seem to have changed significantly during the past few million years.

Migration calculations

The migration distance of uranium has been simulated by Kemakta and RIVM for a number of parameter combinations and model concepts to test whether the observed migration distances could be predicted.

Both teams based their migration calculations on data that had been determined and reported within the ARAP project, e.g. porosity, K_d and rock densities, and the extensive hydrological work that was performed within ARAP as well as within INTRAVAL phase 1.

Kemakta

The Kemakta team focused its work on calculations of the dispersion of uranium and daughter nuclides in the weathered zone. The aim was to test the applicability of rather simple models that generally are used in performance assessment of radioactive waste repositories.

The modelling work was carried out in several iterations with increasing level of complexity. Each iteration included a review of available laboratory and field data, selection of the system to be modelled and a suitable model, and finally a comparison of modelling results with field observations.

In total three iterations were performed in which the following model concepts were applied:

- advection–dispersion model with linear sorption,
- advection–dispersion model with linear sorption and chain-decay,
- advection–dispersion model with linear sorption, chain-decay, α-recoil and phase transfer.

In the first modelling attempt, a simple 1–D advection–dispersion model including linear sorption according to the K_d-concept was used to simulate the dispersion of uranium in the weathered zone. This model may be considered as a simple performance assessment model.

In the second iteration, the system was extended to also include the transport of the daughter nuclides ^{234}U and ^{230}Th. In addition, an attempt was made to consider α-recoil in a very simple manner by assigning a higher K_d-value to ^{234}U than to ^{238}U. Furthermore, the sensitivity of the results to different combinations of groundwater flux and migration time was studied, as well as the effects of alternating periods of flow and no flow.

In the third and final iteration, the model was even further extended by including transfer of radionuclides between different phases of the rock and including a more detailed model description of α-recoil. This model is based on the assumption that radionuclides in the groundwater are sorbed to accessible phases of the weathered rock (amorphous iron oxides and clays) and that sorbed radionuclides are further included in inaccessible crystalline phases of the rock due to recoil effects and recrystallisation of amorphous iron oxides and clays.

Advection–dispersion model with linear sorption

The calculated solid concentration of uranium at different distances from the source together with available small sample data from the depth interval 10–30 m in holes located along the transect 6110 mN are shown in Figure 6. The simulations are based on a migration time of 2 million years. A K_d-value of 10 overpredicts the solid uranium concentration, if the concentration of dissolved uranium at the source is 1000 mg/m^3, but gives a prediction that agrees fairly well with observed values, if the source concentration is 100 mg/m^3. The combination of a K_d-value of 10 and a Darcy flux of 1 m/yr apparently predicts the migration distance fairly well, whereas a higher water flux overpredicts and a lower water flux underpredicts the migration distance.

With this model, the migration distance is dependent on the ratio Darcy flux to the distribution coefficient, and the maximum solid concentration level is dependent on the product of the liquid source concentration and the distribution coefficient. This means that there are an infinite number of combinations of parameter values that will give good predictions. However, the results also show that it is possible to simulate the observed migration distance and concentration levels with a few parameters having values consistent with what has been recommended in independent interpretations.

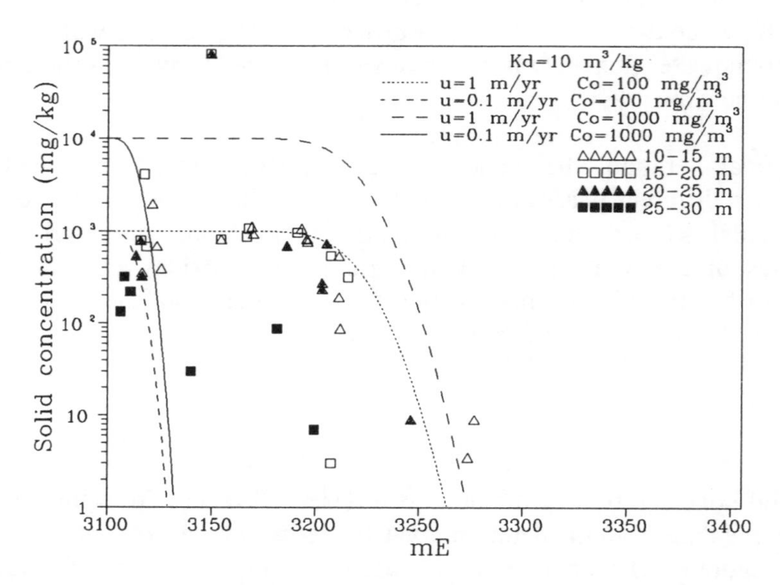

Figure 6. **Kemakta. Predicted and observed uranium concentrations along the transect 6110 mN.**

Advection–dispersion model with linear sorption and chain decay

In the second iteration of the modelling work, the system was extended to also include transport of the ^{238}U daughter nuclides, ^{234}U and ^{230}Th. The observed activity ratios between ^{234}U and ^{238}U indicate a retention of ^{234}U relative to ^{238}U. This could be due to α-recoil of the short-lived ^{234}Th into the solid phase, making ^{234}U, the daughter nuclide of ^{234}Th, more attracted to the solid phase than ^{238}U, the parent nuclide of ^{234}Th. This effect was simulated in a simple manner by assigning a higher K_d-value to ^{234}U than to ^{238}U.

As was found in the first modelling iteration, it is possible to obtain almost identical simulations of the uranium concentration versus distance with different combinations of the values of the distribution coefficient, source concentration, water flux and migration time. However, the activity ratio $^{230}Th/^{234}U$ versus distance from the source is dependent on the migration time, irrespective of the other parameter values. A set of calculations with different combinations of water flux and migration time, all matching the observed uranium concentration profile, were carried out. A comparison between predicted activity ratios and observed data indicates that a migration time of 0.2 million years is too short, and that a migration time in the order of million years gives a better match to the observed data, see Figure 7.

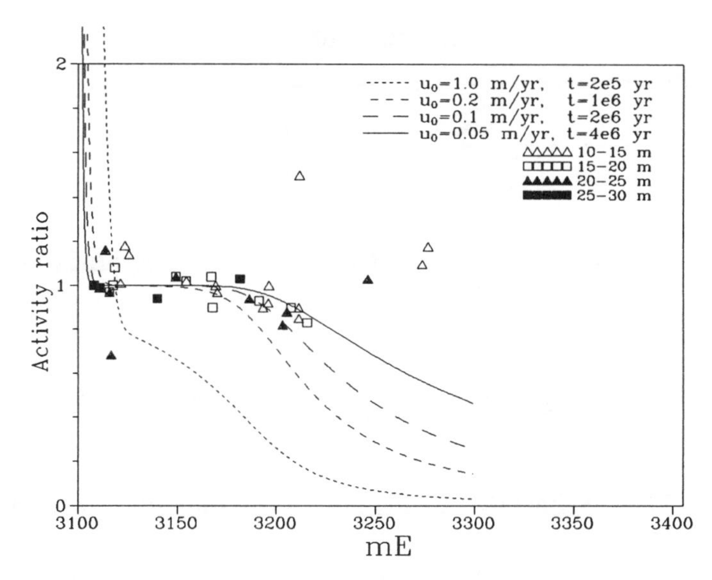

Figure 7. **Kemakta. Predicted and observed $^{230}Th/^{234}U$ activity ratios. Effects of different combinations of water flux and migration time.**

The simulation of variable climatic conditions by assuming alternating periods of flow and no flow showed that the flow conditions during the last 200 000 years are important to the results, given that an average flow rate can be used for times prior to 200 000 years ago.

Advection–dispersion model with linear sorption, chain decay, α-recoil and phase transfer

In this third iteration, the advection-dispersion model was extended in order to investigate the influence of phase transfer and α-recoil on the activity ratios. However, a comparison between calculated and observed data suggests that the results are not significantly improved by including phase transfer and α-recoil in the model, at least not with the combination of phase transfer rates and K_d-values used in the calculations. Adding only the process of α-recoil gave similar results as the simple advection-dispersion-reversible sorption model, while the introduction of phase transfer gave results that were less in agreement with observed data.

RIVM

The contours of solid phase uranium concentrations at different depths suggests that the migration of radionuclides in the fully weathered zone and in the unweathered zone is negligible. RIVM has therefore focused its attention on the transition zone between the highly weathered and the unweathered rock when developing the conceptual transport model. The transition zone was modelled as a two-dimensional region. The main parameters and processes considered were: the velocity distribution, the downward movement of the transition zone, and a few geochemical processes.

The governing equations incorporated the movement of the weathering zone and hence the loss of uranium at the top of the zone and the entering of clean schist at the bottom of the zone (Figure 8). Diffusion of uranium from the highly weathered zone into the transition zone was also included.

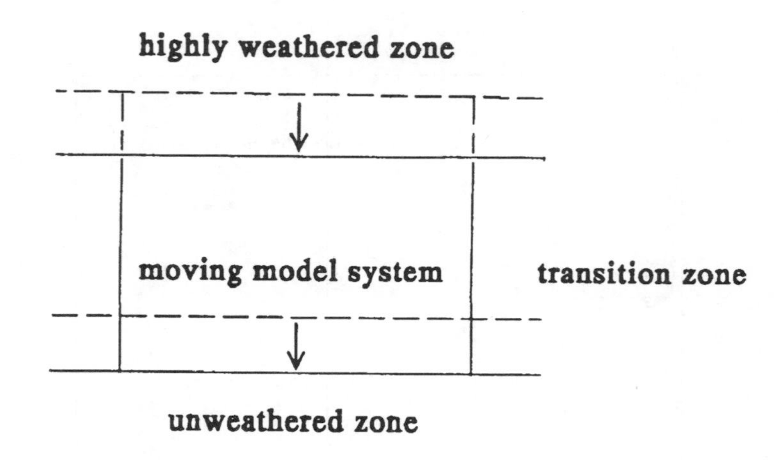

Figure 8. **RIVM. Schematic illustration of the moving transition zone.**

The number of measured values of uranium concentrations in the liquid phase are scare. The calculated values of liquid phase uranium concentrations were therefore converted into solid phase concentrations by using the K_d value in order to be able to compare the modelling results with the present situation.

Rate of downward movement of the transition zone of weathering

Uranium series disequilibria data on the secondary dispersion fan mineralisation indicate that the weathering front reached the top of the ore body sometime in the interval 1–3 million years ago. The lowering of the sea level from 1.8 million years onwards would have increased stream gradients, resulting in increased erosion rates.

If it is assumed that the transition zone has moved about 20 m downward over a time period of 1.8 million years, then a mean velocity of $1.1 \cdot 10^{-5}$ m/y is obtained. However, it cannot be reasonable to assume a constant downward velocity of the transition zone during the past 1.8 million years, since this time period involves cycles of alternative cold and warm climates in combination with advancing and retreating ice sheets, and falling and rising sea levels. Nevertheless, it was considered acceptable to work with an average value of the movement of the transition zone, since data on dry and wet periods and sea levels fluctuations are not available or are too uncertain to provide any reasonably good estimate of the variations in the weathering front velocity. The thickness of the transition zone was assumed to be 5 m, based on a compilation of available field data.

Modelling of uranium transport in the present transition zone - effect of variations in the velocity field

To assess the effect of the groundwater velocity field, three simulations were carried out neglecting the movement of the transition zone. The groundwater velocities were constant in time in the simulations, but both constant and spatially variable velocity fields were applied. A qualitatively fairly good agreement between calculated and measured uranium concentration in the solid phase was obtained for the constant velocity field case and one of the variable velocity field cases. These simulations showed that the uranium concentration distribution pattern is highly dependent on the velocity field. Furthermore, the use of a spatially variable velocity field could not be justified because of the large uncertainties in the boundary conditions.

Modelling uranium transport with a moving transition zone

The boundary conditions used for the calculation of the velocity field are very uncertain. They may be valid for the present transition zone, but are unknown for the past. This being the case, the calculations that included the downward movement of the transition zone were performed with a velocity field that was constant in time and space. The results from one of the simulations is illustrated in Figure 9 at four different depths. A K_d-value of 0.7 m^3/kg, an effective velocity of 1.3 m/y, and a downward movement of the transition zone of $1.1 \cdot 10^{-5}$ m/y were applied in this specific simulation. A concentration distribution pattern that is almost constant with depth was observed in these simulations which is in agreement with the observations at Koongarra as illustrated in Figure 5a-c. One conclusion that can be drawn from these simulations is therefore that the present situation in Koongarra can be simulated by including a moving transition zone. However, the calculated dispersion distance from the ore body was shorter than observed at Koongarra. It is believed that a better "fit" with the measured solid phase uranium concentration distribution could be obtained by changing some of the model parameters like the groundwater velocity, the downward velocity of the transition zone, the adsorption distribution coefficient or the dissolution source term parameters.

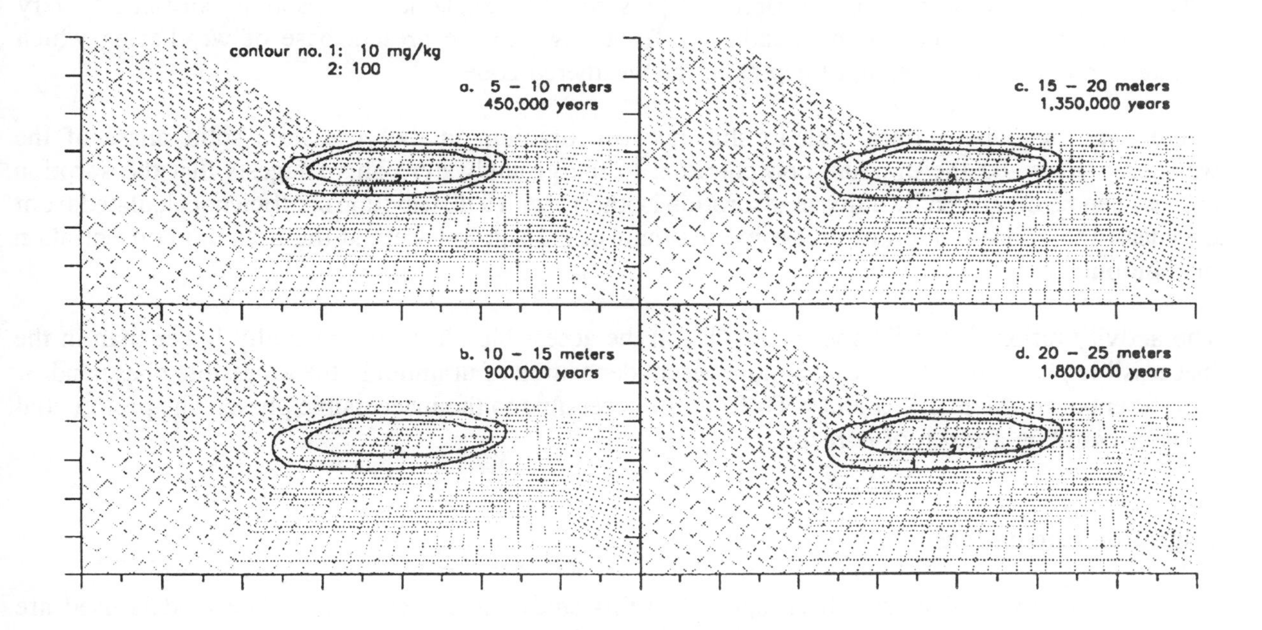

Figure 9. **RIVM. Simulation with a velocity constant in time and space and a downward moving transition zone.**

Discussion

Data review

The review of the available solid uranium concentrations shows that the extension of the dispersed fan in the south direction from the fault is around 350 meters, whereas the extension in the south-east direction is around 100 meters. Information from different depths indicates that the pattern of the dispersion is similar at all depths in the weathered zone. This supports the hypothesis that groundwater movement and dispersion of uranium have taken place mainly in the transition zone. If significant transport had occurred in the fully weathered top layer, then the dispersion distance from the ore body should have decreased with depth. Moreover, the concentration distribution pattern suggests that the direction and magnitude of the average groundwater flow have not changed significantly during the past few million years.

The hypothesis that groundwater movement and dispersion of uranium have taken place mainly in the transition zone is, however, not supported by the observed activity ratio $^{230}Th/^{234}U$. The similarity in this activity ratio profile at different depths in the weathered zone, with values below 1 at the uranium dispersion front, indicate that uranium migration has taken place at all depths in the weathered zone during the last 200 000 years. The similarity in both uranium concentration distribution pattern and $^{230}Th/^{234}U$ activity ratio profile at different depths could be explained by assuming an initially very fast downward movement of the weathering front down to the present base of weathering which initiated uranium migration at all depths in the weathered zone.

There appears to be a linear relationship between uranium in the accessible solid phase of the weathered rock and the uranium concentration in the groundwater. This suggests that the sorption process can be described by a linear adsorption isotherm with a distribution coefficient calculated from the measured concentration of accessible uranium in the solid phase and the uranium concentration in solution.

The activity ratios ($^{234}U/^{238}U$ and $^{230}Th/^{234}U$) in the accessible phase are generally lower than in the inaccessible phase, which indicate a more recent deposition of uranium in the accessible phase and/or a preferential transfer of ^{234}U and ^{230}Th from the accessible to the inaccessible phase. The preferential transfer could be a result of α-recoil.

Migration calculations

The model concepts that have been applied in this study are rather simple. The models used are performance assessment models accounting for advection, dispersion and linear sorption in one or two dimensions. The one-dimension model was extended to include recoil and phase transfer. The vertical movement of the transition zone was included in the two-dimensional model.

One–dimensional calculations

The dispersion distance and solid uranium concentrations calculated with the simplified one-dimensional advection-dispersion model including sorption were in fair agreement with the observed migration distance and the solid uranium concentrations in the dispersion fan. The applied model involves extensive simplifications of the system, but the results showed that it is possible to simulate the observed migration distance and concentration levels with a few parameters having values consistent with what has been recommended in independent interpretations.

The model extended to include chain-decay made it possible to get an independent estimation of the migration time by studying the $^{230}Th/^{234}U$ ratio. This ratio indicated that the migration has continued for a time period in the order of million years.

The introduction of α-recoil and phase transfer allowed a simulation of some observed effects, e.g., a decrease in activity ratios in the bulk rock with distance, and the general trend with most of the observed activity ratios in the accessible phase below 1 and in the inaccessible phase above 1. The extended model did not, however, improve the match between the calculated and observed uranium concentrations in the bulk rock. This model gave a description of the system that was more consistent with what is observed but the number of free parameters was larger, and more data on the system was required, compared to the more simple advection-dispersion-reversible sorption model. This may possibly explain the difficulty in improving the simulation results using the extended model.

Two–dimensional calculations

Calculations of uranium transport in the present transition zone, not considering the moving transition zone, resulted in contour lines of the uranium concentration in the solid phase qualitatively in agreement with the contour lines from the observed concentrations. Taking into account the movement of the transition zone, a dispersion pattern, stationary with depth, is obtained in the weathered zone that is in agreement with the observations at Koongarra. However, the calculated dispersion length was shorter than the one measured. By calibration, a better estimate of the dispersion length could be obtained (e.g., by increasing the effective velocity and/or the dissolution rate of uranium). Thus, the 2-D study showed that it is possible to simulate the present situation in Koongarra using the concept of a moving transition zone.

Results from the ARAP

Modelling of the uranium migration has also been carried out in the ARAP by ANSTO and other organisation. The migration modelling in the ARAP was performed using models ranging from rather simple advection-dispersion-linear sorption models to complex multi-phase models considering recoil and chemical transfer. The advection-dispersion-linear sorption models were found to simulate the observed extension of the uranium migration fairly well. The more complex multi-phase models considering recoil and chemical transfer gave a good description of observed uranium concentrations and activity ratios in the different mineral phases of the rock. The migration modelling carried out in the ARAP as well as by RIVM and Kemakta in INTRAVAL indicated that the uranium migration has continued for a time period of half a million up to a few million years which is in agreement with results from geomorphological investigations.

Validation aspects

The overall aim of this work was to test and, if possible, validate simple and conservative models used in performance assessment of radioactive waste repositories. Although the results from this INTRAVAL study are not enough to validate simple performance assessment models in a strict sense, it has been shown that even simple models are able to describe the present day distribution of uranium in the weathered zone at Koongarra, based on established transport processes and reasonably selected parameter values.

Concluding remarks

Studies of the Alligator Rivers Natural Analogue has demonstrated that the system is very complex. The interaction of many geochemical and geohydrological processes occurring over long times makes it difficult to create a quantitative model of the history of groundwater flow and nuclide transport. The study has shown the importance of a joint interpretation of different types of data and an iterative procedure for data collection, data interpretation and modelling in order to get a consistent picture of the evolution of the site. Furthermore, it was shown that sorption is a major retardation mechanism, that uranium fixation in crystalline phases is a potentially important retardation mechanism in geologic media where significant alteration of the rock is expected, and that α-recoil may have an impact on the distribution of uranium isotopes in the water. Modelling simulations indicated migration times in fair agreement with independent geomorphological information.

A general conclusion from the INTRAVAL study is that rather simple and robust concepts and models seem able to adequately describe the long range migration processes that have occurred. On the other hand, for more detailed modelling and to attain comprehensive understanding the geochemical processes have to be addressed more carefully.

Even though the migration modelling that was carried out within the ARAP has only been briefly addressed in this paper it is encouraging to note that the results from these simulations support the conclusions drawn by Kemakta and RIVM in INTRAVAL.

Acknowledgement

This paper is based on work that was mainly carried out by the two INTRAVAL participants the National Institute of Public Health and Environmental Protection (RIVM) and Kemakta Konsult. The paper was prepared by Kemakta with valuable support by Ms. Rikje van de Weerd at RIVM who supplied us with background material and gave helpful suggestions regarding the content. We would also like to thank Dr. Stig Wingefors at the Swedish Nuclear Power Inspectorate for useful proposals regarding the content.

References

Duerden P., "The Alligator Rivers Natural Analogue", The International Intraval Project, Phase 1, Test Case 8, OECD/NEA, 1992.

Skagius K., Lindgren M., Pers K. and Brandberg F., "The Alligator Rivers Natural Analogue - Modelling of Uranium and Thorium Migration in the Weathered Zone at Koongarra", SKI 94:19, Swedish Nuclear Power Inspectorate, Stockholm, Sweden, 1994.

van de Weerd H., Leijnse A., Hassanizadeh S.M. and Richardson-van der Poel M.A., "INTRAVAL phase 2, test case 8. Alligator Rivers Natural Analogue - Modelling of uranium transport in the weathered zone at Koongarra (Australia)", RIVM Report 715206005, Bilthoven, the Netherlands, 1994.

SESSION III

Chairmen:

H. Umeki
J.-C. Petit

The MEGAS E5 Experiment: A Large 3-D In Situ Gas Injection Experiment for Model Validation

L. Ortiz, G. Volckaert, M. Put
CEN●SCK (Belgium)

M. Impey
INTERA (United Kingdom)

Abstract

For the option of a deep geological disposal facility, several potential sources of gases were identified: i.e. the anaerobic corrosion of iron, degradation of organic materials, generation by gamma-radiolysis, the gas present as such in the waste packages. Of those gases, hydrogen is certainly the gas which can be released in the potentially largest amounts. For the safety evaluation of a repository, it is necessary to assess the effects of gases on the host rocks.

The primary objective of the MEGAS project is to understand the consequences of gas generation in a clay host rock. The final objective of this project is to validate a gas migration model and to confirm our understanding using the MEGAS E5 in situ gas injection experiment.

The MEGAS E5 experiment consists of four multi-piezometers installed in 1992 in the Boom clay formation from the HADES underground research facility. The multi piezometers were installed in a 3-D configuration. A central piezometer is used for the gas injection, while the other piezometers are used to measure the pressure increase caused by the gas injection.

Before the start of the gas injection, hydraulic testing was performed to determine the mean hydraulic parameters in the Boom clay surrounding the piezometers. The results and modelling of this hydraulic testing are also described.

For the modelling of the two phase flow, INTERA has developed the TOPAZ code. The physical and numerical basis of the TOPAZ code are explained.

The results of the gas injection test are presented and compared to the blind predictions made by INTERA using the TOPAZ code.

1. Introduction

The storage of radioactive waste in an argillaceous formation is one of the most important research field at CEN●SCK in Mol, Belgium. One of the phenomena to take into account for the safety assesment of a repository is the corrosion of the waste canisters. This process, among others, will release gases, mainly hydrogen, in large quantities. The effects of gases on the host rock have to be known to assess the performance of a repository.

The study of the possible effects of this release, as well as the modelling of laboratory and in situ gas migration experiments form the basis of the MEGAS project. MEGAS is a multipartner project with BGS, INTERA, ISMES and CEN●SCK. The final objective of MEGAS is the validation of a code modelling the gas migration in clay. This code, TOPAZ, has been developed by INTERA (Henley-on-Thames, UK), which is performing the main modelling work. The CEN●SCK is responsible for gas reaction, diffusion, uniaxial flow and in situ experiments with Boom clay. BGS (Keyworth, UK) is performing gas flow experiments in triaxial cells at room temperature. ISMES (Bergamo, Italy) is performing similar experiments at higher temperature.

MEGAS is part of the CEC umbrella project PEGASUS.

2. The MEGAS E5 experiment

The design of the gas injection experimental set-up consists of four piezonests, installed horizontally from the underground research facility HADES in Mol (Belgium) at 225 meters depth in the Boom clay layer. (see Fig. 1).

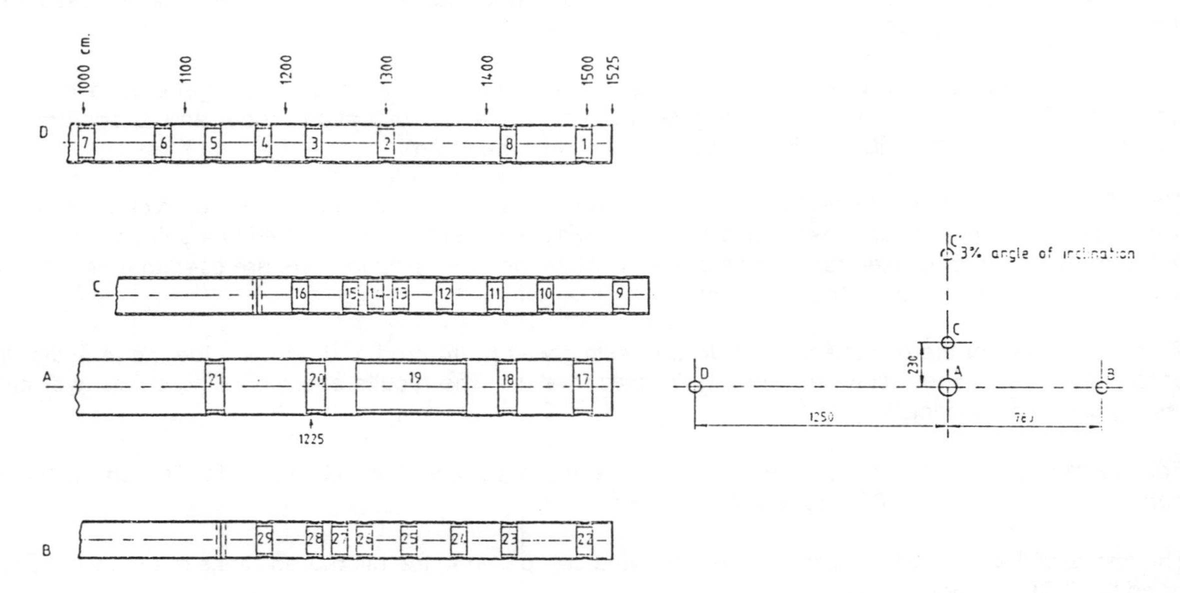

Figure 1. **View of the E5 experimental set-up with the numbering of the filters**

The central piezonest (90 mm diameter) is used for the gas injection. The three peripheral piezonests (60 mm diameter) are used for the detection of local variations in the hydraulic pressure field. One of them is installed in the vertical plane passing through the central piezonest, and is located above it. The two others are located in a horizontal plane, on both sides of the central piezonest, and at different distances. This configuration was chosen, because the previous gas breakthrough experiments realized on small clay cores have shown that gas breakthrough should occur preferentially in a horizontal direction.

Each of the 29 filters is connected to a pressure transducer. A 30 channels data recorder is used for data acquisition. The scanning frequency of this recorder allows us to detect any sudden change in pressure.

After the installation of the four piezonests, a stabilization period of ten months was required to restore the initial ground conditiond before the start of a hydraulic interference test. Therefore, filter 20 of the central piezonest was set to the atmospheric pressure, and the water outflow was measured in function of time. The interstitial pressure decrease in all the other filters was monitored, so it was possible to calculate the mean hydraulic parameters in the vicinity of the experimental set-up. The horizontal hydraulic conductivity K_H , the vertical hydraulic conductivity K_V , and the storage coefficient S_0 were calculated as follows.
In situ Boom clay may be considered as a homogeneous anisotropic water saturated porous medium. Assuming that there is no variation of hydraulic conductivity between the principal directions of the bedding plane, and that these coincide with the x,y principal directions, the equation of continuity can be written as [1]:

$$S_0 \frac{\partial P}{\partial t} = K_H \left(\frac{\partial^2 P}{\partial x^2} + \frac{\partial^2 P}{\partial y^2} \right) + K_V \frac{\partial^2 P}{\partial z^2} \qquad (1)$$

For a source q(t) located at x=0, y=0 and z=0, the pressure decrease P (x,y,z,t) can be calculated by a convolution integral as [2]:

$$P(x,y,z,t) = \int_0^t \frac{q(t') \exp\left(- \frac{(x^2+y^2)}{4k_H\,(t-t')} - \frac{z^2}{4k_V\,(t-t')} \right) dt'}{8\sqrt{\pi^3\, k_H^2\, k_V\, (t-t')^3}} \qquad (2)$$

with k_H and k_V , the horizontal and vertical hydraulic diffusivities defined as $k = K/S_0$.
The time function for the source is developed in a series of exponentials, and the source strength can be written as:

$$S_0\, q(t') = \sum_{j=1}^{N} a_j \exp(-b_j\, t') \qquad (3)$$

The solution of the convolution integral (2) together with (3) gives the pressure decrease P(x,y,z,t) around the source. Calculations for $b_j \geq 0$ gives:

$$P(x,y,z,t) = \frac{\sum_{j=1}^{N} a_j\, F_1(R/\sqrt{t},\, \sqrt{b_j t},\, -b_j t)}{8\pi \sqrt{K_H K_V\,(x^2+y^2) + K_H^2\, z^2}} \qquad (4)$$

with:

$$R^2 = \frac{S_0\,(x^2+y^2)}{4K_H} + \frac{S_0\, z^2}{4K_V} \qquad (5)$$

$$F_1(R/\sqrt{t},\, \sqrt{b_j t},\, -b_j t) = \exp(-bt-2iR\sqrt{b})\, erfc(R/\sqrt{t}-i\sqrt{bt}) + \exp(-bt+2iR\sqrt{b})\, erfc(R/\sqrt{t}+i\sqrt{bt}) \qquad (6)$$

To obtain the values of the parameters S_0 , K_H and K_V , equation (4) has to be fitted to the experimental pressure decrease around the source q(t). A Fortran program has been written for the fitting. The parameters a_j and b_j of equation (3) are obtained through the fitting of a series of exponentials to the measured outflow rate.

The following results were obtained:

Storage coefficient	S_0 =	$8.1 \times 10^{-6} \pm 0.9 \times 10^{-6}$ m^{-1}
Horizontal hydraulic conductivity	K_H =	$5.2 \times 10^{-12} \pm 0.2 \times 10^{-12}$ ms^{-1}
Vertical hydraulic conductivity	K_V =	$2.3 \times 10^{-12} \pm 0.2 \times 10^{-12}$ ms^{-1}

These values are in good agreement with previous results [3]. The fitting was performed for ten data points on each of the 29 filters (i.e. 290 measurement points in total). An example of the fitting for the filter 14 located on the upper piezonest is given in Fig.2. Such a good fitting on the results was obtained for 27 filters. For filters 3 and 4 the fitting was less good, probably due to local variations of the hydraulic parameters.

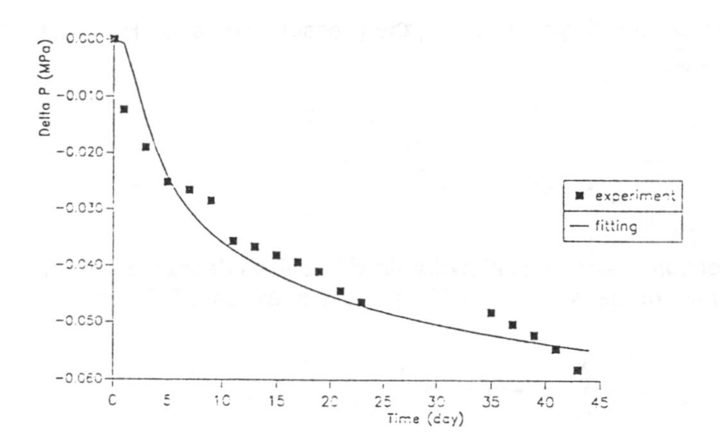

Figure 2. **Pressure decrease in filter 14 together with the fitted curve**

Five months after the realization of the hydraulic campaign in the clay massif surrounding the E5 experimental set-up, helium has been injected through filter 20, i.e. the same filter used as percolation filter during the hydraulic campaign. Helium was chosen as a safe substitute for hydrogen.

The initial interstitial pressure around filter 20 was 1.67 MPa. The gas pressure was first set at 1.77 MPa, and was further weekly increased by an increment of 0.1 MPa until a gas flow was established.

A gas breakthrough to filter 19 was detected after 44 days, at a gas pressure of 2.36 MPa, i.e. 0.69 MPa above the original pore water pressure around the injection filter. The pressure reached in filter 19 was then 2.14 MPa, compared to 1.72 MPa originally, i.e. a breakthrough pressure equivalent to 0.42 MPa. This is a low value compared to the values observed during laboratory experiments (i.e. 1 to 3.1 MPa) [3] and [4].

After the gas entry pressure was increased to 2.38 MPa, the formation of a preferential pathway between filter 20 and filter 19 was established (the pressure in filters 19 and 20 became equal). This phenomenon has also been observed in preliminary experiments [3].

The same phenomenon occurred with filter 18 and filter 21, both located on the injection piezonest after respectively 71 and 72 days of experiment. This happened after the gas injection pressure was increased to 2.53 MPa. The pressure evolution in filters 18,19,20 and 21 is shown on Fig. 3.

A perceptible increase in pore water pressure in the filters of the surrounding piezonests, and especially in those located on the upper piezonest, was observed at the time of gas breakthrough. This sudden pressure increase appearing at distances of more than one meter from the gas injection source can only be attributed to a mechanical effect, and can not be explained hydrodynamically. An example of pressure evolution in filter 12 is shown on Fig. 4 together with the gas injection pressure evolution. Filter 12 is placed at about 1.15m from the gas injection source.

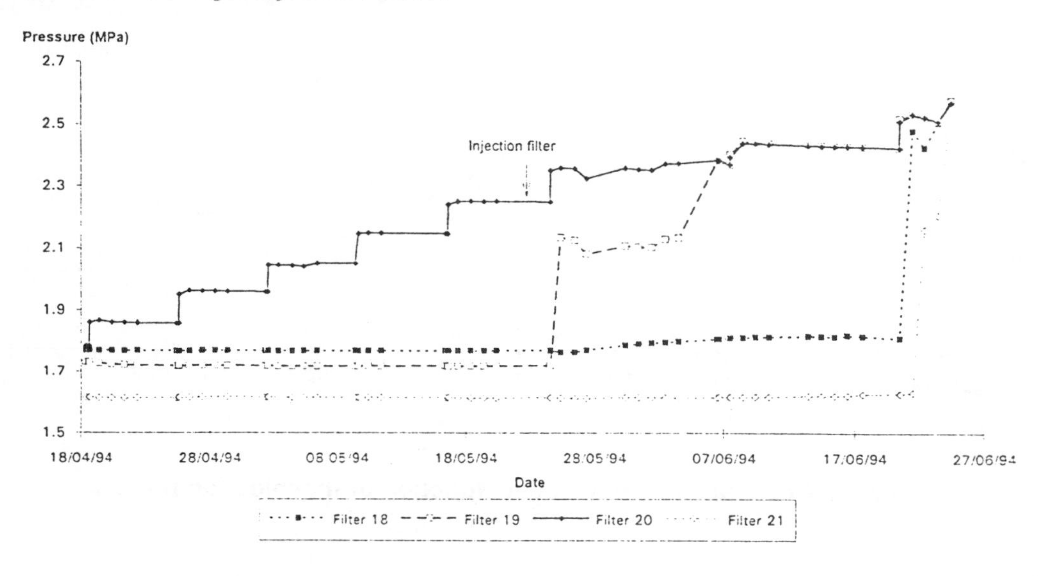

Figure 3. Pressure evolution in filters of the central piezonest during E5 experiment

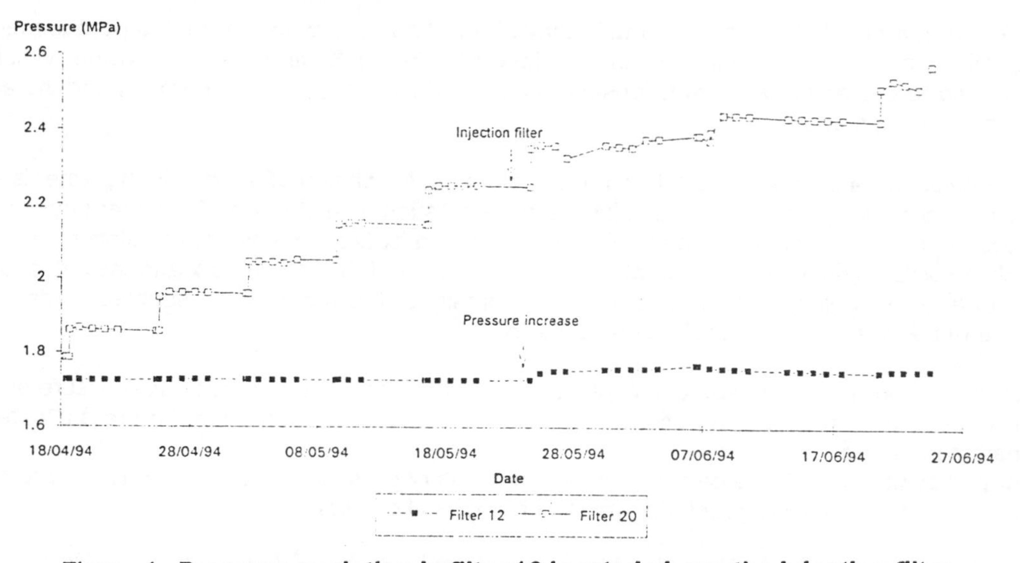

Figure 4. Pressure evolution in filter 12 located above the injection filter

Figure 5 shows the mean gas flow rate evolution in function of the injection pressure. The flow evolution indicates that helium did not propagate by a two phase flow mechanism, but through the formation of preferential pathways along the injection piezonest. This assertion is supported by the observation that beyond a gas injection pressure limit of 2.35 MPa, an immediate increase in gas flow rate followed by an increase in local pressure in the connected filter is perceived. The value of this local pressure becomes equal to the injection pressure.

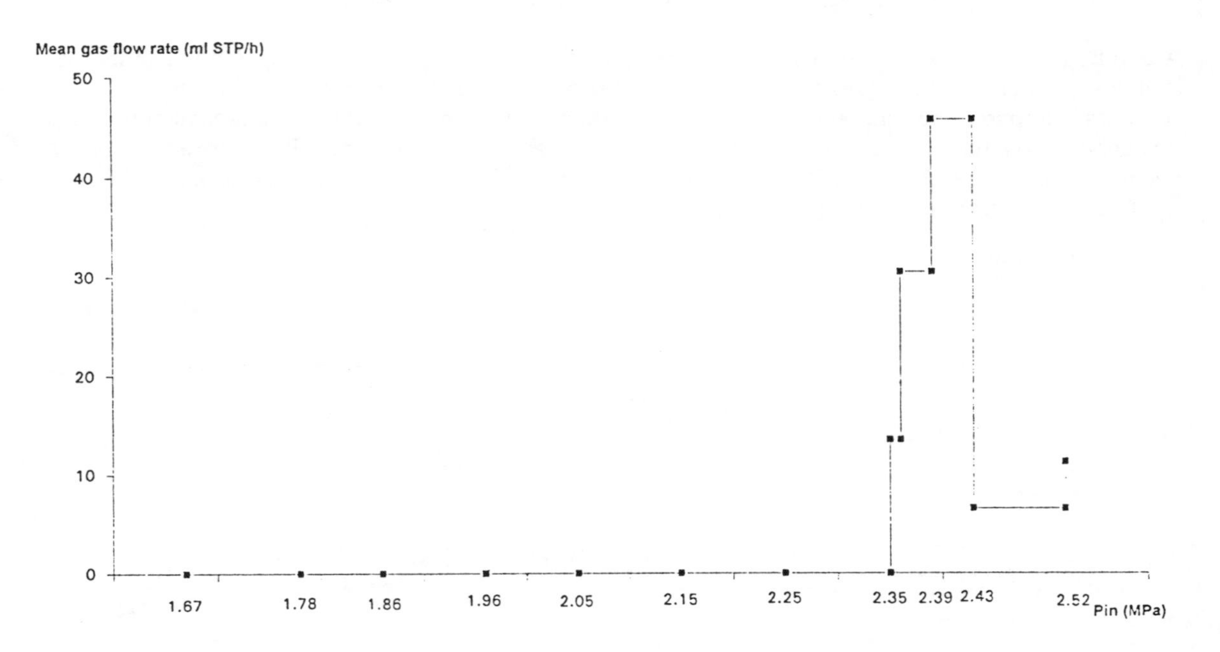

Figure 5. **Mean gas inlet flow rate in function of the injection pressure**

3.Modelling with the two phase flow code TOPAZ

TOPAZ is a numerical model of the simultaneous flow of gas and water through a porous medium [5]. It is essentially a one-dimensional model developed by Intera Environmental Division, which may be extrapolated to two and three dimensions by using cylindrical or spherical geometry and the appropriate symmetry assumptions.

The two fluids are assumed to be totally immiscible, and dissolution of the gas in the water is neglected. This is reasonable for gases with low solubilities in water, such as hydrogen. The water phase is assumed to be incompressible. The flux of each phase is governed by Darcy's law, with a relative permeability for each phase which is dependent on the absolute permeability of the rock matrix and the saturation of each phase in a given region. The permeability and porosity of the medium is assumed to depend on the pressure of the phases, with constant compressibility.

The boundary conditions imposed at the inlet may be either that of a constant gas pressure or a constant gas flux, together with a gas saturation of unity. At the outlet, it is assumed that there is a water reservoir at constant pressure.
The dependences of relative permeabilities and capillary pressure on water saturation are specified in tabular form which does not preclude analytic or empirical solution.

The equations to be solved are the mass conservation for each phase, which are supplemented by some additional constraints and relationships. The mass conservation equations may be written as follows:

$$\frac{\partial \phi \rho_i s_i}{\partial t} = -\nabla . F_i - q_i \qquad (7)$$

where ϕ is the rock porosity, ρ_i is the density and s_i is the saturation of phase i, $\mathbf{F}_i$ is the flux of phase i and q_i is a source or sink term (taking negative and positive values respectively).

The flux $\mathbf{F}_i$ is given by Darcy's law, generalised for two-phase simultaneous flow:

$$F_i = \frac{-kk_i^r \rho_i}{\mu_i}(\nabla P_i - \rho g) \qquad (8)$$

where k is the absolute permeability of the medium, k_i^r is the relative permeability of phase i, P_i is the pressure of phase i, μ_i is the viscosity of phase i, and **g** is the acceleration due to gravity.
The above equations are supplemented by the equation of state for the gas and the relationships between phases saturations and pressures.

In addition, the absolute permeability k and porosity ϕ at pressure P are related to their values k_0 and ϕ_0 at a reference pressure P_0 by the equations:

$$k = k_0(1+\alpha(P-P_0)) \qquad (9)$$

$$\phi = \phi_0(1+\beta(P-P_0)) \qquad (10)$$

where α and β are compressibility constants defined for the clay. This is one of the simplest forms in which stress effects can be introduced. However, this option has not been used for the blind predictions.

The first step in the numerical solution of the above set of nonlinear, coupled, partial differential equations is to write them into a form of difference equations. The finite volume differencing method is used to discretize the spacial parameters, together with an implicit time-stepping scheme. In the finite volume differencing method, the mass conservation equations are first integrated over the domain V of each gridblock, before discretization. This provides mass conservation locally and globally, as mass balance is applied to individual gridblocks. Performing the integration and applying the difference theorem to the flux component, the mass conservation equations become:

$$\int \frac{\partial \phi \rho_i s_i}{\partial t} dV + \int F_i \, dS + Q_i = 0 \qquad (11)$$

where **dS** is the surface area of the gridblock and Q_i is the source or sink term for gridblock i. This set of equations is then integrated over a timestep ΔT using a fully implicit time-stepping scheme, which gives:

$$R_i = M_i(T+\Delta T) - M_i(T) + \Delta T \int F_i \, dS + \Delta T Q_i \qquad (12)$$

where M_i is the mass accumulation term given by $M_i = \int \phi \rho_i s_i \, dV$, and R_i is the residual for gridblock i. $\mathbf{F}_i$ and Q_i are evaluated at $T+\Delta T$.

The solution variables chosen are the gas density and water saturation, so the equations are rewritten in terms of these variables using the equations of state for the gas and the relationships between phase saturations and pressure. The matrix of equations (12) for the residuals R_i, one for each phase and each gridblock, are solved iteratively using the Newton Raphson method. This approach solves the set of nonlinear equations

$$R(\rho_g, s_w) = 0 \tag{13}$$

where **R** is the matrix of residuals. At each iteration, corrections $\Delta(\rho_g, s_w)$ are calculated to the solution variables so that

$$R(\rho_g+\Delta\rho_g, s_w+\Delta s_w) \approx R(\rho_g, s_w)+J\;\Delta(\rho_g, s_w) = 0 \tag{14}$$

where **J** is the Jacobian matrix formed from the partial derivatives of **R** with respect to the solution variables. To find the corrections, we solve the linearized equations:

$$\Delta(\rho_g, s_w) = -J^{-1}\,R(\rho_g, s_w) \tag{15}$$

At each iteration, a test of convergence is made by comparing the absolute sum of the gas and water residuals for each cell with preset tolerance values.

The linearized set of equations above are solved using a direct approach based on the method of singular value decomposition. This is designed to solve systems of equations which may be singular, or numerically very close to singular, i.e. where the solutions for two or more cells are identical, or the residuals are zero.

In transforming the integrated differential equations as difference equations, the mass accumulation and flow terms are rewritten in terms of the nodal values. The nodal positions are interpreted by the code such that the interface between two cells lies at the midpoint between the cell nodes. The mass accumulation terms are obtained from the fluid density at the cell centre solution value and the cell volume. Pressure gradients at the cell interfaces are calculated from the difference in solution pressures at neighbouring cell centres and the internode distances. The flow integrals are replaced by a summation of flux terms into or from neighbouring cells, multiplied by their respective cell interface areas. In changing from a linear to a cylindrical or spherical geometry, it is the cell volumes and the interface areas which are adjusted.

In order to correctly model the displacement of water by a gas, which is analogous to the modelling of a shock wave, it is necessary to introduce some form of numerical diffusion into the solution equations. This is achieved in TOPAZ by the use of a single-point upstream weighting technique to calculate cell mobilities. This means that instead of calculating the mobility of a particular phase at the interface between two cells as a symmetrical combination of the mobilities of the adjacent cells, it is assigned that of the cell from which the phase is flowing.

4. Blind predictions realized with TOPAZ

A series of twenty-five scoping calculations have been performed using TOPAZ, with data supplied from the results of gas breakthrough experiments, realized in laboratory on small dimension cylindrical clay cores. The full details of these calculations are reported elsewhere [6].

These simulations were done eitheir in two dimensions, assuming a cylindrical geometry as it is the case for the in situ gas injection device, and in three dimensions, assuming a spherical geometry. In the two-dimensional calculations, the gas is assumed to act as a line source, with the radius of the inner boundary cell set to 4.5cm, which corresponds to the radius of the gas injection filter. In the three-dimensional calculations, the central boundary cell has a radius of 5 cm, which gives a cell volume approximately equivalent to the one of a cylindrical gas inlet filter of length 10cm and diameter 9cm. The modelled region extends from the inlet cell (radius 4.5 or 5cm) out to a cylindrical or spherical radius of 5m.

Initially, the clay is assumed to be fully saturated with water at hydrostatic pressure, which is set equal to 1.5MPa. The boundary conditions specified at the inlet are a constant gas pressure and a gas saturation of 1. The far boundary cell is assigned a constant water pressure (hydrostatic pressure) and a water saturation of 1. The measurement points in the experiments were located at radii of 0.75, 1.0 and 1.5m, so the far boundary cell is sufficiently distant from the region of interest to have minimal influence on the results. This was checked by repeating selected calculations using a more distant far boundary, which did not alter the results significantly.

The parameters variation and the values used in the various cases were the following:

- for the permeability k, two different values of k were used, namely the permeability in the directions perpendicular and parallel to the bedding plane of the Boom clay (2.1×10^{-19} m² and 4.5×10^{-19} m² respectively)
- two values were chosen for the gas inlet pressure: 3.5 and 4.0 MPa
- for the threshold pressure, three values were used: 1.0, 1.27 and 1.55 MPa. The threshold pressure is defined here to be the capillary pressure which has to be overcome to force gas into clay by dispelling water from the pores with the largest radii
- three different values for the residual gas saturation were used: 0, 10^{-2}, and 10^{-1}
- for the porosity, two values were chosen: 0.36 and 0.15. The first value corresponds to the mean total porosity of Boom clay. The second value is the effective porosity, i.e. the porosity corresponding to mobile water by opposition to the water which is strongly bound to the clay by electrostatic forces [7]

None of the twenty-five performed simulations reflects perfectly the reality of the experimental conditions of E5. Nonetheless, one of these cases, solved in three dimensions, approaches them closely. Table 1 reports the physical parameters taken into account for this specific simulation case.

Table 1. **Physical parameters used in the test case**

Parameter	Value	Unit
Clay porosity	0.36	-
Permeability	4.5×10^{-19}	m²
Residual gas saturation	0	-
Clay compressibility	7.8×10^{-9}	Pa^{-1}
Gas (H_2) viscosity	8.5×10^{-6}	Pa.s
Water viscosity	10^{-3}	Pa.s
Gas inlet pressure	3.5	MPa
Threshold pressure	1	MPa
Hydrostatic pressure	1.5	MPa

The output variables calculated by TOPAZ include the gas inlet flux at the inlet gas pressure of 3.5 or 4.0 MPa, together with the gas pressure (or the sum of water and capillary pressure when no gas is present).

The initial timestep used was 1 day.
The variation of the relative permeability with saturation is shown on Fig. 6. The data points for the gas relative permeability were obtained by CEN•SCK [3], while those for the water relative permeability are taken from Robinet [8]. The theoretical curves fitted to the data are those derived by Fatt and Klikoff [9]. The variation of capillary pressure with saturation used as a tabular input is shown on Fig. 7.

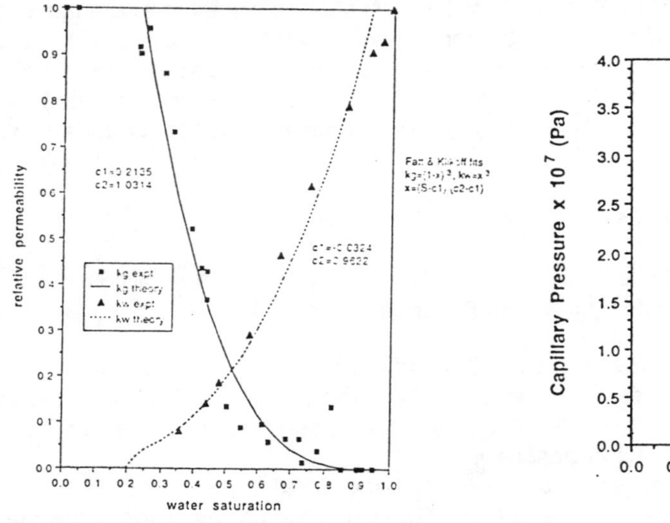

Figure 6. **Variation of gas and water relative permeabilities in Boom clay as a function of saturation**

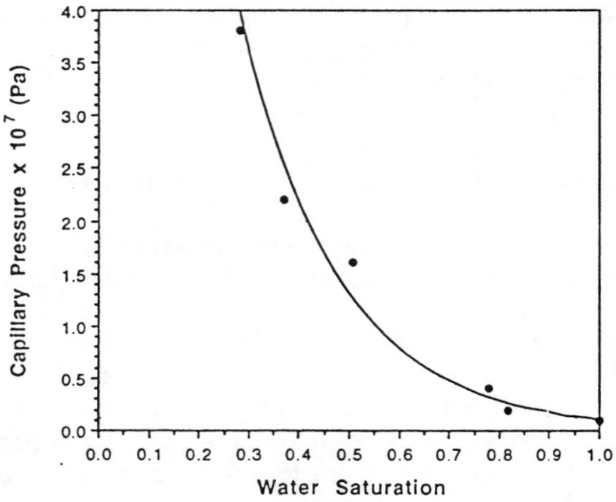

Figure 7. **Variation of capillary pressure in Boom clay as a function of saturation**

The results calculated using TOPAZ include estimates of the gas breakthrough times as well as the gas or total pressure at the monitored positions. In the considered case, the breakthrough times at a distant of 0.75, 1.0 and 1.5 meter from the gas injection source are respectively: 40, 131 and 419 days. Figure 8 shows the gas inlet flux history. The gas inlet flow rate reaches a steady-state value of 1.3 ml per hour after about 200 days.
Figure 9 shows the gas pressure history at 75 cm of the gas source.The figure suggests that the relationship between breakthrough times and distance from the source is non-linear.

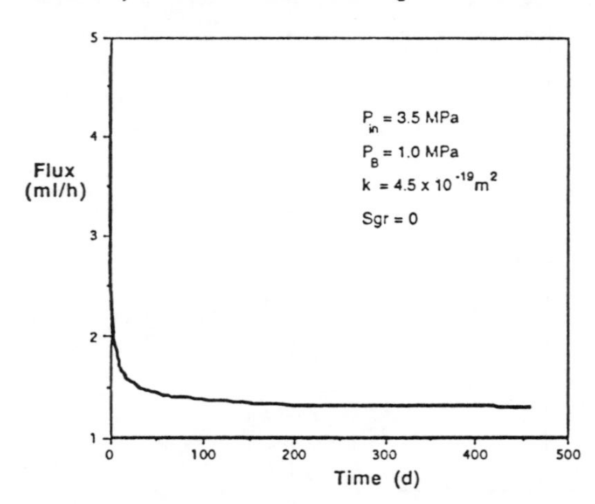

Figure 8. **Gas inlet flux history**

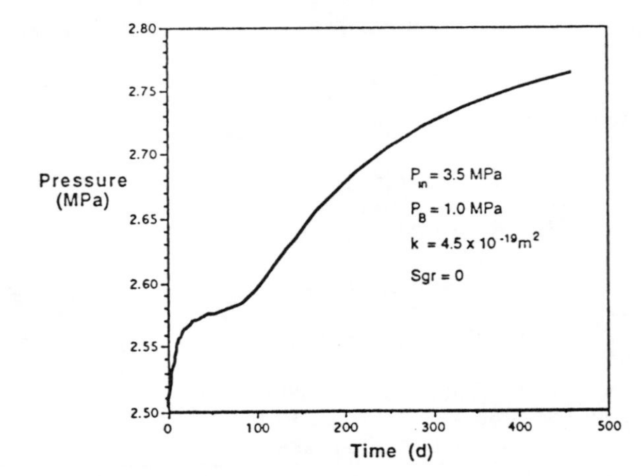

Figure 9. **Gas pressure history at a distance of 75 cm from the injection point**

5.Conclusion

The modelling of the hydraulic test has shown that Darcy's law is valid for water flow in the Boom clay.

The predictions realized with TOPAZ were made before the execution of the hydraulic campaign around the E5 experimental set-up. That explains the divergences regarding the parameters used as input data, such as: gas inlet pressure, hydrostatic pressure, threshold pressure, permeability,...

TOPAZ does not take the anisotropy of the clay into account, but it has been proven that the excess gas pressure necessary to expell water from the largest pores is lower parallel to the bedding plane. Another distinction is that the in situ experiment has been carried out with helium instead of hydrogen for evident safety reasons.

The main difference between the results obtained during the in situ gas injection experiment and the predictions is certainly the gas migration mechanism. The formation of preferential pathways has been observed in situ, while TOPAZ dealt with a two phase flow mechanism for the simulated cases.

A possible explanation for the formation of preferential pathways and the low observed breakthrough pressure value is the influence of the drilling of the borehole on the local geomechanical stress distribution. The drilling of the borehole and the creep of the clay around the experimental device leads to a local reduction in effective stress, which is the global macroscopic stress between the clay particles. In this low effective stress zone, preferential pathways are easily created.

TOPAZ may reproduce in a phenomenological manner the observed migration mechanism by introducing the influence of the effective stress on permeability and porosity (see eq. 9 and 10). Nonetheless, we may not assert that it corresponds with the physical reality.

A gas migration model for a plastic, consolidated clay, should take into account the hydro-mechanical coupling. Wether Darcy's law will be still applicable depends on whether a large number of preferential pathways will be created, or only a few distinct fractures. In the first case, a Darcy type model (equivalent porous medium approach) is applicable, while in the second case it will be necessary to apply a kind of hydro-fracturing model.

6. Acknowledgement

This work has been carried out within the framework of the CEC programme on radioactive waste management and storage (1990 - 1994), task 4, under contract FI2W-CT91-0076. This project is financially supported by the CEC and NIRAS/ONDRAF. This support is gratefully acknowledged.

7. References

[1] Bear J. (1979) Hydraulics of groundwater. Mc Graw Hill

[2] Carslaw H.S. and Jaeger J.C. (1959) Conduction of heat in solids. Oxford Clarendon Press

[3] Volckaert G., Ortiz L., De Cannière P., Put M., Horseman S., Harrington J., Fioravante V., Impey M. (1994) MEGAS: Modelling and experiments on gas migration in repository host rocks. *CEC final report (draft)*, R-3019

[4] Volckaert G., Put M., Ortiz L., De Cannière P., Horseman S., Harrington J., Fioravante V., Impey M. and Worgan K. (1993) MEGAS: Modelling and experiments on gas migration in repository host rocks. *CEC annual progress report*, XII/423/93-EN

[5] Worgan K. (1992) TOPAZ: a model of two phase flow in porous media. *Intera report*, IM2703-3 Version 2

[6] Worgan K. and Impey M. (1992) TOPAZ scoping calculations for the MEGAS experiments in Boom clay. *Intera report*, IM2703-5 Version 3

[7] Geneste P. (1992) Gas pressure build-up in radioactive waste disposal: hydraulic and mechanical effects. *Geostock interim progress report*, GK/Rdd-92/024

[8] Robinet J.C. (1991) Phenomenology of hydrogeological transfer between the atmosphere and underground storage, PHEBUS, FI2 (1-91-11) Euro-Geomat, CEN●SCK

[9] Fatt I. and Klikoff W.A. (1959) Effect of fractional wettability on multiphase flow through porous media. *Pet. Trans. AIME* (216), 426-431

BACCHUS 2: An In Situ Backfill Hydration Experiment for Model Validation

G. Volckaert, F. Bernier
SCK •CEN (Belgium)

E. Alonso, A. Gens
UPC (Spain)

M. Dardaine
CEA (France)

Abstract

The BACCHUS 2 experiment is an in situ backfill hydration test performed in the HADES underground research facility situated in the plastic Boom clay layer at 220 m depth. The experiment aims at the optimization and demonstration of an installation procedure for a clay based backfill material. The instrumentation has been optimized in such a way that the results of the experiments can be used for the validation of hydro-mechanical codes such as NOSAT developed at the University of Catalunya Spain (UPC).

The experimental set-up consists in a bottom flange and a central filter around which the backfill material was applied. The backfill material consist of a mixture of high density clay pellets and clay powder. The experimental set-up and its instrumentation are described in detail. The results of the hydro-mechanical characterization of the backfill material is summarized.

The physical and numerical basis of the NOSATcode are described together with the main model assumptions and the data used for the blind predictions.

The evolution of the stresses, the strain and the degree of saturation in the backfill as obtained from the NOSAT code are compared with the first results of the in situ test. The importance of the model parameters and assumptions will be discussed.

Introduction

When designing a radioactive waste repository in a deep clay layer it is important to know the hydromechanical behaviour of the backfill and sealing materials. Therefore the SCK•CEN has developed, in cooperation with the CEA, the BACCHUS (BACkfilling Control experiment for High level waste in Underground storage) in situ backfill hydration experiments.

In the BACCHUS 1 experiment the applicability of precompacted Ca-bentonite blocks and blocks of highly compacted Boom clay spoil as backfill and sealing material was tested. The BACCHUS 1 experiment was performed between 1988 and 1990. The hydro-thermal behaviour of the backfill and the hydro-thermo-mechanical behaviour of the host rock were studied. The BACCHUS 2 experiment was installed in the HADES underground research facility in 1993 after the retrieval of the BACCHUS 1 experiment.

The BACCHUS 2 experiment aims at the optimization and demonstration of an installation procedure for a clay based backfill material. In this case a granular backfill was applied. The instrumentation has been optimized in such a way that the results of the experiments can be used for the validation of hydro-mechanical codes such as NOSAT developed at the University of Catalunya Spain (UPC).

Description of the BACCHUS 2 experiment

The design of the experiment was discussed as well with the project participants as with the modellers. Their remarks leaded to important modifications in the instrumentation. The resulting design is shown in Fig 1. The experiment consists essentially in a central filter surrounded by a Boom clay pellets-powder mixture on a thickness of about 23 cm. The experiment has been installed at the BACCHUS 1 location between 11 m and 14 m depth from the gallery floor level. The function of the main components is hereafter explained:

- The bottom flange is used as support for instrumentation and as protection for its wiring. The instrumentation implanted in the flange itself consists in total stress and pore water pressure sensors. The transducers have a small sensitive surface (< 1 cm^2) so that the measurements can be regarded as punctual measurements. The flange is also used as support for thin vertical perforated plates. These plates carry total stress and pore water pressure sensors. They allow to perform measurements in the middle of the backfill and at the interface with the natural host rock such as requested by the modellers.

- The central filter (f_{ext} 90 mm) allows the air to escape during the hydration. Once the backfill is saturated, the central filter will allow to record the water pressure at this inner boundary and to determine the hydraulic conductivity of the backfill. The central filter consists of four separated filter sections, which allow to test the homogeneity of the backfill. Between two of the filter sections, total pressure sensors have been installed to measure the swelling pressure exerted on the tube. An inner tube protects the wiring and tubing.

- At the level of the upper two filter sections, a second inner tube is installed. This tube allows to lower a heater.Three "spider webs" (f_{ext} 475 mm) with in total 63 thermocouples have been installed in the backfill material between the 11.5 and 12.5 m levels.They allow to follow the evolution of the radial hydration by measuring the changes in thermal conductivity. The thermal conductivity is derived from the temperature distribution at the end of a heating phase which last maximum five days.

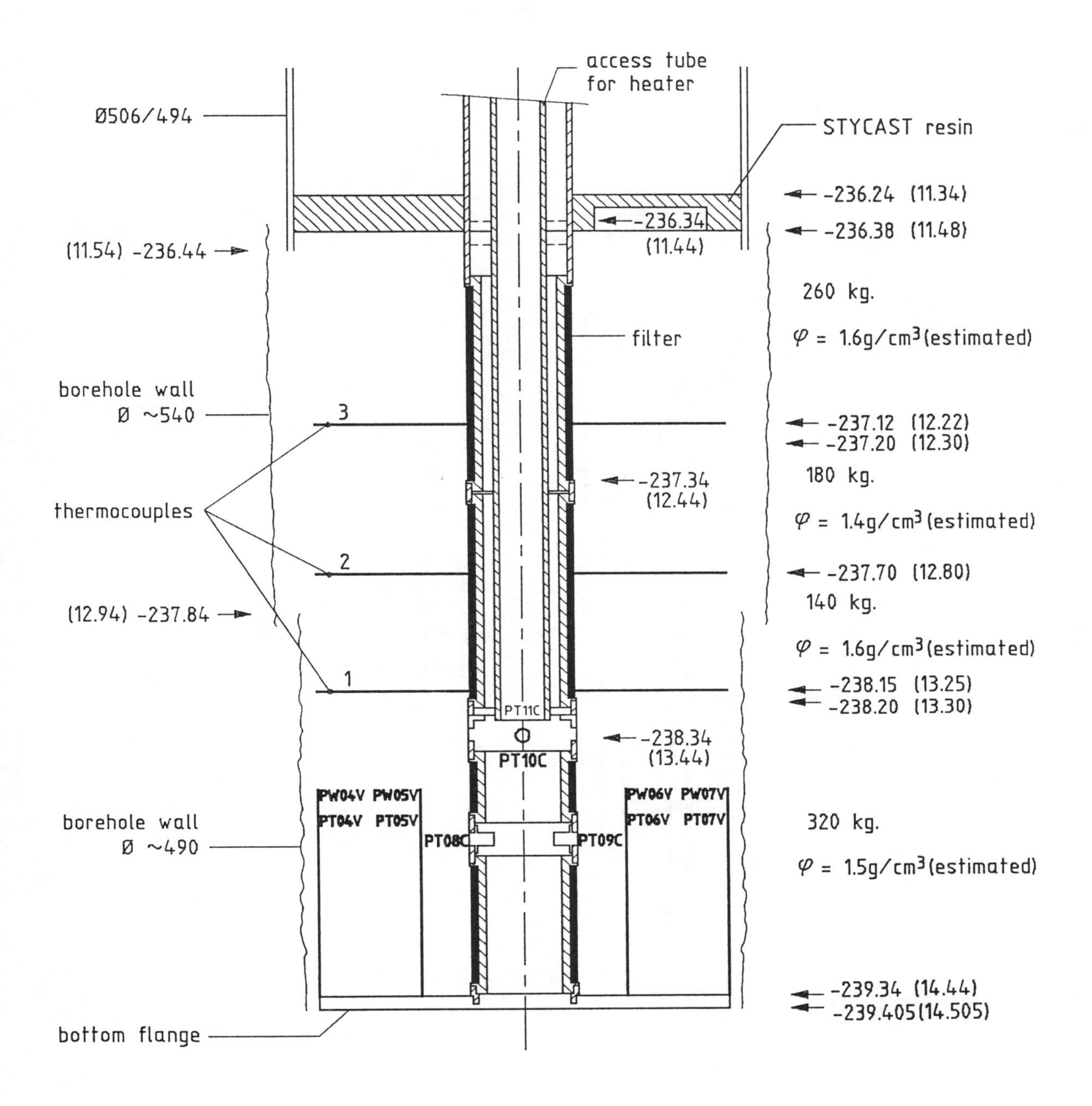

Figure 1 The BACCHUS 2 mock up after installation of the granular backfill and the plates with thermocouples. The estimated backfill density is also indicated.

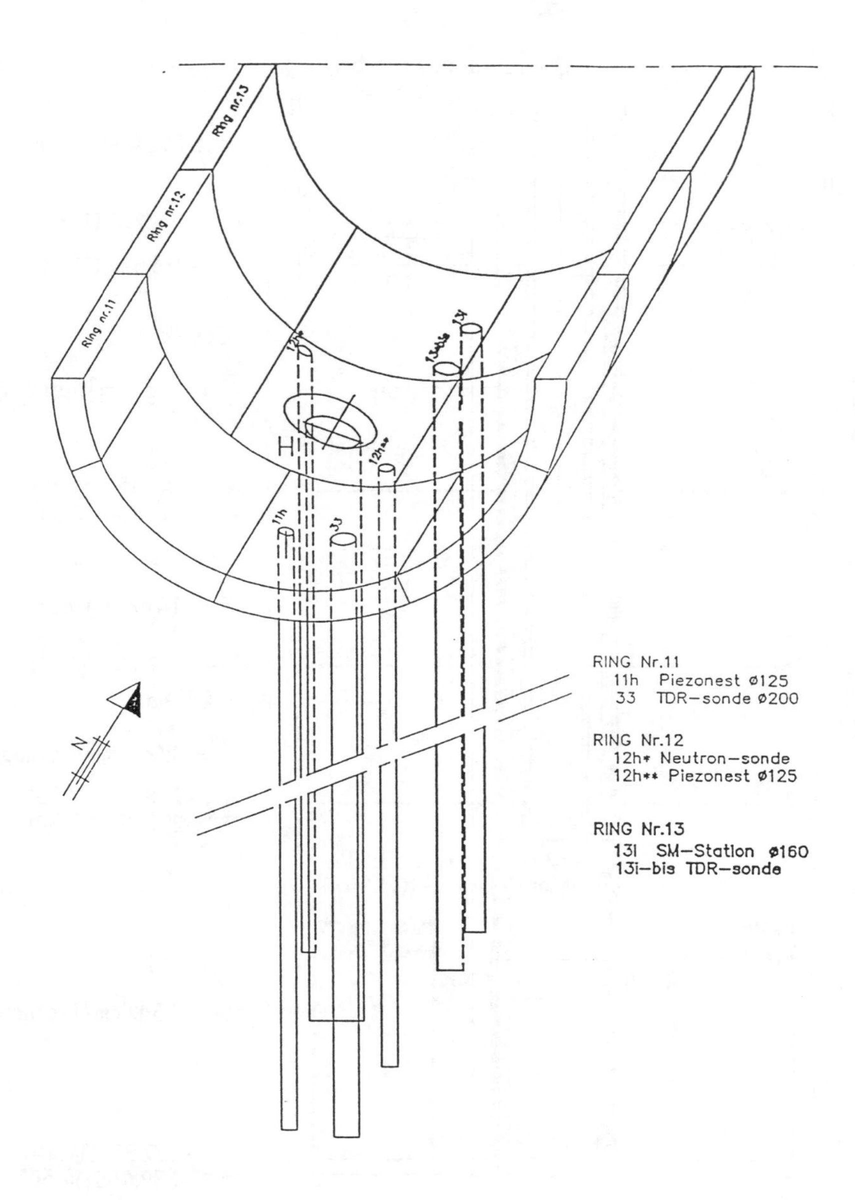

Figure 2: Location of the instrumented boreholes around BACCHUS 2: 3 D view of the gallery

The host rock instrumentation

The host rock instrumentation was installed in 1986-1987 as part of the BACCHUS 1 instrumentation. Tests have shown that it was still operating satisfactorily. It consists of two multi screen piezometers for the measurement of the hydrostatic pressure and a Glötzl stress monitoring station for the measurement of the radial, orthogonal and vertical total stress. Their location is shown on Fig 2. Although the absolute values of the total stress are clearly not reliable, the measured evolution was proven to be reliable so that the stress monitoring station can be further used to measure relative stress changes.

The backfill material and its basic hydro-mechanical properties

The backfill material is a 50/50 mixture Boom clay powder and high density Boom clay pellets. The high density pellets are produced from dry Boom clay powder using industrial equipment. The applied compactor-granulator is of the same type as those used to compact coal or ore powder. This equipment produces pellets with a dry density of about 2.1 g/cm³. The pellets are about 2 cm long and 0.5 cm thick. The mean dry density of the mixture is about 1.5 to 1.6 g/cm³. In vertical boreholes, this mixture is much easier to apply than precompacted high density clay blocks.
The basic idea behind the development of this type of material was that during the hydration the pellets would take up water, swell and compact the powder so that a homogenous low permeability material would be obtained. By applying X-ray tomography during a laboratory hydration experiment, we have shown that the material behaves as expected. Figure 3 shows the same cross-section of the mixture in an X-ray transparent permeameter before and 9 days after the start of the hydration experiment. The tomographs show clearly how the initially inhomogeneous material becomes homogeneous as it takes up water.
Hydration experiments were also performed to measure the hydraulic conductivity and swelling pressure of the mixture. The results given in Table 1 show that the hydraulic conductivity is well reproducible and very sensitive to the mean dry density. For a mixture with a dry density of 1.7 g/cm³, corresponding to the dry density of in situ Boom clay, the hydraulic conductivity is only two to three times larger than the mean in situ hydraulic conductivity. Taking into account the important initial heterogeneity in density of the mixture, this difference in hydraulic conductivity is small. This confirms the X-ray tomography results. At a density of 1.7 g/cm³, the swelling pressure of the mixture is about a factor of two lower than that of the in situ Boom clay (1 MPa).

Table 1 Hydraulic conductivity and swelling pressure as a function of the density of of the granular backfill material

sample number	dry density g/cm³	height cm	max. swelling pressure MPa	hydraulic conductivity m/s
R1D1.6S1	1.59	7.34	-	$3.0\ 10^{-11}$
R1D1.6S2	1.60	7.28	0.35	$1.4\ 10^{-11}$
R1D1.6S3	1.60	7.28	0.22	$1.5\ 10^{-11}$
R1D1.7S1	1.69	6.89	0.6	$7.7\ 10^{-12}$
R1D1.7S2	1.70	6.85	-	$9.9\ 10^{-12}$
R1D1.7S3	1.70	4.34	-	$8.2\ 10^{-12}$
R1D1.7S4	1.70	6.84	0.54	$8.6\ 10^{-12}$
R1D1.8S1	1.80	4.09	-	$6.0\ 10^{-12}$
R1D1.8S2	1.80	6.47	0.95	$6.3\ 10^{-12}$
R1D1.8S3	1.80	6.47	0.82	$5.2\ 10^{-12}$

The compressibility of the dry mixture was determined. An uniaxial compaction pressure of 4 MPa, corresponding to the in situ total stress, leads to an increase of the density with 20 %. We expect the compressibility of the partially saturated mixture to be even higher.

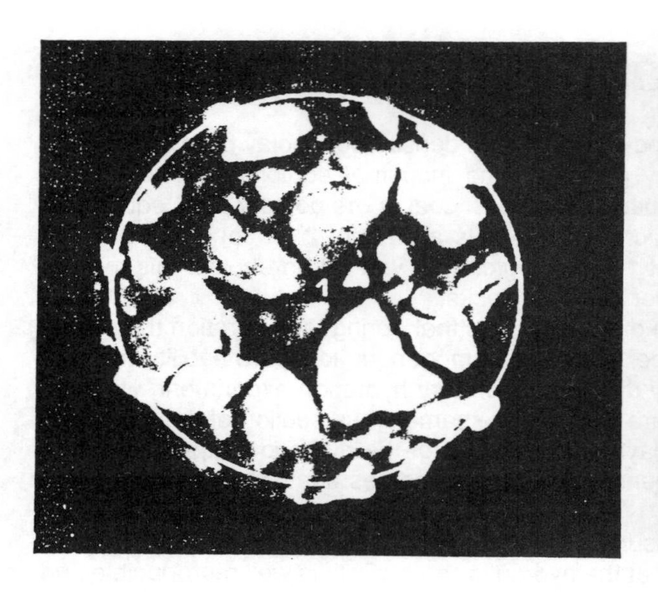
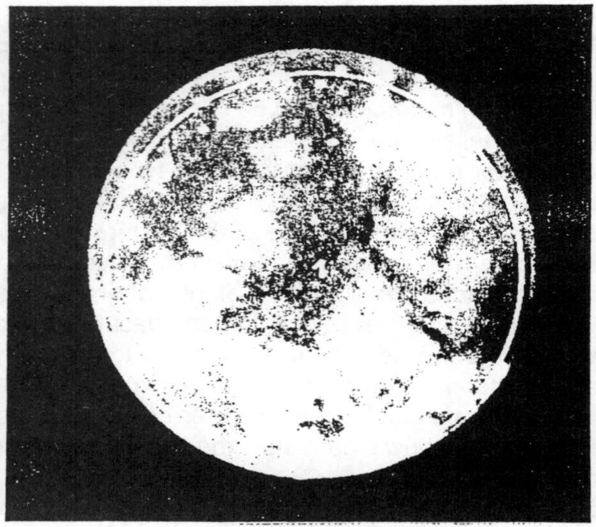

Figure 3 X-ray tomographs of the granular backfill (a cross-section through the X-ray transparent oedometer) before hydration and 9 days after the start of the hydration experiment.

Description of the NOSAT code

Constitutive relations

Two sets of effective stresses are supposed to independently control the mechanical behaviour of an unsaturated soil: the 'net' stress or excess of total stress s over air pressure p_a and the water suction $s=p_a-p_w$, where p_w is the water pressure. In a general way, the mechanical behaviour of an unsaturated soil can be expressed as:

$$ds^* = \boldsymbol{D}(\, de - de_o) \qquad (1)$$

where s^* is the net stress, e the total strain and e_o the strain induced by suction changes. Any constitutive law may be used to define ***D***. In this work, a nonlinear elastic model has been defined through a tangent compressibility coefficient K_t and a tangent shear modulus G_t.

Volumetric strains $(e_o)_v$, have been specified through the concept of state surface (Matyas and Radakrishna, 1968). The following general expression, propose by Lloret and Alonso (1985) for a wide variety of unsaturated soils, has been adopted:

$$(e_o)_v = a'(\, s - p_a) + [b' + c'\, (\, s - p_a)](p_a - p_w) \qquad (2)$$

where $(\, s - p_a)$ is the net mean stress, and a', b' and c' are constants. K_t is obtained from equation (2) by differentiation.

Soil water retention characteristics should also be defined. The concept of state surface is also suitable for these purposes. Experimental results collected by Lloret and Alonso (1985) suggested the following expression for the variation of degree of saturation with suction and net confining stress:

$$S_r = 1 - \{ 1 - \exp[-a''(p_a - p_w)]\}[b'' + c''(s - p_a)] \tag{3}$$

where a", b" and c" are constants.

Flow properties

Generalized Darcy's laws for water and air flow are used. Alonso et al. (1987) made a review of existing empirical relations between air and water permeabilities, water suction, degree of saturation and soil porosity. In the analysis presented later the following relationships suggested by Lloret and Alonso (1985) were used:

$$K_w(e,S_r) = A\,[(S_r - S_{ru})/(1 - S_{ru})]^3\, 10^q \tag{4}$$

$$K_a(e,S_r) = B\,[e(1-S_r)]^c \;\; g_a/m_a \tag{5}$$

where A, S_{ru}, q, B and c are constants, g_a the specific weight of air and m_a the viscosity of air.

Equation (4) combines the change of water permeability with void ratio e, described in Lambe and Whitman (1964) and the variation with degree of saturation proposed by Irmay (1954). Equation (5) has been proposed by Yoshimi and Osterberg (1963).

Field equations

Three basic sets of equations have to be solved: mechanical equilibrium, continuity of air and continuity of water:

mechanical equilibrium

$$\frac{\partial(\sigma_{ij} - \delta_{ij} p_a)}{\partial x_j} + \frac{\partial p_a}{\partial x_i} + b_i = 0 \tag{6}$$

air continuity

$$\frac{\partial}{\partial t}[\rho_a n(1 - S_r + HS_r)] + div[\rho_a(v_a + Hv_w)] = 0 \tag{7}$$

water continuity

$$\frac{\partial(\rho_w n S_r)}{\partial t} + div(\rho_w v_w) = 0 \tag{8}$$

where b_i are body forces, r_a and r_w densities of air and water, n the porosity and H Henry's constant.

The NOSAT code

The above set of differential equations have been discretized using Galerkin's procedure and a finite element computer program, NOSAT, was written to solve this type of unsaturated soil problems. The code allows to specify excavation and construction sequences. In addition to conventional types of boundary conditions, two special types have been implemented for the flow part of the analysis: seepage surface and drain conditions as described in Lloret and Alonso et al. (1994). It can also handle double porosity materials which are a suitable representation for highly expansive clays and rocks (Gens et al., 1993). The results discussed herein were obtained with the basic formulation outlined above. Concerning the mechanical behaviour two types of analysis have been performed: a nonlinear elastic and a no tension version of the preceding case.

Application to the BACCHUS 2 experiment

Conceptual representation

The analyzed geometry is schematically represented in Fig. 4. Radial symmetry and plane strain conditions were supposed to allow a 2D simplification. Figure 5 is a quarter space representation of a cross section through the central filter, which acts as a drain, the granular backfill and the natural clay. As far as the numerical discretization is concerned the outer boundary was located at a radius of 3.5 m. The applied finite element mesh has 135 quadrilateral elements. The mesh was refined in the backfill material and in the vicinity of the natural clay-backfill interface.

Initial stress conditions were obtained through a simulation of the actual installation procedure (see Fig. 5): drilling of the borehole followed by the installation of the backfill. It was supposed that the natural clay was subjected to an isotropic initial compressive stress $s_o = s_{ox} = s_{oy} = s_{oz} = 5$ MPa (Fig. 5a). The drilling of the borehole was then simulated (the shaded area in Fig. 5b is removed. The installation of the probe and the backfill induces a boundary stress s_c (Fig. 5c). This stress corresponds to the assumed initial stress conditions in the backfill material (Fig. 5d): $s_{ox} = s_{oy} = 1$ MPa and $s_{oz} = 0.25$ MPa. A rigid contact was assumed at the probe -backfilling interface: displacements in radial direction towards the centre are impossible.

As far as the initial flow conditions are concerned, the natural clay was fully saturated at a constant pore pressure of 2 MPa. As the backfill was installed at low water content, a relative strong initial suction (15 MPa) was assumed in the compacted backfill.

The outer boundary of the discretized area was assumed impervious to air and water. Along the probe - backfill interface a 'seepage' condition was imposed. The filters of the probe act as a drain for the incoming water.

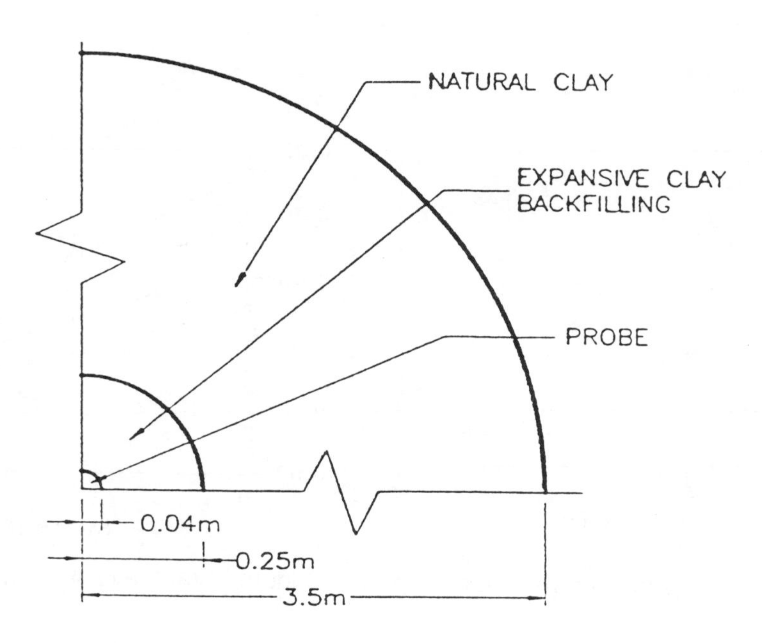

Figure 4 Schematic representation of the analysed geometry.

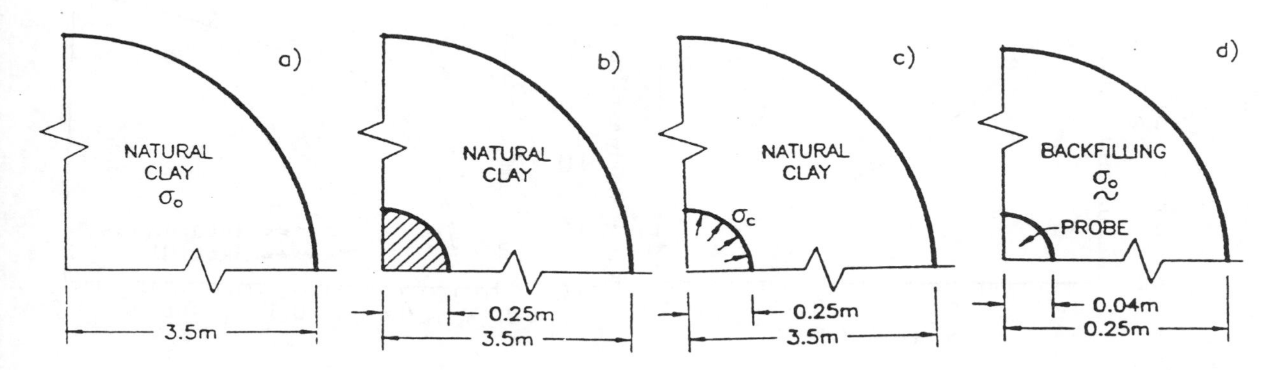

Figure 5 Simulation of the installation of the probe: a) initial state, b) borehole drilling, c) installation of the backfill, d) final state.

Main material properties applied in the simulation

Figure 6 shows the state surface for volumetric deformation adopted for the natural clay. Note that an increase in suction (drying) induces a small volumetric shrinkage, which decreases with applied net confining stress. The equivalent state surface for the backfill is plotted in Fig. 7. Wetting of the backfill is expected to induce considerable swelling for the whole range of expected confining stresses.

A common state surface for the degree of saturation was adopted for the backfill and the natural clay (Fig. 8). The relation between water permeability and saturation is plotted in Fig. 9.

Poisson's ratio was fixed at n = 0.30 for both the natural clay and the backfill material.

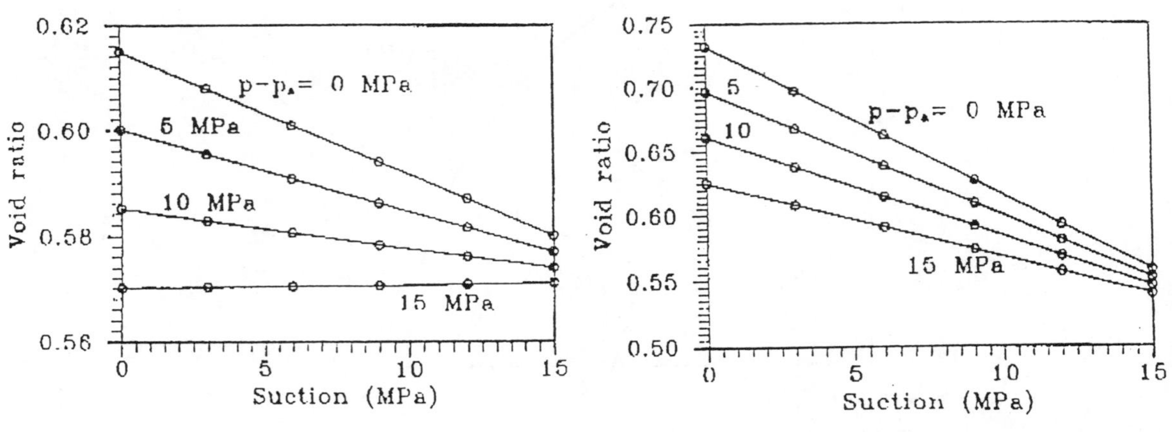

Figure 6 Void ratio state surface of the natural clay.

Figure 7 Void ratio state surface of the backfill.

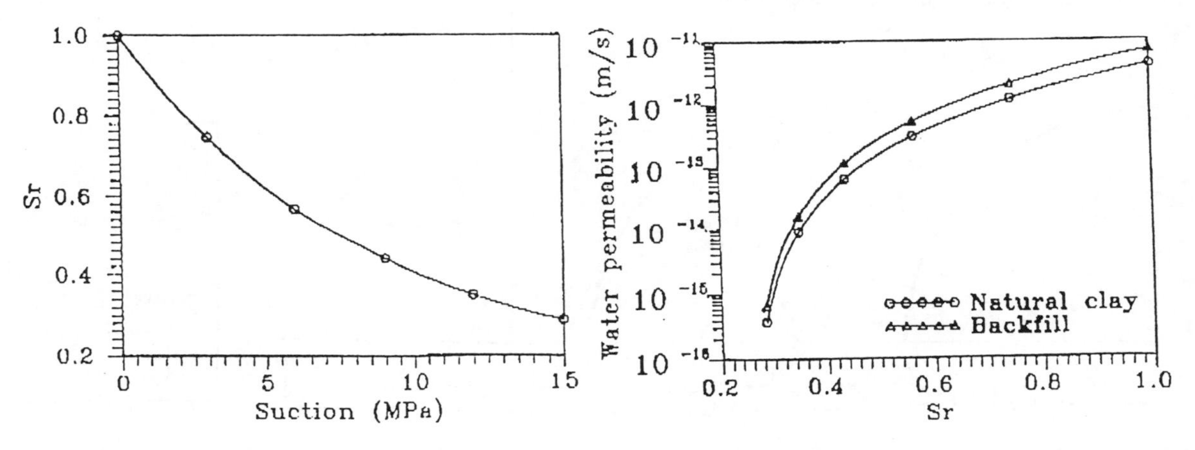

Figure 8 Water retension curve for the natural clay.

Figure 9 Relation between water permeability and saturation.

Results of the simulations

The results of the simulations have been presented as the variation with time of a state variable at different radial distances from the centre.

The induced flow is conveniently visualized through a plot showing the evolution of the degree of saturation (Fig. 10) . The initially dry backfill is wetted as a saturation front driven by the high suction in the backfill, progresses inwards. The high initial suction in the backfill causes a transient drying in the natural clay in the vicinity of the backfill. Steady state conditions are reached after about 150 days. Figure 11a shows the computed radial displacements. Maximum displacements at steady state conditions occur at the backfill-natural clay interface. At the inner boundary a small gap (about 1 mm) was computed after the swelling of the backfill.

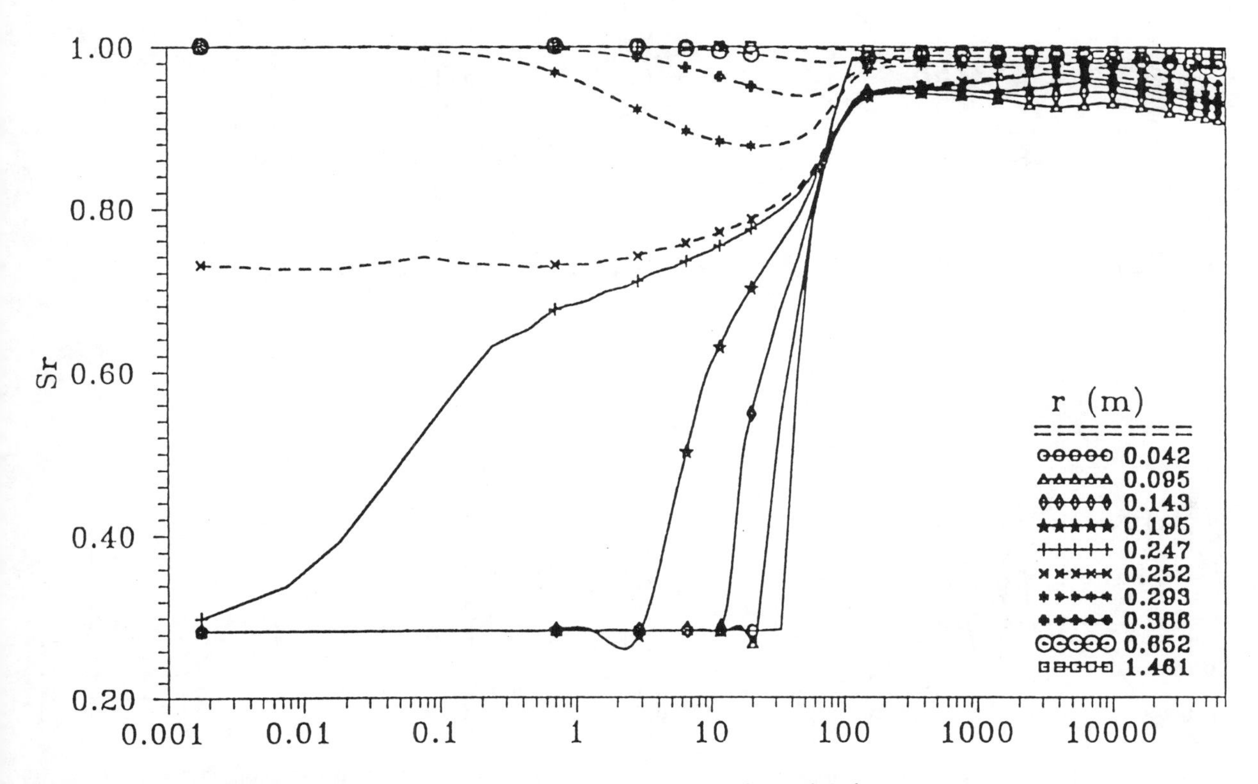

Figure 10 Evolution of the degree of saturation at some selected points

The mechanical effect of the wetting process is clearly demonstrated by the evolution of the circumferential principal stress (see Fig. 12a). The points in the backfill experience an extension-compression cycle as they are reached by the wetting front. At steady state a stable compression is computed in the wetted backfill. On the contrary, the natural surrounding clay is pushed away by the expanding inner backfill and circumferential stresses become tension stresses.

It is clear that the computed tension could not occur in reality in the considered clay soil. Therefore a more realistic no tension analysis was carried out.

The circumferential stresses computed in the no tension analysis are given in Fig. 12b. The hydration of the backfill leads only to compression forces. Although the stress distribution is different, the maximum compression in the backfill is similar to the one obtained in the nonlinear elastic analysis. The computed stresses in the natural clay have decreased to moderate realistic values. Figure 11b shows that the no tension analysis predicts larger overall displacements compared to the elastic analysis and it also predicts a gap at the probe-backfill interface. The calculated flow behaviour in the no tension analysis is very similar to the one obtained in the elastic case.

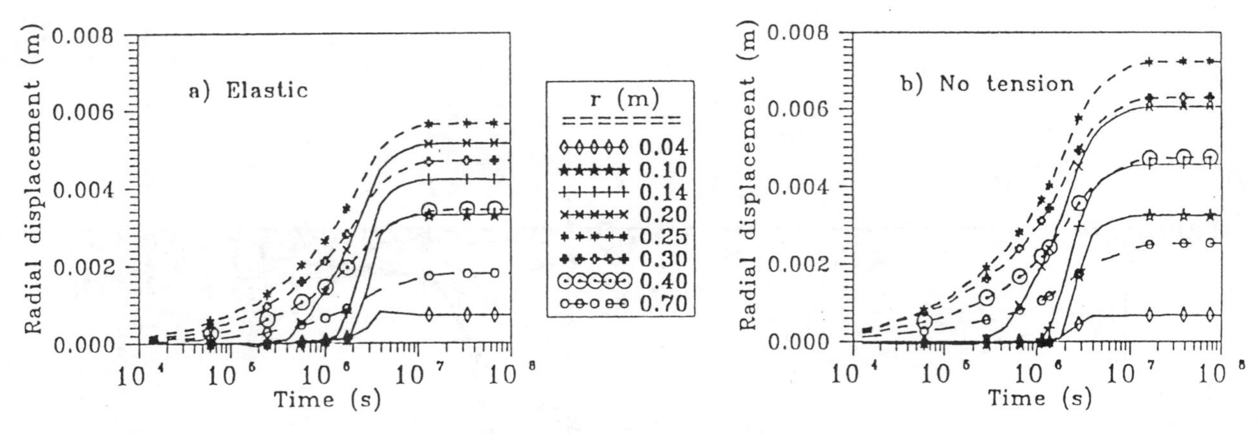

Figure 11 Computed radial displacements: a) elastic analysis, b) no tension analysis.

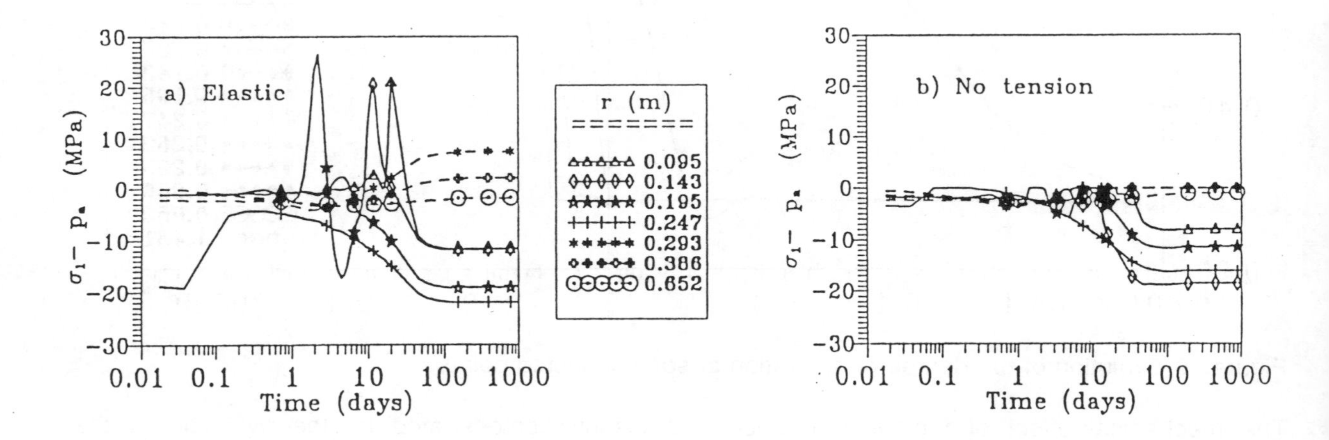

Figure 12 Circumferential stresses: a) elastic analysis, b) no tension analysis.

A further insight into the mechanical phenomena developing during the wetting of the backfill is provided by the stress paths in the (q,p) plane where p is the net mean stress and q the deviatoric stress (Fig. 13). Figure 13a shows the stress paths corresponding to the nonlinear elastic analysis. Loading-unloading cycles with high tension and shear stresses are again clearly identified. It is obvious that these paths cross any reasonable strength envelope which could be assigned to the backfill and natural clay. The stress paths for the no tension analysis plotted in Fig. 13b, still show important shear stresses. Those shear stresses probably reach beyond the shear strength in some cases. These results point out the need for improving the analysis introducing elastoplastic criteria.

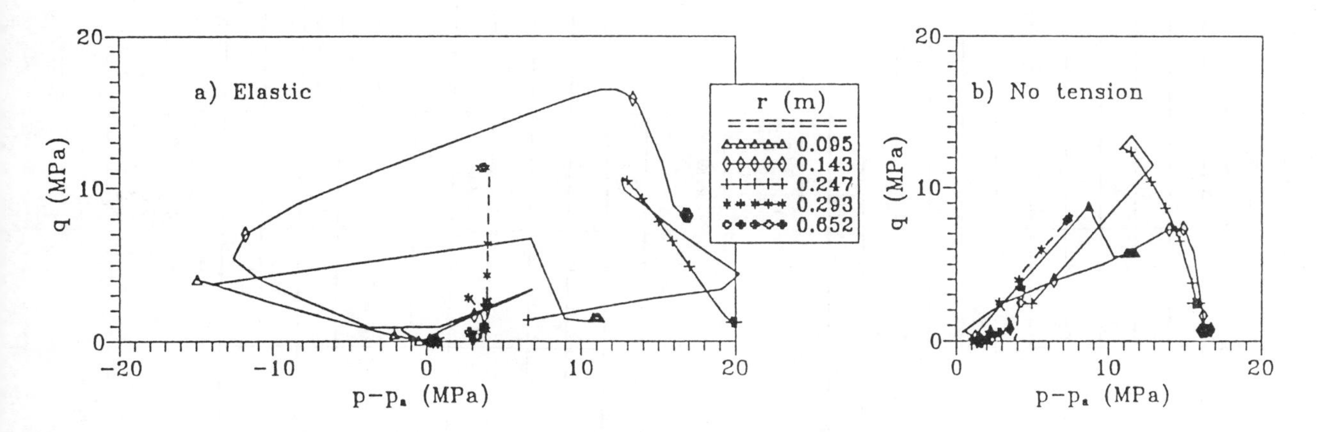

Figure 13 Stress paths in the (p,q) plane: a) elastic analysis, b) no tension analysis.

Experimental results

The BACCHUS 2 experiment is still running, so the experimental results are given until the time of writing of this text i.e. 15 months after the start of the experiment. Only the main experimental results are given below.

Measurements in the backfill material

The evolution of the water content distribution is determined by measuring the thermal conductivity of the backfill. The dry backfill has a thermal conductivity of about 0.6-0.7 W/m°C, while the saturated backfill is expected to have a thermal conductivity close to the natural saturated clay i.e. 1.7 W/m°C. In Fig. 14 the evolution of the thermal conductivity in two different radial directions is plotted. As expected, the thermal conductivities of the backfill increase with time and increase faster close to the natural clay-backfill interface then close to the centre. At the outer interface the backfill is now close to saturation while at the inner boundary it is still fairly dry. No positive pore water pressures have yet been measured what confirms that the backfill is still not saturated. We also observe an inhomogeneity in the hydration process: the conductivity in the middle of the backfill at the south side is higher than at the north side.

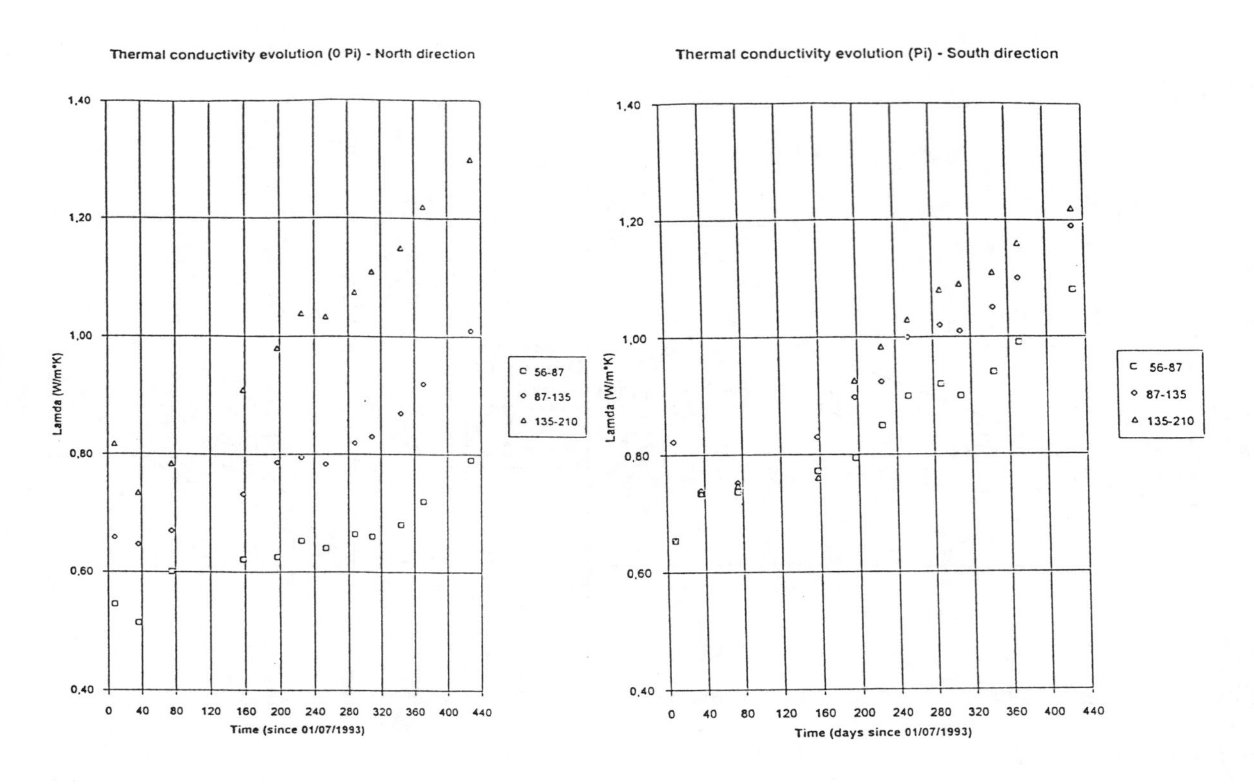

Figure 14 Evolution of the thermal conductivity at the outher boundary (135-210), in the middle of the backfill (87-135) and at the inner boundary (56-87).

This inhomogeneity is also found back in the total stress measurements shown in Fig. 15. This inhomogeneity is probably a consequence of an inhomogeneity in the convergence of the large borehole. At the inner boundary, important total stresses are recorded on the central tube (about 1 MPa). These are possibly a consequence of the consolidation of the dry backfill.

Measurements in the natural clay

Higher pore water pressures are recorded at ring 11 than at ring 12 (see Fig. 16). This difference can be explained by the location of the two devices (see Fig. 2). Indeed, as the piezometers placed at ring 12 are closer to the experiment than those placed at ring 11, the influence of the draining (resulting in lower pore water pressures) caused by the hydration of the backfill material is more important. Also the pressures at the bottom of the experiment are less influenced by the draining. This explains why after the pressure drop caused by the installation of the experiment, only the pore water pressure at the lowest points have immediately started to increase. Up to now, the pore water pressure at the higher levels does not increase because the backfill is not yet saturated.

The total stresses (radial, tangential and vertical) recorded in the host clay at a distance of 1.2 m remain almost constant and seem not to be influenced by the hydration of the backfill.

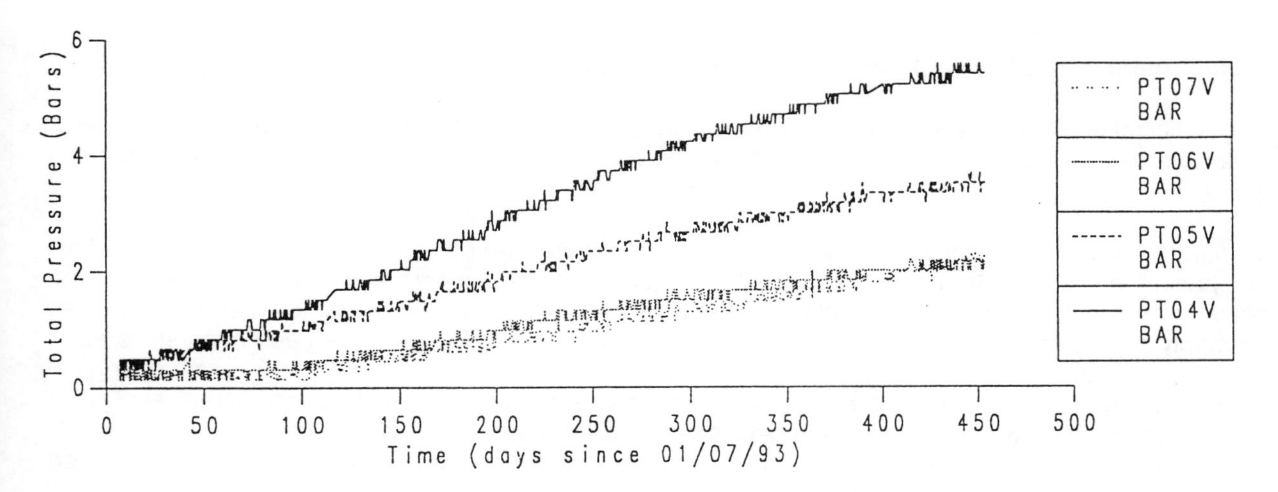

Figure 15 Evolution of the total radial stress in the backfill (sensors PT07V and PT06V south direction, PT04V and PT05V north direction, see also Fig. 1)

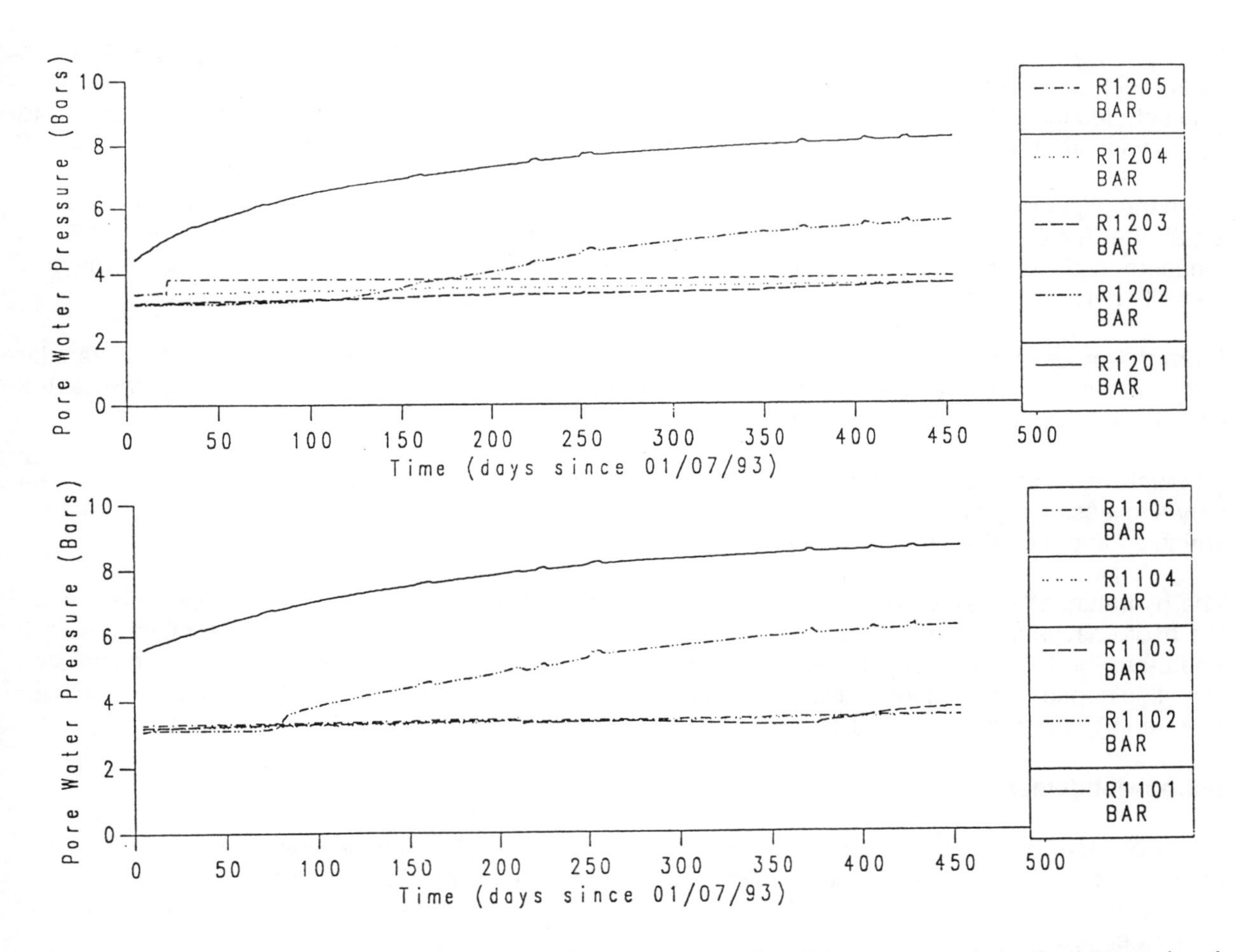

Figure 16 Evolution of the pore water pressure in the natural clay (01 corresponds to the bottom level of the experiment, 05 to the top level and the others are at intermediate levels)

Comparison between computed and experimental results

The calculations were blind predictions performed before a full characterisation of the backfill was available and were mainly intended to show the potential of the code to model coupled hydration process.

Apparently the hydration of the backfill takes much more time than computed: after more than 400 days the backfill is still not saturated while a steady state time of 150 days was predicted. With respect to the saturation time, the hydraulic conductivity of the backfill is a very sensitive parameter whose value has probably been over estimated. Also the initial suction of the backfill might have been overestimated.

The computed total stresses in the backfill are much larger than the measured total stresses. This is most probably due to the adopted parameters that overestimate the swelling properties of the backfill (especially the swelling stress). Up to now no gap has been formed at the inner boundary as indicated by the important pressures recorded at this location.

As computed by the no tension analysis, the influence of the hydration of the backfill on the stress distribution in the natural clay is very small, and this is in accordance with the in situ measurements.

Conclusions

It should be reminded that the aim of the BACCHUS 2 experiment was not only to validate a hydromechanical model but also to demonstrate the applicability of a granular backfill using industrial available techniques. This combination was quite difficult to realize. Yet, it was thought to be useful to try to validate a model in a situation may be close to reality and not idealized.

From the experimental side the main difficulty encountered during the design of the experiment was to install the instrumentation at the locations required by the modellers. It was not always possible to measure in situ all model variables the modellers like to compare with (e.g. displacements in the clay backfill could not be measured).

From the modelling point of view, the main difficulties are: the type of backfill material i.e. a granular material, the estimation of the unknown model parameters and the correct representation of the natural clay- backfill interface.

Although this first attempt to validate a hydromechanical model for unsaturated clay was as such not very successful, it has helped a lot both experimentalists and modellers to point out which parameters, variables and model assumptions are the most critical.

The hydration of a clay backfill in a borehole seems a relative simple problem at first sight. However, the hydromechanical transient analysis of it has shown that complex stress histories might develop. Therefore the NOSAT code will be further extended to include an elastoplastic model for unsaturated clay. It is hoped that this together with an improved characterisation of the granular backfill will allow for a more successful simulation of the experiment.

Acknowledgement

This work has been carried out within the framework of the CEC programme on radioactive waste management and storage Part B HADES under contracts FI2W-CT90-33 and 102. This project is financially supported by the CEC, ANDRA, ENRESA and NIRAS/ONDRAF. This support is gratefully acknowledged.

References

1. Alonso, E., Gens, A., and Hight, D. W. (1987). Special problem soils: general report. Proc. IX I.C.S.M.F.E. Dublin.
2. Gens, A., Alonso, E., Lloret, A. and Batelle, F. (1993). Prediction of long term swelling of expansive soft rocks: a double structure approach. Geotechnical Eng. of Hard Soils-Soft Rocks. Anagnostopoulos et al. eds., Balkema, pp 495-500.
3. Irmay, S. (1954). On the hydraulic conductivity of unsaturated soils. Trans. Amer. Geoph. Union. 35, pp 463- 468
4. Lambe, T. W. and Whitman, R. V. (1968). Soil mechanics. J. Wiley, New York.
5. Lloret, A. and Alonso, E. (1985). State surfaces for partially saturated soils. Proc. XI I.C.S.M.F.E., San Francisco, Vol. 2, pp 557-562.
6. Lloret, A. and Alonso, E. (1994). Unsaturated flow analysis for the design of a multi layer barrier. Proc. XIII I.C.S.M.F.E., New Delhi, Vol 4, pp 1629-1632.
7. Matyas, E. C. and Radhakrishna H. S. (1968). Volume change characteristics of partially saturated soils. Géotechnique, 18, pp 432-448.
8. Yoshimi, V. and Osterberg, J. O. (1963). Compression of partially saturated cohesive soils. Soil Mech. Found. Eng. Div. ASCE 89 (SM4), pp 1-24.

Modelling Hygro-Thermo-Mechanical Behaviour of Engineered Clay Barriers - Validation Phase.

C. Onofrei
AECL Research, Whiteshell Laboratories

(Canada)

ABSTRACT

The Canadian Nuclear Fuel Waste Management Program (CNFWMP) is evaluating the concept of disposal of nuclear fuel waste in an engineered vault at a depth of 500 to 1000 m in the plutonic rock of the Canadian Shield. In common with engineered barrier system designs being developed in other countries, the waste would be contained within durable containers that, in turn, would be isolated from the host rock by clay-based materials.

The objective of the CNFWMP is to develop a disposal concept that will protect human health and the natural environment far into the future. Assessments of the conceptual vault designs are based on system theory in which an attempt is made to correlate experiences with theoretical concepts of planned systems in such a way that the resulting coordination is sound and convincing. By necessity, since experiments with a total disposal system can never be performed, both the design and the performance assessment rely on experiments performed on physical models of vault elements over relatively short times and on information inferred from calculations (mathematical models) that simulate the probable behaviour of the system in the space-time domain of interest. For a simulation model to be successful, that is appied within a real world situation, the model must provide information regarding the behaviour of the system of interest that is clearly better, in some way, than the mental image or other abstract model that would be used instead. The results of a series of tests performed within the activity known as validation serve as tangible evidence regarding the success of a model in representing the system of interest.

This paper focusses on the validation of the models that describe the hygro-thermo-mechanical behaviour of the engineered clay-barriers proposed for application in the Canadian disposal system concept. The strategy being used to address the key issues in modelling to minimize the model error and to maximize the usefulness of the simulation model, based on testing procedures, is reviewed. Finally, a project initiated within the CNFWMP, dedicated to the VALidation of codes/models that describe the Unsaturated behaviour of engineered CLAY barriers (the ***VALUCLAY*** Project), is described.

1. Introduction

The Canadian concept of the repository system is similar to the concepts of a number of other countries and is based on the idea of a multiple barrier system. The multi barrier system consists of: waste packages (containers with heat-generating radioactive waste), other engineered barriers (buffer and backfill, and other sealing materials which will probably be based on clay), and natural barriers (the crystalline rocks of the Canadian Shield).

Although the barrier sub-systems complement each other, the multiple barrier concept implies that each individual barrier type performs effectively over a long period of time. Since experiments with real disposal can never be performed, both the design and the performance assessment of engineered barriers rely on experiments performed on physical models of an engineered barriers sub-system and on information inferred from calculations (mathematical models) that simulate the probable behaviour of the real object in the space-time domain relevant to the safety requirements. Comprehensive assessments of a repository system as well as assessments of individual barriers sub-systems, are based on system theory and an attempt is made to correlate experiences with theoretical concepts of perceived systems in such way that the resulting coordination is sound and convincing.

Within the Canadian Nuclear Fuel Waste Management Program (CNFWMP), one of the more important goals is to determine whether the simulation models used are adequate tools that represent the probable behaviour of the real engineered barriers. There are debates in the literature over the definition of the tests performed to evaluate the mathematical models used to estimate the behaviour of a complex system (Oreskes et al., 1994; Bredehoeft and Konikow, 1993; McCombie and McKinley, 1993). The problem of certainty associated with the output of numerical models is merely a philosophical issue. We all know that it is not possible to describe the complexity of investigated phenomena with certainty. This is not a special case for the nuclear waste management problem, but is generally true for any practical problem that deals with physics, i.e., all natural sciences that base their concepts on measurements that lend themselves to mathematical formulation. The debate centres on the question if there is a difference between the meaning of the scientific concepts and the concepts of the general public. This question is not new and it is generally accepted that the expression of scientific concepts cannot always be made using terms with identical meanings as in ordinary language. This view has been clearly expressed by the scientist philosopher, Albert Einstein (1940) who wrote that "The scientific way of forming concepts differs from that which we use in our daily life, not basically, but merely in the more precise definition of concepts and conclusions ..."

Assessments of engineered clay barriers system based on simulation models make use of a set of scientific concepts, including the concept of validation. This paper presents an overview of the basic concepts used in the analysis of engineered clay barriers derived from system theory in the specific context of a repository system for the disposal of heat-generating nuclear wastes. The next sections address, primarily, three concatenated questions (what models are tested, why are they tested and how do we assess success of a simulation model or a group of models), and answers are provided through an example of a project, recently initiated, dedicated to the VALidation of codes/models that describe the Unsaturated behaviour of engineered CLAY barriers, the ***VALUCLAY*** project.

2. Basic concepts - an overview

For more than a century, system theory has been used successfully to analyze complex, dynamic systems. Fundamental to system theory is a set of related definitions that describe the reality (matter/energy) organized in an integrated fashion (hierarchical) in time and space domains (Miller, 1978).

Basic to the analysis of engineering barrier system are three terms: system, model and simulation model. These terms are used in many different ways. Sometimes the terms system and model are used interchangeably, so that their meanings are often confusing, making the documentation and communication of the model results difficult. To distinguish between system and model the system is often considered to be a part of reality, i.e., an entity, whereas the model is considered to be a simplified representation of reality. These definitions are appropriate in dealing with problems of managing natural (existing) systems. When changes of a natural system are planned, before the decision to change an existing system is implemented, the scientist is asked to facilitate answering "if then" questions of managers and decision-makers. The option employed by scientists from many different branches of science, other than the obvious one of polite evasion, to answer such questions is based on system analysis. This method includes three groups of interrelated activities: 1) order data and information into a logical model that describes the existing system, 2) assure the usefulness of the model by comparing its output with the real system output, i.e., model validation, and 3) include in the model the perceived changes and simulate the behaviour of the system to estimate the outcomes of a direct modification of the real system and/or the probable range of variability of the system's state in both the time and space domains of interest. In engineering work, the objective is often to build a new product. The general procedure used in engineering work is also based on system analysis, but the starting-point of the analysis is a conceptual system, which, by definition, is an abstraction. A simulation model that simplifies the conceptual system is developed, in which numerical values are given to the general statements of size, magnitude and influences. After the simulation model is judiciously tested (validated) it is used as a direction to build the new product, the concrete object. The end product is in a sense a model of the conceptual system, and the simulation model sits between the concept and physical reality. Obviously, a comparison of the simulation model outputs with measurements of the concrete object before the object is built is not possible. Difficulties encountered in the validation of a simulation model are generally resolved by developing physical models of the conceptual system or of components of the system, and testing the model by comparing the outputs from a simulation with the outputs of a physical model.

The following general definitions for the terms system, model and simulation, widely accepted by scientists from various branches of science, are also employed in the system analysis of engineered barriers. System: *"Sets of elements standing in interaction."* (von Bertalanffy, 1956); Model: *"A set of rules and relationships that describe something is a model of that thing."* (Forrester, 1968); Simulation model: *"... the art of building mathematical models and the study of their properties in reference to those of the system."* (de Wit and Arnold, 1976).

Several important corollaries, employed in the analysis of an engineered barriers system, follow from the above definitions:

- The concept of system is always ahead of the concept of model.
- Both system and model can be either an entity or an abstraction.
- Both system and model are defined in terms of elements (components, parts or members), relationships (for dynamic systems the relationships describe processes) and state (condition at an instant time).
- The idea of "set" used in the definitions is fundamental to system analysis and modelling, and has the same meaning as in mathematics (i.e., a collection of objects with common characteristics). This similarity unifies the concepts of system, model and simulation model. The system and the model, however, are not identical objects even when they are of the same type (i.e., when both are entities, the system is a natural system and the model is a physical model); if the two objects are identical, use of the term model is meaningless.
- A system can generate a large number of models.
- Simulation models or, simply, simulations are viewed as a special class of models that display processes in some way comparable with a reference object that operates on real space and time scales. Although desirable, not all processes included in a simulation model can be described mechanistically (i.e., by a description that reflects some concept of the causal mechanism that underlies the relationship). Some correlative descriptions (descriptions that only reflect an observed relationship) are always included in a model to describe a complex system. However, it should be realized that excessive use of the correlative technique can lead to errors. Particularly, errors occur when

the model describes a high hierarchic level of the system and the model is intended to be used for predictive purposes. This technique turns the model from an explanatory type into a correlative one which cannot be used for prediction at all. Keeping in mind the important role that initial and boundary conditions play in determining whether a physically sensible solution can be obtained for a given equation, simulation models can estimate the fundamental mechanisms over large space-time intervals for many practical applications.

- Classical statistical methods are not always the most appropriate techniques to test a simulation model. Often, the models should then be judged, not on an absolute scale that rejects them for failure to describe the true behaviour of the system that is represented. Instead, they are rated on a relative scale, and accepted as useful tools if they successfully match, in some degree, the measurements of real objects, and clarify knowledge and provide insights into the behaviour of the system of interest.

In engineering work there is often an interest to draw inferences about a system at a much higher level of hierarchy than, for example, at the molecular level. A given level, the term used in hierarchy theory, or a system of interest, the term used in system analysis, is perceived to exist within a hierarchical structure where the system is simultaneously a part of a larger unit (a supra-system), an entity by itself that is also a collection of sub-units (sub-systems). Within mathematical models that describe the engineered barrier system, the continuity between different hierarchical levels is preserved. The constraints imposed by the supra-system on the system form the boundary conditions, and some sub-systems are always included in the model in order to describe the dynamics of the system. These sub-systems provide the explanatory characteristics of the model.

The time-like order of activities (objective time) renders dual characteristics to the engineering barriers system. Presently, an engineered clay barrier system for a disposal of heat-generating nuclear waste is a concept (a theoretical viewpoint), i.e., an abstraction. Once a site is selected, a disposal system will include a given geological subsystem, which is a concrete system. The elements (units) of this sub-system are also concrete, and they interact in many different ways (spatial, temporal, spatio-temporal, causal, etc.). As the construction of a repository progresses, the natural sub-system is altered and combined with other concrete objects (waste package and other engineered barriers). At the end of construction, the engineered clay barrier system, as well as the repository system as a whole, turns into a set of entities standing in interaction: this set is itself an entity. The bridge between the concept and the real (physical) object is realized by means of two types of models: simulation and physical. This is shown schematically in Fig. 1.

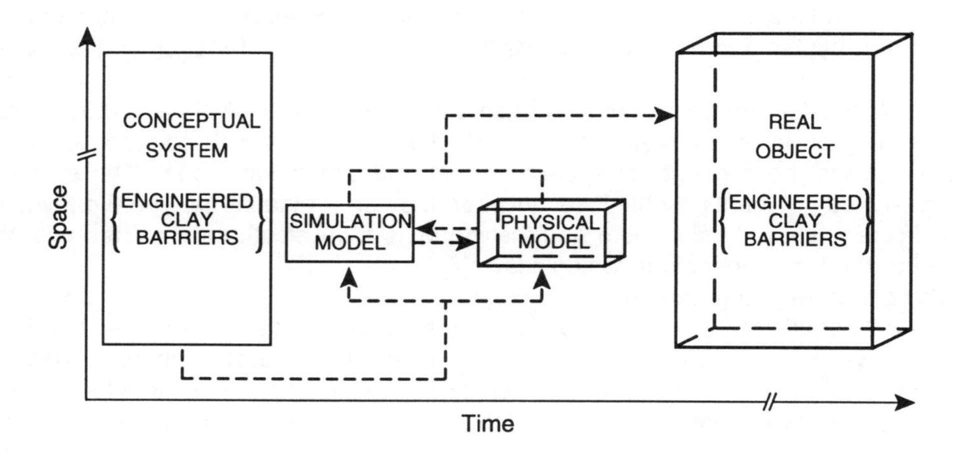

Figure 1. The relationships between a conceptual system, the models and the real object.

Although they are simplifications of the conceptual system, each type of model presents some advantages over the conceptual system. Simulation models, which are also abstractions, convert the qualitative description of the conceptual system into a quantitative description by means of numbers. Physical models are entities that can be used for experimentation and measurement. Experiments performed with physical models yield two benefits. They provide information regarding how a concrete object similar to the object of interest behaves, and they serve as tangible evidence in judging the capability of models to simulate the behaviour of the real object of interest. Neither physical models nor simulation models separately provide enough information regarding the probable behaviour of the real object to be constructed. On one hand, physical models provide information about separate components and about how they function over short space-time intervals. On another hand, mathematical models aimed at the prediction of the behaviour of a real object cannot rely exclusively on information from the theoretical concept. The confidence in the strength of a simulation model to estimate the behaviour of a real object over a long time period depends to a large extent on some demonstration that the model is capable of simulating the function of a physical model similar to the object of interest. The tests of a simulation model versus physical models are performed within the validation stage of the modelling activity. However, a simulation model can never predict the behaviour of a real object exactly, and the simulation always will be to some extent a work of art. Much of the art of simulation of the behaviour of an engineered clay barrier is concerned with choosing the appropriate approximations (mathematical and physical models), and striking a fruitful comparison between the concept and reality.

3. The ***VALUCLAY*** Project

In common within other repository development programs, the CNFWMP is modelling engineered clay barriers such as the buffer that surrounds corrosion resistant waste containers and separates them from the host rock and backfills used to fill the excavations. Immediately after being put into place, the buffer and backfill materials will be unsaturated with water. For some time after emplacement, the temperature of the containers and the hydraulic head in the rock adjacent to the clay barrier undergoes a time-dependent change. Under these conditions, the magnitude of the heat / moisture flow velocity in the clay barriers will vary with time. Subject to thermal loading and the groundwater flux from the rock, swelling pressure and time-dependent stress and deformation, characteristics of buffer and backfill, and complex coupled processes of heat and moisture transfer will take place. It is important to know about these transient processes within engineered clay barriers to design and construct a repository. In addition, transient processes may be important in the long-term performance of the engineered clay barriers and, thus the overall performance of the repository. Being able to estimate the behaviour of engineered clay barriers during the unsaturated stage of the vault (a stage that can last hundreds to thousands of years) is the only tangible evidence to sustain the hypothesis that, from a hydrological and mechanical viewpoint, at the time of radionuclide migration, the transport through the clay barriers takes place only by diffusion. Thus, a project has been initiated to test the mathematical models developed to describe the coupled hygro-thermal-mechanical processes within engineered clay barriers during the unsaturated stage of a vault. This is called the ***VALUCLAY*** Project (VALidation of codes/models that describe the Unsaturated behaviour of engineered CLAY barriers).

3.1 Project scope and postulates

The main objective of the ***VALUCLAY*** project is to determine the degree of plausibility and utility, i.e., to validate, some of the existing models to simulate the probable behaviour of engineered clay barriers over the entire space-time domain relevant to the safety requirements. The project is designed to encourage and facilitate an iterative interplay between theoretical interpretation and experimental work and has the following tasks:

- Review and select models with a potential to describe the unsaturated engineered clay barrier system.

- Create a model-bank and data-bank for the validation phase; complete *in situ* observations and laboratory validation experiments.
- Proceed with the validation. The overall goal of this activity is to establish the relationship between the modelling development and application phase, that is to assess the success of existing models. At the end of the validation phase one of the following decisions will be taken:
  a) progress to the application phase (i.e., draw inferences from simulation of model(s) about the probable behaviour of the clay barrier beyond the range of actual observations), or
  b) return to the modelling development phase (i.e., improve/develop the model(s) in the areas in which it is practical and /or worthwhile).

For disposal of heat-generating nuclear wastes, the performance of the total system, including both the natural and engineered barriers, is of interest. The ***VALUCLAY*** project focuses on engineered clay barriers. Generally, the project rests on two premises: a) interactions between the near-field and the far-field can influence repository performance, and b) processes in the near-field are critical in the design and construction and have the potential to affect the long-term performance of a repository. Consequently, the validation phase focuses on the existing models that have the potential to accommodate the following two general postulates.

1) The structure of the system, commonly termed the near-field, is deemed to include the clay barriers, other fillers and that portion of the host rock that is disturbed mechanically or hydraulically by the rock-mass excavation and influences the performance of the clay-barrier.

2) Hygro-thermo-mechanical processes in unsaturated clay barriers are important processes within the system; these processes are strongly coupled because the thermal and hydrological gradients in the near field during the initial stage of a vault are the highest. The relevant coupled processes assumed in the near field system are shown schematically in Fig. 2.

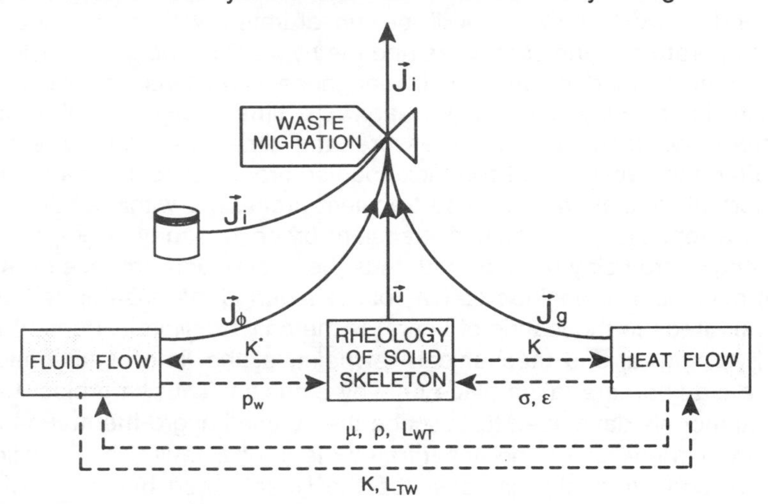

Figure 2. The coupled processes in the near-field relevant to the description of the behaviour of engineered clay barriers; J_i, J_ϕ, J_g is mass density, water and heat flux respectively, u is displacement tensor, K^* is the hydraulic conductivity, K is the thermal conductivity, p_w is pore water pressure, σ is the stress, ε is the strain, μ is viscosity, ρ is density, L_{WT} is water diffusivity due to the temperature gradient and L_{TW} is the thermal diffusivity due to the hydraulic gradient.

3.2 Models included in he *VALUCLAY* project

Many models have been developed (most of them derived from other existing models) in the last decade to describe the major components of a repository system. Some computer programs developed to describe a process or a group of processes, therefore, are potentially capable of describing the behaviour of more than a particular system and are termed codes. For simplicity, the term model will be used throughout this paper for both model and code unless the abbreviated name imposes the use of the term code.

Several models (MOTIF, Chan et al., 1987; HTM, Selvadurai, 1991; TRUCHAM/TISDA, Radhakrishna et al., 1990; and Lau and Radhakrishna, 1992; THAMES, Ohnishi et al., 1987; COMPASS, Thomas, 1992; NAPSAC, Herbert and Lanyon, 1992; FracMan, Deshowitz et al., 1991; TRACR3D, Travis, 1984; FAMOS, Andrews et al., 1986; TOUGH2, Pruess, 1987) have been reviewed. These codes/models are complex and elaborate. However, their basic characteristics can be summarized as follows:

- Initially developed for either scientific research purposes or for specific problems other than to describe the clay barrier system.
- Use different assumptions relative to the system's structure, boundaries and processes.
- Describe, mathematically, transport phenomena of a system at nonequilibrium (i.e., MOTIF - Model Of Transport In Fractured/porous media; TOUGH2 - Transport Of Unsaturated Ground water and Heat, version 2).
- Make use of numerical solutions to calculate net flux and balance equations.
- Are mixed, explanatory / correlative.
- Require validation before use for drawing inferences about the probable conditions of an engineered clay barrier within a real repository.

The rigorous solution of coupled processes that occur in the near-field would require the development of a model based on nonequilibrium thermodynamics. With this approach, both classical laws (flux-force relationships) of known phenomena and cross phenomena are included in the mathematical formulation. Due to the great difficulties in deriving the necessary set of coefficients, a developed and operational model based on this approach is lacking. All existing models make physical assumptions to simplify the solution. These models are based on combinations of equations or derivations from four major groups of equations that can be summarized as follows: a) the fundamental continuity equations, b) the classical laws of known physical phenomena, c) the coupled heat and moisture transfer - using the Philip and de Vries (1957) approach and d) constitutive relationships. Mathematical formulations of the available models vary widely in details. In this paper it is only possible to describe the general theoretical formulation of these models (see Appendix1). In addition to physical assumptions, all models make mathematical simplifications by replacing mathematical expressions with numerical approximations.

None of the available models that have been reviewed are designed or declared to mechanistically describe fully coupled major processes that occur in the engineered clay barriers. The extent to which this limits the usefulness of the models for the prediction of the performance of unsaturated engineered clay barriers will be examined in the ***VALUCLAY*** project. Four models, COMPASS, TRUCHAM/TISDA, HTM and TOUGH2, were selected and included in the validation program. The criteria for selection include the capabilities and assumptions built into models that will permit the postulates mentioned in Section 3.1 to be accommodated, computer requirements and availability. These models are based on different concepts regarding the main processes described. For example, the coupled heat and moisture transport in HTM and TRUCHAM are based on the Philip and de Vries approach. This approach is considered to account for a "weak" coupling of heat and moisture transport. To accommodate high temperature gradients and to describe in a more complete manner the interrelationships between processes, the description of coupled fluid and heat flow in COMPASS and TOUGH2 includes the transport of latent and sensible heat and liquid-vapour phases changes. Table 1 summarizes major characteristics of models included in the project.

Table 1. Summary characteristics of the models included in the ***VALUCLAY*** project

Model	System geometry	Governing equations	Boundary conditions	Numerical method	State-variable* estimated
COMPASS	porous	Mass, Energy balance Equilibrium equation (multi-phase system)	Dirichlet/ Neuman	finite element	T, Ψ (θ), P_a,u
TRUCHAM / TISDA	porous	Philip and de Vries /constitutive rel. (two-phase system)	Neuman	integral finite difference	T, θ, u
HTM	porous	Philip and de Vries	Neuman	finite element	T, θ
TOUGH2	porous / fractured	Mass, Energy balance (multi-phase system)	Dirichlet/ Neuman	integral finite difference	T, h, S_l, P_a, S_g

*where T is temperature, Ψ is water potential (energy/unit mass), θ is the volumetric moisture content, P is pressure, u is displacement, h is the hydraulic head (energy/unit weight), S is the degree of saturation and the subscripts a, l and g stand for air, liquid and gas phase, respectively.

All these models have been tested, to some extent, during the development phase. Although these tests are important they provide a limited confidence in the usefulness of the models. The modelling activity is complex and includes several stages, experiments and tests. The interpretation of the validation concept, however, requires an understanding of the underlying relationships between individual stages (and associated tests) and the larger picture of modelling activity that cements them together.

3.3 Stages in modelling

The validity of any dynamic model that describes the function of engineered clay barriers over a long period of time is always open to question. Yet, model errors are minimized and the usefulness of the simulation model is maximized when several interrelated stages in mathematical modelling, schematically represented in Fig. 3, are undertaken systematically.

a) Objective(s) - It is critical to establish precise objective(s) for a model when the model is indentified to be used to guide the solution to a practical problem. Invariably, when the objective of a model is defined or described in broad terms, decision-makers tend to expect more from modelling results than can be delivered.

b) Postulate - Appropriate postulates regarding the relevant structure (units) and processes (relationships) of the system of interest need to be selected. A postulate is a special assumption (or set of assumptions) required to progress with the solution of a problem. This is the same meaning as generally used in physics and mathematics. The postulates also set the boundary of the system to be modelled. Although to some extent the boundary of a system is arbitrary and is often chosen for convenience, it must be established. Without boundaries the system under study will be continuously moved between hierarchical levels and would be unsolvable.

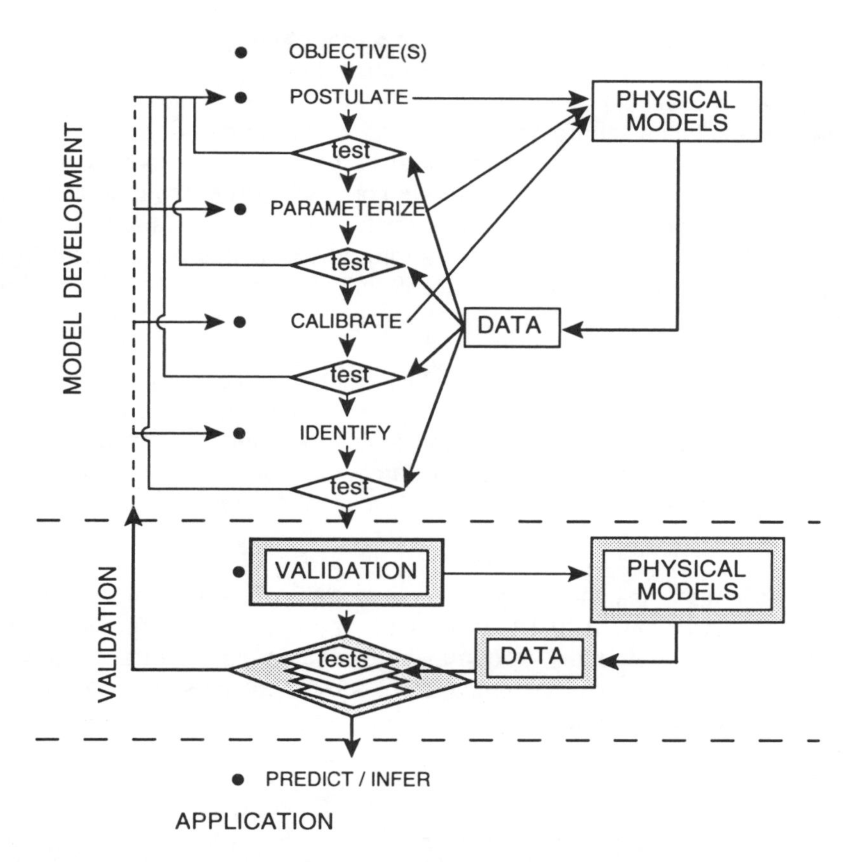

Figure 3. Stages in the modelling activity and their relationships with physical models.

c) Parameterization - For calculation purposes, the characteristics of engineered components that are not true constants within populations have to be parameterized. It becomes evident that the behaviour of engineered barriers would be suitably described by a stochastic model. In this case, parameters should be described in a form of families of probability distribution functions (mass or density functions).

d) Calibration - Certain parameters are particularly poorly known. Sometimes, these parameters are adjusted based on a fitting procedure of model output to the observed data set. This method is called calibration or tuning of the model. Closely related with calibration is the test that is commonly termed sensitivity analysis. This technique is used to determine how much the output of the overall model varies with changes in a given parameter. Generally, this method is used to improve the estimation of relevant parameters (a high sensitivity value of a parameter would suggest a better estimation) and model simplification (lower sensitivity parameters / subroutines can be discarded from the model). Both calibration and sensitivity analysis may be, in some cases, effective testing procedures helpful in the development stage of a model. However, they must be used with caution, particularly when the model is intended to be used for prediction purposes. Because they are based on fitting procedures, the results are highly dependent on the data sets considered; different data sets may give different results. In addition, some parameters are correlated and/or their importance may depend on input variables. The significance, or the lack of it, of a given parameter on the overall output of a model must be carefully appraised.

e) Identification - For calibration and sensitivity analysis to have a practical value, it is necessary to identify if the available data necessary to run the model within the application phase justify the suggested alterations of the model.

f) Validation - This is one of the most important, complex and, often, the most scrutinized phase within a modelling activity. Validation is concerned with the aspects of appropriateness and plausibility of a model to be used in solving practical problems. Results from a series of tests performed within this activity serve as tangible evidence regarding the success of the model in representing the system of interest. If a simulation model is to be successful, that is, used within a real world situation, it must give information or predictions that are clearly better, in some way, than the mental image or other abstracted model that would be used instead. The validation phase is the interface between model development and model application. Results of tests performed within this phase provide the necessary confidence in the strength of a model to provide reliable information regarding the probable behaviour of the system of interest.

g) Application - The final phase in the modelling activity is the application, i.e., the estimation of the probable behaviour of engineered clay barriers by means of simulation. In view of various uncertainties in many areas of the functioning of an engineered barriers system, the models that incorporate random elements and therefore give the outputs in the form of probability distributions, if practically achievable, would be more useful than models that try to predict the state of the engineered barriers deterministically.

It is important to emphasize that stages in the modelling are concatenated and interrelated. The basis for the iteration is always a test that links the theoretical interpretation with experimental data from the physical models at every opportunity. Each test is important. However, the tests performed in the phase of model development are not validation tests; all that is evaluated is the extent to which the model regenerates its own inputs.

3.4 Validation approach

In one sense, there will never be a phase of final synthesis in modelling a complex system. Generally, the completion of many modelling activities represents the starting point for new and improved models. However, to progress towards the application, the final phase of the modelling activity, the focus shifts from the development to the validation phase. Progress in the CNFWMP since 1983 has included the development of models for a engineered barrier system and completion of laboratory and *in situ* experiments. The studies are now focused on the validation phase.

In recent years, within the studies based on simulation modelling, debate has ranged around the use of the terms validation and verification. The concept of validation and use of the term verification for specific tests performed within the validation phase are not products of radioactive waste management research programs. They were developed some time ago and became common terms in many scientific disciplines that deal with system analysis and simulation techniques (Miller, 1978; Gold, 1977; Forrester, 1968; Law and Kelton, 1982; Naylor and Finger, 1967; Schellenberger, 1974; Hermann and Hermann, 1972). In the present study the term validation is used for the activity performed to assess the success of a simulation model. Alternative terms for validation such as "evaluation" (Konikow and Bredehoeft, 1992) or "confirmation" (Oreskes et al., 1994) may also be used. However, use of either term requires a further definition of the concept associated with them.

Validation is not a wholly objective activity that consists of a single test - validation is a larger concept. Model validity is always a matter of degree, affected by the purpose for which the model was designed and by the type of criteria employed in the validation. In the context of engineered barrier systems for nuclear waste management problems, validation consists of a series of tests designed to provide a degree of confidence in the ability of the model(s) to estimate the probable behaviour of the actual engineered clay barriers whose true behaviour we cannot be certain of. Several facets of model validity are considered in the ***VALUCLAY*** project. They are listed below.

Internal validity. - Two tests, properly termed verifications, are performed within this activity. One test is the operation performed to check the computer program of a simulation model that facilitates the translation of a mathematical problem into machine code; the scientific concepts (physical, chemical, biological, geochemical, mathematical, etc.) based on which the model was developed are judged to be true. The test of the computer program versus the scientific theory, the so called program debugging, is a verification process. The second verification test is based on the comparison of a numerical model versus the exact analytical solution; the analytical solution is accepted to be certain. When the problem for which an analytical solution is available is specified properly and the numerical solution matches the analytical solution, the statement that the numerical model verifies on the range considered is a correct claim. These two test types, for which results are either positive or negative, are verification processes. These tests, particularly the debugging, are performed during programming within the model development phase. However, the internal validity of a model is also included within the validation phase. All models selected and included in the ***VALUCLAY*** project are installed on AECL Research computer platforms.

Event validity, i.e., "validation through model testing with experiments" - A model derived from theory, no matter how elegant in itself, is unlikely to be of use in solving a practical problem unless there is a practical way of assessing how well the theory agrees with the observed data. An event in this interpretation is an occurrence related to a concrete object; to each event belongs its place coordinates and time values. The tests included in this stage are based on comparisons of model outputs with the outputs (measurements) of physical models of engineered barriers. In order to secure the integrity of these tests, a completely independent data set, other than data sets used in calibration and sensitivity analysis, are employed. The experiments within this activity are essentially of the same type as those used in the model development phase:

- a) bench laboratory models,
- b) large-scale laboratory models, and
- c) *in situ* models.

However, we realized that the validation through model testing with experiments, does not always allow us to rely on the data as they already exist even when such data sets are unrelated to the development of a given model. Most of the experiments performed in the early stage were designed to provide information regarding the function of the system and/or to provide engineering experience with handling the equipment, materials and geotechnical instrumentation. The main difficulties in using the event validation stage data generated from some of the past experiments can be summarized as follows: a) the initial conditions of the physical model were not properly measured, therefore the starting assumptions in the mathematical simulation cannot be stated explicitly, b) the experimental results were consistent with more than one causal hypothesis; c) a large number of instruments were used in the experiment and this could alter the function of physical model - this cannot be considered in the simulation; and d) some of the major state variables were measured imprecisely due to problems associated with the function of the instruments. Such instances are not uncommon and give rise to arguments over the correct interpretation of the tests performed.

There is an essential interdependence of data collection and data analysis; the validation of particular models requires us to ensure a particular data set. For this reason, the selection of past experiments as well as the set up of the new ones to create an appropriate "data bank" followed the selection of models to be tested, i.e., the creation of a "model bank" in the ***VALUCLAY*** project. Designing appropriate equipment and experiments that will permit a reliable testing procedure of the models considered involves a considerable amount of work and care. Some of the equipment designed and facilities available to the project are represented schematically in Figures 4a, 4b and 4c. Examples of the preliminary event validity tests for COMPUSS model are presented briefly in the next section.

Hypothesis validity.- This activity deals with testing the major hypothesis regarding the processes described in a model. Generally, the hypothesis validity of a simulation increases as its particularity

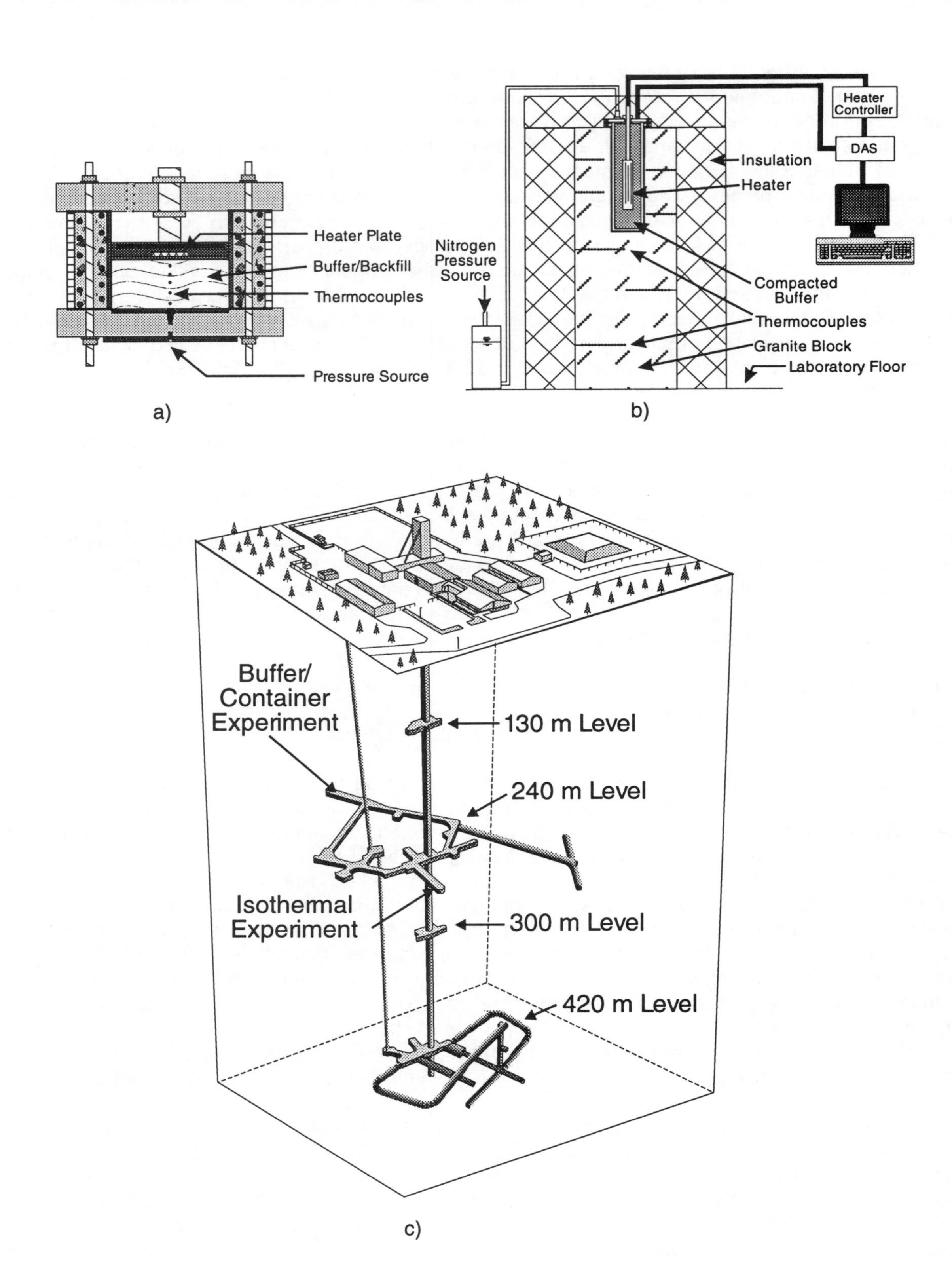

Figure 4. Schematic representation of equipment and facilities available for testing models included in the ***VALUCLAY*** project, event validity phase; a) bench laboratory heating cell used to estimate diffusivity parameters, b) large scale granite block facility and c) the AECL Research Underground Research Laboratory (URL).

increases and as the processes it represents diverge from those of systems that were not considered at the time of model development. However, some models with relatively low hypothesis validity may be more useful to describe a generic repository system if they relate to a larger number of different systems.

Intersimulation validity. - When the simulation progresses beyond the range of actual observations of a physical model, a simulation model output(s) can only be compared with the output(s) of other simulation models developed to solve a similar problem. Comparisons among the outputs of the different models included in the project, all run for a given set of selected benchmark scenarios, will provide valuable information regarding the overall direction and magnitude of a particular state variable of interest.

Face validity. - This is concerned with the impression of realism that the simulation makes on the participants in the validation and application stages. A simulation model, and in general any mathematical model, is an adjunct, a supporter to professional judgement, and never a substitute. Tests performed at this stage are primarily based on the intuition of scientists, therefore they are, to some extent, subjective. However, it is important to emphasize that the results of all tests performed within the validation phase are checked continuously against the scientist's intuition. In this manner, face validity consolidates all other facets of the validation phase. The validation phase was designed to start and to end with face validity. Agreement between mathematics (simulation results), physical confirmation and intuition is regarded as a strong indication that the simulation is successful. Alternatively, disagreement is regarded as a signal for a need for the improvement of the existing models. At the present time, the face validation for the ***VALUCLAY*** project is carried out by a group that includes members of the academic and industrial communities. The expertise of individual members includes geotechnical and civil engineering, applied mechanics, soil physics and mechanics, clay mineralogy, numerical modelling and mathematics.

However, neither modelling development nor validation are definitive activities. As Dormuth (1993) pointed out, the validation and modelling refinement should continue in order to increase the reliability of the estimated behaviour of every component of a repository as the implementation of a repository concept proceeds. Validation is not a phase to be attempted after the modelling activity has already been performed and only if there is remaining time and money. Validation requires considerable amount of effort and care since it is the only insurance against losing the thread of the argument regarding the usefulness of simulation.

3.5 COMPASS - preliminary tests results

COMPASS (COde for Modelling PArtly Saturated Soil) is a set of numerical programs developed to simulate the coupled thermo / mechanical / hydraulic behaviour of unsaturated soil. It consists of a suite of codes developed by Dr H. R. Thomas with a research group at the School of Engineering, University of Wales College of Cardiff. The version included in the ***VALUCLAY*** project accommodates temperature variations (heat transfer by means of conduction, convection and latent heat of vaporisation), moisture content (both the liquid and the vapour phase) changes, air pressure fluctuation (both bulk flow due to pressure gradients and as-dissolved air in the liquid phase) and deformation (Thomas 1992), all of which are processes characteristic of buffer and backfill. The deformation of the solid skeleton is calculated from a constitutive relationship known as a "state-surface" approach. Eight-noded isoparametric elements are employed in the program, which enables irregular/circular boundaries to be readily accommodated. The results from two event validation tests are presented.

In the first test, the model simulated the heat and moisture transfer of a simple physical model, a bench-scale laboratory test on medium sand (Ewen and Thomas, 1989). The sample was compacted into a glass cylinder and heated by an embedded circular heater passing through the centre of the cylinder. Both ends of the glass cylinder were insulated during the experiment. The temperature was recorded during both transient and steady-state conditions. The results of this test are presented in Figures 5a and 5b.

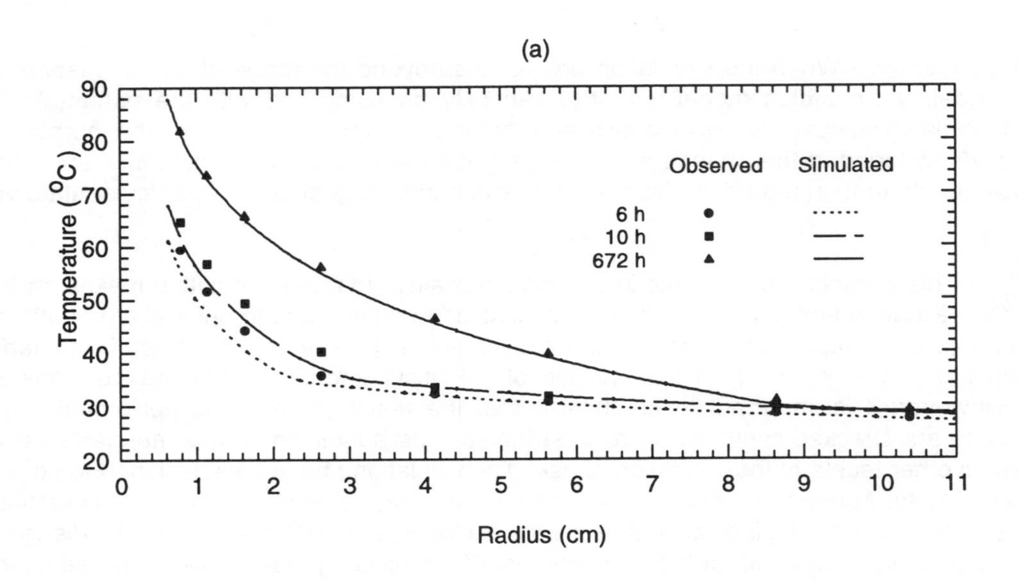

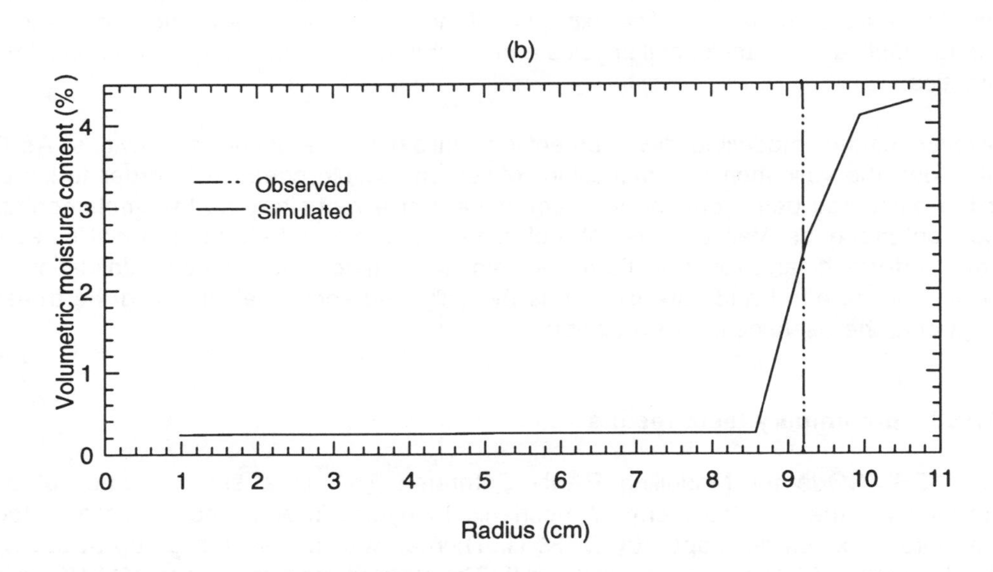

Figure 5. COMPASS simulated heat and moisture transfer for a simple physical model, a bench-scale laboratory test; a) temperature values, observed and simulated, after 6, 10 and 672 hours after the experiment starts, and b) The dry/wet interface front after 672 hours.

The observed and simulated temperature values presented in Fig. 5a indicate a good agreement between measurements and estimate values. Assuming that the absolute error follows a student's t-distribution, there is a 95% probability that, at any distance from the heating source, the simulated temperature value will be within 1.8 °C, or less, about the observed value. Figure 5b shows the steady-state moisture content. The dry/wet interface front was predicted at a radius between 8.57 cm and 9.26 cm (average radius of 8.92 cm), compared with the radius of 9.20 cm, the observed value. Since the radius ranged from r_1=0.635 cm and r_1=10.980 cm, the position of the dry/wet interface is predicted within an error of about 2.71% of the full radius range.

In the second test the model simulated a physical model, the *in situ* test of the Isothermal Experiment installed at the URL. This experiment is being carried out at the 240-m level in crystalline rocks of the Canadian Shield in a 5-m deep, 1.24-m-diameter borehole to examine the water uptake by the buffer under *in situ* boundary conditions, and the buffer's hydraulic and mechanical interactions with the surrounding rock and the concrete restraining plug. This experiment is in progress. Typical results, comparing observed and simulated values of the pore water pressure head, using preliminary data, are presented in Figures 6a and 6b for 100 days and 200 days, respectively, from the start of the experiment.

(a)

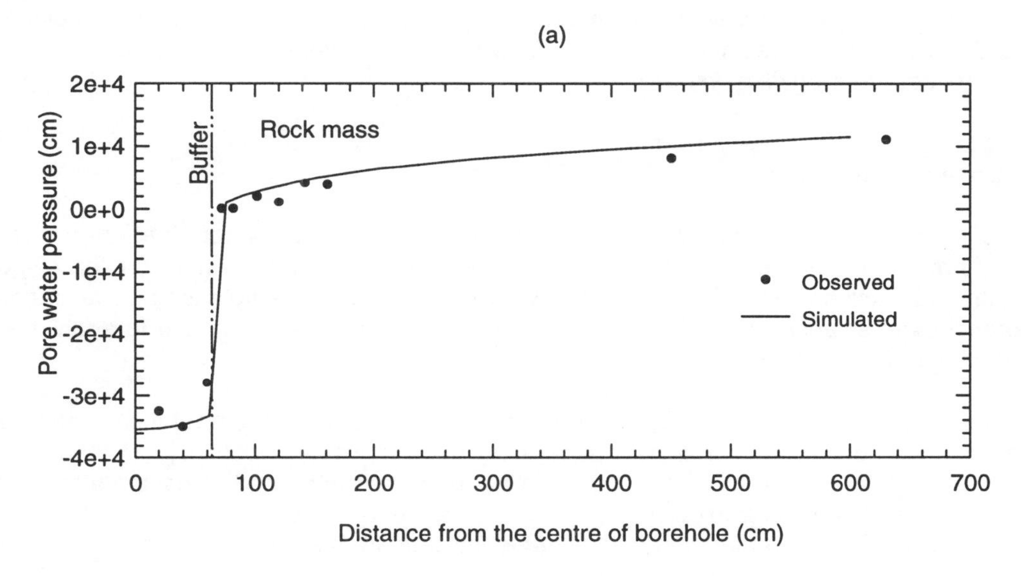

(b)

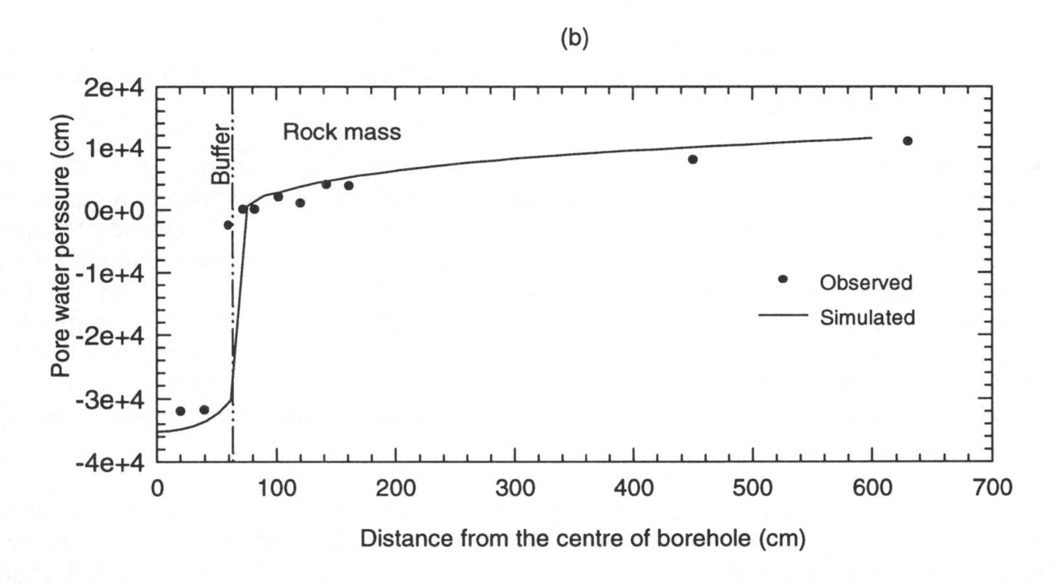

Figure 6. COMPASS simulated pore water pressure for the *in situ* test, Isothermal Experiment - URL, preliminary results; a) after 100 days and b) after 200 days from starting the experiment.

A reasonable agreement between predicted and measured data has been achieved at each time considered. The gradual resaturation of the buffer observed was also indicated by the simulated values. Differences exist, however, between observed and simulated values, particularly regarding the duration of the suction in the rock. The numerical analysis indicates that significant negative pressure heads can be generated in the granite. The magnitude of such pressure heads and the extent of their penetration into the granite then influences the rate of saturation of the bentonite-sand barrier. The simulation results obtained are strongly dependent on the hydraulic properties of the granite. High suction in the granite results in very low hydraulic conductivity in the near-field zone creating a nearly impermeable region. The modelling results serve to highlight the importance of buffer/rock hydraulic interaction and indicate the need for further investigation of the flow characteristics of the rock in the disturbed region near the excavation. Although it is still early in the validation program, COMPASS seems to be a very useful tool that can successfully estimate the behaviour of clay engineered barriers. The model is theoretically sound, includes the major coupled processes that take place in the near-field, and the results of preliminary tests suggest that the simulated values compare favourably with observed values from physical models.

4. Summary and Conclusions

The concepts used in the analyses of the engineered clay barrier system that are derived from system theory have been presented in the specific context of a repository system for the disposal of heat-generating radioactive wastes. The concepts system, model, model simulation and validation used in various branches of science to analyze complex, dynamic systems, are suitable to analyze the behaviour of engineered clay barriers.

Two approaches are used to progress from the concept to the implementation of the decision to build real engineered barriers for a heat-generating nuclear waste repository: 1) investigate the behaviour of the physical models similar to the object to be constructed and 2) estimate the probable behaviour of the future engineered barriers using simulation models. These two approaches are different. However, they are complementary. Neither is included in the other, nor can either of them be reduced to the other, but both of them are necessary, supplementing one another for a fuller understanding of how a real engineered barriers will function. A continuous comparison and matching of the measurements of physical models and the simulated values contributes to the shaping of the necessary confidence that the design and construction of safe engineered barriers for a heat-generating nuclear waste repository is possible.

A knowledge of the transient processes within engineered clay barriers is essential for the design and construction of a repository and for assessing the long-term performance of a repository. In order to determine that the assumptions made regarding the hydrological and mechanical properties of engineered clay barriers at the time when the nuclides start to migrate are indeed reasonable and applicable to a given set of conditions, it is essential to first estimate the behaviour of clay engineered barriers during the unsaturated stage of the vault. The mathematical arguments of a model that simulate the behaviour of a component of a real repository (at any level of hierarchy) and shift the initial conditions, $t \rightarrow 0$, forward to an arbitrary time, is as good as is the support for the starting assumptions. Mathematical work may be right, but when the physical assumptions are doubtful, the final conclusion is always disputed.

Answers to three interrelated questions, considered to be relevant for the activity of validation through model testing with experiments, were provided through the ***VALUCLAY*** project. Section 3.2 described what models are validated; Section 3.4 provided the rationale for validation; and Section 3.5 documented the validation approach used in the ***VALUCLAY*** project. Five types of validation are considered necessary to estimate the strength and credibility of a simulation model to describe the unsaturated behaviour of engineered clay barriers: internal validation, event validation, hypotheses validation, intersimulation validation and face validation. The most convincing tests regarding the usefulness of simulation models are those in which simulated values are compared directly and successfully with observed data from physical models. This largely occurs in the stages referred to as event and hypothesis validation. A balanced validation, however, has to include all types of validation and has to rely on the professional judgement, face validation, in order to explain the underlying relationships between the individual tests performed and the larger picture of model usefulness, i.e., to assess the success in the simulation of the probable behaviour of engineered clay barriers within a heat-generating nuclear waste repository.

ACKNOWLEDGMENTS

I would like to thank to M. N. Gray, K.W. Dormuth, L.H. Johnson and A.G. Wikjord for helpful information and discussions on the modelling validation; K. Tsui, H.R. Thomas, A.G. Wikjord and D. Oscarson for reviewing the manuscript; and D. Hnatiw and A.W.L. Wan for providing data from the isothermal test. The Canadian Nuclear Fuel Waste Management Program is jointly funded by AECL and Ontario Hydro under the auspices of the CANDU Owners Group.

REFERENCES

Andrews, R.W., LaFleur, D.W., Pahwa, S.B., 1986. Resaturation of backfilled tunnels in granite. Nagra, Technical Report 86-27. Baden, Switszerland. 68 pgs.

Bredehoeft, J. D. and Konikow, L. F., 1993. Ground-water models: validate or invalidate. Ground water, 2 (31): 178-179.

Chan, T., Reid, K. and Guvanasen, V., 1987. Numerical modelling of coupled fluid, heat, and solute transport in deformable fractured rock. In Coupled Processes Associated with Nuclear Waste Repositories. Tsang, C.F. (Ed.) Academic Press, Inc. 605-625.

Deshowitz, W.S., Wallmann, P., Geier, J.E., and Lee, G., 1991. Preliminary - Discrete fracture network modelling of tracer migration experiments at the SCV site. SKB Technical Report 91-23.

Dormuth, K.W., 1993. Evaluation of models in performance assessment. Paper presented at SAFEWASTE 93, International Conference on Safe Management and Disposal of Nuclear Waste, Avignon, 13-18 June 1993.

de Wit, T. C. and Arnold, W. G., 1976. Some speculation on simulation. In: G. W. Arnold and C. T. de Wit (Editors), Critical Evaluation of System Analysis in Ecosystems Research and Management. Pudoc, Wageningen.

Einstein, A. 1940. Considerations concerning the fundamentals of theoretical physics. Science, 91 (2369): 487492.

Ewen, J. and Thomas, H.R., 1989. Heating unsaturated medium sand. Geotechnique, 39 (3): 4555-470.

Forrester, W. J., 1968. Principles of Systems. Text and Workbook. Cambridge, Massachusetts.

Gold, J. H., 1977. Mathematical Modeling of Biological Systems - An Introductory Guidebook. John Wiley & Sons, New York.

Herbert, A. and Lanyon, G., 1992. Modelling tracer transport in fractured rock at Stripa. SKB Technical Report 92-01

Hermann, C. F. and Hermann, M. G., 1972. An atempt to simulate the outbreak of World War I. In: H. Guetzkow, P. Kotler and R.L. Schultz (Editors), Simulation in Social and Administrative Science. Prentice-Hall, New Jersey.

Konikow, L.F and Bredehoeft, J. D., 1992. Ground-water models cannot be validated. Adv. in Water Resources, 15, 75.

Lau, K.C. and Radhakrishna, H.S., 1992. Analysis of Buffer/Container Test 1 - Revision 2. Contractor's Report, Technical Record. 36 pgs.

Law, M. A. and Kelton, D. W., 1982. Simulation Modeling and Analysis. McGraw-Hill, New York

McCombie, C. and McKinley, I. 1993. Validation - another perspective. Ground water, 4 (31): 530-531.

Miller, G. J. 1978. Living Systems. McGraw-Hill, New York.

Naylor, T. H. and Finger, J.M., 1967. Verification of computer simulation models. Manage. Sci., 14: 92-101.

Ohnishi, Y., Shibata, H., Kobayashi, A 1987. Development of finite element code for the analysis of coupled thermo-hydro-mechanical behaviours of saturated-unsaturated medium. In: Coupled Processes Associated with Nuclear Waste Repositories. Tsang, C.F. (Ed.) Academic Press, Inc.

679-697.
Oreskes, N., Shrader-Frechette, K. and Belitz, K., 1994. Verification, validation, and confirmation of numerical models in the earth sciences. Science, 236: 641-646.
Philip, J.R. and de Vries, D.A., 1957. Moisture movement in porous materials under temperature gradients. Trans. Am. Geophys. Union 38 (2): 222-232.
Pruess, K., 1987. Tough user's guide. Lawrence Berkeley Laboratory, University of California. 86 pgs.
Radhakrishna, H.S., Lau, K.C. and Crawford, A.M., 1990. Experimental modelling of the near field regime in a nuclear fuel waste disposal vault, Engineering Geology, 28, 337-351.
Schellenberger, R. E., 1974. Criteria for assessing model validity for managerial purposes. Decis. Sci., 5: 544-653.
Selvadurai A.P.S., 1991. Hygro-Thermal behaviour of a clay barrier developed for use in a nuclear waste disposal vault. III. Numerical modelling and comparison with experimental results. Contractor's Report, Technical Record. 50 pgs.
Thomas, H.R., 1992. On the development of a model of heat and moisture transfer in unsaturated soil, Can. Geotech. J., 29, 1107-112.
Travis, B.J., 1984. TRACR3D: A model of flow and transport in porous/fractured media.- LA-9667-MS. 190 pgs.
von Bertalanffy, L. 1956. General system theory. Gen Systems, 1: 3.

Appendix 1.

The reviewed models that describe either the near-field or far-field are based on combinations of equations or derivations from four major groups of equations that can be summarized as follows: a) the fundamental continuity equations, b) the classical laws of known physical phenomena, c) the coupled heat and moisture transfer - using the Philip and de Vries approach and d) constitutive relationships.

a) Continuity equations - The general form of the equation of continuity used to describe transport phenomena within the system with several non-uniform quantities includes three terms to account for the diffusive flux, convective (centre of mass velocity) flux and production/dissipation (p/d) or the so call source/sink term. The common form of the three equations of continuity (mass), energy and motion (momentum) are:

a1) Mass

$$\frac{\partial \rho_i}{\partial t} = -\nabla \cdot \mathbf{j}_i - \nabla \cdot (\rho_i \mathbf{v}) \pm (p/d)_i \tag{1}$$

where $\partial\rho_i/\partial t$ is the rate of change of ρ_i (the partial mass density of a component i, $\rho_i=m_i/V$), t is time, ∇ is the del operator, $\nabla\cdot\mathbf{j}_i$ is the divergence of diffusive flux, $\mathbf{j}_i = \rho_i(\mathbf{v}_i-\mathbf{v})$ where $\mathbf{v}$ is the centre of mass velocity, i.e., the velocity of local centre of mass due to external force, $\nabla\cdot(\rho_i\mathbf{v})$ is the divergence of convective flux and the p/d is the production/dissipation rate (physical/chemical/biological reaction rates, Φ_i, for example).

a2) Energy

$$\frac{\partial\left[\rho\left(E + \frac{1}{2}v^2\right)\right]}{\partial t} = -\nabla \cdot \sum_{i=1}^{n} \overline{H}_i \mathbf{j}_i - \nabla \cdot \mathbf{q} - \nabla \cdot \left[\rho\left(E + \frac{1}{2}v^2\right)\mathbf{v}\right] + \sum_{i=1}^{n} \rho_i \mathbf{v}_i \cdot X_i + \nabla \cdot (\mathbf{v} \cdot \delta) \tag{2}$$

where $\{\partial[\rho(E\frac{1}{2}\ v^2)]\}/\partial t$ is the rate of total energy (specific internal energy and kinetic energy), the first two terms on the right hand side account for conduction of heat flux (conduction of mass, H, enthalpy and pure conduction of heat, **q**), the third term accounts for the convection of mass flux due to the canter of mass velocity and the last two terms account for p/d of energy due to external force (body force, $(\Sigma\ \rho_i\mathbf{v}_i\cdot X_i)$ and due to surface force (shearing upon moving particle, $\nabla\cdot(\mathbf{v}\cdot\delta)$; the p/d term may be also affected by the reaction rates, Φ.

a3) Momentum

$$\frac{\partial(\rho\mathbf{v})}{\partial t} = -/ + \nabla \cdot \delta - \nabla \cdot (\rho\mathbf{v}\mathbf{v}) + \rho\mathbf{x} \tag{3}$$

where $\partial(\rho\mathbf{v})/\partial t$ is the rate of change of momentum, $\nabla\cdot\delta$ is divergence of surface stress tensor, δ, (tensile stress taken as positive) , i.e. momentum diffusion, $\nabla\cdot(\rho\mathbf{v}\mathbf{v})$ is the divergence of momentum flux due to the canter of mass velocity (convection) and $\rho\mathbf{x}$ is the source/dissipation term due to external force.

b) <u>Classical laws</u> - Some of the models describe moisture and heat flow as simple processes using the classical physical laws of known phenomena. The processes are assumed to be uncoupled and the flux is consider directly proportional to a gradient, which in essence is the driving force in the transport process. The most frequent flux - force relationships used in these models are the well known theories Fourier's law, Fick's law, Darcy's law and Richard's equation.

$$\mathbf{q} = -K\nabla T, \quad \mathbf{j}_i = -D_i \nabla C_i, \quad \mathbf{v} = -K^* \nabla h, \quad \mathbf{v}^* = -K^*_{(\phi)} \nabla h \qquad (4)$$

where **q** is the heat flux, K is the thermal conductivity, ∇T is the gradient in temperature, $\mathbf{j}_i$ is the mass flux of component i, D is diffusion coefficient, ∇C is the gradient in concentration, **v** is the water flux within saturated porous media (specific discharge), K^* is the saturated hydraulic conductivity, ∇h is the gradient in hydraulic head, $\mathbf{v}^*$ is unsaturated flow with the provision that K^* is now a function of potential (energy per unit mass), ϕ, i.e., $K_{(\phi)}$ is unsaturated hydraulic conductivity.

Combining eqs. 1 and Fick's law, and assuming **v**=0, i.e., no centre of mass velocity, and $(p/d)_i=0$, the rate of change in mass is expressed using Fick's 2nd law:

$$\frac{\partial \rho_i}{\partial t} = D_i \nabla^2 \rho_i \quad \text{or} \quad \frac{\partial C_i}{\partial t} = D_i \nabla^2 C_i \qquad (5)$$

On substitution of **q**, Fourier's law, into energy transport equation (eqn. 2) and assuming no mass transport and no diffusion, i.e., no centre of mass velocity and no velocity of component i, $\mathbf{v}_i=\mathbf{v}=0$, eqn. 2 reduces to:

$$\rho \frac{\partial E}{\partial t} = K \nabla^2 T \, . \qquad (6)$$

$$\because \; dE = \left(\frac{\partial E}{\partial t}\right)_V dT = C_V \, dT \quad \text{and} \quad k = \frac{K}{\rho C_V} \qquad (7)$$

where C_V is the volumetric heat capacity and k is "Diffusivity" (term given by Kelvin) or "Thermometric Conductivity" (term given by Maxwell), the rate of energy than is expressed by a simple form of eqn. 2, known as Fourier's 2nd law of heat conduction:

$$\therefore \; \frac{\partial E}{\partial t} = k \nabla^2 T \, . \qquad (8)$$

Several models employ equations similar to eqn's. 5 and 8 to calculate the water uptake by the clay barriers and the temperature in the near-field. However, such models strongly simplify the real system. In the liquid subsystem Fick's law does not necessarily hold true, and the energy is transferred by convection and diffusion; the energy transport by conduction flux alone implies the simplest case of a pure solid system.

c) <u>Coupled heat and moisture transfer</u> - Philip and de Vries (1957) developed a model for moisture movement in porous materials under temperature gradients. The general form of governing equations in this model are:

$$\frac{\partial \theta}{\partial t} = \nabla \cdot (D_T \nabla T) + \nabla \cdot (D_\theta \nabla \theta) + \frac{\partial K^*}{\partial z} \qquad (9a)$$

$$C_V \frac{\partial T}{\partial t} = \nabla \cdot (\lambda \nabla T) - L \nabla \cdot (D_{\theta\,vap} \nabla \theta) \qquad (9b)$$

$$D_T = D_{T\,liq} + D_{T\,vap} \qquad (10a)$$

$$D_\theta = D_{\theta\,liq} + D_{\theta\,vap} \qquad (10b)$$

where θ is the volumetric moisture content, D_T is the thermal diffusivity, D_θ is the isothermal diffusivity, the subscript liq and vap refer to the liquid and vapour phases and λ is the thermal conductivity and L is the latent heat of vaporization.

The theory of this model provides a comprehensive basis for estimation of heat and moisture transfer in unsaturated porous materials. Several models were developed based on this approach or on the derivations of Philip and de Vries model. However, the major assumption in this model is that the porous material is incompressible. Consequently models derived from Philip and de Vries model have some limitation when they intend to estimate the behaviour of systems where deformations that take place in the system are of major importance. In addition, this model describes a relatively weak heat-driven flow; under high temperature gradient, increases in the vapour partial pressure and strong forced convection of the gas phase, this model may underestimate the non-isothermal unsaturated flow.

d) <u>Constitutive relationships</u> - Momentum transport problems are very complicated in the multi-components systems in which the liquid is one of the major subsystems and the processes of interest coupled. With the exception of classical Newton's law of stress, there is no theoretical relationship between momentum flux and forces. The interactions between different forces due to cross phenomena can not be included explicitly in the mathematical formulation because only the tensor flux is expressed as tensor force.

Assuming that the momentum flux, $-\delta + pI$, is directly proportional with the gradient of centre of mass velocity of the volume element (i.e., the force that is a tensor) the stress tensor is:

$$\delta = pI - \left[\kappa_1 (\nabla v)_e + \kappa_2 (\nabla v)_d + \kappa_3 (\nabla v)_r\right] \qquad (11)$$

where pI is the pressure tensor (I is a unit tensor), $(\nabla v)_{e,\,d,\,r}$ are the tensor (force) that account for expansion / contraction, deformation and rotation respectively and $\kappa_{1,\,2,\,3}$ are constants of proportionality between flux and forces.

Newton's law of stress relates the momentum flux and forces through the viscosity tensor. The only force that can be associated with momentum flux is $(\nabla v)_d = (\nabla v)^* - \frac{1}{3}\nabla \cdot vI$, where $(\nabla v)^*$ is the symmetric tensor component. For $\kappa_1 = 0$ (i.e., bulk viscosity equal zero), $\kappa_3 = 0$ (i.e., rigid body rotation; no viscous force can be associated with it) and substituting for $\kappa_2 = 2\eta$ [½ in the traceless tensor elements of $(\nabla v)_d$] where η is the coefficient of viscosity, the stress tensor (Newton's law of stress) is:

$$\delta = pI - 2\eta\left[(\nabla v)^* - \tfrac{1}{3}\nabla \cdot vI\right] = pI - 2\eta(\nabla v)^* + \tfrac{2}{3}\eta\nabla \cdot vI \qquad (12)$$

Using the Newtonian stress tensor, the divergence of surface stress tensor, the first term in the momentum transport equation (eqn. 3) becomes:

$$\nabla \cdot \delta = \nabla \cdot pI - 2\eta\nabla \cdot (\nabla v)^* + \tfrac{2}{3}\eta\nabla \cdot (\nabla \cdot vI). \qquad (13)$$

$$\because \nabla \cdot pI = \nabla p, \quad \nabla \cdot (\nabla v)^* = \tfrac{1}{2}\nabla^2 + \tfrac{1}{2}\nabla(\nabla \cdot v) \quad \text{and } \nabla \cdot (\nabla \cdot vI) = \nabla(\nabla \cdot v) \qquad (14)$$

$$\therefore \nabla \cdot \delta = \nabla p - \eta\nabla^2 v + \tfrac{1}{3}\eta\nabla(\nabla \cdot v). \qquad (15)$$

Combining eqs. 15 and 3 and assuming incompressible liquid, the rate of change of momentum (Navier-Stokes equation) is:

$$\frac{\partial \vec{v}}{\partial t} = -\nabla p - \eta\nabla^2 v - \tfrac{1}{3}\eta\nabla(\nabla \cdot v) + \rho\vec{x} \qquad (16)$$

To accommodate the expansion / contraction of skeleton, several stress - strain - time constitutive relationships derived from consolidation theory were developed and included in the simulation models. Deformation of the solid component of the system (the skeleton), both in the far - and near- field is fundamentally based on the relationships between the degree of consolidation and time derived from principle of hydraulics by Terzaghi and Biot. The equation of equilibrium is written in various forms. Some models follow the rules of thermo - elasto - plasticity, others conform to thermo - visco - plasticity. It should be noticed, however, that constitutive models make use of parameters that, often, are difficult to measure and that are not always conservative. Generally, small changes in the parameters of a constitutive model may induce significant differences in the overall simulated results.

Interaction between HLW Glass and Clay: Experiments versus Model

P. Van Iseghem, K. Lemmens, M. Aertsens and M. Put
SCK•CEN, Belgium

Abstract

This paper reviews the approach followed in Belgium to evaluate the long-term performance of high-level nuclear waste glass in Boom clay. The approach is based on a parametric experimental programme on inactive simulated waste glasses in clay media, and on both geochemical and mathematical modelling. In-situ testing of the glass performance in the underground laboratory in clay and laboratory experiments on glasses doped with radionuclides or fully active glasses are also performed, but not discussed in this paper. The experimental conditions discussed include durations until 5 years, and surface area to solution volume ratios of maximum 10000 m^{-1}.

Untill recently decreasing corrosion of the glass in pure solutions due to SiO_2 saturation in solution, and a final, small reaction rate were widely accepted. Subsequent models were developed. We provide evidence that diffusion processes control the long-term glass dissolution. At even further reaction progress, i.e. for boron in solution above 1000 $mg.l^{-1}$, discontinuous excursions in the glass dissolution were observed, related with secondary phase formation. Our models consider the following critical parameters: (1) the forward rate of reaction, (2) the decreasing dissolution rate due to SiO_2 saturation, (3) the diffusion of glass constituents through the glass surface, and (4) the transport properties of SiO_2 through the pore water in clay.

Introduction

NIRAS/ONDRAF, the Belgian authority responsible for nuclear waste management, requests that the compatibility of the conditioned radioactive waste with the geological disposal conditions be investigated. Compatibility may cover different aspects: release of the radionuclides, degradation of the waste matrix, thermo-physical behaviour (e.g., crack formation), impact on the near field (e.g. gas generation). These aspects need to be evaluated for all types of waste forms envisaged for geological disposal, such as glass, cement, and bitumen. The on-going investigations on the compatibility of vitrified waste with the geological disposal selected in Belgium (Boom clay) aim to establish a model for the behaviour of the waste form in the disposal environment, based on experimental data. This model must allow to predict the long-term performance of the waste form, over periods exceeding the laboratory durations. In our programme, we also account for possible changes in the disposal concept, influencing the behaviour of the particular waste forms.

The practical way followed to evaluate the long-term performance of the waste glass is based on three approaches: (1) laboratory interaction tests, (2) conceptual and mathematical modelling, and (3) in-situ verification and demonstration tests. We did not investigate natural analogues in relation to disposal in clay so far. Different types of laboratory experiments have been considered. Experiments on inactive samples are performed to determine the dissolution mechanisms of the glass. The application of a large set of experimental parameters (composition of the solution, concentration and composition of solids in the leachant, pH/E_h, surface area of the sample to solution volume, temperature) must allow to obtain a basic understanding of glass dissolution in various experimental conditions. Experiments on glass samples doped with various radionuclides (Pu, Np, Am, Tc, ...) are performed to investigate the particular leaching behaviour (total release, mobile concentration in the near field) of the radionuclides of relevance in the long term. Experiments on fully active samples are carried out to verify the similarity in dissolution of the inactive simulate samples and the "real" glass.

The underground research laboratory in clay situated beneath the SCK•CEN offers the possibility to add the third approach to the programme [1]. The in-situ experimental conditions offer the most realistic environment. Although interpretation of these experiments can be hampered if, say, the clay rock would be inhomogeneous, these tests are needed to demonstrate the relevance of the laboratory tests. They also differ from our laboratory tests in that they induce glass dissolution in an open system, whereas in the laboratory tests we typically deal with a closed system. The in-situ tests are also very important to increase the public acceptance of geological disposal of nuclear waste packages. The relationship between the different experimental approaches and the on-going modelling efforts is given in Fig. 1.

This paper reviews the inactive corrosion tests, and how they can be used as input for the modelling. The modelling attempts so far are restricted to the leaching of the inactive constituents of the waste glass, because the experimental and theoretical data base for radionuclide leaching is much more limited, and because the radionuclide leaching generally proceeds slower than the leaching of the soluble inactive glass constituents. A review of the radionuclide (Pu, Np, Am, Tc) leaching and of the in-situ tests in Boom clay can be found elsewhere [2,3].

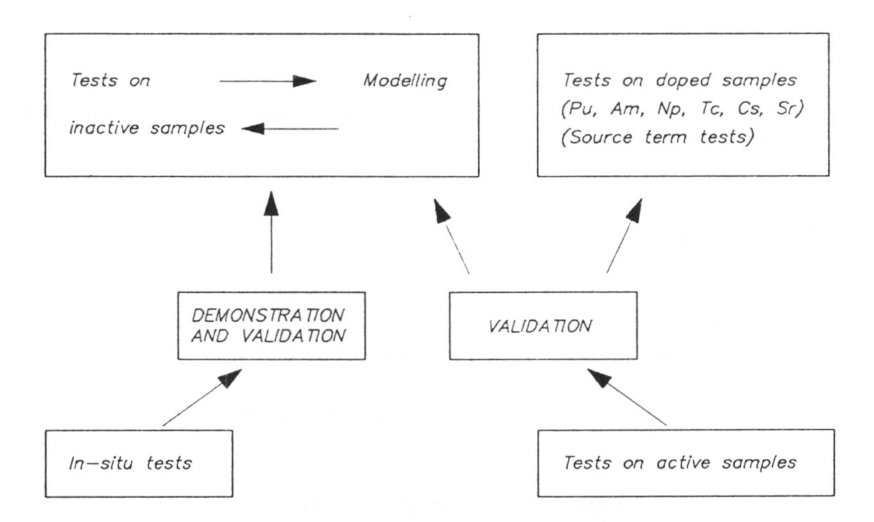

Fig. 1 Relationship between the different experimental and modelling approaches in the SCK•CEN programme

Experimental approach

The goal of the laboratory experimental programme is to study the interaction mechanisms between vitrified waste and the geological disposal surroundings. Because of the existing uncertainties in the scientific understanding of the glass corrosion phenomena [4], even in the simplest system (glass - pure water), we opted for a parametric approach. We considered "simple" two-component systems such as glass-pure water and glass - claywater, as well as more realistic media loaded with solids (clay, with and without additional materials such as container corrosion products and backfill). This approach was supposed to identify the specific role of the solids on the glass dissolution.

We respected the natural pH/E_h conditions prevailing in the anticipated disposal conditions, and static interaction conditions were choosen because of the extremely low hydraulic conductivity in the Boom clay layer (about $4x10^{-12}$ m/s). We applied short (days) and long (until 5 years) interaction durations, because long-term data are of the uppermost importance. With the same objective, we performed experiments at different SA/V (sample surface area to solution volume ratio) values and temperatures. Both parameters are known to influence the interaction kinetics [5].

The experimental parameters are summarized in Table I. In the further discussion, 90 °C is considered as the reference temperature; modelling will be carried out first for the tests at this temperature. The glasses of interest are the DWK/Pamela glasses SM513 (and precursor SM58), SM527 (and precursor SAN60) and SM539, and the Cogéma/R7T7 glass SON68. The composition of these glasses and of the Boom clay can be found elsewhere [6]. The main glass components are listed in Table II. Glasses SM527 and SM539 are rather special glasses, they have Al_2O_3 contents of about 20 wt% (below 5 wt% for the other two glasses). This will be of large importance for the corrosion behaviour of these glasses. Analyses of the experiments is performed with the different techniques available today [7]: besides determining the glass mass loss, the reacted glass surface is investigated by SEM-EDXA, EPMA X-ray mapping, SIMS or XRD. Leaching solutions are analysed by ICP-AES or ICP-MS, after separating the solids or colloids by centrifugation/filtration procedures.

Table I Overview of the experimental parameters considered in the SCK•CEN programme.

medium:	DW*, SIC**, clay slurries: DCSICM***, CCSICM**** (without, with corrosion products and backfill)
temperature (°C):	40, 90, 150, 190
SA/V (m^{-1}):	10, 100, 500, 2500, 10000
E_h:	positive, negative
duration:	days to years

*	DW =	distilled water
**	SIC =	synthetic interstitial claywater (composition, see [6])
***	DCSICM =	diluted mixture of Boom clay with SIC (10 $g.l^{-1}$)
****	CCSICM=	concentrated mixture of Boom clay with SIC (500 $g.l^{-1}$)

Table II Main constituents of the glasses investigated (in wt%)

	SON68	SM58	SM513	SAN60	SM527	SM539
SiO_2	45.48	56.87	52.15	43.41	38.75	35.28
B_2O_3	14.02	12.28	13.08	17.00	21.70	25.58
Na_2O	9.86	8.30	9.05	10.67	8.64	9.21
Li_2O	1.98	3.74	4.18	5.0	3.10	3.49
Al_2O_3	4.91	1.16	3.59	18.09	19.96	19.83
CaO	4.04	3.83	4.54	3.50	3.87	5.05
TiO_2	-	4.45	4.54	-	1.55	0.003
Fission product oxides	13.38	6.06	4.46	1.21	1.58	0.96
Balance	6.33	3.31	4.41	1.12	0.85	0.60

Current situation on glass dissolution mechanisms and modelling

Many summary reports or compendiums have reviewed the dissolution behaviour of glass [5,8,9, 10]. So far there was a general consensus that the glass dissolution in static solutions like pure water, groundwater or brine is mainly controlled by a number of mechanisms, which are briefly recalled here:

- diffusion reactions in the initial stage, corresponding with, e.g. ion exchange, between H^+ and H_3O^+ with alkali (Na, Li) cations from the glass, or with diffusion of water molecules into the glass;
- congruent dissolution, due to the dissolution of the SiO_2 based network;
- saturation phenomena in the leachate, typically of rather insoluble elements such as rare earths, transition metals, but also SiO_2; the saturation of SiO_2 in the solution causes a drop of the corrosion rate to zero;
- long-term dissolution of the glass, at a rate much below the initial dissolution rate.

Our experiments carried out in pure (distilled) water provided evidence of the decrease of the initial glass dissolution due to saturation, and of the "long-term dissolution" of the glass, see Figs. 2 and 3.

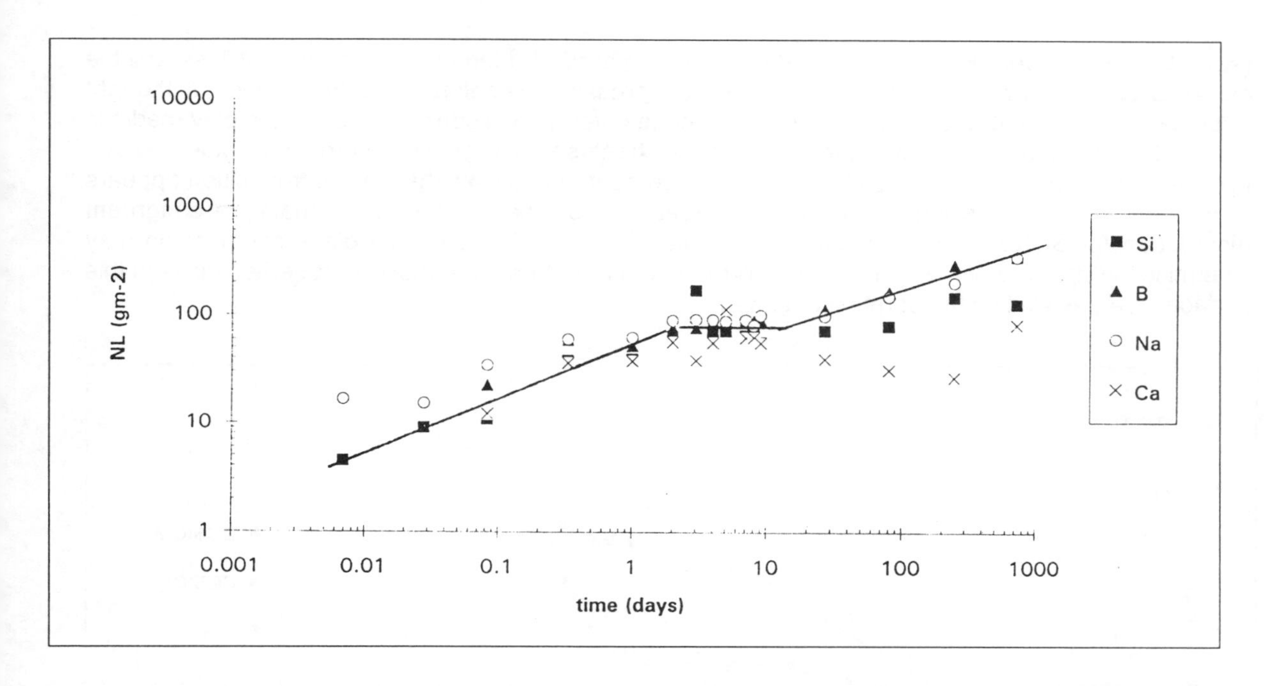

Fig. 2: Normalized losses for B, Si, Na and Ca upon corrosion of glass SAN60 in distilled water at 90°C ($SA/V = 100\ m^{-1}$). The line represents the B data, and is a guide for the eye.

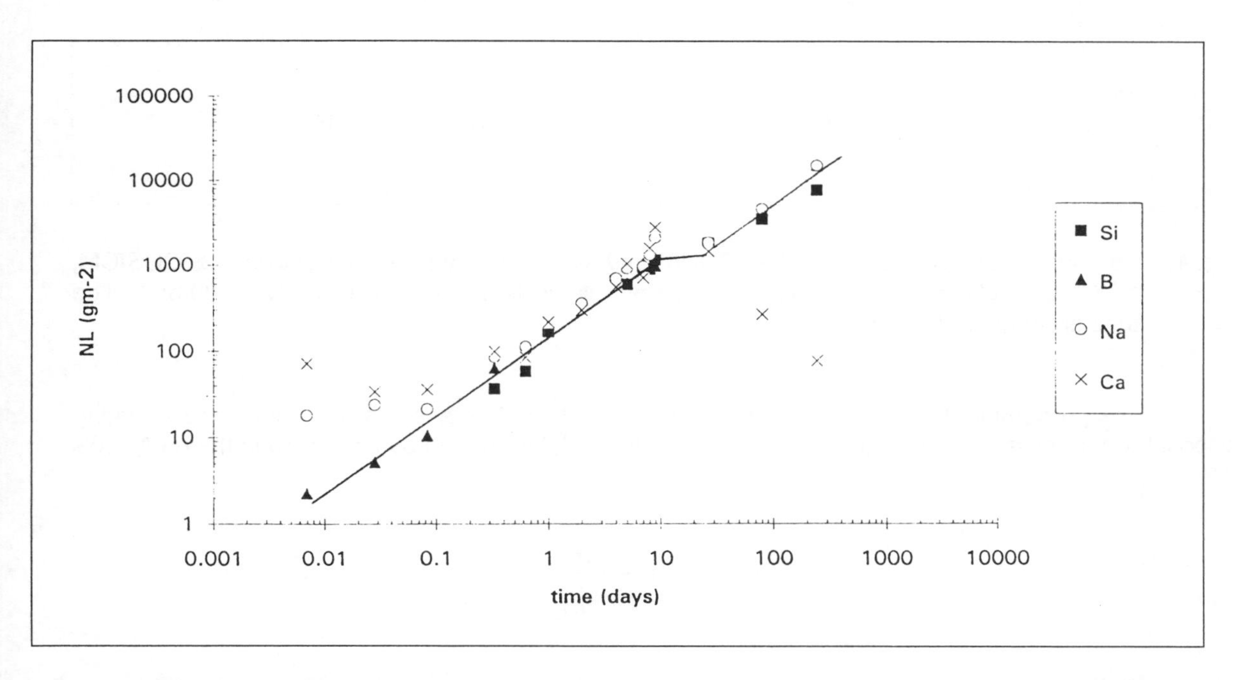

Fig. 3: Normalized losses for B, Si, Na and Ca upon corrosion of glass SM58 in distilled water at 90°C ($SA/V = 10\ m^{-1}$). The line represents the B data, and is a guide for the eye.

The presence of solid materials in the leachant, such as the clay studied in Belgium as the host rock, is known to enhance the glass dissolution mainly due to sorption of less soluble elements, e.g. Si [11]. The strong influence of clay in the leachant on the glass dissolution is shown in Fig. 4, comparing glass dissolution in pure solutions, and solutions loaded with different amounts of clay. Whereas after a few days of interaction mass losses are nearly similar in all media, differences are nearly a factor $x10^3$

(solid Boom clay versus pure water) after about 3 years of interaction. Sorption of less soluble elements, such as Si, Al, Fe, rare earths, on clay, suppressing the saturation of the solution, is thought to be the main phenomenon responsible for the huge difference in corrosion; pH in the clay media is even a little smaller (8.5-9.5) than in the pure water leachant, so no dissolution increase is expected due to differences in pH. Besides the corrosion enhancement due to clay, the glass composition appears to strongly influence the interaction with clay (see Fig. 5); the mechanism is mainly a congruent dissolution for SON68, and is mainly diffusion controlled for SAN60. The glass composition may determine the structure of the surface layer formed, and therefore the transport properties through this surface layer, as well as the sorption onto clay.

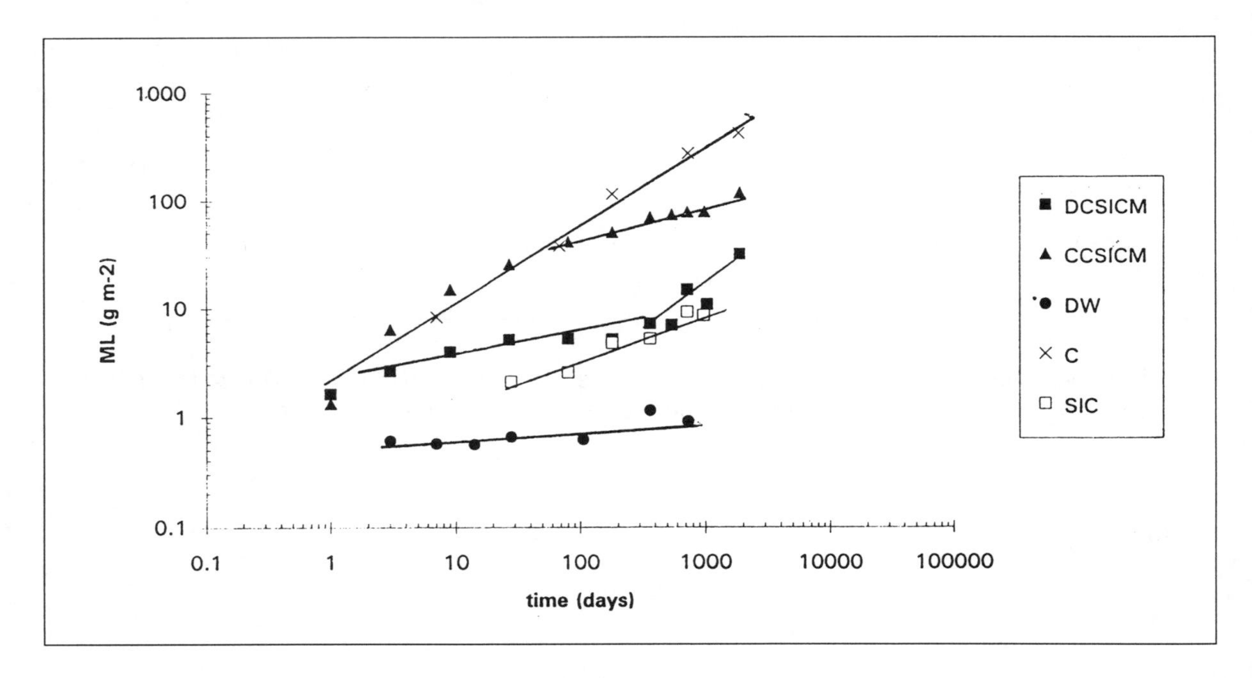

Fig. 4: Mass losses upon corrosion of glass SON68 in: X wet clay, clay-claywater slurries (▲ CCSICM, ■ DCSICM), □ synthetic interstitial claywater and ● distilled water (90°C, SA/V = 100 m^{-1}). The lines are a guide for the eye.

The predominant role of SiO_2 in the dissolution of waste glass in static solutions simulating disposal conditions is acknowledged in the mechanistic model proposed by Grambow in the early '80's [12]:

$$R_M = k_+(1 - \frac{a}{a_{sat}})$$

where R_M the reaction rate
k_+ the forward rate of reaction
a the activity of SiO_2 in the leachate
a_{sat} the activity of SiO_2 in the leachate at saturation

In this equation R_M drops to zero when SiO_2 reaches its saturation concentration. This saturation regime however is a transient to a "final", continued glass dissolution, yielding a final rate of reaction, R_∞ [13].

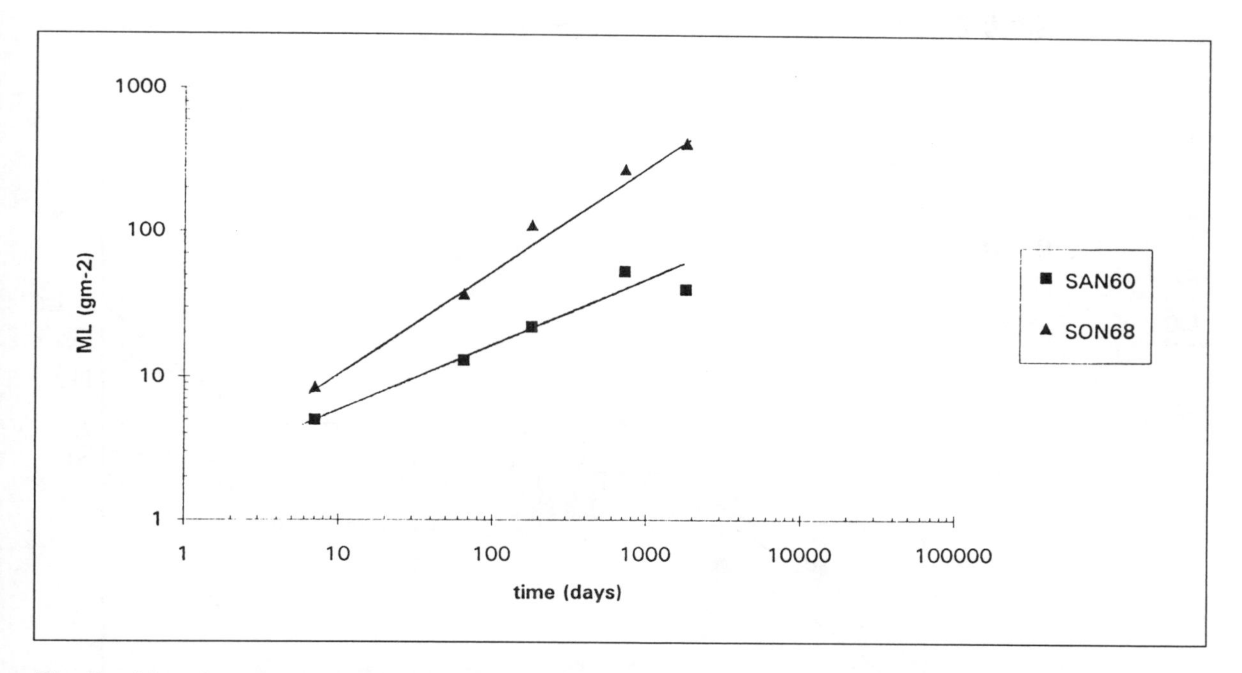

Fig. 5: Mass losses upon corrosion of glasses SON68 and SAN60 in wet clay (90°C, SA/V = 100 m^{-1}) [6]. The lines are a guide for the eye.

This conceptual model is translated to a mathematical model using the "GLASSOL" code [13]. GLASSOL basically calculates the dissolution of the glass matrix (e.g. boron) using the initial (forward) and final reaction (dissolution) rates. Other parameters in this mathematical model are the saturation concentration of SiO_2, the diffusion coefficient of SiO_2 through the surface layer, and the thickness of this layer. The latter two parameters account for the transition to the final reaction rate. The GLASSOL model was successfully applied to our experimental data at 90 °C in pure water, see e.g. Fig 6. B and Li are calculated by GLASSOL to be the soluble elements. The lower releases for Ca, Na, Si and Al were calculated from the PHREEQE geochemical code, which predicts the formation of secondary phases in solution (e.g., $Al(OH)_3$, $NaAlSi_2O_6.H_2O$, and $Ca_4(Si_6O_{15})(OH)_4$) at certain states of the reaction progress. As a consequence, the concentration in solution of these elements is lower than for the soluble elements B and Li.

Attempts to establish mathematical models for the dissolution of glass in the presence of clay containing media are reported in [14,15,16]. Basically the transport by diffusion of SiO_2 through the pore water of clay is added to the glass dissolution kinetics. Curti and Pescatore considered the glass dissolution kinetics developed by Grambow.

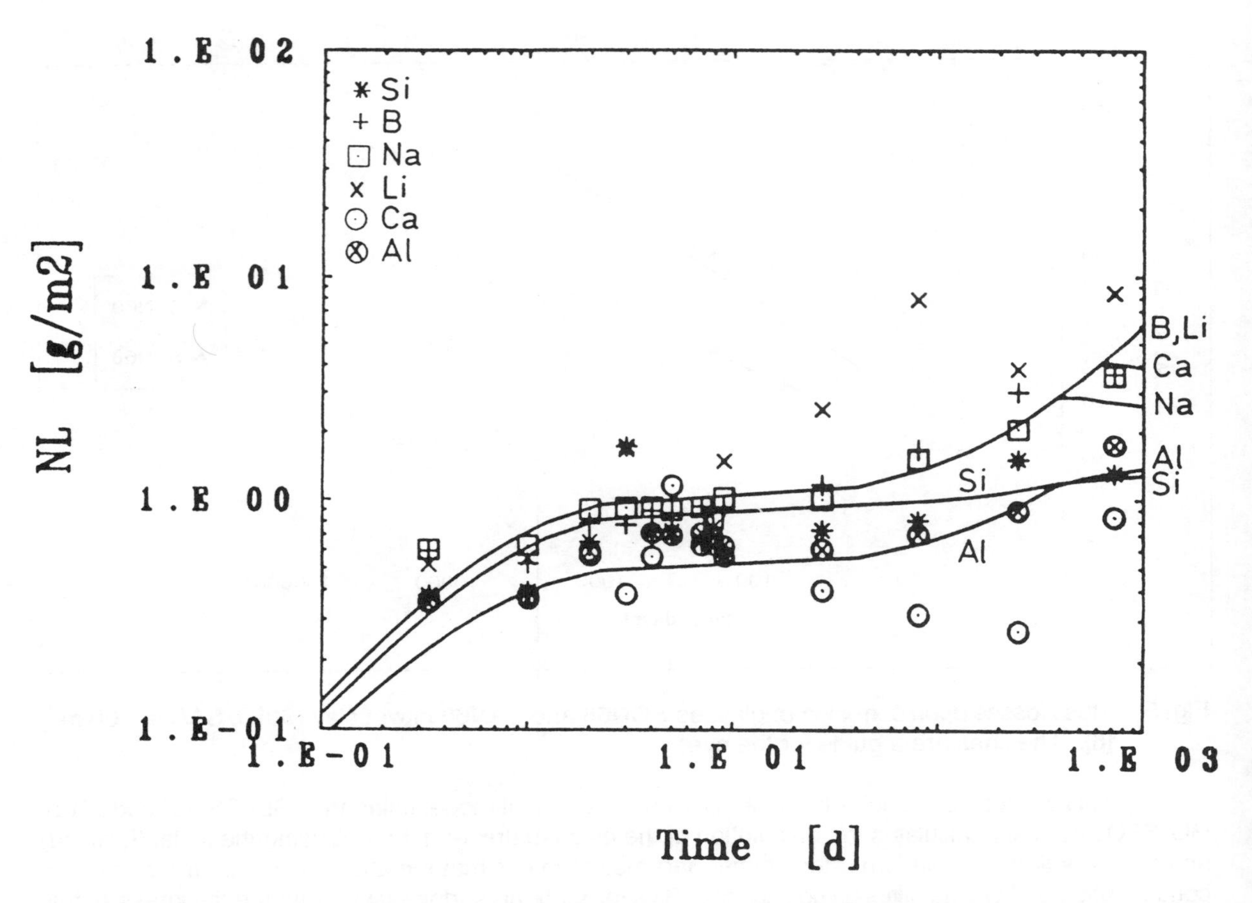

Fig. 6: Experimental normalized losses for glass SAN60 in distilled water at 90°C (SA/V = 100 m^{-1}) and calculated values by GLASSOL [22]

Recent experimental observations

During the past few years, new insights were obtained based on experimental results. The following experiments were in particular relevant:

- long-term duration experiments in various types of media, including clay containing solutions, with durations until five years;
- experiments on powdered glass, to accelerate the accumulation of leached elements in solution;
- experiments to evaluate particular phenomena, such as crystallization at the glass surface, and the role of (Si, Fe, Al) saturation in solution.

Various experiments on powdered glass in pure water were performed [17,18]. Experiments on "simple" glasses with different (SiO_2, Al_2O_3) concentrations were carried out at 90 °C and 150 °C, for SA/V between 100 and 10000 m^{-1} in pure water. In these tests, B concentrations in solution reached final values of about 1000 $mg.l^{-1}$, which is far much higher than in "standard" test conditions (10 m^{-1}). We observe a B release proportional with the square root of time; collecting the data at different SA/V, and plotting all data as a function of the "scaling factor" SA/V x $t^{0.5}$ clearly shows a linear dependence of the B release on SA/V$xt^{0.5}$, for concentrations between, say 100 and 1000 $mg.l^{-1}$ (see Fig. 7). We suggest that the interaction mechanism is diffusion controlled. Effective diffusion coefficients in the order of 10^{-19} to $10^{-20}m^2.s^{-1}$ were calculated. Ion exchange reactions between H_3O^+ or H^+ and Na^+/Li^+ were proposed as the rate controlling process [17]. This suggests that the concept of final rate of dissolution must be changed. Indeed, the final rate of dissolution introduced in [13], should have been existing in this stage of the reaction progress.

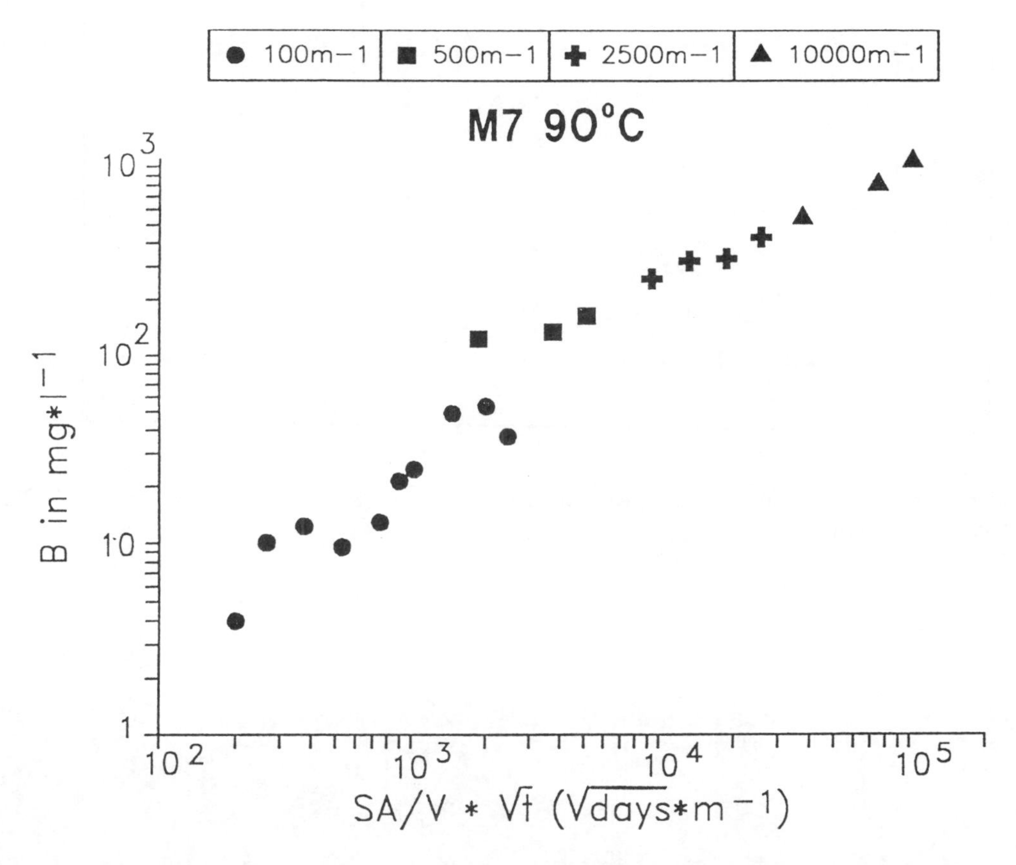

Fig. 7: Boron concentration in solution upon corrosion of glass M7 in distilled water at 90°C. SA/V of 500 m^{-1} and more was achieved with powdered glass. Composition of M7 (in mol%): 40 SiO_2, 6.7 $AlO_{1.5}$, 14.9 $NaO_{0.5}$, 14.5 $LiO_{0.5}$, 21.2 $BO_{1.5}$, 2.7 CaO) [17]

To obtain higher solution concentrations, experiments were carried out at higher temperature. Experiments at 120 °C on glass SAN60 at SA/V = 7000 m^{-1} yielded B concentrations in solution of 10000 $mg.l^{-1}$. As shown in Fig. 8, a very steep increase in dissolution occurs above B concentrations in solution of 1000 $mg.l^{-1}$. This can be correlated with the growth of crystals at the glass surface (see Fig. 9). We believe that this crystal growth enables further glass leaching by suppressing saturation in the solution. Similar observations were reported by other scientists [19]. The steep increase in dissolution cannot be modelled with GLASSOL [20].

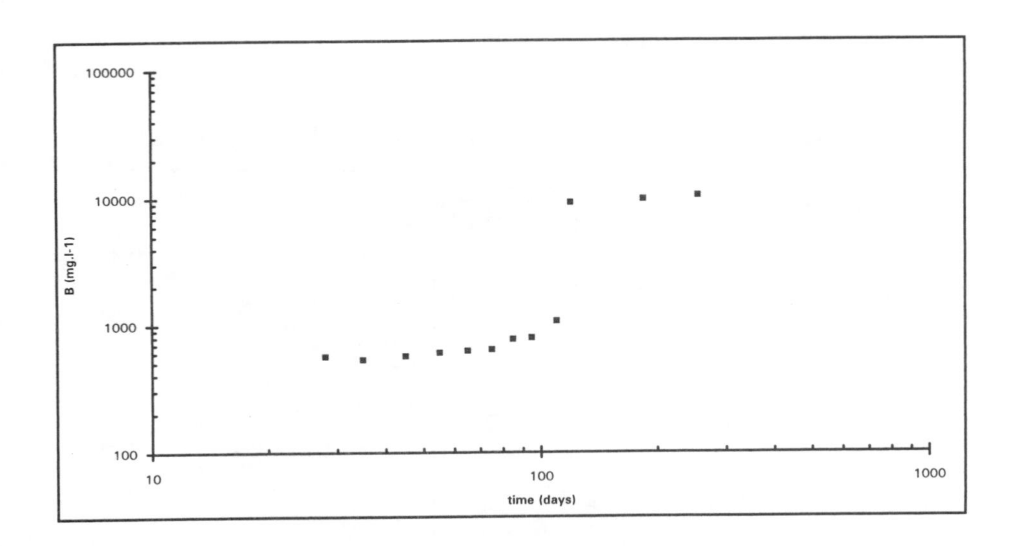

Fig. 8: Boron concentration in solution upon corrosion of glass SAN60 in distilled water at 120°C, with SA/V = 7000 m^{-1}

Fig. 9: SEM surface analysis of glass SAN60, corroded for 185 days in distilled water at 120°C, SA/V = 7000 m^{-1} (see Fig. 8). Different magnifications are shown. The following zeolite phases are observed: analcime (area A), calcium silicate (area B), aluminum silicate (area C), and a calcite (area D).

We have observed similar phenomena in pure claywater and clay/claywater slurries, although the amount of experimental data is smaller compared with pure water. For glass SM527, the corrosion increase is occurring at B concentrations of about 1000 $mg.l^{-1}$, as in pure water. This behaviour was even observed in tests at relatively low SA/V (100 m^{-1}, using glass monoliths), because of the larger glass corrosion in clay media (see Fig. 10) (Si, Al) rich crystals on top of the glass surface were observed as well. Corrosion tests at high temperature (190 °C) in clay/claywater slurries revealed the creation of analcime crystals on top of the glass surface (see Fig. 11). The observations of enhanced dissolution in very far stages of glass reaction progress, associated with crystal growth at the glass/solution interface, therefore are similar in pure water, claywater and clay/claywater slurries. This means that the relevant (for dissolution kinetics) phase formation in clay media does not include organics (humic acids) provided by the clay.

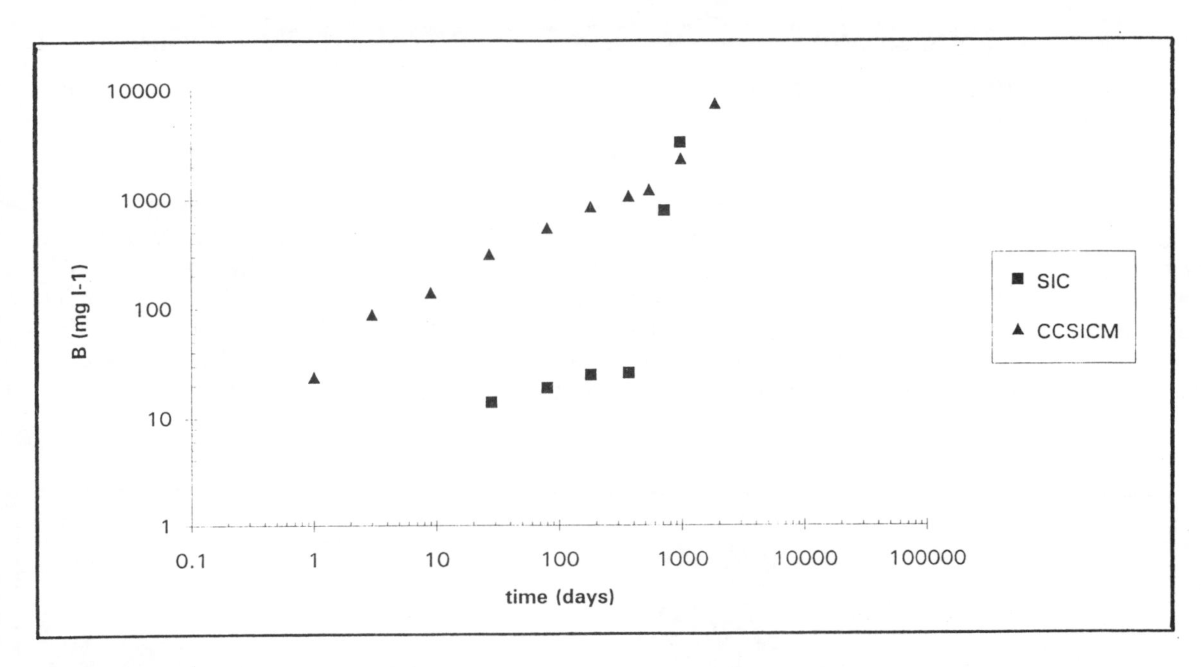

Fig. 10: Boron concentration in solution upon corrosion of glass SM527 in synthetic claywater or a concentrated clay/claywater slurry (90°C, SA/V = 100 m^{-1})

In addition, the strong increase in glass dissolution coupled with crystal growth induces a strongly enhanced leaching of radionuclides such as Pu and Np [6]. The increased leaching is essentially recovered as an insoluble fraction with dimension larger than a few nm.

So far, we have the best evidence of the corrosion enhancing "jump" due to crystal growth from the high Al_2O_3 glass SM527 (and its precursor SAN60). We have only but a weak indication that corrosion enhancement is occurring as well for the other, lower Al_2O_3 glasses SON68 and SM513, for reaction progresses similar as for SM527, i.e. for B concentrations in solution of about 1000 $mg.l^{-1}$ [6].

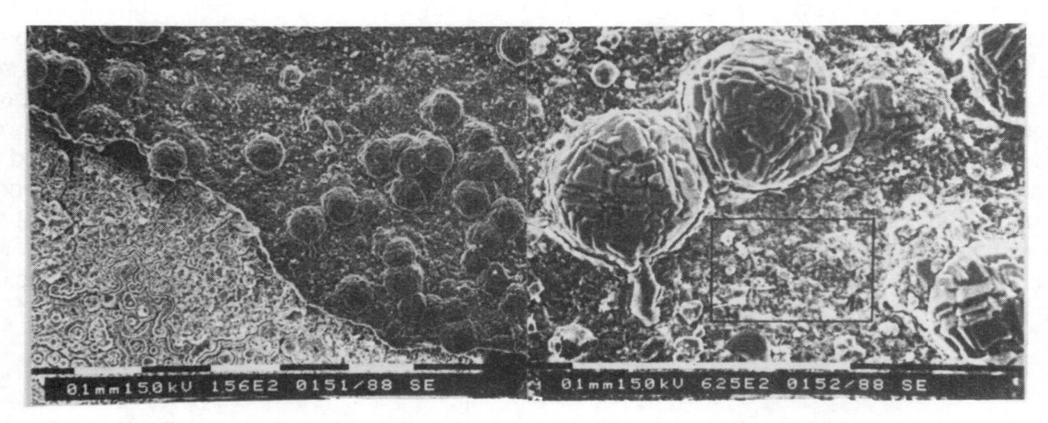

Fig. 11: SEM surface analysis of glass SM527, corroded for 180 days in a concentrated clay slurry (CCSICM) at 190 °C; analcime crystals up to 500 µm can be observed [6]

We also have experimental evidence that SiO_2 saturation in the leachate on itself does not induce a zero leach rate, as would be expected, but yet in agreement with another study [21]. We performed corrosion tests in pure water presaturated with Si, Al and Fe; the latter two elements also have a limited solubility in water, and are important constituents of the glass. The following starting solution for the corrosion test were thus obtained: 103 $mg.l^{-1}$ Fe, 49 $mg.l^{-1}$ Si, and 12 $mg.l^{-1}$ Al. The cumulative leaching for either glass SON68 or glass SM527 within 2 years (90 °C, 100 m^{-1}) were quite the same, based on either the total mass loss or the B release (about 4 g/m^2, see Fig. 12).

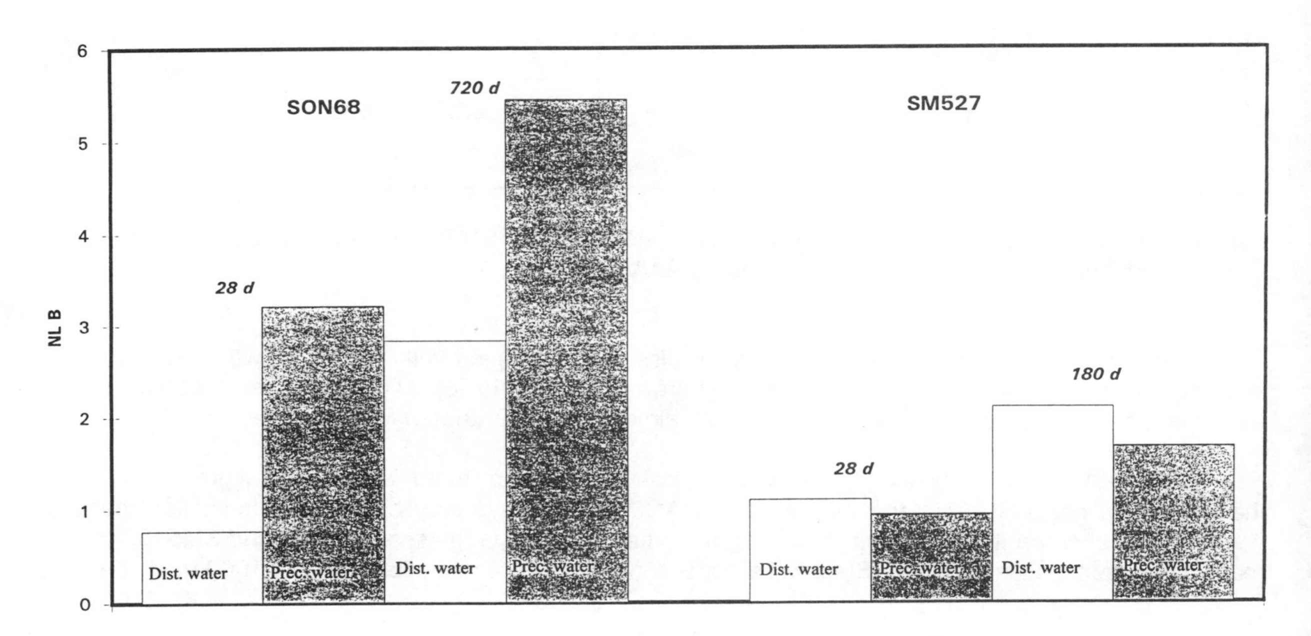

Fig. 12: Comparative B leaching data from glasses SON68 and SM527 in either distilled water or distilled water presaturated in Si, Al, Fe (90°C, SA/V = 100 m^{-1})

Discussion

We have experimental evidence that the mechanistic assumptions in the Grambow model have to be reconsidered. The role of SiO_2 saturation on glass dissolution does not appear to be predominant, and the final "long-term" glass dissolution does not proceed with a constant rate. This is in agreement with observations by Petit [22]. Instead, the long-term glass dissolution in pure solutions is supposed to be controlled by diffusion processes at the glass/solution interface, and secondary phase formation associated with crystal growth may provoke discontinuous excursions of the glass dissolution. The latter phenomenon is occurring in conditions of very high reaction progress, i.e. at B concentrations of about 1000 $mg.l^{-1}$. Both phenomena occur, but possibly to a different degree, for broad glass compositional ranges (i.e., from high Al_2O_3 concentrations to Al_2O_3 concentrations of 5 wt% or less), and in solutions loaded with Boom clay.

Our parametric experimental programme revealed some typical similarities in dissolution behaviour between pure water and the different clay media, especially for the secondary phases formed. Particularly important is the observed correspondence between the dissolution excursion in clay slurries due to secondary phase formation, and an increased leaching of Pu and Np.

We think that the following critical parameters need to be considered in new modelling attempts. Some parameters deal with SiO_2, because of its predominant concentration in the glass, its role as a network former, and its limited solubility in solution compared with other main glass constituents such as B_2O_3. We do not exclude that the role of SiO_2 may be coupled with other less soluble glass constituents such as Al_2O_3 and Fe_2O_3.

* Forward rate of reaction; this is the dissolution rate of the glass in the absence of significant saturation in solution.
* The decrease of the dissolution rate due to SiO_2 saturation; this phenomenon must a priori be taken into consideration, because of the demonstrated occurrence of saturation processes of e.g. SiO_2 in solution [23].

* The diffusion of soluble glass constituents through the glass surface; based on the low effective diffusion coefficients, we expect that diffusion "gradient" zones rather than the less dense "gel" surface layer act as diffusion barriers.
* The transport properties (diffusion and sorption) of SiO_2 through the pore water in clay (in case clay is present); sorption of insoluble elements onto clay is influencing the glass dissolution in clay media; SiO_2 being the main glass constituent is supposed to be most representative and dominant in this respect.

Basically, parameters associated with both the glass and the medium (solution, clay) must be considered as critical parameters for further modelling. Yet, the following experimental data must be known more precisely:

* the SiO_2 saturation concentration for the low Al_2O_3 glasses;
* the role of the other elements (Al, Fe, ...) on the decreasing dissolution rate, as domonstrated by Xing [24];
* the long-term, diffusion controlled reaction rate for the glasses under consideration;
* the diffusion constants through the glass surface;
* the role of secondary phases on the glass dissolution, mainly for the low Al_2O_3 glasses.

Conclusions

Efforts related to the modelling of the interaction between waste glass and clay media are going on, including both experimental and modelling actions. These actions respond to the new insight discussed above. On an experimental basis, a detailed characterization of the corroded glass surfaces is carried out. SIMS (secondary ion mass spectrometry) profiling is expected to provide additional information on the dissolution mechanisms in clay media, for instance the proposed long-term mechanism of diffusion controlled dissolution [25]. Of particular interest is the relatively thin diffusion zone. Diffusion flow-through experiments through Boom clay using ^{32}Si tracer, to determine the diffusion and sorption coefficient of SiO_2 on clay were initiated.

Modelling activities include both mathematical and geochemical modelling. A mathematical model is elaborated, considering a diffusion term for the glass dissolution, and a diffusion term for the transport through clay. Geochemical modelling with the PHREEQE code is expected to provide information on the secondary phases formed during the interaction between glass and clay media.

The development of a model to describe and predict the interaction of waste glass in Boom clay based on both experiments and modelling attempts is advisable. Our experimental programme has provided indications on particular interaction processes, such as the long-term diffusion controlled interaction and the corrosion enhancing secondary crystal growth. The longer durations until 5 years and the difference in glass compositions investigated in particular were very relevant. We believe that the establishment of a reliable model requires the close interference and collaboration of different disciplines, i.e. experimentalists, chemical and physical analysts, geochemists and mathematicians.

Acknowledgements

This work is carried out under contract CCHO-90/123 (KNT 058086) with NIRAS/ONDRAF. We also gratefully acknowledge the technical assistance of B. Gielen and R. Vercauter.

References

[1] A. Bonne, In-situ tests in the HADES facility, in "In Situ Testing of Radioactive Waste Forms and Engineered Barriers", Ed T. McMenamin, Pre-Print of EUR 15629, 1994.

[2] K. Lemmens, P. Van Iseghem and L. Wang, The leaching of Pu, Am, Np and Tc from high-level waste glasses in clay media. Mat. Res. Soc. Symp. Proc. vol 294, Materials Research Society Publ, 1993, 147-154

[3] P. Van Iseghem, In-situ testing of waste glass in clay, Mat. Res. Soc. Symp. Proc. vol 333, Materials Research Society Publ, 1994, 133-144

[4] E. Lue Yen-Bower, D. Clark and L. Hench , An Approach towards Long Term Prediction of Stability of Nuclear Waste Forms. Ceramics in Nuclear Waste Management, Publ by USDOE, CONF-790420, 1979, 41-46

[5] J.E. Mendel comp., Final Report of the Defense High-Level Waste Leaching Mechanisms Program, PNL-5157, 1984

[6] P. Van Iseghem, K. Berghman, K. Lemmens, W. Timmermans, L. Wang, Laboratory and In-situ Interaction between Simulated Waste Glasses and Clay. Final Report 1986-1990, Update 1993, BLG 669

[7] G. Wicks, A. Lodding and M. Molecke, Aqueous Alteration of Nuclear Waste Glasses and Metal Package Components, MRS Bulletin, Sept 1993, 32-39

[8] W. Lutze, Silicate Glasses, in "Radioactive Waste Forms for the Future", Ed W. Lutze and R. Ewing, Elsevier Science Publ, 1988, 3-159

[9] J.C. Cunnane comp.,High-Level Waste Borosilicate Glass. A Compendium of Corrosion Characteristics. DOE-EM-0177, 1994

[10] E.Y. Vernaz and J.L. Dussossoy, Current state of knowledge of nuclear waste glass corrosion mechanisms: the case of R7T7 glass. Applied Geochemistry, Suppl. Issue no1, 1992, 13-22

[11] P. Van Iseghem, R. De Batist and Li Wei-Yin, The Interaction between Nuclear Waste Glasses and Clay. Advances in Ceramics, vol 20: Nuclear Waste Management II. The American Ceramic Society Publ, 1986, 627-637

[12] B. Grambow, A general rate equation for nuclear waste glass corrosion. Mat. Res. Soc. Symp. Proc. vol 44, Materials Research Society Publ, 1985, 15-24

[13] B. Grambow, Nuclear waste glass dissolution: mechanism, model and application. Report to JSS project Phase IV, 87-02, Publ SKB, Stockholm, Sweden, 1987

[14] F. Lanza and A. Saltelli, An experimental and modelling approach of the near-field release and transport processes. Mat. Res. Soc. Symp. Proc. vol 26, Elsevier Science Publ., 1984, 605-612

[15] E. Curti, N. Godon and E. Vernaz, Enhancement of the glass corrosion in the presence of clay minerals: testing experimental results with an integrated glass dissolution model. Mat. Res. Soc. Symp. Proc. vol 294, Materials Research Society Publ. 1993, 163-170

[16] C. Pescatore, The dependence of waste form dissolution on migration phenomena in the host medium. To be published in Radiochemica Acta

[17] P. Van Iseghem, T. Amaya, Y. Suzuki and H.Yamamoto, The role of AL_2O_3 in the long-term corrosion stability of nuclear waste glasses. Journal of Nuclear Materials 190 (1992), 269-276

[18] K. Lemmens and P. Van Iseghem, The long-term dissolution behaviour of the Pamela borosilicate glass SM527 - application of SA/V as accelerating parameter. Mat. Res. Soc. Symp. Proc. vol 257, Materials Research Society Publ. 1992, 49-56

[19] W. L.Ebert and J. Bates, A comparison of glass reaction at high and low glass surface / solution volume. Nuclear Technology, vol 104, 1993, 372-384

[20] J. Patyn, P. Van Iseghem and W. Timmermans, The long-term corrosion and modelling of two simulated Belgian reference high-level waste glasses - part II. Mat. Res. Soc. Symp. Proc. vol 176, Materials Research Society Publ. 1990, 299-307

[21] L. Trotignon, J.C. Petit, G. Della Mea, J.C. Dran, The compared aqueous corrosion of four simple borosilicate glasses: Influence of Al, Ca and Fe on the formation and nature of secondary phases. Journal of Nuclear Materials 190 (1992), 228-246

[22] J. Petit, M. Magonthier, J. Dran, G. Della Mea, Long-term dissolution rate of nuclear waste glasses in confined environments: does a residuel chemical affinity exist? J Mat Science, 25, 3048-3052, 1990

[23] P. Van Iseghem and B. Grambow, The long-term corrosion and modelling of two simulated Belgian reference high-level waste glasses. Mat. Res. Soc. Symp. Proc. vol 112, Materials Research Society Publ, 1988, 631-639

[24] S. Xing, A. Beuchele, I. Pegg, Effect of surface layers on the dissolution of nuclear waste glasses, Mat. Res. Soc. Symp. Proc., Vol. 333, Scientific Basis for Nuclear Waste Management XVII, 1994, 541-548

[25] A. Lodding, H. Odelius and P. Van Iseghem, Belgian HLW glasses after burial in Boom clay: elemental trends in leaching; in "In-Situ Testing of Radioactive Waste Forms and Engineered Barriers", Ed T. McMenamin, Pre-Print of EUR 15629, 1994

Data from a Uranium Ore Body on Release of Dissolved Species: Comparison with a Near Field Release Model

Jinsong Liu and Ivars Neretnieks

Department of Chemical Engineering and Technology
Royal Institute of Technology
Stockholm
SWEDEN

ABSTRACT

In the Swedish repository for spent fuel the canisters containing the uranium oxide fuel are surrounded by compacted bentonite clay which has a low hydraulic conductivity. The transport through the clay will be dominated by diffusion. The water flow in the fractured rock is very small and it is likely that the dissolution rate of the fuel will be solubility limited. Release and transport models based on this assumption have long been used in performance assessments (PA) including that used for the Swedish KBS-3 type repository design. The models basically state that the uranium oxide dissolves to its solubility limit at the surface of the spent fuel and that the dissolved species diffuse out through the clay and into the passing water outside the clay. The rate of transfer is determined by the diffusion resistances in the clay and in the water slowly seeping past. The rate of release and thus dissolution is proportional to the driving force i.e. the concentration difference at the fuels surface and the concentration in the approaching ground water. The proportionality constant is determined by the clay and flow geometry and the material properties.

In the uranium ore body at Cigar Lake there is a striking similarity between the ore body and a repository. The ore is uranium oxide and it is surrounded on the top side by a dense clay which has a much lower hydraulic conductivity than the surrounding rock. We have used the same principles as for the near field release models for the repository to calculate the dissolution rate of uranium and other species at Cigar lake. At this location a large number of measurements on the water composition in and around the ore body have been made. At one location in the ore body itself the first water samples contained a very high concentration of helium, hydrogen and sulphate. This decreased with continued sampling. A departure point in our analysis is the assumption that the first samples represent the steady state concentration of helium in the undisturbed ore. The production of helium by alpha decay of the uranium and thorium chains can readily be calculated and assuming that this is also the steady state release we can calculate the "proportionality factor" i.e. the sum of resistances to release in the model from the decay calculated release and the measured concentrations. This agrees very well with what is obtained from the forward modelling using the geometry, water flowrates and material properties of clay and rock.

INTRODUCTION

The final disposal of spent nuclear fuel in geological formation requires an efficient isolation for very long time of the hazardous radionuclides in the fuel. The processes that can affect the long-term performance of the spent fuel are many and often slow. Safety assessment can therefore not be based solely on short-term experiments and field observations. The experimental results need to be extrapolated to predict the long-term performance. Most of the time a confident extrapolation is difficult to made, because it is usually difficult to assert that the short-term experiments and field observations are still representative to the scenarios of the long-term situation. Some of the mechanisms that affect the long-term performance can simply not be controlled or produced in the short-term experiments and field observations. Models have to be established to take into account the known and possible short- and long-term operating mechanisms. The models should be theoretically self-consistent, and be able to predict short-term behaviour which agrees with experiments and field observations. But this still does not provide a sufficient condition for the models to be correct. The models need also to be validated against the prediction of long-term performance of the spent fuel.

Since long-term performance can be affected by various changes in the repository's future environment, the analysis of future scenarios occupies a central place in the performance assessment. One important question is how to show that no important phenomena or environmental premises have been overlooked in the models (SKB 91). A natural analogue study may provide an excellent case-study for the validation of the various models in this respect. Even though a full validation is still difficult to achieve in a natural analogue study, it can at least considerably increase the confidence of our model predictions.

The Cigar Lake uranium ore deposit in northern Saskatchewan, Canada, is notable as an analogue study site for several reasons (that will be discussed later) (Cramer, 1986a). An international co-operative project was carried out on this site. A large amount of data of hydrology, geochemistry, etc. have been obtained. New models have been developed and existing models have been tested in the studies of the hydrological, geochemical and mass transport processes occurring in the site (Cramer and Smellie, 1993).

In this paper, the mass transport modelling and the model result of the natural analogue study of the Cigar Lake uranium ore deposit are specifically addressed. The model is developed to cope with the specific geometrical features and material properties of the geological formations in the near-filed of the Cigar Lake deposit, but the concepts of the model are not different from those of the existing near-field release models for a final repository (Liu, 1993; Liu *et al.*, 1994). The model results are compared with the known helium production in the ore body. The production of helium from the spontaneous decay of the two uranium series can be calculated based on the well-established knowledge of nuclear reactions. This gives the helium generation rate. As the deposit is a relatively close system and the ore body was formed for such a long time ago (1.3 Ga), it is proper to assume that the present-day release is at a steady-state, i.e., the release rate is equal to the generation rate. The helium concentrations were measured through relatively extensive borehole samplings in the ore body. When representative data are assigned for material properties, the helium release rate can also be calculated from the transport model. The generation rate predicted from nuclear reactions agrees strikingly well with the release rate calculated by the near-field release model. The agreement between the generation rate and release rate suggests that probably no important mass transport mechanisms have been overlooked in the model.

The above finding does have a significant implication to the long-term performance assessment of the final repository. It provides a relatively reliable methodology to validate the existing near-field mass transport models. Helium is chemically inert, and is generated by nuclear reactions which are not affected by the usual hydrogeological and geochemical processes. It is an excellent "tag" against which the release model can be checked. On the other hand, as the Cigar Lake deposit resembles a final repository in many aspects, and the concepts involved are essentially the same in the near-field release model for the Cigar Lake natural analogue study and for the performance assessment of the repository, the agreement of the model results with those observed in the natural analogue study does increase our confidence of our model of the final repository.

Having gained confidence in both the concepts and the actual implementation of the model for the Cigar Lake site we now turn to transport of other species.

Water radiolysis is an important issue in the performance assessment of the spent fuel. It is considered to be the only possible mechanism to change the otherwise reducing near-field geochemical environment to oxidising. When the near-field release model is used to explore the releases of the various redox-related species, it is revealed that the potential oxidising-power-carrying species is sulphate, which is released at approximately the same rate as the dissolved hydrogen. The sulphate is believed to be the oxidation product of some sulphide minerals by the radiolytically generated oxygen and some other oxidising radical species. The formation of the ferric iron (the observed hematised clay zone) and the possible release of dissolved uranium are a few orders of magnitude lower than the release of sulphate and can not be the potential sink of the radiolytically generated oxidants. The preferential oxidation of sulphide helps to stabilise the uranium ore body itself.

The steady-state release rate of sulphate and dissolved hydrogen is the net generation rate of oxidants and reductants by water radiolysis. In this respect, the near-field release model provides a means to estimate the net radiolytic oxidant and reductant production rate. The study of the water radiolysis from the viewpoint of mass transport has not yet been found to be reported in the literature.

The water radiolysis is also addressed from a more fundamental angle of approach (Liu and Neretnieks, 1994). A generic model for water radiolysis is developed. The model concepts are derived from a fundamental level, i. e. the radiation energy deposition in the pore water as well as in the other constituents of the uranium ore body. A maximum possible oxidant production rate can be calculated. When the estimated recombination of the oxidants and the reductants in the system (Christensen and Bjergbakke, 1982) are also accounted for, this calculated oxidant generation rate agrees again with that predicted by the near-field release model. If these are not all just fortuitous, our understanding of the long-term characteristics of mass transport and water radiolysis undergoing in a final repository should be greatly improved.

In this paper, the similarities of the Cigar Lake deposit and a final repository are outlined. The near-field release model(s) for the final repository are reviewed. The near-field release model tailored specifically for the Cigar Lake deposit site is then briefly presented. The concepts, basic assumptions and mathematical representations of the two models are compared where appropriate. Efforts are made to demonstrate that the two models are conceptually essentially the same. The agreement between the model results and the observations in the Cigar Lake deposit does enhance our confidence of the prediction by the near-field release models developed for the final repository.

FEATURES OF A FINAL REPOSITORY AND OF THE CIGAR LAKE DEPOSIT, THEIR RESEMBLANCE AND CONTRASTS

Features of the final repository

The Swedish concept of final disposal of spent nuclear fuel can be briefly described as follows: The fuel, after being decommissioned from the reactor, will first be stored in a temporary storage depot for about 40 years. During these 40 years, much of the residual heat will have been emitted and some of the short-lived fission products will have decayed. The fuel is then encapsulated in metal canisters and buried in a vault excavated in the underground crystalline bedrock. The depth of the vault will be about 500 m. The gap between the canister and the inner wall will be filled with bentonite. The remaining part of the vault and the horizontal shaft along which the vaults are excavated will then be backfilled with a sand/bentonite mixture (KBS-3, 1983). The same repository design principle is also reiterated in a most recently revised disposal concept (SKB 91). The design principles are intended to fulfil the legislative requirement of the Stipulations Act, which is subsequently superseded by the Act on Nuclear Activities, passed by the Swedish Parliament (Riksdagen).

The fuel consists of cylindrical pellets of uranium dioxide enclosed in zirconium ally cladding tubes. This forms the fuel rod. The rods are bound together in fuel assemblies, which are handled as units. The fuel bundles are then encapsulated into copper canisters. The canister will have the dimension of 0.8 m in diameter and 4.5 m long, but can vary depending upon the burnup of the fuel and the reactor types (BWR and PWR in Sweden). The wall thickness of the canister is about 100 mm in KBS-3 and is reduced to 60 mm in SKB 91.

Radioactive elements which have been created in the spent fuel are mainly as follows:

- fission products, such as ^{137}Cs, ^{90}Sr, produced by fission of uranium and generated plutonium;
- isotopes of transuranium elements, such as Np, Pu, Am and Cm, produced by successive neutron captures starting from uranium;
- activation products, like ^{60}Co and ^{59}Ni, produced by neutron capture in the cladding tubes and other parts of the fuel.

The copper canister is supposed to serve as protection during handling in the repository and, its most important function, as a barrier to release of radionuclides from the nuclear fuel. The canister can preserve its integrity for a very long time. Canister failure caused by mechanical stress introduced during the fabrication is well below the ultimate strength of copper (KBS-3). External force exerted due to the swelling of the buffer material and the hydraulic pressure scarcely leads to higher mechanical stresses on the canister. The internal pressure caused by the alpha decay helium will be approximately the same as the external pressure after 10^6 years and will not give rise to stresses that exceed the initial canister stress.

Free oxygen and sulphide are the only possible substances at the disposal environment that can cause copper corrosion. The geochemical environment of the disposal site is reducing. The oxygen that can be entrapped in the pores of the buffer material after the repository is sealed will mostly react with ferrous iron in the buffer material and in the rock. Only a few moles of O_2 will reach the copper canister. This is equivalent to about 0.5 kg of copper to be oxidised (the Swedish Corrosion Research Institute, 1983). Corrosion by oxygen generated by gamma radiolysis is also of relatively limited effect. Two sources of dissolved sulphide are conceivable, one is from direct supply from the inward diffusing groundwater and the other from microbiological reduction of sulphate, but both are of limited quantity. It has been concluded that the canister life for a wall thickness of 60 mm could be over 10^6 years. The more recent conclusion (SKB 91) is that no corrosion penetrations will take place on the canister at a pitting factor (the ratio between pit depth and average material loss) of 2 before more than 100 million years have passed.

The buffer material in the disposal vaults surrounding the canisters is bentonite clay. Its function is to constitute a mechanical and chemical zone of protection around the canister and to limit the inward transport of corrosive substances from the groundwater to the canister surface, and in a later stage, to limit the outward transport of leached radioactive substances from the canister. The backfill in the tunnels and shafts is to preserve the mechanical stability of the excavated space, and to restore the hydrological conditions in the area (KBS-3).

Bentonite is a natural clay formed from volcanic ashes. It is rich in swelling clay minerals, such as smectites, of which montmorillonite is a common variety. The chemical stability of bentonite will exceed more than one million years, provided that the temperature does not exceed 100^oC (Anderson, 1983; Pusch, 1983). Bentonite clay is a good pH buffer. Its pH is about 8-9 in the water-saturated condition. Bentonite has also a good ion exchange capacity, so that any positively charged radionuclides that escape from the canister are retarded.

Bentonite of high density also swells greatly when it absorbs water. The swelling property leads to the self-sealing of most of the pore spaces in the engineered barrier, and prevents water-bearing species from passing through the barrier. Its swelling pressure will make bentonite to penetrate into and seal any fractures that may exist in the walls of the vault.

The density of the compacted bentonite is 2 100 to 2 200 kg m^{-3}. The hydraulic conductivity at this density range is on the order of $5{\cdot}10^{-14}$ m s^{-1}, which means that the material is virtually impenetrable by water after full water saturation. Mass transport through the bentonite is dominated by diffusion. The diffusivity varies with the density of the bentonite slightly, but it is species-specific, i.e., different species have different apparent diffusivities, which is determined by the charge if the species is an ion, the size and the sorbing property of the species. The effective diffusivity is on the order of 10^{-12} m^2 s^{-1}.

The hydraulic conductivity of the bentonite-sand mixture backfill amounts to a maximum of 10^{-9} m s^{-1}.

The final disposal site for Swedish spent nuclear fuel is proposed to be located in some crystalline bedrock at a depth of about 500 m from the surface (SKB 91). Highly fractured zones should be avoided in site selection. From the mass transport point of view, the water conducting channels and fractures, and the micropores within the bedrock are of great importance.

The Finnsjön area, situated in northern Uppland County in Sweden, has been chosen as a calculation example for the safety assessment of the final repository. The bedrock consists of a medium-grain granodiorite (SKB 91). Fracture frequency measured on the outcrops spread out all over the Finnsjön block is 2.9 fractures per metre. At 100 - 200 m depth, the geometric mean value of hydraulic conductivity is 10^{-8} m s^{-1}, at 500 - 600 m depth, it is 10^{-9} m s^{-1}.

Groundwater chemistry in Swedish crystalline bedrocks can be described as follows. Undisturbed groundwaters will almost always contain considerable amounts of carbonate, even in a bedrock without any carbonate minerals at all (KBS-3). The coupling cation is usually Ca^{2+}. A relatively stable pH in the interval 7 - 9 is obtained. Most values of the total carbonate fall within the range 90 – 275 mg l^{-1}. In crystalline granitic or granodioritic rock, the presence of Fe(II) and Fe(III) minerals will determine the redox potential. The groundwater is usually buffered to be reducing by the presence of ferrous iron minerals like pyrite.

Features of the Cigar Lake uranium ore deposit

The Cigar Lake uranium ore body (Cramer 1986b) lies at a depth of about 430 m under the ground surface and occurs as an irregular shaped east-west trending lens (about 2000 m long by 25 to 100 m wide by 1 to 20 m high) inside a 5- to 30-m-thick clay-rich halo in the sandstone. Later studies showed that the thickness of the clay is 3 m on average. Of the 2000 m length of the ore body, about 800 m make up the most prominent part of the ore deposit. The uranium mineralisation is located at the intersection of the unconformity contact of the Athabasca Sandstone Formation and the basement rock. The mechanisms of the ore formation were postulated as hydrothermal. About 1.3 thousand million years ago, reducing hydrothermal fluids were discharged vertically upwards from the basement into the sandstone along the fracture conduits, and mixed with the uranium-bearing oxidising fluid flowing essentially horizontally.

The composition of the uranium ore is relatively simple, containing a high concentration of uranium and a little of other actinides and lanthanides. The uranium minerals are mainly uraninite (UO_2) and coffinite ($USiO_4$). The radionuclides present are thus mostly from the uranium-decay series. Pitchblende, a different structural form of UO_2, also occurs abundantly and is intergrown with the uraninite. The uraninite and pitchblende in the porous part of the ore zone show evidence of surface oxidation in the form of a layer of micro-crystalline U_4O_9 and U_3O_7, which are well rooted into the UO_2 matrix (Sunder *et al.*, 1992). No evidence has been found for the presence of higher uranium oxides, associated with the uraninite and pitchblende, in other parts of the ore zone. Cobalt, nickel, lead and minor amounts of other elements occur in low concentrations in the associated sulphides and arseno-sulphides.

The clay-rich halo consists mainly of illite, kaolinite and quartz. About 1 m of the clay halo, which is 3 m thick on average, has been oxidised to hematised clay. In the massive bleached clay, reducing iron minerals like pyrite and siderite are also identified. In the hematised clay, oxidised iron minerals like hematite and ferric oxyhydroxides are found (Smellie *et al.*, 1991). In general, the clay samples have less SiO_2 but more Al_2O_3, Fe_2O_3 and K_2O; minor increases in TiO_2 are also common. All the trace elements, including U and Th, show increased contents in association with the clay. These chemical trends reflect the variable degree of hydrothermal alteration of the sandstone, which has resulted in the dissolution and removal of quartz accompanied by the formation of clay.

The clay-halo is the least permeable medium of mass transport in the three major lithological and hydrothermal subdivisions in the near-field of the Cigar Lake deposit (the other two are the massive ore body and the overlying sandstone). Its hydraulic conductivity is in the range of 10^{-8} to 10^{-9} m s^{-1} (Smellie *et al.*, 1991). Mass transport through the clay-halo is expected to be dominated by diffusion.

Hydrogeological test and other observations showed that the fractured sandstone above the ore and clay is the main aquifer. The hydraulic conductivity in the altered sandstone is on the order of 10^{-5} m s^{-1} (Smellie *et al.*, 1991). The sandstone immediately out the clay-halo has been bleached grey-white by the removal of hematite during the hydrothermal formation of the ore deposit. Hematite is present in the brick-red unaltered sandstone (Cramer, 1986b). Fractures in the sandstone above the ore body were identified by Cogema, Canada Ltd. (Lavoie *et al.*, 1986). The primary objective of that work was to facilitate mining activity rather than for analogue study. Parameters like fracturing density, fracture aperture, etc., however, are not quantitatively characterised.

Groundwater samples from boreholes throughout the deposit are characterised by their low ionic strength, neutral pH, reducing electrochemistry and low uranium concentrations (Cramer and Nesbitt, 1993). Uranium concentrations from the ore and clay zones vary between 10^{-9} to 10^{-7} mol l^{-1}. The Eh range estimated from the electrochemical couples of CO_2/CH_4, Fe_2O_3/Fe^{2+} and SO_4^{2-}/HS^- is about -0.2 to -0.25 V. Later *in situ* measurements by Au electrode show an Eh range between about -0.3 and 0.3 V.

The U_3O_7 phase appears to be the one mineral that controls the thermodynamic solubilities of uranium in most of the samples. The up-gradient waters in the sandstone to the south of the ore deposit (groundwater flows horizontally from southwest to northeast around the ore body) are undersaturated with respect to this phase. Saturation with respect to U_3O_7 is reached in the ore zone, and the waters appear to be equilibrated with this phase in the sandstones located down-gradient to the north of the deposit (Bruno and Casas, 1993).

In addition to the features described above, another very important feature of the Cigar Lake deposit is that, at the ground surface, there is no direct evidence of the underlying uranium ore body, either radiological, thermal, geophysical or geochemical.

The Cigar Lake deposit as an analogue site of the final repository

In the uranium ore body at Cigar Lake there is a striking similarity between the ore body and a repository. Consequently, it has many features which are valuable for natural analogue studies of the final disposal of spent nuclear fuel. When the features of near-field of the final repository is compared with that of the Cigar Lake uranium deposit, the following points should be noted:

(1) The spent nuclear fuel consists mainly of uranium dioxide, with less than 4% by weight of fission products and other impurities (for BWR with burn-up of 28 000 MWd/tU) (KBS-3, 1983). The main mineral in the Cigar Lake uranium ore is uraninite, with a chemical composition very similar to that of the

spent fuel. The model for dissolution of spent fuel assumes that the stable solid phase for uranium is UO_2. Information from Cigar Lake indicates that UO_2 has persisted as uraninite for $1.3 \cdot 10^9$ years. No neutron-induced fission reactions have occurred to any significant extent at the Cigar Lake deposit. Some radionuclides, however, are generated by *in situ* neutron-capture reactions (Fabryka-Martin, 1993).

(2) Metal canisters and engineered buffering materials will be used in the final repository as barriers to impede radionuclide release. In the Cigar Lake deposit, the ore body is surrounded on top and sides by a clay halo. The clay halo has many properties similar to that of the engineered barrier. Both the illite-rich clay in the deposit and the bentonite as engineered barrier in the repository resist water flow and sorb some dissolved species in the groundwater. There is no equivalent in the Cigar Lake deposit to the canisters in the repository.

(3) Atmospheric oxidants are unlikely to penetrate to the dense bedrock deep under the ground surface. The only potential source of oxidants is radiolysis. In the Cigar Lake uranium deposit, in the 3-m-thick clay halo surrounding the ore body, about 1 m of the clay adjacent to the ore has been hematized with pyrite being oxidised to hematite or iron oxyhydroxide (Smellie et al., 1991). The oxidants might be of radiolytic origin. This case of hematization might provide an analogue for studying the impact of radiolysis on radionuclide transport in the final repository. We are not for sure if such hematisation could occur in the bentonite clay in a final repository, even though the bentonite clay may also contain some ferrous iron minerals. The fate and track of oxidants and hydrogen produced by radiolysis in the deposit can be investigated by mass transport modelling.

(4) The altered sandstone overlying the ore and clay in the Cigar Lake site is much more fractured and permeable than the relatively dense crystalline bedrocks for a final repository. In this perspective, the Cigar Lake site is a worse-case compared to the final disposal site.

(5) Redox conditions in the near-field of the deposit are reducing. The dissolution of uraninite and some other minerals is expected to be solubility limited. The ranges of the pH and Eh values in the groundwater in the Cigar Lake site is about the same as those in the final repository site. Other dissolved species in the groundwaters are also comparable.

NEAR-FIELD RELEASE MODEL DEVELOPED FOR THE REPOSITORY

In order to meet the high demands on safety, the multiple barrier principle is applied in the final repository. For any release of radionuclides from the fuel to be possible, groundwater in the near- and far-field has first to intrude into the repository, i.e. seep inwards through the micropores of the bedrock, or flow through the fractures in the bedrock. It then comes into contact with the engineered buffer material, which, when gets into contact with water, tends to swell and fill up all the cavities left in the material. The hydraulic conductivity of the buffer material will then be at least as low as that of the surrounding rock mass. No water flow of significance can take place in the impervious buffer material. Transport of the dissolved species is dominated completely by diffusion in the engineered barrier (KBS-3).

When the groundwater has diffused inwards through the buffer material, the copper canister then functions as a further barrier. The corrosion rate of the copper canister is quite slow under the environment of the expected groundwater compositions. Only after the canister has been penetrated, the release of radionuclides in the fuel can possibly start. Near-field mass transport models usually do not address the inward flow and diffusion of the groundwater to the spent fuel, but rather assume that, at a certain time (100 000 years in KBS-3), the canisters have been penetrated, or, in the most recently revised final disposal concept SKB 91, the canisters have a certain statistic probability of fabrication fault (it is pessimistically assumed to be 1/1000). The mass transport model starts from when the canister has been penetrated.

The first barrier considered is the spent fuel itself. The release of radionuclides from irradiated fuel in contact with groundwater is the result of two mechanisms (Jonhson *et al.*, 1987):

- release of radionuclides in the gap in the fuel rods, and from grain boundaries in the uranium oxide fuel;
- release of radionuclides due to dissolution or conversion of the uranium oxide matrix.

The former includes the release of some fission products like ^{137}Cs, ^{99}Tc and ^{129}I and some other volatile species. This release is relatively rapid and is of less importance for the long-term performance.

The Fission products and actinides that are released due to matrix dissolution constitute the predominant portion of the fuel's radioactivity (SKB 91). Nuclides that lie embedded in the fuel grains are protected against direct dissolution in groundwater. They can only be released if the fuel matrix is dissolved or converted. In an environment in which UO_2 is oxidised to U_4O_9 or U_3O_7, the quantity of dissolved uranium will be limiting because oxidation to these oxides does not lead to break-up of the crystal lattice.

If the oxidation proceeds further, e.g., to U_3O_8, UO_3 or to some other U(VI) compound, the radionuclides can be released as matrix conversion proceeds, despite the fact that the uranium concentration in solution may still be low (Shoesmith and Sunder, 1991).

Under reducing conditions, i.e. redox potential below about -100 mV, UO_2 is the stable solid phase. Under mildly reducing conditions, one of the phases UO_2, U_4O_9 or U_3O_7 will therefore be stable. Under these conditions, the release of radionuclides will be limited by the dissolution of uranium, since these uranium oxides have the same structure as UO_2 and no re-arrangement of the crystal lattice takes place.

Under reducing conditions with a slow process of mass transfer, the UO_2 dissolution is chemical rather than electrochemical. It can readily achieve equilibrium under disposal conditions (Johnson *et al.*, 1987). At equilibrium conditions, the detailed reaction kinetics can be ignored, because the forward and backward reaction rates cancel each other and give no net contributions to the time variation of the concentrations of the dissolved species. The net rate of dissolution will then not be limited by the rate of chemical reaction, but by the solubility of the solid constituents and the processes of mass transport. This is usually referred to as *solubility-limited dissolution*. Most of the near-field release models for the final repository take the above argument into consideration and make the very basic assumption that the dissolution is solubility-limited. In the release model we developed for the Cigar Lake project (Liu *et al.*, 1994), this is also of central position of the various basic assumptions.

The highly-compacted bentonite clay that surrounds the canister possesses very low permeability to water. The flow of water will therefore be extremely slow and diffusion will be by far the dominant mechanism for the transport of the dissolved substances through the clay.

Most of the nuclides in the near-field exist as cations dissolved in the groundwater. These cations have a strong tendency to be sorbed on the surface of the clay particles. A nuclide that starts diffusing into the clay will migrate very slowly at the beginning, since most of them are sorbed onto the clay surfaces. But when sorption equilibrium has been reached, the nuclides then can migrate further in the pore water. The stage before the sorption equilibrium is established is considered as unsteady-state or transient state. After the equilibrium, steady-state will prevail. In the models for the final repository, both unsteady- and steady-state are considered. In our model specifically devised for the Cigar Lake natural analogue study (Liu *et al.*, 1994), the steady-state is assumed for the reason that the ore has been preserved in a relatively closed system for a geologically long time. Sorption is consequently not considered in the model for the Cigar Lake site.

In steady-state mass transport models for the final repository (Neretnieks, 1978; Andersson *et al.*, 1982), both the diffusion in the clay barrier and the diffusion in the water in water-bearing fractures have been taken into consideration. The diffusion into the rock matrix around the hole is negligible, except during the very first stage. The transport into the clay is described as a three-dimensional diffusion transport from the surface of the fuel to the fracture openings in the rock. The transport into the flowing water is described as diffusion in flowing water between two plane parallel fracture walls. The water flow in the rock is determined to a large extent by regional flow conditions. The transport is affected by the hydraulic conductivity, the fracture frequency of the surrounding bedrock.

The mass transport is represented mathematically by Eq. (1) (KBS-3):

$$N_i = \frac{1}{R}(C_{oi} - 0) \tag{1}$$

where N_i is the quantity of nuclide i that is transported per unit time at a concentration difference of C_{0i} – 0. The parameter R is equal to the sum of the transport resistance in clay and in fractures. All the geometric complexities, the species' diffusivity, and material properties are incorporated into R which can be built up of very complicated functions of the above parameters in certain cases.

Eq. (1) can also be interpreted in the following manner: $1/R$ is the equivalent water flow, Q_{eq} that arrives at the canister with the concentration 0 and leaves it with the concentration C_{0i}, i.e.

$$N_i = Q_{eq} * C_{oi} \tag{2}$$

The near-field steady-state mass transport models for the final repository are conceptually illustrated in Figure 1.

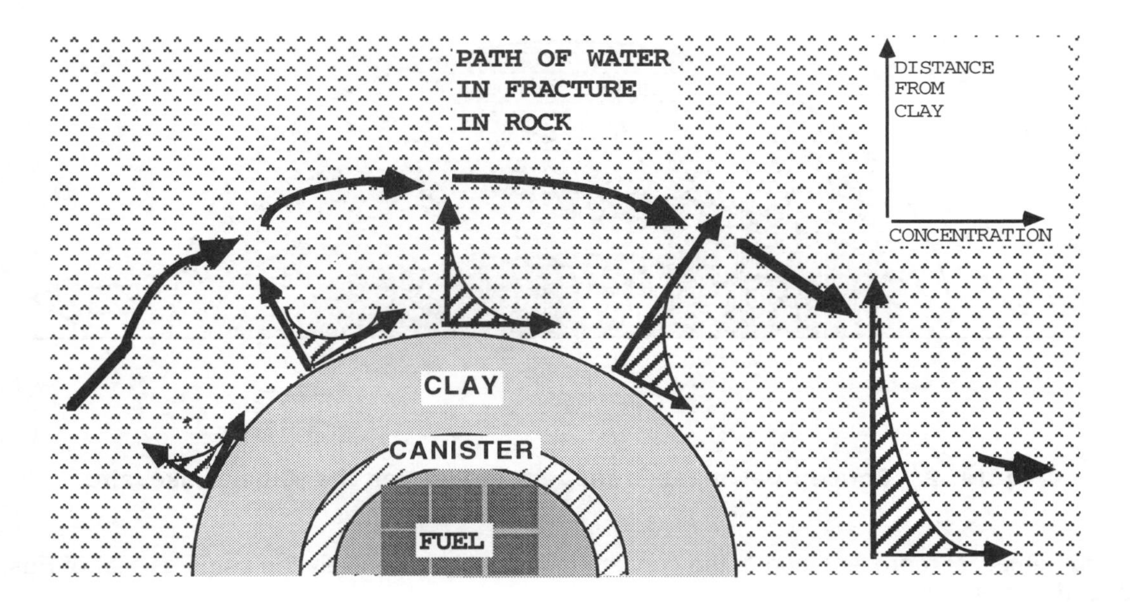

Figure 1. Conceptual illustration of the near-field steady-state mass transport model for the final repository.

The near-field steady-state mass transport model is adapted to cope with the specific geometrical features and material properties of the geological formations in the near-filed of the Cigar Lake deposit, but the concepts of the model are not different from those of the existing near-field release models for a final repository (Liu, 1993; Liu *et al.*, 1994). The model is developed primarily based on the hydrological, lithological and geochemical features of the Cigar Lake uranium deposit. The near-field mass transport media consist of the uranium mineralisation, the covering clay-rich halo, and part of the overlying altered sandstone. The various media are assumed to be homogeneous porous media. Representative central values of mass transport parameters, like porosity, pore diffusivity, are assigned to each of these rock/ore matrices.

The Cigar Lake uranium mineralisation is a naturally occurring, well-preserved system. It is justified to assume a steady-state mass transport in the near-field of this relatively closed system.

The geometry of the ore body is simplified as a regular lens 800 m long, 50 m wide and 10 m high, which is covered everywhere by a clay-rich halo 3 m thick. Outside and above lies the altered sandstone extending to "infinity".

The model is conceptualised as follows (Figure 2) (Garisto, *et al.*, 1991). The groundwater in the ore body is relatively stagnant and a constant aqueous concentration of the transported species, C_0, is assigned in the ore body. The clay-rich halo is the least permeable medium (Winberg and Stevenson, 1993) and groundwater flow in it is neglected. The sole mechanism of release through the clay will be molecular diffusion. In the altered sandstone, groundwater flows in the direction perpendicular to the ore body axis and parallel to the contour of the clay-sandstone interface, and mass transport is therefore of both diffusion in the direction perpendicular to the clay-sandstone interface and advection in the direction of groundwater flow. A constant concentration C_i is assigned to represent the already-existing aqueous species concentration in the approaching groundwater.

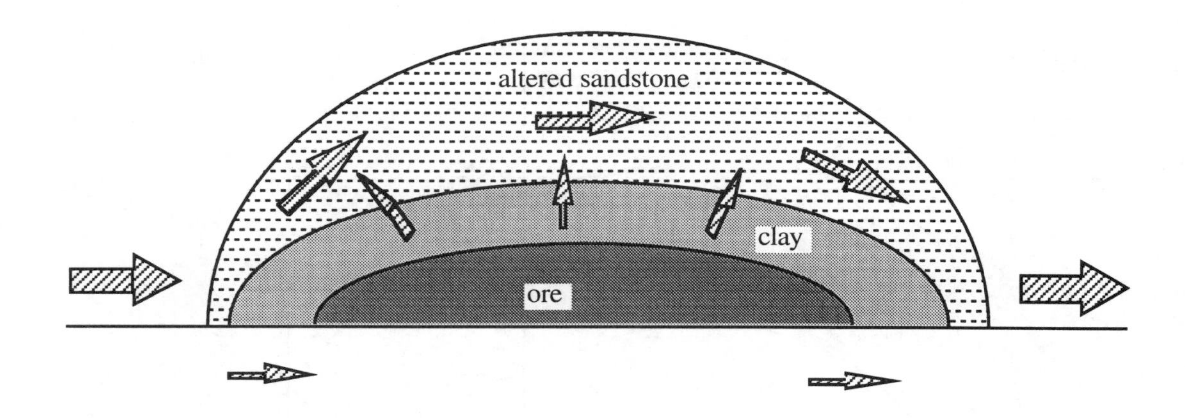

Figure 2. Conceptual mass transport model for the Cigar Lake site.

The mathematical representation of the conceptual model is essentially the same as that of Eqs. (1) and (2) (Liu *et al.*, 1994).

When the transport models for the repository and that for the Cigar Lake site are compared with each other, a striking similarity exists. This is mainly because the Cigar Lake deposit has a similar geometrical configuration as a final repository, and local groundwater flow pattern (the groundwater passing through a repository or the ore and clay zones are made converged with streamlines that follow the out contour of the repository canister or the clay-halo). Major differences in the two models are that, in the Cigar Lake case, porous media are assumed for the rock matrices. In the final repository case, the water flow in the bedrock is mainly in fractures and species also diffuse into the micropores in the rock between fractures.

HELIUM RELEASE RATE AND THE VALIDITY OF THE MODEL

Helium proves to be an excellent species for testing the validity of the mass transport models. Helium is continuously generated by spontaneous decay of the uranium and thorium series and other nuclear reactions. It is geochemically inert. The generation rate can be calculated from the knowledge of nuclear reactions, which is completely independent of the release model. On the other hand, helium release rate can be calculated by the release model, because there exists a distinct concentration gradient of helium from the ore body to the outside. At steady-state, the release rate should be equal to the generation rate if the model is reasonably valid.

The results (Table 1) of the two different approaches agree surprisingly well if the uncertainties in estimating the representative values of the various mass transport parameters are considered.

Table 1. Helium release rate calculated by two independent approaches

	Release rate (mol a^{-1}) from the whole ore body
Calculated from known helium generation	0.46
Calculated theoretically by the near-field release model	0.14 –1.37

TRACK AND FATE OF RADIOLYTICALLY GENERATED OXIDANTS

Uranium minerals are much more soluble when oxidised from tetravalent to hexavalent forms. Atmospheric oxygen is unlikely to penetrate to the deep bedrock formation. A possible source of oxidants is the radiolysis of water in the ore body. Despite the effect of radiolysis the uranium ore has still been stabilised ever since its hydrothermal formation. It is worthwhile exploring, with the steady-state release model, the track and fate of the radiolytically generated oxidants and reductants.

The radiolytically generated reductants are primarily hydrogen molecules. The oxidants can, however, be oxygen molecules, hydrogen peroxide and some oxidising radicals. Dissolved hydrogen is more readily mobile and chemically less reactive; most of the radiolytically generated hydrogen will be expected to be released out of the ore body. According to mass conservation, the same amount of oxidant (in terms of equivalent) is generated in the system and may oxidise some reducing constituents (possibly tetravalent uranium, ferrous iron and sulphides). When oxidised, these species will also be released out of the system if no effective fixation mechanisms exist.

We will attempt to account for the fate of the oxidising and the reducing species. Hydrogen is assumed to be the only reducing species generated by the radiolysis and to leave the system by diffusion. The rate of escape and thus also the rate of formation by radiolysis can then be estimated. The fate of the oxidising species is more complex. There is a very small concentration of oxygen. This will diffuse outward. It is assumed to react with ferrous iron in the clay at a redox front about 1 m into the clay. In addition we account for the transport of the oxidising equivalent which is transported by hexavalent uranium. All escaping uranium is assumed to be in the hexavalent state. This assumption is not well founded but is of practically no consequence as will be seen. The transport of oxidising equivalents by uranium is negligible compared to other means. The only means of substantial loss of oxidising equivalents we have found is by sulphate, which then must be formed by the oxidation of sulphides.

A distinct concentration gradient from within to outside the ore body exists for both the dissolved hydrogen and sulphate. The steady-state release model can thus be applied to calculate their release rate. The release rate is shown in Table 2. It indicates that sulphate and dissolved hydrogen are released at a comparable rate. Total uranium release and oxidant consumption rate by oxidation of reducing iron account for considerably lower values of release rate. This could imply that the sulphides are a potential sink of the radiolytic oxidants. Sulphides are oxidised to sulphates and a concentration gradient from the ore to the outside is built up. The oxidised sulphates are then released from the ore body at a rate that balances the release of hydrogen. The preferential oxidation of sulphides stabilises the tetravalent uranium. This may offer one possible explanation of the stabilisation of the ore body over such a long geological time.

Table 2. Release rate in equivalents per second per cubic metre of ore.

Species	Release rate, N (eqv s^{-1} m^{-3})
Sulphate	$1.72 \cdot 10^{-12}$
Dissolved H_2	$2.34 \cdot 10^{-12}$
Total uranium	$4.74 \cdot 10^{-17}$
Oxidant consumption rate by oxidation of reducing iron	$3.90 \cdot 10^{-15}$

For the sulphate to be produced there is a need of sulphides. There is no known concentration gradient of dissolved sulphide in the water which may lead to transport into the system. There are, however, sulphide minerals present in the ore which may be the source of the sulphides. If the process of generation and transport should have gone on since the ore was formed, there must have been very large amounts of sulphides there initially. The amount of sulphides needed initially is of the same order of magnitude as the ore. This does not seem reasonable because remnants of the sulphide ores in the form of metal oxides would be expected to be present. These are not found in sufficiently large quantities. The source of the sulphides is not yet sufficiently understood.

Table 3. The production rate of oxidant by alpha, beta and gamma radiations

R	(*10^{12} Eqv s^{-1} m^{-3} of ore)
R_{α}	86.1 – 162.7
$R_{\beta,\gamma}$	27.4 – 62.8
Total	113.5 – 225.5

The water radiolysis has also been addressed from a more fundamental level of approach, i.e., the radiation energy deposition in the pore water as well as in the other constituents of the uranium ore body. The geometric dispersion of the various constituents of the ore body is assumed based on field observations. Nuclear radiations are randomly generated from crystals of uranium minerals in the ore body, by Monte Carlo method. The radiation energy is then allowed to deposit into the various constituents of the ore body, according to the interactions of the radiations with matter. The fraction of the total radiation energy absorbed by water is obtained from the model and the extent of water radiolysis can be calculated if the G-value (molecules of water reacted divided by radiation energy absorbed) of water is also known. In this case the calculated maximum possible oxidant production rate is about 100 times larger than the present-day ongoing oxidant production rate. The discrepancy is believed to be due to recombination of radiolytically generated oxidants and reductants. The recombination factor agrees with what other researchers have been projected by other means (Christensen and Bjergbakke, 1982). The calculated maximum oxidant production rate by alpha, beta and gamma radiations are shown in Table 3.

MODELLING OF SHORT HALF-LIFE RADIONUCLIDE RELEASE

In the ore body, neutrons are produced by (α, n) reactions when alpha-particles emitted in spontaneous decay attack light elements with large reaction cross-sections. Neutron-induced fission has not occurred to any significant extent. Some radionuclides are, however, produced by *in situ* neutron-capture nuclear reactions (Fabryka-Martin, 1993). These radionuclides will be released out of the ore body if they do not decay fast enough. The basic concepts of the steady-state release model can be extended to model the mass transport and release of these short-half life radionuclides.

To model the release of short half-life radionuclides, in addition to the diffusive and advective mechanisms of mass transfer, the generation of the nuclides in the ore body, and the decay in the ore body, the clay and the altered sandstone, must be accounted for. The resulting mass conversation equations are solved numerically by the program TRUCHN, an extension of program TRUMP (Edwards, 1972).

The model is based on the following concepts (Liu *et al.,* 1994). The radionuclides are generated homogeneously in the ore. They diffuse outward through the clay surrounding the ore. Decay takes place during the transport. After diffusing through the clay the nuclides diffuse further into the flowing water in the sandstone and are swept away. The assumption for both a homogeneous porous medium of ore body and a fractured ore body are approached. The results are shown in Table 4.

Table 4. Radionuclide release percentage (percent of generated)

	^{3}H	^{14}C	^{36}Cl
Homogeneous ore body	1.5	91.0	97.3
Fractured ore body	1.0	90.0	99.8

The results show that only those nuclides that decay fast enough will decay within the ore and clay zones. Others are mostly released out. Due to the special groundwater flow patterns in the near-field.

The plumes of the released species are confined to a very narrow downstream layer (see Figure 3) and their detection may be very difficult.

Table 5 shows both the calculated and measured concentrations of the three radionuclides. No measured value is available for ^{14}C to compare with. The calculated concentration for ^{3}H is approximately the same as the secular equilibrium concentration without release, obvious due to the reason that ^{3}H decayed almost completely within the ore body. The calculated concentration for ^{36}Cl is about two orders magnitude larger than the measured value. In the calculations, extremely conservative assumption was made that the nuclides generated in the rock and ore would escape completely from the rock to the groundwater without isotopic exchange (dilution). Should the generation in the water alone be accounted for, the calculated concentration would be in the same order of magnitude as the measured value.

Table 5. Concentrations of radionuclides

	^{3}H (TU)	^{14}C (pmC)	^{36}Cl (atoms L^{-1})
Calculated conc. in the ore body	850	350	$6.0 \cdot 10^{11}$
measured concentration	280	-	$4.0 \cdot 10^{9}$

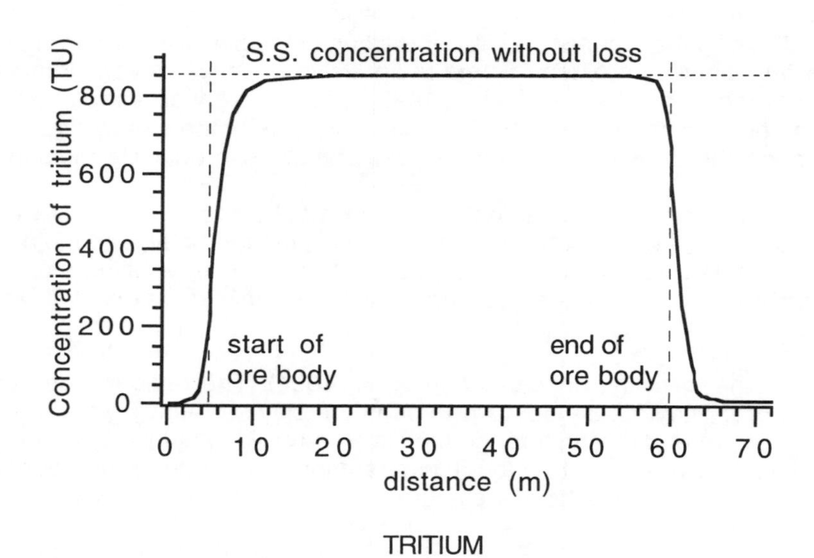

TRITIUM

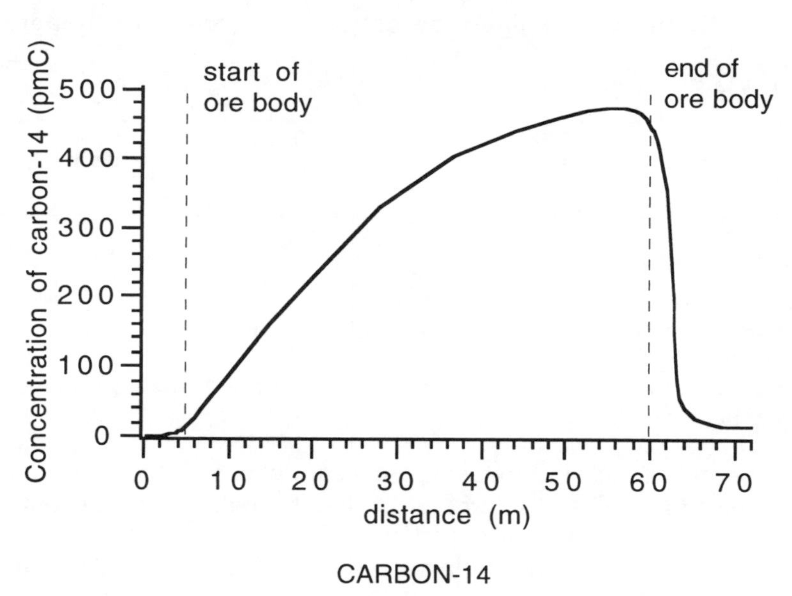

CARBON-14

Figure 3. Concentration profiles of ^{3}H and ^{14}C at the bottom of the ore body.

DISCUSSION AND CONCLUSIONS

The use of the near field model for the release of nuclides from a repository for spent fuel to model the transport of different species at the Cigar Lake uranium ore deposit has given surprisingly good agreement. The calculations for the release of helium were made without any calibration whatsoever. The release calculated based on observed helium concentration in the ore and the known production rate was well within the range of uncertainties. This encourages us that the technique also may be useful for modelling the transport of other species, notably radiolysis products. Also here the production of the oxidising species is equal to that of the reducing species if interpreted by the release rates obtained from the model. Furthermore the magnitude of the oxidant and reductant production found at Cigar Lake agrees very well with calculations we performed with a radiolysis model based on calculating the deposition of the radiation energy in the water and accounting for the partial recombination of the radiolysis products.

Having gained some confidence in the model, further calculations on the fate of ^{3}H, ^{14}C and ^{36}Cl generated by nuclear reactions in the ore indicate that most of the ^{3}H will decay inside the ore whereas the other two nuclides would escape. However, because the streamlines converge to a very narrow stream downstream of the ore body they will hardly be detectable by sampling of the ground water using packers in boreholes. With small packer distances it would be pure luck to find the specific stream containing the nuclides. With a larger packer distance the waters would be so diluted that they could not be found due to high detection limits. These nuclides were not surprisingly not detected outside the ore body.

We have gained considerable confidence in the soundness in our near field model from these investigations.

REFERENCES

Anderson D. M. (1983). Smectite alteration. Proceedings of the Colloquium at State University of New York at Buffalo. May 26-27, 1982. KBS TR 83-03. Feb. 15, 1983.

Anderson G., Rasmusson A. and Neretnieks I. (1982). Migration model for the near-field. KBS TR 82-24.

Bruno J. and Casas I. (1993). Section 3.9. Performance assessment related modelling. In: Final report of the AECL/SKB Cigar Lake analogue study. AECL-10851.

Christensen H. and Bjergbakke E. (1982). Radiolysis of groundwater from HLW stored in copper canisters. KBS Technical Report 82-02.

Cramer J. J. (1986a). Sandstone-hosted uranium deposits in northern Saskatchewan as natural analogs to nuclear fuel waste disposal vaults. *Chemical Geology*. **55**, 269.

Cramer J. J. (1986b). A natural analogue for a fuel waste disposal vaults. Proc. 2nd Int. Conf. Radioactive Waste Management, Canada Nuclear Society. Winnipeg, Canada. Sept. 1986. p.697.

Cramer J. J. and Nesbitt H. W. (1993). Section 3.5. Hydrogeochemistry. In: Final report of the AECL/SKB Cigar Lake analogue study. AECL-10851.

Cramer J. J. and Smellie J. A. T. ed. (1993). Final report of the AECL/SKB Cigar Lake analogue study. AECL-10851.

Edwards, A. L. (1972). TRUMP: A computer program for transient and steady-state temperature distributions in multidimensional systems. UCRL-14754. Lawrence Livermore Laboratory, University of California. Sept. 1972.

Fabryka-Martin J. (1993). Section 3.8. Nuclear reaction product geochemistry. In: Final report of the AECL/SKB Cigar Lake analogue study. AECL-10851.

Garisto N. C., Cramer J. J., Liu J. S. and Casas I. (1991). Modelling of uranium-ore dissolution and uranium migration in the Cigar Lake deposit. In: AECL/SKB/USDOE Cigar Lake Project, Progress Report for the Period May-October 1991. CLR-91-5. Prepared by J. J. Cramer. Nov. 1991.

Johnson L. H., Shoesmith D. W. and Stroes-Gascoyne S. (1987). Spent fuel: Characterisation studies and dissolution behaviour under disposal conditions. Scientific Basis for Nuclear Waste Management XI, *Materials Research Society Symposium Proceedings*, Vol. 112, p. 99.
KBS-3 (1983). Final storage of spent nuclear fuel -- KBS-3. Swedish Nuclear Fuel supply Co. Division KBS. May, 1983.
Lavoie S., Chevalier J., St-Jean R. and Bruneton P. (1986). Cogema Canada Limited, Fracturation in the sandstone above the ore zone, Cigar Lake project, 86-CND-29-04, June, 1986.
Liu, J. (1993). Mass transport and coupled reaction/transport modellings in the near-field of the Cigar Lake uranium ore deposit. Thesis for Licentiate of Engineering. Department of Chemical Engineering, Royal Institute of Technology, Stockholm, Sweden. Feb. 1993.
Liu J., Yu J.-W. and Neretnieks I. (1994). Transport modelling and model validation in the natural analogue study of the Cigar Lake uranium deposit. *J. of Contaminant Hydrology.* In press.
Liu J. and Neretnieks I. (1994). Some evidence of radiolysis in a uranium ore body -- quantification and interpretation. XVIII International Symposium on the Scientific Basis for Nuclear Waste Management. Kyoto, Japan. Oct. 23-27, 1994. *Materials Research Society Symposium Proceedings.* In press.
Neretnieks I. (1978). Transport of oxidants and radionuclides through a clay barrier. KBS TR 79, 1978-02-20.
Pusch R. (1983). Stability of deep-sited smectite minerals in crystalline rock -- chemical aspects. KBS TR 83-16. March 1983.
Shoesmith D. W. and Sunder S. (1991). An electrochemical-based model for the dissolution of UO_2. SKB Technical Report TR 91-63. Dec. 1991.
SKB 91 (1992). Final disposal of spent nuclear fuel. Importance of the bedrock for safety. SKB Technical Report 92-20. May 1992.
Smellie J., Percival J. and Cramer J. (1991). Mineralogical and Geochemical Database for the Major Lithological and Hydrothermal Subdivisions of the Cigar Lake Uranium Deposit, CLR-91-04, November, 1991.
Sunder S., Cramer J. J. and Miller N. H. (1992). X-ray photo-electron spectroscopic study of uranium minerals from the Cigar Lake uranium deposit. Proceedings of the XV International Symposium on the Scientific Basis for Nuclear Waste Management. *Materials Research Society Symposium Proceedings.*, **257**, 449-457.
The Swedish Corrosion Research Institute and Its Reference Group (1983). Corrosion resistance of a copper canister for spent nuclear fuel. KBS TR 83-24. April 1983.
Winberg A. and Stevenson D. (1993). Section 3.4. Hydrogeological modelling. In: Final Report of the AECL/SKB Cigar Lake Analogue Study. AECL-10851. Ed. by J. J. Cramer and J. A. T. Smellie. In press.

SESSION IV

Chairmen:

T. Papp
C. Del Olmo

Influence of Microscopic Heterogeneity on Diffusion for Sedimentary Rocks

Kunio OTA and Hidekazu YOSHIDA

Power Reactor and Nuclear Fuel Development Corporation (PNC),
Tono Geoscience Center, Toki, Gifu, Japan.

ABSTRACT

To build confidence in radionuclide migration models for the performance assessment of geological disposal system of high-level radioactive waste, the limitation of applicability of a diffusion model for sedimentary rocks, has been tested by combination of field observations and relevant laboratory experiments. Comparison of the data between the field observations and an in-diffusion experiment with uranium solution have been carried out. The results of the combination work indicate that microscopic heterogeneity of the flow-path geometry and the distribution of constituent minerals should be taken into account for the further model development. Investigations on the effects of microscopic heterogeneity of rock fabrics must therefore be conducted.

1. INTRODUCTION

Natural geological media are generally considered to be heterogeneous, i.e., they contain spatial heterogeneities in physical and chemical properties on micro- to macro-scale. These heterogeneities can influence the processes of radionuclide migration in such media. The effects of macroscopic heterogeneities have been estimated by site characterisation techniques, for example, in the field of hydrogeology [1]. On the other hand, the effects of microscopic heterogeneities have been mostly tested by stochastic modelling techniques without any field data. In recent years, however, the significance of the detailed geological information has increasingly realised for the confidence in radionuclide migration models [2]. In the matrix of unfractured low-permeable sedimentary rocks, radionuclide migration can be regarded as a diffusional process. The diffusion model which are often over-simplified [3], has been applied to describe the diffusional process. The limitation of applicability of this simplified model should be confirmed.

Power Reactor and Nuclear Fuel Development Corporation (PNC) has been studying the limitation of applicability of a diffusion model for sedimentary rocks at the Tono study site. A framework of the study is the combination of three major fields; field observations, laboratory experiment, and model development. In the Tono study site, natural uranium has been transported by the groundwater, and concentrated in the rock matrix, and then uranium mineralisation has occurred in the low-permeable sedimentary rocks (Fig.1). This evidence therefore can be regarded as an analogue [4] of radionuclide migration in the sedimentary rocks. Field observations of such natural evidence with relevant laboratory experiments will provide useful information for the model testing.

The present paper describes the state-of-the-art in the study for the limitation of applicability of the diffusion model, i.e., it presents the results of the field and laboratory works, and proposes some ideas of future studies for confidence building in the diffusion model.

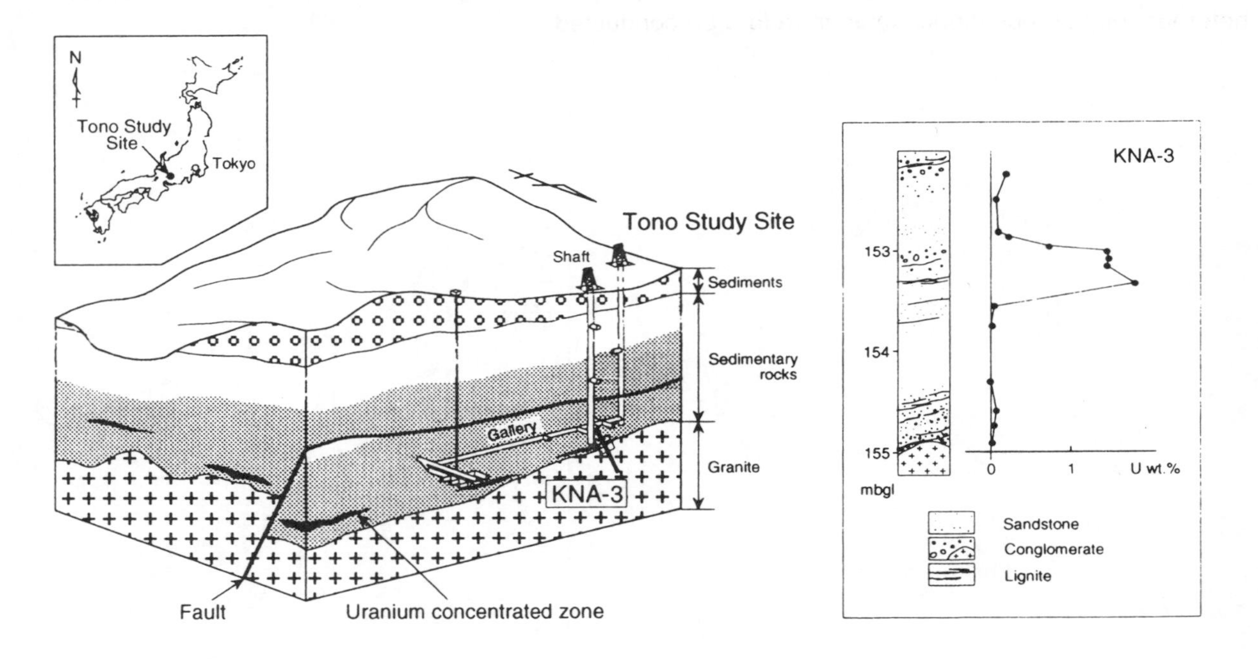

Figure 1. Schematic view of Tono study site and investigated section of KNA-3.

2. METHODS

Field observations

In the Tono study site, compacted sedimentary rocks composed mainly of sandstone and tuffaceous rocks are distributed on the basement granitic rocks. Detailed observations and relevant laboratory experiments have been carried out using the core specimens sampled from the gallery (KNA-3). The sequence of drilled section consists mainly of coarse- to fine-grained tuffaceous sandstone (Fig.1).

To characterise the pore structure and flow-path geometry influencing radionuclide migration, detailed geometrical observations with a scanning electron microscope (SEM), and a flow-path examination by a dye-impregnation method have been carried out with the rock specimens without any natural uranium concentration.

In-diffusion experiment

One-dimensional in-diffusion experiment has been conducted, to test the applicability of the simple diffusion model. Uranium solution, the rock specimens of fine- to coarse-grained tuffaceous sandstone, and an acrylic plastic cell (Fig.2) were utilised for this experiment. The uranium solution was prepared by leaching the uranium concentrated sedimentary rocks. Core specimens saturated with the groundwater in drilled section were fixed in the rock-holding cylinder, and were then dipped into the uranium solution for 14 days. Detailed experimental conditions are shown in Table 1. After the experiment, the core specimens retrieved from the cylinder was sliced with 1mm intervals, then U content in each portion was determined by conventional chemical analysis. Concentration profile of uranium in the core specimens was eventually obtained. Distribution of diffused uranium was also examined by alpha-autoradiography and electron probe microanalysis (EPMA) to find out the correlation to distribution of natural uranium.

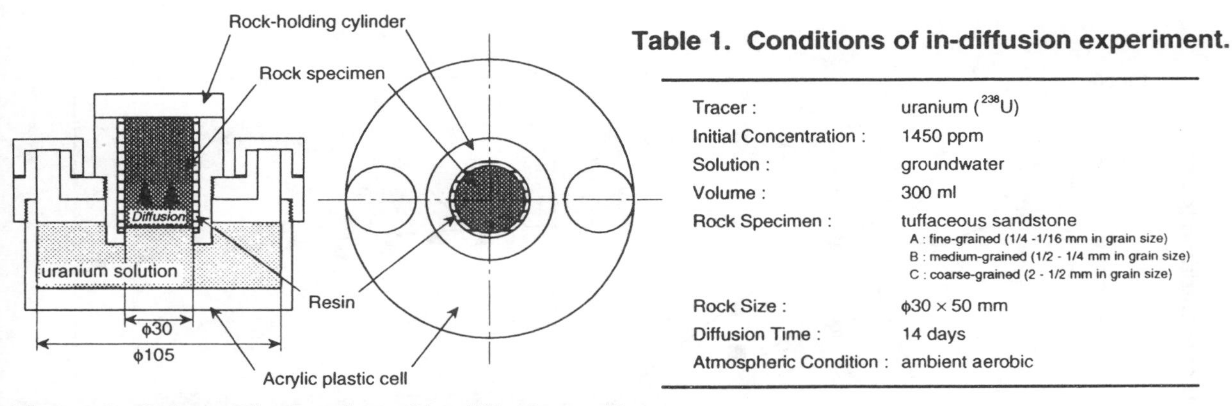

Figure 2. Schematic drawing of in-diffusion cell.

Table 1. Conditions of in-diffusion experiment.

Tracer :	uranium (^{238}U)
Initial Concentration :	1450 ppm
Solution :	groundwater
Volume :	300 ml
Rock Specimen :	tuffaceous sandstone A : fine-grained (1/4 -1/16 mm in grain size) B : medium-grained (1/2 - 1/4 mm in grain size) C : coarse-grained (2 - 1/2 mm in grain size)
Rock Size :	ϕ30 × 50 mm
Diffusion Time :	14 days
Atmospheric Condition :	ambient aerobic

One-dimensional non-steady state diffusion is expressed by the Fick's law by the following equation [6];

$$\frac{\partial C}{\partial t} = Da\frac{\partial^2 C}{\partial x^2}$$

where Da: Apparent diffusion coefficient (m^2/s)
C: Concentration of diffusing substance (ppm)
t: Diffusion time (s)
X: Distance from source (m).

When it is assumed that the substance is diffused from a constant concentration source, ideal concentration profile of the substance can be described as follows;

$$\frac{C}{C_0} = \mathrm{erfc}\left(\frac{x}{2\sqrt{Da\ t}}\right)$$

$$\mathrm{erfc}\ (Y) = 1 - \frac{2}{\sqrt{\pi}} \int_0^Y \exp(-\eta^2)d\eta$$

The value which yielded the best fit can be chosen as the apparent diffusion coefficient.

3. RESULTS AND DISCUSSION

The results of the field observations show that planar splits along (001)-cleavage in the biotite flakes, micro-fractures within quartz and feldspar grains, and pores among detrital grains are microscopically interconnected and permeable, providing further evidence that these pores are accessible to radionuclide migration by fluid movement. The size of the pores range from a few to hundreds of micro-meters, and the distribution of the pores are heterogeneous. Previous studies have revealed that natural uranium has migrated in the pores within detrital grains and in the matrix of the sedimentary rocks within a uranium concentrated zone [6]. This zone is characterised by low permeability (10^{-8} to 10^{-10} m/s) [7]. Because calculated peclet number (Pe) is less than 2, diffusion might be expected as a dominant process of radionuclide migration in the sedimentary rocks in the Tono study site.

The data of uranium concentration, obtained by the in-diffusion experiment, and the profile derived from the model calculations based upon least square method are shown in Figure 3.

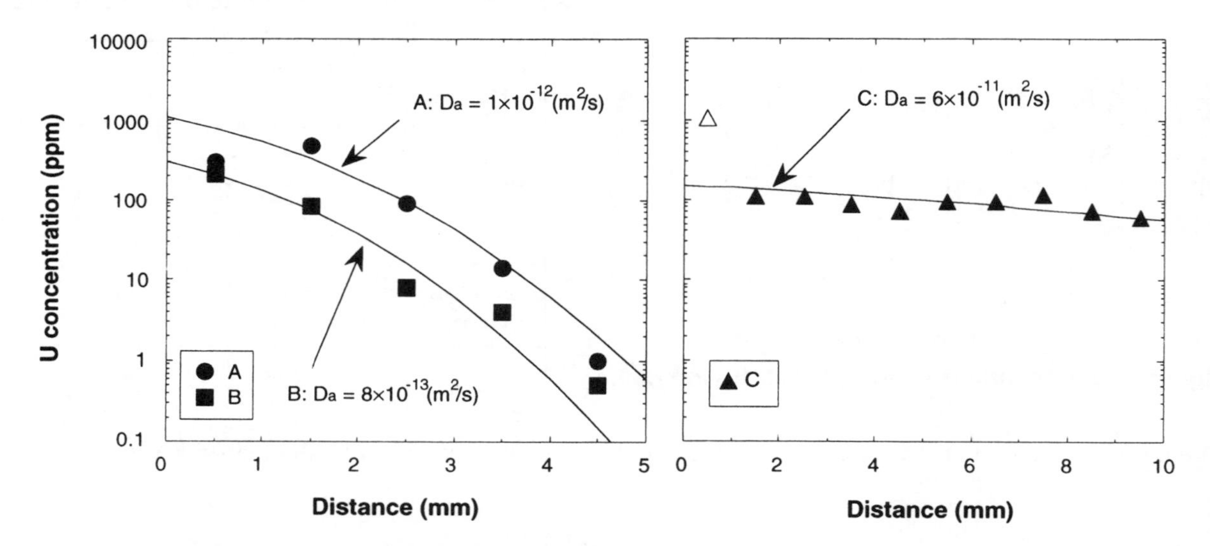

Figure 3. Concentration profile of uranium for each rock specimen.

If the calculated profile fits the measurement data, it will yield the apparent diffusion coefficient of $8x10^{-13}$ to $6x10^{-11}$ m^2/s. These apparent diffusion coefficients of uranium are almost the same values which have been already reported by Conca et al. [8] and Sato et al. [9]. The distribution of uranium is quite similar to that of natural uranium at the Tono study site, described above. Under in-situ conditions and in the in-diffusion experiment, similar (or same) diffusional processes can be expected.The simple diffusion model might therefore be applicable to the description of the diffusional processes in the sedimentary rocks. However, a certain discrepancy is recognised between the measurement data and the calculated profile.

The fluctuation of measured data are considered to be due to the following two effects; experiment itself and diffusion processes. Experimental effects will be derived from analytical error, experimental error, and wall and/or cutting effects. Maximum variation by these effects can be estimated about ±20% for each experimental value. Even if the experimental values vary within ±20% accuracy, fluctuation of the measurement data will not be significantly changed. Therefore, the data fluctuation suggests that the diffusion of uranium has been affected by microscopic heterogeneities, i.e., pore structure, pore connectivity, mineral distribution, and its sorptivity of uranium. Alpha-autoradiographs of uranium within the rock specimens show that uranium can be observed in the matrix of the sandstones, in the pores along grain boundaries and micro-fractures within the detrital grains, and also indicate that uranium has been concentrated in some minerals (Fig.4).

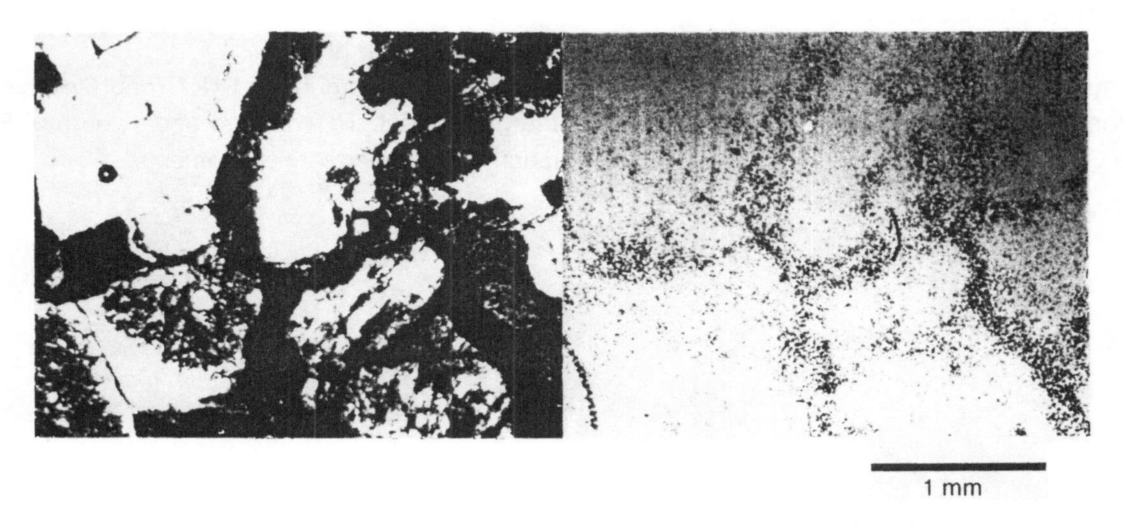

Figure 4. Photomicrograph (left) and corresponding alpha-track pattern (right) on rock specimen B after in-diffusion experiment.

For example, in the section of 1-3mm in the specimen A, uranium has sorbed onto the altered biotite flakes. Therefore, uranium is considered to diffuse into not only matrix of the compacted sandstone but also into well-connected flow-paths within the detrital grains and in the matrix, and to simultaneously interact with constituent minerals such as sorption. For uranium, since some constituent minerals in the sedimentary rocks are efficient as uranium sorbents [12], the differences of sorption capacity of the minerals along the flow-path must also be taken into account for the heterogeneous distribution of the radionuclides.

4. FUTURE WORK

For the specimen A, uranium sorption onto the altered biotite flakes in the section of 1-3mm might cause the fluctuation of the measurement data, as described above. If the measurement values of the section of 1-3mm are excluded from the fitting calculation by the simple diffusion model, the apparent diffusion coefficient will be 2×10^{-12} m^2/s. This value is different from that derived from all the experimental values. The microscopic heterogeneities must therefore be treated with the range of apparent diffusion coefficient.

One idea to estimate the range of apparent diffusion coefficient is that, if the initial concentration of diffusing substance in the rock specimen is known, the range will be expressed by a deviation of the total discrepancy between experimental profile and testing profile, from that between the experimental profile and best fit profile. The other idea is that, if the initial concentration of diffusing substance in the rock specimen is unknown, the range will be simply expressed by minimum and maximum values yielded by a number of probable fitting profiles on the variable initial concentration. However, these ideas should be confirmed by future works with detailed field information. Finally, through these examinations, the limitation of applicability of the simple diffusion model used in the performance assessment might be estimated with confidence.

ACKNOWLEDGEMENTS

The authors should like to acknowledge Dr.Y.Yusa and other colleagues of PNC Tono Geoscience Center for helpful comments and improvement of the manuscript. T.Hanamuro and T.Ando of PNC Tono Geoscience Center are also thanked for preparation on the laboratory experiments.

REFERENCES

[1] Tsang, C.F.: Solute transport in heterogeneous media: a discussion of technical issues coupling site characterisation and predictive assessment. Advances in Water Resources (1994).

[2] Miller,W., Alexander,W.R., Chapman,N.A., McKinley,I.G., Smellie,J.A.T.: Natural Analogue Studies in the Geological Disposal of Radioactive Wastes. Elsevier, Amsterdam (1994).

[3] Oreskes,N., Shrader-Frechette,K., Belitz,K.: Verification, validation, and confirmation of numerical models in the earth sciences. Science **263**, 641-646 (1994).

[4] Yoshida.H.: Relation between U-series nuclide migration and microstructural properties of sedimentary rocks. Applied Geochemistry **9**, 479-490 (1994).

[5] Torstenfelt,B., Kipatsi,H., Allard,B.: Measurements of ion mobilities in clay. Soil Science **139**, 512-516 (1985).

[6] Yoshida,H., Kodama,K., Ota,K.: Role of microscopic flow-paths on nuclide migration in sedimentary rocks. -A case study from the Tono uranium deposit, central Japan. Radiochimica Acta **66/67**, 505-511 (1994).

[7] Yanagizawa,K., Imai,H., Furuya,K., Wakamatsu,H., Umeda,K.: Groundwater flow analyses in Japan. - Part 1: Groundwater flow analyses in central Japan. PNC TN7410 92-019 (1992).

[8] Conca,J.L., Apted,M., Arthur,R.: Aqueous diffusion in repository and backfill environments. In: Scientific Basis for Nuclear Waste Management XVI (1993).

[9] Sato,H., Ashida,T., Kohara,Y., Yui,M., Umeki,H., Ishiguro,K.: Effective diffusion coefficients of radionuclides in compacted bentonite and rocks. PNC TN8410 92-164 (1992).

[10] Ota,K., Kodama,K., Yoshida,H.: Geochemical interactions in relation to natural uranium migration in sedimentary rocks at the Tono uranium deposit, central Japan. In: Abstracts of Scientific Basis for Nuclear Waste Management XVIII (1994).

The Grimsel Radionuclide Migration Experiment - a Contribution to Raising Confidence in the Validity of Solute Transport Models Used in Performance Assessment

Urs Frick
Nagra, National Cooperative for the Disposal of Radioactive Waste, CH-5430 Wettingen, Switzerland

Abstract

An important issue of safety assessment of radioactive waste repositories is to provide confidence that the predictive models utilised are applicable for the specific repository systems. Nagra has carried out radionuclide migration experiments at the Grimsel underground test site for testing of currently used methodologies, data bases, conceptual approaches and codes for modeling radionuclide transport through fractured host rocks. Specific objectives included (i) identification of the relevant transport processes, (ii) to test the extrapolation of laboratory sorption data to field conditions, and (iii) to demonstrate the applicability of currently used methodology for conceptualising or building realistic transport models. Field tests and transport modeling work are complemented by an extensive laboratory program.

Although intense investigations were made to characterise the hydrology of the site, the field experimental activities focused predominantly on establishing appropriate conditions for identifying relevant transport mechanisms on the scale of a few meters, aiming at full recovery of injected tracers, simple geometry and long-term stability of induced dipole flow fields. Laboratory and field experiments were carried out under undisturbed natural chemical conditions and extensive efforts were taken to quantify or rule out any artifacts from instrumental dispersion or equipment sorption. A variety of problems had to be solved and novel technique development was necessary until a satisfactory model calibration was achieved, including model predictions for different tracers and for different flow fields. Tracers utilsed for migration experiments included different non-sorbing anions (uranine, ^{82}Br, ^{123}I), ^{3}He and ^{3}HHO and the reactive tracers $^{22,24}Na^{+}$, $^{85}Sr^{2+}$, $^{86}Rb^{+}$, $^{137}Cs^{+}$, and $^{99m}TcO_4^{-}$.

A relatively simple homogeneous, dual-porosity advection/diffusion model was built with input from a state of the art petrographical characterisation of the water conducting feature (a quasi 2-dimensional, narrow shear zone). It was possible to "calibrate" the model from conservative tracer breakthrough curves. The few model derived parameters were physically reasonable and empirically sound and consistent modeling of a set of tests with different cation-exchanging tracers was possible. The applicability of the model was further demonstrated by quantitatively predicting actually measured concentration vs. time functions for all the employed tracers under different flow conditions. Peak times, peak concentrations and peak shape were accurately predicted over many orders of magnitude. The existence of a significant diffusive component (e.g., "matrix diffusion") was unambiguously identified and independently confirmed by the use of tracers of different diffusion coefficients (e.g. uranine, $^{123}I^{-}$, and ^{3}He).

Currently utilised methodologies for realistic modeling of geosphere transport were shown to be applicable during the time and spatial scale of the Grimsel migration experiment. This lends increased confidence in the used approaches of conceptualising important components of radionuclide transport in a fractured repository host rock.

1. Validation - What does it mean in the Swiss context?

An important aspect of safety assessment of radioactive waste repositories is the predictive modeling of possible doses to humans from potential radionuclide release sometime in the future. For certain scenarios, the model chain is broken up into various components, such as near field processes, geosphere and biosphere transport, etc. As such models are only approximations to real environmental systems, there will always be some uncertainty associated with predictions obtained from their use. Fortunately, it is not critical that these models are "correct", i.e., they are not required to include all the natural details and processes, merely that any uncertainty and simplification results in overestimating of consequences ("conservatism").

A good theory is characterised by the fact that it makes a number of predictions that could in principle be disproved or falsified by observation. Each time new experiments are observed to agree with the predictions the theory survives, and our confidence in it is increased, even if it may be impossible to rigorously prove their validity. Recently such of proof testing has been termed "validation" which, according to IAEA glossaries, has been defined as "... a process, carried out by comparing model predictions with independent field observations and experimental measurements. A model cannot be considered validated until sufficient testing has been performed to ensure an acceptable level of predictive accuracy".

Nagra's work has to be judged according to the updated Guideline HSK-R-21 (1993) from the Swiss Federal Nuclear Safety Inspectorate (HSK), where validation is concisely defined as "Providing confidence that a computer code used in safety analysis is applicable for the specific repository system". The word "validate" no longer connotes absolute truth - which for environmental systems can never be achieved in modeling - but merely acceptability for a specific use. Therefore the process of validation should focus on the level of proof that is needed to show acceptability of a model and how that testing can best be carried out.

The Grimsel migration experiment is one of Nagra's major contribution within the framework of model validation. A number of solute transport experiments, performed with different tracers and in different flow fields, provide a coherent picture which is consistent with a relatively simple model for radionuclide transport in a fractured rock. The confidence in the developed model is substantially raised by successfully predicting additional, largely independent experiments within the same water conducting feature (i.e., "blind" testing). In this paper some important results will be summarised and discussed. It is also shown that considerable effort is required to provide a sound experimental basis for trustworthy solute transport modeling.

2. Fracture flow in Swiss safety assessment concepts

Details of the Swiss safety assessment approach for repositories in a crystalline rock are set out in Nagra (1985, 1993a). Special considerations are given to the regional and local pecularities of Swiss hydrogeology and the long-term, still ongoing, tectonic activities which have caused significant deformation of the crystalline bedrock in Northern Switzerland. The resulting faults are expected to provide the potential flow paths of deep groundwater. A description of the regional geology is set out in Thury et al. (1993). Due to the practically water-saturated, hilly topography, potentially large hydraulic gradients may persist locally down to potential repository depths. It cannot, therefore, be excluded that a small portion of the groundwater may move relatively rapidly in a network of fractures through the surrounding host rock.

Nagra is currently also studying groundwater transport in more ductile, i.e. less "competent", sedimentary rocks (e.g., the marl at the Wellenberg site: Nagra, 1993b). Although such sediments appear generally less permeable than crystalline rocks, groundwater transport in such rocks may also be described as fracture flow.

In all rock formations studied so far for potential HLW and L/ILW repositories, rapid advective flow occurs only in discontinuities which form the key pathways for the performance assessment. Depending upon their explorability, these pathways are generally difficult to characterise due to their local and heterogeneous properties. The key factors for radionuclide transport are the maximum estimated flow velocity in fractures, the distribution of flow, i.e. "channeling" or the flow wetted surface, the diffusive/dispersive penetration of the porespace within the wall rock matrix (= "matrix diffusion"), and sorption on the minerals of the fracture infill as well as in the rock matrix.

Open fractures comprise only a small portion of the water-filled pore space of a rock but they are often up to several orders of magnitude more transmissive than the wall rock. Potentially released radionuclides may be transported by groundwater flowing through such fractures. It is conceivable and demonstrated in certain cases that there exists a certain exchange of solute between flowing water and stagnant pore water which is often termed "matrix diffusion" (Fig.1). This process may lead to a substantial retardation of the solute transport velocity, i.e., the so-called Darcy velocity. Matrix diffusion may play a substantial role in retarding transport of non-sorbing elements, such as iodine. However, the efficiency of this temporary storage is greatly amplified by sorption processes; these, in turn, strongly depend upon nuclide specific properties, water chemistry, cation exchange capacity, mineral surface composition, etc., within the matrix of the wall rock.

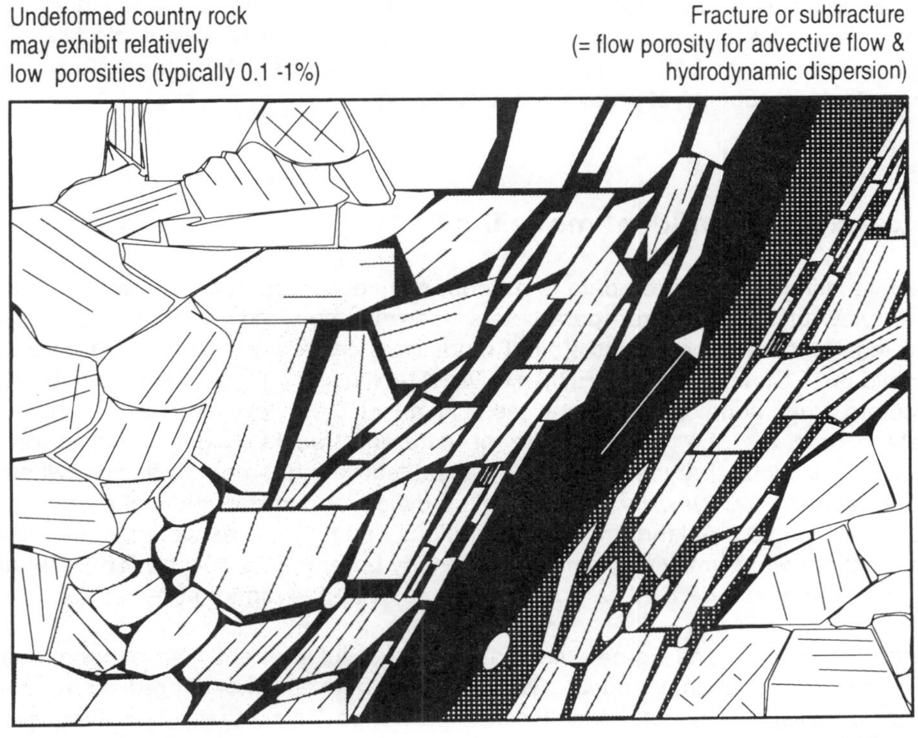
Undeformed country rock
may exhibit relatively
low porosities (typically 0.1 -1%)
Fracture or subfracture
(= flow porosity for advective flow &
hydrodynamic dispersion)
Deformed, sometimes also
altered region of the wall
rock; may exhibit relatively high
porosities (typically 0.5 - 5%)
and/or microfractures (= diffusion porosity)
Fracture infill may exist from
mechanical deformation
(e.g.,fault breccia or fault gouge)
or from hydrogeochemical processes
(e.g., precipitation or mineralization, etc.)

Fig.1 Schematic picture of a water conducting feature in a crystalline rock and its associated porosity distribution. The history of tectonic deformation and mineral alterations cause typical discontinuities in a rock.

In addition, radionuclide transport may also be retarded by sorption on fracture minerals or fracture infill depending upon on the properties and quantity of wetted infill relative to the diffusively penetrated matrix. These two retardation mechanisms have distinctly different effects on solute transport and, therefore, also on the tracer breakthrough characteristics from field experiments.

Retardation during radionuclide transport is an important component of Swiss safety assement models. Even under the conservative assumption of relatively high water flow velocities, the known characteristics of the local flow paths and their surrounding pore space are considered to provide sufficient retardation that allows many important radionuclides of an initially released supply to almost completely decay before they can reach the biosphere.

Model predictions of the geosphere transport from a reference site are based on a set of data which includes the regional and local geological, hydraulic and hydrogeochemical properties, as well as the solubility and sorption data which are applicable for the specific hydrochemical conditions. Although many steps of the modeling may appear highly plausible in view of current knowledge and although the simplifications of the real environmental systems are always aimed to be "conservative" (i.e., pessimistic and leading to higher consequences than a realistic alternative, it is quite demanding to prove the correctness of these steps and to achieve acceptance of the chosen procedures. For this purpose, Nagra decided at a rather early stage, to use a comprehensive approach of field and laboratory experiments as well as geochemical, hydrological and transport modeling to demonstrate the validity of different important components of applied safety assessment models. The Grimsel migration experiment is part of this approach and was planned as a long-term study, attempting to follow ideally an iterative scheme of predictions and testing.

3. How to achieve confident modeling?

Retroactive fitting of measured data is a frequently applied approach to demonstrate the value of modeling. In view of abundant free parameters such attempts must remain questionable, unless predictions of independent experiments ("blind modeling") can be tested by field data. After a few iterative attempts the PSI modelers (HEER & HADERMANN, NTB 94-18) were able to "calibrate" the model parameters from non-sorbing tracer tests in a selected flow field and then succeeded in predicting rather precisely the breakthrough of different tracers under identical or other flow conditions. One might argue that there is still ample subjective judgement to decide, whether or not a prediction is adequate. However, by comparing predicted and measured field data, all important characteristics (i.e., peak breakthrough concentrations, peak time as well as the shape of the breakthrough function) were amazingly similar over many orders of magnitude. The soundness of the applied model and the model-derived parameters seems evident even for non-experts.

The better the predictions agree with results of the subsequent field experiments, the more convincing the chosen model shall appear. Discrepancies between model prediction and experiment either indicate unrealistic concepts, insufficient knowledge of naturally occurring processes or inaccurate assessment of important parameters. Hence, modelers as well as laboratory and field researchers are virtually forced to find about the potential causes.

During the early stages of the migration experiment model predictions repeatedly resulted in rather inaccurate approximations of the naturally occuring conditions or the actually measured tracer breakthrough curves. Subsequent attempts for improving model predictions prompted in a number of additional investigations, field experiments, and equipment modifications.

During later stages, the deviations between predictions and experiments included often rather small details. Nevertheless, these details revealed important clues for the interpretation of the field experiments and resulted in additional fine-tuning of the parameters of the applied model. It should be emphasised, however, that in case of conservative transport modeling during repository safety assessment these would have been unnecessary modifications because such a precision for parameter values is not required for these cases.

4. Overview of the Grimsel migration experiment

Nagra's underground research facility is located near the Grimsel pass road at about 1700 m above sea level in the Central Aare Massif of the Swiss Alps; it was constructed 1983 under about 450m of crystalline rock overburden. The general goals of any work at the Grimsel Test Site (GTS) included the build-up of know-how and the development of necessary methodologies for a radioactive

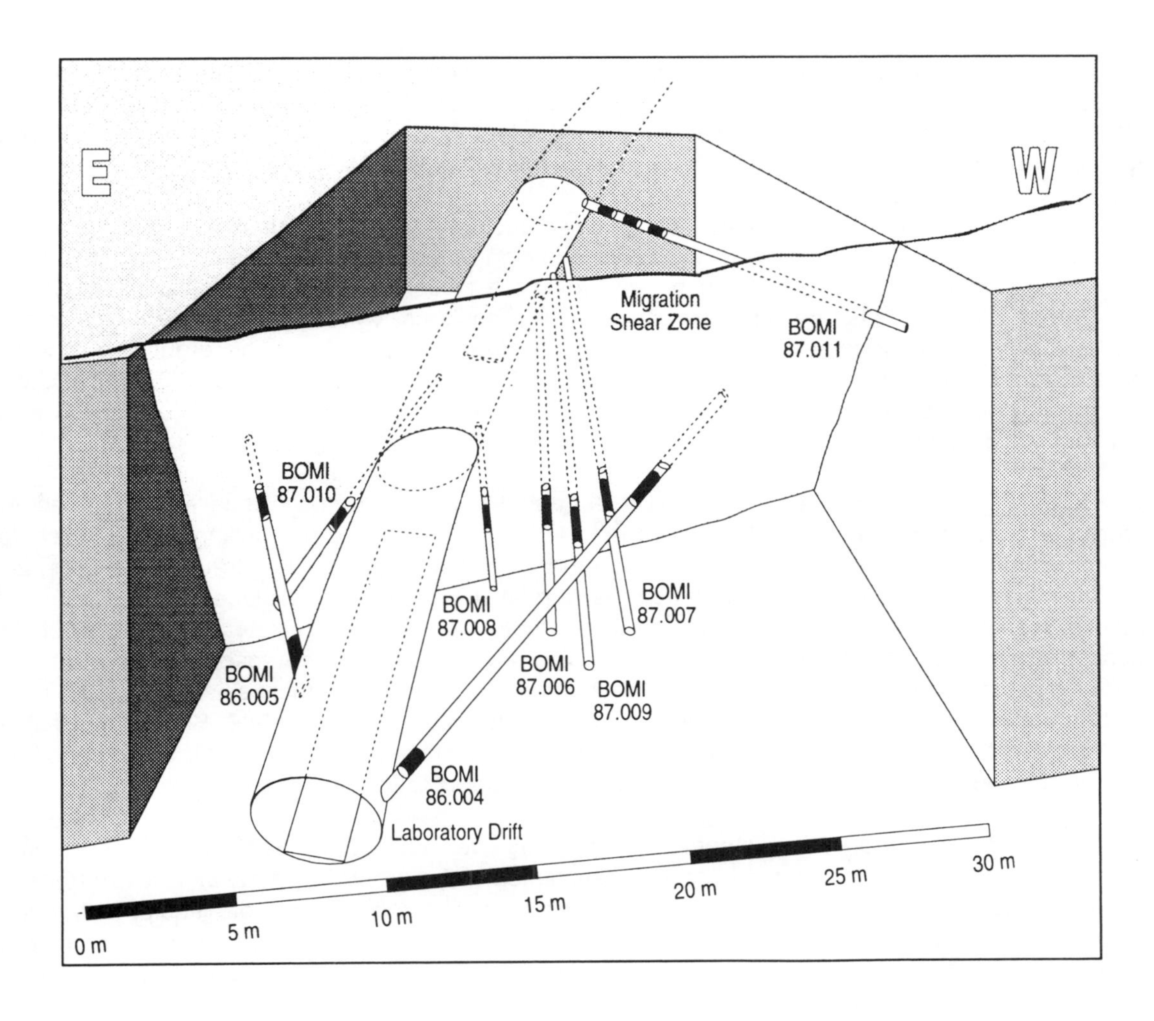

Fig. 2 Overview of the experimental site showing the laboratory drift, the position of the migration shear zone and borehole locations.

waste repository site characterisation. The major aims of the Grimsel migration experiment include field testing of various concepts, data bases, and computer codes which were utilised in the Swiss safety assessment procedures of modeling radionuclide transport. More specifically, the experiments were devised to achieve (i) an identification of the transport processes, (ii) to test the extrapolation of laboratory sorption data to field conditions, and (iii) to demonstrate the applicability of the currently used methodology for conceptualising or building realistic transport models. It was also demanded that laboratory and field experiments had to be carried out under undisturbed natural chemical conditions and that any influence from boreholes or equipment can be neglected or at least quantified.

Investigations started in the laboratory tunnels of the Grimsel test site in 1985 with the hydrogeological exploration of an experimental site and development of suitable equipment. The migration experiments were carried out in a water conducting fault, the so-called "migration fracture" or "migration shear zone" which, on the scale of more than 30 m, can be reasonably described as a 2-dimensional feature. The major flow paths are the product of relatively recent tectonic deformation (Alpine uplift). Brittle processes reactivated older, ductile shear zones and led to the formation of a few cm wide, asymmetrical array of conductive, subparallel openings (subfractures or "flow layers") in an otherwise relatively impermeable, micaceous mylonitei (see also Section 9). The wider subfractures are at least partially filled with highly porous, unconsolidated infill (fault breccia or fault gouge).

Any transport modeling requires a detailed characterisation of the water filled pore space as a necessary basis for a realistic conceptual model and adequate parameter assumption. BOSSART & MAZUREK (1991) have - with the needs of the modelers in mind - provided an extremely useful and easy-to-grasp structural and petrographic description of the migration shear zone.

The main reasons for selecting this particular fracture for the migration experiment were its relatively simple, quasi-planar geometry and its sufficiently large conductivity as indicated by the steady discharge of initially almost 1 L min^{-1}. In order to establish the hydrogeological stability of the system, the migration fracture and the surrounding environs were characterised in terms of hydrochemistry and hydrology prior to performing any further exploration. The groundwater in the migration fracture is anoxic or suboxic, has a high pH (~9.6) and a low ionic strength (0.0012 M); for a more detailed description of the chemistry, see FRICK et al. (1992; Section 3.4), EIKENBERG et al. (1991; Section 3), and EIKENBERG et al. (1994, Section 2.3).

During 1986 and 1987 a total of 8 boreholes, about 6 to 25 m long, were drilled to intersect the fracture plane at distances of 3 to 16 m from the tunnel wall (Fig. 2). Upon drilling, these boreholes were immediately packed off to avoid drainage of the system and to record hydraulic pressures. 2 of the 8 boreholes intersected the migration fracture at relatively impermeable places while subsequent hydraulic testing of the 6 water-discharging boreholes revealed local heterogeneities of the transmissivity from 5.10^{-6} to about 10^{-8} m^2s^{-1} for single-hole tests. On the scale of several meters, however, the fracture appears rather homogeneous with an average transmissivity of 2.10^{-6} m^2s^{-1} based on the results of a large number of cross-hole tests (HOEHN et al., 1989, Section ; FRICK et al., 1992; Section 4). Various hydrodynamic models satisfactorily predict stationary conditions (pressure distribution, flow) but do not adequately reproduce transient responses.

To secure indisturbed and steady hydraulic, as well as stable hydrochemical conditions, required major efforts during field work. Any contamination of the groundwater by air or even transient changes of the water chemistry during drilling, during installation of borehole equipment, during hydraulic testing or later on during tracer injection had to be avoided, as such disturbance could, in principle, lead to dissolution or precipitation phenomena and, thus, shifting hydraulic conductivities or lead to irreversible changes of mineral surfaces.

First pilot tracer tests were carried out in 1988, but a variety of experimental problems had to be solved before a suite of high-quality tracer migration experiments commenced in 1990. In 1994 the

Table 1 Proposed sorption coefficients K_d [in ml g^{-1}] from laboratory experiments as recompiled and modified in by HEER & HADERMANN (1994; see Appendix 6).

Type of Experiment	Size Fraction	K_d [Na] in ml g^{-1}	K_d [Sr] in ml g^{-1}
Radionuclide Batch Sorption	$\leq 63 \mu m$	$1.3 \pm {}^{1.5}_{0.4}$	$41 \pm {}^{39}_{4}$
Radionuclide Batch Sorption	$\leq 250 \mu m$	$0.85 \pm {}^{0.76}_{0.04}$	$25 \pm {}^{22}_{1}$
Radionuclide Batch Sorption[1]	$\leq 2000 \mu m$	$0.43 \pm {}^{0.60}_{0.02}$	$13 \pm {}^{9}_{1}$
Rock-Water Interaction	$\leq 2000 \mu m$	$0.13 \pm {}^{0.13}_{0.02}$	$7.6 \pm {}^{8.2}_{1.7}$

1 fraction is labeled as "loosly disaggregated" to point at the very gentle disaggregation procedures for this fraction (see EIKENBERG et al., 1991: p.44 and 45).

First pilot tracer tests were carried out in 1988, but a variety of experimental problems had to be solved before a suite of high-quality tracer migration experiments commenced in 1990. In 1994 the final field tracer experiments were started. A detailed overview of the investigations 1985-1990 is presented in Frick et al. (1992), including the aims of the project, a description of the regional and local hydrogeological setting; the structural, petrographical and hydraulic characterisation of the migration site; results from complementary laboratory investigations; a description of the the major equipment components and necessary technical developments; a presentation of typical tracer breakthrough data and the results of some preliminary attempts of transport modeling at the Paul Scherrer Institute (PSI).

A thorough presentation of the transport model which has been developed at PSI has just been published (HEER & HADERMANN, 1994). In this paper the calibration of the model by field data, a sensitivity analysis for the involved parameters and subsequent testing of predictions are set out in detail. The complementary documentation of the experimental procedures of the relevant tracer tests is given in EIKENBERG et al. (1994).

6. Laboratory sorption experiments

The main purpose of the laboratory program was to produce sorption data for a variety of tracers under simulated natural conditions (distribution coefficients Kd for sorption and desorption, isotherms, kinetics, etc.). Due to difficulties in obtaining sufficient amounts of fracture infill, different size-fractions of an artificial fault gouge were prepared by gently crushing micaceous mylonite from a nearby shear zone of identical origin. It was shown, that the granulometry, as well as the mineralogical composition of natural and artificial fault gouge were roughly similar (for references, see Section 3 in FRICK et al.,1992). Laboratory rock water interaction and batch radionculide sorption studies were carried out under controlled atmospheric conditions. Detailed documentation of this work is given in BRADBURY et al. (1989) and AKSOYOGLU et al. (1991) and a summary of the laboratory work and the proposed cation exchange model is also included in Section 5 of FRICK et al. (1992). On the basis available mineralogical information, the reported Kd-values for Na^{+} and Sr^{2+} were slightly modified by HEER & HADERMANN (1994, see Appendix 6) and are set out in Table 1. As the modeling by these authors - as discussed later on in this paper - include only non-sorbing anion tracers and the sorbing cations Na^{+} and Sr^{2+}, results of other elements and other laboratory studies are not presented here.

7. The equipment setup for tracer migration experiments

Obtaining reliable experimental data in a field study is a very demanding task. Reasonable model testing can only be achieved when the quality of experimental data and procedures are well established - i.e.. experimental artefacts can reliably be ruled out - and the needs of the modelers are adequately considered. Initial project plans had to be modified repeatedly and the time schedule expanded, as a number of unforeseen experimental problems had to be solved, novel techniques had to be developed, and analytical capabilities had to be improved. In parallel, specific requirements for transport modeling continuously increased with experimental progress. The experimental setup finally used for most of the tracer experiments is set out in Fig. 3; more experimental details are provided in FRICK et al. (1992; Section 6) and EIKENBERG et al. (1994; Sections 2 + 3). The most important experimental implementations are listed here:

- Development of the HPLC pumping technique to provide precise, extremely stable and long-term constant pump rates for injection and withdrawal flow which are not sensitive to pressure fluctuations. Precise pump rates must be known to establish an accurate tracer recovery.
- Development of tracer dosage equipment for pulse or continous injection and to precisely quantify injected tracer mass.
- Construction of low-volume, instrumented packer interval pieces to reduce the effects of unknown instrumental dispersion.
- Increase of analytical range for on-line tracer analyses to allow minimal injection of tracer mass and to enable reproducible and highly sensitive concentration measurements for the breakthrough tail sections that reveal the diagnostic features for process discrimination.
- Sorption-proofing of equipment for the selected tracers to avoid experimental artifacts.
- The development and extensive testing of the helium tracer method for highly sensitive on-line analysis of an inert gas tracer (^{3}He). Needed (i) to check for desaturation phenomena and (ii) to unambiguously identify or independently confirm diffusive processes.
- Dual quartz fiber optical fluorimetry (optrodes), utilising an adjustable argon laser as the excitation light source, was developed to enable continuous, precise and highly sensitive down-hole analyses of uranine. These data are critical for defining the actual tracer input function or to determine the actual instrumental dispersion effects. In addition, fiber optical uranine data can also be obtained in closed-off intervals of passive boreholes; these data provide clues on the rough position or shape of the real dipole flow field in the migration fracture (...but has not been used yet in current modeling).

Sound modeling must include precise knowledge of the tracer input function, that is, the concentration versus time function at the entry of the dipole flow field. Pilot experiments have already shown that unknown equipment dispersion may contribute noticeably or, in case of experiments on the 1.7 m scale, overwhelmingly to the total tracer breakthrough curve. Down-hole fiber fluorometry measurements of uranine concentrations in the injection borehole interval (see Fig. 6-10 in FRICK et al., NTB 91-04 and Figs. 9+22 in HEER & HADERMANN, NTB 94-18) are taken to represent the real valid input function for modeling purposes. By means of equipment tests (i.e., "blanks") it is verified that the measured input function for uranine is also representative for simultaneously injected sorbing tracers which cannot be measured down hole.

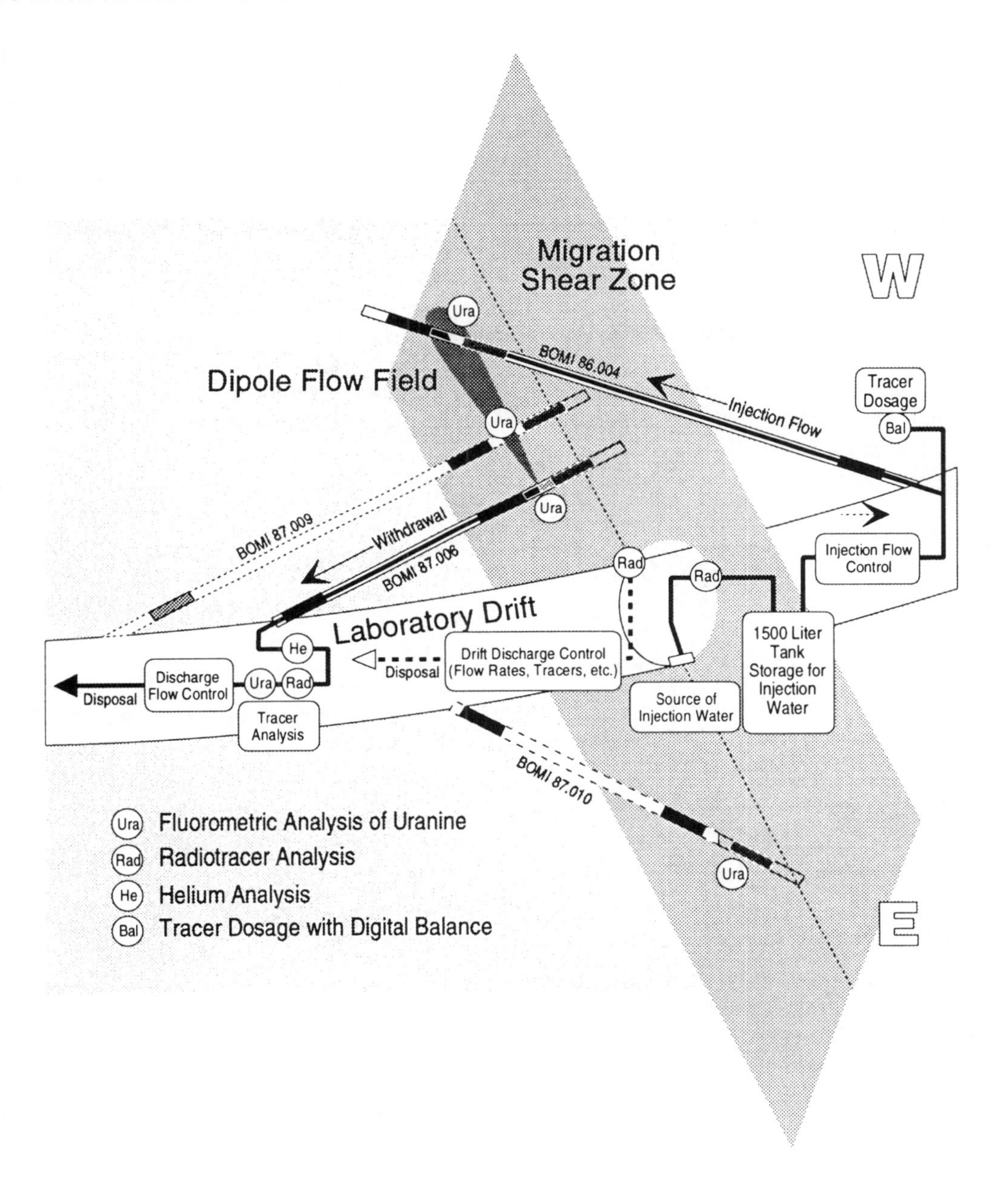

Fig. 3 Schematic diagram of equipment setup for Grimsel tracer migration experiments. A simplified view of a 4.9 m dipole flow field between BOMI 86.004 and BOMI 87.006 is shown; 1.7 m flow distances are achieved between BOMI 86.009 and BOMI 87.006, 14 m flow distances between BOMI 86.004 and BOMI 87.010.

Table 2 Scales of dipole tracer tests 1988-1994, associated injection and withdrawal boreholes, as well as an overview of the employed non-sorbing and sorbing tracers for fracture experiments and equipment tests.

Dipole Distance	Input Borehole BOMI	Withdrawal Borehole BOMI	Non-reactive Tracers	Reactive Tracers
1.7 m	87.009	87.006	Uranine, $^{3,4}He$, $^{82}Br^-$, $^{123}I^-$,	$^{22,24}Na^+$, $^{85}Sr^{2+}$, $^{86}Rb^+$, $^{99m}TcO_4^-$, $^{137}Cs^+$
~4 m	86.004	97.009	Uranine, $^{3,4}He$, $^{82}Br^-$	none
4.9 m	86.004	87.006	Uranine, tritium, $^{3,4}He$, $^{82}Br^-$, $^{123}I^-$,	$^{22,24}Na^+$, $^{85}Sr^{2+}$, $^{86}Rb^+$, $^{99m}TcO_4^-$, $^{137}Cs^+$
14 m	86.004	87.010	Uranine, $^{3,4}He$, $^{82}Br^-$, $^{123}I^-$,	$^{22,24}Na^+$, $^{85}Sr^{2+}$, $^{137}Cs^+$
Equipment Tests			Uranine, ^{4}He, $^{82}Br^-$	$^{22}Na^+$, $^{85}Sr^{2+}$, $^{58}Co^{2+}$, $^{75}SeO_{3,4}^{2-}$, $^{134}Cs^+$

8. The selection of appropriate flow fields

Although intense investigations were carried out to characterise the hydrology of the site, the field experimental activities focused predominantly on establishing appropriate conditions to test transport models and to identify relevant transport mechanisms. For this purpose, very specific flow fields of simple geometry - which allowed full recovery of tracers - were sought for during pilot

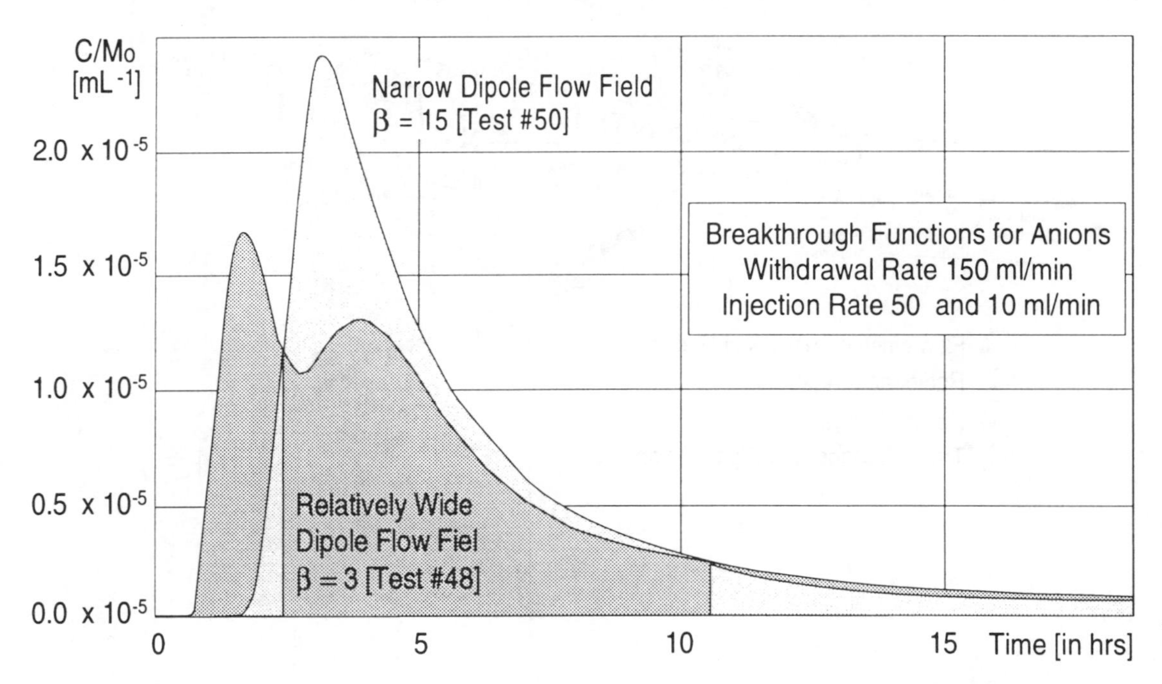

Fig. 4 Measured tracer breakthrough curves after pulse injection of a non-sorbing anion tracer (uranine); the flow distance is 4.9 m, extraction flow rates are 150 ml min^{-1} and the different injection flow rates 10 and 50 ml min^{-1}, corresponding to dipoles of β=15 and β=3, respectively. As a common practice measured tracer concentrations C (in Bq or g ml^{-1}) are normalised to C/Mo, with Mo the total tracer mass (in Bq or g) which is added in a few ml of groundwater to the injection flow at T=0.

experiments with the fluorescent dye uranine (Na-fluorescein). During these experiments, injection and withdrawal rates were varied. After setting constant pump rates, steady-state pressure conditions were reached within days or less. Asssuming negligible background flow the ratio, β, of withdrawal to injection rate determines the shape of the unequal-strength dipole flow field, while the absolute pump rates affect the average local flow velocities. Due to the pressure conditions, the useful range of maximum and minimum withdrawal rates turned out to be rather small, that is, for practical purposes extraction flow rates could only be varied within a factor of about 2.

Fig. 4 depicts the tracer breakthrough curves from experiments with identical withdrawal, but different injection pump rates (i.e, for β=3 and β=15 conditions). The narrow dipole flow field is resulting in a single, sharp peak and subsequent long tail, while the wider dipole produces a multiple breakthrough peak with two concentration maxima, but similar tailing. Obviously, wider dipole fields are more sensitive to heterogeneities of the transmissivity distribution within the fracture and are more prone to result in multiple-peak response to tracer injection (see e.g., MORENO & TSANG (1991). As the description of such heterogeneities requires additional parameters, the set of tracer migration experiments for model calibration and model testing was carried out with narrow dipole fields of β=15. For the narrow dipoles, existing heterogeneities might mainly result in a somewhat prolonged flow path but the simple and smooth breakthrough curve (Fig. 4) indicates that these heteogeneities are effectively averaged out.

In Fig. 5 the position of the laboratory tunnel and borehole intersections with the steeply dipping, quasi planar migration shear zone is shown. Also shown are the schematic shape of flow fields for the different tested distances (1.7, 4, 4.9, and 14 m). Experiments for modeling were all

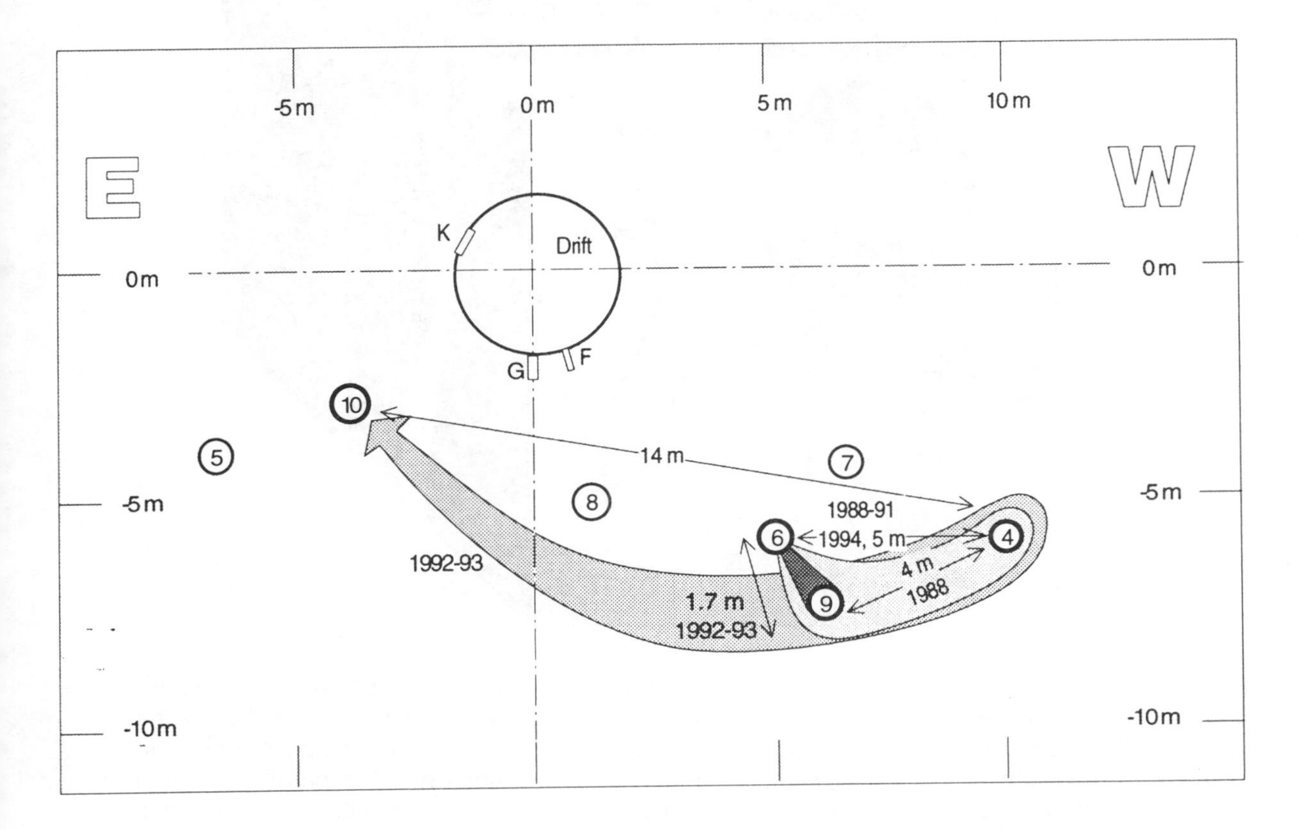

Fig. 5 Position of tunnel and boreholes intersections with the migration shear zone; schematic shape of dipole flow fields over the various tested distances are also depicted.

performed in rather narrow dipole flow fields of β=15. Model calibration and subsequent testing for sorbing tracers was based on 4.9 m distance experiments; further predictions were tested on the 1.7 m flow distance. Experiments from the 14 m distance, mainly with non-sorbing tracers, are not yet modeled. In Table 2 compiled is a summary of the various experiments carried out so far, including the employed tracers, boreholes and dipole flow distances.

9. The PSI transport model

This advection/diffusion transport model was initially developed by HERZOG (1991) and consists of [i] a hydrological part for the calculation of the flow field through a number of streamtubes to account for idealised unequal-strength dipole flow fields in a planar fracture of homogeneous

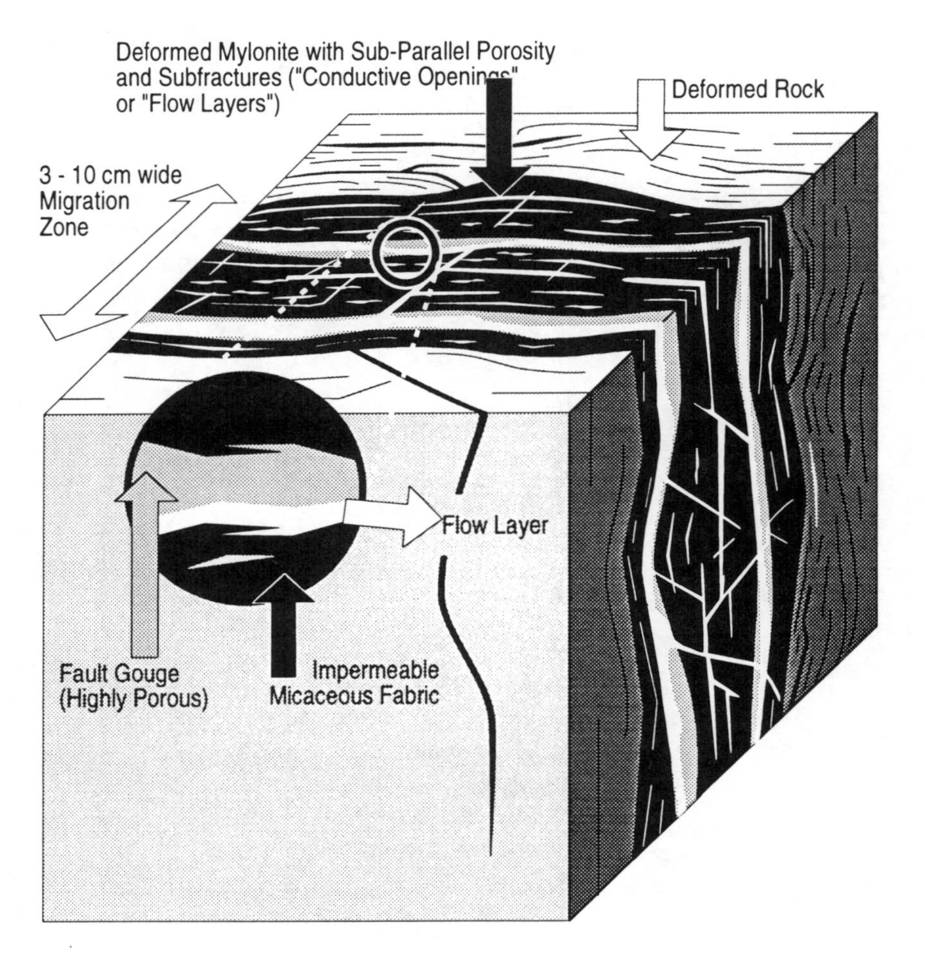

Fig. 6 Simplified view of the migration fracture according to the geological characterisation of the pore space by BOSSART & MAZUREK (NTB 91-12). The shear zone includes the results of ductile and brittle deformation of a granodioritic rock. The mylonite is a few cm wide, mica-rich zone and consists of a parallel oriented network of subfractures with the larger ones at least partially filled with fault gouge. The porosity of the fault gouge is estimated to be around 10-30% while the unfractured, schistose fabric of the mylonite is of relatively low porosity (≈0.1%) and, thus, rather impermeable. The deformed portions of the granodioritic fabric exhibit a relatively large grain-boundary porosity (1-2%).

transmissivity and [ii] a transport part, where tracer transport in each streamtube is calculated by the code RANCHMD (advection/diffusion model) described in JAKOB et al. (1989). Traced water is viewed to flow through a bundle of streamtubes, each being characterised by a different average flow velocity (the actual varying flow velocity depends on the local width of the streamtube). The results of quasi 1-D modeling of transport through each streamtube are then superimposed to obtain the calculated breakthrough curves. In his preliminary attempts HERZOG (1991) did not succeed in matching satisfactorily the tail of the breakthrough curves by his dual-porosity model, but concluded that the field data cannot be interpreted by a single-porosity approach together with a physically meaningful set of parameters.

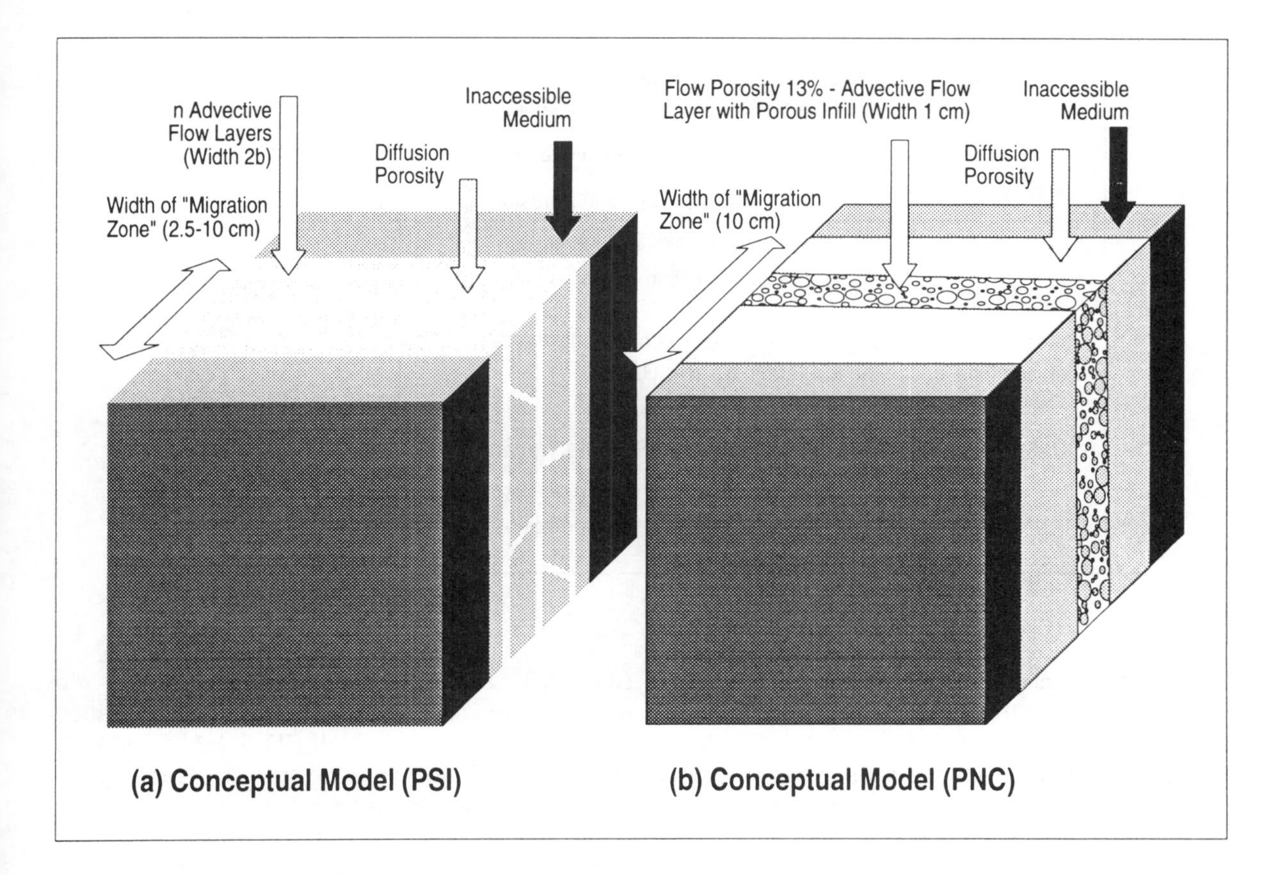

Fig. 7 (a) Schematic illustration of the PSI modeling concepts (HERZOG, NTB 91-31; HEER & HADERMANN, NTB 94-18) with mechanical dispersion (i.e., hydrodynamic dispersion with a negligible component of molecular diffusion) within a number of open interconnected, homogeneously conductive, advective flow layers and molecular diffusion only in the adjacent pore space. This diffusion porosity is thought to represent either the porous fracture infill (fault gouge), the abundant subparallel openings within the mylonite (see Fig. 6), or both.

(b) The Japanese PNC preliminary modeling concept (UMEKI et al. 1994); assumed are a single conductive flow layer with a porous infill (hydrodynamic dispersion, only) and molecular diffusion only in the adjacent pore space.

A priori fixed, assumed and model derived parameters for the different PSI approaches as well as from the PNC work are compiled in Table 4. Perpendicular to its parallel conductive openings, the mylonite is considered practically impermeable which is taken into account by the inaccessible medium, as the limiting boundary for the diffusion porosity in both conceptual models.

Table 3 Compilation of fixed and model-derived parameters of non-sorbing and sorbing tracers for the PSI transport model (HEER & HADERMANN, 1994).

Assumed or Fixed Parameters (from Geological Characterization)	
Width of Migration Zone	a
Number of Flow Layers	n
Density of Porous Matrix	ρ_p
Model-Derived Parameters	
Width of Water-Conducting Openings (Flow Layers)	2b
(b is related to flow porosity = [n·2b]/a)	(ε_f)
Dispersion Length	a_l
Matrix or Diffusion Porosity	ε_p
Pore Diffusion Coefficient	D_p
[1] (Effective Diffusion Coefficient D_e)	($\varepsilon_p \cdot D_p$)
Additional Model-Derived Parameters for Sorbing Tracers	
Matrix Sorption Coefficient	K_d [Na, Sr]
Surface Retardation Coefficient	R_f [Na, Sr]

[1] measured breakthrough curves sometimes allow only the determination of combinations of the two terms ε_p and D_p

With the clarifying geological description of BOSSART & MAZUREK (1991) available - a simplified view of the geologists' characterisation of the water filled pore space of the migration fracture is depicted in Fig. 6 (see also Fig. 3-4 in FRICK et al., 1992) - HEER & HADERMANN (1994) were able to achieve a more realistic conceptualisation of the migration fracture (see Fig. 7). They are able to justify their parameter selection in view of actually observed features, describe in detail the model calibration procedures, including a detailed sensitivity study, and testing of predictions for a selected set of tracer migration experiments with non-sorbing tracers and the sorbing cation tracers Na^+ and Sr^{2+} over the 4.9 as well as the 1.7 m dipole fields.

As the natural pressure gradients were relatively small within the investigated area of the migration shear zone (see e.g., Fig. 3-8 in FRICK et al. NTB 91-04), the radially converging flow to the laboratory tunnel was judged to be negligible for the tests evaluated so far in the PSI models and only artificially induced dipole flow was considered. A simplified view of the associated homogeneous dual-porosity approach is given in Fig. 8 (cf. also Figs. 5 + 6 in HEER & HADERMANN, 1994).

10. PSI Model calibration

The PSI transport model includes 3 fixed parameters, taken from the geological characterisation and, for non-sorbing tracers, a total of 4 model derived parameters. The meaning of these parameters is explained in Table 3. The calibration procedure consists of model-fitting 4 parameters from a reference breakthrough experiment with a non-sorbing (i.e., conservative) tracer (Fig. 9).

Different sections of the fitting breakthrough curve are more or less sensitive to these 4 fitting parameters and from these, values for the flow porosity, dispersion length, diffusion porosity and pore diffusion coefficient can be deconvoluted. Resulting values for dispersion length, flow porosity and the product $\varepsilon_p\sqrt{D_p}$ can be determined almost independently from each other, while variation of the coupled values ε_p and D_p strongly effect fitting of the tail end portion of the breakthrough curve (see Section 4 in HEER and HADERMANN NTB 94-18).

Values for the 4 model-derived parameters of the few centimeter wide "migration zone" (i.e., modelers' definition for the portion of the shear zone, where transport processes occur during the time scale of field experiments) are for the flow porosity $\varepsilon_f = (n \cdot 2b)/a = 0.74\%$; the matrix or diffusion porosity $\varepsilon_p = 6.2\%$ is consistent with the geologists' determinations or observations; the dispersion length $a_L = 25$ cm appears reasonable in view of the observed network of potential flow paths and the pore diffusivity $D_p = 2.5 \cdot 10^{-11}$ m^2s^{-1} is consistent with empirically derived values for the relevant matrix fabric and porosity (c.f., Section 7 and Fig. 28 in FRICK, 1992). For comparison with results from other modeling attempts, these parameters are compiled in Table 4. The latter does not include the uncertainty range for each parameter from HEER & HADERMANN's sensitivity study (see Section 4 and Table 6, therein).

11. Model testing using sorbing tracer tests

11.1 Extrapolation of laboratory sorption coefficients to in-situ field conditions

Predictive radionuclide transport models require the knowledge of average sorption coefficients, Kd, which are representative for the in-situ conditions. Therefore, testing the extrapolation of laboratory sorption data to field conditions was a major aim of the Grimsel migration experiment. Even with the laboratory studies carried out under simulated natural conditions, appropriate sampling and sample preparation remain key problems, as the laboratory material usually consists of artificially disaggregated rock material. Freshly exposed mineral surfaces of crushed fracture material may have different surface properties or somewhat different mineralogy which both effect the uptake capacity (or to be specific to the migration experiment, the cation exchange capacity, CEC). In addition, varying occurrence of trace minerals with high sorption capacities (e.g., mixed layer phyllosilicates, montmorillonite, etc.) may - even for identically prepared size fractions - potentially lead to considerable uncertainties in the laboratory data. Consequently, laboratory samples generally are expected to exhibit a somewhat different sorption behaviour than the actual in-situ flow wetted rock zone. In-situ determinations of the integral cation exchange capacity for a certain flow field were made by a field hydrogeochemical equilibration experiment (EIKENBERG et al., 1991). From such a value, tentative estimates of in-situ Kds can be derived for the different cation exchanging species.

11.2 Model testing with the weakly sorbing tracer $^{22}Na^+$

With the model parameters for conservative tracers being fixed, a field breakthrough experiment, using a weakly sorbing tracer, was predicted and tested against subsequent field observations. In a later step, "improved" predictions are tested first for a more strongly sorbing tracer and, for the same suite of tracers, in a different dipole flow field. A first migration experiment used $^{22}Na^+$ as a weakly sorbing cation. The model prediction was based on the actual flow conditions, the calibrated flow field parameters for conservative tracer runs, and the smallest Kd (i.e., about 0.3 ml g^{-1}) taken from the largest grain-size fraction investigated in the laboratory as proposed by BRADBURY et al. (1989; p. 65; see also Table 5-7 in FRICK et al., 1992).

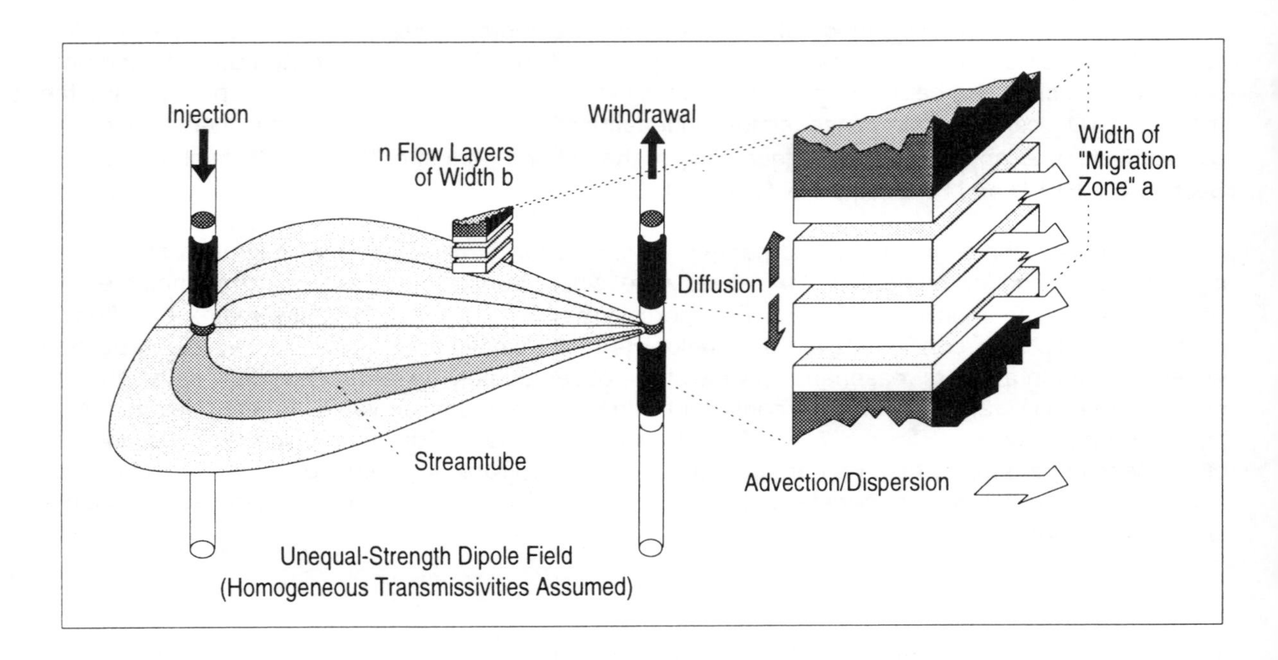

Fig. 8 Simplified presentation of the PSI modeling approach.

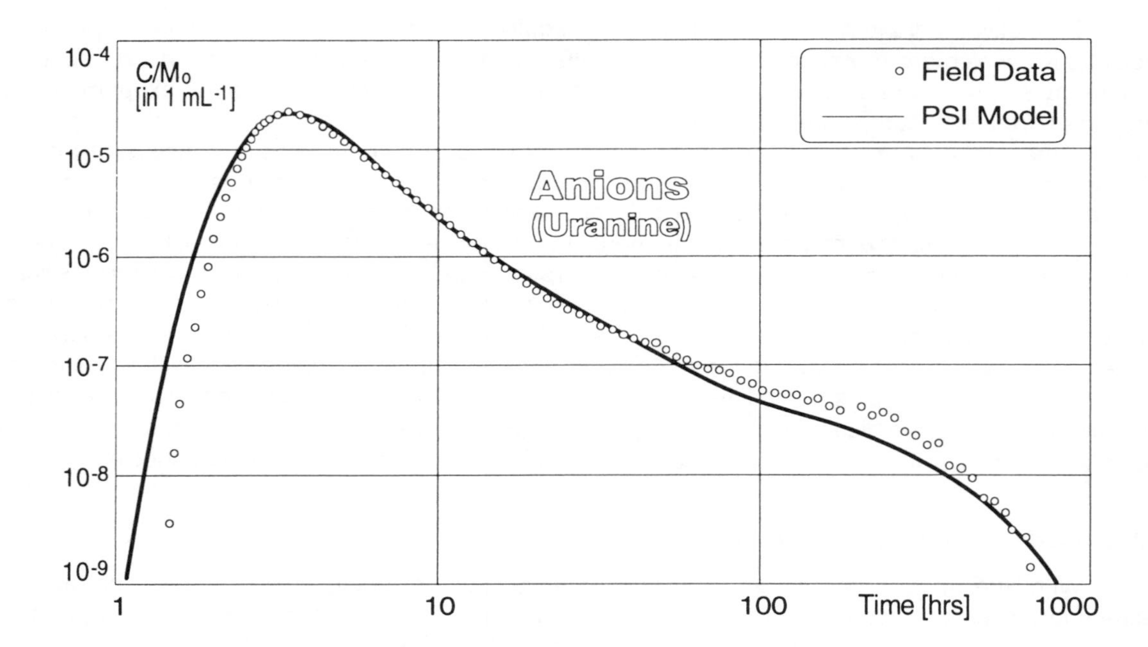

Fig. 9 Comparison of measured data from a field breakthrough experiment with the conservative tracer uranine and the fitted breakthrough curve for the applied flow conditions. This figure also illustrates the necessary analytical range and time scale of field experiments which are both required for adequate process discrimination during modeling of a typical non-sorbing tracer test over about 5 m distance and average flow velocity in the order of 1 m per hour.The good and detailed agreement between field and model data seem to justify a calibration of the required flow field parameters of the model.

Table 4 Comparison of fixed and model-derived parameters from different modeling attempts of 4.9 m dipole tracer tests. Calibration procedures are always based on the results of non-sorbing anions.

		PSI[1]	PSI[2]	**PSI [3] Model**	PNC[4]	[units]
Fixed Parameters (from Geological Characterization)						
Width of Migration Zone	a	1	1	**2.5 - 10**	10	[cm]
Number of Flow Layers	n	10	10	**3 - 6**	1	[-]
Density of Porous Matrix	ρ_p	2.67	2.67	**2.67**	2.70	[g cm^{-3}]
Model-derived Parameters						
Width of Water-Conducting Openings	2b	0.03	0.03	**0.0093**	1	[cm]
flow porosity of migration zone = [n·2b]/a	ε_f	30	30	**0.74**	13	[%]
Dispersion Length	a_l	8	8	**25**	25	[cm]
Matrix or Diffusion Porosity	ε_p	10	1	**6.2**	-	[%]
Pore Diffusion Coefficient	D_p	$5 \cdot 10^{-9}$	$2 \cdot 10^{-10}$	$\mathbf{2.5 \cdot 10^{-11}}$	-	[m^2s^{-1}]
Effective Diffusivity D_e for Anions (Uranine)	$\varepsilon_p \cdot D_p$	$5 \cdot 10^{-11}$	-	$\mathbf{1.55 \cdot 10^{-12}}$	$5 \cdot 10^{-11}$	[m^2s^{-1}]
D_e for Cations ($^{22}Na^+$)	$\varepsilon_p \cdot D_p$	-	$2 \cdot 10^{-12}$	[3]$\mathbf{2.05 \cdot 10^{-12}}$	-	[m^2s^{-1}]
Assumed[5] or Model-Derived[6] Sorption Coefficients						
	K_d [Na]	-	[5]0.3	[6]$\mathbf{0.13 \pm ^{0.16}_{0.07}}$	-	[ml g^{-1}]
	K_d [Sr]	-	-	[6]$\mathbf{21 \pm ^{38}_{14}}$	-	[ml g^{-1}]
Surface Retardation Factor[7] from effectively determined K_d [Na, Sr]	R_f	-	-	[6]**1**	-	[-]

[1] model parameters used for conservative tracer fitting, preliminary attempts by HERZOG (1991)
[2,5] model parameters used for Na tracer fitting, preliminary attempts by HERZOG (1991)
[3,6] for full parameter range from sensitivity analysis, see Table 6 in HEER & HADERMANN (1994)
[4] model calibration from conservative tracer tests (from text and Table 1 in UMEKI et al., 1994)
[7] for definition see Section 3.2 in HEER & HADERMANN (1994)

There was a noticeable difference between predicted and measured breakthrough of $^{22}Na^+$. However, with a Kd about a factor of 3 smaller than the initially proposed value (i.e., Kd = 0.13 ml g^{-1}; see Table 4) model calculations resulted in a good match (not shown here; cf. Fig. 16 in HEER & HADERMANN 1994), similar to the agreement between model and field data as shown in Fig. 9. Possible reasons for this rather minor difference between laboratory Kds and this model derived "in-situ" Kd - the latter may depend to a certain degree upon the used model and its assumed parameters - shall not be discussed further in this paper. Instead, reference is made to BRADBURY & BAEYENS (1992) where the application of laboratory data to the Grimsel field experiment is discussed. Here (Table 2) the effects of different mica contents in samples of natural and artificial fault gouge are considered to be of prime importance. As as result of BRADBURY & BAEYENS (1992), HEER & HADERMANN (1994; Appendix 6) reduced the initially proposed Kd-values by a factor of roughly 3. It is of note that Kd-values for ≤2 mm laboratory samples (see Table 1) agree well with EIKENBERG's (et al., 1991) independent estimate from a hydrogeochemical field experiment (i.e., the inferred Kd for $^{22}Na^{+}$ ranged from 0.1 to 0.3 ml g^{-1}).

These modified values are set out in Table 1 and - perhaps fortuitously so - the largest Kd values for $^{22}Na^{+}$ (from rock-water interacting experiments) now agree well with the model derived Kd. If, instead, the Kd from the ≤63 μm size fraction of radionuclide batch sorption experiments would have been taken as a reference value, a "correction" or "fudge" factor of roughly 10 would have to be introduced to further predict breakthrough tests with reactive tracers which are expected to sorb by an identical mechanism. Regardless of the good model fits over the entire breakthrough curves of anions and $^{22}Na^{+}$, certain model assumption could still be invalid. Further testing was then achieved by utilizing the more strongly sorbing tracer $^{85}Sr^{2+}$.

11.2 Model testing with the moderately sorbing tracer $^{85}Sr^{2+}$

A prediction for Sr was made, using again the calibrated parameters from anion experiments plus a Kd of 10 ml $g^{-1}$for $^{85}Sr^{2+}$ which was roughly consistent with the model fitted Kds for $^{22}Na^{+}$ (see Table 1). The match of peak shape and peak concentration between calculated curves and actual measured field data was already good. It must be mentioned that the goodness of fit was always assessed by visiual inspection (i.e.,"eye-fitting") of calculated and measured breakthrough curves and not by inverse modeling. In Fig. 10 a similar, possibly somewhat better, match is shown for a Kd of 21 ml g^{-1} (see Table 4). The fit may not appear as perfect as the non-sorbing uranine and the weakly-sorbing Na^{+}, however, with the breakthrough of $^{85}Sr^{2+}$ being - relative to anions and Na^{+} - shifted over orders of magnitude to larger peak times and lower peak concentrations, the relatively good agreement again raises confidence for the conceptually simple PSI model.

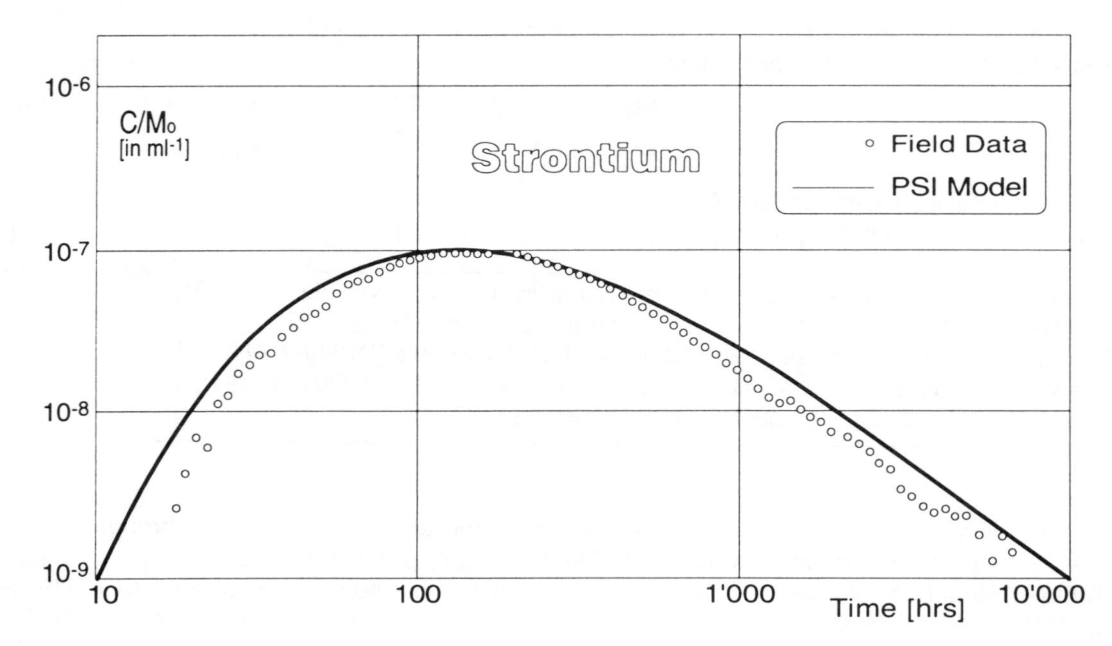

Fig. 10 Comparison of measured data from a field breakthrough experiment for the moderately sorbing tracer $^{85}Sr^{2+}$ in the 4.9 m dipole flow field and the calculated curve, based on fixed parameter values from model calibration and an equilibrium sorption coefficient Kd for Sr (which was a corrected laboratory value; see text). Although $^{85}Sr^{2+}$ exhibits a substantially different breakthrough curve to that observed for non-sorbing anions or the more weakly-sorbing $^{22}Na^{+}$, the good agreement between measured field data and model prediction lends strong support for the chosen model, as well the calibration procedures used.

The breakthrough curve for $^{85}Sr^{2+}$ does not show the typical "advective" peaks of uranine or Na^{+}. The "advective peak" could be viewed to represent mainly transport by hydrodynamic flow and to give a rough indication of effective water flow velocities. This is not shown by the more strongly sorbing $^{85}Sr^{2+}$ which seems to "reside" predominantly on matrix mineral surfaces during its transport through the 4.9 m dipole. Evidently the PSI model is able to reproduce quantively these processes which are- in contrast to the tracers used for model calibration - much more sensitive to the matrix parameters ε_p and $D_{p,}$.

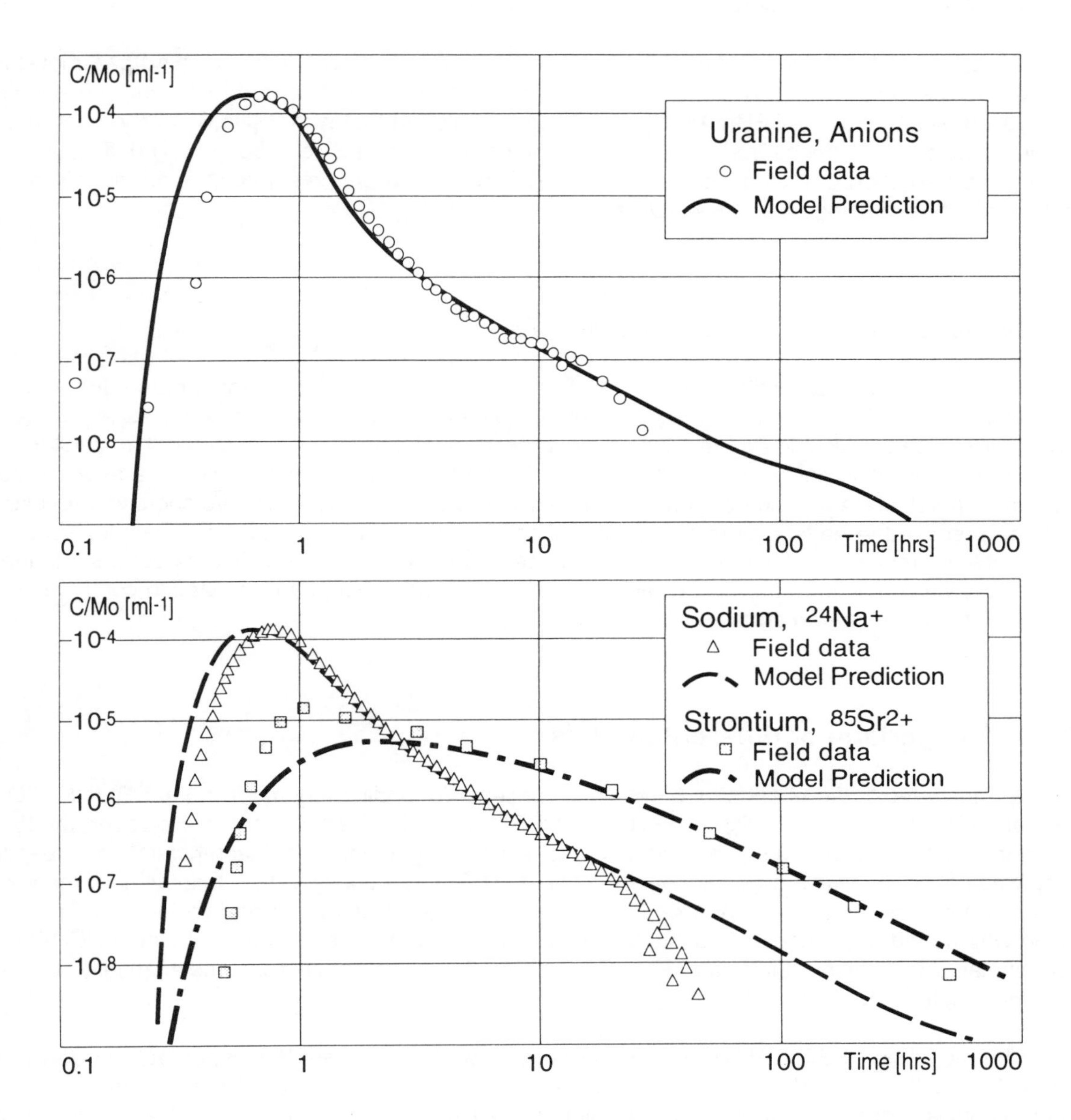

Fig. 11 Comparison of measured and model predicted breakthrough curves for uranine, $^{22}Na^{+}$ and $^{85}Sr^{2+}$ for a different flow field (1.7 m distance; this figure is adopted from HEER & HADERMANN,1994, p.98). The model includes the fit parameters derived from the 4.9 m flow fields (see Table 4) but takes into account the different hydrological flow conditions. For uranine (upper part) and Na (lower part) the agreement between model and experiment is very good, while Sr (lower part) is also adequately matched, although kinetic effects which are not included in the model may be the cause of the poorer match at early times.

Table 4 compiles the fixed and model derived parameters for the 4.9 m tracer tests with anions, $^{22}Na^+$ and $^{85}Sr^{2+}$. It is worth noting the physically and petrographically reasonable range of model derived figures which lends confidence to the applied model. For the cations Na^+ and Sr^{2+}, the model and laboratory derived Kds are roughly consistent, even if a "correction" factor for different effective surfaces accessible during laboratory and field testing were included . Table 4 also comprises the unavoidable uncertainties as obtained from a sensitivity analysis by HEER & HADERMANN (1994). These uncertainties are perhaps important for accurately fitting short-term experiments, however, they would be rather insignificant for safety analytical purposes.

The effects of a potential surface retardation factor R_f, - as an additional fitting parameter for sorbing species which would have a characteristic effect on the breakthrough curve - were also investigated by HEER & HADERMANN (1994). They concluded that with R_f[Na,Sr]=1 the best match between predicted and measured breakthrough curves for Na and Sr can be achieved. This implies that negligible sorption - with respect to the matrix - may occur on the surfaces or on the infill of the water conducting openings (i.e., "flow layers") .

12. Model testing different flow fields

Even more rigorous testing of the model is achieved by fixing all the model parameters for the 4.9 m tests (Table 4) and predicting experiments for different dipole flow conditions. As experiments were shown to be practicable on the 1.7 m scale, breakthrough curves were calculated for similar flow injection and extraction flow rates. As the flow is confined to a much smaller area, the average flow and transport velocities are expected to be substantially higher, thus, advective/dispersive processes would be predicted to have a more dominating effect than the matrix effects. The relatively accurate match between model and field data for all the tracers is shown in Fig. 11. This model testing again provides convincing support for the validity of the PSI modeling concept and enhanced confidence for the associated model parameters.

13. A comparison of different models

As illustrated in Fig. 7 and Fig. 8 the basic concepts for preliminary modeling at PSI (HERZOG, 1991) and model calibration at PSI (HEER & HADERMANN, 1994) are the same. However, the fixed parameters are quite different (see Table 4). As a proper geological characterisation of the pore space was not yet available for preliminary modeling, HERZOG's conservative estimates for the width of the migration zone may have provided insufficient "storage capacity" for retardation - a plausible reason why he did not succeed to achieve acceptable breaktrough tail fitting. In addition HERZOG's best estimates from fitting anion and $^{22}Na^+$ breakthrough curves resulted in a substantially different matrix parameter $\varepsilon_p \cdot D_p$ for both tracers.

The model by HEER & HADERMANN (1994), as well as a recent PNC approach by UMEKI et al. (1994), are both based on BOSSART & MAZUREK's (1991) geological characterisation. Both groups adopt a more realistic width of the migration zone. The key difference is that UMEKI et al. assume only one flow layer with a porous infill, while the PSI group assumes a few narrow, unfilled flow layers (see Fig. 8). In the PNC approach, retardation by sorption is assumed to occur not only in the diffusion or matrix porosity, but also within the flow layer. In contrast, HEER & HADERMANN (NTB 94-18, see Table 6) do not find detectable sorption within their flow layers (i.e., R_f=1) for their approach. A detailed comparisons of predicted and measured breakthrough curves for the 4.9 m as well as the 1.7 m dipole flow fields are currently not available and, hence, a rigorous test of the PNC model is not yet possible. Calculated curves for various Kd-values, however, show reasonable trends for tracer retardation.

A dependable model should also yield physically reasonable, model-derived parameters. The PNC approach (UMEKI et al., 1994) results in a flow porosity of ε_f = 13%; a value for the matrix porosity of $\varepsilon_p \geq 5\%$ can be estimated from $D_e = \varepsilon_p \cdot D_p = 5 \cdot 10^{-11} m^2 s^{-1}$ (see Table 4), assuming an absolute upper limit for a pore diffusion coefficient of $D_{p,} < 1 \cdot 10^{-9} m^2 s^{-1}$ (practically corresponding to diffusion in free water). Although the PNC model shall not be discussed further here, the question may arise, how a 10 mm wide flow layer with a "flow porosity" (or infill porosity) of ε_f=13% can represent mainly advective processes, while the 2 the adjacent matrix layers - each 45 mm wide and of similar porosity, that is with a model derived "matrix porosity" of $\varepsilon_p \geq 5\%$ - are assumed to include no advective flow, but molecular diffusion, exclusively.

14. Retardation and scale of transport

All tests carried out with anions Na, and Sr over the 1.7 and 4.9 m flow distance are plotted in Fig. 12. Also included are field tests with $^{137}Cs^+$, a non-linearly but more strongly sorbing cation - with sorption kinetics (see CONMANS & HOCKLEY 1992) which strongly affect the breakthrough under rapid flow conditions (1.7 m). The Cs experiments shall be discussed in detail elsewhere.

Fig. 12 depicts the different effects of retardation on the 1.7 and on the 4.9 m scale. The sorbing tracers Sr and Cs during experiments with high flow velocities and short residence time in the migration shear zone fracture (1.7 m dipoles) show little peak retardation, but significant tailing with respect to anions. During 4.9 m experiments, however, the cations show - depending upon their sorption coefficient - pronounced shifts in peak times, coupled with considerable dilution of the peak concentration. While anion breakthrough from 4.9 m experiments indicate an average flow velocity of roughly 1 $m \cdot h^{-1}$, the peak time for the mild sorber $^{85}Sr^+$ is around 200 hrs, for the moderately strong sorber ($^{137}Cs^+$) in the order of several 1000 hrs. For longer transport distances, the relative peak retardation and dilution is expected to be even more pronounced.

15. Matrix diffusion - does it really occur?

15.1 Process identification and experimental requirements

It must be noted that heterogeneously transmissive advection models, with molecular diffusion as a component of the hydrodynamic dispersion coefficient, have not been tested yet. Here, the discussion shall be limited to the homogeneous, PSI advection/diffusion model. HEER & HADER-MANN (1994; Section 4.2) have demonstrated that none of the measured breakthrough curves can be modeled by a simple, homogeneous single-porosity approach and that, for large portions of peak tail, the slopes are consistent with analytical solutions for "unlimited" (on the scale of this experiment) diffusion. In order to provide unambiguous evidence for such processes, that is to enable proper transport process identification, there are stringent requirements for the experimental procedures: (i) tracer analyses for a single pulse injection experiment with non-sorbing or weakly sorbing tracers must be extended by a factor of more than 100 over the observed breakthrough peak time and (ii) concentration measurements have to be made over 3 to 4 orders of magnitude between peak and tail end (see e.g., Fig. 9).

As there exists a large set of partially independent experiments which can be coherently modeled with physically reasonable parameters by the PSI dual-porosity approach, there is very strong evidence that matrix diffusion processes indeed can be identified even during the time scale of field experiments. This result could not have been guaranteed during the initial stages of the Grimsel migration experiment, as the suitability of practically achievable field tracer tests for unambiguously identifying matrix diffusion was seriously questioned by many experts.

In many other investigations (see e.g. SMITH et al.,1992 or HIMMELSBACH et al., 1994) matrix difffusion has been alledgedly identified from tracer breakthrough data which can only be reasonably presented - probably due to experimental reasons - on linear scales as exemplified in Fig. 4. Such a presentation does not provide sufficient and unambiguous evidence for process discrimination, as tail fitting on such a rough scale is only of marginally diagnostic value. As HEER & HADERMANN (1994) demonstrate for advectively dominated systems, only prolonged monitoring of the tailing sections provides clear evidence for matrix processes. Typical Grimsel experiments were extended over time spans of in the order of 500 x the peak time for non-sorbing tracers, while in many modeled advection/diffusion experiments elsewhere, tracers were analysed over less than a factor of 10 x the peak time.

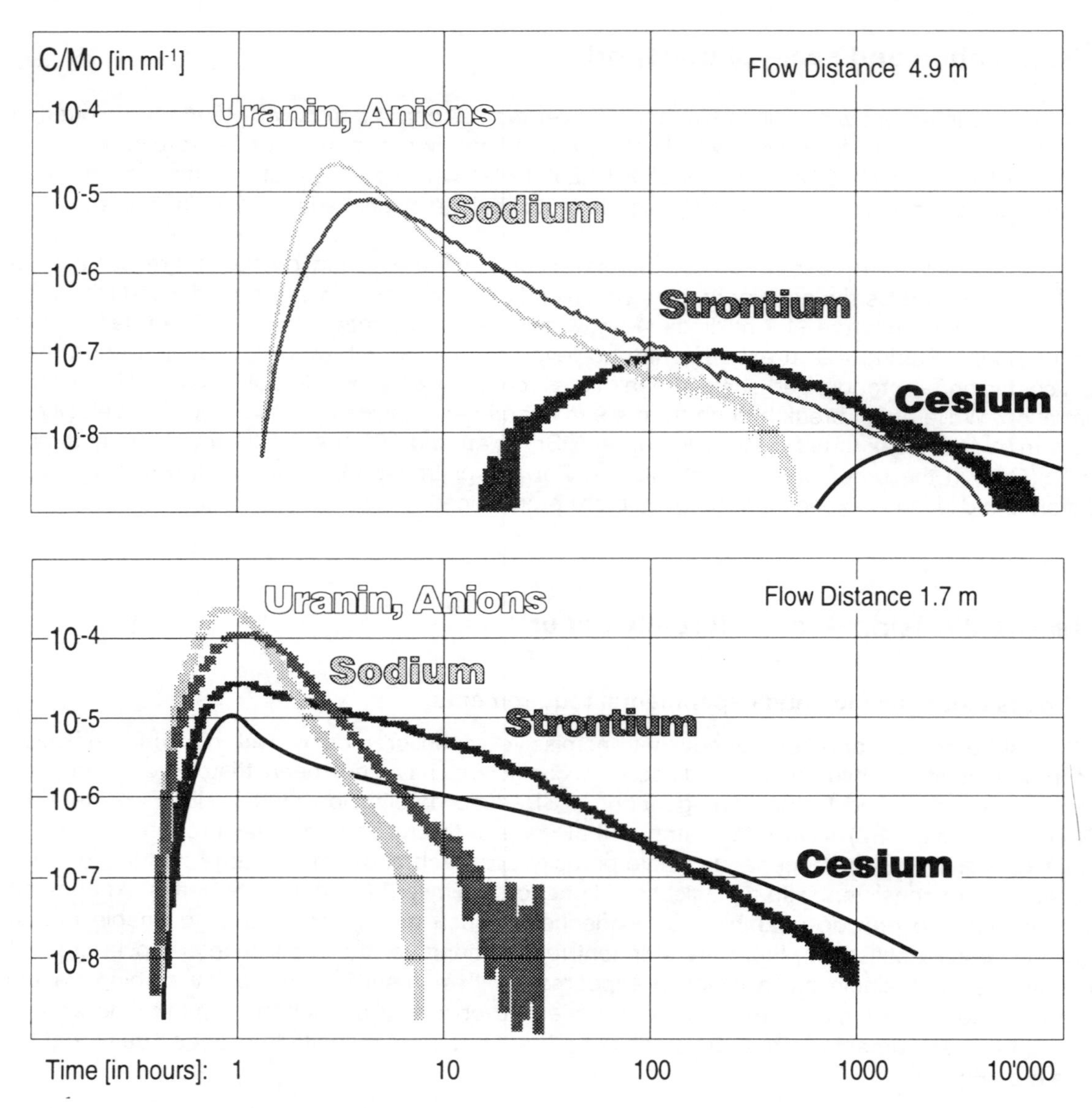

Fig. 12 Retardation for different dipole flow lengths but identical injection and withdrawal pump rates. For small fracture residence times and high flow velocities only minor peak time retardation is measured (1.7 m tests) after pulse injection of anions and the sorbing cations $^{22,24}Na^{+}$, $^{85}Sr^{2+}$, and $^{137}Cs^{+}$. Sorption is mainly evident from more pronounced tailing. Experiments on the 4.9 m scale reveal increasing peak retardation, coupled with strongly decreasing peak concentrations.

15.2 Independent evidence for diffusive processes

There exist independent tracer clues from short-time experiments which cannot be explained by homogeneous single-porosity concepts: a pulse injection test was executed over a 14 m dipole distance (see Fig. 13) with the non-sorbing anion tracers uranine, ^{123}I and the inert, dissolved gas tracer ^{3}He. All of these are characterised by substantially different diffusion coefficients D_w in free water of ~$0.5 \cdot 10^{-9} m^2 s^{-1}$, ~$2 \cdot 10^{-9} m^2 s^{-1}$ and ~$7 \cdot 10^{-9} m^2 s^{-1}$, respectively (FRICK, 1992; p. 20). The corresponding D_p values in a negatively charged mineral matrix are expected to be smaller, while the relative differences between the anions and He are expected to be larger. The increasing retardation (i.e., as seen from lower normalised peak concentrations and more pronounced tailing) for species of larger effective diffusion coefficients (e.g., ^{3}He), strongly supports dual-porosity advection/diffusion concepts. However, these experiments have not been analysed by the existing PSI model yet and it remains to be seen whether or not the observed differences for non-sorbing tracers can be quantitatively interpreted by using the PSI model.

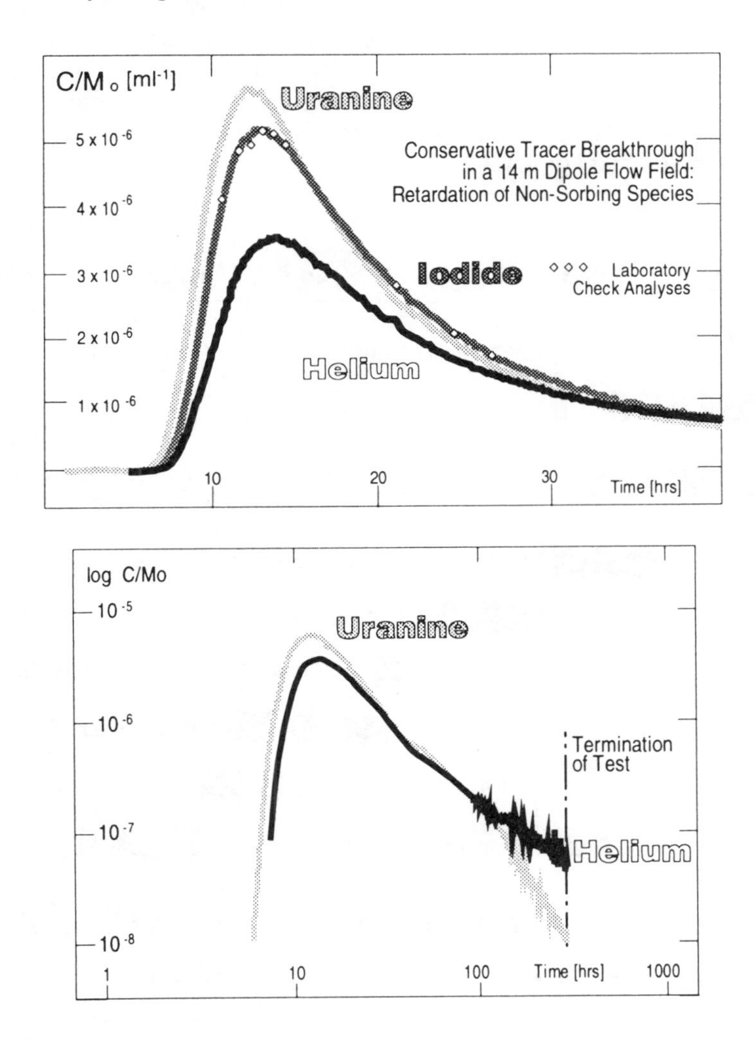

Fig. 13 Breakthrough peaks for the different non-sorbing tracers Uranine, ^{123}I, and ^{3}He. Due to its enhanced diffusion properties (see text), the He breakthrough is noticeably retarded with respect to the anions as seen from its relatively lower normalised peak concentration and more tailing (incompletely shown). The existence of effective molecular diffusion is therefore convincingly demonstrated. Such a result could not be expected in a homogeneous single-porosity flow system.

15.3 Matrix diffusion during experiments and under repository conditions

It can be estimated from the derived diffusion parameters that small portions of non-sorbing tracers may diffuse - within the typical duration of a Grimsel pulse injection experiment of a few hundered hours - no more than a few millimeter into a the porous matrix. This may not be considered as sufficient proof of evidence for effective matrix diffusion processes and dependable retardation during geosphere transport after a potential long-term nuclide realease from a repository. In certain cases, up to a few dm width for the diffusion space would be required (see e.g., Nagra's Kristallin I Safety Assement Report (NTB 93-22; Section 5). Therefore, convincing evidence for such a range of connected porosity perpendicular to a water conducting feature must be obtained from other types of experiments

Isotopic evidence for interconnected porosity extending from water conducting features may be provided by U/Th-decay series disequilibrium studies. For the Grimsel migration shear zone, ALEXANDER et al. (1990; Section 5) have reported activity disequilibrium profiles of some U daughter products which extend roughly 5 cm from an open fissure. Whatever the driving forces were, the range of a geologically rather recent mobilisation clearly indicates a minimum depth of potentially wetted rock matrix. Currently, additional experiments are planned: (a) excavation and post-mortem analyses of the migration shear zone and (b) connected porosity studies extending from water conducting features of the Grimsel granites.

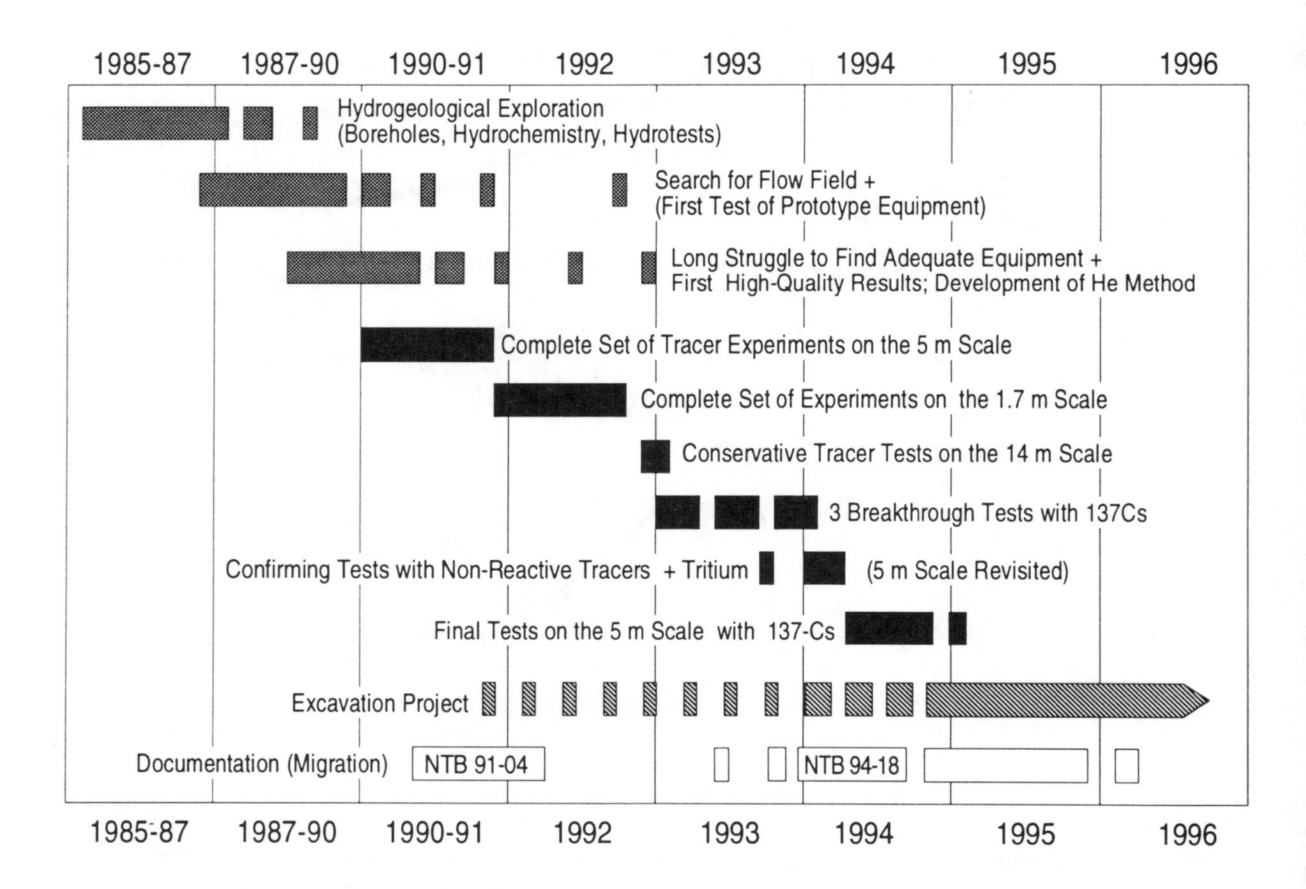

Fig. 14 Simplified time plan of the field activities during the Grimsel migration experiment (investigations during complementary laboratory or modeling programs are not included)

16. Conclusions

16.1 Successful model testing confirms crucial concepts of radionuclide transport

Recent attempts of modeling an extensive set of Grimsel field experiments, carried out with a variety of tracers and under different flow conditions, have convincingly demonstrated the applicability of relatively simple dual-porosity concepts and that the extrapolation of laboratory sorption values to field conditions - if adequate care is taken - can be achieved. Model-testing included successful and accurate predictions of peak concentration, peak time, and shape of breakthrough curve for many related as well as independent experiments. In addition, actually observed parameters from geological characterisations were included for the conceptualisation and model-derived parameters are physically reasonable and consistent in view of the current knowledge or empirically sound. Confident model testing can only be achieved when a number of requirements are met:

- Number of free variables: a simplified, but realistic transport model should comprise a minimum of free parameters. This can be achieved by aiming at simple experimental conditions, consideration of simple geometry and chosing a relatively simple conceptual approach.
- Choice of fixed or assumed parameters: any model includes implicitly or explicitly a number of underlying assumptions and limiting parameters. A state of the art petrographical characterisation of the water conducting features and their adjacent pore space is an absolute necessity for a meaningful conceptualisation of geosphere transport.
- High quality experiments for modeling: with experiments devised for transport process discrimination or identification, extremely high demands are set for the chemical and hydraulic stability of the artificially induced flow fields and the long-term stability and sensitivity of the tracer analyses. As monitoring the tail sections of the tracer breakthrough curve provides important clues for matrix processes, typical experimental periods of more the 100 x the peak time or in the order of 1000 hrs for a single experiment with non-sorbing tracers are required. Experimental artifacts from equipment sorption must be ruled out and instrumental dispersion effects must be quantified.
- Choice of tracers: (i) the injection of tracer should not disturb the natural hydrogeochemical conditions; this is ideally met by using carrier-free, radioactive tracers; (ii) application of different conservative tracers (e.g., fluorescent dyes, radioactive anions such as $^{82}Br^-$, tritiated water THO, ^{3}He) provides important clues for different particle size- or charge-related mechanisms during aqueous transport; (iii) conservative and sorbing tracers must be modeled by a consistent set of parameters; (iv) different tracers with various sorption coefficients, but undergoing similar sorbing mechanisms, are needed to further constrain the degree of freedom during modeling. To widen the more general understanding of retardation, it would be desirable to use tracers with a range of different sorption mechanisms as initially planned for the migration experiment.
- Choice of flow fields for tracer testing: (i) at least for non-sorbing tracers, recovery of injected tracers should be close to 100% (bounded system) within reasonably achievable time scales; (ii) narrow unequal-strength dipoles or quasi 1-D flow fields are much less sensitive to heterogeneities of the 2-D aquifer and, thus, enable a more sensitive transport process identification.
- Complementary laboratory program: laboratory sorption experiments must be executed under controlled atmospheric and simulated natural conditions. They are aimed to identify the specific sorption mechanisms for selected tracers (e.g., cation exchange for Na, Sr, Cs), to determine the corresponding equilibrium sorption coefficients and to demonstrate the reversibility of the sorption processes. Relevant rock material must be sampled, disaggregated and characterised in terms of grain-size distribution, surface properties and mineralogy. Conditioning of tracer solutions is also carried out and

backup analyses of aliquotes samples are made for accurate calibration of the on-line tracer detection.

- Hydrochemical modeling: is needed to infer the pH- and Eh- buffer capacity of the rock/ groundwater system, to estimate potential effects of precipitation or dissolution phenomena on certain tracers, to quantify possible rock-water interaction processes, to identify "invisible" solid phases, etc. Application of such codes aid establish or define adequate experimental conditions for tracer injection, to infer the stability of certain equipment components under the given chemical conditions and to estimate the sensitivity of the natural hydrogeochemical system to potential disturbances.

- Time and costs: It is quite obvious that such a multidisciplinary venture for performing unambiguous process identification and model-testing on an adequate level requires substantial time and resources. For future planning of similar projects, time scale and evolution of the Grimsel project are illustrated in Fig. 14. As much of the extension of the timescale relative to early plans was due to equipment problems, and development or modification of an appropriate experimental setup, the practical experience built up at Grimsel could greatly help to speed up subsequent studies elsewhere.

16.2 Confidence in the applied methodology

Different tracer tests at the Grimsel test site were done at groundwater velocities of in the order of 1 m per hour, over distances of about 2 to 14 m, with the tracers being injected as short-time pulses of a few minutes, and several weeks to months required for full recording of a single test. For practically achievable experiments, a small percentage of a non-sorbing tracer may show residence times of up to 100 hrs. The corresponding maximum range for molecular diffusion in the pore water of the mineral matrix is in the order of mm. Time and spatial scales for groundwater transport after a potential release from a waste repository are quite different; the groundwater flow velocities will be certainly far smaller and the range of diffusive processes must be considerably larger (say ~1 dm or more for current Swiss concepts to result in the required retardation following of a long-term nuclide release). With the Grimsel experiments Nagra has demonstrated that - at least for geosphere transport modeling, as an important component of ongoing safety assessment efforts - the current methodology of hydrological and geological characterisation of water conductive features, followed by their abstraction for relatively simple conceptual approaches, and applying standard numerical solutions for advective/diffusive transport, are all indeed applicable steps for confidently or conservatively predicting solute transport through a host rock. In a sense it is hoped that the "validity" of Nagra's methodological approach has been shown..

17. Acknowledgements

The author gratefully acknowledges the invaluable efforts and contributions of a large number of individuals from many different institutions who have contributed to this successful multidisciplinary project. Particularly appreciated is the dedicated experimental support of the on-site radiotracer analyses by E. Reichlmayr (Institute for Hydrology GSF Munich), for development and operation of the superb field testing equipment by the creative staff of SOLEXPERTS, especially T. Fierz. Special thanks are given to E. Hoehn (EAWAG) for his hydrogeological expertise, P. Bossart (Geotechnical Institute) and M. Mazurek (GGWW, University of Berne) for their clarifying geological missions and to J. Eikenberg (PSI), for his operational laboratory input and radiotracer analyses. Careful investigations during the laboratory program and valuable advice was provided by the staff of the PSI Hotlab (M. Bradbury, B. Baeyens, S. Aksoyoglu, and many others). The fruitful interaction between experimentalists and modelers was obtained from the PSI transport modeling group, especially

W. Heer. Special thanks to I. McKinley (Nagra) and R. Alexander (GGWW, University of Berne) for their competent input during all stages of the experiment and their constructive criticism for improving this paper.

The collaborating staff is aware that the Grimsel migration experiment was an unusually long-time and costly venture. Nagra, the Japanese Power Reactor and Nuclear Fuel Development Corporation (PNC), and the Paul Scherrer Institute (PSI Villigen) generously provided the facilities, manpower and funding to produce an excellent experimental setup, a high-quality, complementary laboratory programs and careful modeling.

18. References

AKSOYOGLU S., BAJO C., & MANTOVANI M. (1991): Grimsel Test Site: Batch sorption experiments with iodine, bromine, strontium, sodium and cesium on Grimsel mylonite. Nagra Technical Report NTB 91-06, Nagra, Wettingen, Switzerland (also published as PSI Report Nr. 83, Paul Scherrer Institute, Villigen, Switzerland).

BOSSART P. & MAZUREK M. (1991): Grimsel Test Site: Structural geology and water flow paths in the migration shear zone. Nagra Technical Report NTB 91-12, Nagra, Wettingen, Switzerland.

BRADBURY M.H., editor (1989): Grimsel Test Site: Laboratory investigations in support of migration experiments. Nagra Technical Report NTB 88-23, Nagra, Wettingen, Switzerland.

BRADBURY M.H.,and BAEYENS B. (1992): Modelling the sorption of Cs: application to the Grimsel migration experiment. PSI Annual Report 1992, Annex IV, PSI Nuclear Energy Research, Progress Report 1992, p. 59-64; Paul Scherrer Institute, Villigen, Switzerland.

CONMANS R.N.J. and HOCKLEY D.E. (1992): Kinetics of cesium sorption on illite. Geochim. Cosmochim. Acta Vol. 56, p.1157-1164.

EIKENBERG J., BAEYENS B., & BRADBURY M.H. (1991): The Grimsel migration experiment: a hydrogeochemical equilibration test. Nagra Technical Report NTB 90-39, Nagra, Wettingen, Switzerland (also published as PSI Report Nr.100, Paul Scherrer Institute, Villigen, Switzerland).

EIKENBERG J., HOEHN E., FIERZ Th. (1994): Grimsel Test Site: Preparation and performance of migration experiments with radioisotopes of sodium, strontium and iodine. Nagra Technical Report NTB 94-17, Nagra, Wettingen, Switzerland (also published as PSI Report Nr. 94-11, Paul Scherrer Institute, Villigen, Switzerland).

FRICK U., ALEXANDER W.R., BAEYENS B., BOSSART P., BRADBURY M.H., BÜHLER Ch., EIKENBERG J., FIERZ Th., HEER W., HOEHN E., McKINLEY I.G., and SMITH P. (1992): Grimsel Test Site: The radionuclide migration experiment - overview of investigations 1985-1990. Nagra Technical Report NTB 91-04, Nagra, Wettingen, Switzerland.

FRICK U., (1992): Beurteilung der Diffusion im Grundwasser von Kristallingesteinen - ein Beitrag zur Kristallinstudie 1993. Nagra Internal Report of limited distribution, is available on request (in German, 130 p., including appendices by K. Skagius and W.R. Alexander). Nagra, Wettingen, Switzerland.

HEER W. & HADERMANN J. (1994): Grimsel Test Site: Modeling radionuclide migration field experiments. Nagra Technical Report NTB 94-18, Nagra, Wettingen, Switzerland (also published as PSI Report Nr. 94-13, Paul Scherrer Institute, Villigen, Switzerland).

HERZOG F. (1991): Grimsel Test Site: a simple transport model for the grimsel migration experiments. Nagra Technical Report NTB 91-31, Nagra, Wettingen, Switzerland (also published as PSI Report Nr. 106, Paul Scherrer Institute, Villigen, Switzerland).

HIMMELSBACH Th., HÖTZL H., & MALOSZEWSKI P. (1994): Forced gradient tracer tests in a highly permeable fault zone; Applied Hydrogeology 3/94; p. 40-47.

HOEHN E., FIERZ Th., & THORNE P. (1990): Grimsel Test Site: Hydrogeological characterisation of the migration experimental area. Nagra Technical Report NTB 89-15, Nagra, Wettingen, Switzerland (also published as PSI Report Nr. 60, Paul Scherrer Institute, Villigen, Switzerland).

HSK (1993) Guideline HSK-R-21/e: Protection Objectives for the Disposal of Radioactive Waste; Swiss federal Nuclear Safety Inspectorate (HSK) and federal Commission for the safety of Nuclear Installations (KSA); Revised Version November 1993; CH-5232 Villigen-HSK, Switzerland

JAKOB A., HADERMANN J. & ROESEL F. (1989): Radionuclide chain transport with matrix diffusion and non-linear sorption. Nagra Technical Report NTB 90-13 Nagra, Wettingen, Switzerland (also published as PSI Report Nr. 54, Paul Scherrer Institute, Villigen, Switzerland).

MORENO L. & TSANG C.F. (1991): Multiple peak response to tracer injecxtion tests in single fractures: a numerical study. Water Resources Research. Vol.27, Nr.8; pp.2143-2150

NAGRA PG85 - Project Gewähr 1985; Vols.1-8, Vol.9 (English Summary; Nagra Gewähr Report Series NGB 85-01/09), Nagra, Wettingen, Switzerland.

NAGRA NTB 93-22 (1993a): Kristallin I - Safety Assessment Report. Nagra Technical Report, Nagra, Wettingen, Switzerland.

NAGRA NTB 93-26 (1993b): Beurteilung der langzeitsicherheit des Endlagers SMA am Standort Wellenberg - Endlager für kurzlebige schwach- und mittelaktive Abfälle (Endlager SMA). Nagra Technical Report, Nagra, Wettingen, Switzerland.

THURY M., GAUTSCHI A., MÜLLER W.H., NAEF H., PEARSON F.J., VOBORNY O., VOMVORIS S. & WILSON W. (1994) Geologie und Hydrogeologie des Kristallins der Nordschweiz. Nagra Technical Report NTB 93-01, Nagra, Wettingen, Switzerland.

UMEKI H., HATANAKA K., ALEXANDER W.R., McKINLEY I.G., & FRICK. U.(1994): The Nagra/PNC Grimsel Test Site radionuclide migration experiment: rigorous field testing of transport models; Preprint submitted to the Proceedings of the XVIII International Symposium on the Scientific Basis for Nuclear Waste Management, October 23-27, 1994, Kyoto, Japan.

Redox Processes in Disturbed Groundwater Systems: Conclusions from an Äspö-HRL Study

Steven Banwart
The Royal Institute of Technology

Peter Wikberg
The Swedish Nuclear Fuel and Waste Management Company (SKB)

(Sweden)

Abstract

The Swedish concept for disposal of high level waste includes isolation of spent fuel in copper canisters, buried deep within granite bedrock. Deep repository construction will lead to a highly disturbed hydraulic system in the surrounding aquifer. In addition to opening the deep environment to atmospheric conditions, there is concern that dissolved oxygen may enter vertical fracture zones with the increased surface water recharge. Oxygen can lead to canister corrosion, and enhanced mobility of redox-sensitive radionuclides; Technetium, Plutonium, Neptunium, Uranium. This study addresses the possible inflow of oxygenated surface water into vertical fracture zones during construction of the Äspö Hard Rock Laboratory (HRL).

Entrance tunnel construction at the ÄHRL opened a conductive vertical fracture zone at a depth of 70 meters on March 13, 1991. Three weeks later a sharp dilution front, corresponding to 80% shallow water inflow to the previously saline fracture zone, arrived at the entrance tunnel depth. The predicted outcome of the experiment was that reaction with Fe(II)-bearing fracture minerals would retard oxygen breakthrough. This oxidizing alteration would change the geochemical character of the migration barrier provided by fracture-filling material. The observed outcome is that almost no oxygen breakthrough ever occurred.

In spite of the large inflow of shallow water, the fracture zone remains persistently anoxic. There is increased input of organic carbon, rather than molecular oxygen, with the surface water inflow. Microbial activity is important in the upper bedrock environment studied here. Because of the inflow and microbially-mediated oxidation of organic carbon, this fracture performs as a conduit for additional reducing capacity into the deep environment, rather than as a path for penetration of dissolved oxygen.

INTRODUCTION

The SKB concept for disposal of high activity nuclear waste includes isolation of spent fuel in copper canisters buried several hundred meters in granite bedrock. The most critical safety aspect is the design of engineered barriers; i.e. canisters and backfill material. Safety assessment must consider eventual failure of these barriers. Exposure of the biosphere to long-lived radionuclides then depends on hydrology and radionuclide adsorption and solubility. Of special concern are the long-lived isotopes of Neptunium, Plutonium, Technetium, Iodine, and Cesium. Of these elements, Technetium and the actinides form sparingly soluble solid phases when reduced but are highly soluble under oxic conditions.

Performance assessment issues related to groundwater redox chemistry are:

1. The presence and fate of molecular oxygen, which can corrode copper metal, in the deep environment at the time of repository closure.

2. Production of hydrogen sulfide, which can corrode copper metal, in the deep environment after repository closure.

3. Dissolved oxygen in vertical fracture zones due to increased surface water inflow during construction and operation of the repository.

The Äspö Redox Experiment in Block Scale addresses the performance of vertical fracture zones as conduits for dissolved oxygen transport from the surface during repository construction.

On March 13, 1991 construction of the access tunnel to the Äspö Hard Rock Laboratory (HRL) intersected a vertical fracture zone at a depth of 70 meters. The fracture zone hydrochemistry was then monitored through time. Groundwater sampling and analysis before intersection by tunnel construction, and examination of drillcores taken from the fracture zone early in the experiment, provided a reference state against which to compare subsequent evolution of groundwater conditions. The shallow groundwater was dilute while the native groundwater is extremely saline. Dilution of chloride ion was a good indicator of surface water intrusion.

OBJECTIVES

The objectives of this experiment are:

1. to determine the extent of surface water intrusion induced by opening the fracture zone at a depth of 70 meters,

2. to observe whether molecular oxygen transport from the surface or from the tunnel can create oxic conditions in the fracture zone,

3. to assess dominant transport and reaction processes controlling the fracture zone geochemistry under the disturbed conditions.

PREDICTION OF OXYGEN FRONT BREAKTHROUGH

We predicted the breakthrough times for surface water and dissolved oxygen to the 70 meter depth expected upon intersection of the fracture zone (Banwart and Gustafsson, 1991). The objective of these predictions was to list possible influences on oxygen transport and consumption, and to understand the relative importance of hydraulic and chemical parameters; porosity, wetted surface area, reducing capacity. We predicted surface water breakthrough to occur between 6 hours and 21 days after opening the fracture zone by tunnel construction.

We predicted retardation of the oxygen front by reaction with the reducing capacity of the fracture zone which was quantified by the concentration of dissolved organic carbon, dissolved Fe(II)- and Mn(II)-species and the abundance of reducing mineral surfaces in contact with the wetted flow path. Because of the large uncertainty in wetted surface area, there was an associated uncertainty in the reducing capacity. We assumed that recharge to the fracture zone contained dissolved oxygen in equilibrium with the atmospheric reservoir. We predicted oxygen breakthrough to occur between 36 hours and 5 years.

OBSERVED RESULT OF ENHANCED SURFACE WATER INFLOW

A sharp dilution front arrived in the access tunnel 21 days after intersection of the fracture zone. A short-lived redox breakthrough was observed between days 25 - 50, as evidenced by the temporary disappearance of dissolved iron in the inflow to the access tunnel during this period. The breakthrough is expected based on the conceptual model upon which the predictions are based. This short-lived breakthrough occurred within the wide range of predicted breakthrough times. However, the return to reducing conditions after day 50 showed that, in fact, the boundary conditions chosen for our predictions were incorrect. We assumed a continuous input of dissolved oxygen with the recharge, and a finite reservoir of reducing capacity within the fracture. Our main conclusion from this project is that rather than adding molecular oxygen, increased inflow of surface water adds reducing capacity to the fracture zone in the form of very young organic carbon.

This increases rates of anaerobic respiration within the fracture zone as shown by elevated $PCO_{2(g)}$, ^{13}C signatures in dissolved inorganic carbon that are consistent with input of biogenic $CO_{2(g)}$ and significant increases in the per cent modern carbon in both dissolved organic and inorganic carbon.

Three years after the initial disturbance by tunnel construction, the fracture zone remains persistently anoxic. Molecular oxygen is apparently transport-controlled by slow diffusion processes. Any traces of O_2 entering the fracture zone are apparently consumed in the most shallow reaches of the overburden, and in a relatively thin reactive "skin" around the access tunnel.

PROJECT ORGANIZATION

This project was not a blind compilation of results from a variety of field investigations. We planned and carried out this work as a field research experiment. We defined our initial hypothesis by predicting the surface water and oxygen breakthrough times in the tunnel. We designed field investigations to systematically test the conceptual model upon which we based the predictions. In addition to identifying dominant hydrologic and chemical processes, we needed to quantify element mass balances and process rates. Figure 1 represents the sequence of activities that developed during the project.

Figure 1. Project stages from initial predictions to development of a site model for coupled hydrologic, chemical, and microbiological processes

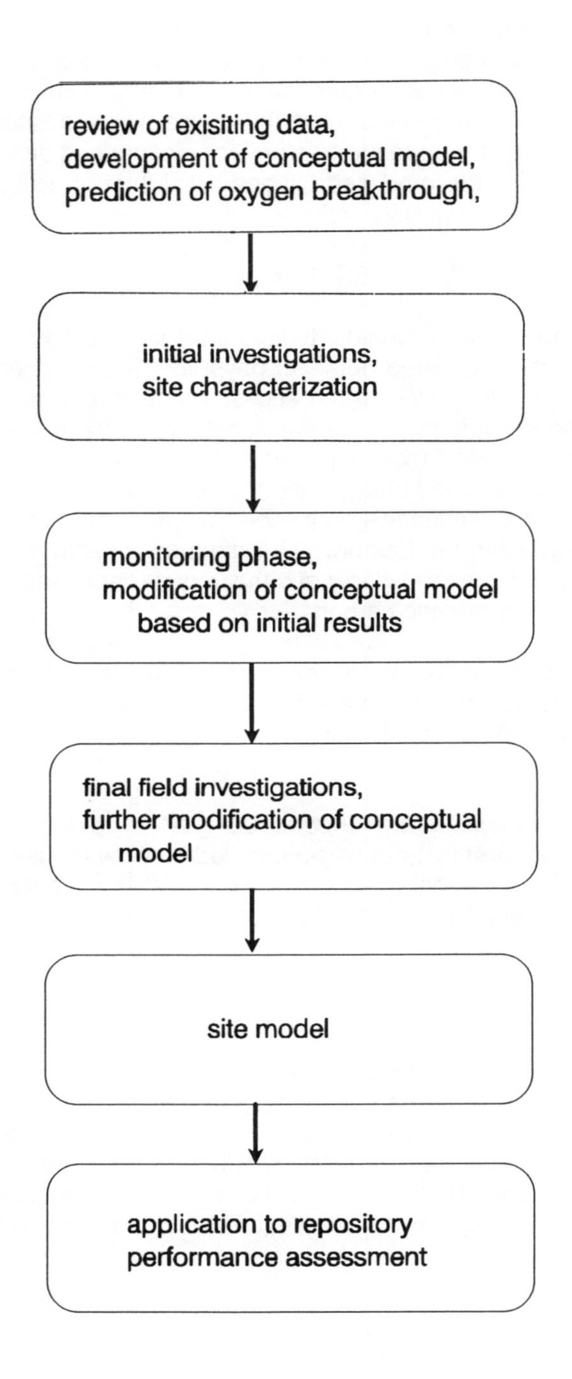

After the first 7½ months of investigations we compiled the results as a site characterization (Banwart et al.,1992a). During the second year of the project, we monitored the fracture zone hydrochemistry and also re-evaluated our predictions and the initial conceptual model (Banwart et al.,1992b; Banwart et al.,1994a). The revised model was then tested by a final series of specific field investigations during 1993. Based on these results, we developed a site model for coupled hydrologic, geochemical and microbiological processes (Banwart et al., SKB report in preparation).

The scientific coordinator (S. Banwart) had the responsibility to synthesize project results and conclusions, and to report them formally to SKB. Interim results and conclusions were reported at the tertial meetings of the SKB groundwater chemistry program where decisions on experimental planning were taken. Figure 2 shows a schematic representation of the project organization.

It was apparent from early in the project that investigation techniques would need to be combined from a variety of disciplines; hydrochemistry, hydrology, mineralogy, colloid chemistry, microbiology, stable isotope geochemistry. Synthesis of a model for coupled hydrological, geochemical, and microbiological processes at the site was a collaborative effort between investigators. This required project participants to learn basic theoretical concepts, and to understand field investigation techniques, from all of these disciplines.

Figure 2. Project organization for the Äspö redox experiment in block scale

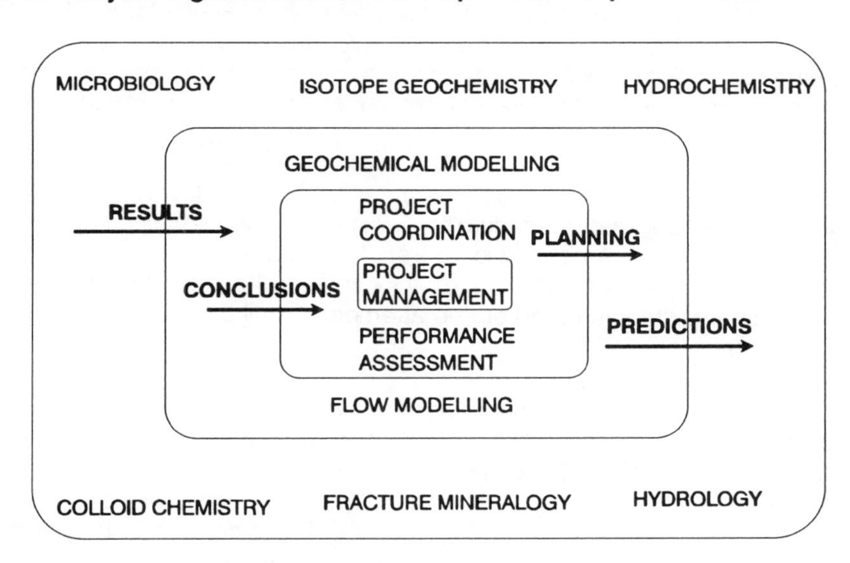

SITE DESCRIPTION

The Äspö Hard Rock Laboratory (HRL) is located on the Baltic coast near the town of Oskarshamn in S. E. Sweden. The access tunnel to the HRL starts at the coastline and proceeds approximately 1 km out under the sea floor and terminates under the island of Äspö. The experimental site for this project is on the small island of Hålö lying between the coastline and Äspö. Figure 3 shows the location of the HRL and the fracture zone studied in this project. The access tunnel to the HRL is shown as a heavy black line starting at Simpevarp on the Baltic coast and passing under the sea floor to the island of Äspö. The laboratory is housed in the tunnel that spirals down to a depth of 500 meters below Äspö.

The island of Hålö comprises a slightly undulating topography of well exposed rock 5 - 10 m above sea level. The geology is characterized by a red to gray porphyritic granite-granodiorite known locally as "Småland" granite, belonging to the vast Transscandinavian Granite-Porphyry Belt with U-Pb intrusion ages between 1760 - 1840 Ma, i.e., late- to post-orogenic in relation to the Svecofennian origin (1800-1850 Ma). The major fractures and fracture zones control recharge, discharge and ground water flow through the island(s) (Smellie and Laaksoharju,1992).

The fracture zone studied here transects the island of Hålö and appears from the surface as a small ditch in the exposed granite. The depression is 2-5 m wide, 2-3 meters deep, and extends laterally across the northern tip of the island (220 m). The topography along the fracture zone defines a catchment area, on Hålö itself, of 10,000 m^2. Considering average annual precipitation (550-675 mm) and evapotranspiration (500 mm) leaves 50-175 mm each year for runoff and groundwater recharge. This corresponds to a maximum recharge of 1.0 - 3.3 L min^{-1} from Hålö alone. Prior to intersection of the fracture zone by tunnel construction, the water table above the tunnel was approximately 0.5 below the ground surface and approximately 1.5 m above the surrounding level of the Baltic Sea. There was only a small drawdown in the water table during the experiment, i.e., the fracture zone remained hydraulically saturated.

Within the depression, directly above the tunnel, there is a 0.2 m deep layer of soil. This soil overlies a zone of re-worked sand and gravel extending to 0.5 meters, below which a layer of glacial clay extends to at least 1 meter depth. A percussion borehole drilled from the surface showed a granite base at a depth of 5 meters. A moraine layer exists between the glacial clay and the granite, as indicated by loose moraine fragments found during attempts to drill a borehole into the fracture zone at a depth of only 3 meters.

The plane of fracture is approximately vertical and is clearly visible from the interior of the access tunnel as a band of water-bearing fractured rock with a nominal width of 1 meter. The tunnel intersects the vertical fracture plane 513 meters from the tunnel mouth.

Figure 4 summarizes the fracture zone mineralogy and the Fe(II) content of host rock and fracture minerals based on characterization of drillcores take from the three boreholes in the side tunnel. The unaltered host rock is a typical granite composed of 30% quartz, 30% plagioclase, 30% K-feldspar, and 10% biotite. Upon approaching the altered section of the drillcores, there is increasing (ancient) oxidation of the rock. Complete oxidation of magnetite to hematite and extensive alteration of biotite to chlorite characterize the altered layer. Micro-fractures, filled with what appears to be iron oxyhydroxide, appear when approaching the fracture surface. Chlorite, calcite and epidote are the most abundant fracture minerals. Clay minerals, iron oxyhydroxides, and some hematite occur.

Based on the groundwater classification for Äspö (Smellie and Laaksoharu, 1992) and the location of this fracture zone within the Baltic Archipelago, we identified three dominant water types at this site; dilute shallow recharge water ([Cl^-]=10 mg L^{-1}), saline native groundwater ([Cl^-]=5000 mg L^{-1}), and brackish water from the Baltic Sea([Cl^-]=2000-4000 mg L^{-1}). Table 1 lists characteristic compositions of these three dominant hydrochemical resevoirs.

Figure 3. Location of the Äspö Hard Rock Laboratory (HRL), the fracture zone studied on Hålö island and the location of the sampling boreholes (see results section)

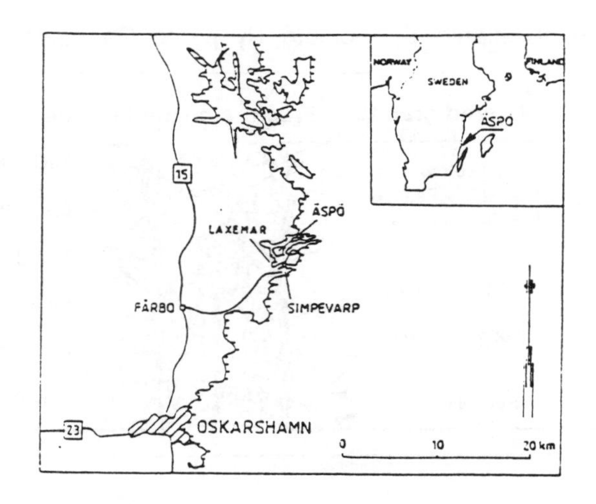

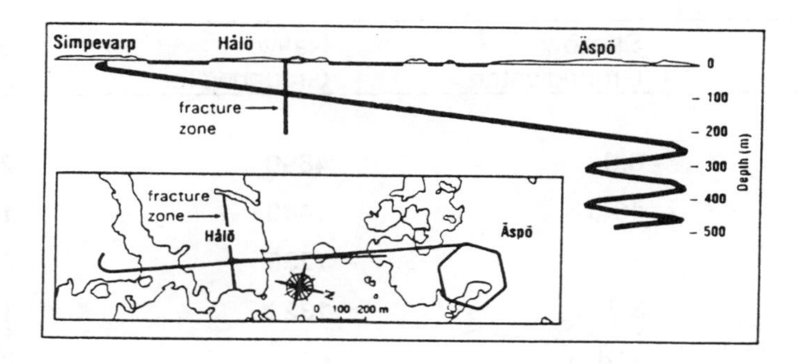

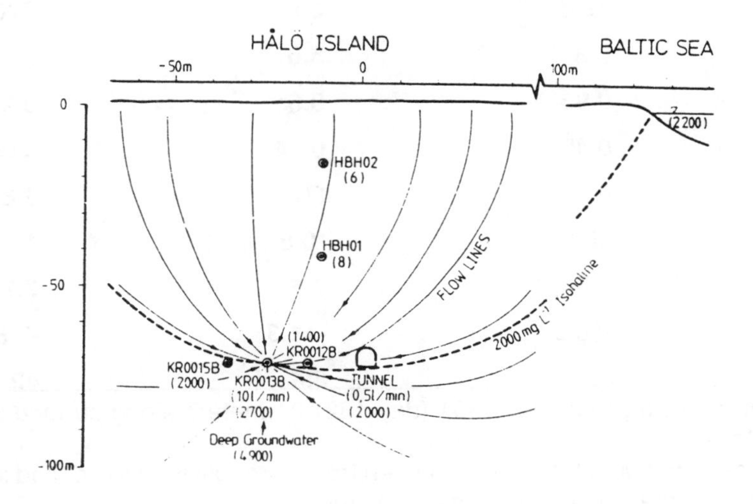

Figure 4. Changes in fracture zone mineralogy going from fresh granite (at left) to highly altered fracture filling material

(Chl=chlorite, Ep=epidote, Py=pyrite, Ca=calcite)

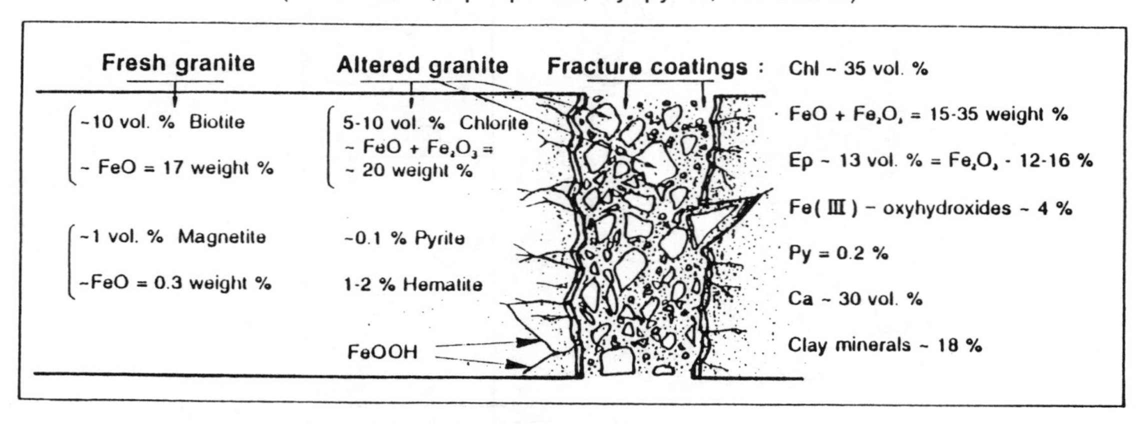

Table 1. Chemical composition of shallow and native groundwater and Baltic Sea Water

Variable	Shallow Groundwater	Native Groundwater	Baltic Sea water
$[Cl^-]$, mg L^{-1}	**6.0**	**4890**	**2180**
$[Na^+]$, mg L^{-1}	**10.3**	**1480**	**1160**
$[Ca^{2+}]$, mg L^{-1}	**43**	**1250**	**61**
$[Mg^{2+}]$, mg L^{-1}	**3.3**	**132**	**140**
$[HCO_3^-]$, mg L^{-1}	**114**	**42**	**22**
$[SO_4^{2-}]$, mg L^{-1}	**19.5**	**60**	**320**
$[Si]_{total}$, mg L^{-1}	**6.8**	**5.6**	**2.6**
$[Fe]_{total}$, mg L^{-1}	**0.95**	**0.6**	**0.2**
$[Mn]_{total}$, mg L^{-1}	**0.43**	**0.94**	**0.05**
[Br-], mg L^{-1}	**0**	**30**	**3.6**
TOC, mg L^{-1}	**20.0**	***0.5**	**16**
pH	**6.6**	**7.5**	**7.7**
$\delta^{18}O$, o/oo SMOW	**-10.2**	**-11.3**	**-7.8**
^3H, TU	**60**	**8.4**	**42**

Shallow Groundwater: sampled Sept. 12, 1991 (day 182) 24 meters along surface borehole HBH02 (15m depth from surface).

Deep Groundwater: sampled March 12, 1991 (day -4) from borehole KA0483 drilled into fracture zone at 70 meter depth prior to intersection by tunnel construction.

Baltic Sea water: sampled May 14, 1992 (day 428) from 2 m depth between Äspö and Hålö.

* TOC (total dissolved organic carbon): was not analyzed in the sample taken from KA0483, but was estimated by comparing data for similar groundwaters in the Äspö area.

EXPERIMENTAL METHODS

Experimental methods are described in detail in interim SKB reports (Banwart and Gustafsson,1991; Banwart et al., 1992a,b) and in two peer-review publications (Banwart et al.,1994a,b). Of particular interest here is application of a flow-through measurement cell connected on-line to the borehole (KR0013B) that remained open for continuous discharge during the experiment, except for the first 7 months. This on-line cell measured redox potentials using inert electrodes of differing material; glassy carbon, platinum and gold. Measured potentials agreed well and remained stable between calibrations. Because redox electrodes are extremely sensitive to dissolved oxygen, stable measurements of low potentials (<-100 mV, relative to the standard hydrogen potential, observed here) over long periods of time, with good agreement between electrodes, are a strong indicator of prevailing anoxic conditions.

MAIN RESULTS

Banwart et al. (1994a,b) report hydrochemical data that is summarized here. Figure 5 shows changes in chloride and dissolved iron with time during the experiment. The sampling locations are shown in Figure 3.

Figure 5. Changes in dissolved chloride ion and iron(II) concentrations during the experiment

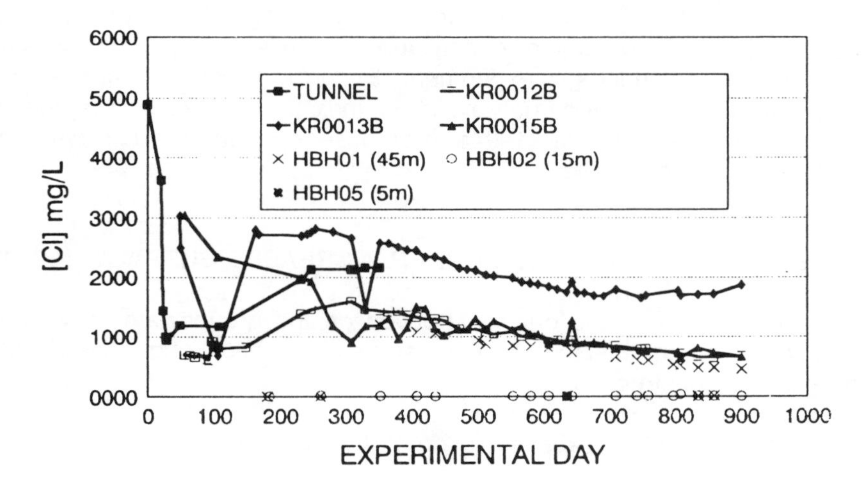

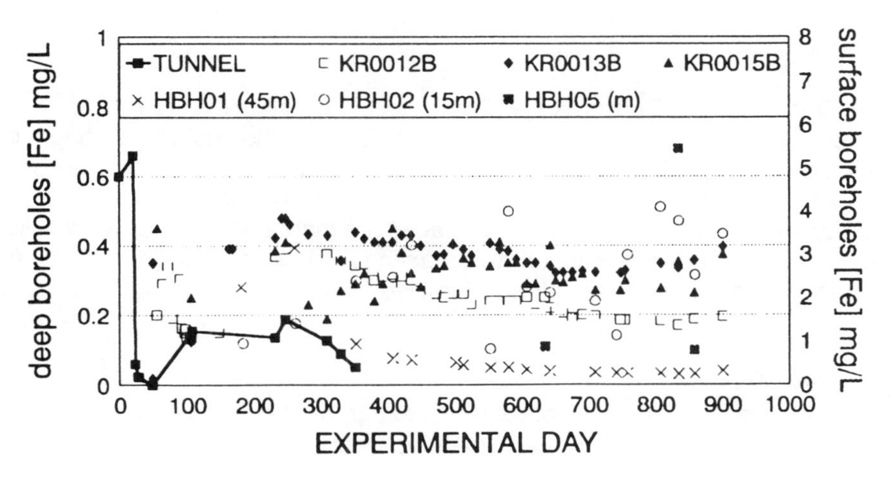

The sharp dilution front is seen at day 20, with a subsequent decrease in dissolved iron between days 20 and 50. This decrease in iron may represent a short-lived oxygen breakthrough at the 70 meter depth. More importantly, iron concentrations then increased and remained high during the remainder of the experiment.

Redox potentials measured in the flow-through cell connected on-line to the discharge borehole ranged between -100 mV to - 150 mV., decreasing slowly during the experiment. The bicarbonate concentration (not shown) increased during the same time period, starting at 200 mg/L and finally leveling off at 300 mg/L in the discharge borehole by the end of the experiment. The pH remained nearly constant throughout the experiment, varying only within a narrow range between pH 7.6 and 8.0. Gas sampling and inorganic speciation calculations agreed well and showed a partial pressure of carbon dioxide near $PCO_{2(g)} = 10^{-2}$ atm. Results from carbon stable isotope analysis of dissolved inorganic carbon, and carbon-14 dating of dissolved organic and inorganic carbon confirmed ongoing respiration of organic carbon as the dominant input of inorganic carbon to the groundwater.

Banwart et al. (1994a,b) propose the following stoichiometric reaction to represent the dominant pathway for oxidation of organic carbon during the experiment.

$$4Fe(OH)_{3(s)} + CH_2O + 8H^+ = 4Fe^{2+} + CO_{2(g)} + 11H_2O \quad (1)$$

This process is also supported by results from identification and viable counts of bacteria sampled from the groundwater (Pedersen et al., SKB technical report in preparation). An important result was that no hydrochemical evidence of sulfate reduction, nor microbiological evidence of viable sulfate-reducing bacterial populations, were found during this experiment. Because dissolved iron(II) is not mobilized on a scale anticipated due to the degree of carbon oxidation, indicated for example by the large increase in biogenic inorganic carbon, Banwart et al. (1994a,b) propose that iron(II) is incorporated into fracture minerals; iron-carbonates, -oxides or -silicates.

THE PERFORMANCE OF FRACTURE ZONES UNDER INCREASED SURFACE INFLOW

We consider the effect of repository construction on the groundwater environment. In this experiment there are two dominant effects from increased surface inflow under the disturbed hydraulic conditions created by tunnel construction.

1. increased surface water inflow causes dilution of the saline groundwater

2. increased surface water inflow adds reducing capacity in the form of organic carbon to the groundwater

Dilution Effects On Fracture Zone Performance

Groundwater salinity affects aqueous speciation and colloid stability through ionic strength effects. Changes in aqueous speciation affect the solubility of mineral phases and the sorption behavior of trace elements. Decreasing ionic strength favors colloid stability and thus transport of colloids and associated sorbed trace elements. Ion exchange surfaces are increasingly selective for divalent cations upon increasing dilution. Groundwater dilution and reaction with ion-exchange minerals may therefore result in lower calcium/sodium ratios in solution. This also favors colloid stability and can affect calcite solubility.

Conditions for Prevailing Anoxia

Prevailing anoxic conditions are favored in this experiment by the input of organic carbon from the surface, and also by the abundance of Fe(II)-bearing fracture minerals. Iron reduction by organic carbon can produce an even larger reservoir of Fe(II) fracture minerals in the flow path. If organic carbon input from the surface were to stop, this reservoir of mineral reductants would still provide a barrier to oxygen intrusion. Although it is primarily beneficial to maintain anoxic conditions in the groundwater, a secondary concern for increased input of organic carbon to sulfate-rich groundwaters is the possibility of stimulating sulfate reduction at depths that studied at this site. This problem is addressed in a separate project "Microbial Sulphate Reduction in the Äspö HRL Tunnel" (Pedersen et al., SKB technical report in preparation).

Anoxia results in this experiment because there is a net reducing capacity in the subsurface, i.e., more organic carbon and Fe(II) than oxygen, and because reaction of oxygen with these reductants is more rapid than oxygen transport into the fracture zone.

Conditions that favor prevailing anoxia over time scales on the order of this experiment are:

1. a diffusion barrier to atmospheric oxygen; saturated soil conditions and long diffusion paths (thick soils/sediments),

2. no preferential flow paths that short-circuit organic-rich soil or sediment layers,

3. an abundance of Fe(II)-bearing oxide, carbonate, silicate or sulfide minerals in contact with vertical flow paths.

Over longer time scales, the size of the organic carbon reservoir, and the relative rate of biomass production and decomposition must be considered:

1. sufficient biomass production to maintain a reservoir of organic carbon,

2. sufficient decomposition of complex biomass (plant material) to dissolved or colloidal forms that can be hydrodynamically transported,

3. sufficient infiltration to transport organic carbon into the subsurface

At a specific site, one or a few of these conditions and processes may be controlling.

CONCLUSIONS

An assessment of this project shows that integration of the various scientific disciplines was critical to development of a conceptual model for dominant hydrologic, geochemical and microbiological processes at the site. It was necessary that the initial hypothesis for a site model, represented by the predictions of surface water and oxygen breakthrough into the tunnel, was a collaborative effort between a geochemist and a hydrologist. The result of persistent anoxia resulted in the initial conceptual model to be modified to include microbiological processes and a better understanding of the boundary conditions for input of oxygen and organic carbon.

Our initial hypothesis was that oxygen could penetrate rapidly and extensively into the subsurface. Our best defense against wrongly rejecting this hypothesis as a general result for all fracture zones is to assess conditions favoring anoxia. The experiment was designed at a site with little overburden and in a region of the bedrock that was near the surface (depth < 100 meters). Because such conditions should favor penetration of oxygen into the subsurface, we have an accordingly greater confidence in rejecting our initial hypothesis of rapid and extensive penetration of the subsurface. A more quantitative assessment of conditions favoring oxygen penetration or rapid downward transport of organic carbon requires coupled modeling of transport and geochemical processes. Such modeling is planned within the Äspö Geochemistry Workshop modeling exercises.

ACKNOWLEDGMENTS

This project was a collaborative effort by: E. Gustafsson, Geosigma AB (Uppsala); M. Laaksoharju, GeoPoint AB, (Sollentune); Ann-Chatrin Nilsson, The Royal Institute of Technology (Stockholm); Karsten Pedersen, University of Gothenberg; Eva-Lena Tullborg, Terralogica AB (Gråbo); Bill Wallin, Geokema AB (Lidingö). Geochemical and hydrogeological modelling was carried out collaboratively with TVO and collaborators; Margit Snellman, TVO; Petteri Pitkänen, VTT; Aimo Hautajärvi, TVO. Soil and sediment sampling and analysis was carried out in collaboration with the SKB biosphere project; Sverker Nilsson (SKB): and the field investigation team of Björn Sundblad, Ove Landström, Ingrid Aggeryd, Lena Mathiasson, Studvik Ecosafe AB (Nyköping). Katinka Klingberg and Karl-Göran Nederfeldt did much of the hydrochemical sampling and laboratory analysis.

REFERENCES

Banwart S. and Gustafsson E. (1991).The large scale redox experiment: prediction of surface water and redox front breakthrough, The Swedish Nuclear Fuel and Waste Management Company (SKB), PR-25-91-06.

Banwart S., Laaksoharju M., Nilsson A.-C., Tullborg E.-L., Wallin B. (1992a). The large scale redox experiment: initial characterization of the fracture zone, The Swedish Nuclear Fuel and Waste Management Company (SKB), PR-5-92-04.

Banwart S., Gustafsson E., Laaksoharju M., Nilsson A.-C., Tullborg E.-L., Wallin B. (1992b). The large scale redox experiment: redox processes in a granitic coastal aquifer. The Swedish Nuclear Fuel and Waste Management Company (SKB), PR-25-93-03.

Banwart S., Gustafsson E., Laaksoharju, M., Nilsson A.-C., Tullborg E.-L., Wallin B. (1994a). Large-scale intrusion of shallow water into a vertical fracture zone in crystalline bedrock. initial hydrochemical perturbation during tunnel construction at the Äspö Hard Rock Laboratory, S.E. Sweden. Water Resources Res.,30:6, 1747-1763.

Banwart S., Gustafsson E., Laaksoharju, M., Nilsson A.-C., Tullborg E.-L., Wallin B. (1994b). Organic carbon oxidation induced by large-scale shallow water intrusion into a vertical fracture zone at the Äspö Hard Rock Laboratory. Proceedings of the conference "Migration '93", Charleston (USA), Dec. 17-22, 1993, Radiochimica Acta, in press.

Smellie J. and Laaksoharju M. (1992). The Äspö Hard Rock Laboratory: Final evaluation of the hydrochemical pre-investigations in relation to existing geologic and hydraulic conditions. The Swedish Nuclear Fuel and Waste Management Company (SKB), TR 92-31.

Results and Experiences from the Äspö Hard Rock Laboratory Characterization Approach

G Bäckblom
Swedish Nuclear Fuel&Waste Management Co (SKB)

Gunnar Gustafson
Chalmers University of Technology

I Rhén
VBBVIAK

R Stanfors
R Stanfors Consulting

P Wikberg
Swedish Nuclear Fuel&Waste Management Co (SKB)

(Sweden)

Abstract

SKB plans to start deposition of spent fuel within about 15 years from now. The design work for a canister encapsulation plant is under way. Feasibility studies for repository siting are in progress in cooperation with two (September 1994) municipalities.

An extensive research, development and demonstration programme will support the detailed design and licensing of the encapsulation plant and the deep repository.

Construction of the Äspö Hard Rock Laboratory down to 450 m below the surface will be complete in early 1995. The last portion of the facility was excavated by fullface boring. Prior to excavation, the results of the pre-investigations of the bedrock were analyzed and modelled on several geometric scales. For each scale, key issues were selected. Several hundred parameter values were then predicted. During construction these predicted values are being compared to the situation in the tunnel and in the boreholes. Between 100 - 200 borehole sections have been monitored on-line.

The prediction-outcome matches are now under evaluation and the final report will be published 1995. Many important results have been obtained and valuable experience has been gained.

The paper summarizes important findings which will influence the detailed planning of the site characterization programme for the Swedish deep repository.

BACKGROUND

SKB has decided to construct the Äspö Hard Rock Laboratory for the main purpose of providing an opportunity for research and development in a realistic and undisturbed underground rock environment down to the depth planned for the future deep repository in Sweden.

One of the stage goals is to "Verify pre-investigation methods" - e.g. to demonstrate that investigations on the ground surface and in boreholes provide sufficient data on essential safety-related properties of the rock at repository level.

A comprehensive site characterization programme was carried out prior to the start of construction in 1990. During the excavation work down to the 450 m level comparisons have been made between the predictive descriptions based on the pre-investigations and the actual outcome as documented in conjunction with the construction of the laboratory. The facility is depicted in Figure 1.Evaluations of model and method capability are under way and will be completed in 1995.

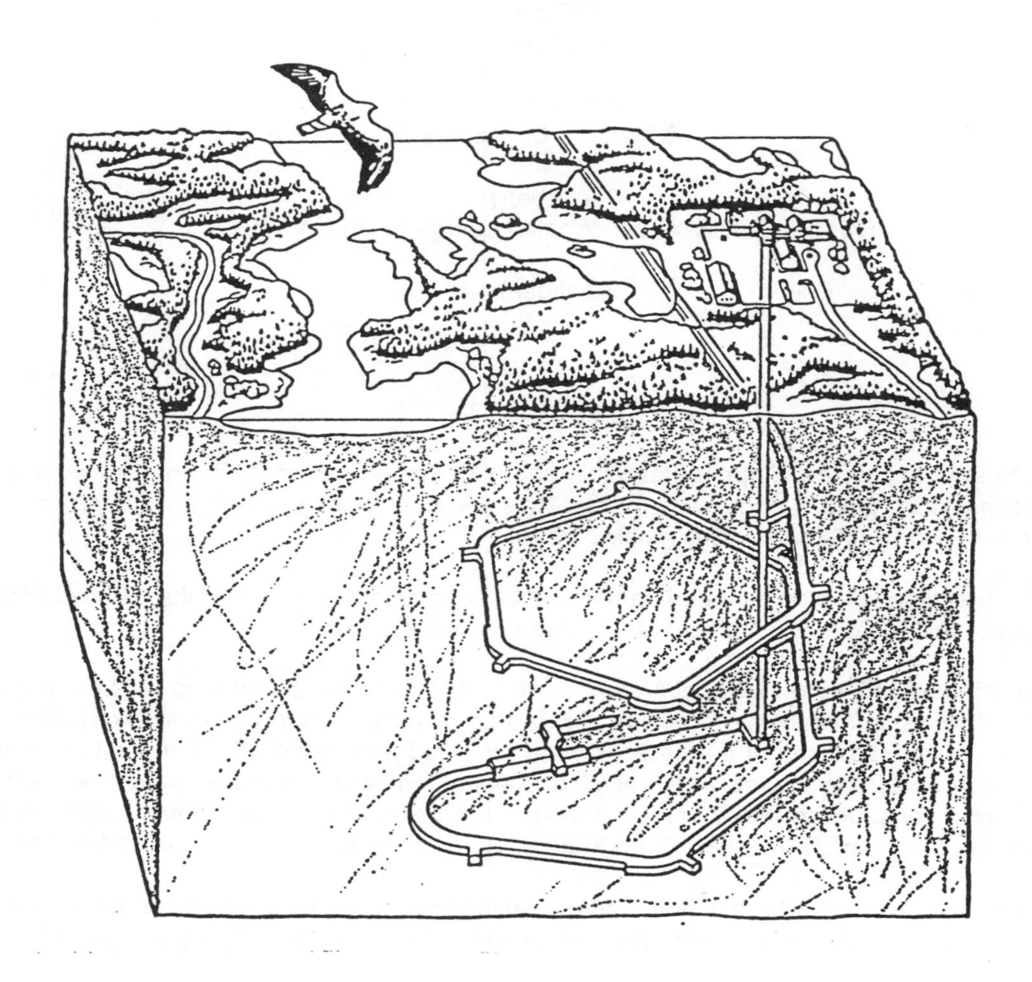

Figure 1. Sketch of the Äspö Hard Rock Laboratory

There is now a strong international interest in the work with participation from Canada (AECL), Japan (PNC, CRIEPI), Finland (TVO), France (ANDRA), Switzerland (NAGRA), the United Kingdom (NIREX), and the United States (USDOE).

HOW WILL THE SITE INVESTIGATION RESULTS BE USED?

The purpose of pre-investigations or site investigations is to:

- demonstrate whether a site has suitable geological properties
- obtain data and knowledge concerning the bedrock on the site so that a preliminary emplacement of the repository in a suitable rock volume can be done as a basis for constructability analysis
- obtain the necessary data for a preliminary safety assessment, which in Sweden shall serve as support for an application under NRL (the Act Concerning the Management of Natural Resources) to carry out detailed site characterization
- obtain data for planning of detailed site characterization, which is done in conjunction with the construction work down to repository level.

It is thus important to show that pre-investigations provide reasonable and robust results.

The reliability of the pre-investigations can be tested during the detailed characterization. The work at Äspö is thus a "dress rehearsal" for the site investigations and the detailed characterization for the Swedish deep repository.

WHAT SHOULD BE ACHIEVED IN SITE CHARACTERIZATION?

Site characterization shall provide an understanding of the site to:

- Demonstrate the existence of properties and their importance
- Justify division of the properties into domains
- Make parameter estimations as an input to model realizations, to support division into domains and to prove the existence or non-existence of properties

Demonstrating or rejecting the existence of properties is of the utmost importance in the characterization programme. Properties may have to do with be lithology, chemical processes, structural features etc. At the outset of characterization it is important that the programme be broad so that unexpected properties can also be detected.

Division of properties into domains can be done for e.g. lithological bodies and groundwater composition such as fresh or saline water, or be applicable to the geometry of structural features.

Parameter estimation can be done with alternative evaluation models. The safety assessor's desire for "alternative models" is warrented for parameter estimation purposes.

The relevance of each property, domain and parameter should be assessed by both the engineer and the safety assessor.

SAFETY ASSESSMENT SHOULD ANALYZE THE REPOSITORY "AS BUILT"

From the view point of the safety assessor it is important that site characterization be so reliable that the engineering and construction work can be done in a reliable way. The reason is as follows:

The engineer basically divides the rock into two families. "Good rock" implies that the rock can be excavated without any sealing or support work. "Bad rock" means that the rock needs grouting or reinforcement by steel and/or concrete.

It is argued that this very simple division is of great importance. Rock bolting with steel bars provides a connection between the backfill of a deposition drift and a possible disturbed zone; shotcreting will interfere with the required close interaction between the backfill and the host rock over thousands of years due to the deterioration of the concrete leaving a gap between the backfill and the rock; pre-grouting of deposition tunnels will make it difficult to obtain accurate values of transmissivity at canister positions.

The chemical inventory for a deep repository could include some 1000 tonnes of cement, maybe some 100 tons of nitrogen compounds (residues from blasting agents) and some minor inventory of oil, cellulose etc. The backfill material will include huge amounts of oxygen to be consumed in the closed repository.

Based on the site investigations a constructability analysis /1/ should be performed so that the safety assessment can give an account of the repository "as built" even prior to construction.

The engineer's wish to avoid "bad rock" is an advantage both for the engineering and for the safety of the repository. It will also simplify the safety assessment. The engineer's wish to find "good rock" can be an advantage. However there is a need to check for stress or thermomechanical problems that may appear in very good (unfractured) rock. Thus, ordinary engineering practice will be of the utmost importance in the licensing process for a deep repository. Utilizing centuries of knowledge supports confidence building for the site characterization and engineering of the repository.

The safety assessor sometimes asks for alternative models to be presented. This is not always acceptable to the engineer responsible for the layout of a facility. He must e.g. be certain that no major water conducting fracture zones are present at the underground facility; he thus needs a "best estimate" where the safety assessor would prefer alternative models.

THE APPROACH TO SITE CHARACTERIZATION AT ÄSPÖ

A basic description of the site characterization approach can be found in /2/. Site characterization is a multi- and inter-disciplinary task that necessitates integration in data acquisition, evaluation and presentation of results.

In order to facilitate such integration, four basic decisions were made for the site characterization.

- to divide the characterization and interpretation into distinct stages

- to devise models on several geometric scales (regional >> 1000 m, site scale 100 - 1000 m, block scale 10 - 100 m and detailed scale 1-10 m)

- to select key issues related to the geological-structural model, groundwater flow, groundwater chemistry, transport of solutes and mechanical stability

- to integrate and facilitate integration of data collection and data interpretation

For each of the three stages, summary evaluations were reported /3--5/.

The last stage was the "Prediction Stage" /6/ where the rock descriptions were put into predictions to be checked during the construction of the laboratory.

The integration of all data was to some extent facilitated by having few principal investigators with broad responsibilities.

VERIFICATION OF PRE-INVESTIGATION METHODOLOGY

Both a "validation" and a"verification" strategy have been devised.

The validation strategy /7/ comprises three basic elements

- a systematic comparison of prediction and outcome

- a judgement of whether the predictions are good enough

- a scrutiny of underlying structures and processes

The Äspö Principal Investigators thought it best that the judgement of the outcome as a ""success" should be made by other interested parties. Such judgements have been passed by the Äspö Scientific Advisory Committee and by Swedish authorities in their review of the work. Publishing of the results is part of this validation strategy.

The "verification strategy" is depicted in Figure 2.

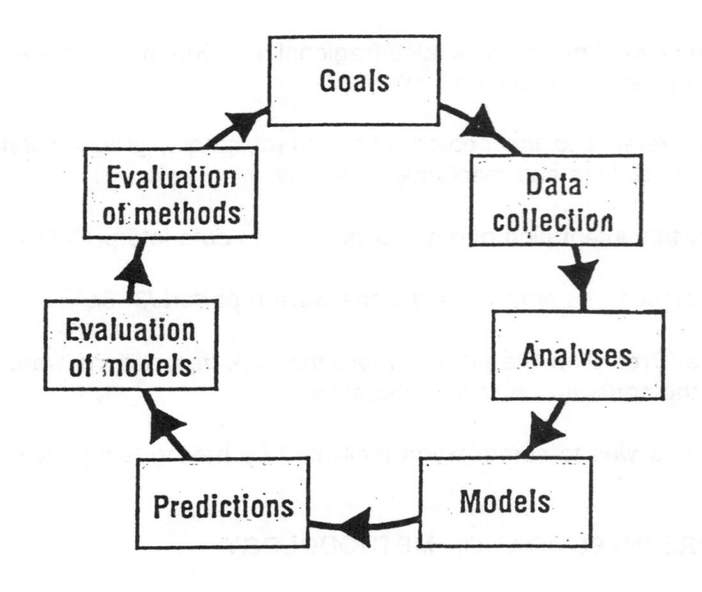

Figure 2. Strategy for verification of pre-investigations.

Evaluations of models and methods are under way and will be finalized in 1995.

The evaluation work will e.g. comprise evaluation of:

- ability to prove the existence of properties and non-existence of properties
- ability to divide properties into domains
- reliability of parameter estimations

The evaluation will deal both with the actual results and the experiences gained using the characterization approach.

RESULTS FROM COMPARISONS OF PREDICTIONS AND OUTCOMES

The evaluations of the pre-investigations and associated predictions were presented prior to the start of the excavation work /5, 6, 8, 9/. Some supplementary work has also been reported /10, 11/.

Based on evaluations to date some of the results are discussed below.

Existence of properties

The bedrock at Äspö is lithologically heterogeneous. Based on evaluation of significance the rock is basically divided into five different lithologies. One of these, the fine-grained granite, is much more fractured than other rock types and was assumed to be more water-onducting and thus would be of interest both to the engineer and the safety assessor.

The existence of the fine-grained granite was predicted and has been confirmed.

The existence of several major fracture zones was predicted. They have all been confirmed. No other major zones exist. To avoid confusion a nomenclature for fracture zones and reliability (Certain, Probable, Possible) was developed /12/.

One subhorizontal zone classified as "Possible" has not been found. The existence of minor fracture zones was predicted. Their existence has been confirmed

The existence of saline water was predicted and its existence has been confirmed.

A process that was not predicted is oxygen consumption (reducing conditions) due to bacterial activity /13/.

Division into property domains

Due to the hererogenity of the lithology it was considered inappropriate to predict the loacation of fine-grained granite. This proved to be a wise decision. Even after construction work, such a prediction is not appropriate. However, the average volume (%) of fine-grained granite was predicted.

The dip and strike of the most important fracture zone NE-1 was predicted and the outcome showed small deviations.

One major fracture zone. NE-2 has a complicated dip. Parts of the zone dip north, other parts of the zone dip south. As the zone was considered to be of minor importance (a true prediction!) few investigations were made to determine the exact geometry of the zone. The outcome is that the zone dips south instead of north as predicted.

The strike of the minor fracture zones were predicted, but not their exact location. The strikes and dips of the minor fracture zones trending NNW have been confirmed to conform to the predictions. Their exact locations were determined in the detailed characterization in connection with excavation.

The groundwater chemistry was predicted to be composed of different bodies with higher salinity at greater depth. This has been confirmed.

Parameter estimation

Evaluation of the predicted parameter estimations is under way. The average amount of e.g. fine-grained granite was predicted and was very close to the outcome.

Transmissivity evaluations for different rock types and for fracture zones are in progress.

The outcome of groundwater chemistry is outside the range of the predictions. Water from greater depths is sampled earlier than predicted. The underlying assumptions made in the prediction work are being examined.

EXPERIENCE FROM THE SITE CHARACTERIZATION APPROACH

The site characterization approach as described above has been useful. The idea of dividing the investigations into different stages, scales and key issues has simplified the evaluation work considerably and further use of this approach is advocated.

Due to the lithological heterogeneity, it seems that block scale predictions were not as useful as site and detailed scale predictions.

Constructability analysis should be added as a key issue to ascertain a good coupling between site characterization and the engineering work.

Based on a general site investigation strategy the site specific strategy shall not be fixed until some preliminary field work has been evaluated, remembering e.g. that major fracture zones can be low-conductive and minor fracture zones can be highly conductive.

In the pre-investigation phase it was difficult to determine:

- The exact position and orientation of minor subvertical fracture zones (e.g. NNW-structures at Äspö).
- The detailed character and importance of subvertical fracture zones.
- The relative importance of the different subhorizontal fracture zone indications.
- The location and distribution of minor rock units at depth (e.g. greenstone lenses and veins of fine-grained granite).
- The validity of the theoretical model of the scale dependency of hydraulic properties.
- The hydraulic properties of minor rock types.
- The absolute water pressures in boreholes at great depth due to varying salinity with depth.
- The groundwater chemistry in low conductive rock masses.

There is a need for a more stringent methodology for assessment of the accuracy of measurement and evaluation methods.

The need for appropriate classification systems should not be underestimated. The fracture zone classification used has been useful /12/.

The nomenclature for theories, models and conceptual models was found out to be confusing. A nomenclature for SKB work has now been established /14/. The basic elements of the nomenclature are depicted in Figure 3.

MODEL NAME/DEFINITION	
Model scope or purpose Specify the intended use of the model	
Process description Specification of the processes accounted for in the model, definition of constitutive equations	
CONCEPTS	DATA
Geometric framework and parameters	
dimensionality and/or symmetry of model specification of what the geometric (structural) units of the model are and the geometric parameters (the ones fixed implicitly in the model and the variable parameters)	specify size of modelled volume specify source of data for geometric parameters (or geometric structure) specify size of units or resolution
Material properties	
specification of the material parameters contained in the model (should be possible to derive from the process and structural descriptions)	specify source of data for material parameters (should normally be derived from output of some other model)
Spatial assignment method	
specification of the principles for how material (and if applicable geometric) parameters are assigned throughout the modelled volume	specify source of data for model, material and geometric parameters as well as stochastic parameters
Boundary conditions	
specifications of (type of) boundary conditions for the modelled volume	specify source of data on boundary and initial conditions
Numerical tool Computer code used	
Output parameters Specify computed parameters and possibly derived parameters of interest	

Figure 3. Basic elements of the SKB model nomenclature.

Three major drilling campaigns were carried out at Äspö prior to construction. The first three holes basically showed all the important properties to be found at Äspö. These three holes also provided a good basis for parameter estimation of both average values and the variance of the parameters. The second and third campaigns (11 deep cored holes) were basically useful for increasing confidence in the geometry of major fracture zones and learning more about the minor fracture zones at Äspö. The geometry of the major fracture zones was very important for the layout of the Äspö facility, to ensure that it was located in good rock.

CONCLUDING REMARKS

Evaluation of the Äspö work is under way. This evaluation concerns the scientific results as well as the experiences gained.

Altogether the work will be of great interest for the implementation of deep repositories in Sweden and elsewhere.

Some concluding comments are:

- Dividing the investigation strategy in to stages, scales and key issues was useful

- Predicitions useful for engineering were adequate

- Constructability analysis should be a key issue

- Evaluation of usefulness of data for safety assessments in progress

ACKNOWLEDGEMENT

The Äspö work has been executed with several experts from Sweden and from abroad. Their dedicated efforts have been most appreciated and have provided inspiration and enhanced the quality of the work in many ways.

REFERENCES

/1/ Bäckblom, G, Leijon B, Stille H, 1994. Constructability analysis for a deep repository - some thoughts on possibilities and limitations. Proc 2nd Int Workshop on Design & Construction of final repositories, Winnipeg, February 15-16, 1994. AECL Research TR (in print)

/2/ G Bäckblom, G Gustafson, R Stanfors and P Wikberg, 1991. Site characterization for the Swedish Hard Rock Laboratory. Proc NEA/SKI Symposium Stockholm, 14-17 May 1990. OECD, Paris.

/3/ Gustafson G, Stanfors R, Wikberg P, 1988. Swedish Hard Rock Laboratory. First evaluation of 1988 year pre-investigations and description of the target area characterization. SKB Technical Report, TR 88-16.

/4/ Gustafson G, Stanfors R, Wikberg P, 1989. Swedish Hard Rock Laboratory. Evaluation of 1988 year pre-investigations and description of the target area, the island of Äspö. SKB Technical Report, TR 89-16.

/5/ Wikberg P, Gustafson G, Rhén I, Stanfors R, 1991. Äspö Hard Rock Laboratory. Evaluation and conceptual modelling based on the pre-investigations 1986-1990. SKB Technical Report, TR 91-22.

/6/ Gustafson G, Liedholm M, Rhén I, Stanfors R, Wikberg P, 1991. Äspö Hard Rock Laboratory. Predictions prior to excavation and the process of their validation. SKB Technical Report, TR 91-23.

/7/ Bäckblom G, Gustafson G, Stanfors P, Wikberg P, 1990. A synopsis of predictions before the construction of the Äspö Hard Rock Laboratory and the process of their validation. SKB PR 25-90-14, Stockholm.

/8/ Stanfors R, Erlström M, Markström I, 1991. Äspö Hard Rock Laboratory. Overview of the investigations 1986-1990. SKB Technical Report, TR 91-20.

/9/ Almén K-E, Zellman O, 1991. Äspö Hard Rock Laboratory. Field investigation methodology and instruments used in the pre-investigation phase, 1986-1990. SKB Technical Report, TR 91-21.

/10/ Rhén I, Svensson U (eds), Andersson J-E, Andersson P, Eriksson C-O, Gustafsson E, Ittner and Nordqvist R, 1992. Äspö Hard Rock Laboratory. Evaluation of the combined longterm pumping and tracer test (LPT2) in borehole KAS06, SKB Technical Report, TR 92-32.

/11/ Smellie J, Laaksoharju M, 1992. Äspö Hard Rock Laboratory. Final evaluation of the hydrogeochemical pre-investigations in relation to existing geologic and hydraulic conditions. SKB Technical Report, TR 92-31.

/12/ Bäckblom G, 1989. Guidelines for use of nomenclature on fractures, fracture zones and other topics. SKB Technical Document.

/13/ Steven Banwart, Erik Gustafsson, Marcus Laaksoharju, Ann-Chatrin Nilsson, Eva-Lena Tullborg and Bill Wallin. Large-scale intrusion of shallow water into a vertical fracture zone in crystalline bedrock: Initial hydrochemical perturbation during tunnel construction at the Äspö Hard Rock Laboratory, southeastern Sweden. Water Resources Reasearch, Vol 30, No 6, pages 1747-1763, June 1994.

/14/ Olle Olsson, Göran Bäckblom, Gunnar Gustafson, Ingvar Rhén, Roy Stanfors, Peter Wikberg, 1994. The structure of conceptual models with application to the Äspö HRL Project. SKB Technical Report 94-08.

Efforts Toward Validation of a Hydrogeological Model of the Asse Area

E. Fein, K. Klarr, C. von Stempel
GSF-Institut für Tieflagerung
Theodor-Heuss-Straße 4
D-38122 Braunschweig

Abstract

The Asse anticline near Braunschweig is part of the Subhercynian Basin which is bounded in the south by the Harz Mountains, in the east and the north by the river Aller, and in the west by the river Oker which flows into the river Aller. The Asse structure ranges lengthwise and breadthwise approximately eight and three km, respectively. It is oriented in NW-SE direction.

In 1965 the GSF Research Center for Environment and Health acquired the former Asse salt mine on behalf of the FRG in order to carry out research and development work with a view to safe disposal of radioactive waste.

To assess long term safety it is important to predict groundwater flow and radionuclide transport in the vicinity of the salt anticline. In order to show the reliability of these predictions they have to be proved by means of validation. For this purpose an experimental program was carried out with the twofold intention to set-up and validate hydrogeological models of the overburden of the Asse salt mine and to provide these with data.

In the course of about 25 years five deep boreholes from 700 m to approximately 2,250 m below surface, and four geological exploration shallow boreholes were drilled in order to improve the knowledge about the geological formation of the Asse area. Moreover, 19 hydrogeological piezometers and 27 hydrogeological exploration boreholes were sunk to perform pumping and tracer tests and yearly borehole loggings. In the end about 50 boreholes and wells, 25 measuring weirs and about 70 creeks, drainages and springs were available to collect hydrological data and water samples. It turns out that the water movement within the mesozoic rocks of the overburden of the Asse anticline as deep as 200 m below surface is influenced by exogenic (e.g. morphological and meteorological) factors, whereas below 700 m the movement is almost stagnating.

The different experiments and their evaluations as well as different hydrogeological models are presented and discussed. Since there is no suitable numerical model available which is able to take into account the effects of variable density due to salinity for large and complex geological structures like the Asse up to now all models neglect salinity.

1 Introduction

The Asse salt anticline is located about 20 km southeast of the City of Braunschweig (fig. 1). It covers an area of about 25 km^2 and is oriented in northwest-southeast direction.

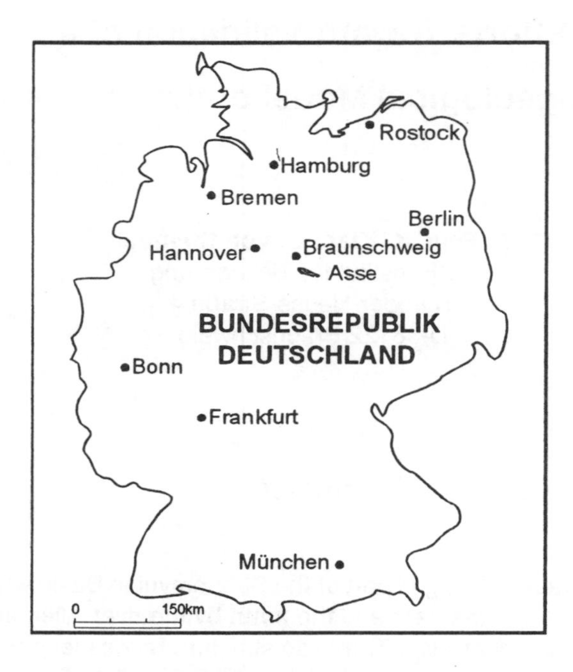

Figure 1. **Location of the Asse anticline**

Salt mining began at Asse with the sinking of the Asse 1 shaft at Wittmar between 1899 and 1901. Due to inappropriate mining in the uppermost section of the salt anticline and in ignorance of the hydrogeological conditions, water inflow occurred in 1906, as a result of which the shaft was flooded in 1906. After the abandonment of the first shaft, the Asse 2 shaft was sunk, some 1.5 km to the east near Remlingen between 1906 and 1908. Initially potash salt was mined exclusively, but in 1916 mining of rock salt began. At the end of 1925 the mining of potash salt was discontinued and all future mining was restricted to rock salt. For economic reasons on 31st March 1964 even this was discontinued and, since then, no exploitation activities whatsoever have been carried out in the Asse salt formation.

In 1965 the GSF - Research Center for Environment and Health acquired the former Asse salt mine on behalf of the FRG in order to carry out research and development work with a view to safe disposal of radioactive waste. Between 1967 and 1978 a total of around 125,000 canisters containing low level waste and around 1,300 canisters containing intermediate level waste were emplaced.

In 1976 the fourth amendment to the Atomic Act provides for a so-called "Planfeststellungsverfahren" as a licensing procedure for final disposal of radioactive wastes. Consequently disposal of radioactive waste was stopped and since 1978 the Asse mine has served as underground laboratory.

2 Site Characterization

2.1 Geology

Geologically the Asse Salt Anticline belongs to the Subhercynian Basin, which is limited in the south by the Harz Mountains, in the east and the north by the river Aller and in the west by the Oker, a tributary of the Aller.

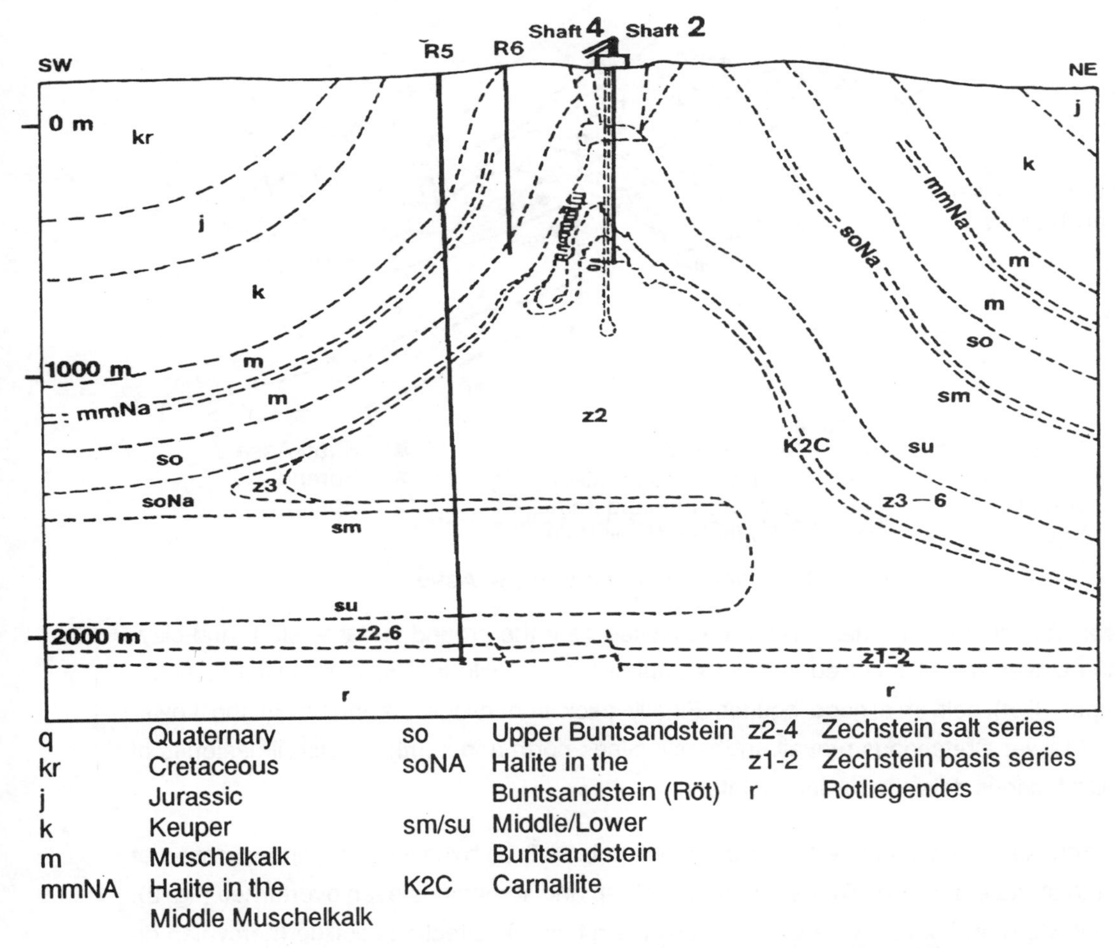

Figure 2. **Geological cross-section through the Asse Salt Anticline**

Morphologically the Asse forms a range of hills of about 8 km length and 3 km breadth oriented northwest-southeast parallel to the northern edge of the Harz Mountains. Whereas the impermeable salt belongs to the Zechstein formation, the overburden rocks consist of dragged up layers from Buntsandstein to Jurassic age (fig.2 and fig. 3). With the exception of a karstified and therefore permeable oolithic limestone in the Lower Buntsandstein the strata in the Buntsandstein consists of sand-, silt- and claystones behaving generally as an aquitard. Only the Upper Buntsandstein forms an aquiclude. In the Lower and Upper Muschelkalk limestones form aquifers, whereas marlstones and evaporitic layers (gypsum and rocksalt) of the Middle Muschelkalk

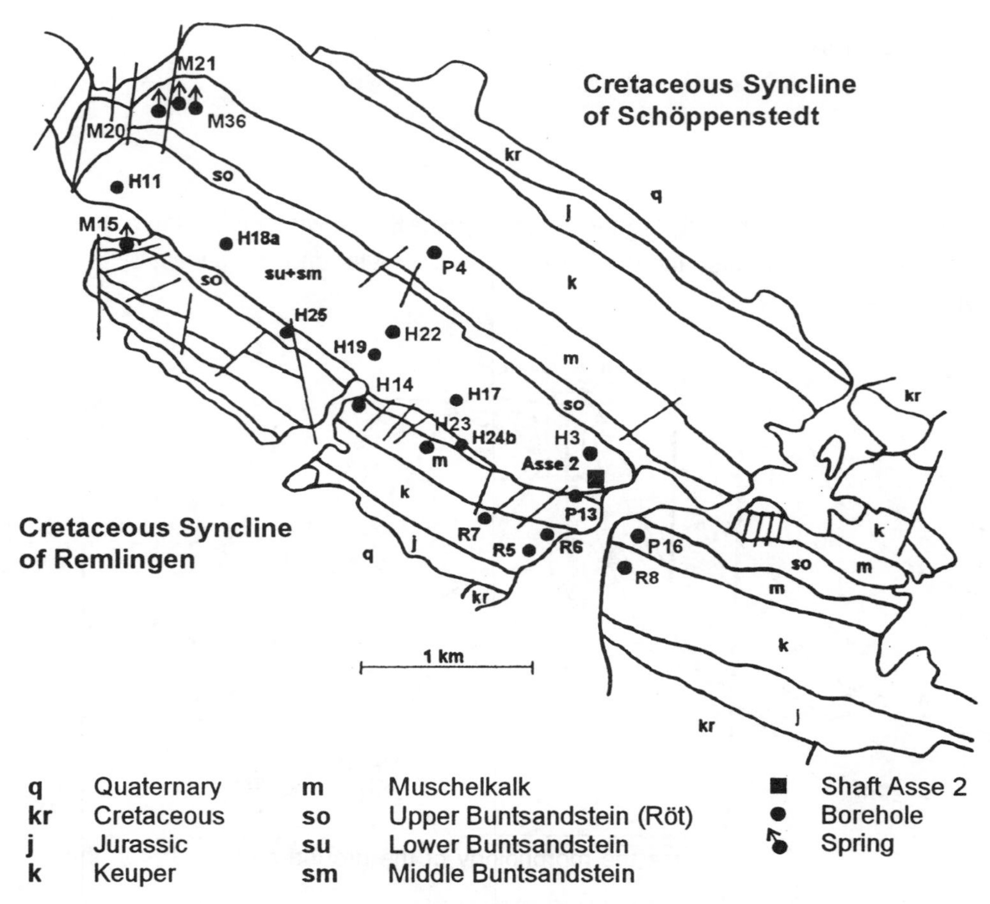

Figure 3. **Hydrogeological Map of the Asse**

behave as an aquiclude in the depth but are karstified near the ground surface. Marl- and claystones in the Lower and Middle Keuper act as aquitards to aquicludes, while sandstones in the Upper Keuper (Rät) deliver a good aquifer. Finally thick impervious claystones in the Lower Jurassic and Lower Cretaceous prevail. In the synclines north and south of Asse limestones of the Upper Cretaceous deliver excellent aquifers.

While the northeastern flank of the Asse Salt Anticline is built up by layers with an inclination of about 50°, the strata of the southwestern flank are steep and sometimes even overturned (fig. 2). At the northwestern end of the Asse structure two main tectonic directions collide (hercynian direction NW-SE and rhenanian direction NNE-SSW). This created a multitude of faults leading to the outlet of many springs [1]. Two major faults, one in the middle of the structure near the village of Wittmar and one in the southeast between Groß Vahlberg and the shaft of mine Asse II cross the Asse structure and partly affect the groundwater movement which in the depth generally follows the striking of the strata from southeast to northwest. The latter fault crosses the anticline in a more diagonal direction and has a horizontal and vertical slip of several hundred meters. In the central part of the structure the salt table ("Salzspiegel") representing a more or less plain surface up to which groundwaters were able to dissolve the salt, is situated about 200 m below the ground surface. The mostly faulted rocks lying above this salt table belong to a collapse structure at the top of the salt anticline.

2.2 Hydrogeology

As the Asse is a morphological elevation consisting of three hill ranges, the northern and southern belonging to the hard limestones of the Lower Muschelkalk and the middle one built up by the oolithic limestone of the Lower Buntsandstein, there are altogether 13 groundwater basins having each one its own hydrogeological characteristics.

The mean precipitation sum per year is 600 mm (165 mm in the winter months november until february and 435 mm in the summer months march until october). The mean annual temperature (1968 to 1990) is 9.1° C. Within the Asse there exist two rain gauges and one meteorological station which records the air temperature, the relative air humidity, the radiation balance and the wind velocity and direction. These parameters will permit to obtain a general value for the evapotranspiration which is especially important in summer due to the forest cover occupying the hill ranges nearly entirely.

For the hydrogeological investigations and groundwater monitoring altogether 48 boreholes were sunk within the Asse area. 27 of them called "hydrogeological investigation boreholes" have depths of 6 up to 390 m with a diameter large enough to perform pumping tests. 18 boreholes with a small diameter served as piezometers and injection boreholes and 4 deep boreholes allow the sampling of formation waters down to 890 m depth. All these boreholes were supplied with automatic water level gauges. Once a year profiles of temperature, electrical conductivity and density in most of these boreholes were measured.

The hydraulic head follows near the surface the morphology of the ground surface, but in the depth it is usually inclined from southeast to northwest following the general striking of the geologic formations, in which the different aquifers exist. This direction of flow can also be observed in the brine-filled flumes at the salt table in the northwestern part of the Asse anticline. Within this confined aquifer-system, which behaves completely different from the other aquifers above, there is a quick pressure response on a distance of about 2.8 km between the region around shaft Asse I (borehole H 17) and the naturally outflowing salt water in the northwest with the artesian borehole H 11.

For the observation of the surface waters 25 measuring weirs were constructed. Despite of many difficulties in recording automatically the yields of these 70 springs, creeks and drainages at least the measuring values of three of them could be interpreted.

In all the hydrogeological investigation boreholes pumping tests were performed. Because of the limited groundwater reserves steady state conditions could be observed in hardly any pumping test. The influence of drawdown effects could only be observed in two boreholes situated mostly nearby the pump wells, in the karstified oolithic limestone of the Lower Buntsandstein and in the system of brine filled flumes between the Zechstein salt and the cap rock at the northwestern end of the structure. This is also the reason why the storage coefficient could in most cases not be determined. In aquifers with very low yields the emptying effect of the borehole in the pumping test was too strong. Therefore the transient response method delivered in those special cases better results than with pumping tests [4]. This method relies on the fact that seismic waves produced by earthquakes influence the groundwater system and that under special conditions the transmissivity of the aquifer can be deduced from the resulting oscillations of the groundwater level [3]. The hydraulic conductivities ranged between 10^{-5} to 10^{-7} m/s in the overburden rocks up to 160 m depth, they reached about 10^{-3} m/s in the brine filled flumes (204 to 241 m depth) and about 10^{-4} m/s in good aquifers on the flank (11 to 31 m depth). In the four deep boreholes only packertests could be performed to obtain values of the hydraulic conductivity which ranges between 10^{-14} up to 10^{-5} m/s. Fig. 4 shows the range of these values and the influence of the depth below ground surface on them. As expected the hydraulic conductivities generally decrease with increasing depth. While in a long-time pumping test on the northeastern flank the drawdown in the pump well H 22 could be observed in the borehole H 3 in 1.5 km distance, another pumping test lasting as long on the southwestern flank showed no influence between the pump well H 23 and an observation borehole (H 14) in only 750 m distance. This shows how heterogeneous the hydraulic behaviour is in such fissured aquifers as exist in the Asse area.

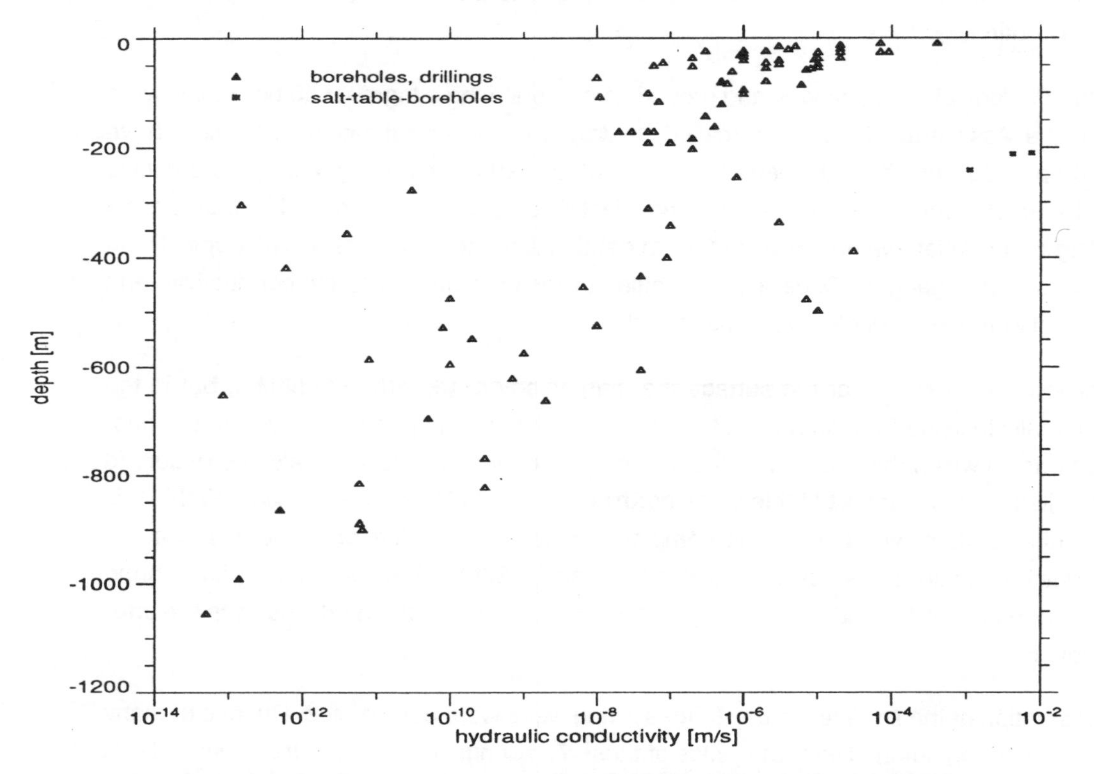

Figure 4. **Dependence of the hydraulic conductivities with the depth**

Water samples from all boreholes at different depths and surface waters were regularly sampled and analysed. Nearly all the groundwaters at the Asse can be considered as mineral waters because of their high mineral contents (> 1 g/l). The sweet groundwaters are mostly of the calcareous type ($CaHCO_3$), gypsum-type $CaSO_4$ with influences of Mg, Na and Cl. All deep waters are saline (NaCl-type). In two deep boreholes (R 5 and R 7) these waters are nearly saturated with NaCl. Because the salt origin may originate from Zechstein, Upper Buntsandstein or Middle Muschelkalk, it is not obvious where these formation waters come from. New interpretation methods are in the state of development to obtain reliable information on their origin also with the help of isotopes.

By analysis of isotopes like ^{2}H (Deuterium), ^{18}O, ^{3}H (Tritium) and the relation of $^{32}S/^{34}S$ a classification of three groundwater zones could be made [2]:

- a zone near the surface up to 150 to 200 m depth contains young groundwater with tritium and a relation of $^{2}H/^{18}O$ cnaracteristic for recent precipitations.
- up to a depth of about 700 m the groundwaters contain no more tritium but have still a relation $^{2}H/^{18}O$, which is characteristic for near-surface-waters.
- below 700 m the waters show no tritium any more and a relation $^{2}H/^{18}O$ completely different from that of the upper zones. This water probably originates from a warmer climate than today.

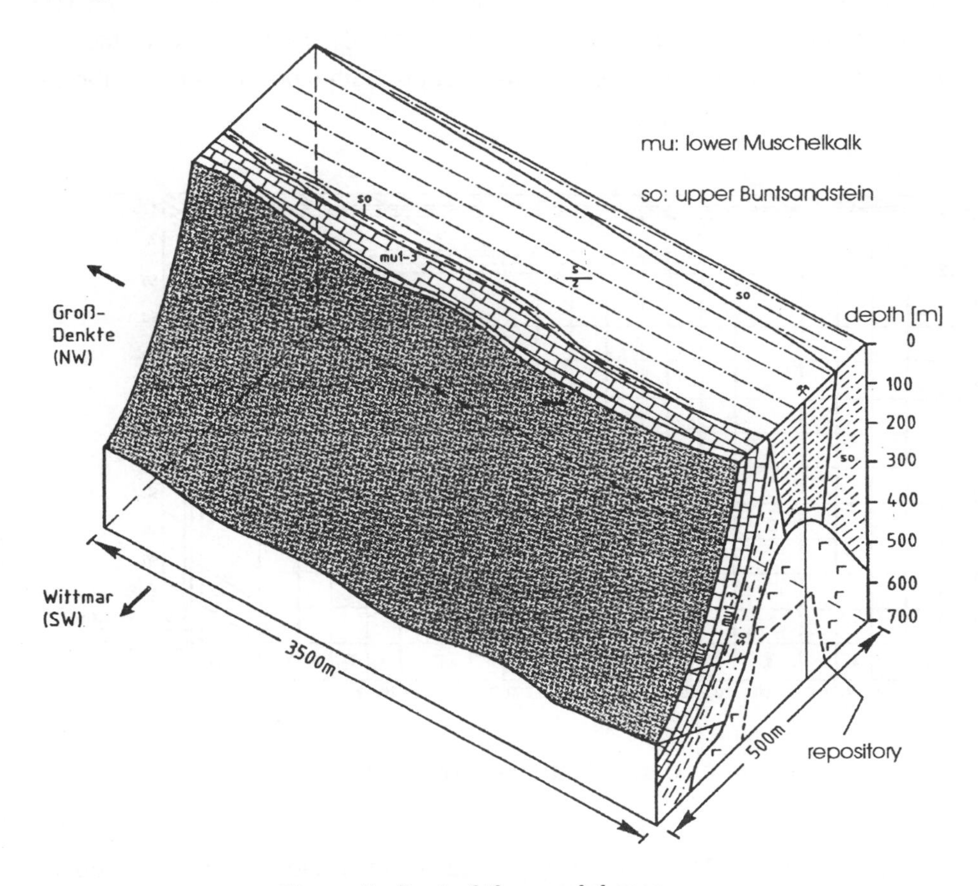

Figure 5. **Part of the model area**

Finally altogether 5 tracer experiments were carried out. First the zone of the brine-filled flumes at the salt table in the northwestern part was marked with LiCl. After 114 days the tracer was found in the artesian borehole H 11 and a saltwater spring (M 63), showing a maximum groundwater movement of 8.8 m/day. Then uranine was injected into a piezometer (P 4) in the Middle Muschelkalk on the northeastern flank. The tracer reappeared after 8.5 months in a spring of the Upper Muschelkalk at a distance of 2.25 km showing a similar maximum groundwater velocity of 8.7 m/day. Calculations show the importance of matrix diffusion. Without any additional knowledge dispersion and matrixdiffusion is not separable. Therefore the dispersion coefficient cannot be determined uniquely. Three other tracer experiments on the southwestern flank with uranine, eosine and pyridine showed no results after four years up to now. If any reappearance of the tracer still occurs in the future, it is probable that it will not be measurable because of the high dilution.

3 Modelling

After two precursor attempts in two-dimensional modelling of the groundwater flow in the Asse area in 1979 by Intera and in 1983 by TUB (Technical University of Berlin) in 1987 GSF and TUB together developed a three-dimensional model HYGAM1 (**HY**dro**G**eologisches **A**sse **M**odell). Later in 1990 the two dimensional model HYGAM2 was set up. Both models start from the assumption that the overburden of the Asse anticline (fig. 5) can be modelled in the porous medium approach taking into account layers as confined aquifers. Both the simulations of the two- and the three-dimensional model are performed by use of the computer code SWIFT [5].

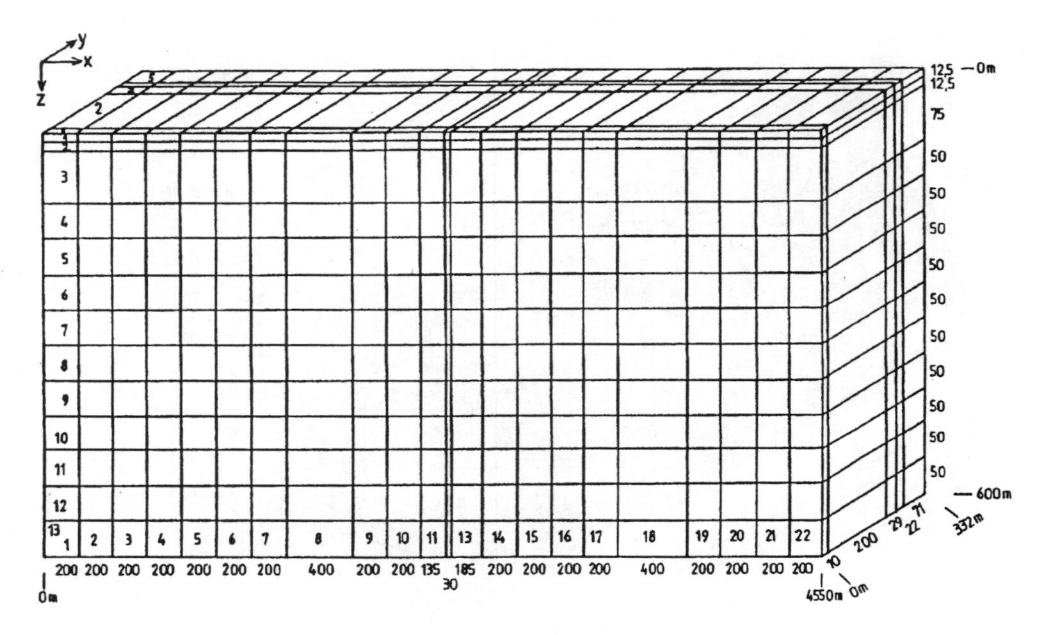

Figure 6. **Discretization of HYGAM1**

3.1 The Three-Dimensional Model HYGAM1

In this approach [6] only the range of the Muschelkalk at the southwestern flank of the Asse anticline is modelled as a vertical stratum (fig. 6) with a thickness of about 330 m. This stratum itself is divided in the relatively high permeable Lower Muschelkalk between two slices of Middle Muschelkalk and Upper Buntsandstein acting as an aquitard and an aquiclude, respectively. Both hydraulic conductivities and porosities decrease with depth below surface (fig. 4). The model is bounded in the northeast by the Upper Buntsandstein which serves as an impermeable boundary. In the southwest Keuper is adjoining the Muschelkalk with lower permeabilities. Therefore this side is taken into account as a permeable boundary. The boundary at the southeast is totally impermeable, whereas in the northwest only the most upper part is permeable. Due to the lack of a suitable numerical model, which is able to take into account the density dependence of the groundwater flow and since the water movement below is almost stagnating, the sharp interface between sweet water and highly saline water at a depth of about 600 - 700 m below surface is used as impermeable bottom of the model. The top of the model is given by interpolated density corrected groundwater levels. As result of our simulation we obtained a flow field directed from southeast to northwest. Moreover pathlines reach the surface at the northwestern boundary. But it turns out that using the measured hydraulic conductivities of the upmost layers results in a much too high recharge rate. Consequently in order to get realistic recharge data, the hydraulic conductivities at the top of the model were reduced by three to four orders of magnitude. One

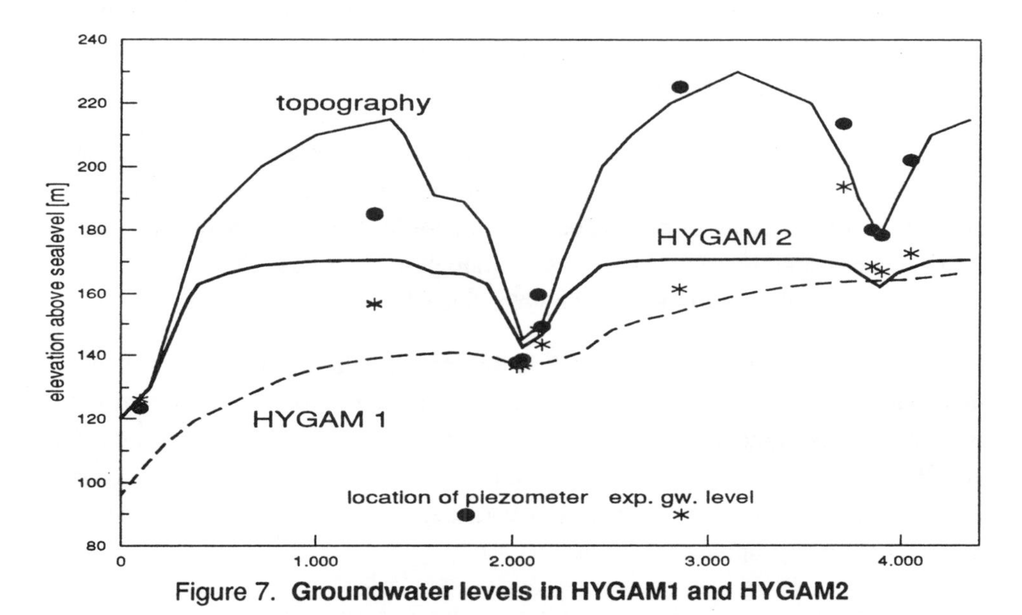

Figure 7. **Groundwater levels in HYGAM1 and HYGAM2**

can imagine a possible increase of the measured conductivity by means of a disaggregation of the rocks due to the drilling. But it is hard to understand the extent of enlargement.

3.2 The Two-Dimensional Model HYGAM2

In contrast to the above described model HYGAM2 [6] is two-dimensional and takes into account only the high permeable stratum Lower Muschelkalk. This time again as well hydraulic conductivity as porosity decreases with increasing depth but the conductivities at the top of the model are not reduced. Instead of interpolating groundwater levels the dependence of the levels on the elevation of the piezometers above sea level is used to fix the boundary condition of the top of the model. In figure 7 the groundwater levels for the two different models are depicted together with the profile of the topography of the Lower Muschelkalk. The main difference is the enhancement of the groundwater level in the northwestern part of the model area. This results in a totally different flow field. In HYGAM1 pathlines reach the surface at the northwestern boundary of the model, whereas in HYGAM2 they reach the surface at the transverse fault zones near Wittmar or at the creek Ammerbeek. But in contrast we get realistic recharge data of about 150 mm/a without reducing hydraulic conductivities. On the other hand this flow field can be interpreted as an indication of the inappropriateness of two-dimensional modelling.

4 Validation Aspects

In the second edition of the IAEA Radioactive Waste Management Glossary [7] validation is defined as "a process carried out by comparison of model predictions with independent field observations and experimental measurements." In principle two measurements independent of the calibration of our model are available to test validity. On one hand these are the velocities measured in boreholes and on the other hand it is the averaged recharge rate. All other field data from hydrochemistry, isotope chemistry, etc. are useful in a qualitative way to improve the understanding of our model, but they cannot be used for validation purposes.

From our point of view in our case the comparison of measured to predicted velocities is only theoretically a method to compare reality to our model and is only mentioned here for completeness sake. On one hand in the vicinity of boreholes not only the hydraulic conductivity is disturbed but the flow field is changed too. In addition the method of measurement itself is affected with large imperfection. Thereby the agreement of the velocities will only be by chance.

Common to both approaches HYGAM1and HYGAM2 is the fact that the available data do not suffice to calibrate perfectly the model. Due to our little knowledge about the water table at the top there remains an ambiguity in modelling. Either one has to reduce the hydraulic conductivities in order to predict the estimated recharge rate or one has to change the groundwater levels at the top of our model. Anyway in our case the recharge rate is no longer an independent measurement. Therefore it is not an appropriate tool to prove validity.

Since there is no suitable numerical model available to take into account the effects of variable density due to salinity by that time we completely neglect these effects of salinity. In using an appropriate numerical model by simulating density driven flow fields we will be able to compare predicted density profiles with measurements. Dependent on the experience made in Intraval Phase 2 [8] with the salt test cases we are aware of the additional difficulties and degrees of freedom which will be introduced by taking into account density effects. But anyway we are sure we will not validate our model in a rigorous way, if this in general is possible. But we hope to increase confidence in our model in obtaining the present density distribution by choosing proper initial conditions.

5 Conclusions and Recommendations

In our case the field data which are available from site evaluation are not sufficient to validate our model. Additionally to date no suitable numerical model is on hand to simulate density driven flow fields for large and complex structures like the Asse area. For that reasons the present status of our effort toward validation is unsatisfactory. But we hope we will be able to improve the situation in the near future when we will be provided with an better numerical model.

For future developments it is very important to enhance the co-operation between people who perform experiments for the purpose of site evaluation and the people who access long term safety. The setting up of a model has to coincidence with the site evaluation program in such a way that it is possible to feed back findings from the theoretical to the experimental side and vice versa. After calibration of the model at least one independent field experiment has to be performed for the sake of validation. This experiment has to be well designed in order to be sensitive in properties of the model which can be predicted by the model. It is obvious that it might be useful to complete the program with additional laboratory experiments.

6 References:

[1] BATSCHE, H. & STEMPEL, C. v.: Hydrogeological investigations in the covering rock-strata of the pilot waste repository Asse (FRG). - Proc. Internat. Sympos. in Orléans 7-10 June, Documents du B.R.G.M., 160 (1988) 113-125, Orléans

[2] BATSCHE, H., KLARR, K. & STEMPEL, C. v.: Hydrogeologisches Forschungsprogramm Asse - Abschlußbericht. - Abteilungsbericht IfT 4/94 (1994) 459 p., Braunschweig

[3] KRAUS, I.: Die Bestimmung der Transmissivität von Grundwasserleitern aus dem Einschwingverhalten des Brunnen-Grundwasserleiter-Systems. - J. Geophys., 40 (1974) 381-400, Berlin

[4] SCHÖNFELD, E.: Die Grundwasserbewegung im Deckgebirge und am Salzspiegel des Salzstocks Asse. - GSF-Bericht, 26/86 (1986) 107 S., Neuherberg

[5] HOSSAIN, S., ARENS, G., FEIN, E.: SWIFT: Intera Simulator for Waste Injection, Flow and Transport. Version GSF 2. GSF-Bericht, 28/90 (1990), Neuherberg

[6] STORCK, R., private communication 1987

[7] ARENS, G., private communication 1990

[8] International Atomic Energy Agency: Radioactive Waste Management Glossary, 2nd edition. IAEA-TECDOC-447, Vienna, 1988

[9] Intraval Phase 2: Studies on Saline Groundwater Movement in an Erosional Channel Crossing a Salt Dome. Working Group Report: Working Group 3 (in preparation)

SESSION V

Chairmen:

B. Vignal
A. Van Luik

Characterization and Tracer Tests in the Full-Scale Deposition Holes in the TVO Research Tunnel

Aimo Hautojärvi & Timo Vieno
VTT Energy, Finland

Jorma Autio, Erik Johansson & Antti Öhberg
Saanio & Riekkola Consulting Engineers, Finland

Jukka-Pekka Salo
Teollisuuden Voima Oy, Finland

Abstract

Three wells of the size of the deposition holes (diameter 1.5 m, depth 7.5 m) in a KBS-3 type repository for spent fuel have been bored with a new fullface boring method in the TVO Research Tunnel. The Research Tunnel lays in the crystalline bedrock at the depth of 60 metres in the VLJ Repository at Olkiluoto. The wells are 6 metres apart. Comprehensive pre- and post-characterization of the rock and the wells has and will be performed. The investigation programme includes geological mapping, geophysical measurements and rock modelling as well as hydrological measurements and tracer tests. The aim is to evaluate the feasibility of the deposition hole boring method and to study properties of fractures and groundwater flow in the rock and in the disturbed zones around the deposition holes and below the floor of the tunnel.

Characterization and tests in the bored wells and evaluation of the results are in progress. So far, the investigations have revealed the following: Fullface boring is a practicable method to make deposition holes. The inflow rate of groundwater in two of the three wells is 6 and 18 litres/hour. The well in the middle is almost dry (inflow 0.05 litres/hour) as expected on the basis of the pre-characterization. Visual inspection of the inflow and tracer tests give indications of flow in sparse and narrow channels. The observed non-Fickian dispersion is thought to be caused by velocity differences in the channel or by diffusion into stagnant pools in the fracture filling. The inflow rate of groundwater into the wells is almost as high as the total inflow rate into the Research Tunnel which is 50 metres in length and 7 metres in height. The inflow into the tunnel takes place at few spots, too. Below the floor of the tunnel, to the depth of about one metre, rock is clearly more fractured than deeper in the holes. There seems to be a skin around the tunnel which prevents inflow of water.

The suitability of a location for a deposition hole in a repository may be checked by means of a small-diameter investigation borehole. Highly conductive fractures are revealed by visual inspection of the drill core and an inflow measurement. In our opinion, the suitability of a location for a deposition hole at the depth of 500 metres is questionable if the inflow rate into the investigation borehole exceeds 10 litres/hour which corresponds to a flow rate of the order of 100 litres/year in the steady state after the sealing of the repository.

1 INTRODUCTION

The TVO Research Tunnel lays in the crystalline bedrock at the depth of 60 metres in the VLJ repository at the Olkiluoto nuclear power plant. The repository consists of two silos for low and medium level reactor waste (Figure 1). It has been in operation since 1992. In winter 1994, Teollisuuden Voima Oy (TVO) and Swedish Nuclear Fuel and Waste Management Co (SKB) in co-operation carried out a boring experiment to study the feasibility of a new fullface boring technique to make deposition holes for canisters in a KBS-3 type repository for spent fuel. The novel boring technique is based on rotary crushing of rock and removal of muck by air vacuum suction through the drillstring. The boring equipment was composed of a standard raiseboring machine, frame, drillstring, cutterhead and vacuum suction system (Figure 2).

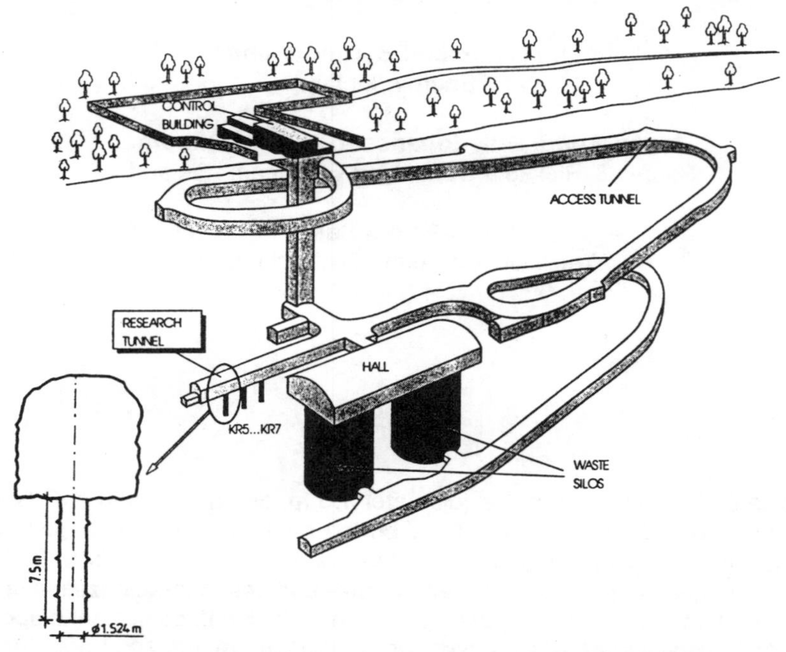

Figure 1. TVO Research Tunnel in the VLJ Repository.

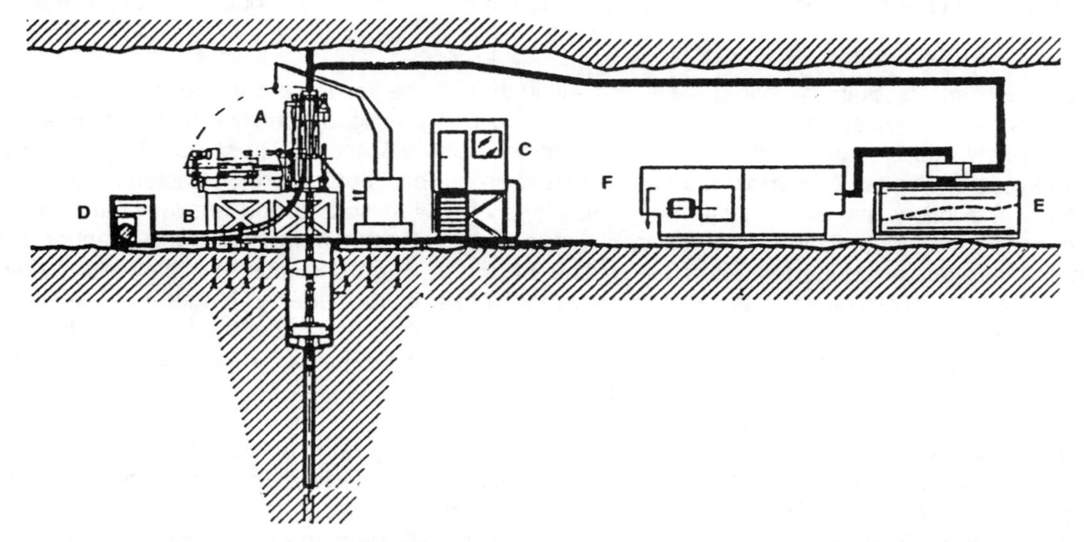

Figure 2. Fullface boring system for deposition holes: a) boring machine, b) frame, c) control room, d) power pack, e) muck tank, f) vacuum unit.

The raiseboring machine used to create the torque and thrust necessary for the cutting of the rock was an old smaller size standard machine with a maximum thrust of 50 - 60 tonnes. The cutterhead was an used Sandvik CBH-5 blind hole head designed for boxhole boring with a hole diameter of 1524 mm (Figure 3). The head was furnished with 8 roller button cutters and 4 gauge rollers. A pilot bit with a diameter of 311 mm was used for boring a pilot hole and guiding the cutterhead. Two nozzles, which were located in the cutterhead between the cutters were used to suck up the muck. The muck was sucked through the cutterhead and drillstring to the suction line which was composed of transport pipes, a muck tank and a vacuum unit with a power of 200 kW.

Figure 3. Cutterhead above bored well.

A test programme was carried out to find optimal operating parameters (Autio & Salo 1994). The main interests in the technical performance are the penetration rate and efficiency of the vacuum suction system. As concerns the quality of the hole, the most interesting features are roughness, microfracturing and porosity of the rock surface. Muck samples were taken during the boring to find out the effect of different thrust forces and rotation speeds on the quality of muck.

Three wells with a diameter of 1.5 m and a depth of 7.5 m were bored. The holes are 6 metres apart. The boring system performed well. A section of one well was bored without a separate pilot hole. The highest rate of penetration achieved during the test programme was about 1 metre/hour with a total maximum thrust of 73 tonnes and rotation speed of 8 rpm. With a more powerful boring machine and suction system, penetration rates over 3 metres/hour could be achieved for the tonalite rock type of the TVO Research Tunnel.

The boring tests were accompanied by comprehensive pre- and post-characterization of the rock and the bored wells. The investigation programme includes geological mapping (Äikäs & Sacklén 1993), geophysical measurements (Johansson & Autio 1993, Cosma 1994) and rock modelling (Johansson & Hakala 1994) as well as hydraulic (Hautojärvi et al. 1993, Öhberg & Sacklén 1994) and tracer tests (Hautojärvi 1994, Viitanen 1994). Characterization and tests in the deposition holes and evaluation of the results are still in progress.

2 GEOLOGICAL AND GEOPHYSICAL CHARACTERIZATION OF THE NEAR-FIELD

The VLJ repository has been thoroughly investigated before and after the construction. The repository has been excavated in a tonalite formation surrounded by miga gneiss. The rock types found in the Research Tunnel are tonalite and pegmatite. Tonalite is usually slightly foliated, medium grained, massive and sparsely fractured. Pegmatite is non-foliated, coarse grained and sparsely fractured. The Research Tunnel is 47 metres in length and 4 - 7 metres in height. It was excavated using conventional drill and blast technique. The tunnel is located above a gently dipping hydraulically conductive zone (RF) and is intersected by a vertical zone (RV1) (see Figure 4).

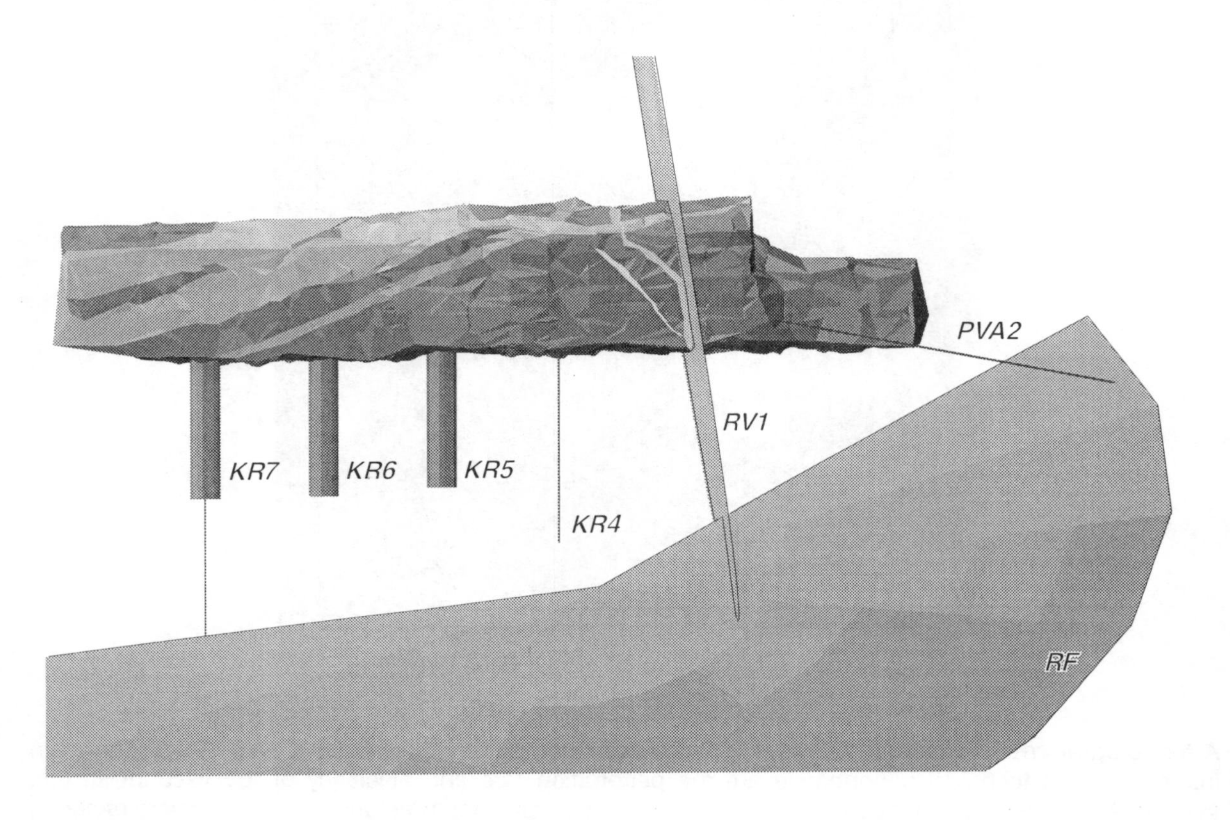

Figure 4. Deposition holes bored in the Research Tunnel. Hydraulically conductive fracture zones in the vicinity of the tunnel are also shown.

The fracturing of the rock at the locations of the deposition holes and in the periphery of the tunnel was investigated by various geoscientific methods before and after the boring of the holes (Table 1). Core drilling was used to assess the subhorizontal fractures and continuity of the features found on the walls of the tunnel. Geophysical measurements were carried out to evaluate the existence of subsurface features and to confirm the geometry of distinct structures, and to estimate their continuity. Water conducting fractures were identified by means of hydraulic and tracer tests. The results of the investigations were compiled in a structural model which was updated in the different phases of the boring and investigation programme.

Table 1. Investigations carried out before and after boring of the full-scale deposition holes.

In the tunnel before boring of the investigation boreholes
- detailed geological mapping of the surface with a cut-off length of 1 m
- surveying of tunnel surface, location of fractures, and half barrels of blast holes
- mapping of excavation damage on half barrels of blast holes
- determination of mechanical properties of rock
- ground penetrating radar on the floor of the tunnel
- inflow mapping
- structural modelling

In the investigation boreholes before boring of the deposition holes
- surveying of direction
- mapping of cores
- high frequency seismic reflection survey
- seismic tomography in holes KR4 and KR5
- hydraulic conductivity and head measurements
- inflow measurement
- tracer test between KR4 and KR5
- modelling of flow paths
- structural modelling

After the boring of the deposition holes
- surveying of geometry
- detailed geological mapping
- high frequency seismic reflection survey with a high resolution equipment
- ground penetrating radar
- inflow measurement
- mapping of inflow points
- colour penetration mapping
- core samples from wall
- structural modelling

Core drilling gave information on the location of horizontal or gently dipping filled fractures. Still, many of the fractures detected on the walls of the full-scale holes were not observed in the drill cores and were probably partly misinterpreted as mechanical breaks of core samples and foliation. In the upper parts of the deposition holes, to the depth of 1.5 metres below the floor of the tunnel, the fracture intensity is in average 4.5 fractures/m^2 which is over ten times higher than the average fracture intensity in the lower parts of the holes. The extent of the disturbed zone corresponds to the relatively high charge density (1.1 kg-dyn/m) used in the blasting. The investigation of the perturbation caused by the drill and blast excavation of the tunnel and by the fullface boring of the holes is still in progress.

The natural starting point for assessing fracturing of bedrock in the vicinity of deposition holes in a repository is data from the tunnel in which the holes will be bored. If the conductive fracturing forms a cubic type system like in the Research Tunnel, the locations of vertical conductive fractures can be found on the basis of traces and leaks on the walls of the tunnel. Subhorizontal fractures can be detected by core drilling and supplementary geophysical methods. Hydraulic characteristics of the rock and fracture connections can be evaluated by means of hydraulic measurements and tracer tests which are discussed below. 3-D modelling of the fracture observations facilitates assessing of the connections in the fracture network.

3 HYDRAULIC MEASUREMENTS AND MODELLING

Hydraulically the dominant feature in the vicinity of the Research Tunnel is the fracture zone RF which dips gently below the tunnel. The hydraulic head in RF at the depth of the Research Tunnel is measured in an adjacent borehole. The head in RF is about -3 metres in relation to sea level and thus the head difference between RF and the open tunnel at the level of -60 metres is about 57 metres. A concept of two flow resistances between the three conductors (fracture zone RF, off-packed borehole, tunnel) was considered to give a rough idea of the hydraulic coupling (Figure 5). In the modelling it was assumed that head boundary conditions (-3 metres in RF and -60 metres in the tunnel) were at a distance of 10 metres for both resistances in the case of every borehole. Exact distances depend, of course, on the detailed geometry of the fracture and flow systems.

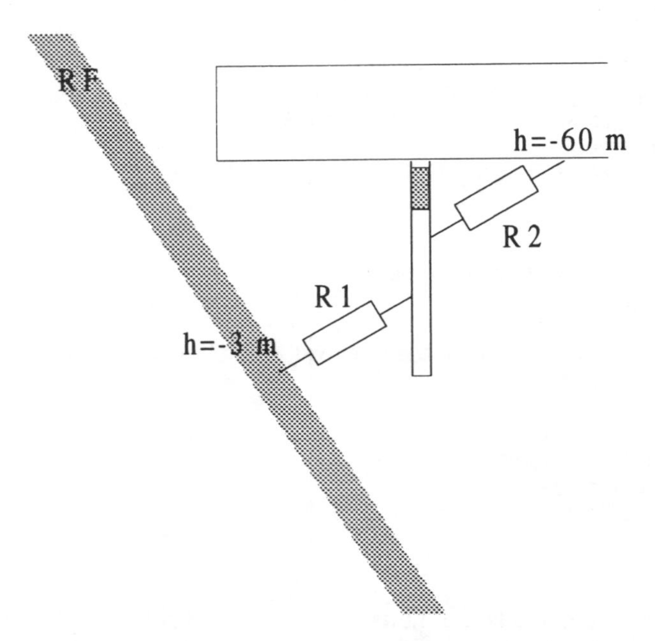

Figure 5. Concept of two flow resistances, R1 and R2, between fracture zone RF, borehole, and tunnel.

In the measurements simple methods feasible also during the work in the tunnel were employed. Flow rates against atmospheric pressure were determined by collecting water typically a couple of minutes into a bottle and weighing the collected amount. Later on the travel time between markers in an hose was also used to determine the flow rate and thus only a stopwatch was needed for flow measurements. The inflow measurements in the boreholes give directly an estimate of the resistance R1. In order to determine the resistance R2 additional measurements were needed. First the ends of the outflow hoses were lifted as high as possible up to the ceiling of the tunnel which is about 5 metres. A reduction due to the back pressure was systematically observed. More back pressure was needed, however, to determine the value of the resistance R2. This was arranged by means of pressurized nitrogen flask as back pressure.

On the basis of the measured head in RF, it was a priori estimated that the inflow rate into the investigations boreholes would be of the order of 0.001 - 0.1 l/h, if there is no hydraulic connection to RF, and 1 - 100 l/h if a good connection exists. The flow rates 2 - 20 l/h from three of the four investigation boreholes were a slight surprise because the bedrock is very sparsely fractured and its overall hydraulic conductivity is rather low. The fourth investigation borehole KR6 had a low flow rate of 0.05 l/h. The total flow rate into the boreholes was almost as high as the total inflow into the Research Tunnel which was measured to be about 30 l/h. A good hydraulic connection seemed to exist between boreholes KR4 and KR5 because opening either of them caused a clear pressure response in the other.

Results of the back pressure measurements also indicated that a skin surrounds the tunnel diminishing the amount of water leaking into the tunnel. The resistance R2 is typically much higher than R1. In the case of hole KR6 the transmissivity to the tunnel is somewhat greater than the transmissivity between the hole and fracture zone RF but both values are so small that the inflows are anyhow extremely small.

The inflow rates of water into the large holes 1, 2 and 3 bored in this order at the locations of the investigation boreholes KR5, KR6 and KR7 were correspondingly 18 l/h (47 l/h), 0.05 l/h and 6 l/h (14 l/h), the estimates made on basis of the measurements in the investigation boreholes shown in the parenthesis. The hole 2 in the middle was almost dry as expected on the basis of the behaviour of the investigation borehole KR6. The inflow rates into the full-scale (diameter 1.5 m) holes were about the same as into the small (56 mm) investigation holes.

According to the observations and measurements the hydraulic behaviour can be explained by a few discrete fractures. Mapped fractures from the cores are shown in Figure 6 as elliptic discs centred at the observation points. The two lowest lying fractures in KR4 and the lowest one in KR5 are plotted with larger dimensions because they are considered to be mainly responsible for the hydraulic behaviour of these boreholes and to form the connection between the holes.

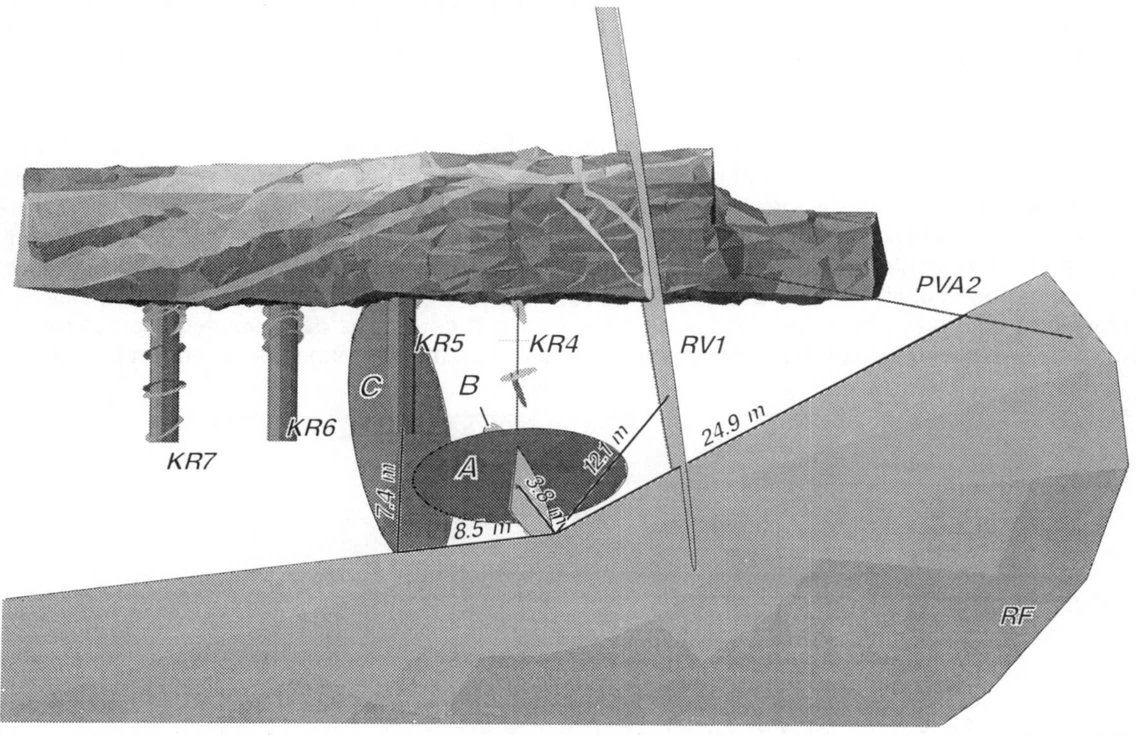

Figure 6. Mapped fractures in the boreholes and hydraulically significant structures RV1 and RF. The shortest distances between boreholes and intersections of fractures/structures are also shown.

4 TRACER TESTS

Three tracer tests have been performed between the borehole KR4 and the borehole (and later the full-scale hole) at the location of KR5. The first test was run after all four investigation holes were drilled. The observed good hydraulic connection between these boreholes was assessed to make a successful tracer test possible. The purpose of the first test was to investigate transport in single fractures before boring of full-scale holes. The second tracer test was run after all three full-scale deposition holes were bored. In that test the main target was to study the mass outflow pattern from the wall and floor of the large hole but some samples were collected as well to get some idea of the intensities of the mass flows. In the third tracer test the water level was allowed to rise above the observed outflow areas and two plates were installed to cover the main tracer outflow areas, one on the wall and one on the floor. Two break-through curves were measured in this test. The water flow rate through the injection borehole KR4 could be measured accurately during all three tests by using the dilution probe technique.

First test

The first test was run between the investigation boreholes KR4 and KR5 being 6 metres apart from each other. Dye (Uranine) and radioactive tracer (Br-82) were injected and recovered successfully in short packed off sections containing only single fractures in each of the boreholes. The injection and discharge sections were only 30 cm long and both were located at the bottom of the holes.

The break-through curve was analyzed by using various techniques of model fitting and convolution or deconvolution. In the present case the injection procedure is such that the deconvolution problem is stable and all of the techniques give essentially similar results. This is not always the case. There is no unique explanation for the obtained impulse response based on transport mechanisms. Much more experiments would have been needed to draw conclusions on the transport processes in detail but the main objectives of this test was to get an idea of the connectivity of fractures and flow conditions in the near-field of a deposition hole as well as to develop tracer experiment techniques for single fractures having relatively low transmissivities.

It is quite clear that an advection-dispersion model with Fickian dispersion does not apply in the present case. The break-through curve could be explained alternatively by diffusion from a very narrow channel (a few millimetres) into the fracture filling or otherwise stagnant areas in the fracture plane or by a wider velocity field (a few centimetres) where both very slow and higher velocities exist and the transport time is too short for molecules to diffuse across the velocity field and cause mixing. There is no possibility to see convincingly matrix diffusion and to extract parameters of it from such an experiment before the mechanisms causing hydrodynamic dispersion are known in detail. The deconvoluted impulse response derived by direct inversion of the Toeplitz matrix formed from the injection pulse is shown in Figure 7.

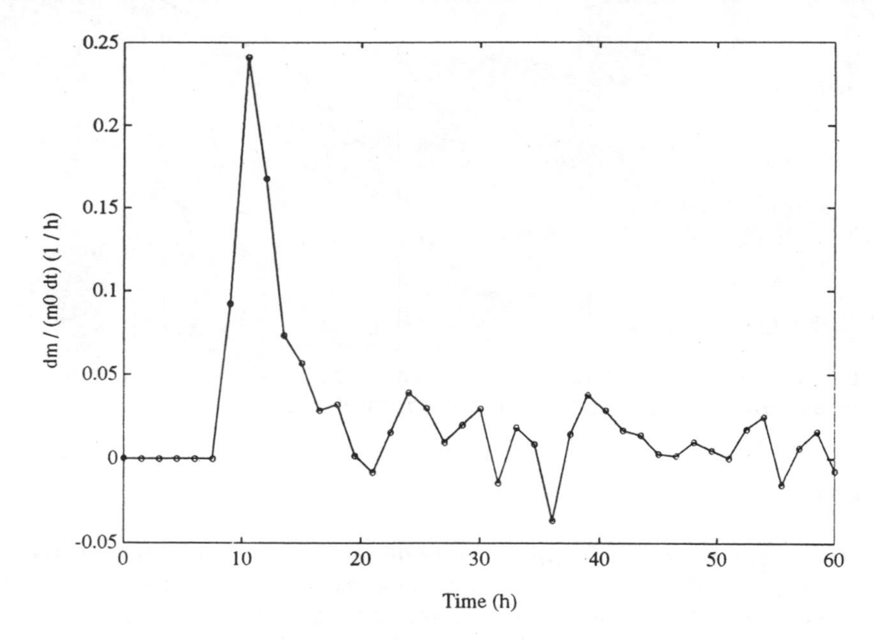

Figure 7. The impulse response of the transport path from KR4 to KR5 determined by direct inversion of the Toeplitz matrix formed from the injection pulse.

The flow path cannot be deduced unambiguously but some possibilities can be given. The path might lead from KR4 via fracture B to RF and from there along fracture C into KR5 (see Figure 6). This would be a somewhat complicated transport path and it is not easy to understand the flow rate 18 ml/h through KR4 in that configuration. Another possibility is that fracture A connects fractures B and C where the tracer was injected and recovered, respectively. Fracture A would then carry also the water to KR4 either from RV1 or RF. The uncertainty about the transport path makes the transport path length also uncertain which affects the other transport parameters in the models. It can be concluded, however, that the sharpness of the break-through curve indicates a well defined single transport path from KR4 to KR5 via a few fractures although it might be difficult to find out where this path is.

Second test

In the second test tracer injection was done in the same way as in the first test. Small improvements allowed, however, a better control of the tracer entering the rock. The biggest difference was that a visual observation was possible in the discharge hole, which was now the full-scale deposition hole 1 bored at the former location of KR5. It had been observed earlier that water entered the full-scale hole from four spots: two at the opposite sides on the wall at a height of about one metre from the bottom and two on the floor in the same steeply dipping fracture.

During this test the flow rate through the same injection section as in the first test was a factor of two higher being now 35 ml/h. The first arrival of tracer into the full-scale hole 1 occurred 5.5 hours after the injection start.

All tracer that could be observed came into the hole 1 from the southern points of water leak and from the fracture line connecting them, although very little water came from the fracture except the strong point leaks. Some samples were taken from the point leaks and break-through curves were determined roughly. Due to a much shorter injection period (7 hours) than in the first test (83 hours) the break-through resembles as such closely the impulse response. A noticeable difference was that the first arrival took place 2 - 3 hours earlier than in the first test. This was anticipated because the flow rate through the injection section had been increased. Otherwise the break-through curves of the first two tests are reasonably similar.

Third test

In the third test the aim was to determine more accurately the break-through curves from the wall and from the floor of the full-scale hole 1. Two plates were assembled to cover the leak areas and water was pumped from the middle of each plate. Thus all of the tracer arriving between the hole wall or floor and the plate could be collected and sampled. Two break-through curves were observed showing a similar behaviour but with different intensities. In this test the tracer arrived still in a shorter time and the flow rate through the injection section was highest of the three tests (47 ml/h). The analysis of the break-through curves is still in progress but the same conclusions as from the first and second test apply.

5 DISCUSSION

The boring tests proved that deposition holes in a KBS-3 type repository for spent fuel can be bored efficiently by using a boring machine based on fullface rotary crushing and removal of muck from the bottom of the hole by direct air vacuum suction. The test boring without a pilot hole succeeded well and the deposition hole cân be made as well without it.

Visual inspection of the inflow and tracer tests give indications of flow in sparse and narrow channels. The non-Fickian behaviour indicate transport processes where the dispersion is not in steady state or in other words the dispersion coefficient is time dependent. There are two interpretations. The tracer may be transported in a channel where there exist strong variations of velocities over the channel width. Velocities being practically zero and maximum velocities may occur within a distance of a few centimetres over the channel width. This could cause the observed transport behaviour. Another possibility is that the flow channel is just a thin pipe through the fracture filling and tracer molecules diffuse from the pipe into the fracture filling or other stagnant areas around the pipe. In this kind of in-situ experiments, it is not possible to distinguish the effects of matrix diffusion from other phenomena causing dispersion (Hautojärvi 1993, Ilvonen et al. 1994). With non-sorbing tracers, the ratio of the flow wetted surface (channel width × channel length) and the flow rate should be larger by at least two orders of magnitude to reveal effects of matrix diffusion.

From the point of view of performance analysis, interesting results are expected also from the characterization of the disturbed rock zones around the deposition holes and below the floor of the tunnel. Rock is clearly more fractured in the upper parts of the holes, to the depth of about 1.5 metres below the tunnel floor. The extent of the disturbed zone corresponds to the relatively high charge density used in the blasting. Hydraulic measurements indicate that there is a skin around the tunnel which prevents inflow of water.

The suitability of a location for a deposition hole may be checked by means of a small-diameter investigation borehole. Highly conductive fractures are revealed by visual inspection of the drill core and an inflow measurement. In the Research Tunnel, which lays 60 metres below the surface, the observed inflow rates into the investigation boreholes and full-scale deposition holes range from 0.05 to 20 litres/hour. In a repository at a depth of 500 metres, the gradients during the construction and operation phase are typically ten times higher. On the other hand, the deposition holes in the Research Tunnel were deliberately located near to a hydraulically conductive fracture zone. In a repository for spent fuel, a respect distance should be left between deposition holes and a highly conductive fracture zone. If the distance from the deposition hole to a major fracture zone is 50 metres and the head difference is 500 metres, we have a gradient of about 10 when the repository is open. In the steady state after the sealing of the repository the gradient is typically less than 0.01. An inflow rate of 20 litres/hour into an investigation borehole at a potential location of a deposition hole would thus mean that the post-closure flow rate around the deposition hole is of the order of $0.01/10 \times 20$ litres/hour ≈ 0.02 litres/hour ≈ 200 litres/year. The mass flow rates in the deposition hole are effectively limited by the impermeable bentonite buffer. The diffusive mass flow rates through the bentonite correspond to an equivalent flow rate of less than 10 litres/year irrespective of the flow rate of groundwater in the rock around the deposition hole (Vieno et al. 1992, SKB 1992). Nevertheless, in accordance to the multibarrier principle, we think that the flow rate around the deposition hole should not be much higher than 100 litres/year in the post-closure phase. Accordingly, the suitability of a location for a deposition hole is questionable if the flow rate into the investigation borehole exceeds 10 litres/hour. More accurate limits have to decided in the construction phase on the basis of site-specific data.

REFERENCES

Autio, J. & Salo, J.-P. 1994. Boring of full-scale deposition holes in the TVO-research tunnel. Äspö Hard Rock Laboratory International Workshop on the Use of Tunnel Boring Machines for Deep Repositories. Äspö, June 13 - 14, 1994.

Cosma, C. & Honkanen, S. 1994. High frequency seismic survey in the research tunnel. Teollisuuden Voima Oy, Work Report TEKA-94-05.

Johansson, E. & Autio, J. 1993. Rock mechanical properties of intact rock in TVO's research tunnel. Teollisuuden Voima Oy, Work Report TEKA-93-05.

Johansson, E. & Hakala, M. 1994. 3D-modelling of TVO's research tunnel. Teollisuuden Voima Oy, Work Report TEKA-94-02.

Hautojärvi, A. 1993. Intraval Phase 2: Do break-through curves in field tests reveal matrix diffusion ? Teollisuuden Voima Oy, Work Report Safety and Technology 93-03.

Hautojärvi, A., Ingman, M. & Öhberg, A. 1993. Hydraulic characterization of the near-field of the boreholes in TVO's research tunnel. Teollisuuden Voima Oy, Work Report TEKA-93-10.

Hautojärvi, A. 1994. Evaluation of the tracer test in Olkiluoto research tunnel. Teollisuuden Voima Oy, Work Report TEKA-94-09.

Ilvonen, M., Hautojärvi, A. & Paatero, P. 1994. Intraval Project Phase 2: Analysis of Stripa 3D data by a deconvolution technique. Nuclear Waste Commission of Finnish Power Companies, Report YJT-94-14.

SKB 1992. SKB 91 - Final disposal of spent nuclear fuel. Importance of the bedrock for safety. Swedish Nuclear Fuel and Waste Management Co (SKB), Technical Report 92-20.

Vieno, T., Hautojärvi, A., Koskinen, L. and Nordman, H. 1992. TVO-92 safety analysis of spent fuel disposal. Nuclear Waste Commission of Finnish Power Companies, Report YJT-92-33.

Viitanen, P. 1994. Combined use of radioactive and inactive tracers in a borehole experiment. Nuclear Geophysics 8(1994), pp. 335 - 341.

Äikäs, K. & Sacklén, N. 1993. Fracture mapping in the research tunnel. Teollisuuden Voima Oy, Work Report Research Tunnel 93-01.

Öhberg, A. & Sacklén, N. 1994. Mapping of inflow points on the floor of the research tunnel. Teollisuuden Voima Oy, Work Report TEKA-94-03.

Joint ANDRA/Nirex/SKB Zone of Excavation Disturbance Experiment (ZEDEX) at the Äspö Hard Rock Laboratory

A J Hooper
United Kingdom Nirex Ltd (UK)

O Olsson
Conterra AB (Sweden)

Abstract

The excavation of access shafts and tunnels and of the disposal areas of a waste repository will cause a disturbance in the surrounding rock mass with possible alterations to rock mass stability and hydraulic properties. For a number of disposal concepts this disturbance may be important for the operational and/or post-closure safety of the repository. Furthermore the disturbance may extend over time as a consequence of processes such as stress relaxation.

The sponsors of ZEDEX, namely ANDRA, Nirex and SKB, are interested in developing the ability to produce reliable models of the disturbed zone that will develop around large cross-section excavations in fractured hard rock masses that are initially water saturated. Various models have been developed to calculate the important characteristics of the disturbed zone in such rock masses as a function of parameters related to the rock mass quality and the geometric description of the excavation.

ZEDEX was initiated in the Äspö Hard Rock Laboratory in April 1994 with the drilling and instrumentation of boreholes running alongside the planned extension of the spiral access ramp and a planned parallel experimental tunnel. The experiment seeks to exploit the extensive database that has been established for the bedrock through the site characterisation programme at Äspö. A range of geophysical measurements are to be carried out in the instrumented boreholes with the objective of determining the characteristics of the disturbed zone resulting from subsequent excavation. The characteristics will be expressed in terms of parameters that are output from various models applied to the rock mass and which are also applicable in performance assessment.

ZEDEX has been designed to generate information for alternative methods of excavation. The extension to the spiral ramp is to be made by tunnel boring whereas the parallel experimental tunnel will be excavated in part by "normal" blasting and in part by smooth blasting. The objective is to build confidence in the modelling of the disturbed zone to support the selection of excavation methods for repository construction.

1. INTRODUCTION

The excavation access of shafts and tunnels and of the disposal areas of a waste repository will cause a disturbance to the rock surrounding the excavation. The character and the magnitude of the disturbance is due to the existence of the air-filled void represented by the excavated feature and the method of excavation used to construct it. The properties of the disturbed zone around excavations is of importance to long-term repository performance in that it may provide a preferential pathway for radionuclide transport or may affect the efficiency of plugs placed to seal drifts. The character and magnitude of the Excavation Disturbed Zone (EDZ) is due to:

- the air-filled void represented by the excavated feature,
- the method of excavation.
- the value of certain in-situ parameters such as frequency and orientation of discontinuities, rock mass properties, and stress.

To obtain a better understanding of the properties of the disturbed zone and its dependence on the method of excavation ANDRA, Nirex and SKB decided through their participation in the international Äspö Hard Rock Laboratory Project to perform a joint study of disturbed zone effects. The project is named ZEDEX (Zone of Excavation Disturbance EXperiment).

The method of excavation used for the Äspö HRL spiral was changed in the summer of 1994 from drill and blast to tunnel boring (TBM). This afforded an opportunity to study the effects of different excavation methods on the rock surrounding the tunnel.

The ZEDEX project is expected to contribute to the understanding of the measurement and development of the EDZ, thereby providing a firmer basis for selecting or optimizing the construction method or combination of methods for a deep repository and subsequent sealing.

The project is, at the time of writing, at a relatively early stage and no results are available. The paper focuses upon the design philosophy for the experiment.

The objectives for the project are:

- to understand the mechanical behaviour of the Excavation Disturbed Zone (EDZ) with respect to its origin, character, magnitude of property change, extent, and its dependence on excavation method,
- to perform supporting studies to increase understanding of the hydraulic significance of the EDZ, and
- to test equipment and methodology for quantifying the EDZ.

The project does not include, as a major aim, the study of possible changes in hydraulic properties in the disturbed zone caused by the two excavation methods.

2. RATIONALE

2.1 RELEVANCE TO REPOSITORY PERFORMANCE

The properties of the disturbed zone must be considered in the design of a repository and in the assessment of its long term safety. In addition, the data collected in drifts, shafts or tunnels to access the repository will be used for detailed characterisation of the repository host rock and hence used in performance assessment. For these reasons it is important to understand how the method of excavation affects the properties and extent of the disturbed zone and the ability to characterise the rock. This knowledge is essential in deciding what excavation method or combination of methods to use for a future repository. The ultimate aim in constructing vaults for the emplacement of radioactive waste is to do it in a manner that ensures that the EDZ does not have a significant impact upon repository performance.

This test has relevance to repository construction and performance in that it will attempt to relate physical measurements of displacement and disturbance to excavation method and geological conditions.

2.2 PREVIOUS EXPERIENCE

ANDRA were involved in a smooth blasting experiment carried out in 1991 at the URL in Canada. The main result was to prove that it was possible to reduce the extension of the damaged zone significantly (from >1 m to 10 or 20 cm) using gas-energy explosives combined with appropriate blast designs.

The Norwegian Geotechnical Institute (Nirex's Geotechnical Consultants) have previously been involved with international projects on EDZ estimation at the Canadian URL and Stripa. Professor R Paul Young and the Applied Seismology Group (Nirex's Consultants) are responsible for acoustic emission (AE) studies. They developed AE technology for the URL in Canada and have extensive experience in the application of AE in rock mechanics and underground mining. It is intended to build on existing experience and to develop confidence in current technology for the measurement of the EDZ.

During 1991 SKB performed an experiment to study the extent and character of the disturbed zone at the Äspö HRL. The aim of this experiment, the "Blasting Damage Investigation", was to study the extent of the zone damaged by blasting for three different drill and blast schemes. The experiment showed that the damage in the floor of the drift was more extensive than in the walls for all drill and blast schemes used.

2.3 JUSTIFICATION FOR THE EXPERIMENTAL WORK

There is a need to demonstrate that we can quantify the EDZ and relate this to its impact on hydraulic conductivity (especially in the vicinity of plugs) in support of safety assessment.

The change in the method of excavation of the tunnel spiral of the Äspö HRL from drill and blast to tunnel boring provided an opportunity for a direct comparison of the disturbance to the rock induced by the different excavation methods. A detailed comparison of these effects for different excavation techniques has previously not been performed at realistic repository depths.

There is a need to demonstrate that the extent of the EDZ can be reduced by the selection of the optimum drilling pattern for a blasted drift. In this instance this experiment is intended to add to and complete similar experiments carried out at the Canadian URL on two aspects:

- Industrial aspect: after defining the optimum blast design, we should obtain predictable and consistent results.
- Measurement aspect: the tools that the Bundesanstalt für Geowissenshaft und Rohstoffe (BGR) intends to use in support of the ANDRA project should be more accurate than the ones used previously by ANDRA at the URL.

There is a need to select and develop equipment for the measurement of EDZ properties and extent and gain experience and confidence in its use before application to the planned Nirex Rock Characterisation Facility (RCF).

Äspö offers the opportunity to test out equipment and methods under similar conditions before application to build further confidence within the Nirex RCF and the planned ANDRA underground laboratories.

3. EXPERIMENTAL CONCEPTS

3.1 LOCALIZATION OF EXPERIMENT

The test site for a comparative study of excavation disturbance has to be located somewhere along the main TBM drift as this is the only place where TBM excavation can be performed at reasonable cost. A dedicated TBM drift for these tests would have incurred large extra costs. The flexibility in locating a drill and blast excavated drift is relatively large and it could in principle be placed anywhere along the TBM drift. However, a study of excavation induced disturbance to the rock requires drilling of test boreholes at appropriate locations before excavation commences. As the TBM drift was to be excavated into virgin rock volumes the only possibility to drill test boreholes for the TBM drift was at the very beginning of the drift. Hence, the most appropriate location of this experiment was where the excavation method was changed by SKB from drill and blast to TBM. This determined that the experimental drifts will be located adjacent to the TBM Assembly Hall (Figure 1).

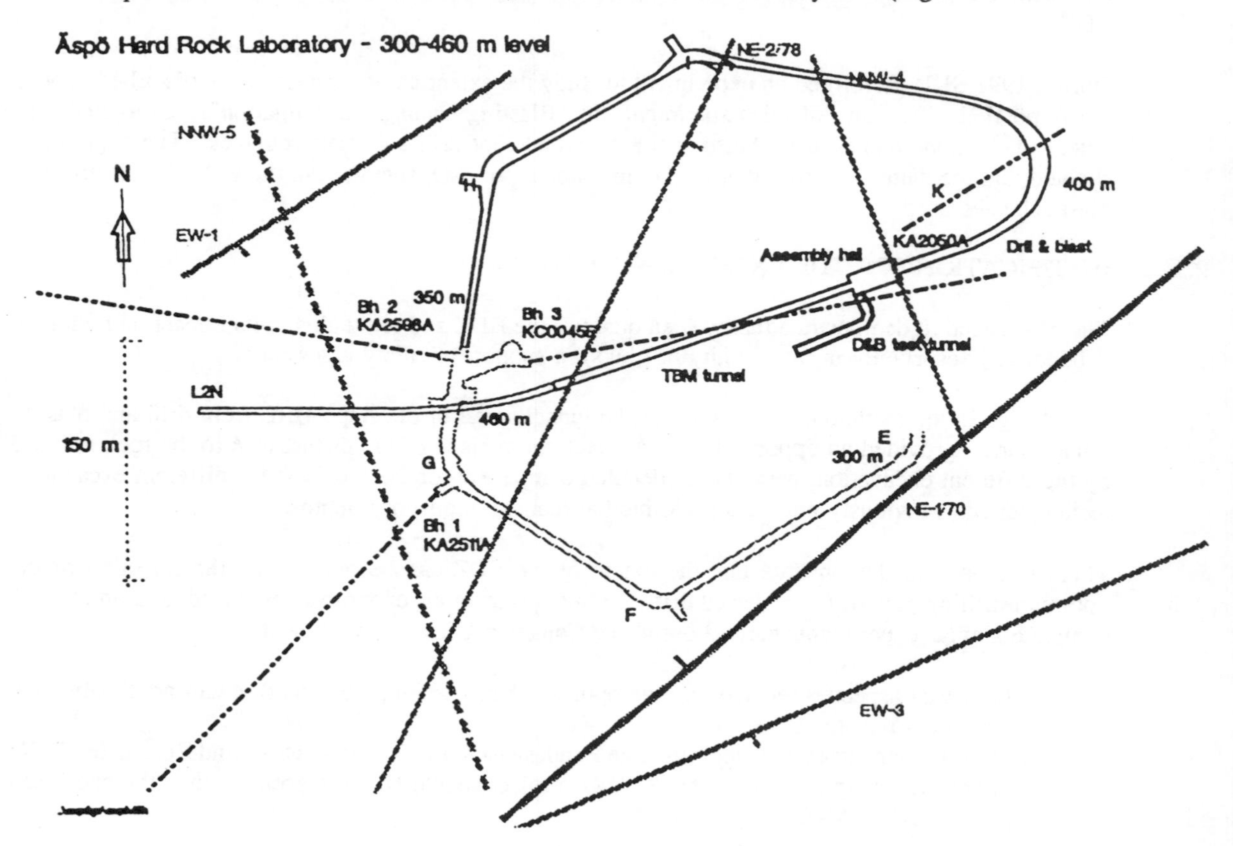

Figure 1. Projection of the Äspö HRL 300-460 m level showing the test drifts located adjacent to the TBM Assembly Hall of the HRL tunnel.

There will be two test drifts, one to be excavated by drill and blast and one already excavated by tunnel boring. The test drifts will be parallel and located approximately 25 m apart. The intention is that the test drifts should be located in relatively homogeneous Äspö diorite so that the geological conditions at the two drifts are similar to facilitate a meaningful comparison. The test drifts will be located at an approximate depth of 420 m below ground surface and the orientation of the drifts will be roughly perpendicular to the main horizontal stress direction.

3.2 TESTED HYPOTHESIS

The disturbance to the rock depends on both the excavation method used and the combined effects of the existence of a void and the geological and rock mechanical properties of the excavated rock (excavation method independent disturbances). The hypothesis is that near field (< 2 m) disturbance can be reduced by application of an appropriate excavation method (smooth blasting or tunnel boring). It is hoped the test will confirm that smooth blasting limits the extension of the damaged zone without affecting the blast productivity. The hypothesis is that far field disturbance (> 2 m) will be essentially independent of excavation method as it is caused by stress redistributions, discontinuity geometry, and mechanical properties of the rock.

The ZEDEX Project will include tests of the following excavation techniques:

- "normal" blasting, similar to that used for excavation of the Äspö HRL tunnel to a depth of approximately 430m,
- smooth blasting based on the application of low shock explosives and an optimised drilling pattern, and
- tunnel boring.

The test should determine displacements, micro-crack movements, and geophysical profiles in an EDZ during mine-by, which can then be related to hydraulic conductivity changes.

3.3 CONFIGURATION

The TBM test drift constitutes a part of the main access tunnel of the Äspö HRL and is located shortly after the TBM Assembly Hall. The Drill & Blast test drift will be parallel and located approximately 25 m to the south of the TBM tunnel as shown in Figure 2. An access drift (12 in Figure 2) will be excavated from the end of the Assembly Hall to access the Drill & Blast test drift. The purpose of the first round in the Drill & Blast test drift is to reduce the effects of the anomalous stress field caused by the drilling niches. The next four rounds will be used for testing of a smooth blasting technique based on low-shock explosives, and the following five rounds will be used to study effects of normal blasting.

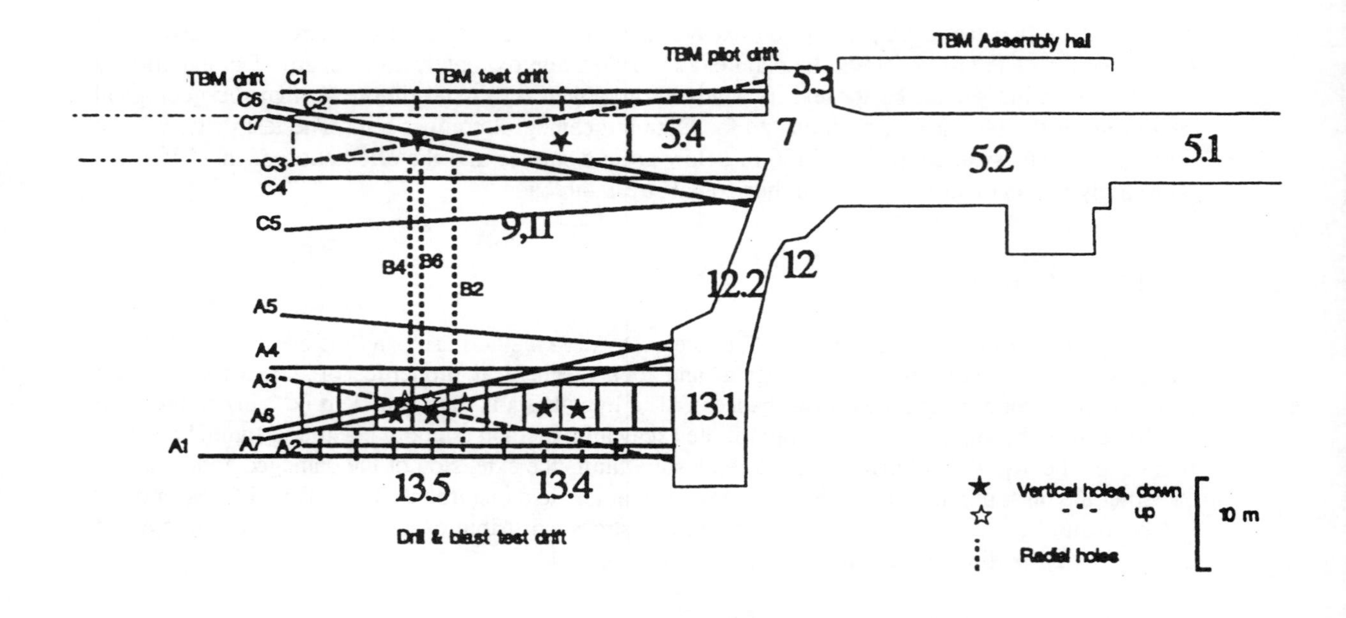

Figure 2. Proposed experimental configuration for the ZEDEX project. Numbers in the figure other than borehole nos. A1-7 and C1-7 refer to major project items.

The shape of the blasted drift should be rounded (circular with a flat floor) and the diameter of the blasted drift should be about the same as for the TBM drift, i.e 5 m.

There will be a number of boreholes drilled axially and radially relative to the test drifts to assess the properties and extent of the EDZ. The location of the boreholes in plan and vertical section is shown in Figures 2 and 3, respectively. A borehole for accelerometer measurements (diameter 86 mm) has been drilled parallel to and at a distance of 3 m from each test drift (boreholes A1 and C1, respectively). At each drift six boreholes (three to the side, two above and one directed below the drift, boreholes A2-7 and C2-7, respectively) with a length of 40-50 m have been drilled to facilitate acoustic emission, directional radar, seismic tomography, and hydraulic conductivity measurements before and after excavation of the drifts.

After excavation of the drifts a number of short (3 m) radial boreholes will be drilled in each drift to assess the extent of the disturbed zone in the near field. There will also be a set of longer boreholes extending
radially from the centre of the blasted test drift to investigate properties of the disturbed zone at a larger distance from the wall. Three of these holes will be drilled from the TBM drift towards the Drill & Blast drift
and will be used to measure displacements and changes in rock properties at the Drill & Blast drift before and after excavation.

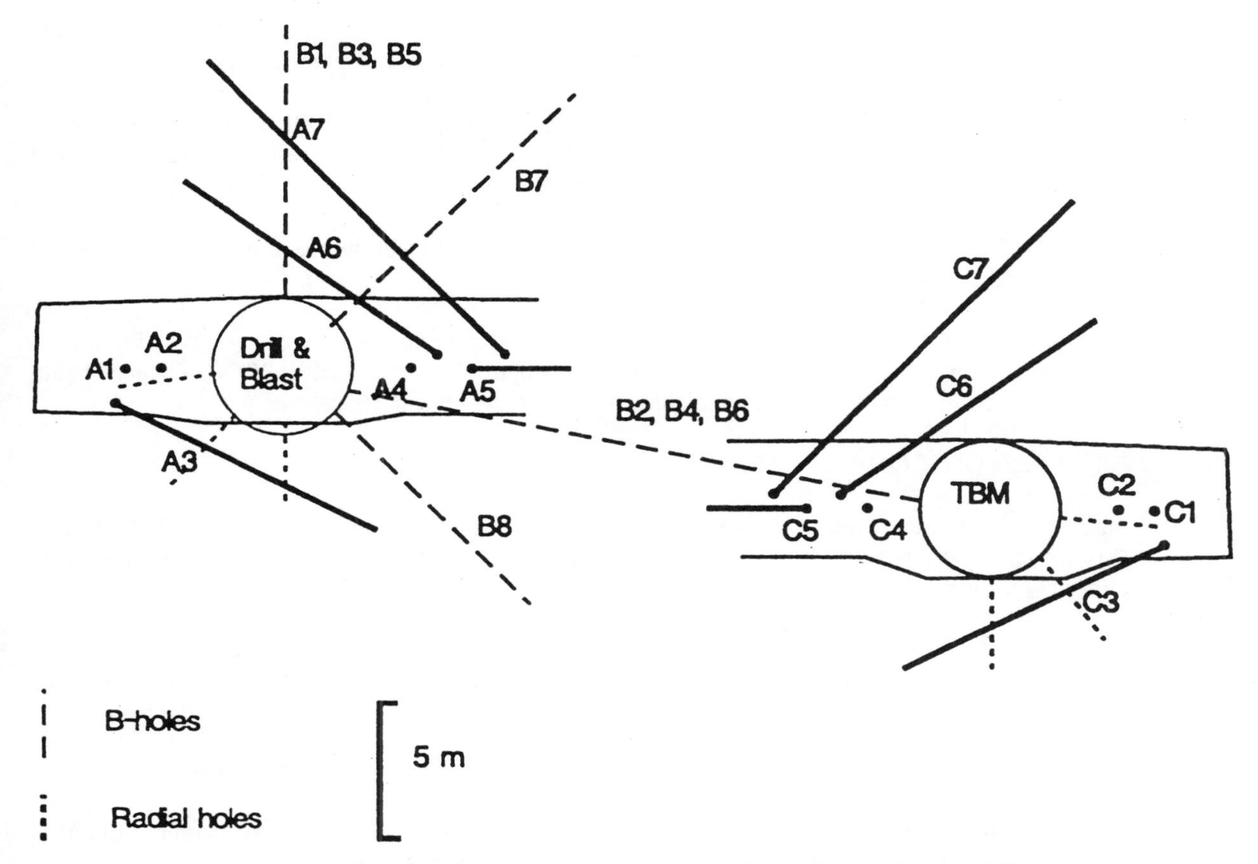

Figure 3. Vertical section showing location of boreholes in relation to the test drifts.

3.4 PARAMETERS QUANTIFIED

Excavation disturbance has to be expressed in terms of measurable quantities where "disturbance" is defined as a change in properties from the natural state before excavation. The ZEDEX Project will include quantifying the following parameters:

- acceleration
- displacement
- P-wave velocity.
- hydraulic conductivity
- natural and induced fracturing
- temperature
- acoustic energy release
- stress state

The following remote sensing techniques will be applied for characterisation of the EDZ:

i) P wave seismic velocity tomography and its variation across the EDZ. Variations of seismic velocity with the direction of wave propagation relates to the preferential directions of open joints.

ii) The attenuation of seismic waves obtained by tomographic techniques for estimating the general effects of stiffness of the joints and/or their aperture, conductive length, and frequency.

iii) Acoustic emission monitoring to detect the location of micro-crack events. This should provide information on the extent of the zone where joint movements occur.

iv) The velocity of electro-magnetic waves (radar) obtained from reflection measurements in the tunnel and boreholes along the tunnel yielding information on the presence and orientation of fractures and on the average water content of a given rock volume.

v) Measurements of acceleration during excavation to estimate the magnitude of force applied to the rock.

Only one test for each excavation method is planned, but measurements will be made with several methods at multiple locations along the tunnels. The length of each test drift is 30-40 m which should facilitate discrimination of excavation disturbances from geological heterogeneity.

3.5 EXPECTED OUTCOME OF EXPERIMENT

The experiment is expected to improve understanding of the extent and character of the EDZ and its dependence on the method of excavation used. Specifically, the experiment is expected to provide information on the following items:

- Mechanical damage to the rock for the excavation methods used, this includes crushing of rock (at blast holes or at TBM grippers) and induced fracturing, both on macro and micro scale.

- The magnitude of force (vibrations) applied to the rock mass during excavation.

- Comparison of quality and information content obtained from geological mapping (number of fractures, fracture characteristics) for the two excavation methods.

- Magnitude of excavation disturbance as a function of distance from drift perimeter measured in terms of fracture frequency (induced fractures), micro fracturing, acoustic emissions, radar and seismic velocities, temperature changes, and hydraulic conductivity.

- Character of tunnel inflow for the two excavation methods (channelling etc).

- A test of a system for measuring the total extent of the EDZ by using geophysical techniques and rock mechanics assessment in order to quantify parameters on the configuration of the damaged zone required for safety assessment.

- A basis for selection of optimized methods of excavation.

3.6 PROBLEM AREAS

A major problem is to identify a suitable experiment area with appropriate rock conditions (sufficient similarity in conditions at the two test drifts) considering the constraints imposed by tunnel construction. The flexibility in selecting a site is limited and it is not possible to guarantee ideal experimental conditions in advance. SKB has performed a comprehensive investigation program in borehole KC0045F (Bh 3 in Figure 1) in order to provide information to select the location of the test drifts.

The coordination of the experimental work with construction work will be a major endeavour. This may cause delays, increase costs, and possibly compromise the quality of experimental results if not managed properly. If there is malfunction of equipment it may not be possible to perform the planned measurement within the allotted time and in a worst case to perform it at all.

Integrating near EDZ and far EDZ data sets into a total assessment of the whole EDZ is an important but challenging problem. This should be overcome by good cooperation between the different groups involved and close management of the project.

To ensure optimization of smooth blast designs, the special explosives and detonators should be in sufficient quantity to be able to adapt the next blast design to the results obtained with the previous blast.

Proper function of measurement tools during the experiment is essential. Two differing tools will be used by BGR to measure the extension of the damaged zone. These tools were tested in a mine in Germany in August 1993 prior to the experiment in Sweden.

4. DATA ANALYSIS AND REPORTING

An integrated analysis will be made of the results obtained from the Drill & Blast and TBM drifts, respectively. These results from each drift will be compared and evaluated with respect to the expected outcome of the excavation method used. The results will be presented in the final report of the ZEDEX Project which will be published as an Äspö Project International Cooperation Report.

An important objective is the publication of papers in the peer-reviewed scientific literature.

The in-situ tests should, according to the current schedule for excavation, take place during the period May 1994-May 1995. Tests in the TBM drift are expected to be completed in October 1994. Tests in the Drill & Blast test drift are expected to begin in November 1994.

Integrated analysis of results will commence shortly and is expected to be completed in early 1996 and finally reported in mid 1996.

5. ACKNOWLEDGEMENT

The authors wish to thank their colleagues at ANDRA, Nirex and SKB and in particular A Cournut and K Ben Slimane (ANDRA), D Mellor and N Davies (Nirex) and G Bäckblom (SKB) for their invaluable contributions to the design and implementation of the ZEDEX Project.

Integrated Modeling and Experimental Programs to Predict Brine and Gas Flow at the Waste Isolation Pilot Plant

R.L. Beauheim, S.M. Howarth, P. Vaughn, S.W. Webb, and K.W. Larson
Sandia National Laboratories

(USA)

Abstract

Evaluation of the performance of the WIPP repository involves modeling of brine and gas flow in the host rocks of the Salado Formation, which consist of halite and higher permeability anhydrite interbeds. Numerous physical, chemical, and structural processes, some of them coupled, must be understood to perform this modeling. Gas generation within the repository, for example, is strongly coupled to the amount of brine inflow to the repository because brine aids in the corrosion of metals and associated generation of hydrogen gas. Increasing gas pressure in the repository, in turn, decreases the rate of brine inflow. Ultimately, the gas pressure may exceed the brine pressure and gas may flow out of the repository. The initial models used by WIPP Performance Assessment (PA) were simplified because of a lack of WIPP-specific data on important processes and parameters. Relative-permeability curves and a correlation between threshold pressure and permeability taken from studies reported in the literature were used in PA models prior to being experimentally verified as appropriate for WIPP. In addition, interbed permeabilities were treated as constant and independent of effective stress in early models. Subsequently, the process of interbed fracturing (or fracture dilation) was recognized to limit gas pressures in the repository to values below lithostatic, and assumed (and unverified) relationships between porosity, permeability, and pore pressure were employed. Parameter-sensitivity studies performed using the simplified models identified important parameters for which site-specific data were needed. Unrealistic modeling results, such as room pressures substantially above lithostatic, showed the need to include additional processes in the models. Field and laboratory experimental programs have been initiated in conjunction with continued model development to provide information on important processes and parameters. Current field experiments are aimed at determining the permeability of anhydrite and halite beds under undisturbed conditions, the pressure-dependence of anhydrite fracture permeability, and the threshold pressure of anhydrite fractures. Laboratory experiments are being performed on anhydrite core samples to determine capillary-pressure curves, relative-permeability curves, and permeability and porosity as a function of applied stress. The current PA models built upon these experimental data are considerably more realistic and credible than the initial models used and, in some cases, predict significantly different consequences.

Introduction

The Waste Isolation Pilot Plant (WIPP) is a repository for transuranic wastes located 655 m below ground surface in bedded evaporites of the Permian-age Salado Formation. The WIPP site is located near Carlsbad, New Mexico, USA (Figure 1). One scenario of concern to the performance of WIPP is the generation and migration of gas from the repository. While gases are not themselves of direct regulatory concern, they may serve to mobilize (facilitate the transport of) volatile organic compounds (VOC's) present in the waste. Transport of VOC's across the WIPP site boundary is subject to regulation.

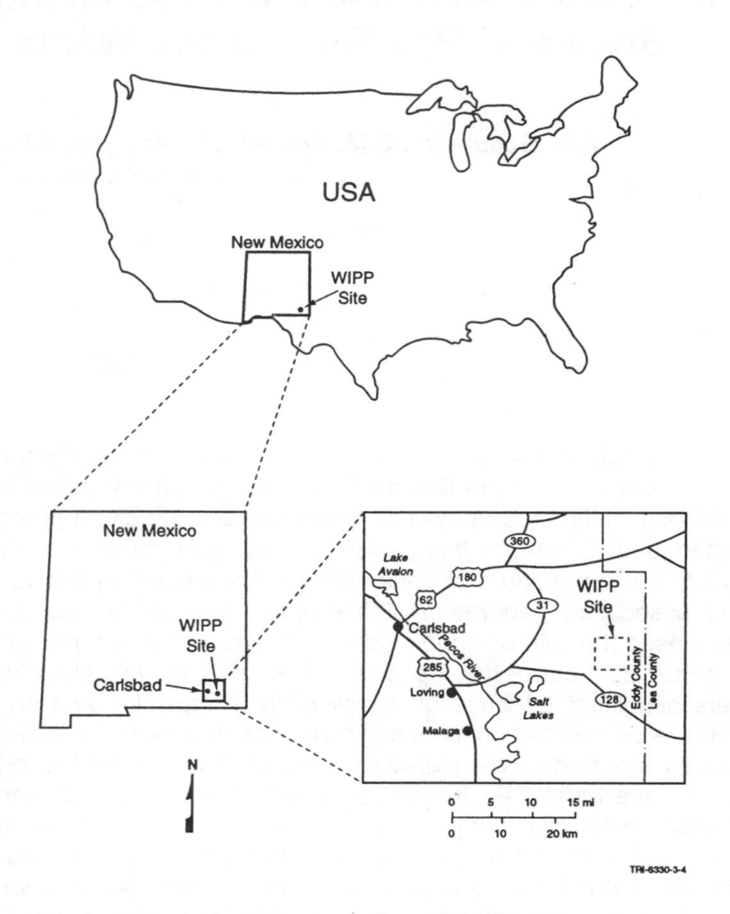

Figure 1. **Location of the WIPP site.**

The gas generation and migration scenario involves a number of interrelated processes. Among these is brine inflow caused by pressure disequilibrium between the repository, initially at atmospheric pressure, and the fluid in the surrounding rock, which because of the plasticity of halite, is at a pressure approaching lithostatic. The majority of the brine entering the repository comes not through the halite layers that make up most of the stratigraphy, but through thin anhydrite interbeds containing naturally occurring fractures. The brine entering the repository reacts with metals present in the waste and drums to form hydrogen gas. Other gases are generated by microbial degradation of organic materials present in the waste. As gases accumulate and the surrounding salt creeps inward, the pressure within the repository rises. Eventually the gas pressure in the repository exceeds the brine pressure in the surrounding rock. At this point there are two limits on the ability of the gas to leave the repository. First, the rock may have a threshold pressure, which is the pressure increment above the brine pressure that gas must achieve before it can begin to displace brine from the pore spaces in the rock and establish continuous flow paths. Second, the permeability of the rock, either intrinsic or relative, may be too low to allow the gas to flow away as fast as it is being generated. If this occurs, the pressure in the repository may continue to rise.

Just as they provide the primary pathway for brine to enter the repository, so will the anhydrite interbeds provide the most likely pathway for gas to leave the repository. Because of their higher permeabilities relative to halite, the interbeds have lower threshold pressures and admit gas more readily than halite. Once gas enters the interbed fractures, the two-phase flow characteristics (capillary pressure and relative permeability) of the rock will control the rate at which the gas can flow away. If this rate is lower than the rate at which gas is being generated, the pressure in the interbed fractures will rise. As this occurs, the fractures may dilate, increasing both in permeability and storage capacity. This dilation may continue until the enhanced permeability and storage can accommodate the gas being generated without additional increase in pressure.

Modeling of the gas generation and migration scenario requires knowledge (or assumptions) of the parameters and processes involved. The intrinsic permeability and far-field pore pressure in the anhydrite interbeds have been characterized by an *in situ* testing program (Beauheim et al., 1991, 1993). Information on two-phase flow properties, threshold pressure, and pressure-dependent permeability (fracture dilation) is less easily obtained. The strategy employed at WIPP for setting experimental priorities is to first make assumptions about properties and processes that bound the expected ranges, and then perform model calculations to determine if the uncertainties affect evaluation of repository performance. If uncertainty in a particular area can affect whether or not calculated performance complies with regulations, an experimental program to reduce the uncertainty may be warranted. In some cases, an iterative sequence of initial models, initial experiments, new models, and new experiments is required to resolve areas of uncertainty adequately. This paper describes how three modeling and experimental programs have been integrated to address the WIPP gas generation and migration scenario.

Two-Phase Flow Properties

Characteristic curves for capillary pressure and relative permeability are used to determine the pressures and rates at which gas is able to leave the repository and enter anhydrite interbeds. The capillary-pressure function (Figure 2a) describes the pressure at which gas can first enter a brine-saturated medium (defined here as the threshold pressure) and the pressure disequilibrium that occurs between brine and gas as the gas saturation of the medium increases. The relative-permeability function (Figure 2b) describes how the relative permeabilities of the medium to brine and gas change as gas saturation increases and brine saturation decreases. Both capillary pressure and relative permeability must be measured at known saturations to define the characteristic curves. No method is known to determine saturations under field, or *in situ*, conditions. Characteristic curves, therefore, must be defined from tests on core samples in a laboratory where saturations can be controlled and measured.

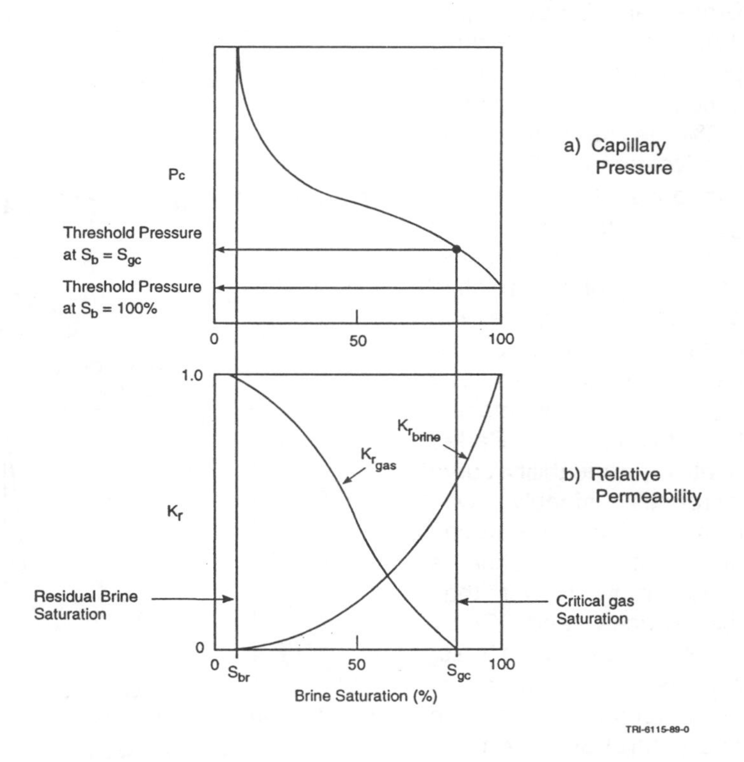

Figure 2. **Relationship between (a) capillary pressure and (b) relative permeability.**

Before undertaking these measurements, modeling studies were performed using different sets of characteristic curves with widely different forms. Webb (1992a) studied the effect of the Brooks and Corey (1964) curves and the Sandia Functions (Pruess, 1987), which are based on van Genuchten (1980). The two-phase characteristic curves can have a dramatic effect on the gas-migration distance. Subsequently, similar studies were performed using the Brooks and Corey and van Genuchten/Parker curves (WIPP Performance Assessment Department, 1992). The van Genuchten/Parker curves are similar to the Sandia Functions except the nonwetting phase relative permeability is given by Parker et al. (1987). Again, the gas-migration distance was found to be sensitive to the two-phase characteristic curves.

The forms of the two-phase curves employed are shown in Figures 3 and 4. As shown in Figure 3, the Brooks and Corey and van Genuchten capillary-pressure curves have a similar shape except approaching full liquid saturation. The Brooks and Corey curve (Figure 3a) has a threshold pressure on the order of 1 MPa for WIPP materials, while the van Genuchten relationship (Figure 3b) has zero threshold pressure. Figure 4 presents the relative-permeability curves. The wetting-phase curves for both models are similar, although the Brooks and Corey value is always higher than that of van Genuchten. The nonwetting-phase (gas) relationships are dramatically different. For Brooks and Corey (Figure 4a), the gas relative permeability is zero until a critical gas saturation (20%) is reached. The relative permeability slowly increases with increasing gas saturation. In contrast, the Parker nonwetting-phase relative permeability (Figure 4b) does not have a critical gas saturation, so any amount of gas is mobile. In addition, the Parker relative-permeability curve increases steeply with increasing gas saturation and at any saturation is considerably higher than the Brooks and Corey curve. Webb (1992b) discusses the effect of characteristic curves on gas-saturation profiles, including both Brooks and Corey and van Genuchten/Parker curves. Both the Brooks and

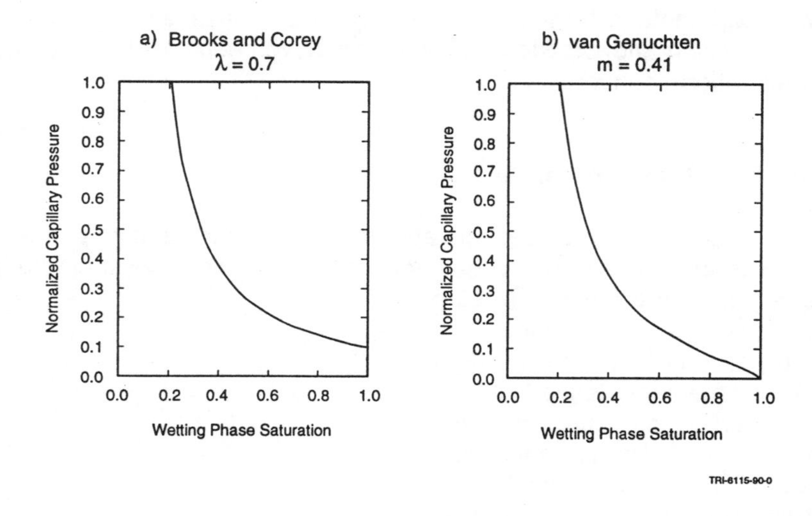

Figure 3. **Example capillary-pressure curves used by WIPP PA.**

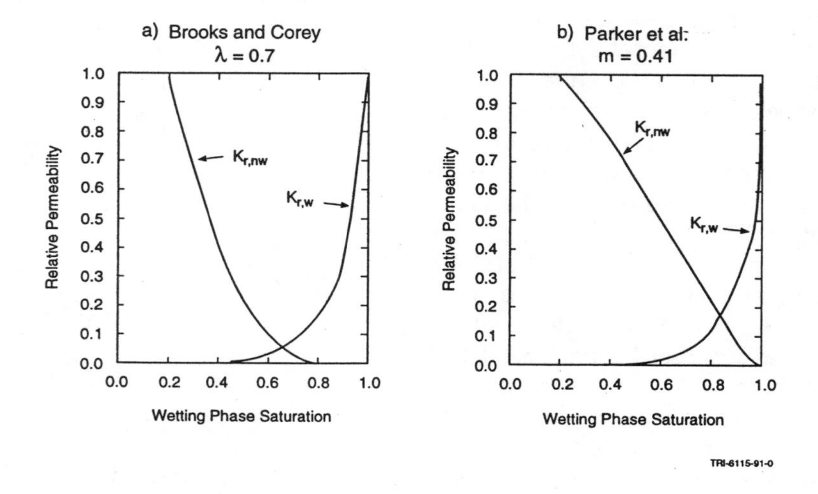

Figure 4. **Example relative-permeability curves used by WIPP PA.**

Corey and van Genuchten/Parker curves used by WIPP PA were developed from studies of porous media. Their applicability to fractured media is, therefore, uncertain, but no characteristic curves developed specifically for fractured media are known.

Modeling of gas migration from the repository was performed assuming both flat-lying (horizontal) stratigraphy and a 1° dip more representative of actual conditions at the WIPP. The modeled stratigraphy included single anhydrite interbeds above and below the repository horizon. In all simulations, gas-migration distances were greater in the upper interbed than in the lower interbed. The modeling shows that the selection of Brooks and Corey or van Genuchten/Parker characteristic curves has a large effect on the calculated gas-migration distance. The Brooks and Corey curves, because they can include a threshold pressure greater than zero, a critical gas saturation of 20%, and low relative permeability to gas at low gas saturations, tend to restrict gas migration, keeping it within about 1 km of the repository in the upper interbed over a 10,000-yr period in the simulation shown in Figure 5a. Use of van Genuchten/Parker curves in an otherwise identical simulation causes gas to migrate over 5 km from the repository in the upper interbed over 10,000 yr (Figure 5b). Gas saturations within the interbeds, however, are much lower using the van Genuchten/Parker curves because those curves include neither threshold pressure nor a critical gas saturation. The effect of 1° dip on the simulations is shown in Figures 6a and 6b. Using Brooks and Corey curves, gas has a slight tendency to move preferentially up dip, but is still confined within a radius of about 1 km over 10,000 yr.

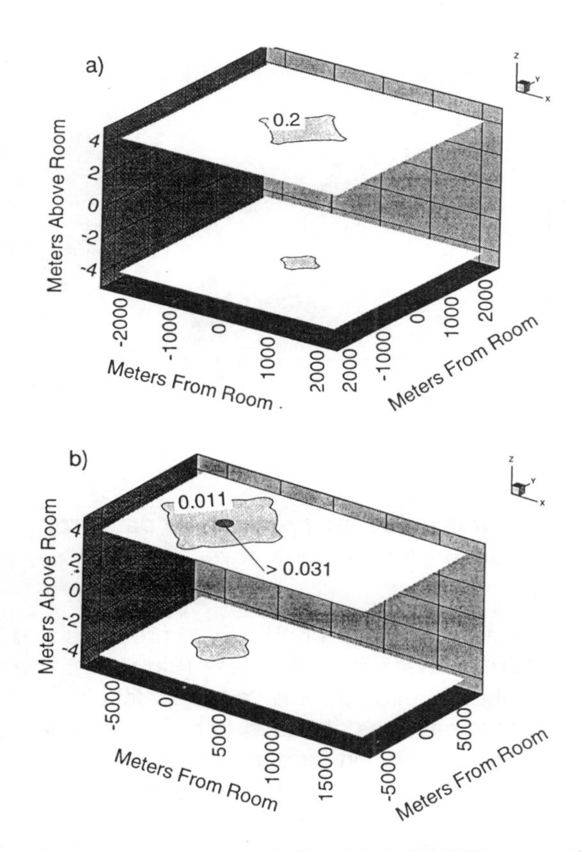

Figure 5. **Gas saturations at 10,000 years in horizontal anhydrite interbeds using (a) Brooks and Corey characteristic curves and (b) van Genuchten/Parker characteristic curves.**

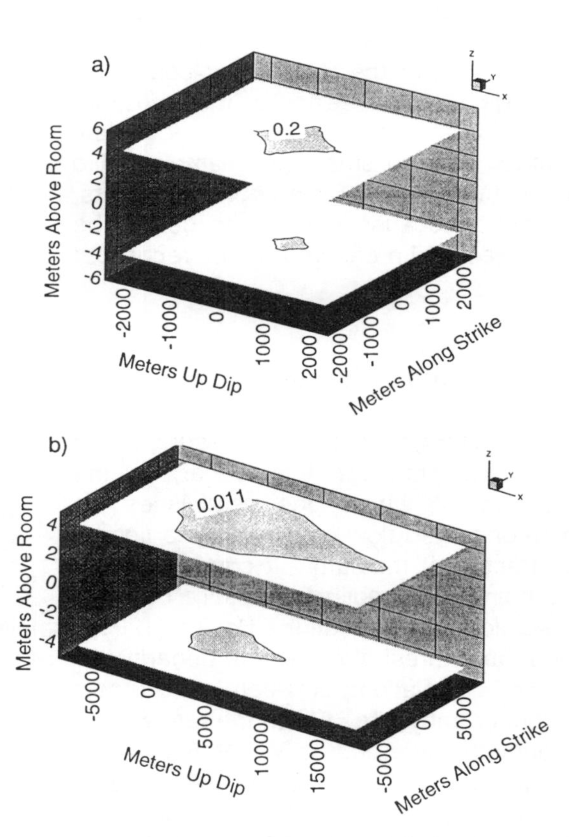

Figure 6. **Gas saturations at 10,000 years in anhydrite interbeds with 1° dip using (a) Brooks and Corey characteristic curves and (b) van Genuchten/Parker characteristic curves.**

Preferential up-dip movement is much more pronounced using van Genuchten/Parker curves (Figure 6b). The simulated gas-migration distance in the upper interbed increases to about 15 km over 10,000 yr, relative to the horizontal simulation (Figure 5b).

These simulations showed that knowledge of the true nature of the characteristic curves for anhydrite interbeds is needed to perform realistic modeling of gas migration. A laboratory test program is currently underway to provide the needed two-phase flow characteristics. Initial experiments included determination of capillary-pressure curves for 12 core samples under no confining pressure using centrifuge and mercury-injection techniques (Figure 7). The capillary-pressure results from the oil-air centrifuge tests and the mercury-air mercury-injection tests were corrected for air-brine conditions for WIPP applications. These data could be fit with either a van Genuchten/Parker model or a Brooks and Corey model with a low threshold pressure. Additional measurements are planned in 1995 of capillary pressure and relative permeability under a range of confining pressures to provide characteristic curves that may be used directly in future simulations of gas migration. Reliable extrapolation of data from the laboratory scale to the repository scale is always of concern because of uncertainty about whether or not the features or conditions controlling two-phase flow on the repository scale are, or can be, adequately characterized in small core samples. No alternative to this extrapolation exists, however, because saturations cannot be controlled and measured on the repository scale.

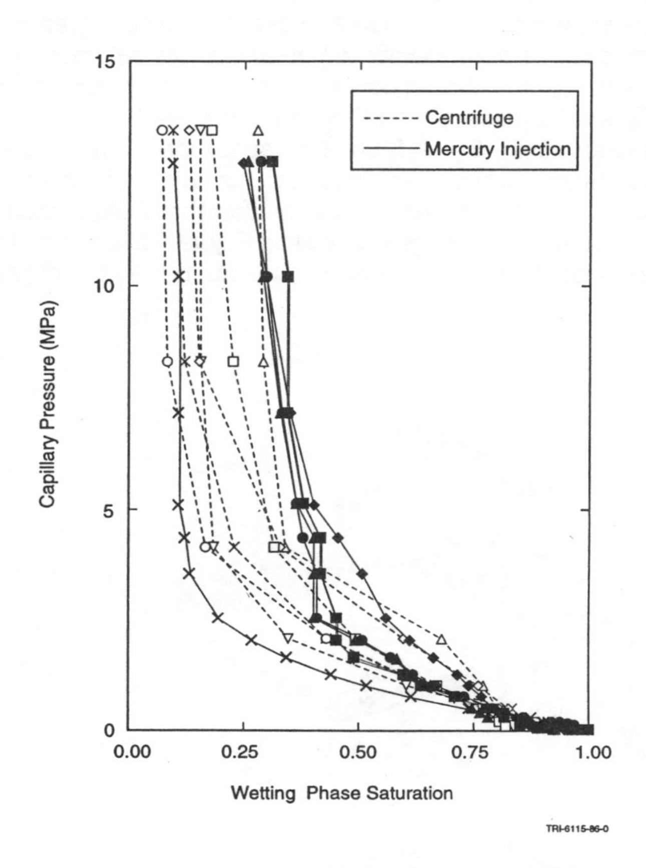

Figure 7. **Laboratory capillary-pressure data from tests on anhydrite core.**

Gas Threshold Pressure

The gas threshold pressure, in the usage adopted for this paper, is the pressure at which gas establishes an interconnected flow path through a porous (or fractured) medium. It corresponds to the point on a capillary-pressure curve at the critical gas saturation. Davies (1991) compiled literature data available on threshold pressure and found that threshold pressure generally increases as intrinsic permeability decreases (Figure 8). He defined correlations between intrinsic permeability and threshold pressure for a variety of lithologies, including anhydrite. Given the range of vertically averaged permeabilities obtained from *in situ* testing of anhydrite interbeds at WIPP (6×10^{-20} to 1×10^{-18} m^2), Davies' correlation for anhydrite indicated that the gas threshold pressure of these beds might lie between 0.1 and 5 MPa. If threshold pressures truly were as high as 5 MPa, then hydraulic (or pneumatic) fracturing of the host rock surrounding the WIPP repository might occur before gas would enter the rock.

Laboratory threshold-pressure experiments were performed to determine if the threshold pressures of Salado anhydrites were really so high as to make hydraulic fracturing a possibility. Threshold pressures ranging from 0.016 to 0.003 MPa were determined from six mercury-injection tests performed on anhydrite core samples having permeabilities between 4×10^{-19} and 4×10^{-18} m^2. These threshold pressure values are one to two orders of magnitude lower than predicted by Davies' correlation. The tests, however, were conducted under no confining pressure, which may have contributed to the low values. Additional laboratory tests are planned under a range of confining pressures to provide results more representative of *in situ* conditions.

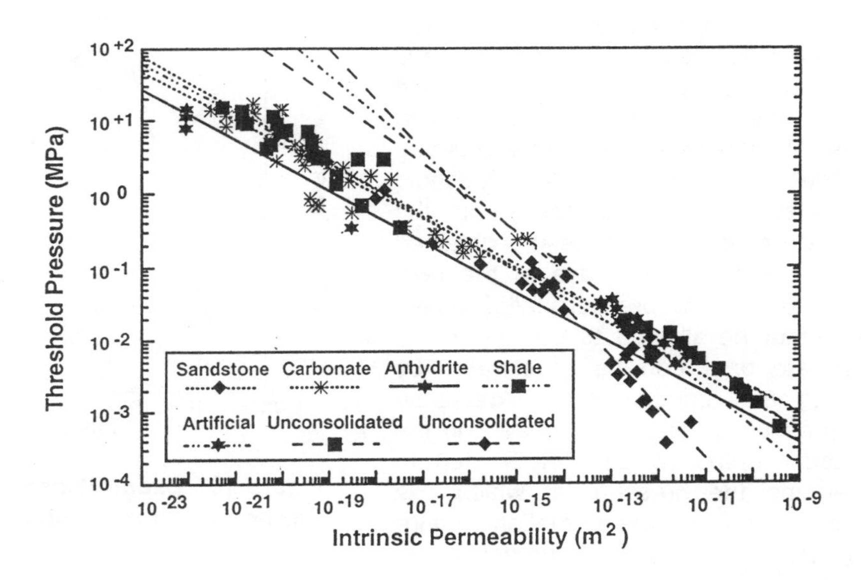

Figure 8. **Literature data and correlations compiled by Davies (1991) between intrinsic permeability and threshold pressure.**

Alone among two-phase properties, threshold pressure is amenable to measurement through field tests because it does not rely on determination of relative saturations. Provided that a medium is known to be fully brine saturated, its threshold pressure can be determined through a gas-injection test. Three *in situ* threshold-pressure tests were conducted in anhydrite interbeds above and below the WIPP repository in which permeability tests had already been completed. These three tests provided estimates of threshold pressure ranging from 0.27 to 0.01 MPa for permeabilities ranging from 7×10^{-20} to 3×10^{-18} m^2. The optimal testing procedure determined from these tests is as follows: 1) Isolate the interbed to be tested with packers and wait for the brine pressure to stabilize. 2) Inject gas (nitrogen) at the top of the test zone while withdrawing brine from the bottom, perturbing the pressure as little as possible. 3) After the brine in the test zone is completely replaced by gas, allow the test-zone pressure to stabilize. 4) Inject gas at a constant mass rate (1 to 3 cm^3/min at STP) while monitoring the pressure increase in the test zone and plotting pressure change and its derivative with respect to log time versus elapsed time on a log-log plot. Threshold pressure is reached when the pressure-change data and, more clearly, the pressure-derivative data deviate from the unit-slope line on the plot characteristic of wellbore storage (Figure 9). That is, until gas begins to penetrate the formation, the pressure in the test zone will rise linearly as gas is injected. Once gas begins to move out in the rock, the rate of pressure increase in the test zone will decrease.

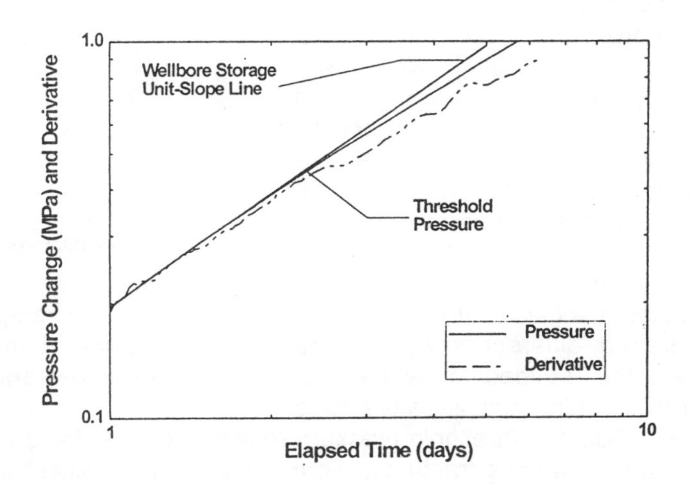

Figure 9. **Determination of threshold pressure from pressure and pressure-derivative responses during gas injection.**

Additional information can be obtained by shutting in the test zone after the threshold pressure is surpassed and observing the subsequent pressure falloff as gas continues to flow from the test zone into the formation. The pressure decline should cease at the threshold pressure, providing a check on the estimate of threshold pressure obtained during the injection phase of the test. Operationally, the pressure-falloff phase can either be allowed to proceed at its own rate or it can be accelerated by venting small amounts of gas to decrease the pressure in steps and observing the pressure response after each step to determine the pressure at which gas stops flowing into the formation (Figure 10). Above the threshold pressure, the immediate response after each pressure drop will be an increase in the test-zone pressure as the local equilibrium between the test zone and the surrounding rock is restored, followed by a decrease in the test-zone pressure as outward flow resumes (Figure 11a). After a drop below the threshold pressure, the test-zone pressure will increase as gas flows back from the formation to the test zone (Figure 11b), ultimately to stabilize at the threshold pressure. Additional information might also be obtained by repeating the gas-injection test at different injection rates, but saturation uncertainties become more problematic as the test increases in complexity.

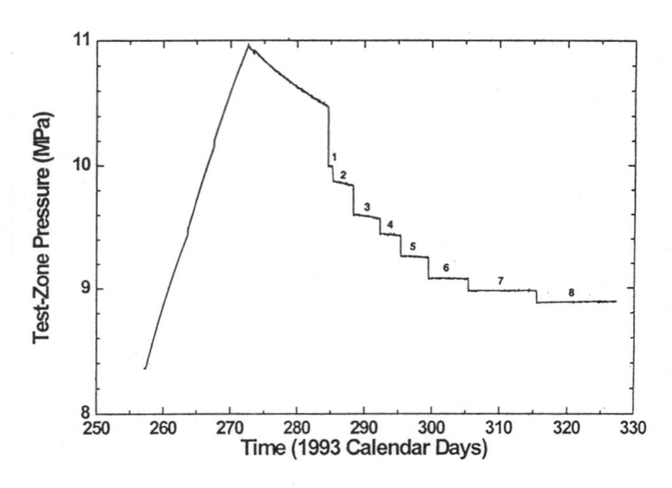

Figure 10. **Example of stepwise pressure-falloff approach to determining threshold pressure.**

As shown on Figure 12, both the laboratory and field values of threshold pressure are lower than the correlation of Davies (1991) for anhydrite would predict. Several factors might contribute to this apparent discrepancy. First, the literature data (from Ibrahim et al., 1970) are sparse: only seven values are given, two of which are presented as upper bounds only and one as a lower bound only. Thus, the correlation is not well defined. Second, the literature data are from a salt dome caprock which may have a different sedimentological structure and, therefore, different flow properties than the bedded anhydrites at WIPP. Third, the literature data may reflect the threshold pressure required to enter the anhydrite matrix whereas at WIPP we believe the gas is entering fractures. Fourth, the "permeabilities" used to plot the WIPP data on Figure 12 do not directly represent the fractures that gas is probably entering, but instead represent the average permeabilities of the entire thicknesses of anhydrite tested (2.5 cm to 1 m). The true fracture permeabilities must be higher than the thickness-averaged values, but exact values cannot be determined.

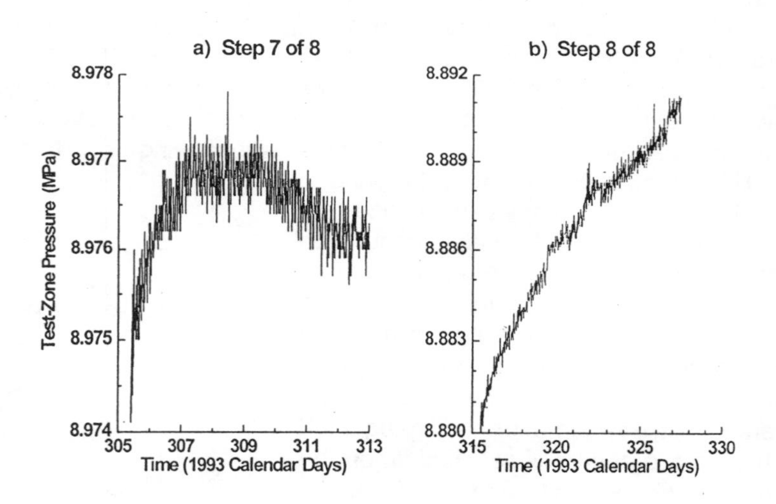

Figure 11. **Example of bracketing threshold pressure between pressure-falloff steps (a) above threshold pressure and (b) below threshold pressure.**

Regardless of how well the WIPP data agree with the correlation of Davies, the laboratory and field experiments completed to date have shown that the threshold pressures of the anhydrite interbeds are low enough that gas may enter the interbeds at pressures well below the hydraulic-fracturing pressures. Thus, further field experimentation and model development to examine the effects of high threshold pressures in anhydrite interbeds are not warranted.

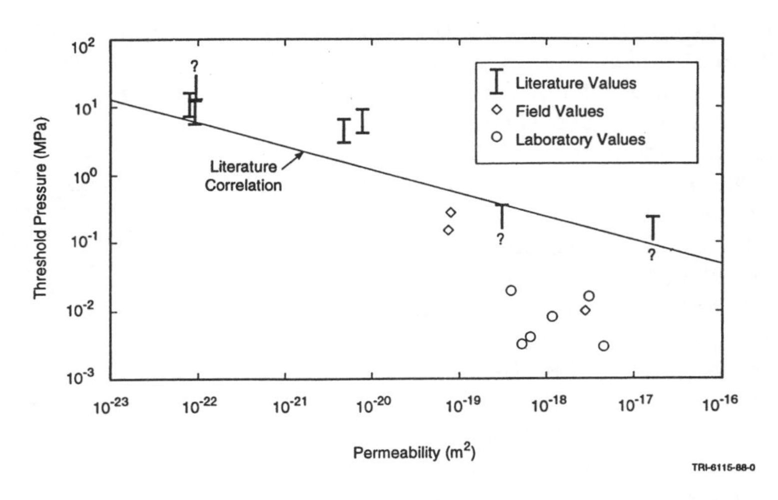

Figure 12. **Comparison of literature and measured values for anhydrite threshold pressure.**

Pressure-Dependent Permeability

Early modelers (e.g., Davies et al., 1992; Webb, 1992a) examining the effects of gas-generated pressures on repository performance assumed that the permeability and porosity of anhydrite interbeds would remain constant regardless of how pressures changed. They found that under some conditions, pressures in the repository could exceed lithostatic (Figure 13) because gas was generated more rapidly than it could flow away through the interbeds. Pressures several MPa greater than lithostatic were not considered realistic, however. Mendenhall et al. (1992) argued that either existing fractures in the interbeds would dilate or new fractures would be created as the pressure in the repository increased until the resulting enhanced permeability and porosity were adequate to accommodate the gas being generated. One potentially undesirable side effect of this enhanced permeability was an increase in the distance to which gas would migrate from the repository.

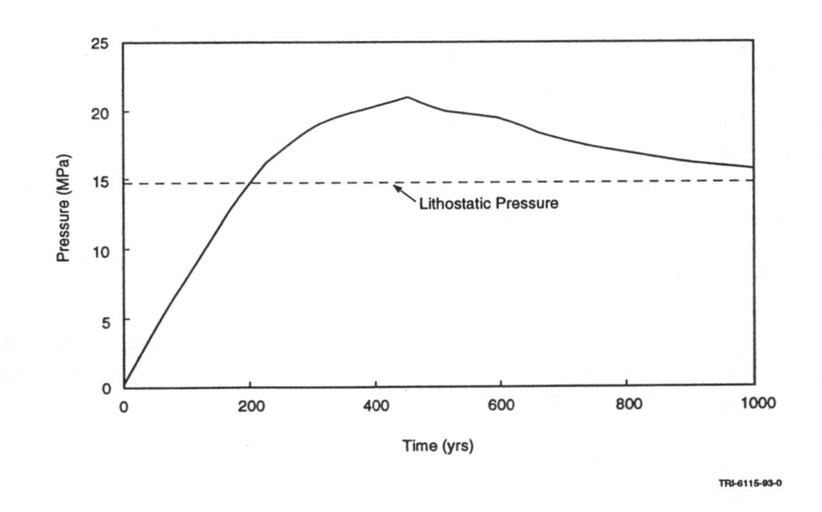

Figure 13. **Example calculation of repository pressurization in the absence of pressure-dependent hydraulic properties.**

Early scoping studies were conducted assuming interbed permeability increased above a defined "fracture" pressure while the porosity was assumed to remain constant. Results indicated that the repository pressure was easily capped by increasing the interbed permeability. However, the increase in gas-migration distance was large. Subsequently, WIPP Performance Assessment (PA) modelers began to consider more sophisticated ways of incorporating pressure-dependent interbed permeability and porosity in their models at the same time as field and laboratory experiments began to study the phenomenon. Laboratory tests on eight core samples showed that interbed permeabilities decreased by factors ranging from 1.6 to 5.3 as the net confining pressure was increased from 2 to 10 MPa, with 52 to 92 percent of the reduction occurring as the confining pressure was increased from 2 to 6 MPa (Figure 14). This behavior would translate to an increase in interbed permeability around the repository as pore pressures in the interbeds increased, thereby decreasing the effective confining pressure. One experiment was performed to examine the effect of pore pressure on permeability in the field. This experiment was run with individual constant-pressure flow tests at four different pressures, one test at a pressure below the ambient pressure in the interbed and three tests at pressures above ambient. These tests showed a threefold increase in permeability as the test pressure was increased from about 7.5 to 10.8 MPa, with the rate of permeability increase accelerating as the pressure increased (Figure 15). Thus, both the laboratory and field measurements showed pressure-dependent permeability to be a plausible mechanism to relieve high gas pressures generated within the repository.

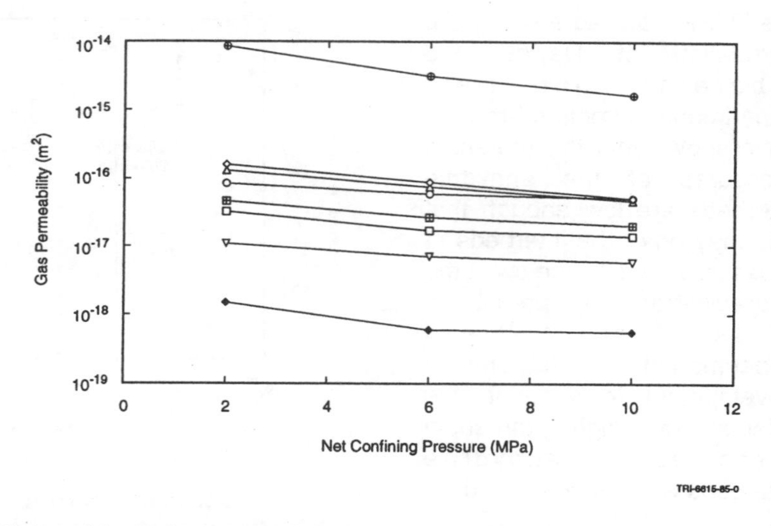

Figure 14. **Laboratory data showing reduction in permeability as confining pressure increases.**

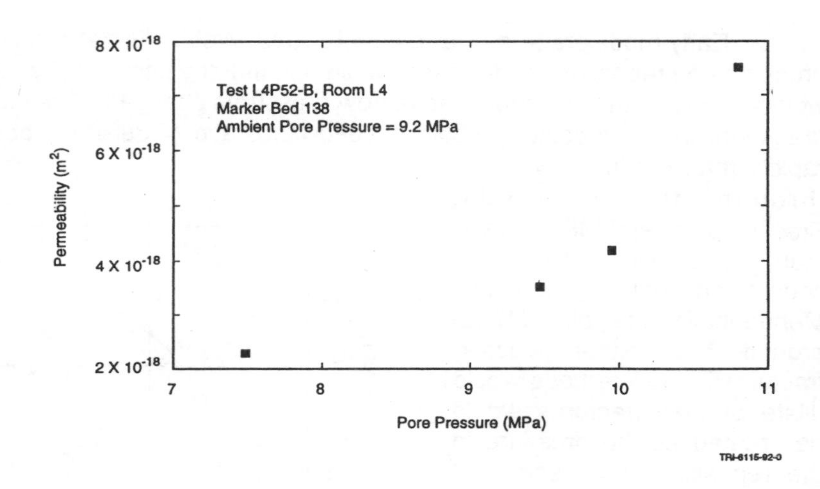

Figure 15. **Field data showing increase in permeability as induced pore pressure increases.**

For PA purposes, a relatively simple, single-continuum model was developed that relates interbed porosity and permeability to pressure by first assuming a pressure-dependent compressibility. The pressure dependence is assumed to be initiated at a pore pressure slightly in excess of the estimated far-field pressure. Below this "initiating pressure (P_i)," anhydrite compressibility and permeability are treated as constant. For pressures above the initiating pressure, compressibility is assumed to increase linearly with pressure up to a so-called "fully altered pressure (P_a)," above which no further property alteration occurs. PA related porosity (ϕ) to pressure (P) through a definition of compressibility (C) given by:

$$C = \frac{1}{\phi}\frac{d\phi}{dP} \qquad (1)$$

Porosity as a function of pore pressure is obtained by integration of the compressibility equation. This results in the porosity function shown in Figure 16 where the ratio of porosity to intact porosity is a function of pore pressure. By analogy to the parallel plate approach to determining fracture permeability, in which permeability is proportional to a power of fracture aperture (or porosity), altered permeability is related to altered porosity by:

$$\frac{k}{k_i} = \left[\frac{\phi}{\phi_i}\right]^J \qquad (2)$$

where: k = permeability of altered anhydrite,
k_i = permeability of intact anhydrite,
ϕ = porosity of altered anhydrite,
ϕ_i = porosity of intact anhydrite, and
J = an empirical parameter.

This model relies on the specification of "fully altered" values of porosity (ϕ_a) and permeability (k_a). PA handles uncertainty in these values by sampling each of them from ranges determined to be reasonable based on available geologic and hydrologic information. The empirical parameter J given above takes whatever value is necessary to provide the sampled value of fully altered permeability given the sampled value of fully altered porosity. Figure 16 shows conceptually how permeability increases much more rapidly than porosity as the pore pressure increases.

Preliminary calculations using this model showed that it was effective at relieving gas pressure in the repository; the fully altered pressure was not reached for any sampled combination of parameters. However, the calculated gas-migration distance was found to be highly sensitive to the sampled values. High porosity ranges coupled with low permeability ranges led to low migration distances and low porosity ranges coupled with high permeability ranges led to high migration distances. Without data to narrow the sampled ranges, gas-migration distances would remain highly uncertain.

Because the exact form of the relationship between pore pressure and permeability is unknown, a second model relating the two was developed. In contrast to the so-called "porosity" model developed by PA, the second model, termed the "aperture" model, assumes that increasing pore pressures cause

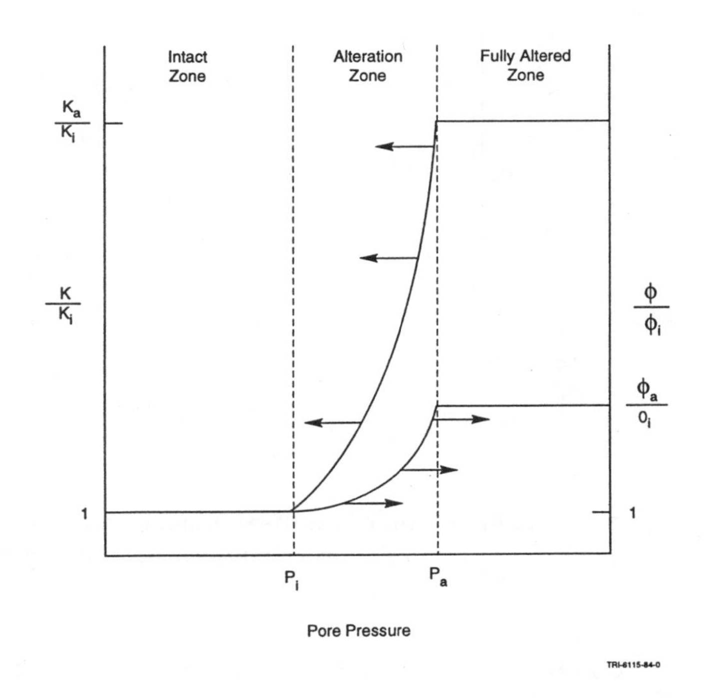

Figure 16. **PA "porosity" model for pressure-dependent porosity and permeability.**

fracture apertures to increase and calculates the permeability of the dilated fractures using the cubic law. The overall porosity (aperture) change is related to pressure change through Eq. 1 above and then permeability is calculated from:

$$k = \frac{nb^3}{12h} + k_m \quad \textbf{(3)}$$

where: n = number of fractures among which the porosity is distributed,
b = average fracture aperture,
h = thickness of rock matrix, and
k_m = matrix permeability.

A drawback associated with this equation is that the parameter n (number of fractures) must be either known or estimated.

For a given porosity change, the porosity and aperture models predict greatly different permeability changes. Figure 17 shows how calculated permeabilities differ between the two models for a one percent porosity change, assuming an initial (unaltered) permeability of 10^{-19} m^2 and ranges of values of the parameters J and n in Eq. 2 and 3, respectively. The initial condition used for the aperture-model calculations shown in Figure 17 may be somewhat unrealistic because, by starting with an overall permeability equal to the assumed matrix permeability, the model effectively assumes that no fractures exist initially. Regardless of the reasonableness of this starting assumption, however, the aperture model shows that permeabilities calculated using a cubic-law expression such as Eq. 3 are several orders of magnitude higher than those predicted by the porosity model for a given porosity (total fracture aperture).

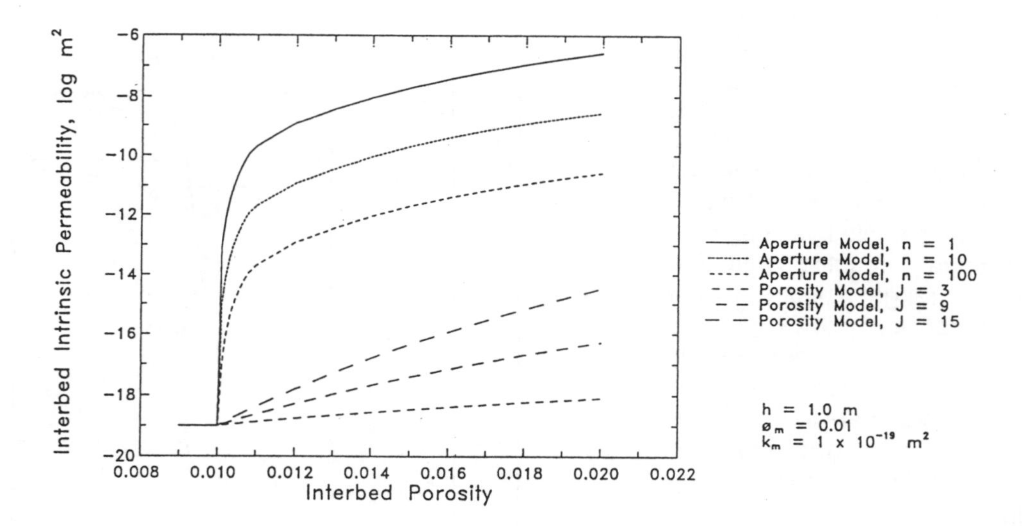

Figure 17. **Comparison of permeabilities calculated by porosity and aperture models for a one percent porosity change.**

Use of the porosity and aperture models to predict gas-migration distances also results in significantly different consequences. Figure 18 shows how the predicted porosity changes directly above a waste-storage room over a period of 10,000 years differ between the two models. The porosity model predicts much greater increases in porosity, particularly within the first 2000 yr, than the aperture model. Increasing the porosity has the effect of decreasing the gas-migration distance, as shown in Figure 19. To access the storage volume needed for gas pressure to reach equilibrium with the far-field brine pressure, gas must travel a much greater distance under aperture-model assumptions than under porosity-model assumptions.

These calculations have shown the importance of understanding the functional relationships among pressure, porosity, and permeability changes in the anhydrite interbeds if accurate calculations of gas-migrations distances are to be performed. An effort was made to fit the porosity and aperture models to the field data already collected relating permeability to pore pressure (Figure 15) to determine if either or both was inconsistent with those data. By modifying values of parameters such as initiating pressure, both models could be made to fit the data, as shown in Figure 20.

To distinguish between the two models (or develop a different, more appropriate model), not just permeability change but also porosity change needs to be measured as a function of pressure. Consequently, a new test tool has been developed to measure the dilation of the overall thickness of anhydrite interbeds during permeability testing over a range of pressures. The tool is conceptually similar to the Pac-ex system developed at the Canadian Underground Research Laboratory (Thompson et al., 1989), but is adapted to a corrosive high-pressure environment. The tool uses linear variable-differential transformers (LVDT's) to measure changes in the separation of anchors set above and below interbeds during testing (Figure 21). This tool will provide information only on the total amount of dilation that occurs, not on the number of fractures dilating. Nevertheless, the data can be used to determine the appropriateness of the porosity model

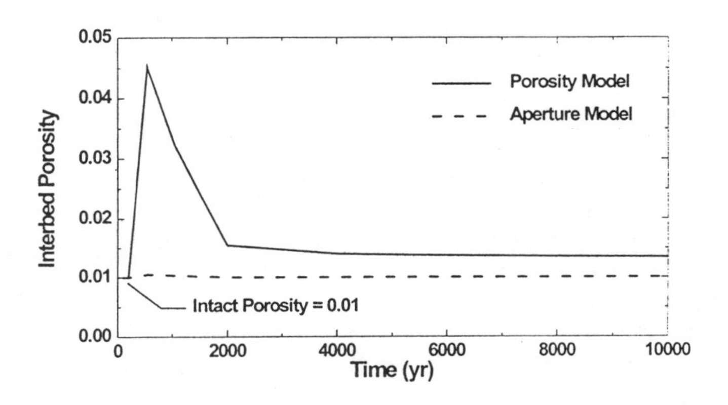

Figure 18. **Comparison of interbed porosities above disposal room calculated using porosity and aperture models.**

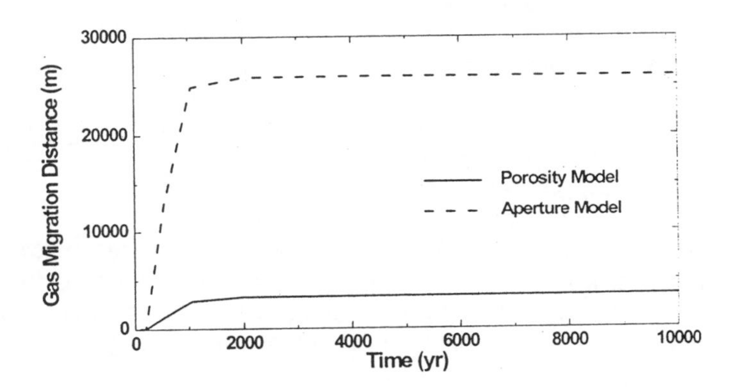

Figure 19. **Comparison of gas-migration distances calculated using porosity and aperture models.**

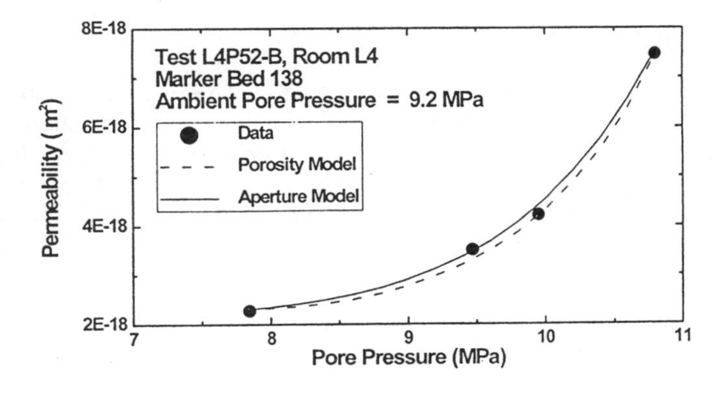

Figure 20. **Example matches of porosity and aperture models to available field data.**

and the aperture model (assuming a "reasonable" range for the number of affected fractures).

Summary and Conclusions

The initial models used by WIPP PA to model the gas generation and migration scenario were simplified because of a lack of WIPP data on important processes and parameters. Two sets of two-phase characteristic curves taken from studies reported in the literature were used in PA and other models to evaluate their effects on gas release. Calculated gas-migration distances were found to be strongly dependent on the nature of the characteristic curves used, motivating laboratory experiments on core samples to define WIPP-specific characteristic curves for anhydrite. The information obtained from the laboratory program may be used directly in future simulations of gas migration, with appropriate upscaling. A correlation between anhydrite permeability and threshold pressure derived from literature data, combined with *in situ* permeability data, suggested that gas might not be able to enter anhydrite interbeds around the WIPP repository at pressures below the hydraulic fracturing pressure. Both laboratory and field experiments have shown that threshold pressures in the interbeds are much lower than predicted by the literature correlation, and reaching the hydraulic fracturing pressure before gas entry is no longer of concern. Early models treating fracture apertures and permeabilities as constant and independent of effective stress allowed repository pressures to reach unrealistically high levels above lithostatic pressure because gas could not leave the repository as fast as it was generated. Field and laboratory experiments showed that anhydrite fractures would dilate and increase in permeability as pore pressures increased, but did not provide enough information to define the functional relationship between pressure and fracture aperture and permeability. Two models relating pressure to permeability and aperture in different ways were developed and were shown to predict significantly different gas-migration distances. A field experimental program has been developed to provide direct measurements of permeability and aperture change as functions of pressure to allow development of a realistic model coupling rock mechanics with fluid flow.

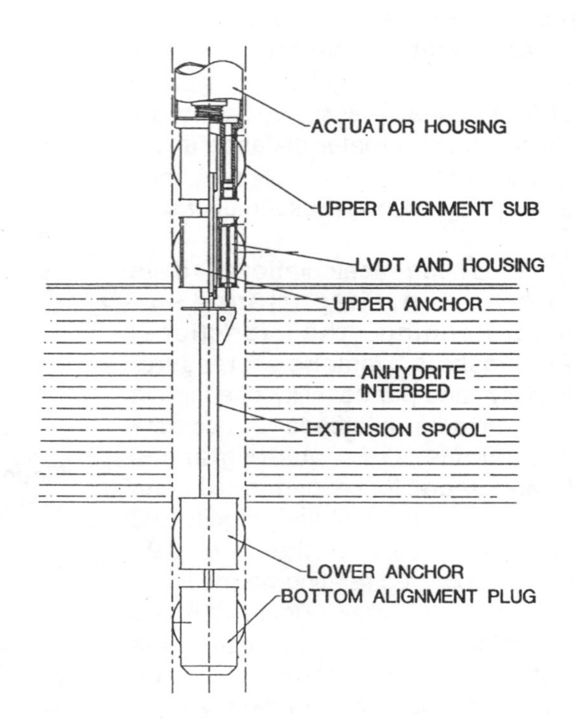

Figure 21. **Schematic of tool to measure interbed dilation.**

These examples illustrate the WIPP project approach of performing iterative modeling and experimental activities to, first, determine the importance of different processes and parameters and, second, reduce uncertainty in those areas that have the largest effect on the calculated performance of the repository. This approach reduces expenditures in areas that have little impact on regulatory compliance and provides assurance that the most important parts of the system receive adequate attention.

References

Beauheim, R.L., R.M. Roberts, T.F. Dale, M.D. Fort, and W.A. Stensrud. 1993. *Hydraulic Testing of Salado Formation Evaporites at the Waste Isolation Pilot Plant Site: Second Interpretive Report.* SAND92-0533. Albuquerque, NM: Sandia National Laboratories.

Beauheim, R.L., G.J. Saulnier, Jr., and J.D. Avis. 1991. *Interpretation of Brine-Permeability Tests of the Salado Formation at the Waste Isolation Pilot Plant Site: First Interim Report.* SAND90-0083. Albuquerque, NM: Sandia National Laboratories.

Brooks, R.H., and A.T. Corey. 1964. *Hydraulic Properties of Porous Media.* Hydrology Paper No. 3. Ft. Collins, CO: Colorado State University.

Davies, P.B. 1991. *Evaluation of the Role of Threshold Pressure in Controlling Flow of Waste-Generated Gas into Bedded Salt at the Waste Isolation Pilot Plant.* SAND90-3246. Albuquerque, NM: Sandia National Laboratories.

Davies, P.B., L.H. Brush, and F.T. Mendenhall. 1992. "Assessing the Impact of Waste-Generated Gas from the Degradation of Transuranic Waste at the Waste Isolation Pilot Plant (WIPP): An Overview of Strongly Coupled Chemical, Hydrologic, and Structural Processes," *Proceedings: Workshop on Gas Generation and Release from Radioactive Waste Repositories, Aix-en-Provence, France, September 23-26, 1991.* Paris, France: OECD Nuclear Energy Agency. 54-74.

Ibrahim, M.A., M.R. Tek, and D.L. Katz. 1970. *Threshold Pressure in Gas Storage.* Project 26-47 of the Pipeline Research Committee, American Gas Association at The University of Michigan. Arlington, VA: American Gas Association, Inc.

Mendenhall, F.T., B.M. Butcher, and P.B. Davies. 1992. "Investigations Into the Coupled Fluid Flow and Mechanical Creep Closure Behavior of Waste Disposal Rooms in Bedded Salt," *Proceedings: Workshop on Gas Generation and Release from Radioactive Waste Repositories, Aix-en-Provence, France, September 23-26, 1991.* Paris, France: OECD Nuclear Energy Agency. 282-295.

Parker, J.C., R.J. Lenhard, and T. Kuppusamy. 1987. "A Parametric Model for Constitutive Properties Governing Multiphase Flow in Porous Media," *Water Resources Research.* Vol. 23, no. 4, 618-624.

Preuss, K. 1987. *TOUGH User's Guide.* NUREG/CR-4645, SAND86-7104, LBL-20700. Washington, DC: US Nuclear Regulatory Commission.

Thompson, P.M., E.T. Kozak, and C.D. Martin. 1989. "Rock Displacement Instrumentation and Coupled Hydraulic Pressure/Rock Displacement Instrumentation for Use in Stiff Crystalline Rock," *Proceedings: Workshop on Excavation Response in Geological Repositories for Radioactive Waste, Winnipeg, Canada, April 26-28, 1988.* Paris, France: OECD Nuclear Energy Agency. 257-270.

Van Genuchten, M.T. 1980. "A Closed-Form Equation for Predicting the Hydraulic Conductivity of Unsaturated Soils," *Soil Science Society of America Journal.* Vol. 44, 892-898.

Webb, S.W. 1992a. "Sensitivity Studies for Gas Release from the Waste Isolation Pilot Plant (WIPP)," *Proceedings: Workshop on Gas Generation and Release from Radioactive Waste Repositories, Aix-en-Provence, France, September 23-26, 1991.* Paris, France: OECD Nuclear Energy Agency. 309-326.

Webb, S.W. 1992b. "Steady-State Saturation Profiles for Linear Immiscible Fluid Displacement in Porous Media," *Heat and Mass Transfer in Porous Media*, ASME HTD, Vol. 216, 35-46.

WIPP Performance Assessment Department. 1993. *Preliminary Performance Assessment for the Waste Isolation Pilot Plant, December 1992, Volume 4: Uncertainty and Sensitivity Analyses for 40 CFR 191, Subpart B.* SAND92-0700/4. Albuquerque, NM: Sandia National Laboratories.

Development of Coupled Models and Their Validation against Experiments - DECOVALEX Project

O.Stephansson and L. Jing
Royal Institute of Technology, Stockhol

(Sweden)

C-F. Tsang
Lawrence Berkeley Laboratory
1 Cyclotron Road, Berkeley, CA 94720

(USA)

F. Kautsky
Swedish Nuclear Power Inspectorate
(Sweden)

Abstract

DECOVALEX is an international co-operative research project for theoretical and experimental studies of coupled thermal, hydrological and mechanical processes in hard rocks. Different mathematical models and computer codes have been developed by research teams from different countries. These models and codes are used to study the so-called Bench Mark Test and Test Case problems developed within this project. Bench-Mark Tests are defined as hypothetical initial-boundary value problems of a generic nature, and Test Cases are experimental investigations of part or full aspects of coupled thermo-hydro-mechanical processes in hard rocks. Analytical and semi-analytical solutions related to coupled T-H-M processes are also developed for problems with simpler geometry and initial-boundary conditions. These solutions are developed to verify algorithms and their computer implementations. In this contribution the motivation, organization and approaches and current status of the project are presented, together with definitions of Bench-Mark Tests and Test Case problems. The definition and part of results for a BMT problem (BMT3) for a near-field repository model are described as an example.

There are currently nine Funding Organizations, one Funding Party, one Participating Party and Two Observers in the project, representing nine countries plus CEC. The Funding Organizations support fourteen Research Teams working in parallel on three Bench-Mart Test and six Test Case problems. Three phases of the project have been successfully conducted and a possible extension of the project is under planning.

1 INTRODUCTION

The rock mass response to storage of radioactive waste and spent nuclear fuel is a coupled phenomenon involving thermal (T), hydrological (H), mechanical (M) and chemical (C) processes [1]. The term "Coupled Processes" implies that one process affects the initiation and progress of another and vice versa (see Fig. 1). Therefore, the rock mass response to waste storage cannot be predicted with confidence by considering each process independently. Traditional disciplines in geoscience and geoengineering (for example, structural geology, geohydrology, mining and civil engineering,. geophysics, etc.) have to be integrated interactively together with applied mathematics and physics to gain a proper understanding of coupled behaviour of rock masses in which radioactive waste repositories are built. Mathematical models and computational methods are specially important to understand the coupled behaviour of rock masses since the processes involved are very complex, of long-term, and of large scale.

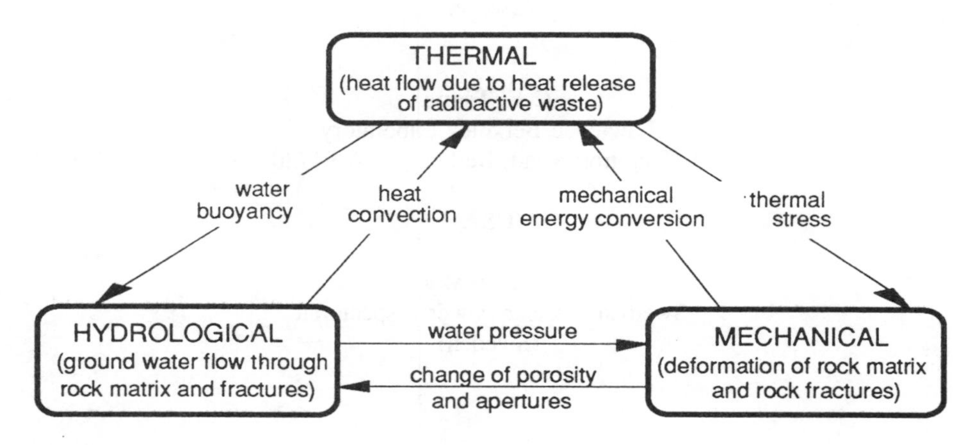

Figure 1 Coupled thermo-hydro-mechanical processes in a fractured rock

Though the importance of coupled thermo-hydro-mechanical (T-H-M) processes has been recognized for some years, the current capability of modelling such processes is still very limited, partially because of lack of applicable and realistic test cases (experimental studies) to establish conceptual models and to validate mathematical models and computer codes. Considerable work is needed to develop robust and reliable computer codes capable of modelling fully coupled T-H-M processes in geologic systems. To this end, an international project, **DECOVALEX** (acronym for **DE**velopment of **CO**upled models and their **VAL**idation against **EX**periments in nuclear waste isolation) has been established to support the development of mathematical models of coupled thermo-hydro-mechanical processes in the geologic media and their applications and validation against experiment in the field of radioactive waste isolation [2]. The overall goal of DECOVALEX project is to increase our understanding of the various aspects of coupled T-H-M processes of importance in the release and transport of radionuclides from a repository to the biosphere and how these processes can be described by mathematical models. The objectives of the DECOVALEX project can be summarized as following subjects:

- to support the development of computer codes for T-H-M modelling
- to investigate and to apply suitable algorithms for T-H-M modelling
- to investigate the capabilities of different computer codes to describe recent laboratory experiments and to perform code verification
- to compare theory and model calculations with results from field experiments
- to design new experiments of coupled T-H-M processes for further code development

2 ORGANIZATION

The organization of project DECOVALEX is illustrated in Fig. 2 and Table 1. The Steering Committee is formed by funding organizations and has the overall responsibility for the project. The committee is assisted in administrative and technical matters by the project Secretariat. The research teams are financially supported by their respective funding organizations to perform experimental or modelling works defined in the project.

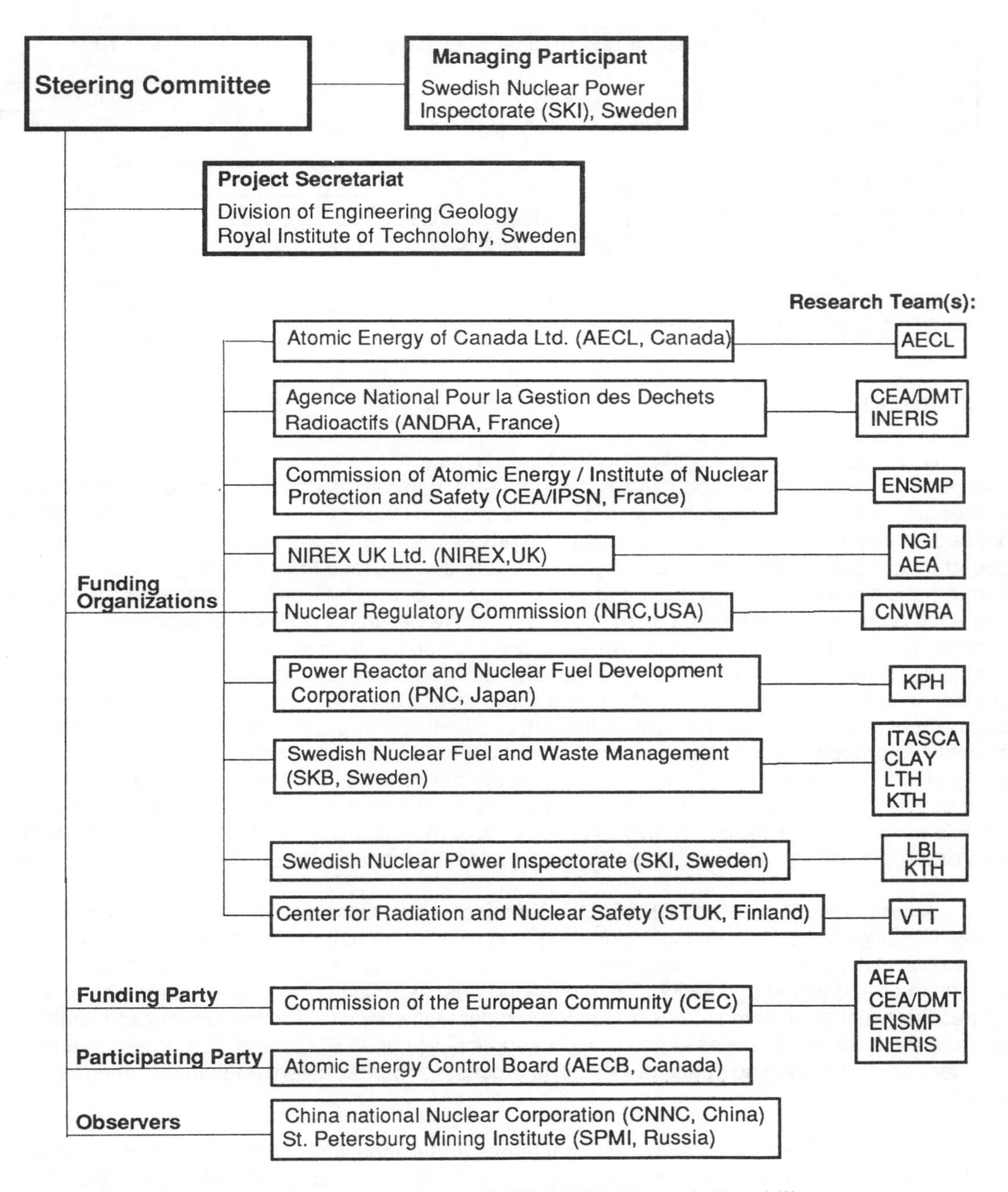

Figure 2 Organization of DECOVALEX, Phase I, II and III

Table 1 Research Teams in DECOVALEX, Phase I, II and III

Full name	Acronym
Applied Geoscience Branch, Whiteshell Laboratories, AECL, Canada	AECL
Commissariat à l'Energie Atomique (CEA/DMT), Department de Mecanique et de Technologie, France	CEA/DMT
INERIS/Laboratoire de Mécanique des Terrains Parc de Saurupt, France	INERIS
Ecole Nationale Supérieure des Mines de Paris, France	ENSMP
AEA Technology, UK	AEA
Kyoto University, Power Reactor and Nuclear Fuel Development, and Hazama Corporation, Japan	KPH
Norwegian Geotechnical Institute, Norway	NGI
Center for Nuclear Waste Regulatory Analysis, Southwest Research Institute, USA	CNWRA
Road, Traffic and Geotechnical Laboratory, Finland	VTT
Lawrence Berkeley Laboratory, University of California, Berkeley, USA	LBL
Royal Institute of Technology, Stockholm, Sweden	KTH
Clay Technology AB, Lund, Sweden	CLAY
Lund University of Technology, Lund, Sweden	LTH
ITASCA Geomechanics AB, Sweden	ITASCA

3 APPROACH

A number of problems, called Bench-Mark Tests (BMT) and Test Cases (TC) are defined or selected for research teams to analyze using different mathematical models and computer codes. The former are hypothetical problems and the latter are actual laboratory or field experiments. Both are defined as initial-boundary value problems with proper thermal, hydraulic ana mechanical initial-boundary conditions. The aim is to compare, validate and improve the predicting capacities of the mathematical models and computer codes. The number of the BMT and TC problems are limited so that each of them can be studied by multiple research teams in parallel and at successive stages or phases. Therefore, lessons learned from early phases can be used to improve later analyses. The results are presented and compared at regularly held workshops and the differences and similarities are discussed in detail. In this way, the models and results of each research team will undergo detailed peer review by all other teams, leading to significant in-depth exchange of information and scientific knowledge from different disciplines represented in the task forces. Out of these studies and discussions, new experiments with more reasonable designs are proposed to provide more rational tests of the models, and to advance the state-of-the-art of both our understanding and mathematical models of coupled processes. Analytical and semianalytical solutions to the coupled problems are also developed whenever possible in the project to assist model verifications and code validations.

A great advantage of the DECOVALEX project is the successful integration of different disciplines of geoscience and geoengineering, and of different national research teams. Represented in the project are physicians, hydrogeologists, researchers and practitioners in rock mechanics, civil and mining engineering and national waste management agencies. They all work together in both their respective national research teams and at the regular workshops to discuss, understand, defined and analyze the scientific soundness and practical applicability of all BMTs and TCs, from scientific, engineering and managing points of view. The process is therefore very educational and mutual beneficial.

Besides the integration about national teams and scientific disciplines, a special integration in the DECOVALEX is about different numerical methods applied to coupled T-H-M processes, including the finite element method (FEM), the finite difference method (FDM), the distinct element method

(DEM) and discrete fracture network method (DFN). The FEM and FDM are based on continuum approach and DEM treats the rock mass as an assemblage of deformable blocks. The FDN, on the other hand, considers the fracture space only, see Fig. 3. The mathematical models developed for these different numerical tools are wide spread and lead to different results. The uncertainties and differences in results and interpretations naturally lead to active exchange of conceptions, models and algorithms, which remain to be the frontier of the current research.

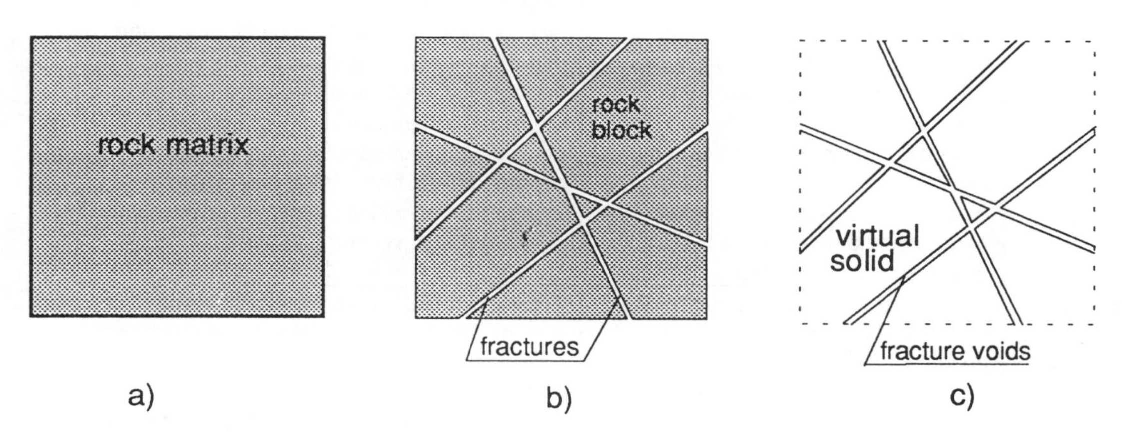

Fig. 3 Geometrical representations of a fractured medium by a) continuum approach (FEM and FDM, b) discontinuum approach (DEM), and c) discrete fracture network approach (DFN)

4 BMT/TC PROBLEMS AND COMPUTER CODES

Table 2 lists a brief description to the physical processes involved in BMT and TC problems studied in the first three phases of the DECOVALEX project. The reasons why these problems are defined or selected are also briefly explained. The laboratory and field experiments are conducted in parallel with computer modelling so that prediction from the numerical simulations can be examined with experimental measurements, except TC2 which was completed before DECOVALEX started. The computer codes and their characteristics used in DECOVALEX are listed in Table 3.

A detailed description of all these BMTs and TCs is beyond the scope of this contribution. A relatively detailed definition of BMT3 is given below as an example of a BMT problem in DECOVALEX.

BMT3 is a near field repository model set up as a two-dimensional plane-strain problem in which a tunnel with a deposition hole is located in a fractured rock mass. The model is 50 x 50 m in dimension and 500 meters below the ground level (Fig. 4). The fracture pattern of the problem area, shown in Fig. 5a, is called the reference fracture network. It is a two-dimensional realization of a realistic three-dimensional fracture network model of a rock mass from the Stripa Mine, Sweden. The total number of fractures (discontinuities) in the reference network is 6580, consisting of six sets and with a mean density of 2.632 fractures per square meter. The problem is set up as a fully coupled Thermo-Hydro-Mechanical problem in a fractured medium, representing the near field behaviour of a repository. The thermal effect is caused by the radioactive waste in the deposition hole (the heater). The heat source decays exponentially with time. The rock matrix is assumed to be isotropic and linearly elastic, and its mechanical properties do not change with temperature variations. The thermal conductivity and expansion of the rock matrix are also assumed to be isotropic. The fractures are assumed to consist of parallel, planar, smooth surfaces at the macroscopic level with an effective hydraulic aperture. The initial and boundary conditions for the mechanical, thermal and hydraulic effects are shown in Figs. 5b and Fig. 6, respectively. Three loading sequences are specified for

Table 2 Physical processes studied in BMT and TC problems of the DECOVALEX project

Problem	Physical Processes	Reasons for selection
BMT1	A model of repository at 500 m depth and in fractured rocks with two orthogonal sets of persistent discontinuities. Domain size 3000 x 1000 m, a 2-D far-field problem with thermal, hydraulic and mechanical boundary/initial conditions	To examine the capabilities of computer codes for handling large scale problems of coupled T-H-M processes in rock masses containing a large number of discontinuities; Test the behaviour of different constitutive laws of rock matrix and discontinuities, and different numerical methods.
BMT2	A model of rock mass with four intersecting discontinuities and nine blocks in the vicinity of a heat source. Domain size 0.75 x 0.5 m, a 2-D near-field problem with thermal, hydraulic and mechanical boundary/initial conditions	To examine the capabilities of different computer codes for handling discrete discontinuities of specific constitutive behaviour under combined thermo-hydro-mechanical conditions with special emphasis on the heat convection
BMT3	A model of repository at 500 m depth in rocks with a realistic fracture network of 6580 fractures, Domain size 50 x 50 m, 2-D near-field problem with thermal, hydraulic and mechanical boundary/initial conditions	To examine the capabilities of different computer codes for handling large number of and randomly distributed discrete discontinuities of in a relatively small area surrounding a repository and to test different methods for fracture representation, homogenization and heat convection
TC1	A laboratory test of coupled shear - flow test of a single rock joint without thermal loading, performed at NGI	To test the performance of constitutive laws in the computer codes and their capacities to handle coupled H-M processes
TC1:2	Same set up as TC1 with improved problem geometry and different joint properties, performed at NGI	To test the performance of constitutive laws in the computer codes and their capacities to handle coupled H-M processes
TC2	A field experiment for coupled T-H-M processes in fractured rocks, 3-D problem. Domain size 10 x 10 x 5 m, conducted at Fanay-Augères (France)	To test the predicting capacities of computer codes for temperature, stresses and displacements measured during experiments in three-dimensional conditions
TC3	A large scale laboratory (Big-Ben) experiment of coupled T-H-M processes in engineered buffer materials, conducted at Tokai by PNC, Japan	To test the predicting capacities of computer codes for coupled T-H-M processes in both saturated and unsaturated situations in three-dimensional or axisymmetric conditions
TC4	A laboratory, true triaxial experiment of coupled normal stress-flow processes for a single rock discontinuity, designed at VTT, Finland	To test and establish physical laws governing coupled T-H-M processes for a single rock discontinuity and validate computer models in different codes
TC5	A laboratory test of coupled shear-flow experiments of a single rock joint on a direct shear machine, conducted at CNWRA, USA	To test the coupled hydro-mechanical behaviour of natural rock discontinuities, determine material properties and validate constitutive laws with emphasis on roughness characterization
TC6	An in situ borehole injection experiment at 150 m depth in fractured rock, conducted at Luleå, Sweden	To test predicting capabilities of different computer codes for coupled hydro-mechanical processes under in situ conditions

Table 3 Computer codes and their characteristics used in DECOVALEX

Code	User	For	Major Characteristics
MOTIF	AECL	BMT2 TC1 TC1:2 TC6	FEM solution of transport problems in porous/fractured media for 3-D problems of transient and steady-state groundwater flow, heat transfer and quasi-static T-H-M processes
THAMES	KPH	BMT1 BMT3 TC3 TC5	FEM solution of coupled T-H-M processes for porous/fractured media with crack tensor approach
ADINA-T & JRTEMP	VTT	BMT2	2-D FEM solutions of heat transfer (ADINA-T) and deformation analysis (JRTEMP) without fluid flow
ROCMAS	LBL KTH	BMT2 TC1 TC1:2 TC6	FEM solution of coupled T-H-M problems with special joint elements and a simple differentiation of a non-incremental formulation
CHEF HYDREF VIPLEF	ENSMP	BMT1 TC2	FEM solution of coupled T-H-M 2-D problems for porous or/and fractured media with special joint elements
TRIO-EF & CASTEM-2000	CEA/DMT	BMT1	FEM solution of coupled T-H-M 2-D & 3-D problems for porous media
NAPSAC	AEA	BMT3 TC6	Discrete fracture network solution for ground water flow
FRACON	AECB	TC6	FEM solution with joint elements for coupled H-M processes of continua
ABAQUS	CNWRA CLAY	TC3 TC5	FEM solution of coupled T-H-M problems for continua
FLAC (2D)	ITASCA	BMT3	Finite difference solution of coupled T-H-M problems of continua
UDEC	CNWRA, NGI INERIS ITASCA VTT	BMT1 BMT2 BMT3 TC1 TC1:2	DEM solution of 2-D, quasi-static, problems of coupled T-H-M processes for discrete, deformable block assemblages
3DEC	INERIS	TC2	3-D distinct element solution for coupled T-M problems of discontinua

BMT3, see Fig. 7a. Sequence 1 is the loading of the computational model prior to excavation of the tunnel under initial and boundary conditions of hydro-mechanical effects. An initial hydro-mechanical equilibrium should be achieved at the end of this loading stage. Sequence 2 is the excavation of the tunnel, executed at the end of sequence 1. The initial hydro-mechanical equilibrium is disturbed and a new pseudo-steady state of hydro-mechanical equilibrium should be achieved at the end of this stage, denoted as time $t=t^*$. Sequence 3 begins the thermal loading, starting at time $t = t^*$ and maintained for 100 years.

Figure 7b shows the output specifications for BMT3. The output required at eight points, two lines (profiles), five segments and nine regions are:

- Point A - H: Temperature and components of normal stresses and displacements.
- Line I and II: Temperature and stress components normal to the lines.

- Segments ABCD, EF, FG, GH, and HE: Water flux.
- Regions 1 - 9: Cumulative histograms of porosity, aperture and conductivity.

This Bench-Mark Test problem was set up to examine the capabilities of mathematical models and computer codes to handle a large number of, and randomly distributed, discontinuities in natural rocks. Particular interest was placed in the schemes of homogenization applied by continuum based numerical methods and simplification techniques applied by the discrete element method. Eight research teams have studied this problem (AEA, NGI, INERIS, CNWRA, VTT, KPH, CEA/DMT, ITASCA). Eight different computational models were built, which leaded to different results, especially regarding the stresses and water flux. Uncertainties in homogenization and simplification schemes remained and have been subjects of intense discussions.

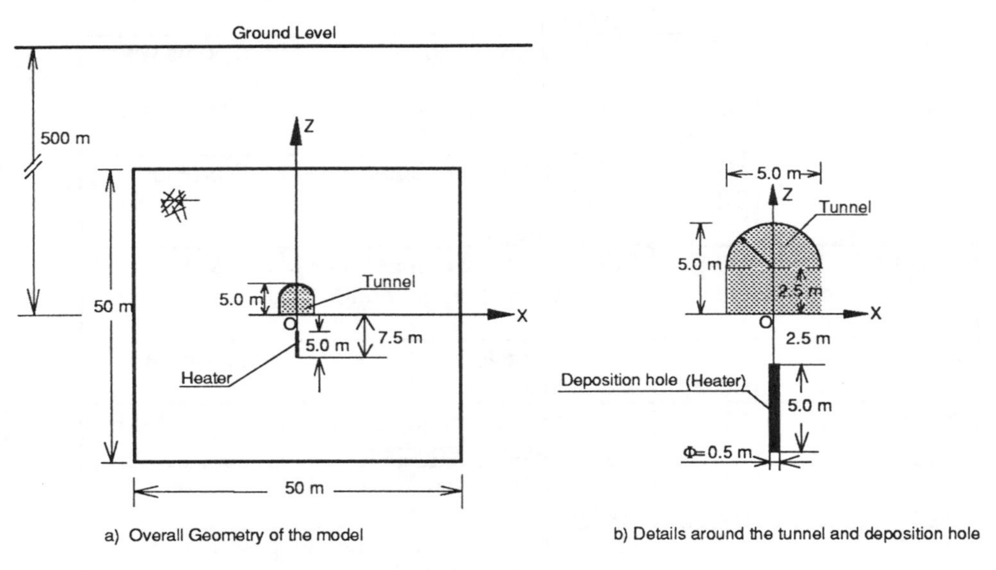

Fig. 4 Problem geometry of BMT3

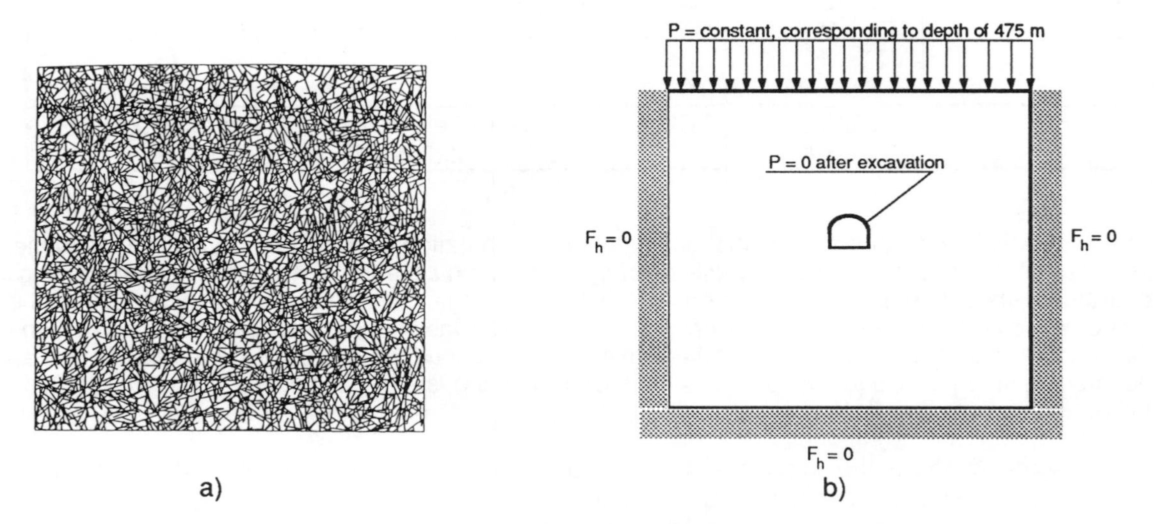

Fig. 5 The Reference Fracture Network (a) and the initial and boundary conditions of the hydraulic effects (b), BMT3. P - water pressure, F_h - water flux.

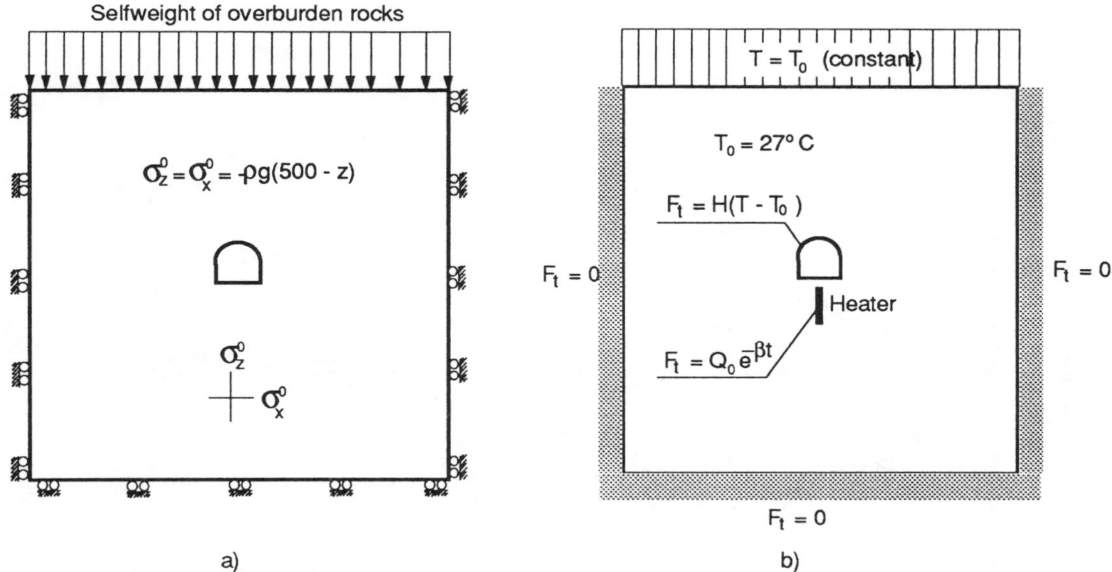

Fig. 6 The initial and boundary conditions for the mechanical and thermal effects, BMT3. T - temperature, T_0 - initial temperature, F_t - heat flux, H - surface heat transfer coefficient (= 7 $W/m^2 °C$), Q_0= 470 W/m^3 , β = 0.02/year.

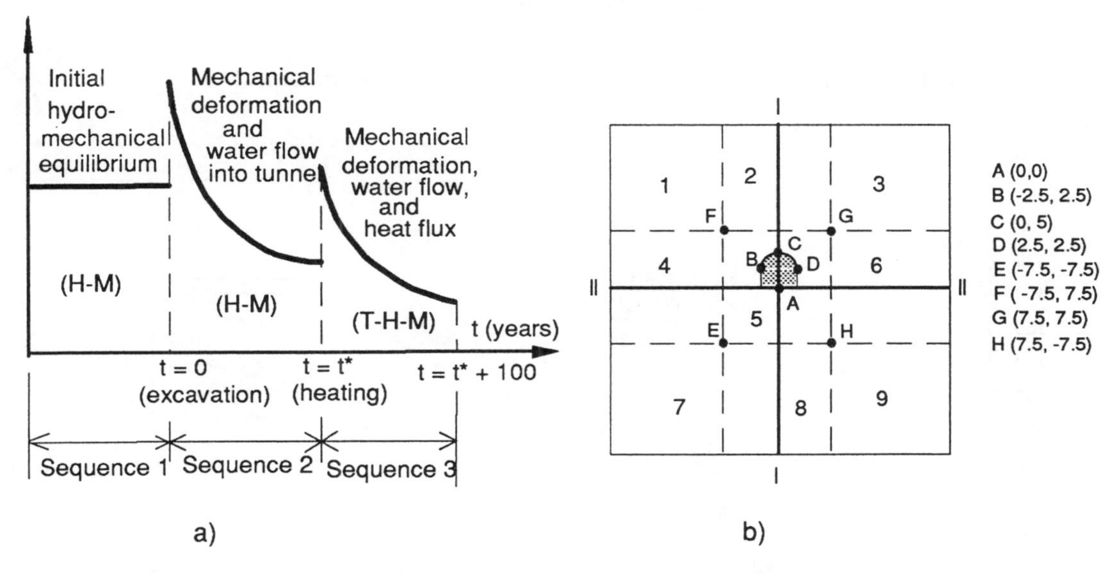

Fig. 7 Loading sequences (a) and output specifications (b) for BMT3.

Figures 8 - 10 show the results of the horizontal stress (Sxx) and temperature along profile I at t = 4 years (when the maximum temperature at heat source is reached), and the water flux into the tunnel (whose surface is represented by segment ABCD) versus time, respectively. Good agreement is obtained for temperature, but great discrepancies appeared for stresses and water flux, due to either different approaches (continuum vs. discontinuum), different homogenization schemes (between CEA, and KPH models) and different simplification and internal discretization techniques (among VTT, CNWRA, INERIS, and NGI, all using UDEC code).

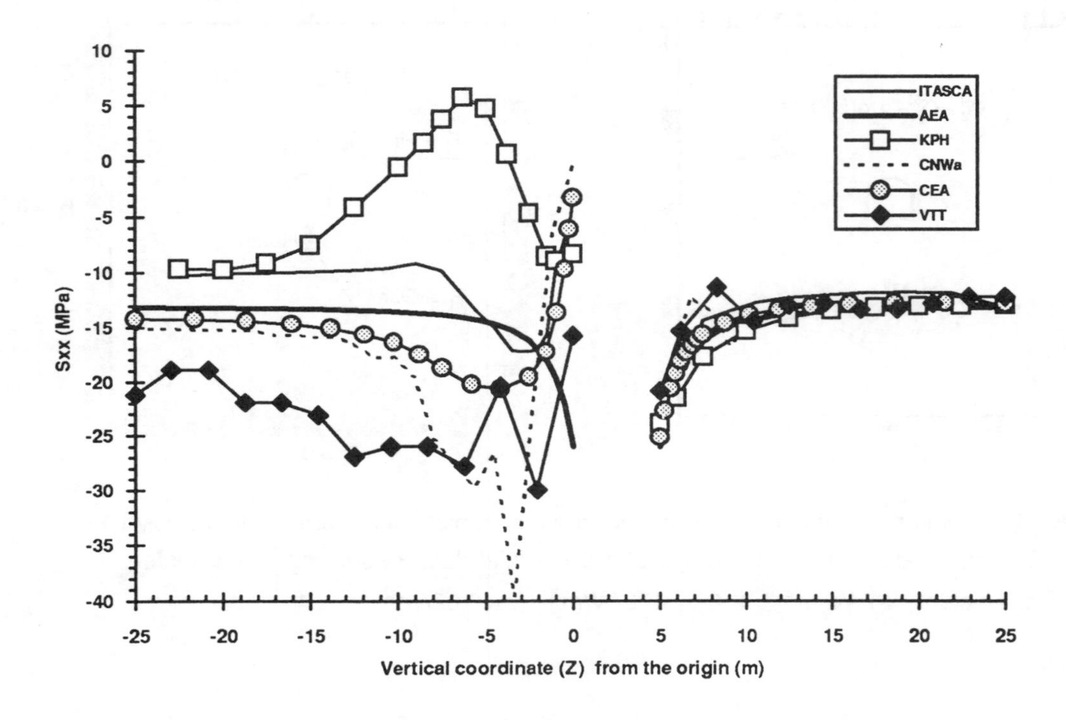

Fig. 8 Horizontal stress (Sxx) along profile I at t = 4 years for BMT3

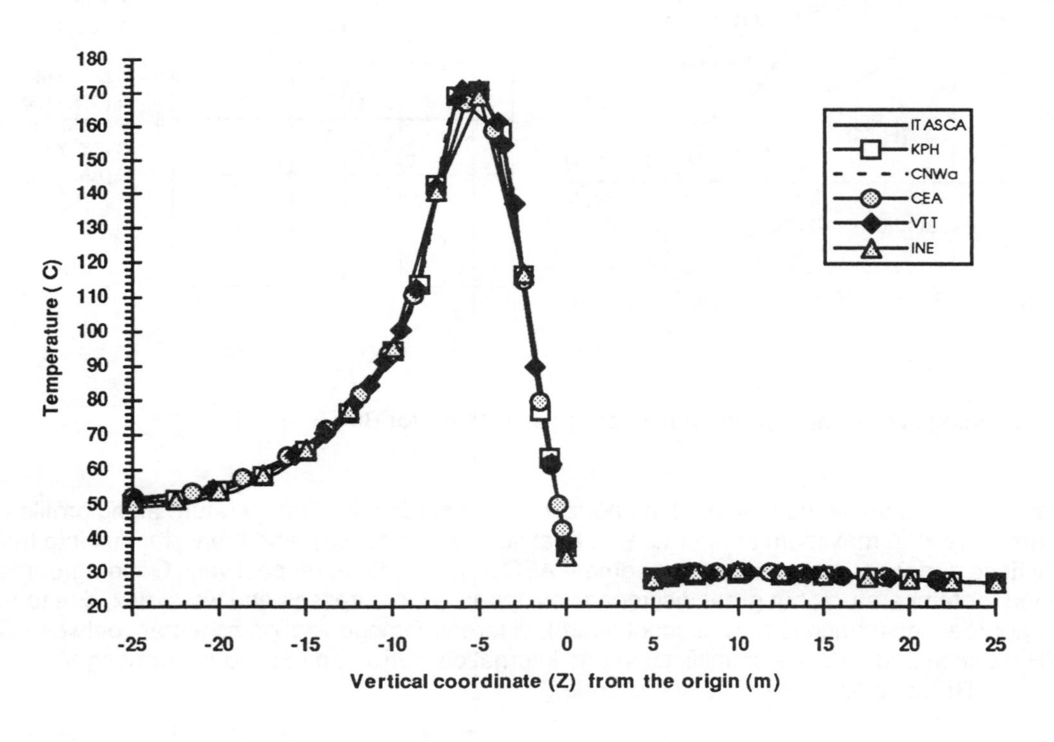

Fig. 9 Temperature along profile I at t = 4 years for BMT3

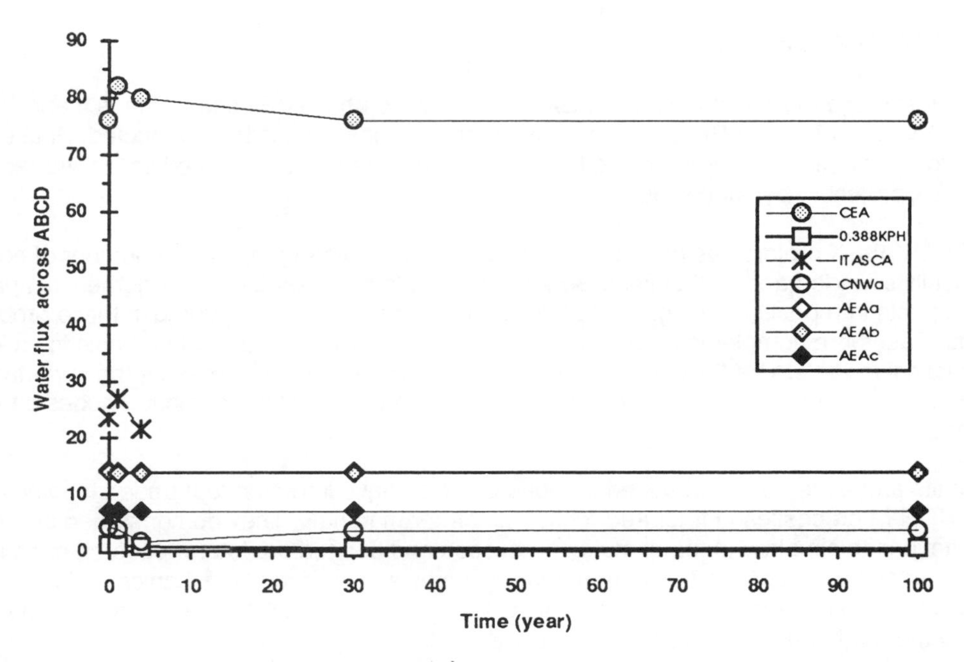

Fig. 10 Water flux ($\times 10^{-8}\,m^3/s$) across segment ABCD (tunnel surface) vs. time for BMT3

5 CURRENT STATUS OF THE DECOVALEX PROJECT

Three phases (I, II and III) have been successfully conducted with progress in both model and code improvement and overall understanding of coupled T-H-M processes in fractured rocks. The compositions of the first three Phases of DECOVALEX project and their implementation stages with series of workshops are illustrated in Fig. 11. For these three phases, the DECOVALEX project is aimed to validate the mathematical models and computer codes against either Bench-Mark Test problems or small scale laboratory or field Test Case problems. To continue the validation process, planning for possible extension of DECOVALEX is currently underway, with emphasis on simulation of, and integration with, one or two ongoing large scale underground in situ experiments of coupled T-H-M processes in fractured rocks and buffer materials.

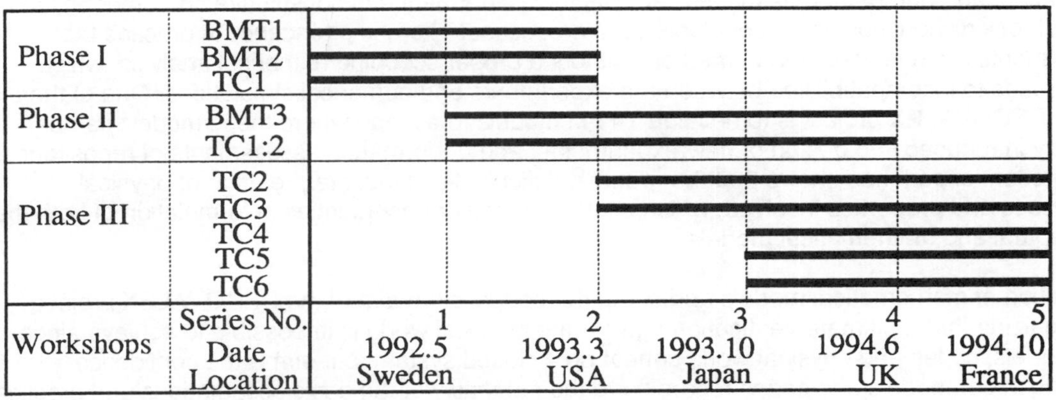

Phase I	BMT1 BMT2 TC1					
Phase II	BMT3 TC1:2					
Phase III	TC2 TC3 TC4 TC5 TC6					
Workshops	Series No.	1	2	3	4	5
	Date	1992.5	1993.3	1993.10	1994.6	1994.10
	Location	Sweden	USA	Japan	UK	France

Figure 11 Phases of DECOVALEX project, BMT and TC problems and Workshop series

6 CONCLUDING REMARKS

The first stage of project DECOVALEX includes three phases: Phase I, Phase II and Phase III. To date (October, 1994) both Phase I and Phase II have been successfully conducted. Phase III is also close to its completion. Summarizing the results and experiences obtained so far, we regard the DECOVALEX project to be successful.

1) The DECOVALEX project has been successful in its effort to bring together a number of both scientifically sounded and well defined Bench-Mark Tests and Test Cases for numerical analyses. Broad aspects in repository design and performance assessment are covered in these problems. The Test Cases are completed or ongoing experiments which provide a sound basis to validate the mathematical models. The Bench-Mark tests are carefully defined to represent the important aspects of a repository *in situ*, so that computational capabilities of the computer codes can be examined.

2) The results and experiences obtained in these exercise have improved our understanding of the coupled T-H-M processes in fractured rocks and buffer materials. They demonstrated our abilities and limitations to simulate these complex processes and pointed out future research directions with many uncertainties becoming forefront research subjects in both rock mechanics and hydrogeology. As a result of these exercises, the computer codes applied in the DECOVALEX have undergone a relatively thorough validation process in many aspects.

3) Integration of different geoscience and geoengineering disciplines has been implemented. The disciplines involved include: physics, applied mathematics, geohydrology, hydraulics, engineering geology, soil and rock mechanics, and civil and mining engineering. The integration of these disciplines are covered in the formulation of BMTs and TCs, evaluation of models and results, and estimation of uncertainties. The national waste managing agencies are also actively involved in every aspect of the project with their respective input on practical needs and problem scopes.

4) Integration of different numerical methods with both continuum and discontinuum approaches has been a special aspect of DECOVALEX and proves to be very fruitful and educational for both camps. Convergence rather than divergence appears to be the general trend to these different approaches to improve the understanding of capabilities, limitations, and uncertainties in their respective formulations. This is also reflected in the joint scientific publications from people with different background.

5) Some uncertainties and difficulties remain in the project. The lessons learned are currently under evaluation, and one of them is the less satisfactory performance of constitutive and other physical laws of rock discontinuities for coupled T-H-M processes. There is a tendency in practice that experiments are designed and carried out without a proper scooping numerical analysis, which often leads to incompatibilities between the experiments and numerical simulations. One of the aim of the DECOVALEX project is to develop T-H-M models to a stage where these models can be actively introduced into design of new experiments, and performance assessments of repositories. Our present experience shows that the greatest difficulty lie in the interpretation of physical processes and properties involved, which causes the major uncertainties in formulation of both the conceptual and mathematical models.

6) Validation of mathematical models and computer codes is the aim of the DECOVALEX project. Recognizing that a ultimate verification against the physical world is impossible to achieve since the models, codes and physical world cannot be a closed system, our aim is the confidence building in our models and codes regarding to the rationality of the basic assumptions and practical needs. The term validation is therefore always relative in terms of degrees. Recognizing also the importance of models and codes for the design and performance assessments of repositories, the

validation of the models and codes will remain necessary and important for the radioactive waste repositories until a relatively high level of confidence is achieved.

The key factor which makes the DECOVALEX project fruitful in advancement of scientific knowledge concerning coupled processes is the spirit of close co-operation among the different national organizations responsible for radioactive waste management and among the individual researchers from different disciplines involved in the project. Since our understanding of coupled behaviour of geomaterials for radioactive waste isolation is still at an early stage, free exchange of knowledge, in-depth discussion and friendly interactions among the international teams provide the much needed catalysts which are required for more reliable safety assessment of radioactive waste repositories.

7 ACKNOWLEDGMENT

The authors of this contribution, on behalf of the Steering Committee of the DECOVALEX project, would like to thank all research teams and individuals for their scientific contributions to the project, and Funding Organizations and Funding Parties for their financial support. The general spirit of scientific cooperation and constructive discussions have made all the efforts very fruitful.

8 REFERENCES

1. Tsang, C-F. (ed.), Coupled processes associated with nuclear waste repositories, Academic Press Inc., 1987
2. Jing, L., Stephansson, O., Rutqvist, J., Tsang, C-F., and Kautsky, F., DECOVALEX - mathematical models of coupled T-H-M processes for nuclear waste repositories, Phase I report, SKI Technical Report 93:31, June, 1993.
3. Jing, L., Stephansson, O., Rutqvist, J., Tsang, C-F., and Kautsky, F., DECOVALEX - mathematical models of coupled T-H-M processes for nuclear waste repositories, Phase II report, SKI Technical Report, 1994 (in press).

The Potential Use of Natural Analogue Studies in Radioactive Waste Disposal: A Review.

J.A.T. Smellie
Conterra AB
Box 493
751 06 Uppsala

F. Karlsson
Swedish Nuclear Fuel and Waste Management Company
Box 5864
102 40 Stockholm

B. Grundfelt
Kemakta Konsult AB
Box 12655
112 93 Stockholm

(Sweden)

Abstract

Radioactive waste must be disposed of in such a way that it remains safe for periods of time which are often well beyond our ability to guarantee control and surveillance. To ensure adequate protection to the biosphere, a series of barriers are erected to prevent the widespread dispersal of potentially dangerous radionuclides. Having designed such a multibarrier system, the next stage is to demonstrate its safety. The performance of the repository cannot, of course, be demonstrated practically nor by experiment because of the long time scales involved. Model predictions should be valid for thousands of years and, in the case of very long-lived radionuclides, even hundreds of thousands of years. Consequently, safety analysts over the past 10 years have recognised that the study of natural systems (or natural analogues) provides opportunities to test, by observation and measurement, many of the geochemical processes that are expected to influence the predicted reliability of radioactive waste containment over realistically long periods of geological time. In addition to the time scale factor, these analogue studies attempt to understand the multiprocess complexity of the natural system by an interdisciplinary approach, which contrasts with the limitations of the laboratory.

This paper reviews the application of natural analogue studies to demonstrate the safety of deep radioactive waste disposal; emphasis has been put on the Swedish disposal concept as an example. Examples of such applications are described and finally the main performance assessment objectives are discussed. Future areas of improvement are also addressed.

THE POTENTIAL USE OF NATURAL ANALOGUE STUDIES IN RADIOACTIVE WASTE DISPOSAL: A REVIEW

1. Introduction.

Radioactive waste is produced mainly by nuclear power plants with minor quantities from research facilities, industry and hospitals. The waste, low-, intermediate- and high-level radioactive forms, must be disposed of in such a way that it remains safe for periods of time which are often well beyond our ability to guarantee control and surveillance. To ensure adequate protection to the biosphere, a series of barriers are erected to prevent the widespread dispersal of potentially dangerous radionulides. In general, these barriers comprise a leach resistant waste form, stable encapsulation and backfill materials, and finally the massive volume of host rock that separates an underground repository from the biosphere.

Having designed a multibarrier system, the next stage is to demonstrate its safety. The performance of the repository cannot, of course, be demonstrated practically nor by experiment because of the long timescales involved. However, through the employment of dynamic models, based on sound experimental and field data, certain predictions can be made regarding the retardation of radionuclide transport to the biosphere, at least long enough for the radioactivity to decay to acceptable levels. Model predictions should be valid for thousands of years and, in the case of very long-lived radionuclides, even hundreds of thousands of years.

It has been increasingly recognised by safety analysts over the past 10 years that the study of natural geological systems (or natural analogues) provides opportunities to test, by observation and measurement, many of the geochemical processes that are expected to influence the predicted reliability of radioactive waste containment over realistically long periods of geological time. Their use in model testing therefore has the potential to describe the long-term development of the repository system in an appropriate way. In addition to the timescale factor, these analogue studies attempt to understand the multiprocess complexity of the natural system by an interdisciplinary approach, which contrasts with the limitations of the laboratory.

In general, natural analogue studies of contrasting geological environments are used to support repository performance assessment by: a) identifying the major geochemical processes which are and have been active in the formation and evolution of the geological environment being studied, b) selecting those geochemical features and processes most pertinent to repository performance assessment, c) introducing an iterative protocol to test and further develop those conceptual models used in performance assessment, and d) planning, collection and implementation of relevant field data to satisfy the above-mentioned criteria.

Natural analogue studies also encompass examples taken from anthropogenic sources, including archaeologic metal artefacts which have been used as indicators to canister material stability.

A significant volume of data, potentially applicable to the problems of predicting repository performance assessment over long periods of time, are now available in the open literature as a result of studies pertaining to natural analogues and geological systems (e.g. see reviews and references therein by Chapman et al., 1984; Pearcy and Murphy, 1991; Brandberg et al., 1993; Miller et al., 1994).

In addition to the obvious scientific value of natural analogue studies, they also play an increasingly important public relations role. There are clear indications from the informed scientific and technological communities, in addition to the less informed public, that confidence in the reliability of a certain feature or process increases when presented with evidence from natural geological systems. Consequently, reference to natural analogue studies and the resulting data are often used as information material to the media. The dependability of analogues in building confidence is based on the fact that they have developed over timescales comparable to the predicted duration of a repository system. Furthermore, since they have not been influenced or controlled by man, unexpected events and processes can be identified and observed.

This paper reviews some applications of natural analogue data to demonstrate the safety of deep radioactive waste disposal; emphasis has been put on the Swedish disposal concept as an illustration.

2. Natural Analogue Studies and Disposal Concepts.

Many countries will dispose of their high-level waste deep in a saturated bedrock environment, ideally in an area of groundwater recharge. Recharging flow paths penetrate deep into the bedrock, through the repository, and eventually to the biosphere, usually facilitated by large-scale conducting fracture zones (e.g. crystalline rock) or more porous horizons (e.g. clay). In order to assess the long-term safety performance of a repository, a series of scenarios have been considered which may result in a failure of the engineered barrier system eventually releasing radionuclides to the near-field. It is customary to select a rather pessimistic "reference scenario" approach whereupon radionuclides are released to the near-field. Although these will migrate along the flow paths, it is assumed they will undergo dilution prior to reaching the biosphere. To demonstrate that radioactive dosages calculated from the reference scenario case will fall within specified public safety limits, is thus an important performance assessment assignment.

Some countries, for example the USA, intend to locate their disposal facility in the unsaturated zone. This means that only sporadic amounts of water will come in contact with the repository, but even these amounts are assumed to eventually lead to a breakdown of the engineered barrier system and the release of radionuclide material to the saturated zone. Once in the saturated zone, radionuclide migration and dilution processes will be similar to those described for the case above. Despite the fact that this, and other disposal concepts are fundamentally different from each other, they share many similarities when individual near- and far-field geochemical processes are considered, especially in the far-field. It is these processes which are identified in geological systems and studied as natural analogues to waste disposal.

The key geochemical processes relevant to most high-level waste disposal concepts are shown schematically in Figures 1 and 2; these figures illustrate an idealised repository intersected by hydraulically conducting fracture zones. Figure 1 represents the intact canister scenario; the bedrock is assumed to be reducing and the potential long-term influences on the repository system are listed. The bentonite overpack serves as a mechanical buffer to changes in rock pressure and rock displacements (suitable plastic properties), as a hydraulic barrier (low permeability) to groundwater flow and as a chemical buffer to pH and O_2. Figure 2, the penetrated canister scenario, shows a whole series of processes and reactions resulting from canister penetration; radiolysis, redox front propagation, waste dissolution and radionuclide mobilisation and dispersion from the near- to the far-field along hydraulically conducting fracture zones extending to the biosphere. At various locations along this flow path, radionuclide migration may be enhanced by the influence of colloids, organics, microbes and gases, and retarded by processes such as mineral sorption, precipitation and matrix diffusion.

3. Natural Analogue Sites

That geological systems can demonstrate geochemical processes analogous to those expected to occur within and around a repository during long-term isolation is beyond dispute. The main problem remaining is to identify a suitable geological system with the required geometry and well constrained physico-chemical boundary conditions to make the study worthwhile.

It is important to bear in mind that there is no single analogue site that provides a complete analogy to a repository system. A single analogue site may be selected to try and study as many of the near- to far-field geochemical processes as possible, as in the case of most of the major international studies of uranium deposits such as at Poços de Caldas, Brazil (Chapman et al., 1992), Alligator Rivers, Australia (Duerden et al., 1994), Cigar Lake, Canada (Cramer and Smellie [Eds.], 1994), Peña Blanca, Mexico (Murphy and Pearcy, 1994), El Berrocal (Astudillo, 1994) and Oklo, Gabon (Toulhoat et al., 1994). These may be termed large-scale multiprocess analogues. Contrastingly, one or more small-scale analogues may be selected to focus on single processes, e.g. matrix diffusion (Smellie et al., 1986; Alexander et al., 1990), near-field transport and retardation in a clay buffer (Hooker et al., 1985) and far-field mass transport in sediments (Read et al., 1991), or single components of a repository system, e.g. bentonite stability (Pusch, 1983; Pusch and Karnland, 1988), canister stability (Johnson and Francis, 1980; Hallberg et al., 1988) and concrete stability, mainly applicable to low- to intermediate-level waste facilities (Khoury et al., 1992; Alexander et al., 1992).

Uranium deposits are chosen as they are usually well characterised, either from prospecting and/or subsequent mining activities. This facilitates the location of strategic profiles for detailed analogue studies, e.g. along groundwater flow paths from the deposit into the host rock. Such studies are invaluable in that they represent the multiprocess complexity of any geological system, and by analogy any repository site. An obvious disadvantage to such sites is that mining activities have often perturbed the present groundwater system.

Natural analogue studies need not focus only on uranium mineralisation; other possibilities include carbonatite deposits (rare-earth concentrations, chemically analogous to several radionuclides in the waste form) and other concentrations of trace elements of particular relevance to high-level waste, e.g. Mo. Se, Zn, Sn, Pb, Cr, Cu, As etc.

4. Natural Analogues and Performance Assessment

Large-scale analogue studies normally generate large volumes of analytical data and substantial time and effort is often spent in the compilation, evaluation and modelling of these data to understand the nature and geochemical evolution of the sites under investigation. This is, of-course, quite essential, but unfortunately a full appraisal of these data with respect to repository performance assessment (PA) generally tends to be frustratingly inadequate.

The limited use of analogue data in performance assessments may be partly explained by historical reasons, whereupon performance assessors tend to represent the engineering and chemical engineering fraternities and, as such, have concentrated on the performance of the near-field technical barrier systems which can be conveniently tested under laboratory conditions. With the onset of natural analogue and geological system studies, however, the reliability of laboratory data gave way to the uncertainity of complex field-derived data. As a result, existing models were either too simple or not robust enough to cope. On the other hand, often field-derived data are inadequate for modelling purposes due to poorly analysed samples or inadequately defined boundary conditions etc., thus reflecting a failure of the geoscientist to recognise the type and quality of information required for performance assessment purposes. There is much room for improvement on both sides.

Natural analogue data fall generally into three categories: quantitative (input data to performance assessment), qualitative (confidence building by demonstration) and public awareness (seeing is understanding); the last mentioned is dealt with separately. Because of the complexity of geological systems, most analogue data fall under the qualitative category, where important geochemical processes can be demonstrated to occur, but the boundary conditions cannot or have not been well enough defined to be measured. Despite this, such demonstrations are valuable in illustrating the concepts and safety of geological disposal.

However, some quantitative data have emerged from analogue studies, for example, corrosion rates of canister materials, the stability of the bentonite overpack and penetration depths of matrix diffusion processes etc. This is reflected in the inclusion of such data, albeit limited, in the Swedish and Swiss national programmes. In Sweden reference to natural analogues has been made in safety performance assessments such as KBS-1 on vitrified waste and KBS-2, KBS-3 and SKB-91 on spent fuel. In these assessments natural analogue applications have been mostly concerned with the near-field environment, e.g. the stability of the copper canister and the bentonite overpack. These and other applications are documented in the literature by Smellie (1989), Brandberg et al. (1993) and Karlsson et al. (1994) and summarised in Table 1. In Switzerland reference to analogues have been made in the Project Gewähr (Nagra, 1985) and Kristallin-I (Nagra, 1994; McKinley et al., 1992) performance assessments; further analogue applications are discussed by, for example, Alexander and McKinley (1992), McKinley and Alexander (1992) and Pate et al. (1992). More recent reference to general analogue applications in performance assessment have been presented by Miller et al. (1994) and McKinley and Alexander (1994).

Natural analogues should not be seen as an alternative to laboratory studies; rather their function should be seen as a complement, bridging the gap between the field and the laboratory, i.e. providing realistic time-scales to reactions and processes, allowing comparison of a complex vs simple systems and comparing similar reactions under the controlled, well defined boundary conditions of the laboratory, with the uncontrolled conditions of geological systems. The link-up of analogue and laboratory studies with reference to model testing/validation and performance assessment is shown in Figure 3 after Miller (1994).

4.1 Groundwater Redox Conditions: A Pragmatic Approach to PA

Central to the long-term behaviour of a repository is precise knowledge of groundwater redox conditions. This is essential to our understanding of geochemical processes operating in the bedrock, as redox and pH constitute two of the major master variables used to model such geochemical processes. Traditionally, the geochemical approach to groundwater redox conditions has been based on guarded theoretical sceptism. The performance assessment approach, in contrast, is based more on practicalities, in particular there are two main requirements: a) that the groundwaters at the depth of disposal are oxygen-free (measured Eh <0 mV), and b) that potentially dangerous redox-sensitive nuclides such as U, Np and Tc remain in their reduced states (measured Eh <-250 mV). Quite simply, reducing groundwater conditions are critical to the long-term safety and stability of the repository system. This is illustrated in Figure 4 which underlines the central role played by groundwater redox conditions in influencing several of the major processes which determine repository stability.

In the near-field the presence of oxygen can result in: a) the stability of the canister being affected by localised corrosion attack (i.e. pitting), b) an increase in the availability of Ca^{2+} and K^{+} resulting in a deterioration in the physico-chemical properties of the bentonite backfill material, e.g. reduction in the swelling and plastic properties and a reduction in the redox buffering capacity, c) the waste form being subject to greater solubility and mobility, e.g. oxidation of spent fuel initially resulting in a change of the UO_2 crystal structure and subsequently in the conversion of U(IV) to U(VI); this leads to the "matrix release" of uranium and other radionuclides otherwise trapped in the UO_2 crystal lattice, and d) stimulated growth of microbes (i.e. bacteria oxidising Fe^{2+}).

Oxidation results in an increase in the mobility and transportation of radionuclides away from the near- to the far-field. This is facilitated by the presence of high oxidation states of U, Np and Tc, e.g. as uranyl, neptunyl and pertechnetate complexes respectively, increased colloidal material, e.g. polysulphides and Fe-oxyhydroxides, and accelerated microbial activity. Oxidation also serves to minimise radionuclide retardation effects due to the decreased tendency of actinides at higher oxidation states to sorb unto fracture surfaces, and a general decrease in the redox buffering capacity of the geosphere system (e.g. by oxidation of Fe^{2+} to Fe^{3+}) available to the migrating radionuclides.

4.1.1 Determination of Redox Potential in Groundwater

Precise determination of redox potential in natural waters will increase the understanding of systems that govern groundwater redox conditions; this is essential for the credibility of performance assessment. For example, within the Swedish programme considerable attention has been given to field measurements of redox potential, both recorded at the surface and downhole (e.g. Wikberg, 1987; Wikberg, 1991). These field measurements have been compared with calculations based on different redox-determining couples, in particular the Fe(II)/Fe(III) couple (Grenthe et al., 1992). This model was subsequently extended to include the redox potentials calculated for the different solid ferric mineral phases which characterise the system, based on the hypothesis that different ferric minerals may determine the redox potential at different sites (Brandberg et al., 1993). This showed that redox conditions could be characterised by the simple analysis of Fe(II), pH, HCO_3 and temperature in groundwater, in combination with field observations of the age and crystallinity of fracture filling iron minerals in contact with the groundwaters. Essentially, it should be possible to simply calculate groundwater redox measurements by omitting direct measurement of Eh, provided that the other groundwater parameters are accurately determined.

These redox studies are a good example of model development partly driven by the demands of natural analogue studies to derive reliable input redox values to test geochemical codes and databases used in the performance assessment modelling of trace element solubility and speciation, and geochemical equilibrium models of water-rock interaction.

4.1.2 Radiolysis as a Source of Oxidation

From the preceeding discussion, all indications seem to support the premise that natural groundwater systems at repository depths will be sufficiently reducing to ensure long-term stability. Other sources of oxidation, however, particularly resulting from radiolysis reactions in the near-field, should not be overlooked. Analogue studies show that radiolysis processes do occur in nature influencing the "near-field" redox chemistry which can result in localised oxidation and radionuclide mobility (see discussion below).

4.2 Testing Geochemical Codes and Databases

In recent years, some of the most valuable applications of natural analogues to performance assessment have been in testing the geochemical models (i.e. the geochemical codes and databases) used to calculate radionuclide solubility and speciation in groundwaters, and their sorption behaviour in contact with fracture mineral assemblages. Chemical thermodynamic models provide two main types of data for repository assessment purposes: a) solubilities of particular elements or radionuclides (this parameter places an important constraint on release rates from, e.g. spent fuel), and b) speciation of such elements in solution (allows their transport properties to be estimated).

To test such models, a series of solubility "blind" predictions have been carried out, notably at Oman, Poços de Caldas, Cigar Lake and Maqarin. In the Poços de Caldas study five different groups participated in the exercise (Bruno et al., 1992), at Maqarin four groups (Alexander et al., 1992); Oman (Bath et al., 1987) and Cigar Lake (Casas and Bruno, 1994) each involved one. Selected major element analyses of groundwaters were used as input to predict the solubility, speciation and limiting solid phase for a number of elements considered important in repository performance assessment (e.g. U, Th, Pb, Ba, Cu, Sr, Mo, As, Sn, Se, Mn, Ni and Ra). The predictions were then compared with observed aqueous concentrations, speciation and trace mineralogy. From all sites there was reasonable agreement between the predicted solubilities and the measured trace element concentrations, but discrepancies occurred in the predicted aqueous speciation. This was particularly obvious for uranium, indicating the large uncertainties which still exist in the solute speciation databases. To help tackle this problem, experimental field in situ speciation techniques were attempted at Poços de Caldas, but unfortunately with only limited success.

Further work is needed in this area; such "blind" predictive exercises should ideally involve the comparison of several codes and databases thus facilitating an integrated approach to identifying areas of uncertainty and/or lack of data within different thermodynamic databases. Additionally, more effort is required to develop reliable in situ speciation techniques and continued effort is required to identify fracture mineral phases known to be in contact with the sampled groundwaters (Brandberg et al., 1993).

4.3 Radionuclide Chemistry and Transport in the Near-field

Current performance assessment models used to describe spent fuel dissolution assume that oxidising species are formed by alpha-radiolysis of the groundwater as it comes in contact with the spent fuel uranium pellets subsequent to canister penetration. This can lead to changes in redox conditions, creating an oxidising groundwater environment around the waste causing oxidation of UO_2. Oxidation of UO_2 beyond U_3O_7 may cause a change in the crystal lattice leading to the release of radionuclides enclosed in the uranium oxide matrix. Further oxidation will increase the solubility of uranium considerably (hexavalent form). Of course dissolution of the uranium oxide matrix will also result in radionuclide release. Excessive radiolysis may also result in the transport of uranium and other oxidising species in association with a propagating redox front away from the spent fuel, through the bentonite overpack, and out into the surrounding rock. However, the natural redox buffering capacity of the bentonite and rock (e.g. mainly due to Fe^{2+} contained in the mafic mineral phases) is expected to restrict the movement of the redox front.

The near-field formation and propagation of redox fronts resulting from oxidation processes (e.g. radiolysis), accompanied by radionuclide transport, have been studied at Oklo (Curtis and Gancarz, 1983; Menet et al., 1992), Cigar Lake (Cramer and Smellie (Eds.), 1994; Karlsson et al., 1994) and Poços de Caldas (MacKenzie at al., 1992). These data have been compiled and discussed by Brandberg et al. (1993) and the following description is based wholly on that work.

Studies show that radiolysis processes do occur in nature influencing the "near-field" redox chemistry which can result in localised oxidation and radionuclide mobility. Model calculations at Cigar Lake confirm that the radiolysis products of water oxidise Fe(II) to Fe(III) which may then precipitate due to low solubility. This might account for some of the characteristic red colouration due to the presence of trivalent iron (as Fe-oxyhydroxides) evident at the ore/clay boundary. Mineralogical evidence in the orebody also points to some oxidative effects, i.e. the presence of higher mixed uranium oxides; furthermore hydrogen gas (known to result from radiolysis/water reactions) and sulphate (known to result from sulphide oxidation) has been measured in the groundwater from the orebody.

Radiolysis evidence at Oklo is believed to have resulted in a net reduction of iron to its divalent form, probably accompanied by a net increase in total iron, associated with the reactor zone rocks; the aureole rocks showed less extreme reduction features. Furthermore, some of the fissionable products are believed to have undergone oxidation and subsequent migration away from the reactor zone. However, it must be remembered that most of the radiolysis reactions at Oklo have occurred during hydrothermal conditions so that hydrogen has been reactive and participated in the redox reactions (Fe(III) reduction).

Taken collectively, present results support the fact that radiolytic oxidation in the near-field of the exposed spent nuclear fuel, within the concept of an underground repository, cannot be ruled out. However, studies at Cigar Lake indicate that the net changes are considerably overestimated by models currently used by performance assessment (Karlsson et al., 1994).

The outward propagation of redox fronts from the radiolytic source at the canister/bentonite interface, and their migration rates, depends on the waste, the design of the repository, the nature of the host rock etc., and may involve distances from centimetres to hundreds of metres per Ma. Low temperature redox front migration is best exemplified in the Poços de Caldas study. Based on estimated geomorphological net erosion rates of the Poços de Caldas plateau, and assuming that the downward migration of the redox fronts was dependent on the rate of erosion, the average rate of redox front movement was calculated to be around 10 metres per Ma which is roughly comparable to the rates expected to occur around a repository.

As predicted by the near-field model, redox-sensitive radionuclides are also observed to precipitate as they pass in solution through the redox front. This is particularly convincing at Poços de Caldas (for uranium and its decay daughters) and also at Oklo (for several fission end products); however further detailed studies are needed at Cigar Lake to confirm similar processes. At Poços de Caldas it has been shown that reduction/precipitation mechanisms, as well as scavenging by coprecipitation and sorption, take place at the redox fronts. Some of the substances reduced (e.g. UO_2^{2+} and SO_4^{2-}) are found in the order of centimetres downflow of the front. It has been postulated that the reduction of these species may be kinetically controlled. The models used, however, cannot be considered predictive as they greatly simplify the system being described.

Some of the elements enriched at the Poços de Caldas redox fronts may have resulted from sorption or coprecipitation with, for example, ferric iron hydroxides. This may very well have important performance assessment implications as there exists the potential for interaction between easily soluble radionuclides and canister corrosion products.

Processes relating to coprecipitation and solid solution of radionuclides in Fe-oxyhydroxides cannot presently be simulated by current near-field models. Their role has not been quantitatively included in performance assessment models due to a lack of appropriate data. Neglecting this phenomena may be conservative, but then again this might lead to an overestimation of concentrations by many orders of magnitude. The observed magnitude of this phenomenon was unexpected before the investigations, thus serving to identify an important feature that had been overlooked in early performance assessments. This is a good example of how natural analogue studies can help to recognise processes which are non-linear and therefore unpredictable in a precise and deterministic fashion. Data do exist, for example from Poços de Caldas, that may be used for testing simple co-precipitaton models such as solid solution models.

Observations at both the Cigar Lake and Oklo sites, further supported by coupled mass transport calculations at Cigar Lake, endorse the efficiency of the clay halo as a barrier to radionuclide and trace element migration. At Cigar Lake, the massive clay (metres) contained within and surrounding the ore, effectively shields the orebody hyrologically and also acts as a buffer to geochemical reactions and release processes. Models used to calculate mass transport in the near-field of a repository using a clay buffer have been successfully applied to explain the release of helium generated in the orebody (Liu et al., 1994). At Oklo, although the clay-rich alteration rim around the reactor is contrastingly very narrow (centimetres), its retardation properties at hydrothermal temperatures have been impressive. These studies therefore strongly support the integrity of the bentonite overpack intended for waste disposal, despite the fact that illite, rather than the smectite considered in disposal concepts, characterises both the Cigar Lake and Oklo sites.

4.4 Radionuclide Chemistry and Transport in the Far-field

Central to any repository performance assessment is an estimation of the mass transport rate of radionuclides from the near-field engineered overpack out into the geosphere and ultimately to the biosphere. As a general rule, mass transport models try to simulate the current spatial distribution of some natural tracer from its source. The use of uranium (and its decay daughters) as a source term has many advantages: a) it constitutes a natural groundwater tracer, b) it conveniently occurs as concentrations in rock at depths and in hydrogeological environments comparable to many disposal concepts, c) it is the major component of spent nuclear fuel, d) it is a chemical analogy to other actinides in the tetravalent and hexavalent state, and e) its redox behaviour helps to predict the geochemical redox reactions of other actinides such as Np(V) to Np(III) (redox probe).

Far-field processes have been mainly restricted to the large-scale, multiprocess analogue sites, such as Poços de Caldas, Alligator Rivers, Cigar Lake and more recently Oklo. Of special importance is an understanding of the present (and past) hydrogeological groundwater flow environment of the sites and, specifically, the nature of groundwater interaction with the radioactive source term, i.e. usually a uranium orebody. Ideally, groundwater transport paths to and from the orebody should be confined and rock and groundwater sampling points restricted to these paths.

Movement and retardation of radionuclides through the geosphere has been dealt with under the general headings of transport (advection/dispersion), channelling, colloidal influence and retardation mechanisms such as matrix diffusion.

Radionuclide mass transport

Transport models of the type commonly used in repository performance assessment have been used to simulate radionuclide migration in the shallow weathered zone of the Koongarra uranium deposit in the Alligator Rivers region (Golian and Lever [Eds.], 1993). These different model attempts give a fairly consistent picture of uranium migration in the weathered zone, in the order of 1 Ma. However, palaeohydrogeological studies are still necessary to accomodate climatic variations which are known to have influenced this region within the last million years.

In the Cigar Lake study, due to an initial lack of suitable performance assessment mass transport models that could be directly applied, a series of three iterative models were subsequently developed to describe the system (Liu et al., 1994). In contrast to Alligator Rivers, where uranium mobiliation and transportation have been measured, the Cigar Lake site shows very little field evidence of uranium mobilisation and migration. All the models were therefore based on the assumption that uranium was being mobilised from the orebody. Two of the models, one involving molecular diffusion and advection with a rapid transfer of uranium without interaction/retardation effects, and the other coupling uranium migration with possible geochemical processes, showed that uranium movement did not extend beyond the massive clay halo surrounding the deposit.

A further model, however, considered the mass transport of in situ nuclear capture products (^{36}Cl, ^{14}C and ^{3}H) from the ore through the clay to the porous altered sandstone; enhancements of both ^{14}C and ^{3}H have been measured in groundwaters from the altered sandstone close to the deposit. Transport processes were considered to be by diffusion (through the clay) and advection (within the sandstones). The results showed that the released plume of nuclear products through the clay halo (most of the products are contained by the clay) would tend to be restricted to the bottom of a very narrow downstream layer, and that concentrations would be very low. These findings essentially supported the field observations and the conceptual hydrogeological model for the Cigar Lake deposit.

Channelling

The flow of groundwater through a fractured crystalline rock via a network of anastomosing channels has long been accepted by the geoscientific fraternity. Recently the performance assessment implications for far-field transport (i.e. rapid transport pathways) and retardation mechanisms (i.e. less fracture surface area available) have been recognised and attempts have subsequently been made to quantify this phenomenon. However, field evidence is rare; any evidence is furthermore difficult to quantify. A semi-quantitative attempt was made in the Poços de Caldas study (Romero et al., 1992). Although of limited value, this study helped to underline the importance of certain parameters such as the distribution of flow-rate and the interconnecting pattern network of the channels.

In conclusion, the character, distribution and dynamics of channelling in crystalline rocks is still not well known with confidence; more studies are needed before a quantitative appraisal can be conducted.
Natural analogues should be important in this context as they represent processes that have been going on very slowly for long periods of time and under conditions undisturbed by stress redistributions and artificial hydraulic gradients typical for man-made tracer experiments. Potential studies could involve the staining of fracture surfaces by coloured precipitates or other clearly detectable tracers capable of resolving areas of water-rock contact.

Matrix diffusion

Matrix diffusion is potentially a very effective retardation process to far-field radionuclide transport and, if verified, has important implications on performance assessment. The major implications are that sorbing radionuclides are not just limited to the fracture surfaces, by diffusing through the fracture coatings they have access to a much greater rock matrix volume.

Natural analogue studies are probably more reliable than laboratory data in most respects by virtue of realistic time-scales and that the measured radionuclide profiles represent processes which have been occurring under natural groundwater and overburden pressure conditions. A considerable amount of laboratory and field data have now been compiled and reviewed (see Brandberg et al., 1993). Natural analogue studies, mainly focussing on diffusion penetration depths, have been conducted in a variety of hydrogeological environments and some kind of pattern seems to be emerging. Most penetration depths range from 1 mm up to 50 mm (values of 50 cm have also been mentioned); the higher range values are most commonly observed in fractures which have undergone alteration (usually early hydrothermal) and the lower range values from relatively "fresh" fractures.

Unfortunately few field data exist at the moment for deriving diffusivity coefficients; this is often due to practical considerations such as sample size restrictions etc. which inhibit the fine measurement of physical parameters. Moreover, considering those data available, a confusingly large range of values for granites have been reported (10^{-19} to 10^{-10} m^2/s). At the moment, therefore, emphasis must still put on laboratory-derived values; it should be pointed out that such data should be treated with caution (as experimental artefacts tend to produce overestimations) and should be considered as non-conservative.

In general the basic theory of matrix diffusion has been clearly verified from natural analogue studies, at least for fractured crystalline rocks, with connected porosities extending well into the rock matrix. Furthermore, the combination of observed penetration depths of uranium and estimated time-scales are in good agreement with predictions with a simple diffusion model using apparent diffusivities based on laboratory measurements. However, more quantitative field studies are required to support this statement.

The role of colloids

A major interest in performance assessment is the potential far-field transport of trace elements distributed between groundwater, the fracture surfaces and the colloid or particulate content in the groundwater. To date no model exists to quantitatively accomodate these processes. The role of colloids is usually regarded as unimportant, colloids being considered immobile, produced and present in too small concentrations, and that interaction with elements is restricted to rapid and reversible sorption mechanisms. In principle, colloids can cause dissolution of radionuclides above their solubility limits and prevent sorption on rock mineral surfaces. If this occurs to any notable extent, then the colloids would act as a "short circuit" to the important barriers of solubility and sorption.

Colloids have been studied from several of the analogue sites; Alligator Rivers (Duerden et al., 1994), Poços de Caldas (Morro do Ferro; Miekeley et al., 1992), Cigar Lake (Vilks et al., 1991), Broubster (Longworth et al., 1989) etc. In most cases the separated colloid size fractions have been chemically analysed (major and trace element ions, uranium decay series and REEs) and physically classified (size, shape, mineralogical character etc.). In general, the total colloid concentrations were found to be quite low, for example at Morro do Ferro they were <1 mg/L, and that colloid transport of radionuclides was found to be relatively unimportant. Uranium migrates mostly as dissolved species, whereas thorium (+ actinium and REEs) is mostly associated with larger, relatively immobile particles (>1.0 μm) such as amorphous Fe-oxyhydroxides and clays which reflect the local fracture mineralogy. At Morro do Ferro the uptake of three- and tetravalent elements by the colloidal particles was stronger than anticipated. There is some evidence from Broubster that fulvic acid can contribute to keep hexavalent uranium in solution thereby promoting transport.

The natural analogue observations generally support the assumption that sorption of uranium, thorium and lathanides onto naturally occurring colloids in the groundwater can be either reversible, or, colloidal particles are immobile. It is also likely, from the observations, that the colloids are efficiently filtered by the geological medium at the studied sites (e.g. Morro do Ferro; Broubster); there is no major evidence of massive transport of colloid-bound radionuclides or other trace elements.

Unfortunately, the application of analogue-derived data to performance assessment is very sparse due to the complexities of sampling, separation and sensitivity to groundwater chemistry (e.g. mixing). Many of these problems remain to be resolved.

4.5 The Role of Microbes

It is an established fact that microbes have been identified in deep groundwaters down to 1 km depth (e.g. Pedersen, 1989), and that they play an important role in the evolution of groundwater chemistry (e.g. by influencing iron oxidation and sulphate reduction), although their efficiency very much depends on the supply of nutrients and the absence of extreme conditions such as high temperature, high pH and radiation effects etc. The repercussion of microbial activity on repository performance assessment (both high- and low-level wastes) centres around gas generation, canister corrosion, redox reactions, formation of complexing agents and radionuclide groundwater transport.

It should be pointed out, however, that this work is still at its infancy; reliable microbial sampling techniques from deep groundwater sources and associated rocks still need to be improved. This is not only necessary to characterise the "natural" microbial content of deep groundwaters, but knowing this will also provide a reference microbial "population" to assess the amount and type of contamination expected to occur during repository construction and sealing.

Although the role of microbial activity in producing gas, generating organic acids that form strong complexes with many metal ions, and retaining dissolved radionuclides in groundwater is of importance in repository performance assessments, little analogue data exist that illustrates these mechanisms. Most analogue evidence centres around canister corrosion and redox reactions.

Corrosion

The presence of aqueous sulphide is a potential threat to the integrity of the metal materials (mostly steel and copper) used in repository construction. Bacterial influences (mainly sulphate-reducing forms) on canister corrosion have recently received widespread attention in connection with low- to high-level radioactive waste repositories. As an example, sulphate-reducing bacteria have been associated with the corrosion of steel oil-well casings (McKinley et al., 1985). It is generally expected that aerobic corrosion will dominate until air trapped in the sealed repository has been consumed, whereupon anaerobic corrosion will take over. Limiting factors to microbial acitivity are the availability of nutrients and energy; specifically for sulphide-reducing bacteria, the major limiting factor is the supply of electron donors from sources such as organic material and hydrogen and methane gas.

With regard to pitting corrosion, analogous occurrences exist in nature of microbial activity which serve to illustrate the localised nature of these phenomena. For example, reduction spheres (cm to dm in size and often containing sulphide ore phases) occur in an otherwise oxidised and hematite-rich host rock. It has been conjectured (Hofmann, 1990) that these spheres may have resulted from sulphide-reducing bacteria, although an external source of reductants would also have been necessary. Of particular concern to disposal concepts therefore, is the localised nature of the activity (i.e. pitting), the production of sulphides and the unknown external source of reductants. On the brighter side, trace elements, including uranium, are retained within the spheres and, moreover, the spheres take a long time to develop.

Microbial activity and oxic corrosion of canister and reinforcement materials (from residual trapped oxygen) have also been addressed in performance assessment analyses. Pitting, observed in domestic copper water pipes, has been attributed to microbial activity (Bremer and Geesey, 1991); furthermore, this has been reproduced under laboratory conditions. Contrastingly, no major microbial activity has been associated with the well preserved bronze Swedish canon from the 17th century warship, Kronan. Even though residing in brackish waters, albeit partially submerged in clay sediments, the rate of corrosion was estimated to be less than a cm during a 100 000 year period. However, microbial activity may explain the reduction of surface tenorite (CuO) corrosion products to copper sulphides.

Redox reactions

Bacterial reactions, in addition to mediating metal corrosive effects, also serve to lower the solubility of some radionuclide elements by decreasing the redox potential and generating sulphide; in the absence of bacteria, sulphate reduction would not occur at ambient temperatures. Bacteria may also play an important role in catalysing reactions such as the oxidation of Fe(II) and sulphide minerals and the reduction of U(VI). For example, oxidation of iron and pryrite contained in bentonite results in a reduced system prohibiting further oxidation. However, these reactions may not require the presence of microbes. This is preferable from a performance assessment viewpoint as it increases reliability, predictability and therefore confidence in repository behaviour. Reliance on bacterial action in bentonite would, at best, be dubious.

Some evidence can be gleaned from the Poços de Caldas redox front analogue studies (West et al., 1992) which shows that microbial activity is not always necessary for pyrite oxidation. In this study the oxidation of pyrite was shown to be remarkably fast when compared with diffusion. Moreover, sulphur-oxidising bacteria were not found at the redox front. This is not proof, but a rather strong indication that pyrite oxidation at the redox front will proceed as a series of pure inorganic reactions when similar circumstances are maintained.

Reduction of dissolved U(VI) at a redox front is another interesting reaction from the performance assessment point of view. This reaction is predicted to be sluggish and, as shown by the Poços de Caldas studies, U(IV) is not precipitated immediately but a few centimetres ahead of the redox front on the reducing side. Uranium is precipitated as pitchblende nodules, often together with secondary pyrite. The role of microbes in these reactions was also investigated at Poços de Caldas but no clear answer was found.

4.6 Construction Materials

4.6.1 High-level waste

The near-field isolation system (see Figs 1 and 2), which constitutes the major component for safety performance assessment, essentially comprises the engineered barriers which are in contact with, and close to, the high-level radioactive waste. The waste itself, whether in the form of UO_2 spent fuel rods or matrixed as a borosilicate glass, is also considered an integral part of the barrier system. The system consists of a canister or package containing the waste, a bentonite clay packed around a canister, and a sand/bentonite backfill used to seal the access shafts and galleries when disposal is complete. The functions of the barriers are to: a) isolate the radioactive waste from contact with the groundwater, b) contain the waste until the level of radioactivity has dropped to acceptable levels, c) in the event of canister corrosion and radionuclide dissolution into the groundwater, ensure that the transport out of the near-field is limited, and d) the retardation in the bedrock mass will ultimately hinder transport to the biosphere.

The materials comprising the engineered barrier system have important physical properties, the suitability of which must be demonstrated over the very long time-scales (hundreds to thousands of years) required to assess the safety performance of a repository. Various models are used to extrapolate the laboratory characterisation of barrier behaviour. However, to test or validate these models, which relate to short-term material degradation over hundreds to thousands of years, is beyond conventional engineering experience. Validation, even though being only partial, is therefore sought by the natural analogue approach.

Waste Form

The features of the waste form which are of interest in performance assessment essentially relate to the variability of the physico-chemical properties of the waste with time, i.e. chemical, thermal, mechanical and radioactive properties. Chemical alteration of the matrix will lead to the release of radionuclides; this may occur by the dissolution of the UO_2-matrix or by oxidation of UO_2 to form higher uranium oxides with contrasting crystalline structures. Elements bedded in the original matrix can thus become released and come in contact with the aqueous phase. Mechanical fracturing of the waste can increase the surface area available for alteration, and radioactive effects can lead to radiation-induced dissolution transformation to other solid phases by corrosion reactions.

Spent nuclear fuel is the preferred waste form for several high-level radioactive waste programmes. The major waste component in spent fuel is UO_2 (>95%), together with matrix impurities such as nuclear fission products, actinides and actinide daughters. Naturally occurring uraninite or the less pure variety of pitchblende are commonly considered analogous to spent fuel, containing in some cases up to 90% UO_2, but important dissimilarities do occur. There are also important differences in morphology between UO_2 minerals and spent fuel. Despite these differences, however, the study of natural uraninite ore and its surroundings provides important clues to the predicted long-term behaviour of spent fuel in a repository environment.

Natural analogues have been used to address the stability of uraninite/pitchblende under normal geological and repository conditions (e.g. Cigar Lake; Cramer and Smellie [Eds.], 1994). The main conclusion reached is that as long as conditions remain reducing, the uranium is practically immobile; measured concentrations in groundwater are around 1 ppb, similar to UO_2 solubilities measured in the laboratory.

In the context of spent fuel stability, much attention has been given to the Oklo natural reactors which underwent criticality some 2.0 Ga ago resulting in the concentration of nuclear fission products, actinides and actinide daughters associated intimately with uraninite which crystallised from pitchblende during fission. The general stability of the uraninite since criticality to the present time has been demonstrated (see Brookins, 1990 and Jakubick, 1986). However, important differences do exist when the uraninite from the old reactor zones is compared to modern spent fuel: a) uraninite from the Oklo reactors contains lower concentrations of fission products, b) maximum temperatures reached in a modern reactor far exceed those experienced by the uraninite in the zones of criticality at Oklo; i.e. less fractionation in uraninite, and c) Oklo uraninite is better preserved than spent fuel because of a less complete nuclear reaction, thus rendering the reaction products less susceptible to mobilisation.

Canister

The function of the canister is to isolate the spent fuel from groundwater contact for designated time periods ranging from thousands to hundreds to thousands of years depending on waste composition (reprocessing waste or spent fuel) and design goal. The life span of the copper canister (e.g. as considered in the Swedish concept) depends on metal purity and the quality of the engineering (e.g. canister sealing), together with the composition of the groundwater (e.g. sulphide content) which it will ultimately come in contact with. Corrosion rates (0.025-1.3 μm/a) calculated from both historical and archaeological copper artefacts found in a wide range of wet/dry and oxic/anoxic environments demonstrates the long-term durability of copper, consistent with thermodynamic considerations (Chapman et al., 1984). However, owing to the predicted low supply of corrodents expected to reach the canister (because of reducing repository conditions), only local corrosion attack (i.e. "pitting") is considered important. This would be the only way for a limited amount of corrodents to breach the canister. Safety assessment calculations for the KBS-3 concept involved a pitting factor of 5 (as observed from archaeological materials) and an unfavourable case of 25 (as observed from the exposure of copper in soil). All available evidence from natural occurrences and archaeological artefacts showed that a pitting factor of 5 was more realistic, concluding that the proposed copper canisters should exceed their designed lifetime of 10^5 a by at least one order of magnitude.

Contrastingly, natural analogues for steel (e.g. as considered in the Swiss concept), which is a relatively recent technological alloy, are mostly limited to geological or archaeological artefacts (Hellmuth, 1991). Iron meteorites and archaeological artefacts comprising iron or iron alloys have been categorised (Johnson and Francis, 1980), native iron occurrences have been studied at Disko Island, Greenland (Bird and Weathers, 1977), and archaeological artefacts such as a horde of Roman iron nails discovered at Inchtuthil, Scotland, are described (Pitts and St. Joseph, 1985). Although no useful analogue information on localised corrosion (i.e. pitting) of iron or steel is available, the most appropriate data indicate a total corrosion time to be anything from 10 000 to 250 000 a (Nagra, 1985).

In summary, there would appear to be adequate archaeological and geological analogue evidence to support the stability of copper over the large time-scales necessary for reliable performance assessment predictions, but is it enough? More examples to support the steel canister concept are desireable. It is recommended that the quest for geological and archeological metal analogues should continue to locate examples (such as the Kronan canon; Hallberg et al., 1988) within specific hydrogeological and hydrogeochemical environments where the boundary conditions can be precisely defined.

Bentonite Buffer

The function of the buffer material is to: a) constitute a mechanical and chemical zone of protection around the canister, b) limit the inward migration of potentially corrosive substances from the groundwater to the canister surface, c) filter fine particulate and colloidal material that may form during canister reactions, and d) in the event of canister corrosion, to limit the dispersion of leached radionuclides from the canister out into the bedrock.

One of the major problems regarding the use of bentonite is the conversion of smectite clay (which includes Na-bentonite) to illite clay during increases in temperature, pressure and water circulation. Analogue studies of bentonite stability have therefore involved long-term naturally occurring scenarios of clay alteration in different geological and thermal environments (see Miller et al., 1994 and references therein). For example, bentonite stability under laboratory conditions, within the range of 100-130°C (i.e. the maximum temperature range expected to result from thermal loading during spent fuel disposal), is not well established. Natural analogue studies of naturally occurring bentonite showed that even having experienced temperatures of 110-120°C over a period of 10 Ma, there was no evidence of any adverse effects on the buffer properties. At higher contact metamorphic temperatures (105-240°C) the results indicated that even though there was substantial illitisation at temperatures up to 110°C (maintained for several hundreds of years), there was no obvious loss in plasticity or expansion properties on hydration. These analogue observations thus play an important role when stating that the thermal pulse predicted in the deposition hole will not result in any deterioration of the clay barrier.

A possible disadvantage with bentonite is its load-bearing capacity, i.e. the possibility of the canister sinking with time, reducing the bentonite insulation thickness between the canister bottom and the underlying bedrock. A further problem is the generally low thermal conductivity in contrast with the surrounding bedrock; even the bentonite/sand backfill material has higher thermal conductivity. Both of these aspects have not yet been fully addressed by suitable analogue studies.

In the context of hydraulic and elemental retardation properties, it is worth mentioning observations from the Cigar Lake uranium deposit and the Oklo uranium reactor occurrence, both of which are surrounded by an envelope of clay, mostly illite in composition. These clay layers have demonstrated their effectiveness, both as a hydraulic barrier against water circulation (Cigar Lake and possibly also Oklo) and a barrier to fission product migration (Oklo). Other examples of hydraulic integrity include the clay preserved tree forest at Dunarobba, Italy (Ambrossetti et al., 1990).

In summary, analogue studies of bentonite-type clays, representing a wide range of physical and chemical properties, show a consistent picture of alteration rates below those assumed for the safety analysis. At ambient repository temperatures (less than 80°C), they indeed suggest that the final maximum alteration will be around 30%, and hence the assumption in the safety analysis that the bentonite properties will remain constant for at least 1 Ma, seems to be well justified.

4.6.2 Low- to Intermediate-level Radioactive Waste

These wastes are packaged in steel drums and in concrete or steel containers. In some of these wastes, for example, ion exchange resins from the primary circuits, are solidified in a cement or bitumen matrix. Following final sealing and saturation of the repository, the near-field groundwater chemistry is predicted to become very alkaline (due to concrete) and reducing. In this environment, the stability of concrete/steel constructions, the infuence of alkaline solutions on the near-field rock and backfill barriers (i.e. bentonite clay), the durability of the cement and bitumen matrices solidifying the wastes, and the nature of the breakdown of organic products that may possibly enhance the solubility and subsequent transport of radionuclides, are therefore critical factors to the performance safety assessment of such repositories.

Concretes and Cements

The long-term integrity of repository construction materials such as cement, steel and bitumen in a highly alkaline groundwater environment is largely surmised only from laboratory-based modelled simulations. In the short-term, two analogue approaches have been used to study the durability of concretes and cements: a) archaeological building materials (hundreds to thousands of years), and b) industrial building material (tens to hundreds of years). The major problem with the archaeological material analogy is that modern concretes are composed of portland cement (mainly calcium silicates with little free lime) with chemical and physical properties much more extensive than ancient lime-based cements. However, some of the ancient cements also contain CSH (Calcium Silicate Hydrate) compounds, formed due to specific conditions of the cement preparation, which are the main hydration products of modern-day portland cement. These CSH compounds have consequently served to stabilise and preserve these ancient cements for up to 2000 years. It follows that with to-day's technology, CSH compound-bearing concretes should exhibit a durability even better than indicated from archaeological materials.

The stability of cement binders has been demonstrated by studies from Hadrian's Wall in northern England and from the Gallo-Roman baths in southwestern France. These studies confirm that the long-term carbonation of hydrosilicates is the dominant alteration mechanism of the cement binders (see Petit, 1990). Furthermore, in more modern times (the last 100-150 years or so), industrial portland cements have shown that the CSH compounds have not exhibited any form of degradation, although hydration of the cement is incomplete.

Other anthropogenic analogues presently being studied include the cement lining of a water tunnel (cast in 1914) in northern Sweden (Grudemo, 1982). Another Swedish example is a steel tank for drinking water dating to 1906 which has an inner lining of concrete to protect the steel from corrosion. Inside a carbonation front of about 5mm the mortar is remarkably well preserved even though it has been in contact with soft water for almost 90 years; there is no clear sign of leaching of portlandite and the steel has been well protected (Lagerblad and Trägårdh, 1994).

Clay (used as backfill material) in contact with concrete is of importance for repository performance. Concrete pore water, particularly the hyroxide, calcium and potassium, sulphate and carbonate ions, is expected to have an influence on the bentonite clay constituents. Contrastingly, the sulphate and carbonate ions from the clay are expected to influence the concrete causing ettringite formation and carbonatisation. In this respect old construction concretes embedded in clay have been studied but little change has been noted (Anderson and Fontain, 1981).

Natural analogue studies which can be directly applicable to a low- to intermediate-level cementitious repository are presently being conducted in the Maqarin region of N.W. Jordan. The chemistry of the source rock is defined by the spontaneous combustion of hydrocarbon-rich marls, followed by low temperature hydration of the high temperature mineral assemblage, eventually forming cement minerals such as portlandite, ettringite and thaumasite. Thus, in addition to the high pH groundwater environment (by interaction of normal groundwaters with the cement minerals), the likely sequence of source leachates (NaOH and KOH followed by $Ca(OH)_2$) is precisely that predicted by various models for cement degradation in groundwater.

The main disadvantage of the Maqarin site is that so far only the reactions occurring under oxidising conditions have been studied. However, this is not expected to influence the typical cement reactions.

Bitumen

In low- to intermediate-level waste programmes bitumen is often used as an alternative to cement as a matrix material to solidify the wastes. Both geological and archaeological occurrences of bitumen and bitumen-related hydrocarbons have been used as analogues to ascertain its stability and preservation properties in varying geochemical environments (see Hellmuth, 1989 and Miller et al., 1994). However, very little information is available for relevant high pH conditions (McKinley and Alexander, 1992).

Long-term, radioactive waste isolation properties of bitumen-related hydrocarbons under assumed reducing conditions is presently being studied at Oklo (Nagy et al., 1991). Initial results suggest that the Oklo reactor uraninites, which incorporated fission products during criticality, where held immobile within resolidified hydrocarbons until geologic events at 1 Ga following criticality resulted in some mobilisation of U and Pb. Even so, the uraninite encased in solid graphitic matter in the organic-rich reactor zones lost virtually no fissiogenic lanthanide isotopes during this later event.

Short-term preservation properties of bitumen can be observed from archaeological artefacts; even in Babylonian times (1300 B.C.) it had been long used to preserve organic material (e.g. wooden boats, roofing, baskets etc.). In almost all cases were artefacts have been found coated in bitumen, they have been well preserved except when mechanical damage to the bitumen has occurred, due to a hardening of the bitumen coatings with time (Hellmuth, 1989).

In general, natural and archaeological analogues studies presently support the stability, preservation and immobilising properties of bitumen and related hydrocarbons under the temperature and redox conditions predicted to prevail in a repository situation. However, the absence of high pH groundwater conditions associated with many of the examples documented is a point of concern. The project at Maqarin, Jordan (Khoury et al., 1992; Alexander et al., 1992), is presently trying to remedy this situation by studying the interaction of high pH groundwaters with a Bituminous Marl Formation. The hydrocarbons are rich in many trace elements, including uranium, considered important to repository safety assessments. Initial results indicate that whilst certain trace elements may be preferentially mobilised, the hydrocarbon matrix is relatively resistant. However fossil reactions, similar to those presently occurring at Maqarin, and some hundreds of thousands years old, exist in Central Jordan. The disturbing fact is that no hydrocarbon material apparently remains; more studies are required in this region.

The major disadvantages in the study of natural analogues to understand the behaviour of bitumen are: a) the general absence of bitumen; many of the hydrocarbons studied are a form of kerogen, b) the absence of interacting high pH solutions (exception of Maqarin), and c) little information exists on the bitumen stability in the presence of saline water.

5. Natural Analogues and Public Awareness

Natural analogue studies are playing an increasing major role in heightening public awareness and understanding in the disposal of radioactive waste. Analogue studies, such as Cigar Lake, Poços de Caldas, Oklo etc, are being widely used to convey the principles of waste disposal and illustrate the safety of such concepts by a "seeing is believing" approach. This is proving to be a very effective means of informing, not only the layman, but also much of the scientific fraternity, about the long-term safety of radioactive waste disposal. To these ends, numerous brochures have been produced for domestic distribution, and recently international co-operation has produced an analogue video (entitled "Traces of the Future") for world-wide public consumption.

6. Conclusions

Natural analogues are widely used to demonstrate the long-term safety of a repository in addition to increasing the credibility of the performance assessment as such. This demonstration is not only geared towards the informed scientific community, but also to the general public. The former tends to search for confirmation of the occurrence and magnitude of a process or processes anticipated to occur in various performance assessment scenarios. Communication with the general public is more geared to showing that the processes occurring in the repository can be easily understood in natural systems over long timescales, are not unique to radioactive waste disposal; this serves to strengthen the predictability of repository performance into the future.

The value of natural analogues to the scientific community is at least two-fold: 1) to identify and confirm processes occurring in a repository system, and 2) testing of predictive models describing those processes. In this present review the indentification of processes has been illustrated by numerous examples including phenomena relating to the stability of construstion materials. In at least one of the cases, co-precipitation of radionuclides with ferric oxyhydroxides, the magnitude of the phenomenon observed in the analogue study was surprisingly large. This led to the conclusion that the resistance against leakage of radionuclides from breached canisters may have been underestimated in previous performance assessments. Therefore, there may be reason to develope and test models for this phenomenon.

Model testing using data from natural analogues has generally been more successful for equilibrium models than for dynamic models of long-term radionuclide migration. One reason for this is that the equilibrium models only need present-daya data whereas the dynamic models require data on initial conditions and material properties. The current understanding of these historical data are usually interpretations of present-day observations using models to be tested. The situation is thus one of "circular evidence".

Presently there exists a considerable volume of natural analogue data but unfortunately a full appraisal of these data with respect to repository performance tends to be frustratingly inadequate. Some applications have been described in this review, but additional time and resources are required to re-evalute some of the existing analogue data in an attempt to derive full benefit for performance assessment. It is recommended that this situation be addressed in the near future. Further recommendations include the pooling of international resources to continue with large-scale analogue studies which reflect the complexity of natural systems which ultimately will host the repository. Furthermore, such studies require a multidisciplinary approach which can serve as a "dry run" for repository site-specific investigations. Other suggestions centre around closer integration of laboratory studies with process-specific analogues (e.g. comparing laboratory and field radionuclide diffusion profiles in similar media) and more thought for *in situ* experiments, i.e. to influence rather than observe (e.g. downhole leaching of uranium from an orebody under controlled conditions).

7. References

Alexander, W.R., MacKenzie, A.B., Scott, R.D. and McKinley, I.G., 1990. Natural analogue studies in crystalline rock: The influence of water-bearing fractures on radionuclide immobilisation in a granitic rock repository. Nagra Tech. Rep. (NTB 87-08), Baden.

Alexander, W.R., Dayal, R., Eagleson, K., Hamilton, E., Linklater, C.M., McKinley, I.G. and Tweed, C.J., 1992. A natural analogue of high pH cement pore waters from the Maqarin area of northern Jordan. II: Results of predictive geochemical calculations. J. Geochem. Explor., 46, 133-146.

Alexander, W.R. and McKinley, I.G., 1992. A review of the application of natural analogues in performance assessment: Improving models of radionuclide transport in groundwaters. J. Geochem. Explor., 46, 83-115.

Ambrossetti, P., Bascilici, G., Gentili, S., Biondi, E., Cerquaglia, Z. and Girotti, O., 1990. La Foresta Fossile di Dunarobba. Ediart (Publ.), Todi, Italy.

Anderson, D.M. and Fontain, J., 1981. Investigation of the chemical stability of clays employed as buffer materials in the storage of nuclear waste materials. SKBF/KBS Prog. Rep. (AR 82-05), Stockholm. (Available from SKB).

Astudillo, J., 1994. Characterisation and validation of natural radionuclide migration processes under real conditions in a fissured granitic environment. 5th CEC/NAWG Meeting and Alligator Rivers Analogue Project (ARAP) Final Workshop. In: Proceedings of an International Workshop, Toledo, October 5-9th, 1992. EUR 15176, Brussels.

Bath, A. H., Christofi, N., Philip, J.C., Cave, M.R., McKinley, I.G. and Berner, V., 1987. Trace element and microbiological studies of alkaline groundwaters in Oman, Arabian Gulf. A natural analogue for cement pore waters. Nagra Tech. Rep., (NTB 87-16), Wettingen.

Bird, J. and Weathers, M., 1977. Native iron occurrences of Disko Island, Greenland. J. Geol., 85, 359-371.

Blomqvist, R., 1994. Natural analogue studies in Finland. Proceedings of the 6th CEC-NAWG Workshop Meeting, Sante Fe, USA (Sept. 12-16), 1994 (In press).

Bremer, Ph.J. and Geesey, G.G., 1991. Laboratory based model of microbiologically induced corrosion of copper. Appl. Environ. Microbiol., 57, 1956-1962.

Brookins, D.G., 1990. Radionuclide behaviour at the Oklo nuclear reactor, Gabon. Waste Manag., 10, 285-296.

Brandberg, F., Grundfelt, B., Höglund, L-O., Karlsson, F., Skagius, K. and Smellie, J., 1993. Studies of natural analogues and geologic systems - Their importance to performance assessment. SKB Tech. Rep. (TR 93-05), Stockholm.

Bruno, J., Cross, J.E., Eikenberg, J., McKinley, I.G., Read, D., Sandino, A. and Sellin, P., 1992. Testing models of trace element geochemistry at Poços de Caldas. J. Geochem. Explor., 45, 451-470.

Casas, I. and Bruno, J., 1994. Testing of solubility and speciation codes. In: J.J. Cramer and J.A.T. Smellie (Eds.), Final Report of the AECL/SKB Cigar Lake Analog Study. AECL Tech. Rep. (AECL-10851), Pinawa, and SKB Tech. Rep. (TR 94-04), Stockholm.

Chapman, N.A., McKinley, I.G. and Smellie, J.A.T., 1984. The potential of natural analogues in assessing systems for deep disposal of high-level radioactive waste. SKB Tech. Rep. (TR 84-16), Stockholm and Nagra Tech. Rep. NTB 84-41), Baden.

Chapman, N.A., McKinley, I.G., Penna Franca, E., Shea, M.E. and Smellie, J.A.T., 1992. The Poços de Caldas Project: An introduction and summary of its implications for radioactive waste disposal. J. Geochem. Explor., 45, 1-24.

Cramer, J.J. and Smellie, J.A.T., 1994. The AECL/SKB Cigar Lake Analog Study: Some implications for performance assessment. 5th CEC/NAWG Meeting and Alligator Rivers Analogue Project (ARAP) Final Workshop. In: Proceedings of an International Workshop, Toledo, October 5-9th, 1992. EUR 15176, Brussels.

Cramer, J.J. and Smellie, J.A.T. (Eds.), 1994. Final Report of the AECL/SKB Cigar Lake Analog Study. AECL Tech. Rep. (AECL-10851), Pinawa, and SKB Tech. Rep. (TR 94-04), Stockholm.

Curtis, D.B. and Gancarz, A.J., 1983. Radiolysis in nature: Evidence from the Oklo natural reactors. SKBF/KBS Tech. Rep. (TR 83-10), Stockholm.

Duerden, P., Lever, D., Sverjensky, D.A. and Townley, L.R., 1994. Summary of findings. Alligator Rivers Analogue Project Final Rep. Vol. 1. ISBN Tech. Rep. (0-642-59941-6), DOE/HMIP Tech. Rep. (RR/92/072), SKI Tech. Rep. (TR 92:20-2).

Golian, C. and Lever, D. (Eds.), 1993. Alligator Rivers Analogue Project - Final Report, vol. 14, Radionuclide Transport: OECD/NEA, Paris (In press).

Grenthe, I., Stumm, W., Laaksoharju, M., Nilsson, A-C. and Wikberg, P., 1992. Redox potentials and redox reactions in deep groundwater systems. Chem. Geol., 98, 131-150.

Grudemo, Å., 1982. X-ray diffractometric investigation of the state of crystallization in the cement matrix of a very old concrete. Nordic Concrete Res., 1, 1-22.

Hallberg, R.O., Östlund, P. and Wadsten., T., 1988. Inferences from a corrosion study of a bronze canon, applied to high-level nuclear waste disposal. Appl. Geochem., 3, 273-280.

Hellmuth, K-H., 1989. Natural analogues of bitumen and bituminised waste. Finn. Cent. Rad. Nucl. Safety, Tech. Rep. (STUK-B-VALO 58), Helsinki.

Hellmuth, K-H., 1991. The existence of native iron - implications for nuclear waste management. Part I: Evidence from existing knowledge. Finn. Cent. Rad. Nucl. Safety, Tech. Rep. (STUK-B-VALO 67), Helsinki.

Hofmann, B.A., 1990. Reduction spheres in hematitic rocks from northern Switzerland. Implications for the mobility of some rare elements. Nagra Tech. Rep. (NTB 89-17), Helsinki.

Hooker, P.J., MacKenzie, A.B., Scott, R.D., Ridgway, I., McKinley, I.G. and West, J.M., 1985. A study of natural and long term (10^3 - 10^4 years) elemental migration in saturated clays and sediments. part III. BGS Tech. Rep. (FLPU 85-9); CEC Waste Management Series (EUR 10788/2), CEC, Luxembourg.

Jakubick, A.T., 1986. Oklo natural reactors: Geological and geochemical considerations - A review. At. Ener. Control. Board Res. Rep. (INFO-0179), Ontario.

Johnson, A.B. and Francis, B., 1980. Durability of metals from archaeological objects, metal meteorites and native metals. Battelle Pacific Northwest Lab. (PNL-3198).

Karlsson, F., Smellie, J.A.T. and Höglund, L-O., 1994. The application of natural analogues to the Swedish SKB-91 safety performance assessment. 5th CEC/NAWG Meeting and Alligator Rivers Analogue Project (ARAP) Final Workshop. In: Proceedings of an International Workshop, Toledo, October 5-9th, 1992. EUR 15176, Brussels.

KBS-1, 1978. Handling of nuclear fuel and final storage of vitrified high-level reprocessing waste. KBS Final Rep., Vols. I-V, Stockholm.

KBS-2, 1978. Handling and final storage of unreprocessed spent nuclear fuel. KBS Final Rep., Vols. I-II, Stockholm.

KBS-3, 1983. Final storage of spent nuclear fuel. SKBF/KBS Final Rep., Vols. I-IV, Stockholm.

Khoury, H.N., Salameh, E., Clark, I.D., Fritz, P., Bajjali, W., Milodowski, A.E., Cave, M.R. and Alexander, W.R. 1992. A natural analogue of high pH cement pore waters from the Maqarin area of northern Jordan. I: Introduction to the site. J. Geochem. Explor., 46, 117-132.

Lagerblad, B. and Trägårdh, J., 1994. Conceptual model for concrete long time degradation in a deep site nuclear waste repository. Swedish Cement and Concrete Res. Inst. Rep. (No. 94022), Stockholm. (In press).

Liu, J., Neretnieks, I. and Yu, J-W., 1994. Mass transport modelling. In: J.J. Cramer and J.A.T. Smellie (Eds.), Final Report of the AECL/SKB Cigar Lake Analog Study. AECL Tech. Rep. (AECL-10851), Pinawa, and SKB Tech. Rep. (TR 94-04), Stockholm.

Longworth, G., Ivanovich, M. and Wilkins, M.A., 1989. Uranium series disequilibrium studies at the Broubster analogue site. UK DOE Tech. Rep., (DOE/RW/89.100), London.

MacKenzie, A.G., Scott, R.D., Linsalata, P. and Miekeley, N., 1992. Natural decay series studies of the redox front system in the Poços de Caldas uranium mineralisation. J. Geochem. Explor., 45, 289-322.

McKinley, I.G., West, J.M. and Grogan, H.A., 1985. An analytical overview of the consequences of microbial activity in a Swiss HLW repository. Nagra Tech. Rep., (NTB 85-43), Wettingen.

McKinley, I.G. and Alexander, W.R., 1992. A review of the use of natural analogues to test performance assessment models of a cementitious near-field. Waste Manag., 12, 253-259.

McKinley, I.G., Alexander, W.R., McCombie, C. and Zuidema, P., 1992. Application of results from the Poços de Caldas project in the Kristallin-I HLW performance assessment. Int. Con. High-level Waste Mang., (April 12-16, 1992), Las Vegas.

McKinley, I.G. and Alexander, W.R., 1994. The uses of natural analogue input in repository performance assessment: an overview. Proceedings of the 6th CEC-NAWG Workshop Meeting, Sante Fe, USA (Sept. 12-16), 1994 (In press).

Menet, K., Ménager, M-T. and Petit, J-C., 1992. Migration of radioelements around the new nuclear reactors at Oklo: Analogies with a high-level waste repository. Radiochim. Acta, 58/59, 395-400.

Miekeley, N., Coutinho de Jesus, H., Porto da Silveira, C.L. and Degueldre, C., 1992. Chemical and physical characterization of suspended particles and colloids in waters from the Osamu Utsumi mine and Morro do Ferro analogue study sites, Poços de Caldas, Brazil. J. Geochem. Explor., 45, 409-437.

Miller, W.M., Alexander, W.R., Chapman, N.A., McKinley, I.G. and Smellie, J.A.T., 1994. Natural analogue studies in the geological disposal of radioactive wastes. Nagra Tech. Rep. (NTB 93-03), Wettingen.

Miller, W.M., 1994. The value of natural analogues. Proceedings of the 6th CEC-NAWG Workshop Meeting, Sante Fe, USA (Sept. 12-16), 1994 (In press).

Murphy, W.M. and Pearcy, E.C., 1994. Performance assessment significance of natural analog studies at Peña Blanca, Mexico, and at Santorini, Greece. 5th CEC/NAWG Meeting and Alligator Rivers Analogue Project (ARAP) Final Workshop. In: Proceedings of an International Workshop, Toledo, October 5-9th, 1992. EUR 15176, Brussels.

Nagra, 1985. Projekt Gewähr, 1985. Nuclear waste management in Switzerland - Feasibility studies and safety analysis - Summary. Nagra Tech. Rep. (NGB 85-09), Baden.

Nagra, 1994. Kristallin-I, 1994. Safety Assessment Report. Nagra Tech. Rep. (NTB 93-22), Wettingen.

Nagy, B., Gauthier-Lafaye, F., Holliger, P., Davis, D.W., Mossman, D.J., Leventhal, J.S., Rigali, M.J. and Parnell, J., 1991. Organic matter and containment of uranium and fissiogenic isotopes at the Oklo natural reactors. Nature, 354, 472-475.

Pate, S.M., McKinley, I.G. and Alexander, W.R., 1992. Use of natural analogue test cases to evaluate a new performance assessment TDB. 5th CEC/NAWG Meeting and Alligator Rivers Analogue Project (ARAP) Final Workshop. In: Proceedings of an International Workshop, Toledo, October 5-9th, 1992. EUR 15176, Brussels.

Pearcy, E.C. and Murphy, W.M., 1991. Geochemical natural analogs: Literature review. Center for Nuclear Waste Regulatory Analyses, San Antonio, Texas. Int. Rep. CNWRA 90-008.

Pedersen, K., 1989. Deep groundwater microbiology in Swedish granitic rock and its relevance for radionuclide migration from a Swedish high-level nuclear waste repository. SKB Tech. Rep. (TR 89-23), Stockholm.

Petit, J-C., 1990. Design and performance assessment of radioactive waste forms: What can we learn from natural analogues? Proceedings of the 4th CEC-NAWG Meeting and Poços de Caldas Project Final Workshop, Pitlochry, Scotland. Tech. Rep. (EUR 13014), Brussels.

Pitts, L. and St. Joseph, A., 1985. Inchtuthil Roman legionary fortress excavation, 1952-1965. Allan Sutton.

Pusch, R., 1983. Stability of deep-seated smectite minerals in crystalline rock: Chemical aspects. SKBF/KBS Tech. Rep. (TR 83-16), Stockholm.

Pusch, R. and Karnland, O., 1988. Geological evidence of smectite longevity: The Sardinian and Gotland cases. SKB Tech. Rep. (TR 88-26), Stockholm.

Read, D., Bennett, D., Hooker, P.J., Ivanovich, M., Longworth, G., Milodowski, A.E. and Noy, D.J., 1991. The migration of uranium into peat rich soils at Broubster, Caithness, Scotland. In: 3rd International Conference on Chemistry and Migration Behaviour of Actinide and Fission Products in the Geosphere (Migration '91), Jerez de la Frontera, Spain, Oct. 1991.

Romero, L., Neretnieks, I. and Moreno, L., 1992. Movement of the redox front at the Osamu Utsumi uranium mine, Poços de Caldas, Brazil. J. Geochem. Explor., 45, 471-502.

SKB-91, 1992. Final disposal of spent nuclear fuel. Importance of the bedrock for safety. SKB Tech. Rep., (TR 92-20), Stockholm.

Smellie, J.A.T. and Rosholt, J.N., 1984. Radioactive disequilibrium in mineralised fracture samples from two uranium occurrences in northern Sweden. Lithos, 17, 215-225.

Smellie, J.A.T., MacKenzie, A.B. and Scott, R.D., 1986. An analogue validation study of natural radionuclide migration in crystalline rock using uranium-series disequilibrium studies. Chem. Geol., 55, 233-254.

Smellie, J.A.T., 1989. Some Swedish natural analogue studies. A review. Proceedings of the 3rd. CEC -NAWG Meeting, Snowbird, Salt Lake City, June 15-17, CEC Tech. Rep. (EUR 11725 EN), Brussels.

Toulhoat, P., Gallien, J-P., Louvat, D., Moulin, V., l'Henoret, P., Ledoux, E., Gurban, I., Smellie, J.A.T. and Winberg, A., 1994. Present-time migration around the Oklo natural reactor. Evidence of uranium isotopic anomalies in groundwaters close to reaction zones. Migration '93, Charleston, S. Carolina (1993).

Vilks, P., Cramer, J.J., Bachinski, D.C. and Miller, H.G., 1991. Studies of colloids and suspended particles in the Cigar Lake uranium deposit. Conference on Chemistry and Migration Behaviour of Actinide and Fission Products in the Geosphere (Migration '91), Jerez de la Frontera, Spain, Oct. 1991.

West, J.M., McKinley, I.G. and Vialta, A., 1992. Microbiological analysis at the Poços de Caldas natural analogue sites. J. Geochem. Explor., 45, 439-449.

Wikberg, P., 1987. The chemistry of deep groundwaters in crystalline rock. Ph.D. Thesis. Department of Inorganic Chemistry, Royal Institute of Technology, Stockholm, Sweden (TRITA-OOK-1018; ISSN 0348-825X).

Wikberg, P., 1991. Laboratory E_h simulations in relation to the redox conditions in natural granitic groundwaters. In: Radionuclide Sorption from Safety Perspective. Proceedings of an NEA Workshop in Interlaken, Switzerland, Oct. 16-18, 1991.

Table 1: Key processes and analogues studied within the Swedish radioactive wast programme

Near-field Processes	*Natural Analogue*	*Performance Assessment*
Canister stability	7	KBS-3
Bentonite stability	8 (3,4,6)	KBS-3
Concrete stability	6,11,12	
Spent fuel stability	3,4,5,9	KBS-2; SKB-91
Radiolysis	3,4	KBS-3; SKB-91
Redox front propagation	1	
Radionuclide solubility and speciation	1,3,6	
Radionuclide transport processes		
- colloids	1.3.6 (4)	
- organics	1,3,6 (4)	SFR
- microbes	1,3,6 (4)	
Far-field Processes		
Radionuclide solubility and speciation	1,3	
Radionuclide transport processes		
- colloids	1,2,3,5,6 (4)	SKB-91
- organics	3 (4,5,6)	
- microbes	3 (4)	
Radionuclide retardation processes		
- absorption	1,2 (4,5)	
- co-precipitation	1,2,3,6 (4,5)	
- matrix diffusion	3,5, (4,6)	SKB-91

1. Poços de Caldas (Chapman et al., 1992)
2. Alligator Rivers (Duerden et al. [Eds.], 1994)
3. Cigar Lake (Cramer and Smellie [Eds.], 1994)
4. Oklo (Curtis and Gancarz, 1983; Toulhoat et al., 1993)
5. Palmottu (Blomqvist et al., 1994)
6. Maqarin (Khoury et al., 1992; Alexander et al., 1992)
7. Copper stability (Hallberg et al., 1988)
8. Bentonite stability (Pusch, 1983; Pusch and Karnland, 1988)
9. Uraninite stability (Smellie and Rosholt, 1984)
10. Matrix diffusion (Smellie et al., 1986)
11. Porjus (Grudemo, 1982)
12. Uppsala fresh water resevoir (Lagerblad and Trägårdh, 1994)

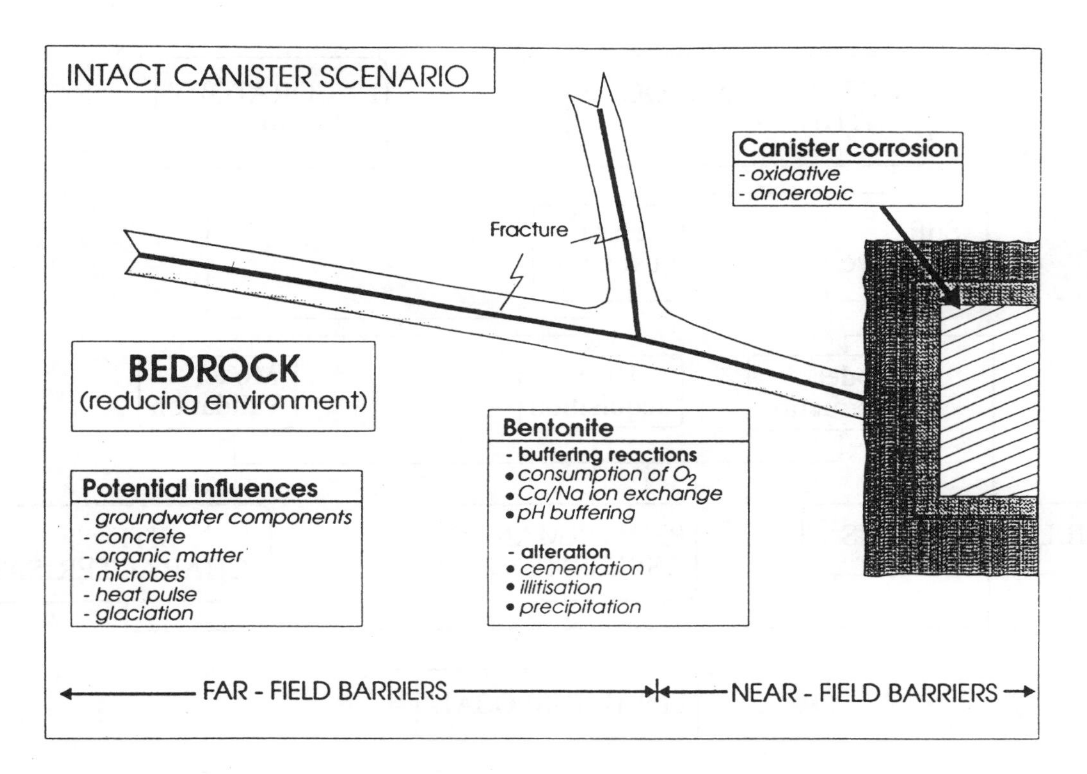

Figure 1. Schematic representation of an intact canister scenario.

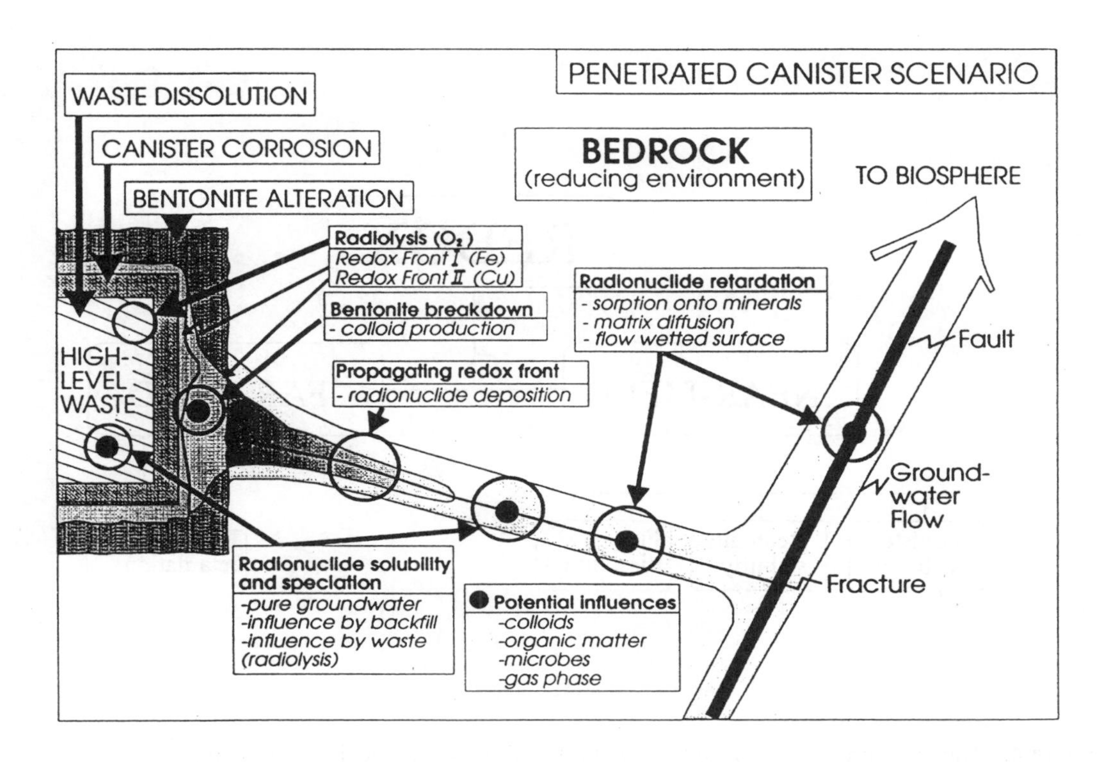

Figure 2. Schematic representation of a penetrated canister scenario.

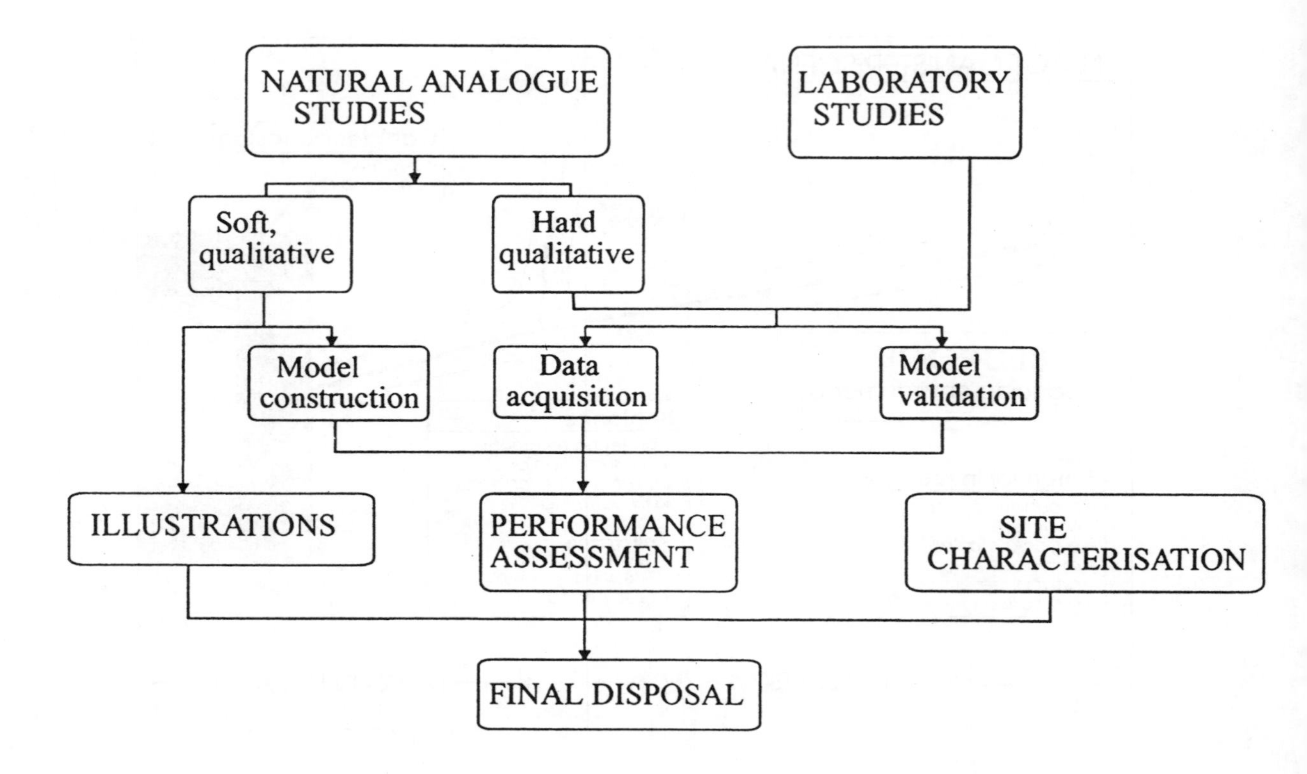

Figure 3. Waste repository safety case; the role and interrelationships of natural analogues (Miller, 1994).

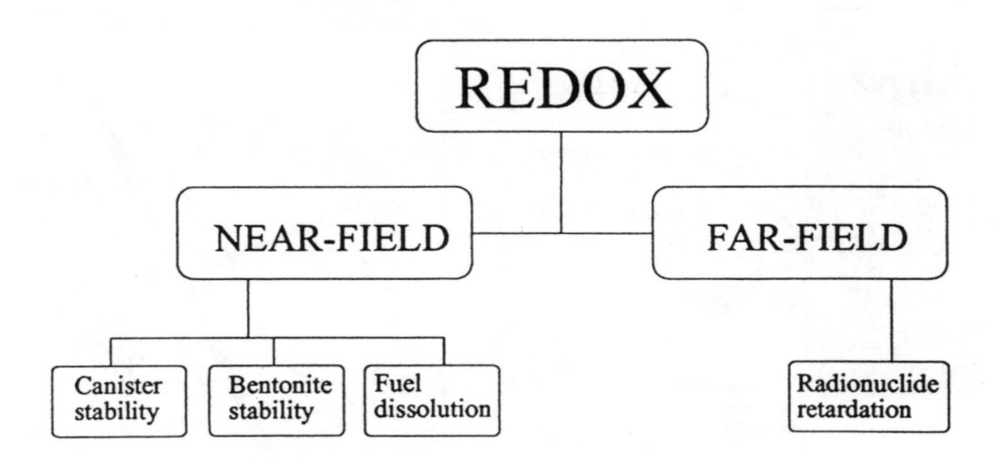

Figure 4. The central role of groundwater redox conditions on repository stability.

SESSION VI

Chairmen:

A.G. Duncan
J. Vigfusson

The Approach to Model Testing and Site Characterisation Studies Undertaken by United Kingdom Nirex Ltd

A J Hooper
United Kingdom Nirex Ltd (UK)

Abstract

U K Nirex Ltd is currently characterising a site at Sellafield in West Cumbria to determine its suitability for the deep disposal of intermediate-level and some low-level radioactive wastes in a repository located within hard fractured basement rock: The Borrowdale Volcanic Group. A programme of drilling has been underway at the site since 1989 and by mid 1994 some eighteen deep boreholes had been drilled and extensive geophysical surveys had been carried out on both regional and site scale. Geological evaluation of the site was carried out, using existing information, prior to 1989 to inform the process leading to the selection of the site for investigation.

The paper records the process by which conceptual models for the site were evaluated and successively rejected or revised as an integral part of the site characterisation programme. The process has become increasingly specific as the programme has developed. Initially the objective of the site characterisation programme was to confirm the availability at Sellafield of the Basement Rock Under Sedimentary Cover (BUSC) concept that performed well in the assessments carried out to inform site selection, and to confirm that rock properties lay within the range that had been employed in these assessments. Thereafter, the site characterisation programme has been directed to address in a systematic manner the most significant uncertainties in the conceptualisation of the site. These uncertainties have included the hydrogeological characteristics of faults that were detectable by geophysical methods and the nature of the controls on the effective hydraulic conductivity of the fractured BVG.

The paper notes that as the issues to be addressed in the site characterisation programme become more specific, the relationship between the various models used to interpret and evaluate field data becomes increasingly important.

Finally the paper notes the requirement for site characterisation to be carried out on an appropriate scale and discusses the role of a large scale pump test planned for late 1994 as a major development of the overall strategy. The strategy for further development from surface-based characterisation to underground experiments in the proposed Rock Characterisation Facility is also discussed briefly.

Coordinated Site Characterization and Performance Assessment - an Iterative Approach for the Site Evaluation

T. Papp, L. O. Ericsson, C. Thegerström

SKB, Swedish Nuclear Fuel and Waste Management Co.

K-E Almén

KEA GEO-Konsult

(Sweden)

Abstract

SKB planning for siting a deep repository involves feasibility studies in 5-10 municipalities, surface based characterization and drilling on two canditate sites and detailed characterization of one site including a shaft to proposed repository depth. The selection of a site or the detailed layout of the repository defines characteristics that might influence safety in a broad sense. There is a strong link between the safety, technical (engineering) and functional aspects.

At an early phase of the investigations only limited information is available for safety evaluation. The site selection will be based on general geoscientific information, ie mechanical stability, ground-water chemistry, slow ground-water movements and complicating factors like high potential for mineralization.

Site characterization provides input to more quantified safety considerations. The general layout of the repository in the actual geological stucture of the site must be done with regard to a number of guidelines, eg to hydraulically separate the parts of the repository containing the spent nuclear fuel from those for other types of longlived waste and to separate the two stages of the spent fuel repository so they can be handled separately in the licensing process.

When the various parts of the repository have been tentatively located the consequence of the multiple barrier principle is that the layout of the various parts should be made with the aim to utilize the available natural barrier system at the site as well as possible.

Siting and design of nuclear waste repository is an iterative process with a gradual knowledge build-up by means of data collection, conceptualization, analysis and renewed proposals for data collection, etc. Safety and design aspects follow the site characterization process in parallell.

COORDINATED SITE CHARACTERIZATION AND PERFORMANCE ASSESSMENT - AN ITERATIVE APPROACH TO SITE EVALUATION

PREMISES FOR THE SITING WORK

Since the mid-1970s, SKB has carried out investigations of the geological conditions at depth in the Swedish bedrock. Extensive studies have particularly been conducted in the Stripa Mine and are currently being conducted in the Äspö HRL. Furthermore, SKB and other organizations have conducted a number of detailed safety assessments for final disposal in the environment existing in the Swedish bedrock.

SKB's siting strategy has recently been elaborated in a supplementary report to the R&D programme 92 (1). The siting strategy is based on the conviction that it is possible to find a site that meets stringent environmental and safety requirements at the same time as a local understanding for the establishment of the deep repository is sought. Figure 1 shows ongoing and planned activities for the siting and construction of a deep geologic repository in Sweden. Safety evaluations at different levels are an important basis for the decissions taken.

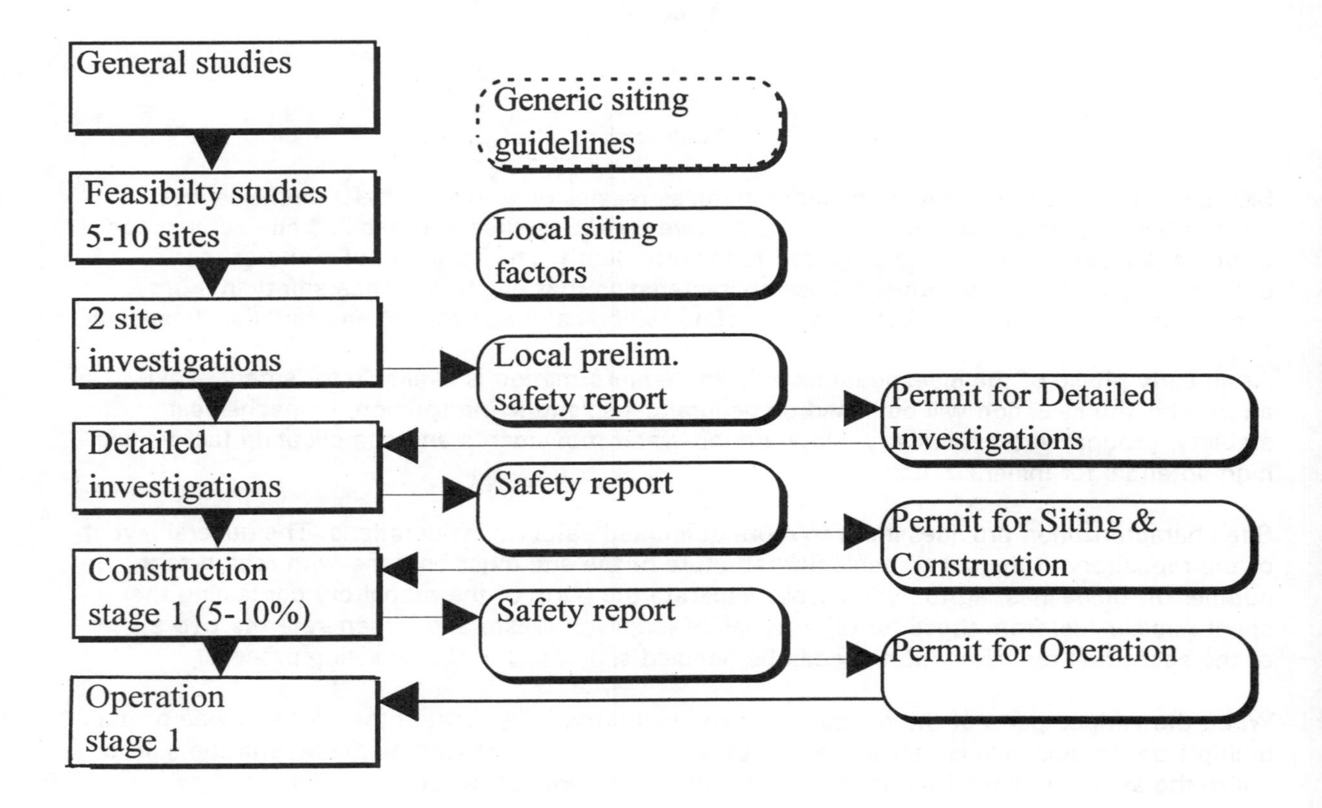

Figure 1: Activities for siting and construction of a deep geologic repository for spent nuclear fuel in Sweden

There are fundamental requirements that must be met by a deep repository. They primarily have to do with long-term safety and other environmental impact. These requirements are set forth in laws and regulations issued by the authorities. Regardless of how the site has been selected, it is the results of a broad and deep assessments of safety and environmental impact which ultimately decides whether the deep repository may be built on the site in question.

In 1993, the Nordic radiation protection and nuclear safety authorities published the report "Disposal of High Level Radioactive Waste, Consideration of some Basic Criteria" /2/. It describes the fundamental requirements for disposal of high-level waste, with an emphasis on long-term safety aspects in a deep geological repository. According to the document, site specific factors affecting the siting can be divided into three main groups: Factors related to the geological medium, the environment and the society. SKB employs a classification that covers in the three main groups but ties in more strongly with the functional requirements on the deep repository:

Safety — Siting factors of importance for the long-term safety of the deep repository.

Technology — Siting factors of importance for the construction, transport, performance and safe operation of the deep repository.

Land and environment — Siting factors of importance for land use and general environmental impact

Societal aspects — Siting factors connected to political considerations and community impact.

A comprehensive assessment of long-term safety requires access to site-specific data on the bedrock conditions on the site. Such data can only be obtained by means of extensive investigations on sites which must be selected on the basis of incomplete information. This distinguishes siting of underground facilities in general and a deep repository in particular from other industrial sitings (surface facilities) where knowledge of all important factors is relatively easily accessible. This in turn affects the strategy and organization of the siting work and the way of working with siting criteria. The process is started with the identification of

- fundamental safety requirements on a deep repository.
- generally favourable conditions for the possibility of siting and building a safe deep repository.
- disqualifying factors for the possibility of siting and building a safe deep repository.

Based on this, siting factors have in turn been identified.

SAFETY-RELATED SITING FACTORS

The basic safety principle for the deep disposal system planned by SKB is to completely contain and isolate the spent nuclear fuel in tightly sealed canisters deposited at a depth of about 500 metres on the selected repository site. This isolation shall be preserved over very long periods of time so that the radioactive materials decay inside the canister and cannot be released. This means that the most important safety-related function of the rock is to guarantee stable chemical and mechanical conditions for the engineered barriers over a long period of time.

The safety strategy for a deep repository is based on the multiple-barrier principle. This means that safety must not be solely dependent on one single barriers' functioning as planned. Another important safety-related function of the bedrock on a deep repository site is to retain the radionuclides or retard their transport if the engineered barriers should be damaged. Finally, it is in principle favourable to have biosphere conditions that ensure that only very small quantities of any radioactive materials that are released will ever reach man.

Table 1 summarizes the safety functions and the siting factors that are linked to them.

Table 1. Safety functions related to the site characteristics at a deep repository

Safety function - Required	Related site factors
Level 1: Ensure longterm stable conditions for canister and bentonite clay so waste is isolated	- chemical environment for bentonite/canister; - mechanical stability of rock; - transport in rock of corrodants - risk of future intrusion
Level 2: Ensure low dissolution of exposed fuel and slow transport of released radionuclides through the rock	- chemical environment for fuel; - transport in rock or radionuclides
- Desirable	
Level 3: Ensure favourable discharge area conditions	- biosphere conditions

Level 1 provides complete isolation of the waste. Level 2 counteracts release and transport of radionuclides if there are damaged canisters. Level 3 contributes to low individual doses if the safety functions on levels 1 and 2 should not perform satisfactorily. The influence of the site factors on the repository system is commented below and some specific parameters are given in table 2.

Mechanical stability: The repository shall be situated in parts of the rock that do not consist of zones of fractured rock in which future significant fault movements could preferentially be released.

Chemical environment: Long-term stable reducing conditions shall prevail in the groundwater in the selected rock regime. The groundwater must have properties that contribute to:
- preservation of the properties of the bentonite;
- low corrosion rate for the canister material;
- low dissolution rate of the fuel (uranium dioxide);
- low mobility and good sorption properties for the radionuclides.

Ability of rock to limit transport: The rock shall constitute a safety barrier by sorbing and retaining any released radionuclides so that their transport through the rock will be slow.

Intrusion: A repository should not be located so close to valuable or potentially valuable natural resources that a future exploitation of these resources would entail damage to the repository's barrier system.

Discharge area conditions: Conditions (dilution, salt/freshwater, accumulation) in possible discharge areas for deep groundwater in the biosphere should be taken into account in a comprehensive assessment and comparison between different sites.

Table 2: Parameters for the Chemical and Mechanical siting factors

MECHANICAL STABILITY		
Geology, Structures	**Mechanical Parameters**	**Processes**
Distribution of rock type Fracture geometry Zones, Lineaments	Rock stresses Properties of the rock Properties of the fractures/zones	Plate tectonics, Glaciation, Aseismic changes, Seismisity, Induced disturbancies

CHEMICAL ENVIRONMENT			
Canister	**Bentonite**	**Spent Fuel**	**Nuclide transport**
Redox buffering Sulphides Chlorides Micro organisms	Calcium Potassium Chlorides Sulphates	Redox potential pH Carbonates	Mineralogy Humic substanses Micro organisms Colloids, Geogas

OTHER SITING FACTORS

The siting of the deep repository must also take into account the technical options available for transportation and for design/construction of the facilities. As a rule the technical solutions are flexible and can be adapted to varying site conditions and bedrock characteristics. This means that evaluations of different technical factors for alternative site choices could be made in economic terms.

The site of the deep repository shall offer:

- *Bedrock conditions that permit construction of stable shafts, tunnels and rock caverns so that safety requirements during construction and operation are met.*
- *Good options for carrying out all transports to the site and all activities in the deep repository facility in a safe manner.*

Site selection and design of the facilities shall be done so that conflicts with opposing interests are minimized. Consideration shall be given to the environment, cultural monuments, recreation, hunting, fishing, other outdoor activities, important natural resources, agriculture and forestry, existing and planned land use. In brief, the requirements can be formulated as follows:

The site for the deep repository shall have
- *few opposing interests for land use;*
- *good prospects for being able to build and operate the facilities in compliance with all environmental protection requirements.*

Socio-economic consideration are important for both site selection and design of the facilities on the selected site. Establishment and operation of a deep repository will have different impacts on the locality and the region. These include e.g. impact on employment, the local business community and local services.

Siting of a deep repository shall be carried out so that
- *investigation activities in different stages, construction and commissioning and operation, are firmly rooted in a democratic decision process.*
- *social and socio-economic consequences are taken into account.*

OUTLINE OF SKB SITING WORK

The purpose of the siting work is to obtain all the data necessary to choose a site and obtain a permit to commence detailed geoscientific characterization. The siting work entails general studies, feasibility studies and site investigations and will cover siting factors on different scales. Figure 2 shows a schematic illustration of the variation in scale that is involved. The studies in different scales should be relevant to the geologic understanding of the site and the safety of the repository. However, it is not easy to convert the generalized level of descriptive regional geology into parameters used in the model chain of safety assessments. As the scale changes the use of siting factors will change from being a general guidance on the regional scale, to an actual use of the parameters in the design optimization and safety assessment modelling during the actual construction of the repository.

General studies - shall provide an understanding of the regional setting of the site.

Feasibility studies - shall examine the prospects for a deep repository in potentially suitable and interrested municipalities. A feasibility study is performed mainly on the basis of existing material.

The following questions are dealt with:
- What are the general prospects for siting a repository in the municipality?
- Within which parts of the municipality might there be suitable sites for a deep repository, considering geoscientific and societal aspects?
- How can the repository be designed with respect to local conditions?
- How can transportation be arranged?
- What are the important environmental and safety issues?
- What are the possible consequences (positive and negative) for the environment, the local economy, tourism etc. in the municipality and the region?

Feasibility studies of specific municipalities, together with general studies of the entire country, are to provide the necessary background data and the breadth of alternatives that are required to identify suitable sites for site investigations. SKB believes that a reasonable number can be between 5 and 10 feasibility studies of municipalities in different parts of the country. Two such studies are currently in progress in the municipalities of Storuman and Malå.

General studies		Feasibility studies	Site investigations	
Countrywide	Regional	Municipality	Area	Site

1 000 km x 1 000 km --- 100 km x 100 km --- 10 km x 10 km --- 1 km x 1 km

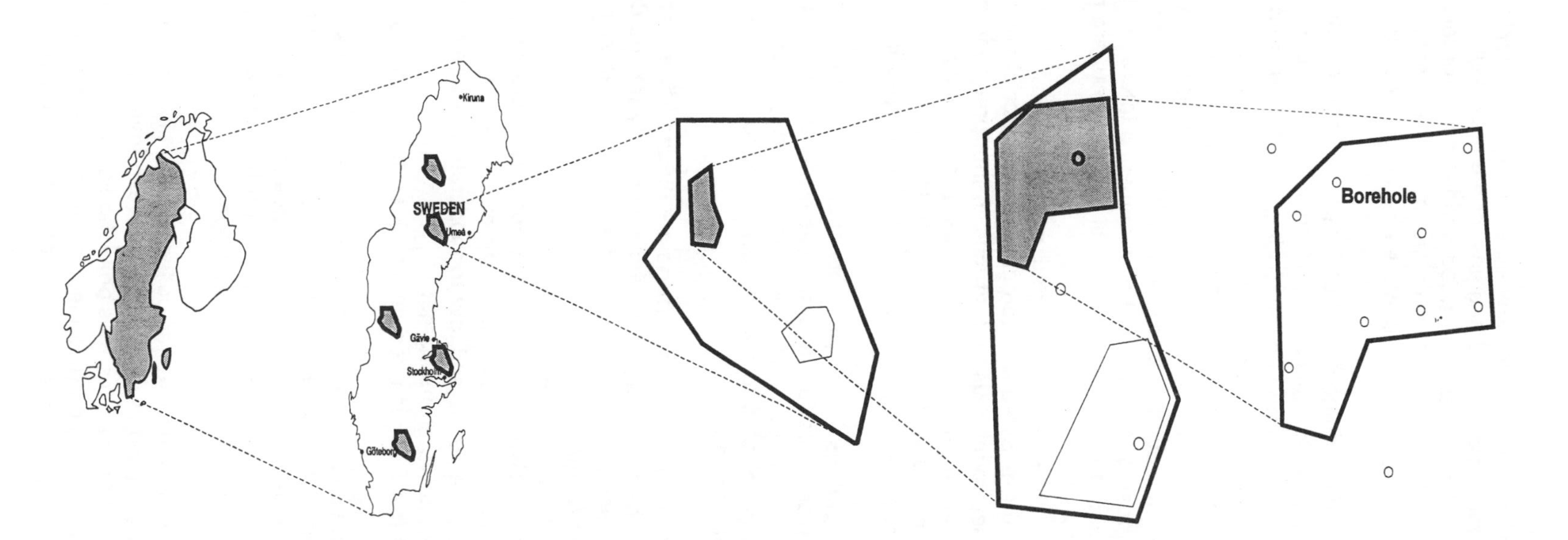

Figure 2. A scematic picture of the scales involved in the siting work. The areas marked on the map are fictive.

Key questions in the initial siting phase are:
- Which sites have particularly good chances of meeting the requirements with regard to safety, technology, land and environment as well as societal aspects?
- Which of these sites offer good opportunities for later carrying out a reliable characterization of, above all, the important environmental and safety factors'?
- How can these sites be identified based on existing material?

The following conditions are thereby primarily favorable (give a good prognosis):
- A common rock type without interest for other utilization of natural resources. Gives prospects for a good understanding of safety-related bedrock conditions and reduces the risk that the area will be of interest for other use.
- A large site with few major fracture zones. Provides flexibility in the utilization of the site and improves the prospects to construct a repository with ample room for the canister positions in sound rock and with a high level of safety.
- Few opposing land-use and environmental interests. Good prospects for adapting the facilities so that the environmental requirements are met in a satisfactory fashion.

Conditions that should be avoided are
- abnormal (for Swedish bedrock) groundwater chemistry;
- highly heterogeneous and difficult-to-interpret bedrock;
- known deformation zones and postglacial faults;
- pronounced discharge areas for groundwater;
- rock types that might be of interest for prospecting.

Site investigations - shall yield site-specific data for safety assessments and environmental impact assessments for sites in areas found to be of interest in the feasibility studies and in the light of the general studies. The purpose is to gather material for a preliminary determination of whether it is possible to build a deep repository on the site that can meet all environmental and safety requirements.

During a site investigations efforts are directed firstly at clarifying the conditions that will prevail for the repository as a whole, secondly for the different repository parts, and finally during detailed investigations for the individual canister positions. The following conditions are particularly favourable:
- reducing groundwater chemistry;
- normal (for Swedish bedrock) groundwater chemistry in other respects;
- homogeneous and easy-to-interpret bedrock;
- few fracture zones and low to moderate fracture density;
- low groundwater discharge;
- normal (for Swedish bedrock) rock stresses, strength properties and thermal conductivity.

Conditions which can cause a site to be abandoned are:
- extreme groundwater chemistry, e.g. oxidizing groundwater;
- valuable ores or minerals in the repository area;
- several closely-spaced water-bearing fracture zones;
- extreme rock-mechanical properties.

The site investigations entail collection and analysis of field data. The work is done step by step and iteratively in interaction between investigations, performance assessments and constructability analyses (design). Should the initial studies indicate favourable conditions, the investigations will be broadened to complete site investigations. The investigations will be conducted in accordance with a detailed investigation programme, which will be presented before the initial site investigations start. The general structure of theprogram is discussed below.

SITE INVESTIGATION PROGRAMME

Site investigations consist primarily of geoscientific surveys of a specific site from the ground surface and in boreholes. The results of the site investigation work are compiled in assessments for site-specific environmental impact assessment, long-term safety, and constructability. More detailed determinations of the parameters will be made in the next stage of the siting work, the detailed investigations, where investigations from tunnels and/or shafts within the potential repository rock volume will be included.

The programme for data collection during the site investigation will be focused on the factors of importance which were discussed in the previous sections. However, in order to understand these related parameters a general understanding of the site geology must be gained. Therefore the site specific data collection will be carried out as a geoscientific characterization of the whole rock volume, and include all parameters and factors defined as important for the deep repository. This general understanding of the site is also regarded as necessary for the acceptance, among expertise as well as among the common public, of results and future decisions based on the site investigations.

Strategy of the site investigations

The site investigation programme will primarily be based on knowledge and experiences gained during previous site investigations in Sweden, i.e. more than 10 Study Site investigations, the Stripa Project, the Äspö Hard Rock Laboratory and other field investigations, but also on experiences from site investigation programmes abroad.

For efficiency in the fieldwork and evaluations, the site investigation will be carried out in stages, permitting a successive development of the structural model of the site with regard to geology, geohydrology and hydrogeochemistry. The planning, the field measurements and the evaluations will require a high degree of integration. During the course of the site investigations, iterative performance assessments in different scales and site specific revisions of layout/design will guide the further field investigations. For each iteration of investigations and interpretations, and evaluation of safety and constructability more information will be avialable for the assessments, and better guidance can be given on what are the most important data, and what areas or structures will be most important to investigate to reduce the uncertainties (Figure 3).

A possible sequence would be

- Identify the regional fracture zones and the general groundwater circulation in the area.
- Locate the various parts of the repository on the site with regard to the wish to utilize the site as well as possible and to separate the repository for the spent fuel from that for the other waste types.
- Select areas for access tunnels or shafts and for ventilation shafts.
- Identify outflow areas of the deep groundwater from the area.
- Select depth of the repository parts and respect distances from dominating transport routes for groundwater.
- Identify the access shaft/tunnel for the detailed geoscientific investigations.

A site investigations will continue for a period of about three years. The selection of areas for investigations will be based on the earlier described siting factors addressed during the

Figure 3

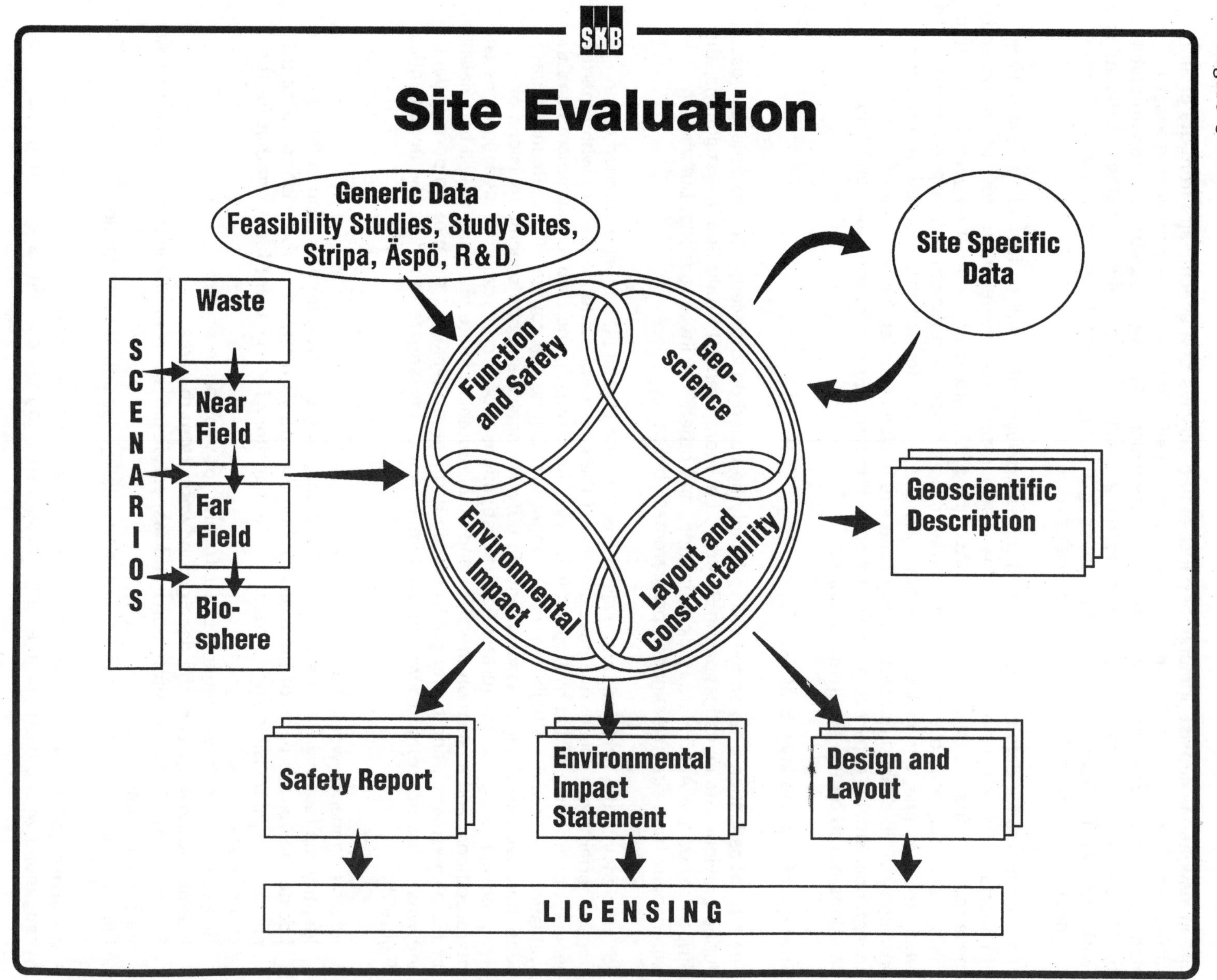

SKB
Site Evaluation
Generic Data
Feasibility Studies, Study Sites,
Stripa, Äspö, R & D
Site Specific Data
SCENARIOS
Waste
Near Field
Far Field
Bio-sphere
Function and Safety
Geo-science
Environmental Impact
Layout and Constructability
Geoscientific Description
Safety Report
Environmental Impact Statement
Design and Layout
LICENSING

feasibility studies. One or more areas might be available in a selected municipality. Since no field measurement have been made during the feasibility studies the site investigation programme will start with a first stage called initial site investigations.

Initial site investigations

An initial site investigation is a limited effort to address specific issues that have not been well enough resolved in the feasibility studies, but might be of a discriminating nature, eg. rock-type contacts with high potential för mineralizations or extreem rock stresses. The initial investigations will also have the goal to map the natural status of groundwater and surface waters of the area. If such unfavourable conditions exists and dominate above the favourable conditions the site will be abandoned from further investigations. A decision to carry out a complete site investigations will accordingly not be taken until the initial site investigation at a site have been performed and the favorable conditions assumed when selecting the site have been substantiated as far as can be understood from the limited set of data.

The initial site investigation starts with geological and geophysical mapping, from the ground surface and from the air.The extent depends on the amount of already existing information. Important issues are to characterize the surrounding in a regional scale, with regard to lineament structures and the general geology. Structure analysis and rock type classification gives indications on the most interesting areas of homogeneous bedrock, maybe surrounded by major fracture zones, but with few fracture zones within the area.

This investigation stage will contain the first site-drillings, including a few relatively shallow holes for controlling the orientation of dominating fracture zones. An important borehole is a deep cored hole in the central part of the area of interest. The hole will be drilled down to a depth of at least 1000 m. Major efforts will be made to minimize the contamination of the groundwater as the chemistry of the groundwater is one of the most important factors for the long term function of the deep repository.

As subhorizontal fracture zones might have strong influence on the ground water flow around the repository the existence of such zones as well as other larger fracture zones in the area are looked for by reflection seismics from the surface and by Vertical Seismic Profiling from the deep borehole. The drill-cores from the hole will be used to confirm the frequency of fractures and fracture zones, the homogeneity of the bedrock and the potential for ore bodies or valuable minerals.

Topografic data, the geometry of the larger fracture zones and the first data on hydraulic conductivity will be input to the first crude modelling of the groundwater flow in the area, its general directions and pathways.

The site investigations

While the initial site investigation addresses specific factors that can be of prohibitive character, the broader scope of the full site investigation calls for extensive data collections of many different factors and parameters. This work is a step by step procedure where often several alternative interpretations of the collected dataset have to be analysed and tested in the process of establishing confidence in the descriptive model. During this process the close interaction between the various assessments, as has been shown in Figure 3, will be important for a balanced view of what importance to ascribe to remaining uncertainties.

Several boreholes will be drilled, measured and hydraulically tested, using a variety of methods (single hole and cross-hole) which are found useful in the Äspö HRL program or elsewhere in similar rock conditions. For example cross-hole hydraulic tests and cross-hole tracer tests are very important for the construction of the (geohydrological) structure model and for the determination of groundwater pathways and transport parameters. The hydrogeochemistry will be one important tool for the determination and the understanding of the groundwater flow, not less in the regional scale. Rock stress measurements will be conducted in sections of boreholes selected for optimizing the representability of the data.

At this stage of the site investigations the site model will include deterministic location of the major structures which are regarded as important for the functioning of the repository, and for safety during construction and operation of the repository. Minor structures are characterized with regard to frequency, spacing and dominant orientation, and handeled probabilistically in the modelling.

Based on this structural model a preliminary layout will be made, including depth and position of the various parts of the repository and the ways of access to the repository.

The iterative character of site investigations means that new investigations will be used to test the validity of the old site model and to build a revised model. Bedrock composition, fracture frequencies, existence and orientation of fracture zones are examples of parameters to be predicted in the site model, see also "Results and Experience fom the Äspö Hard Rock Laboratory Characterization Approach, Göran Bäckblom et al at session 4. Hydraulic modelling of flow, groundwater pressure drawdown and tracer arrival are parameters to be predictive modelled before large scale pumping tests will be performed.

All data collections, analyses, evaluations and modellings will be conducted in the framework of a quality assurance programme. The most importance aspect of the QA programme is the need of traceability of data in all stages and levels of the data management.

When two complete site investigations have been carried out, all relevant material from the siting work is compiled in an application for a permission to carry out detailed characterization on one of the two sites. The reasons for the choice of site are presented, along with all background material in the form of data, analyses, investigations, arguments for and against and judgements.

REFERENCES

1. FUD-PROGRAM 92, Kompletterande redovisning, SKB, aug 1994 (in translation to english)

2. Disposal of High Level Radioactive Waste, Consideration of some basic criteria; Radiation Protection and Nuclear Safety Authorities in Denmark, Finland, Iceland, Norway and Sweden, 1993.

The SKI SITE-94 Project Approach to Analyzing Confidence in Site-Specific Data

Björn Dverstorp and Johan Andersson
Swedish Nuclear Power Inspectorate (SKI)

(Sweden)

Abstract

The ongoing SKI SITE-94 project is a fully integrated performance assessment based on a hypothetical repository at 500 m depth in crystalline rock. One main objective of the project is to develop a methodology for incorporating data from a site characterization into the performance assessment. The hypothetical repository is located at SKB's Hard Rock Laboratory at Äspö in south-eastern Sweden. The site evaluation in SITE-94 uses data from the pre-excavation phase that comprised measurements performed on the ground and in boreholes, including cross-hole hydraulic and tracer experiments.

Interpretation of site specific data involves several steps of model testing, each associated with different sources of uncertainty. These uncertainties affect the final performance assessment results to a varying degree. Uncertainties related to measurement technique, equipment and methods for interpretation were evaluated through a critical review of geohydraulic measurement methods and a complete re-evaluation of the hydraulic packer tests using the generalised radial flow (GRF) theory. Groundwater chemistry samples were analyzed for representativity and sampling errors.

A wide range of site models within geology, hydrogeology, geochemistry and rock mechanics has been developed and tested with the site characterization data. Key features of the model testing include assessment of reasonableness through simple scoping calculations, development of alternative conceptual models on different spatial scales - each consistent with field data. The question of consistency between geologic, hydrogeologic, rock mechanical and geochemical models of the site is also being addressed.

The multiple interpretation approach makes it possible to illustrate conceptual uncertainties regarding the description of processes and structures. Integrated interpretation of geologic, hydraulic and geochemical data using 3D CAD tools together with multivariate analyses are expected to give information on possible correlations in the site data and to aid in the discrimination between the alternative conceptual models. Remaining unresolvable conceptual and other uncertainties will be propagated through the assessment and quantified in the subsequent calculations of radio nuclide release and transport to determine their importance. Attempts are also made to illustrate the effects of errors introduced when estimating effective parameters used in the performance assessment models. Although experiment planning and design are beyond the scope of SITE-94, suggestions for improved measurement strategies is another expected outcome from this site evaluation exercise.

INTRODUCTION

The ongoing SKI SITE-94 project is a fully integrated performance assessment using site specific data from crystalline rock at SKBs Hard Rock Laboratory site at Äspö in southeastern Sweden. The assessment is based on a KBS-3 type repository for spent nuclear fuel, hypothetically located at 500 m depth at the southern part of the Äspö island. SITE-94 is structured around five sub-projects: *Scenario development, Site evaluation, Engineered barriers, Radio nuclide transport, and Quality assurance.* One of the main objectives of SITE-94 is to develop a methodology for analyzing site specific data for the performance assessment.

The objectives of the site evaluation in SITE-94 are to deliver parameters for calculation of stability of the engineered barriers, radio-nuclide release and transport in the near-field rock and for calculation of radio-nuclide transport through the geosphere under different scenarios. Above all, it should develop an understanding of the site's structures and processes and their evolution in time to render credibility in the predictions made with the site models and the performance assessment results. A key to such confidence building is a systematic and transparent evaluation of uncertainties and their impact on the performance assessment.

DATA USED

The site evaluation in SITE-94 uses site characterization data from the SKB Äspö Hard Rock Laboratory project (Stanfors et al., 1991, Almén and Zellman, 1991, Wikberg et al., 1991, and Gustafson et al., 1991). The data comprise a wealth of measurements performed on the ground and in boreholes drilled from the surface, such as:

- various types of maps and areal photos (satellite photos, digital terrain models, geologic maps, geophysical maps etc),
- local ground geophysical measurements,
- core logs (fracturing, mineralogy etc),
- borehole geophysical logging (radar, neutron, salinity, natural gamma, etc),
- hydraulic test data (single-hole packer tests, hydraulic interference tests),
- a long-term pumping and tracer test,
- groundwater chemistry samples and,
- rock stress measurements.

A total of 14 cored boreholes with depths down to 1000 m and 20 shallow percussion boreholes were drilled on the Äspö island. Data is also available from boreholes on the surrounding islands and the mainland. It should be noted that the objective of the site characterization at Äspö was not to produce a data base for performance assessment. Still, the type and amount of data is similar to what can be expected from the first phase of a site characterization at a candidate site for a nuclear waste repository in Sweden. No data from the construction phase of the Äspö project, i.e. measurements performed from the underground excavation, were used in SITE-94.

Despite the wealth of data gathered in a site characterization like the one performed at Äspö the actual volume of rock that is represented by measurements is relatively small. Figure 1 illustrates the Äspö island and the locations of cored boreholes at the surface and horizontal cuts at different depths below surface. The circles around the boreholes indicate the maximum radius at which a fracture can be detected with the borehole radar. Note that there is only one borehole below repository depth in the southern part of Äspö where the hypothetical SITE-94 repository has been located.

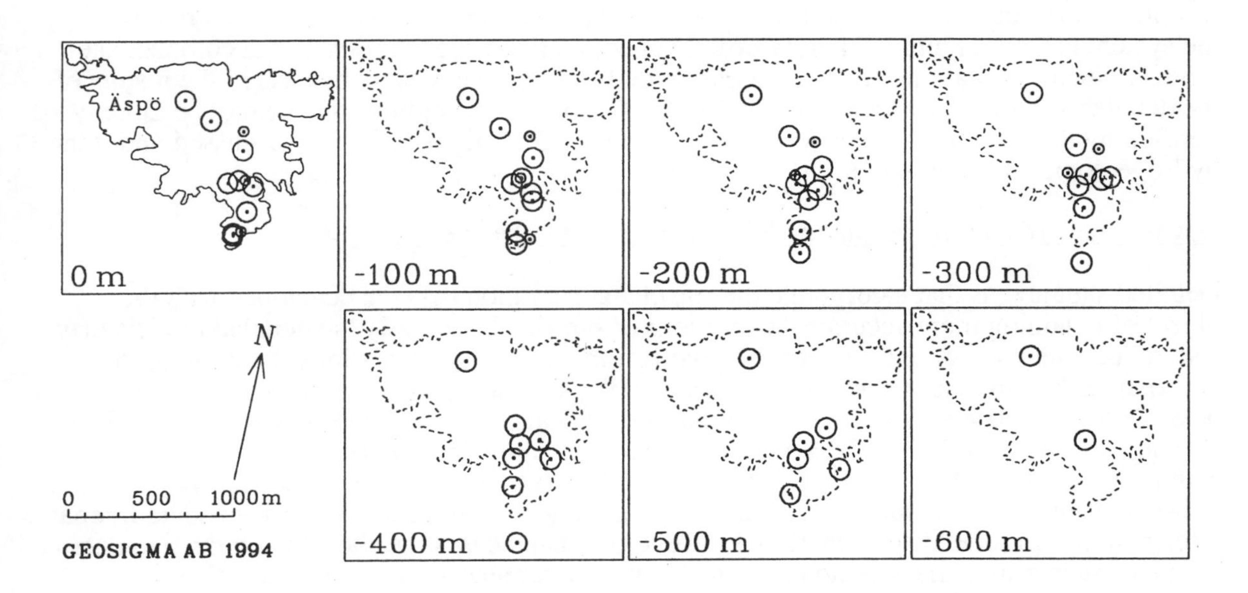

Figure 1. **Locations of cored boreholes at different depths at the Äspö site (see text for explanation).**

STRATEGY FOR ANALYZING SITE DATA

The site evaluation comprises analyses of a large amount of site specific data and development and testing of interdisciplinary *site models* within geology, hydrogeology, geochemistry and rock mechanics. Site model refers to an integrated description of the site within a particular discipline and may consist of a variety of sub-models for different processes and structures. The subsequent sections give some examples of this model development within the disciplines of geology, hydrogeology and geochemistry. The overall strategy is summarized in the following.

Assessment of uncertainties in the site characterization data

Site characterization data are associated with several sources of potential errors and uncertainty regarding equipment, measurement technique, representativity etc. Many of the parameters used in the site evaluation, e.g. hydraulic conductivity, cannot be directly measured but need to be estimated indirectly using interpretation models. Assumptions associated with such interpretations introduce additional uncertainties. Description of data uncertainties is part of the quality assurance and should ideally be an integrated part of the site characterization data base.

The activities in SITE-94 comprise a critical review of the site characterization methods employed at Äspö and detailed analyses of a few selected parameters in the site characterization data base, including hydraulic conductivity and groundwater chemistry.

Simple scoping calculations

Simple scoping calculations and sensitivity analyses are means to estimate wide bounds of the system behaviour and to identify critical uncertainties to be addressed with complex site models. Such calculations are also valuable for assessing the reasonableness of the results of the detailed site models which often build on conceptual assumptions that cannot be directly verified by the site data. The use of simple scoping calculations is further discussed under the hydrogeologic site evaluation.

Development of alternative site models consistent with the site data

Detailed site models that incorporate the site characterization data are developed with the object of obtaining more accurate descriptions of the site. Potentially the detailed models may also reduce the wide bounds determined with simple scoping calculations. Because of the pronounced heterogeneity and the limitations in the site data discussed above it is often possible to formulate alternative conceptual models that fit with the site characterization data. To some extent the validity of the alternative conceptual models can be evaluated by calibration on one data set and validation on another data set. Analysis of consistency between different site models e.g. within hydrogeology and geochemistry may add additional information for model discrimination. However, in many cases the site characterization data do not contain critical information for conclusive discrimination between alternative conceptual models. The envelope of alternative models is a measure of conceptual uncertainties in the description of the site.

The strategy adopted in SITE-94 for evaluation of conceptual model uncertainties is to develop alternative conceptual site models consistent with the site characterization data and propagate the results through the assessment. Conceptual uncertainties related to the interpretation of fracture zones and site geology will be addressed by evaluating and comparing three independently developed three-dimensional geologic structure models. Site hydrogeology is analyzed with alternative conceptual models, including discrete fracture network and stochastic continuum models, that have been calibrated with the existing hydrological data. The former model is based on the discrete representation of fractures in the geologic structure models whereas the latter is based on a geostatistical description of rock heterogeneity. The alternative geologic structure models are also used for analyzing the mechanical stability of the site.

Integrated interpretation of interdisciplinary site data

Integration activities play an important role for confidence building in the site evaluation. Integrated interpretation of interdisciplinary data gives information on correlations and provides information for discrimination between alternative site models. The main integration activity in SITE-94 is an integrated interpretation of geologic, geophysical, hydrogeologic and geochemical data based on the SITE-94 geologic structure model. By means of 3D-CAD technique it is possible to visualize large and complex data sets and test different hypotheses on correlation between e.g. geologic features and hydrologic and geochemical properties.

Assessment of long-term evolution

Predictions of long-term evolution involves extrapolation of site models for hydrology, geochemistry and rock mechanics to other boundary conditions and temporal scales than those for which the models were developed. In SITE-94 a sequence of time dependent boundary conditions has been specified in a special climate evolution scenario that includes glaciation cycles, sea level changes and permafrost events over the nearest 120, 000 years.

Special large-scale hydrologic models have been developed to simulate the long-term evolution of the groundwater flow field and distribution of fresh and saline waters. The site's mechanical response to the external loads caused by repository heat and later by glaciation is analyzed with the rock mechanical site models. Within the geochemical site evaluation efforts are made to analyze indirect evidence of the site's historic evolution and the long-term processes that has led to today's conditions to build confidence in the predictions made into the future. This work includes evaluation of groundwater chemistry and development of models for the origin of different groundwaters.

Propagation of uncertainties

Eventually the site models should provide parameters for the consequence calculations that comprise calculation of stability of the engineered barriers and calculation of radionuclide release and transport in the geosphere. The ambition is to propagate all conceptual and other uncertainties that could not be resolved by the site characterization data and evaluate their impact on repository safety. However, there is not a direct pipeline between site evaluation and the consequence calculations. An important step is therefore to compile and organize results of the site evaluation in such a way that different sources of uncertainty can be evaluated. This involves book keeping of uncertainties, simplifications and assumptions made throughout the site evaluation and a systematic approach to defining calculation variants for the consequence calculations. Another important aspect is to ensure consistency between parameters both within a discipline and between disciplines, e.g. between hydrology and geochemistry.

The radionuclide release and transport calculations will in turn give feedback regarding the importance of different uncertainties in the site models and required accuracy in the predictions made. Only then it can be judged whether the conceptual uncertainties are acceptable. Judgements of validity of sub-models or certain process models of the site will be arbitrary as long as their relative impact on the overall repository performance is unknown.

DEVELOPMENT OF AN INTEGRATED GEOLOGIC STRUCTURE MODEL

The classical approach to developing an integrated geologic site model in crystalline rock is to start with a deterministic description of major geologic structures and units, here referred to as a geologic structure model. The basic assumption is that the site's properties and behaviour, from the performance assessment perspective, can be related to identifiable features such as extensive fractures, fracture zones, lithological units, veins and dykes etc. In this respect the geologic structure model should be seen as a conceptual framework for synthesis of a variety of site characterization data from the disciplines of geology, geophysics, hydrogeology, geochemistry and rock mechanics.

The specific objectives of the geologic structure model developed in SITE-94 are:

- to serve as a geometrical framework for the development of an integrated site description using all available data,
- to provide a simplified conceptual model of fractures and fracture zones for one of the main hydrogeologic site models and for calculations of rock mechanical stability and,
- to serve as a basis for placing a repository in the studied rock portion.

Development of a geometric description of site structures

Integration of spatial scales

The SITE-94 geologic structure model was developed through a stepwise analysis of data from regional to local site scale. The result is a series of increasingly detailed descriptions of geologic structures going from regional scale to the local site scale. Table 1 summarizes the type of data used and conceptual representation of structures developed on different spatial scales.

Spatial scale	Type of structure model	Information/data used
Southeastern Sweden	Description of geologic and tectonic evolution	Existing geological literature, and maps etc.
Regional: 35 * 25 km^2	2D surface model of major fracture zones (lineament model)	Areal photos, satellite photos, topographical maps, digital terrain models, nautical charts, geological maps, geophysical maps etc.
Semi-regional: 12 * 10 km^2	2D surface model of major fracture zones. Estimates of dip, width and structural characteristics	As above.
Local site model covering the Äspö island: 2 * 2 * 1 km^3	3D-CAD model of extensive fractures and fracture zones	Borehole radar and geophysical measurements, fractured sections in drill cores, surface structures observed on Äspö, ground geophysical measurements (refraction seismics, magnetic measurements, VLF)

Table 1. **Spatial scales of the SITE-94 geologic structure model**

Integration of spatial scales and good knowledge of the geologic setting on a regional scale is a key to the development of an understanding of local site geology. First, the rock forming and major tectonic processes are by nature large scale processes which must be evaluated on a scale much larger than the investigation area. Secondly, many of the structural features that can be observed at the site are part of larger features that are not sufficiently characterized by the local site data. For example, many of the site scale structures observed at Äspö can be interpreted as being part of a larger regional fractures zones extending several tens of kilometres away from the site. Similarly, the formation of hydraulically important veins and dykes observed at the site can be related to the intrusion of a granitic body outside the site area.

Three-dimensional site model of geometric structures

The 3D structure model of the Äspö site was developed through 3D-modelling of fractures and fracture zones using 3D-CAD technique. The first step was to identify intersections between fractures/fractures zones and the boreholes using core data and indirect geophysical indications of fracturing. Estimates of the orientation of the structures were obtained from borehole radar measurements and oriented cores. In the next step each structure was extrapolated in space, assuming extensive and planar geometry, and tested in other boreholes that it ought to intersect. By extrapolating an identified structure to the surface it could be tested whether it would coincide with a surface lineament observed on the Äspö island. Three criteria were in most cases fulfilled for the adopted structures:

- Maximum 15 degrees deviation of relative orientation at intersected boreholes
- Similar fracturing characteristics at the intersected boreholes (e.g. fracture frequency and occurrence of crushed rock)
- Maximum 25 m deviation between predicted surface outcrop and observed surface lineament (depending on depth of data points).

The resulting geometric model consists of a set of 52 extensive fractures and fracture zones.

Verification of geologic structures through integrated interpretation

The geometric structure model was tested through incorporation of all relevant geologic, geophysical, hydrological and geochemical site data. The basic idea was to assign properties to each of the postulated fracture planes and test whether these properties could be correlated between the borehole intersection points. Another important objective was to evaluate possible correlations in the data, e.g. between lithology, fracturing and hydraulic properties. The testing procedure involves both visual inspection in the 3D-CAD model and statistical analysis of a large amount of interdisciplinary data.

Preliminary results indicate that almost 60% of the flow-points (spinner or geophysical indications of flow in the boreholes) coincide with the interpreted fractures. The question remains as to what would explain the remaining flow indications. More fractures or other kinds of structures that could not be identified from the data ? Despite this difficulty, the identified sub-set of water conducting fractures in the structure model explained nearly all of the transmission of early drawdown observed during all interference tests performed at Äspö. Another interesting result is a strong correlation between flow-points in the boreholes and occurrence of aplite dikes indicating that a good understanding of lithology is valuable in determining the site's hydrologic behaviour. The local measurements of hydraulic conductivity (3-m packer tests) turned out to be rather poor indicators of locations of flow. More than half of all points with high hydraulic conductivity had no measurable flow at those positions in the rock. This confirms that connectivity is more important than local measurements of hydraulic conductivity in describing flow through fractured rocks. The identified conductive fractures on the average crossed 2.5 borehole sections with measured flow (or 75% of all intersections with the boreholes), which gives some support to the assumption of spatially extensive and hydraulically connected features.

Preliminary inferences made from the incorporation of groundwater chemistry data suggest that a few water conducting fractures that outcrop in the water south of Äspö contain water with the chemical characteristics of seawater. Deeper fractures contain shield brine and isolated fractures at shallower depths contain water that may be glacial meltwater. These type of results will be used to build an integrated understanding of the hydrogeologic evolution at the site.

Alternative geologic structure models

The development of a geologic structure model is typically a non-unique process because it involves assessment of a large amount of interdisciplinary data and several steps of subjective expertise judgement (geologists' interpretations). In order to address such interpretation uncertainties three different structure models of the Äspö site are to be evaluated and compared: 1) the SITE-94 structure model, 2) SKB's own interpretation (Gustafson et al., 1991) and 3) a model developed by an independent analysis group (SKN, 1992). Preliminary comparison between these structure models indeed confirms that different groups will arrive at different models both in terms of number of structures and their properties.

HYDROGEOLOGICAL SITE EVALUATION

Multiple interpretations is a key feature of the hydrogeologic evaluation in SITE-94. The hypothesis is that conceptually different models that can be calibrated to fit the site characterization data do not necessarily give the same results when extrapolated to the spatial scales and boundary conditions to be evaluated in the performance assessment. Table 2 summarizes all hydrogeological conceptual models that are being tested, some of which are described more in detail below.

Spatial Scale	Conceptual model
Sweden scale (southern Sweden)	• 2-dim continuum surface model • 2-dim continuum vertical section with density dependent flow
Regional scale (length scale < 10 km)	• 2D continuum vertical section
Site scale (length scale < 5 km)	• Simple 1D application of Darcy's law • 3D deterministic discrete feature site model • 3D stochastic continuum site model • Lithofacies model (feasibility study)
Near-field rock (length scale < 50 m)	• 3D stochastic discrete fracture networks • 3D variable aperture discrete fracture network model

Table 2. **Hydrogeologic conceptual models tested in SITE-94**

Evaluation of uncertainties in the site characterization data

Because SKI has no previous experience of making field measurements the hydrogeologic evaluation started with a critical review of the geohydraulic measurement methods employed at Äspö with the objective to describe the type of errors and uncertainties that are associated with the site characterization data base (Andersson et al., 1993). This review clearly demonstrated that site characterization data cannot be taken for granted. Errors and uncertainties associated with instrumentation, measurement technique, and not the least measurement interpretation in many cases may affect estimated parameters by more than an order of magnitude. A pre-requisite for analyzing such data uncertainties is a careful documentation of measurement procedures and any assumptions made in the interpretation of the measurements.

As a basis for the further model development SKI initiated a complete re-evaluation of all hydraulic packer tests performed at Äspö using an alternative interpretation model based on the Generalized Radial Flow (GRF) theory (Barker, 1988 and Doe and Geier, 1990). This interpretation model produces estimates of hydraulic conductivity but also gives information on flow dimension which can be related to the small-scale hydrologic structure of conductive features. The objectives were to derive an alternative data base for hydraulic conductivity and transmissivity and to assess the suitability of particular conceptual models for specific geologic units within the site.

Preliminary results show that assumptions made with regard to flow geometry may affect the hydraulic conductivity estimates by several orders of magnitude. Figure 2 shows a scatter-plot of log hydraulic conductivities estimated with the GRF-model versus log conductivities interpreted by SKB using an interpretation model based on two-dimensional flow geometry, for one of the boreholes at Äspö. The numbers represent estimates of flow dimension. Both sets of conductivities were evaluated and compared in the hydrological site models.

Simple evaluation of groundwater flux

As an alternative to the detailed hydrogeological site models developed in SITE-94 the hydrologic behaviour was assessed with simple scoping calculations of groundwater flux based on one-dimensional application of Darcy's law and simple assumptions of flow field structure and bounding values of driving forces. The objectives were to estimate ranges of the hydrologic behaviour and to identify critical factors for the determination of groundwater fluxes. The analysis considers the Äspö site and seven other study sites in crystalline rock in Sweden and employs SKB's interpretation of structures and hydraulic conductivities.

An important inference made from the simple evaluation is that connectivity and spatial structure of hydraulic conductivity are the critical uncertainties for the determination of groundwater flux. The variability of hydraulic conductivity estimates at the studied sites, including Äspö, is very large (up to 5 orders of magnitude) both in fracture zones and in the less fractured parts of the rock. However, limited or no direct information is available in the existing site characterization data regarding how conductive elements in the rock mass and fracture zones connect in space as most hydraulic measurements represent only a small volume of rock. The properties of a calibrated flow model will thus to a large extent depend on conceptual assumptions that cannot be directly supported from the site characterization data.

As a consequence of the large variability of the hydraulic conductivity the simple evaluation produced a wide range of darcy velocities (up to 7 orders of magnitude). The question to be addressed in SITE-94 is whether a more comprehensive use of detailed site characterization data will reduce this prediction uncertainty or if this is a true variability of the site.

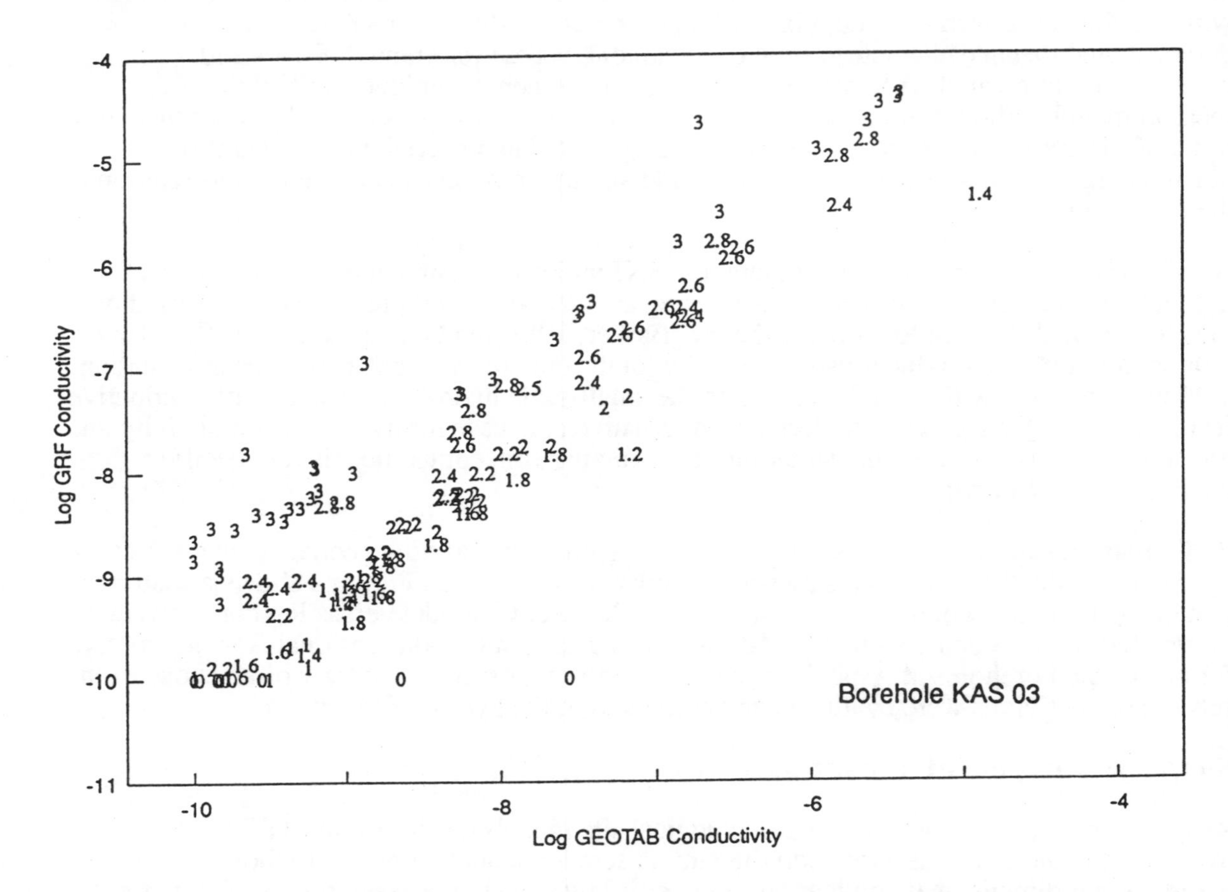

Figure 2. **Scatter-plot of log hydraulic conductivities estimated with Generalized Flow Geometry (GRF) model versus SKB estimates based on two-dimensional flow geometry. Numbers represent GRF estimates of flow dimension.**

Regional groundwater flow model

The regional flow system was analyzed through semi-generic simulations in a two-dimensional cross-section over Äspö (Boghammar and Grundfelt, 1993). The cross-section was 7 km long and 1600 m deep. Location and properties of major fracture zones were taken from SKB's interpretation of the site (Gustafson et al., 1991). The main objectives of the regional model were to determine boundary conditions for the site models and to determine an appropriate size of model domain for the detailed hydrologic site models.

Sensitivity analyses in the regional model indicated that local topography on the Äspö island governs the flow down to repository depth , i.e. 500 m depth. At greater depth small regional driving forces control the flow. This pattern was not significantly changed even when introducing an extensive sub-horizontal fracture zone at repository depth. The water that flows through the repository area discharges via fracture zones to the strait SE of Äspö.

Because the simulations did not account for density effects (increasing salinity with depth) the conclusions regarding relative importance of local versus regional driving forces at the site could be questioned. This issue is further analyzed in the hydrogeological evaluation for the climate evolution scenario.

Development of detailed site models

The hydrogeologic evaluation comprises two main conceptual site models; the discrete feature site model that integrates the geologic interpretation of structures and the stochastic continuum site model that relies primarily on hydraulic test data.

Discrete feature site model

The discrete feature model is based on a three-dimensional deterministic representation of the fractures identified in the SITE-94 geologic structure model and a stochastic representation of small scale fractures in the near-field rock. Figure 3 is a schematic illustration of the discrete model and the hypothetical repository. Note that the representation of fracture zones and fractures becomes more detailed towards the repository.

Initial estimates of hydraulic properties of the deterministic structures were obtained from the GRF-analysis and from type-curve analysis of hydraulic cross-hole tests. In the next step the initial estimates were refined through transient simulations of the hydraulic cross-hole tests. Finally the validity of the calibrated model was tested by predicting the draw down during a long-term pumping test. The goodness of fit was judged in terms of the transient response in packed-off sections of the responding boreholes.

The development of the stochastic near-field rock model started with a statistical analysis of fracture data from cores and outcrop fracture mapping. A separate geometric discrete fracture network (DFN) model was derived for each of the main rock types found at Äspö. Alternative models were considered for each of the following network properties; location (or clustering), fracture transmissivity, orientation and size. The hydraulic properties of the different DFN models were estimated through transient simulations of 3 m packer tests. The validity of each DFN model was assessed in a statistical sense by comparing the distributions of the interpreted transmissivity and flow dimension with those estimated in the GRF analysis of the actual tests. Two alternative models representing the dominating rock types at repository depth (Äspö diorite and Småland granite) were selected for analysis in the integrated discrete feature model.

In addition to the SITE-94 geologic structure model, the two independently developed geologic structure models (see above) are being evaluated hydraulically using the discrete feature concept. Because the objectives of the different structure models were not identical no attempt is made to determine which model that best describes site hydrology. However, the hydraulic evaluation of these alternative structure models will allow us to quantify the impact of geologic interpretation uncertainties on hydrologic predictions.

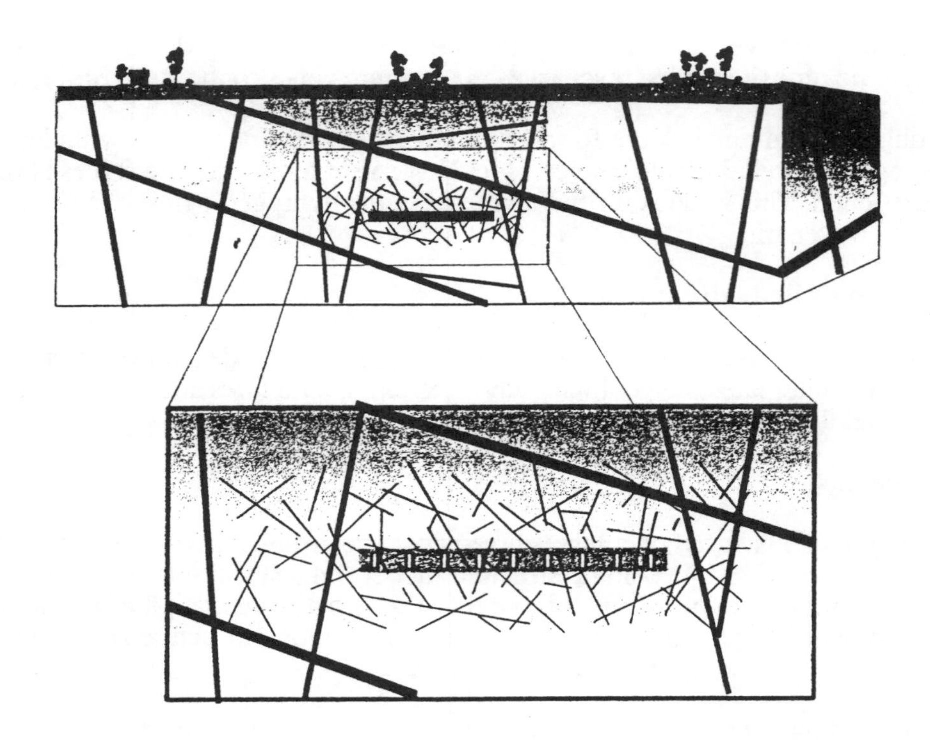

Figure 3. **Schematic view of the SITE-94 discrete feature hydrologic site model and a repository deposition tunnel.**

Stochastic continuum site model

The second main hydrologic site model is based on the stochastic continuum concept. This model generates three-dimensional realizations of the hydraulic conductivity field based on a geostatistical description of hydraulic packer test data. By means of conditioning each simulation reproduces the point values of hydraulic conductivity estimated from the packer tests.

The stochastic continuum model is calibrated through transient simulations of the hydraulic interference tests. However, because the stochastic continuum model is not heavily constrained by geologic data (or interpretations) it is possible to formulate different conceptual models of the hydraulic conductivity field that are equally consistent with the hydraulic test data. The conceptual variants to be evaluated in SITE-94 include different types of correlation structures, parametric (gaussian) versus non-parametric transmissivity distributions and alternative interpretations of hydraulic conductivities.

A non-parametric indicator simulation model (Gómez-Hernández and Srivastava, 1990) makes it possible to assign different correlation lengths to different classes of hydraulic conductivity. In this way it is possible to incorporate "soft" geologic information on fracture zones in terms of very long correlation for the highest class of hydraulic conductivities. The orientation of these geostatistically simulated "fracture zones" were taken from the geologic structure model of the site.

Evaluation calculations - preliminary results

Eventually the calibrated hydrogeologic site models will be used to develop predictions of hydrologic parameters for the evaluation of the performance of the engineered barriers and for the radio nuclide transport calculations, i.e. the consequence calculations. The procedure for estimation of these parameters involves the following steps:

1) Solving steady-state flow for specified boundary conditions

2) Sampling of calculated darcy velocities, fracture apertures, fracture spacing and other parameters in the near-field rock

3) Performing particle tracking calculations of releases from specified canister positions in the repository

4) Estimation of far-field transport parameters for releases from each canister position by fitting the 1D-ADE solution to the mass breakthrough at the interphase with the biosphere (far-field darcy velocities, dispersivities etc)

Somewhat surprisingly the results of the first set of evaluation calculations were rather consistent between the discrete feature and stochastic continuum models in terms of estimated ranges of darcy velocities and dispersivities. Moreover, the use of the alternative geologic structure models did not significantly change the hydrologic predictions made with the discrete feature site model. However, the hydrology models differed in terms of connectivity properties and correlation between near-field and far-field darcy velocities. In the discrete feature site model only 20% of all canister positions were connected to a flowing structure whereas the stochastic continuum model is hydraulically connected everywhere. In contrast to the discrete model there is only a weak coupling between near and far-field fluxes in the stochastic continuum model. These are examples of conceptual uncertainties that cannot be directly resolved from the site characterization data.

The more comprehensive use of data in the detailed site models reduced the prediction uncertainty of darcy velocity with several orders of magnitude compared to the simple evaluation. In both hydrologic site models the variability between realizations (ensemble variance) is small in comparison to the variability in a given realization. This implies that the remaining variability in the detailed site models is a true property of the site rather than a conceptual uncertainty.

GEOCHEMICAL SITE EVALUATION

The objectives of the geochemical site evaluation in relation to the performance assessment is to develop predictions of groundwater chemistry and geochemical conditions in the repository and along major groundwater flow paths from the repository to the biosphere, both for present-day conditions and for the site's future evolution. The approach to the geochemical evaluation is to analyze all available geochemical and isotopic information and build a general understanding of site geochemistry and its historic evolution through integration with geologic and hydrogeologic data.

Specifically the geochemical site evaluation should determine:

- origin of different groundwaters,
- the processes responsible for the chemical evolution of the Äspö site,
- residence time estimates (age) for the Äspö groundwaters,
- the structural and hydrologic implications of differences in groundwater chemical and isotopic composition, and
- the implications of the mineralogic record on past groundwater environments.

The preliminary results of the geochemical characterization suggest that five distinct groundwater types are present at Äspö: (1) recent waters, (2) deep saline waters, (3) glacial melt water, (4) waters with seawater imprint, and (5) groundwater with the chemical characteristics of the Baltic. These results gives indirect information on the climatic and hydrologic evolution in the Äspö region. The structural implication of these results will be evaluated by correlating different groundwaters to the structures in the geologic site model. However, such integration is not always unambiguous. For example the groundwaters with the chemical characteristics of sea water appear in structures that according to the hydrologic site models discharge into the Baltic.

A critical issue for the geochemical evaluation is to determine the representativity of the groundwater chemistry samples. At Äspö several groundwater sampling procedures were employed, including sampling during drilling, sampling during pumping tests and complete chemical characterization which involves continuous pumping of a borehole section prior to sampling. Because porosity in fractured rock is very low even extraction of small water volumes will result in withdrawal and mixing of groundwaters from large volumes of rock. Consequently the representativity of groundwater samples will depend on sampling procedure and borehole activities such as hydraulic testing performed prior to the sampling. In SITE-94 special efforts have been made to make use of ”first strike data”, i.e. samples taken in connection with drilling in order to determine undisturbed groundwater chemistry.

Another issue of importance, both for interpretation of the observed variability of groundwater chemistry and for prediction of radionuclide behaviour, is the ability to account for heterogeneities in mineralogy. Preliminary results of geochemical modelling indicate that the observed sets of fracture minerals are consistent with the analyzed groundwater compositions. Geochemical modelling has also been applied to describe the spatial variability in groundwater chemistry of importance for radionuclide behaviour. On the other hand, conclusive evidence on correlation of geochemical properties (e.g. in terms of fracture mineralogy) along certain groundwater flow paths has not yet been found. One explanation for this is the sparse sampling of fracture minerals due to the large spacing between boreholes. Another explanation could be that the possible changes in flow direction and recharge groundwater composition have not given a sufficiently well defined imprint on the observed fracture mineralogy.

DISCUSSION

The more stringent conclusions from the Site Evaluation sub-project in SITE-94 are yet to come. Such conclusions require feedback from the consequence analyses currently underway. Still, already at this point some observations can be made.

The preliminary results of the site evaluation in SITE-94 support the hypothesis that model development based on site characterization data from heterogeneous crystalline rock is highly non-unique. The multiple interpretation approach adopted in SITE-94 is a means to quantify such conceptual uncertainties. There is no guarantee that the analyzed models cover all types of critical uncertainties. The success of the approach will thus be restricted by the limited number of conceptual models that actually have been tested and the extent and accuracy of the available data. Still, all models tested strongly support the pronounced spatial variability of flow and other safety related properties of the rock at Äspö. Despite the principal differences between the applied models, all models predict fairly similar performance of the site.

Integrated interpretation of interdisciplinary site data helps constraining the possible conceptualizations of the site and gives confidence to the site evaluation. On the other hand, inconsistencies between different site models should be taken as a warning sign that our understanding of the site is insufficient in the first place, not that one or the other model is more or less valid.

A potentially important source of uncertainty not discussed in this paper is the transfer of results from the complex and detailed site models within hydrogeology and geochemistry to the often simplified performance assessment models. For example, in estimating effective hydrologic transport parameters such as average darcy velocities and dispersivities much of the complexity in terms of mixing processes, parameter variability and scale dependencies may be lost. Some aspects of these problems will be analyzed by testing different methods for parameter estimation, including whole curve and multiple peak analyses of the breakthrough curves predicted by the hydrological site models. In addition, a special simulation suite has been set up for the radionuclide transport model with the objective to compare chemical transport based on stochastic groundwater travel time distributions with corresponding deterministic simulations using effective values of fluxes and dispersion.

ACKNOWLEDGEMENTS

Most data analyses and modelling for the Site Evaluation in SITE-94 are performed by experts outside SKI, whereas integration, results compilation and synthesis is performed within SKI. Because SITE-94 is an ongoing project and most of the supporting documents are in draft form it has not been possible, at this stage, to acknowledge all our contractors through references. We therefore would like to acknowledge the following persons whose involvement in SITE-94 made this paper possible: Clifford Voss (USGS), Pierre Glynn (USGS), Joel Geier (Golder), Yvonne Tsang (LBL) and Sven Tirén (Geosigma).

REFERENCES

Almén K.-E., and O. Zellman, Äspö Hard Rock Laboratory. Field investigation methodology and instruments used in the pre-investigation phase, 1986 - 1990, SKB Technical Report 91-21, Swedish Nuclear Fuel and Waste Management Company, December 1991.

Andersson, P., J.-E. Andersson, E. Gustafsson, R. Nordqvist, and C. Voss, Site characterization in fractured crystalline rock. A critical review of geohydraulic measurement methods, SKI Technical report 93:23, March 1993.

Barker, J.A., A generalized radial-flow model for pumping tests in fractured rock, Water Resources Research, Vol. 24, pp. 1796-1804, 1988.

Boghammar, A., and B. Grundfelt, Initial two dimensional groundwater flow calculations for SITE-94, SKI Technical report 93:25, May 1993.

Doe, T.W., and J.E. Geier, Interpreting fracture system geometry using well test data, Stripa Project Technical Report 91-03, SKB, Stockholm, 1990.

Goméz-Hernández, J.J. and R. M. Srivastava, ISIM3D; An ANSI-C three-dimensional multiple indicator conditional simulation program, Computers and Geosciences Vol. 16, No. 4, p. 395 - 440, 1990.

Gustafson, G., M. Liedholm, I. Rhén, R. Stanfors, and P. Wikberg, Äspö Hard Rock Laboratory. Predictions prior to excavation and the process of their validation, SKB Technical Report 91-23, Swedish Nuclear Fuel and Waste Management Company, June 1991.

SKN, SKNs fortsatta undersökningar av förundersökningar och prognoser avseende Äspölaboratoriet, SKN report 61, Statens Kärnbränslenämnd, June 1992 (in Swedish).

Stanfors, R., M. Erlström, and I. Markström, Äspö Hard Rock Laboratory. Overview of the investigations 1986 - 1990, SKB Technical Report 91-20, Swedish Nuclear Fuel and Waste Management Company, June 1991.

Wikberg, P., G. Gustafson, I. Rhén, and R. Stanfors, Äspö Hard Rock Laboratory. Evaluation and conceptual modelling based on the pre-investigations 1986 - 1990, SKB Technical Report 91-22, Swedish Nuclear Fuel and Waste Management Company, June 1991.

Model Validation From A Regulatory Perspective: A Summary

N. Eisenberg and M. Federline
U.S. Nuclear Regulatory Commission
Washington, DC 20555

B. Sagar and G. Wittmeyer
Center for Nuclear Waste Regulatory Analyses
6220 Culebra Rd.
San Antonio, TX 78238-5166

(U.S.A)

J. Andersson and S. Wingefors
Swedish Nuclear Power Inspectorate
Box 27106
S-102 52 Stockholm, Sweden

(Sweden)

Abstract

Building confidence in the mathematical models used in safety assessments of HLW repositories is necessary if such models are to be applied successfully to the support of regulatory decisions. Conventional comparisons of model predictions to actual performance are not possible for such models over the temporal and spatial scales of interest. Uncertainties inherent in describing and modeling complex natural and engineering systems, as well as their interactions, make it difficult to discriminate between inadequacies in the mathematical models and the inadequacies of input data. A successful regulatory strategy for model validation, therefore, must attempt to address and carefully document these difficulties. Regulatory model validation efforts should seek to provide a documented enhancement of confidence in the model in so far as the model is necessary to support regulatory decisions. This document summarizes an approach jointly developed by members of the staff of the U.S. Nuclear Regulatory Commission and of the Swedish Nuclear Power Inspectorate for the validation of models used to assess the performance of geologic repositories for high-level waste disposal. The terms "validation" and "confidence building" are used interchangeably throughout this summary. "Confidence building" reflects a recognition that full scientific validation of models for performance assessment of the repository system may be impossible and that the acceptance of mathematical models for regulatory purposes should be based on appropriate testing which will lead to a reasonable assurance that their results are acceptable.

INTRODUCTION

Performance assessment (PA) using mathematical models is a key component of the evaluation of the long-term radiologic safety provided by a geologic repository. There is general agreement in the international community that a repository for high-level radioactive waste (HLW), including spent nuclear fuel, will consist of multiple barriers, and that these barriers will each contribute to the safety of the overall system by providing some level of redundancy. The long-term performance of the overall repository system and its components will be demonstrated by using mathematical models. Before such models can be used in the regulatory process to demonstrate compliance with safety standards, some measure of credibility and confidence in these models must be demonstrated in a process transparent to the parties to the regulatory process. The terms "validation" and "confidence building" are used interchangeably throughout this summary. "Confidence building" reflects a recognition that full scientific validation of models for PA of the repository system may be impossible and that the acceptance of mathematical models for regulatory purposes should be based on appropriate testing which will lead to a reasonable assurance that their results are acceptable. In the regulatory context, the level of confidence required for a given model is necessarily a function of the importance of that model's results to safety decisions.

The literature on the validation of models used to assess the performance of HLW repositories is vast and often contradictory, clearly reflecting the disparate views of those scientists, engineers, and policymakers involved in HLW management. This document summarizes a proposed regulatory approach, jointly developed by staff members of the U.S. Nuclear Regulatory Commission (NRC) and the Swedish Nuclear Power Inspectorate (SKI), to the validation of models used to assess the performance of geologic repositories for HLW disposal. The results of this collaborative effort represent the views of the authors and will be presented in more detail in a separate report (U.S. NRC and SKI, in preparation). Repository developers should not view this summary or the more detailed document as guidance, but merely as a report on current thinking with a view to developing future guidance.

REGULATORY CONTEXT

Although the United States and Swedish regulatory structures are different, both rely on predictive models to demonstrate compliance, and both nations' regulatory programs share a common interest in the development of procedures for demonstrating the validity of PA models.

Under current NRC regulations[1], models used to predict future conditions and changes in the geologic setting of a repository must be supported by a combination of field and representative laboratory tests, along with monitoring data and natural analogue studies. NRC regulations also require a performance confirmation program, through which the adequacy of modeling assumptions and performance predictions is to be verified, to the extent possible. Compliance must be demonstrated for both the overall environmental standard established by the U.S.

[1] In the U.S., the Environmental Protection Agency (EPA) is responsible for establishing generally applicable environmental standards for the management and disposal of high-level radioactive wastes including spent nuclear fuel. Under the Energy Policy Act of 1992, the U.S. Congress directed EPA to promulgate health-based environmental standards applicable to a proposed repository at Yucca Mountain, Nevada, consistent with the recommendations of a National Academy of Sciences' study currently underway. The NRC is required to conform its technical regulations to final EPA standards.

Environmental Protection Agency (EPA), and with the quantitative criteria established by the NRC for the performance of key repository subsystems.

As yet, similar regulations have not been issued in Sweden. However, criteria under development are expected to follow the recommendations presented jointly by the nuclear safety and radiation protection authorities in the Nordic countries (Nordic, 1993). These recommendations call for models used in safety assessments to be validated as far as practicable based on laboratory data and field measurements from the HLW repository site and natural analogue experiments.

It is anticipated that U.S. EPA standards applicable to disposal of HLW at Yucca Mountain will be probabilistic in nature and require that uncertainties be considered explicitly in performance estimates. But, even when regulatory standards are deterministic (e.g., the Swedish standard is expected to be expressed in terms of dose without attached probability; similarly the NRC's subsystem performance requirements are deterministic), regulators require that the uncertainty in model predictions be estimated and presented either as a range or, more commonly, as a probability distribution. This means that even where conservatism is invoked, it is necessary that estimates be made of uncertainties introduced due to model structure and assignments of certain preferred values to parameters.

Under both U.S. and Swedish regulatory regimes, the long-term performance of a repository will be assessed using quantitative PA and modeling techniques. Evaluation of the adequacy of these assessments will not only check whether estimated radioactive releases comply with specified criteria, but must also ascertain whether essential physical and chemical processes and their interactions have been identified, described adequately, and addressed.

COMPONENTS OF PA MODELING

Modeling for assessing repository performance is closely tied to site characterization and repository design. Data from site characterization and design features are crucial, not only to the development of appropriate conceptual models, but also to the extraction of parameter values that are employed to obtain numerical estimates of repository performance. Development of the conceptual model or models is the first step in PA modeling. Conceptual model development includes determining the governing equations, the geometry of the system, initial conditions, appropriate boundary conditions, and level of detail. For most natural and many engineered systems, formulation of a single, acceptable conceptual model is difficult, if not impossible, to achieve. In most cases, several classes of conceptual models are derived that satisfy the known constraints to varying degrees. Formulation of conceptual models for the natural system introduces problems that may not be encountered for engineered systems. Engineered systems, within limits, can be designed to meet prescribed performance criteria; whereas, the geologic system may only be explored and characterized. Complete characterization, however, is never possible for large and complex natural systems. Because tests may perturb the very properties being measured and because of the possibility that destructive testing could impair the barrier properties of the site, site testing may be further constrained. The site conceptual model, therefore, is based on considerable extrapolation of sparse quantitative and qualitative data, which can give rise to large conceptual and parameter uncertainty. In view of this, it is imperative that alternate site models be formulated and tested to account for possible biases in conceptual model formulation.

Once the requisite conceptual models have been developed, mathematical models representative of each conceptual model must then be formulated and usually are implemented using computer codes. Mathematical models of the total system, and their corresponding computer codes, which include realistic details of all system components and which treat parameter and future states uncertainty, can become very complex and computationally onerous. Under such circumstances, it is logical and appropriate to perform modeling using a hierarchy of models (Figure 1). The very detailed, and more realistic, models of individual processes comprise the first level of this hierarchy and are useful for understanding the sensitivity of a process to parameter variations and external forces. These first-level models also may be used to demonstrate the conservatism of assumptions and to provide a basis for second-level models in the hierarchy. In the second level, a limited number of the detailed models, with some simplifications, are coupled with one another, to gain some understanding of the interfaces among processes. In the third and final level, all component models are further simplified and coupled to formulate a "total system PA (TSPA) model." These are the fast, efficient models required for a probabilistic treatment of performance. However, it must be kept in mind, that if the coupling among the detailed models is strongly nonlinear, then it may not be easy to ascertain whether assumptions for conservatism are valid for the coupled system. In addition, not all processes are reduced to the third level of simplicity for inclusion in the system model; some processes have such a strong effect on the final result that they must be included in full detail. When such a hierarchy of models is used in demonstrating compliance with performance requirements, then all parts of the hierarchy need to be evaluated to build the required confidence, even though the type and amount of testing for each level in the hierarchy will be different.

The NRC (U.S. Nuclear Regulatory Commission, 1992) and SKI (Swedish Nuclear Power Inspectorate, 1991) have conducted PA iterations applicable to the current disposal concepts under consideration in their respective countries. In SKI's Project-90 and NRC's Iterative Performance Assessment (IPA), the repository and the neighboring rock are termed the "process system" or "repository system." Classes of events and processes external to the repository system are referred to as "external events" or "external environment." External events or the external environment acting on the repository system gives rise to scenarios. The initial system description, at the time the repository is sealed, is called the undisturbed-, base-, or nominal-case. Future disturbances of this system as a result of either natural or human-initiated external events or ongoing processes may alter the boundary conditions on the system or may be so profound as to modify the system description and necessitate corresponding changes to the underlying conceptual model. Scenarios may be defined as "physically plausible sequences of events and processes [occurring in the external environment] that could lead to release and transport of radionuclides from a repository" [Nuclear Regulatory Commission, 1992]. However, all HLW programs do not define scenarios in the same manner. For purposes of model validation, a uniform definition of scenarios is less important than the fact that each scenario is associated with a conceptual model. Each of these conceptual models will require some degree of validation. Because the effects of each disruptive scenario on the repository may be manifest in a number of different ways, it may be prudent to thoroughly exercise the models that implement these scenarios to ensure that the resulting total system performance assessments are reasonable.

REGULATORY APPROACH TO MODEL VALIDATION

The primary objective of this proposed approach is to provide a means to establish the adequacy of a model for its intended use in the regulatory context. Although it has been prepared by regulators, it is intended primarily for application by repository developers.

HLW regulators will be responsible for determining compliance of a proposed repository with environmental standards and implementing criteria. For regulators in both the United States and Sweden, the test of compliance with regulatory criteria is "reasonable assurance." This concept recognizes that absolute assurance of compliance is neither possible nor required. Instead, an applicant must provide such information as may be necessary to convince a reasonable decision-

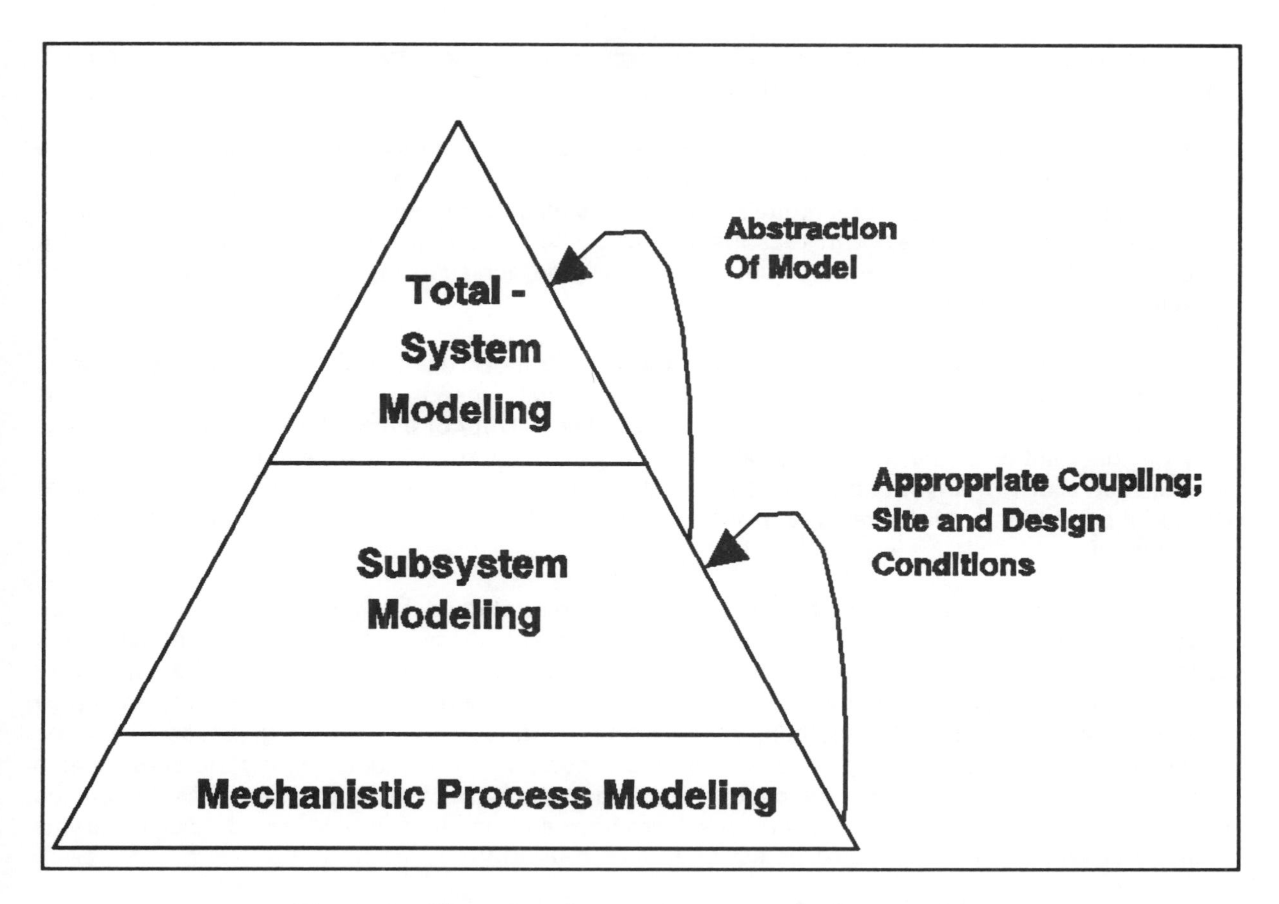

Figure 1. Hierarchy of performance assessment models

maker that compliance with regulatory criteria will be achieved. For these reasons, regulatory expectations for model validation are based on an applied science approach and differ from those appropriate to a purely scientific approach to developing and testing models. For example, a purely scientific approach compels pursuit of complete and detailed explanations for all observed phenomena independent of any particular model application. The regulatory approach requires only an adequate description of the phenomena for a given purpose (e.g., for the licensing of a repository) (Davis et al., 1991).

If, in the regulatory context, one assumes "validation" to mean demonstration that a model is sufficiently accurate for the purpose for which the model is used, there can be no standard answer to the question "How much validation is enough?" Rather, the answer will depend on the model itself and on its specific application. This does not imply that regulatory validation is entirely subjective. It is possible to envision a process or strategy whereby a repository developer and the regulator could reach an agreement on the degree of validation needed for each model used in a repository performance assessment.

Goals of a Validation Strategy

In devising such a strategy for validation of PA models, it should be made clear that the overall goals of validation are to: (i) establish whether the scientific basis is adequate for each model, and

(ii) demonstrate that each model is sufficiently accurate for the purpose for which the model is used by assuring that the scientific basis is correctly applied.

If a model is to be used as part of a demonstration of repository safety (or to challenge the projections of a model used for that purpose) it must be shown to have an adequate scientific basis. Speculative or conjectural models that have no plausible theoretical foundation or empirical basis will not demonstrate, with reasonable assurance, that repository performance will meet regulatory criteria. Thus, the minimum threshold to be achieved in validating PA models is to establish their scientific credibility.

Additionally, it must be demonstrated that any model used in a safety assessment is sufficiently accurate for the purpose for which the model is used. Implicit in this second goal is the need to validate each application of the model. The validity of a model prediction depends not only on the validity of the model, but also on the validity of the input parameters used with the model, the validity of any numerical implementation of the model, and the validity of interpretation of model projections. However, the strategy presented here focusses primarily on validating the model itself.

The repository developer should prepare a validation strategy describing the plans for validation of each model to be used as part of a repository performance assessment. A principal goal of this validation plan is to establish in a transparent fashion the process by which the repository developer will demonstrate a level of confidence in models consistent with their contribution to repository performance. In addition, the validation strategy will help to guide or focus the repository developer on formulating site characterization plans and in determining the performance goals for the components of the overall repository system. Validation strategies should be established in the early phases of a disposal program and updated as warranted. For programs well under way, this issue should be assigned a high priority.

For certain components of the repository system, a regulator may elect to develop his own PA models and may, therefore, need to establish an independent strategy for their validation. However, since the purpose of such models is not to demonstrate the safety of the repository system, but to probe and evaluate the projections of the developer's models, the goals of the regulator's validation strategy may be less ambitious than those of the developer. In many cases, the regulator only needs to establish the scientific credibility of its models so that the projections of those models can be compared to the projections of the developer's models. It should be recognized, however, that the regulator, in addition, will have to develop competence and procedures for review of the licensees compliance demonstration in this area. Any guidelines or rules for a validation strategy will have to be decided by the regulating body before the last phase of the licensing procedure.

A Strategy for Developing Confidence in Models

Shown in Figure 2, is a strategy for developing confidence in models that are used in HLW PA to demonstrate compliance with safety criteria. Implementation of this approach requires certain decisions, feedback, and iteration between a number of steps as follows:

(1) Define a compliance demonstration strategy that allocates performance of modeled systems and defines goals for model validation

(2) Determine the documented support that exists for each model necessary to demonstrate compliance

(3) Compare the validation goals to the existing support

(4) Decide whether to revise or retain the compliance strategy

(5) Obtain additional support for the model

Each of these steps is discussed briefly below and developed more fully in the a joint NRC/SKI white paper currently in preparation.

Define Compliance Demonstration Strategy

The confidence building process should begin with the development of an overall strategy for demonstrating compliance with the regulations. This strategy should identify the quantitative post-closure performance objectives and include plans for demonstrating that a repository will meet these objectives. The strategy is developed by taking into account repository design information and site information.

The quantitative performance objectives for a repository should be stated in terms of specific performance measures. A performance measure is a physical quantity, which depends on the long-term behavior of the repository and which indicates how well the repository isolates the radioactive waste from the environment or how well the environment is protected. Examples of performance measures include the concentration of radionuclides in groundwater or the dose to the maximally exposed individual. Minimum or maximum allowable values of the performance measure are identified as performance limits. Sometimes the performance measure may be estimated by a suite of linked computer codes, which represent models for various components of the repository or the environment. The resulting estimate of a given performance measure can then be compared to the performance limit for that component.

The repository developer may choose to include only certain components in a demonstration of compliance, either because (i) the components are very robust and drive the system performance measure well below the maximum limit allowed, or (ii), the components do not significantly add to the uncertainty in the estimate of performance. These choices of which components to include in models for demonstrating compliance comprise the compliance demonstration strategy. The degree to which each component is necessary for demonstrating compliance constitutes the allocation of performance to that component. This allocation of performance determines the level of validation required for a particular model. Models representing those components central to a compliance demonstration necessarily will be subject to a greater degree of validation than those representing more peripheral components. Although performance allocation can be described quantitatively, it is doubtful that a universally applicable, quantitative measure of model validity can be devised. Performance allocation should be used to establish semi-quantitative goals for the desired level of validity of various models. These goals might be expressed as the rank ordering of importance of the models or as a small number (2-5) of categories representing smaller or greater need for validation.

An iterative process is best suited for determining acceptable levels of performance for each component of an overall repository system. Based on the safety standards for the performance of the entire repository system, an initial design is made before site characterization begins. This initial design should describe: (i) the level of performance expected for the natural barriers, (ii) the level of performance to be obtained from engineered barriers, (iii) the level of confidence anticipated for each projection of performance, and (iv) the safety factors, margins for error, or redundancy among barriers (if any) to be incorporated into the design of the repository system. Descriptions of the desired level of confidence for projecting performance should include a

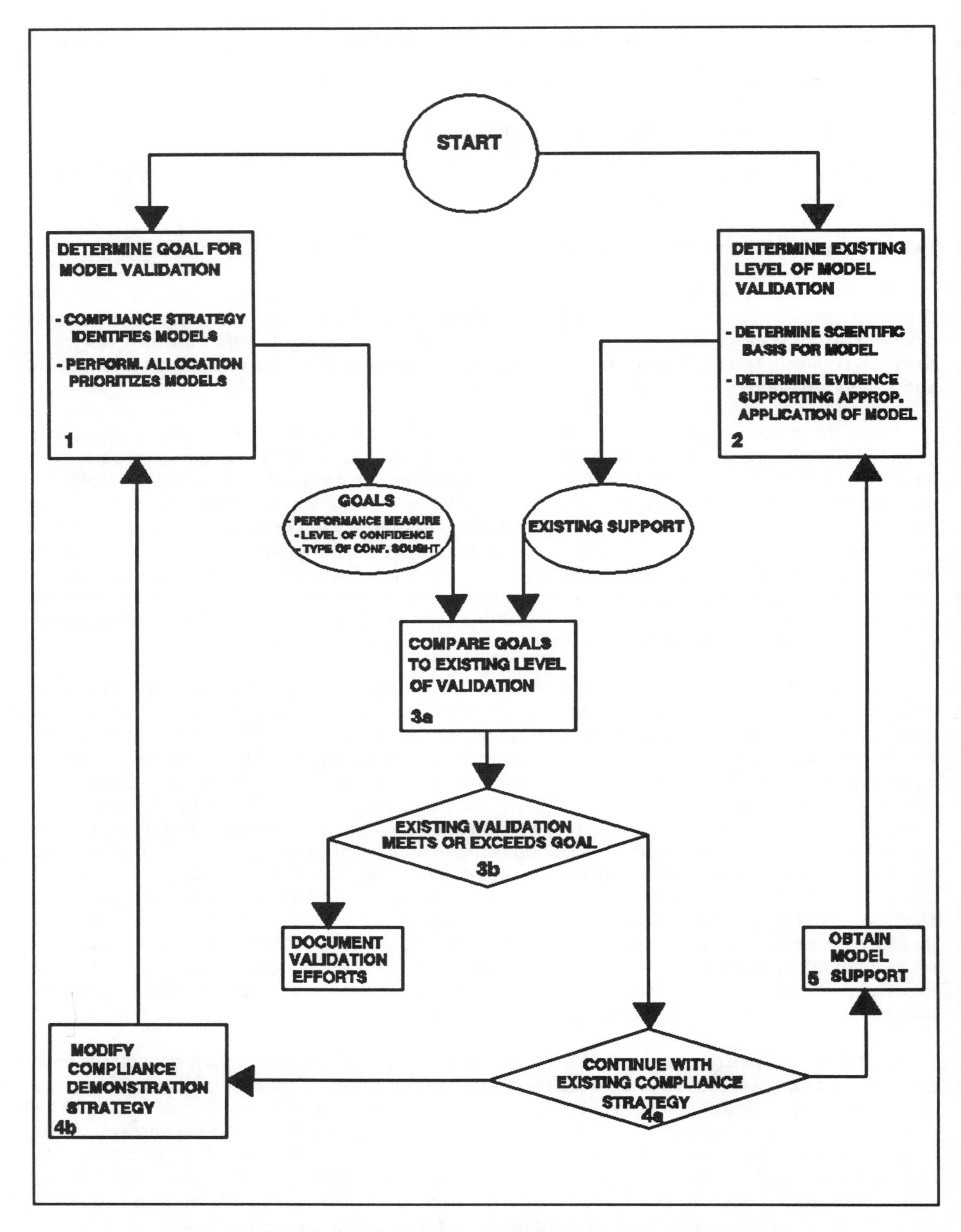

Figure 2. Regulatory strategy for developing confidence in models

discussion of the repository developer's plans for validating any models employed as well as for obtaining peer and regulator review of the results. Each IPA should be followed by a systematic review resulting in updated judgments of the relative importance of various submodels and assumptions.

Determine the Existing Level of Support for the Model

Models used in PA must be supported by a sound and well-documented scientific basis if regulators and the public are to have confidence in modeling results. It is essential to identify which aspects of a given model are based on accepted science, and which lack credibility or are potential sources of uncertainty. Having identified the models and performance measures of interest, an effort should be undertaken to review and assimilate the scientific literature relevant to the use of the models for the applications of interest. Because many studies might have employed a particular model, but may not have applied it in the same manner or for the same purpose, it may be necessary to reevaluate and/or recompute experimental or theoretical results, in order to apply the model to the compliance measure in question and subsequently arrive at a judgment as to what level of model support is exists.

In order to establish scientific support for a given model, it is necessary to both examine its theoretical basis and to evaluate the application of scientific principles in the model to assure the application is appropriate. Normally, the validity of the application of principles is achieved by comparing the model predictions against empirical information. For models used in HLW disposal, the evaluation of the application of scientific principles and the comparison of model results to empirical data is limited, because each site is unique. This limits the modelers' ability to extrapolate from one site to another. Another limitation on the use of empirical data is that experimental data collected over the time scales of interest are not available. Evidence from field data and natural analogue studies are available for longer time and spatial scales, but there is considerable uncertainty about the environmental conditions prevailing for the system over these long times.

Compare the Validation Goals to the Existing Support

If the existing support for the model exceeds that required to demonstrate compliance, then no further validation activities are needed. If the existing support is insufficient, then it is either necessary to reduce the reliance placed upon the modelled component and revise the compliance strategy accordingly, or to acquire additional support for the model. Reducing reliance on one component, in general, can be expected to require increased reliance on one or more other components.

Decide Whether to Revise Compliance Strategy

Most proposed repository systems should have sufficient margins-of-safety or redundancies such that several different combinations of system components assigned different relative priorities may be included in a successful demonstration of compliance. Overall safety criteria may be fixed, but there may be several legitimate means of demonstrating safety with those criteria, each of which may be associated with some degree of uncertainty. If it is decided to revise the compliance demonstration strategy, a new set of components must be selected, or new priorities assigned to the old ones. This will require, in turn, that new models with different performance measures be identified, or different priorities assigned to the old models. In either case, the evaluation and comparison with the existing level of support will need to be revisited. If the original compliance strategy is to be retained, then additional technical support must be acquired.

For example, if the Swedish HLW program were to decide that descriptions of far-field migration are very uncertain, it could elect to place greatest reliance on the long-term stability of the canister. However, if subsequent study of the validity of canister stability models cast doubt on their validity, more confidence might then be placed on the retarding mechanisms of the far-field (i.e., change the compliance strategy). Alternatively, it might be decided to conduct further research on canister corrosion phenomena (i.e., obtain further model support).

A similar hypothetical example, applicable to the proposed U.S. repository at Yucca Mountain, could be postulated such that only transport in the unsaturated zone would be modeled because the cumulative releases of radionuclides from the unsaturated zone to the saturated zone can be shown to meet the regulations. Conversely, if water transport in the saturated zone was shown to be slow, it could be assumed that radionuclides are released directly into the saturated zone, and the more complex modeling of the unsaturated zone transport could be avoided.

Modification of a compliance demonstration strategy may reduce the need for additional model support. However, in most cases, additional support will be required for adequate decision making.

Obtain Additional Model Support

The support for a given model may be bolstered by theoretical advances, the acquisition of additional laboratory or field data, or from further study of appropriate natural analogues. Generally speaking, some measure of theoretical support is required for all models used in PA and, in some cases, theoretical support may be used to substitute for experimental evidence. Virtually all the models used in PA are based on well-established scientific principles, such as conservation of mass, momentum, and energy. Difficulties arise in applying these principles to complex situations, such as the flow of water in heterogeneous, partially-saturated, fractured rock. Nevertheless, extensive theoretical analyses, with evaluative experimental studies, are available on topics and systems relevant to nuclear waste disposal. To the extent that the scientific basis is well-established for both the fundamental theory and the application of that theory to processes, phenomena, or systems related to nuclear waste, this information (which has incorporated previously obtained empirical results) may be substituted for experimental support for validation. This type of evidence, when presented in a logical fashion, may be especially useful in supporting claims that a particular model is conservative in a given application. Some models, for example, very simple "models," that consist solely of correlations of variables in experiments of limited scope, require further theoretical substantiation, if model results are to be extrapolated to times or conditions not encompassed by the original data.

Progress in validation may also require additional experimental evidence. This can be the case for generic issues, like coupled near-field phenomena, as well as for cases where the validation issue is to show that a particular process or structure is applicable to a specific site. Additional experimental evidence may be used to add confidence to either the scientific basis for the model or the particular application of the model. In the latter case, it may be necessary to show how well characterization measurements actually characterize the site. This will generally require experiments both at the actual site, as well as at other sites, in order to confirm the reliability of the site characterization techniques. In addition, confirmation of the application of established principles may be directed to the interpretation of site characterization data, especially as they pertain to conceptual models. Experimental evidence is more likely to be needed to support the scientific basis for a model for areas in which the theoretical and empirical bases are incomplete or still evolving (e.g., models of the formation and migration of colloidal contaminants in the geosphere).

Experimental data may not be necessary for validation of PA models in all cases. Applying the concept of model hierarchy discussed earlier, the theoretical and empirical basis for more detailed models may be sufficiently well-established, such that one may be able to use theoretical arguments, based on the detailed models, to confirm the validity of the abstracted models. This is limited by the degree to which the model of interest is coupled with other models. Such an approach may be especially useful for conservative or bounding assumptions used for the modeling of engineered components of the repository. In such cases, the support sought may be gained by development of a deeper understanding of the process in question than is actually needed in the PA model. Such an understanding can be attained simply by more detailed and careful modeling of important processes.

Due to scale and time limitations associated with conducting laboratory and field experiments, their usefulness in model validation is limited mainly to understanding the processes at work in the real system; however, if a model fails to agree with experiments conducted over limited scales, chances are small that it will be satisfactory at larger scales. Comparing PA subsystem and complete system models to laboratory and field test data is a much more difficult task, since the function of these models is to predict system performance over large spatial and time scales. In general, there will be more confidence in a model which compares favorably to several types of evidence covering a range of spatial and time scales, such as laboratory tests and natural analogues.

Laboratory experiments are useful because they can be performed in a controlled environment that minimizes uncertainty in initial and boundary conditions, and experiments can utilize samples that exhibit relatively little geometric variability or whose variability can be measured. However, the use of laboratory experiments in validation efforts is limited by the inability to perform tests on the long time and large spatial scales representative of a HLW repository, difficulties in testing some coupled processes, and the possibility that the systems used are not representative of *in situ* conditions (e.g., samples damaged in collection, not enough samples collected to characterize spatial variability, laboratory conditions inconsistent with field conditions which may produce phenomena that do not actually occur in nature) (Davis et al. 1991).

Field experiments overcome, to a degree, the problem of whether data is representative and some of the scale problems that plague laboratory experiments. To a certain extent, field experiments can be direct surrogates of repository performance (e.g., field heater tests and tracer tests). However, the usefulness of field experiments is limited by uncertainties in initial and boundary conditions and, to a large degree, by the possible conceptual misunderstanding of field conditions (Davis et al. 1991). Nevertheless, field experiments are necessary tools for site characterization and a thorough understanding of their potential usefulness and limitations is certainly warranted.

Lastly, evidence from appropriate natural analogues have value for increasing the temporal and spatial scales available for experimental study. In some sense, nature could be considered to have initiated experiments that could be used for validation. The transport of radionuclides from uranium deposits and the transport and deposition of minerals along fractures are two examples. These "experiments" have the advantage of having taken place on temporal and spatial scales that are comparable to those relevant to HLW repository systems. In addition, coupled processes are often involved that are difficult to reproduce in either the laboratory or the field. Uncertainty in initial conditions, boundary conditions, and the temporal evolution of the physical system, however, limit the usefulness of natural analogues in validating models (Davis et al. 1991). Another potential drawback of natural analogues is their complexity. It is difficult to demonstrate a sufficient understanding of an analogue, with inferred historic evolution, in order to support claims that the studied analogue is directly relevant to the system or subsystem modelled.

When planning experiments or field studies to support a model, it should be stressed that the experiments should be planned based on a systematic analysis of their potential for resolving the identified problems. In such planning there are some "good practices" which may prove helpful. These practices, which can be applied to the collection of laboratory, site, and natural analogue data, include the following:

(1) Identify potential alternative models and then design tests that will discriminate among the various alternatives and the preferred model, if there is one.

(2) Design experiments that will enhance the fundamental understanding of important processes included in a model. A suite of experiments carried out on different scales, if achievable, will add confidence.

(3) To the extent practical, design experiments that test models over the type and range of conditions for which the models will be used. When it is impractical to test over the full range, as will usually be the case for repository models, means to expand the data base (accelerated testing) or to scale the data (e.g., by using dimensionless numbers) should be used with great caution. Since many of the phenomena of interest are dependent on scale and/or experimental conditions, simple relationships for scaling or extension of data may be unusable. Tests should be designed to identify the conditions for which model results will be invalid.

(4) Various scenarios may be represented by different boundary conditions applied to existing models or, in the case of profound changes produced by the scenario, completely different models. A complete program for building confidence in models must treat these different conditions and/or models to the extent that they contribute to the overall measure of safety of the repository.

(5) If a model intended for predictions over long times and large spatial scales fails to predict processes accurately over shorter time and smaller spatial scales, this is strong evidence that the model may not be valid.

(6) When analyzing the results of an experiment it is recommended that subsets of data not be excluded for arbitrary reasons; all relevant data should be used to evaluate the accuracy of the model and to evaluate the potential for errors and biases introduced by the experimental technique. However, only the data relevant to the predictive model under study need be used; and only the phenomena and variables of interest need be explained by the model.

(7) Agreement with an experiment is insufficient, alone, to validate a model; the scientific basis for the model must be supported and scrutable. One must ensure that generally accepted scientific principles (e.g., those describing flow of groundwater through a porous matrix) are applied to model the conditions anticipated for a specific repository.

(8) In general, if a single model is divided into two or more submodels, the degree of confidence imparted by evaluating the submodels individually will not be as great as the degree of confidence achieved by evaluating the submodels linked together. Therefore, it is desirable to perform additional tests designed to validate the combination of submodels, as the combination will be used for repository performance assessments. In the absence of the practical ability to perform tests for combinations of submodels, careful theoretical evaluation is called for.

(9) Data used to develop or calibrate a model cannot be used alone to validate that model. Model calibration is performed to demonstrate that the model is consistent with the system being modeled. Validation on the other hand, is the testing of the models ability to simulate the same system under different conditions. Thus, at least two data sets (or a partitioned set of data) are required for model validation.

(10) If a model is intended to be conservative rather than realistic (i.e., to overestimate potential repository impacts), tests, or proofs, should be designed to verify that the model is, in fact, conservative.

(11) Accurate records of model development and testing should be maintained, subject to periodic peer reviews during the development and testing process. These records should include the analyses and rationale supporting the decision to accept or reject the plausibility of various conceptual models.

With all their limitations, long-term field experiments and natural analogue studies nonetheless make it possible to study complex, coupled systems. This constitutes an invaluable check that no essential process or coupling effect has been omitted. Furthermore, even a qualitative fit between standard PA models and results from long-term experiments provides increased confidence in the model.

DOCUMENTING STATEMENTS OF VALIDITY

Experiments, derivation of models and assumptions, compliance demonstration strategies, and other reasoning should be openly and clearly documented in order to make possible a comprehensive review of the validation strategy. Thorough and systematic review, possibly including international cooperative efforts, is fundamental in judging validity. However, it should be emphasized that review efforts will be productive only when the material to be reviewed is supported by quantitative analyses of experiments or other proofs, or is derived from interpretation of natural analogues or more detailed modeling.

In a license application, the documentation of the validation strategy will be of great importance in judging the credibility of safety analysis calculations. The overall objective of this documentation should be development of a framework which facilitates the acceptance (or rejection) of models, based on transparent and logical reasoning.

SUMMARY

Model validation has been a topic of debate in HLW management for several years. There is considerable disagreement within the technical community about what constitutes model validation and how to achieve it. Furthermore, the process for validating PA models is likely to be a lengthy one, requiring laboratory and field experiments and natural analogue and theoretical investigations. It is important that PA be viewed as an applied science, rather than a pure science, and that the models used need only be valid enough to provide predictions useful for their intended application. An approach to model validation has been presented which outlines an iterative process of defining and redefining compliance demonstration strategies based upon the extent of scientific support available or reasonably attainable throughout the characterization of a potential repository. A number of good practices are advocated that may assist the repository developer in building confidence in PA models and in their interpretation of their results.

REFERENCES

Davis, P.A., N.E. Olague, and M.T. Goodrich. 1991. *Approaches for the Validation of Models Used for Performance Assessment of High-Level Nuclear Waste Repositories*. NUREG/CR-5537, SAND90-0575. Washington, DC: U.S. Nuclear Regulatory Commission.

Environmental Protection Agency. 1993. Environmental radiation protection standards for management and disposal of spent nuclear fuel, high-level, and transuranic radioactive wastes. Code of Federal Regulations 40, Part 191. Washington, DC: Environmental Protection Agency.

Nordic Document. 1993. *Disposal of high-level radioactive waste. Consideration of some basic criteria*. The Radiation Protection and Nuclear Safety Authorities in Denmark, Finland, Iceland, Norway, and Sweden.

Swedish Nuclear Power Inspectorate. 1991. *SKI Project-90*. Stockholm, Sweden: Swedish Nuclear Power Inspectorate: SKI TR 91:23.

U.S. Nuclear Regulatory Commission. 1983. *10 CFR Part 60; Disposal of High-Level Radioactive Wastes in Geologic Repositories—Technical Criteria.* Federal Regulation 48: Washington, DC: U.S. Nuclear Regulatory Commission: 28,194–28,229.

U.S. Nuclear Regulatory Commission. 1983. Final Technical Position on Documentation of Computer Codes for High-Level Waste Management. NUREG-0856. Washington, DC: U.S. Nuclear Regulatory Commission.

U.S. Nuclear Regulatory Commission. 1992. *Initial Demonstration of the NRC's Capability to Conduct a Performance Assessment for a High-Level Waste Repository.* NUREG-1327. Washington, DC: U.S. Nuclear Regulatory Commission.

U.S. Nuclear Regulatory Commission and Swedish Nuclear Power Inspectorate. *Model Validation from a Regulatory Perspective: A White Paper*, in preparation.

Evolution of flow and transport conceptual models at the waste isolation pilot plant, 1978 to 1984

Peter Davies
Tom Corbet

Geohydrology Department
Sandia National Laboratories

Abstract

Flow and transport conceptual models have evolved as the result of interactions among specific model concepts, field studies, and numerical simulation studies. This paper examines major steps in that evolution process beginning with the relatively primitive models used in the original Environmental Impact Statement in the late 1970's. Major iterations in field testing and model revision have occurred over the past 15 years. This review illustrates the evolution from relatively little conscious consideration of underlying conceptual models to significant focus on the assessment of alternative conceptual models, both numerically and in the field. Major flow conceptual issues include degree and importance of flow confinement in the Culebra Dolomite and the existence and importance of long-term transients. A major transport conceptual issue is characterisation of the degree of fracture-matrix interaction in the transport process.

Validation Strategies for Licensing Nuclear Waste Repositories in Salt Formations in Germany

P. Bogorinski
Gesellschaft für Anlagen- und Reaktorsicherheit (GRS) mbH

K. Schelkes
Bundesanstalt für Geowissenschaften und Rohstoffe

R. Storck
GSF - Forschungszentrum für Umwelt und Gesundheit GmbH

J. Wollrath
Bundesamt für Strahlenschutz

(Germany)

Abstract

The Gorleben salt dome in Germany has been selected as a potential site for a radioactive waste repository to be constructed within a salt dome. Licensing of such a nuclear waste repository requires the demonstration of the post-closure safety of the facility. This involves extensive use of models to assess the potential future evolution of the site with respect to thermal-geomechanical effects, behaviour of the waste, release of radionuclides and their migration through the geosphere and uptake by man. Since observations of the relevant processes will not be available over the long time period involved, confidence-building measures are needed to convince the authorities as well as the public, that the repository is safe. Part of this process is model testing, usually referred to as validation, which requires the design of appropriate experiments and data sampling, both to be carried out during site characterization as well as in the underground facilities to be constructed.

Validation has to be done within the framework of German regulations which require compliance with deterministic dose limits.

With rock salt as host formation, a number of specific phenomena and processes have to be considered. Models for the creep of salt and compaction of backfill material and seals due to heat and pressure as the most important processes in the near-field have to be tested against in-situ experiments and supporting laboratory experiments. Appropriate laboratory experiments are also carried out to validate geochemical models used to determine the degradation of the waste form and the solubility of the radionuclides under the conditions of a brine environment. Groundwater movement and hence radionuclide migration in the strata overlying the salt dome is strongly influenced by density changes caused by the dissolution and transport of salt. The observed current salt distribution in the aquifer system can be used as a natural analogue for modelling the corresponding processes in the far-field. Sorption processes in a brine environment require special attention.

Since mechanisms in the near-field are independent of those in the far-field, validation work on these areas can be performed separately. However, the scope of the work in these fields should be consistent and concentrate on the relevant issues. Use of natural analogues helps to build up confidence in analyses carried out over long time periods.

1. Introduction

The Gorleben salt dome has been selected as a potential site for the German radioactive waste repository. It is located in the rural district of Lüchow-Dannenberg near the community of Gorleben in the northeastern part of Lower-Saxony. The Gorleben salt dome has a length of approx. 14 km and a width of up to 4 km and reaches from a depth of approx. 3,500 m up to 260 m below the surface. It is overlain by a sequence of Tertiary and Quarternary sedimentary layers forming a heterogeneous system of aquifers and aquitards. The repository will be located at a depth of approx. 850 m which is approx. 600 m below the top of the salt dome. It is planned to dispose of there all kinds of solid or solidified radioactive wastes arising in Germany.

Licensing of such a nuclear waste repository requires the demonstration of the post-closure safety of the facility. This involves extensive use of models to assess the potential future evolution of the facility with respect to thermal-geomechanical effects, behaviour of the waste, release of radionuclides and the transport through the geosphere and uptake by man.

Most of the surrounding salt formation will not be affected by the radioactive waste repository. In case of the normal evolution of the disposal system due to the low permeability of the host rock transport of radionuclides is limited to the very near field of the waste containers, where small amounts of brine from inclusions in the surrounding rock could get contaminated. This contaminated brine would then penetrate into backfill material and sealings as well as the surrounding rock up to a few meters but will never be released from the salt formation. The accidental intrusion of brine from outside the salt formation into the backfilled repository is being considered as a disturbed evolution of the disposal system. One possible natural pathway between the repository and the aquifer system overlying the salt dome may be an anhydrite layer, where water can flow from the overburden into the repository and at later times back into the overburden driven by the convergence of the salt formation. This brine intrusion scenario was the basis for existing post-closure safety assessments and is the basis for further discussion of validation strategies for radionuclide transport phenomena.

The principal requirements for the safety assessment of a radioactive waste repository in Germany are laid down in the Safety Criteria for the Disposal of Radioactive Wastes in a Mine [14], which are set up by the Federal Government due to the recommendation of the Reactor Safety Commision. A tenet of the approach to post-closure safety is to provide the same level of protection to future generations as it is afforded to people today. The safety assessment has to be done on a site-specific basis using state of the art data and models. The need of a site-specific safety assessment implies that a site is only suitable if the analyses show that the protective goal defined in the Radiation Protection Ordinance [17] is reached. The main requirement laid down by the Federal Government is expressed as follows: movement of radioactivity from a nuclear facility should not lead to a significant increase in the dose rate naturally occurring in Germany. This is reached by limiting the dose rates an individual is exposed to to the range of variation of the naturally occuring dose rate. More specifically compliance with a deterministic dose limit of 0.3 mSv/a is required [17].

With rock salt as host formation for a radioactive waste repository a number of specific phenomena and processes have to be considered such as creep of salt and compaction of the backfill material due to heat and pressure, the effect of limited brine intrusion on the degradation of the waste forms and the solubilitiy of radionuclides, and the influence of density changes due to salt dissolution on the groundwater flow in the overburden which all effect the release of radionuclides from the waste into the biosphere. To evaluate the post-closure safety of the overall disposal system the radiological consequences of the brine intrusion scenario are being estimated. Several safety analyses have been carried out over the last years using repository designs of that time and up to date information from the site investigation programme (e.g. [15]). These studies increased the knowledge about the feasibility of the site. The results are beneficial in steering the site investigation and scientific research programmes. They help to show the important features and processes as well as the gap of information and data.

To investigate the geology and hydrogeology of the Gorleben site an investigation programme was carried out between 1979 and 1985 consisting of drilling numerous boreholes and piezometric wells, pumping tests, geoelectrical and geothermal studies, gravimetry, seismology, geochemistry and micropalaeontology [2]. The investigation of the site from the ground surface will be continued for a few more years and includes studies on the territory of the former GDR north of the River Elbe. Especially these results will help to reduce uncertainties in the hydrogeological description of the area. The objective of the future exploration from underground is to acquire all information needed to evaluate the safety of the planned repository. The stratigraphy and structure of the salt dome will be investigated by geological and geophysical methods. The mechanical behaviour of different parts of the salt dome will be investigated by extensive laboratory and in situ tests. All this work is complemented by numerous laboratory experiments and in situ experiments in the Asse salt mine. It gives the basic information to perform validation exercises.

2. Validation Strategies

2.1 General Remarks

The term validation usually refers to testing how accurately models describe observable natural processes. In the context of waste management several organizations have formulated definitions of the term validation. These definitions generally indicate that validated models should give a good representation of the real system and that the validation process should be performed by comparing modelling results with experimental results or other empirical observations [6]. This is a generic statement which holds also for salt. Therefore, this definition is also usable for the safety assessment of the planned radioactive waste repository located in the Gorleben salt dome.

One basic difficulty in validating models for use in post-closure safety assessments is that predictions have to be made for the far future. Another difficulty is the proof of the transferability of results of laboratory or in situ experiments on the scale and conditions of the planned repository and the extrapolation in time. It is not possible to test such predictions directly. Hence, confidence must be built in the adequacy of the extrapolation and transferability of the models.

Validation is important in reducing model uncertainties by providing assurance that models adequately represent reality and indicating where models require further development. Validation cannot be expected to result in precise agreement between observations or measurements and model predictions. But they are useful in building confidence in the safety assessment. The results of such safety assessments could be seen as one main safety indicator of the disposal system [11]. Therefore, the use of validated models might give an essential contribution to confidence building. They are needed to convince the authorities as well as the public, that the repository is safe. Another important aspect of confidence building in the safety assessment is peer review through the international scientific community. For this purpose German organizations responsible for development and licensing radioactive waste repositories take part in international projects such as PAGIS [15] or INTRAVAL [6].

The German deterministic safety criteria requires principally a validated model for an integrated safety assessment of the overall system. Because this is not possible, the German approach is to validate models sufficiently for parts of the system or distinct processes. Due to the fact that the processes in the salt formation, the groundwater movement and the radionuclide transport in the overburden are decoupled from each other, all processes influencing this three main topics of the safety analyses could be handled seperately. In Chapter 2.2 the important processes in the salt formation are identified and the state of the validation efforts is pointed out, Chapter 2.3 deals with the main processes influencing the groundwater movement and Chapter 2.4 concerns to the radionuclide transport in the overburden of the salt dome.

2.2 Radionuclide Transport in the Salt Formation

Considering the brine intrusion scenario, the transport of radionuclides in the salt formation is affected by the intrusion process itself, the contamination of the brine after contact with the waste, and the transport of contaminated brine through the salt formation. Due to the low permeability of undisturbed rock salt the transport of radionuclides takes place through backfill and sealings and through disturbed zones of the salt formation. The major disturbed zone to be considered is the intrusion pathway which is assumed to be an anhydrite vein which opens after mechanical deformations caused by the thermal input into the salt formation.

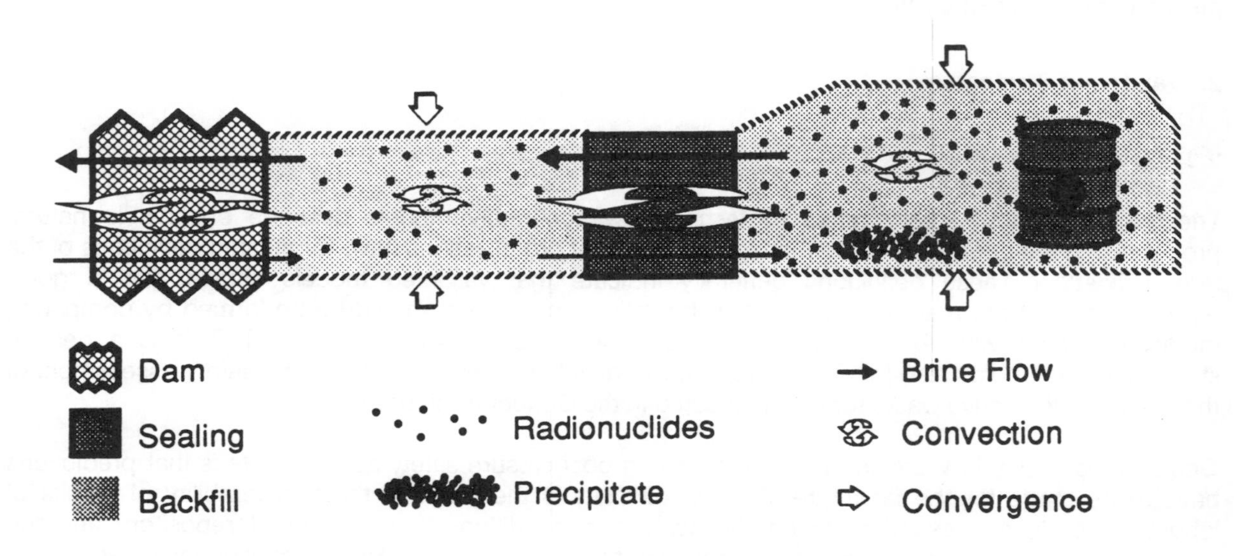

Figure 1. **Schematic overview of transport phenomena in the salt formation**

Models and data for relevant radionuclide transport phenomena in the salt formation were included in all performance assessment studies up to now. A description of the transport phenomena is given with most of the assessments (e.g. [15]) and a schematic overview on the phenomena is given in figure 1. The transport phenomena within the salt formation are also called near-field phenomena. Although the models and data for the near field are not validated in a strong sense, they are considered to have a sufficient degree of confidence to draw conclusions from the results on the importance of phenomena and data for the long-term safety of the repository. Near-field phenomena ranked by decreasing importance are listed in the following:

- convergence rate of the salt formation
- permeability of dams
- volume of brine pockets
- permeability of the backfill
- time of brine intrusion
- solubility limits

As can be seen from the list, the mechanical phenomena are of higher importance compared to the chemical phenomena. This is a result of the mobile radionuclides which are clearly dominating the peak doses in all performance assessments. Nevertheless, the validation efforts are devoted to mechanical phenomena as well as chemical phenomena. The mechanical phenomena are included due to their high importance. The chemical phenomena are included because the lower importance of the less mobile radionuclides is a result of chemical retardation phenomena. The activities and strategies for the validation of near-field phenomena are discussed in the following.

Open cavities in rock salt are being closed by the pressure of the overburden and the plastic characteristics of the rock salt. The closure rate of open cavities, the so-called *convergence rate*, has been observed in existing mines over many years [5]. At the planned repository depth and at corresponding rock temperatures the convergence rate in a homogenous medium varies between 0.5 and 1.0%/year and is well based.

Convergence rates could seriously be affected by heterogeneous structures of the salt formation. Salt domes with massive anhydrite layers show convergence rates of one or two orders of magnitude lower. This site-specific and also location-specific phenomena can only be investigated at the given site. Hence, an extensive site investigation or an on-site confirmation programme has to be performed, which covers the dependencies of the convergence rate on the location and also the depth.

The convergence rate of rock salt is also seriously affected by the rock temperature. This phenomena has been investigated by several heater tests in the Asse mine attended by extensive rock mechanics modelling work. Since the disposal of heat-producing waste is foreseen into predominating homogenous rock, the effect of increasing temperatures on the convergence rate can be predicted sufficiently with respect to the long-term safety of the repository.

The convergence process in backfilled cavities will result in a *consolidation of the backfill*. If crushed salt is used for backfilling, the consolidation will continue up to the characteristics of the undisturbed salt formation are reached. This full reconsolidation will require several centuries assuming undisturbed rock temperatures. The compaction process has not been observed over longer time periods especially not at lower porosities. The model for the reconsolidation of crushed salt is therefore mainly based on theoretical considerations.

Modelling of the backfill compaction will require greatest efforts in the future. This includes laboratory experiments especially at lower porosities for dry material as well as for different levels of brine content. To get enough information on the full time scale of this process, the compaction process has to be accelerated without modifying the type of backfill consolidation [19], [16].

The validation of backfill compaction models also requires the arrangement of in situ experiments. While laboratory experiments are limited to the backfill material itself and to smaller set ups, the in situ experiments could include the rock matrix and therefore the interaction between the rock salt and the backfill material as well as they consider the influence of the real scale. Both, the laboratory experiments and the in situ experiments are planned within the Debora project of the European Community [9].

The low *permeability of the dams* hinder the inflowing brine to reach the waste packages. At the same time the convergence/compaction process especially at higher temperatures isolates the disposed waste within the rock mass. Therefore, most of the waste packages will never be reached by the inflowing brine. This is the reason for the high importance of the dam permeability and flow resistances in general for the long-term safety of a salt repository.

Unfortunately, the data for the dam permeability used in safety assessments are not based on real measurements or experiences from existing mines but are based on expert opinion. Hence, there is a strong need to investigate the behaviour of dams especially over long time frames. The individual components of a dam such as the mechanical abutment, the long-term seal and the surrounding rock have to be investigated by laboratory experiments so that a prediction of the overall performance of the dam system can be made and subsequently validated by the results of an in situ experiment.

Waste matrixes which have been considered up to now are the vitrified high level waste, the spent fuel itself and cemented low and medium level waste. In all cases a full *degradation of the waste forms* has to be assumed under the saturated brine conditions already after a few centuries. Container under discussion are thin-walled containers for transportation purpose only and thick-walled containers which last for several hundred years.

All these time scales are short compared to the available time periods for the transport of contaminated brine of several thousand years. Because of this, minor importances of the waste packages in long-term safety analyses have been observed [1]. This would reduce the need for a pretentious validation exercise. Only, if the time requirements for container failure or the waste degradation are much higher, additional efforts should be made for validation of waste degradation models. However, the extrapolation of data from short-scale experiments to the long-term environment with maybe changing chemical conditions has to be born in mind.

The reason for the low importance of all the actinides and the less mobile fission products is the *solubility-limited release* of radionuclides, especially of the transuranic elements, from the disposal locations. The solubility limits used in existing safety assessments are based on expert opinion with some background from laboratory experiments. This is by no means adequate for a license application. Hence, a well founded data base for solubility limits has to be established [10].

The development of well based solubility limits requires in a first step a variety of laboratory experiments under different chemical conditions to get sufficient thermodynamical data. In a second step geochemical models have to be used for interpretation of laboratory experiments and the development of simplified models and data. These are then to be included in safety assessment codes to handle the solubility-limited releases from disposal vaults with different waste contents and chemical environments.

2.3 Groundwater Movement in the Geosphere

The groundwater flow in the sediments covering the Gorleben salt dome depends on the distribution of the permeabilities and porosities, and on the hydrogeological situation, which is given by groundwater levels or the recharge and discharge distribution and locations of the hydraulical boundaries. Especially in an environment around a salt dome like Gorleben or other salt structures, the density and viscosity of the fluid, whose changes are caused by salt content, temperature and pressure, as well as salt dissolution processes and diffusion and dispersion effects, which influence the salt distribution, have to be considered, too [18].

The hydrogeological system surrounding the salt dome consists of Tertiary and Quarternary sediments, which form a multiple-aquifer system of a porous medium. Only the caprock of the salt dome has to be considered as a fractured medium or transferred into an equivalent porous medium. One characteristic feature is an Elsterian subglacial erosion channel filled with sands and gravels, which crosses the salt dome from south to north and forms the lowermost aquifer above the salt dome.

Like in other sedimentary areas, the groundwater flow is widely determined by the distribution of the hydrogeological units, characterized by their permeabilities. From laboratory and field experiments permeabilities as well as porosities were determined. Together with transferrable data from other investigation areas and hydrogeological expertise a data base was built up for the Gorleben area, which can be used for the modelling of the groundwater flow system. This data base will be enlarged by results of the ongoing investigation program.

Permeability values, which were measured or calculated from measurements on samples often differ very much from values measured in field experiments. The anisotropy and inhomogeneity of the system have an important impact on these values. Therefore, the variability in the hydrogeological parameters has to be taken into account. Field experiments like pumping tests or tracer tests can be used as calibration experiments with respect to the hydrogeological structure of the system. Calculations on the basis of pumping tests were carried out for the Gorleben area under the simplifying assumption that density effects on the flow field could be neglected. The results gave a good picture of the heterogeneous structure and information on anisotropies of the aquifers up to some distance of the pumped well [2]. Geological and geophysical field work as well as a new pumping test will give a better understanding of the hydrogeological structure especially about the area of the former GDR. It has to be taken into account that new field experiments should be planned and carried out in such a way that they are also useful as validation exercises for groundwater flow models.

To investigate the behaviour of the overall groundwater system, a deterministic modelling approach with a sensitivity analysis for the parameters, which, based on hydrogeological expertise, may have an impact on the flow field, was used during the last years in the German approach [3]. All calculations in the first deterministic groundwater models were carried out assuming groundwater with constant water density [3], [2]. But the groundwater movement in the surrounding layers of a salt dome, thus also throughout the investigation area of the Gorleben salt dome, is normally strongly influenced by the variable density of the water caused by the variations in the salt content. Here, the freshwater body is underlain by rather saline water. The salt content usually increases with depth. It reaches that of saturated brine near the top of the salt dome [3]. Therefore, the variable density has to be taken into account. Otherwise it has to be proven that the assumption of a constant water density is conservative with respect to contaminant transport, which could be difficult, sometimes also not possible.

Connected to the influence of the salt content is the fact that for the calculation of the groundwater flow data were used for the above-mentioned porosity distribution, the diffusivity and tortuosity and for the dispersion effects. Other influencing factors like temperature changes or short time variations of the groundwater level are of minor importance for the groundwater regime under natural conditions in this area. For most of these data a sufficient data base is available except for the dispersion coefficients. For the dispersion, the Scheidegger model with constant dispersivities is used in the German approach, although it is known that for very high density gradients the value will change [6]. But such high gradients were not observed in the field until now. Additional studies on the theoretical basis of the dispersion concept could be significant. Basic scientific research has also to be done for the description of the subrosion processes and especially their mathematical description for the use in numerical models.

Dispersion coefficients as defined above, which have been calculated from laboratory experiments, are generally not transferrable to a regional scale. In addition to the use of internationally available data on regional dispersion lengths special experiments, e.g. tracer experiments, are necessary to give information on this transport parameter. Such tracer experiments are very difficult to carry out under natural conditions in a heterogeneous groundwater system with very low flow velocities of less than one meter per year and experimental depth of more than one hundred meter. They have to be designed carefully. Possibilities for such experiments are given in combination with a planned pumping test or as a direct tracer experiment in lower depth with forced groundwater flow.

In general, a validation of constant density groundwater models with respect to very long times of prediction seems to be difficult. It led to controversial discussions in the scientific community during the last years (i.e. [7]). This holds much more, if transport processes have to be taken into account as it has to be done in the case of groundwater flow with varying density. As a great difficulty, groundwater models with variable density, which are mathematically based on coupled nonlinear partial differential equations, need much more computer space and computing time. Therefore, such calculations for the Gorleben site have been carried out until now with two dimensional models only and coarse grids [18], [12]. Because it is necessary to investigate the density dependent groundwater movement in more detail, research projects are going on to develop computer codes which will use modern mathematical methods and which are therefore much faster and more efficient in modelling such systems than it was possible with the existing codes.

However, the salt content and the density distribution in the groundwater give the possibility for a validation of the groundwater models in a paleohydrogeological approach (fig. 2). In this approach the actual salinity distribution will be used as a target for a time dependent modelling of the transient flow system. The idea is to start from a definite initial situation in the past, e.g. at the end of an ice age, and calculate the present status of the groundwater system with the observed salinity distribution. This procedure is also used in natural analogue studies. For the Gorleben site, the results on the behaviour of the sodium chlorid concentration with time can be seen as a natural analogue for the transport of radioactive tracers. The greatest benefit of this approach can be seen in its degree of confidence building with regard to the possibility of a calculation of the groundwater flow processes at the site over long time periods. The knowledge about the system behaviour in the past will allow a better extrapolation into the future.

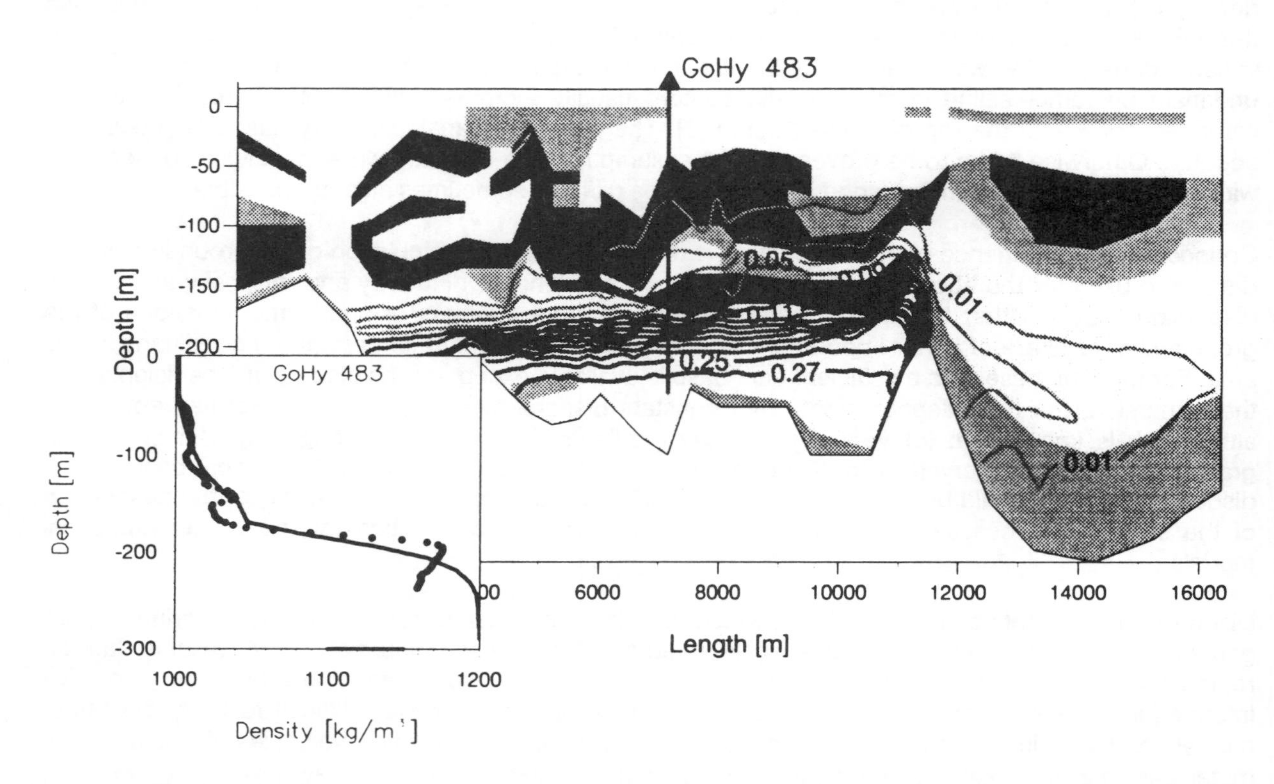

Figure 2. **Hydrogeological cross section** (permeabilities from 10^{-9} m/s (dark grey) to 10^{-5} m/s (light grey)) **with an example of the density distribution** (isolines of the salt content in kg/kg) **10,000 years after the end of an ice age with a definite initial density distribution Intercomparison of calculated** (straight line) **and measured** (dotted line) **density data as a function of depth at the location of observation well GoHy 483**

The initial studies with regard to this approach [18], [13] have shown some difficulties, which have to be investigated furtheron. It became clear that the arrangement of the hydrogeological units has a major impact on the regional groundwater flow patterns and the associated salinity distribution which makes it necessary to look in more detail in the hydrogeological settings. It could be shown that the initial salinity distribution and the length of the time simulated have a profound influence on the intermediate simulation results. The hydrogeological conditions in the past, e.g. the density distribution during ice ages, appear to affect the present density distribution and the associated groundwater system.

Therefore, investigations on the paleoclimatic conditions have to be carried out to give the initial and time dependent boundary conditions for the modelling of the density-dependent groundwater flow system [13]. This will be another very difficult part of the definition of this system, because the paleoclimatic considerations will probably not produce a unique initial situation as well as changes in the surface boundary conditions.

2.4 Radionuclide Transport in the Geosphere

The most likely pathway for the release of radionuclides from a nuclear waste repository to the environment is their transport with flowing groundwater through the natural barrier system of the geosphere. It involves advective transport with the flowing groundwater, dilution processes due to dispersion, and retardation mechanisms like sorption and chemical reactions.

Retardation encompasses all processes in which transport of substances solved in the groundwater is delayed due to interaction between solutes and surrounding rock. It plays an important role in limiting the consequences of potential releases of long-lived radionuclides from the disposal facility. Assuming substantial retardation these radionuclides have sufficient time to decay due to the inherent delay in their travel time. Therefore, these processes must be studied carefully.

The relevance of the various mechanisms depends on the nature of the system in which the transport takes place as well as on the type of the transported radionuclides. Considering the potential repository site of Gorleben the following processes play a major role in limiting migration of radionuclides through the geosphere:

- sorption as a combination of various physical and geochemical processes,
- dispersion, already dealt with in the previous section,
- diffusion into stagnant water usually referred to as matrix diffusion.

Processes like dispersion or matrix diffusion also effect the migration of non-reactive tracers, i.e. those radionuclides which do not interact with the solid phase or with other groundwater constituents. However, sorption including geochemical processes are among the most effective retardation mechanisms [10].

Current computer resources do not allow the use of complex geochemical models in the frame of performance assessments, when a large number of simulations will have to be carried out. But even if this could be resolved will it be almost impossible to acquire all data needed for detailed retardation models considering the spatial variability involved. Therefore, only simple retardation models are normally applied in performance analyses which mostly uses one single parameter, the distribution coefficient k_D, to describe the ratio of radionuclide mass sorbed at the solid phase to the radionuclide mass in solution. These models assume the process to be in equilibrium at any time and to be fully reversible. It is evident that such models do not account for complex physical or chemical mechanisms in a rock-water system. This holds also for even more detailed sorption models which assume the system to behave non-linearly, i.e. where sorption depends on the concentration of the radionuclides assuming saturation of the surface of the solid phase at higher concentrations. Surface complexation models seem to be the most appropriate approach taking into account the relevant influences of the solid and the liquid phase. However, thermodynamical equilibrium models require detailed knowledge of the geochemical system which may not be available at real sites [8].

The main means for studying retardation processes is by performing laboratory experiments, either dynamic column experiments or static batch experiments to study sorption processes only. Such experiments should be well controlled and the data acquired should have small uncertainty ranges. The relevance of these experiments primarily depend on the technique used in the experiments, their duration, and how representative the geochemical and other parameters are for real-field situations.

However, non-standard experimental procedures contribute to the difficulties in the interpretation of the results. Nevertheless, the techniques are well known and if the experiments are carried out carefully they should provide data for the purpose of simple retardation models. To date the value of performing such experiments is therefore in acquiring data for a specified set of conditions. But even with such well controlled experiments it will be difficult to separate the various processes contributing to retardation and therefore it will also be difficult to test models encompassing only single retardation mechanisms against those experiments. Sorption data are sufficiently well known to date to provide a satisfactory picture of radionuclide migration in the geosphere at Gorleben for the purpose of safety assessments, however these have to be supported by more detailed geochemical analyses and modelling.

Field experiments are normally carried out to obtain the whole qualitative and quantitative picture of solute transport in natural systems. However, field experiments are normally not as well characterised and controlled as laboratory tests. Furthermore it will also be almost impossible to study the various retardation processes separately in real systems. This is especially true for a complex site like Gorleben with its sequence of sandy and gravelly aquifers containing clay and silt lenses and clayey aquitards with sandy interbeds. As a starting point for radionuclide transport field studies an appropriate hydrologic model of the investigaion site is essential. Heterogeneities will have a big influence on this interpretation as will the varying salt content of the groundwater. The same site characteristics make the proper modelling of radionuclide transport difficult: Heterogeneities will contribute to large-scale dispersion, solutes will diffuse into clayey material and will get retarded by matrix diffusion and additionally by sorption, the salt content in the system will influence the geochemical reactions with varying conditions throughout the system. These processes have not yet been studied in detail by in situ tests. Only such tracer tests complemented by appropriate laboratory experiments will provide sufficient information to validate the models needed in safety assessments for the Gorleben site.

Both laboratory experiments as well as field studies lack the spatial and temporal scale involved in performance assessments. One of the means to cope with this scientific challenge of extrapolation into the far future is the study of natural analogue systems. These are usually sites where migration of elements from ore bodies occured under natural conditions over long time periods involving hydrogeological and geochemical processes. Climatic changes and consequently variations in the hydraulic system during the past contribute to uncertainty. The unknown history of the system as well as its unknown initial conditions may be the source of doubts about the quantitative validity of the results of natural analogue studies. However, careful investigation of how the system evolved in the past by using information from all fields of geosciences may overcome these doubts to a certain degree. The ability to simulate the history of a natural analogue system with respect to hydrogeology and solute transport using scientifically based assumptions with models developed for performance assessment will result in confidence building in the results of long-term safety analyses. Furthermore, natural analogue systems demonstrate the retention potential of the geosphere qualitatively. At Gorleben transport of salt through the system in the past may be looked at as a natural analogue for how the system may behave in the future. The same holds of naturally occurring isotopes like ^{14}C or ^{18}O and others. This will have to be studied more carefully in the future.

3. Conclusions

The validation process needs a long term stragegy [6]. It is an iterative process between performing experiments and model calculations. In some parts of the performance assessment this process has come to results [4], [12], whereas in other parts this process is still at the beginning. The need for additional experiments taking into account site-specific phenomena as well as validation-specific requirements is obvious.

Models and data for the near field are not validated in a broad sense. Utilizing the results of existing safety assessments, the validation effort can be directed to the most important areas. Special attention should be devoted to the site-specific influences on the convergence rate, the compaction resistance of the backfill material, the permeability of dams and sealing systems and the solubility limitations for the very near field.

The data base for the calculation of the groundwater movement is sufficient for most of the parameters. It will be enlarged during the ongoing field work. Special attention has to be given to the problems connected with the modelling of the dispersion of constituents. New field experiments and accompanying laboratory experiments have to be planned with respect to their use as validation exercises. But, looking at the validation process, for the groundwater movement only the short-time behaviour can be validated by such site-specific experiments. The results of these exercises help in confidence building in the groundwater flow models and their hydrogeological data basis. They can only be used to a certain extent as an argument for the discussion on the long-term behaviour of the system. The paleohydrogeological approach is one possibility to show the site-specific applicability of groundwater models for the long-term behaviour and gives a scientific basis for the use of such a model in prediction calculations. Validation studies on natural analogues, and in this case on the site itself, seem to be one very important instrument for confidence building in the models used for long-term predictions.

To date most of what can be expected from retardation and sorption experiments on laboratory scale using site-specific sedimentary material as well as groundwater from the site has been achieved satisfactorily. Additional geochemical studies may prove to be useful to back the result especially when extrapolating the results to different conditions e.g. to account for complexation caused by naturally occurring humic acids. One of the major unresolved issues in radionuclide migration as in other scientific areas of concern is the upscaling from laboratory to repository scale in space and time. Therefore, in the future one should put emphasis on the design of appropriate field experiments and geochemical studies to account for the relevant site conditions and their variation. These may also result in the need for new laboratory experiments. Having built confidence in the models on this scale, studies on migration of naturally occurring radionuclides during the past using appropriate interpretations from paleogeological information may help to build up confidence in the capabilities to estimate the behaviour of radionuclides potentially leaking from the repository in the future.

References

[1] Buhmann, D.; Brenner, J.; Storck, R.: The Efficiency of Source Terms in Performance Assessments for Salt Repositories. Proc. 1994 Int. High-Level Radioactive Waste Management Conf., 22-26 May 1994, Las Vegas, 1994

[2] Bundesamt für Strahlenschutz: Fortschreibung des Zusammenfassenden Zwischenberichts über bisherige Ergebnisse der Standortuntersuchung Gorleben im Mai 1983. Report ET-2/90, Salzgitter, 1990

[3] Fielitz, K.;Giesel, W.; Schelkes, K.; Schmidt, G.: Calculations of Groundwater Movement and Salt Transport in the Groundwater of the Sedimentary Cover of the Gorleben Salt Dome. Proc. Int. Symp. on Groundwater Resources Utilization and Contaminant Hydrogeology, Vol. II, Montreal, 1984, pp. 383-392

[4] Glasbergen, P. (ed.): Experimental Study of Brine Transport in Porous Media, INTRAVAL Phase 1, Test Case 13. Swedish Nuclear Power Inspectorate and OECD/NEA, Paris, 1992

[5] Heijdra, J.-J.; Prij, J.: Convergence Measurements in a 300 m Deep Borehole in Rock Salt. ECN-Report ECN-C-92-016, Petten, 1992

[6] INTRAVAL: The International INTRAVAL Project, Phase 1, Summary Report. Swedish Nuclear Power Inspectorate and OECD/NEA, Paris, 1994

[8] Koß, V.: Stand des Wissens zur Sorptionsmodellierung für die Bewertung der Radionuklidmigration in der Sicherheitsanalyse - Statusbericht. GSF-Report 39/91, GSF - Forschungszentrum für Umwelt und Gesundheit GmbH, Braunschweig, 1991

[7] Konikow, L.F.; Bredehoeft, J.D.: Ground-Water Models Cannot Be Validated. Advances in Water Resources 15(1992), No. 1, pp. 75-83

[9] Prij, J.; Rothfuchs, T.; Spies, T.: Sealing of HAW-boreholes in Salt Formations: Objectives and First Results of the DEBORA Project. Proc. PEGASUS Meeting 11-12 June 1992, EUR 14816 EN, Commission of the European Communities, Brussels, 1992

[10] Read, D.: Status Report on Geochemical Modelling. GSF-Report 42/91, GSF - Forschungszentrum für Umwelt und Gesundheit GmbH, Braunschweig, 1991

[11] Röthemeyer, H.: Langzeitsicherheit von Endlagern - Sicherheitsindikatoren und natürliche Analoga. In: Atomwirtschaft 39 (1994), Nr. 2, pp. 136-139

[12] Schelkes, K. (ed.): Studies on Saline Groundwater Movement in an Erosion Channel Crossing a Salt Dome, INTRAVAL Phase 2, Gorleben Test Case, Report (in prep.)

[13] Schelkes, K.; Vogel, P.: Paleohydrogeological Information as an Important Tool for Groundwater Modeling of Gorleben Site. In: Paleohydrogeological Methods and their Applications for Radioactive Waste Disposal, Proc. of an OECD/NEA Workshop, OECD/NEA, Paris, 1992, pp. 237-250

[14] Sicherheitskriterien für die Endlagerung radioaktiver Abfälle in einem Bergwerk (Safety Criteria for the Disposal of Radioactive Wastes in a Mine). In: Bundesanzeiger 35 (1983), Nr. 2

[15] Storck, R.; Aschenbach, J.; Hirsekorn, R.-P.; Nies, A.; Stelte, N.: Performance Assessment of Geological Isolation Systems for Radioactive Waste (PAGIS): Disposal in Salt Formations. EUR 11778 EN, GSF-Report 23/88, Commission of the European Communities, Gesellschaft für Strahlen- und Umweltforschung mbH Munich, Brussels, Luxembourg, 1988

[16] Stührenberg, D.; Zhang, C.L.: Untersuchungen zum Kompaktionsverhalten von Salzgrus. Salzmechanik, 13 (Festschrift Langer), BGR, Hannover, 1993

[17] Verordnung über den Schutz vor Schäden durch ionisirende Strahlen - Strahlenschutzverordnung vom 13.10.1976, BGBl. I, p. 2905, in der Neufassung vom 30.6.1989, BGBl. I, p. 1321, zuletzt geändert durch die Strahlenschutzregisterverordnung vom 3.4.1990, BGBl. I, p. 607

[18] Vogel, P.; Schelkes, K.; Giesel, W.: Modeling of Variable-Density Flow in an Aquifer Crossing a Salt Dome - First Results. In: Custodio, E.; Galfore, A. (eds.): Study and Modelling of Salt Water Intrusion. Proc. 12th Salt Water Intrusion Meeting, CIMNE, Barcelona, 1992, pp. 359-369

[19] Zhang, C.-L.; Heemann, U.; Schmidt, M.-W.; Staupendahl, G.: A Constitutive Model for Description of Compaction Behavior of Crushed Salt Backfill. In: Proc. ISRM Int. Symp. EUROCK '93, Lisboa, Portugal, 21-24 June 1993

Evaluating Models Used for the Postclosure Assessment of a Nuclear Fuel Waste Disposal Concept

K.W. Dormuth and A.G. Wikjord
AECL Research
Canada

Abstract

The evaluation of models for long-term performance assessment of geological disposal systems plays a vital role in establishing the reliability of the assessment. A model may be evaluated by testing the underlying assumptions and hypotheses or by comparing model results with observations. Evaluations do not in themselves allow the conclusion that the use of the models in an assessment is valid. Rather, the validity of the assessment must be discussed in the overall context of data, models, assumptions, model results, interpretation of results, and conclusions. In assessing the safety of waste disposal systems, it may not be necessary or practicable to model a particular process in detail. Conservative and unrealistic assumptions are often made which lead to an overestimate of impacts. Examples from the assessment of the Canadian disposal concept illustrate a framework for evaluation of performance assessment models.

1 INTRODUCTION

Environmental assessments or license applications for nuclear waste disposal facilities are often based on a comparison of estimates of long-term impacts with a quantitative criterion. Mathematical models must be used to make the required estimates, and the validity of the safety case will depend to a large degree on the adequacy of these models for the time frames and space domains of interest.

In the simulation literature, a mathematical model is considered adequate for a particular purpose once it passes a "validation" test in which the behaviour of the mathematical model is compared with the behaviour of the system being simulated to determine whether the two are in reasonable agreement. The criterion by which reasonable agreement is determined is a matter of judgment.

Use of a validated model does not in itself imply that the conclusion drawn is valid, because the models must generally be applied beyond the range of the data values used for validation. As a textbook example, Murthy *et al.* [1] constructed a regression model to predict the maximum speed of aeroplanes in the year 2000. They determined the model parameters using historical data for the years 1900–1940. They then validated the model by demonstrating that its estimates for the years 1941–1965 agreed with historical data for those years within their predefined criterion. Having thus validated the model, they used it to forecast the maximum speed of aeroplanes in the year 2000 to be 16 000 m.p.h. In spite of the use of a validated model, the result should be subjected to tests of reasonableness. It is the result and any conclusion drawn that must be judged valid, not the model employed [2].

Unfortunately, some have argued that to say a model has been *validated* implies it should make reliable predictions without qualifications [3]. In part to avoid this confusion over usage, we shall use the broader term *model evaluation* to refer to the testing and analysis by which the adequacy of a model is inferred.

The role of model evaluation in the performance assessment process is to provide information to justify the use made of the model to arrive at conclusions important to the assessment. The evaluation indicates that the model will have a certain accuracy or that it will give a conservative result; or the evaluation may provide only indirect evidence through supporting an underlying hypothesis. In any case, the analyst uses the information to support the reliability of calculations used in the assessment. Judgment enters into the use of the information from the model evaluation by both the analyst and the reviewer .

We discuss below the evaluation of the postclosure assessment models used for a long-term safety case study for the Canadian nuclear fuel waste disposal concept.

2 Postclosure Assessment Models

The Regulatory Policy Statement R-104 of the Canadian Atomic Energy Control Board (AECB) requires that for ten thousand years following closure the estimated probability that an individual of the critical group will incur a fatal cancer or serious genetic effect must be less than $10^{-6}/a$ [4]. A system model has been constructed to make the estimates required for comparison with this risk criterion for a disposal system consisting of nuclear fuel waste emplaced deep in plutonic rock.

To ensure that the model represents a realistic situation and is based on a consistent data set, it has been constructed using site- and design-specific information for a case study [5]. The site features are based on the geological characteristics of AECL's Whiteshell Research Area in Manitoba, and a surface environment representative of the Canadian Shield. The design features are based on a conceptual engineering study [6].

The system model pertains to a hypothetical array of disposal rooms, about 4 km^2 in area, at 500 m depth in a large granite batholith. The rooms and all the access tunnels and shafts are backfilled and sealed. Long lasting metal containers with waste are emplaced in the rooms surrounded by sealing material. The batholith was formed over 2.5 billion years ago and is several kilometres deep with and exposed surface over 60 km long and 20 km across at its widest part. The model was designed to be applicable in the absence of a major disturbance such as glaciation, i.e., about 10^4 a or more.

The system model consists of three major components:

- The vault model calculates the release of contaminants to groundwater and migration through the sealing materials near the waste containers into the pore spaces and fractures of the surrounding rock [7].

- The geosphere model calculates the movement of contaminants through the pore spaces and fractures of the rock to the surface environment [8].

- The biosphere model calculates the movement of contaminants through the surface environment and the resulting dose to an individual of the critical group [9].

Because the results of the model are to be compared to the risk criterion, it is not necessary that the system model provide an accurate estimate of the dose to the individual. Rather, the model should provide a high estimate. Therefore, many conservative assumptions (assumptions that lead to higher estimated dose than any reasonable alternative assumptions) are employed in the models. For example, containers are assumed to provide no protection for the waste form when a defect of any size occurs in the container wall.

In the application of the system model, both deterministic and probabilistic approaches were used. The probabilistic approach is employed to examine the variability in model results resulting from variability in the input data [10].

3 Framework for Evaluation of Performance Assessment Models

Comparison of the system model with an actual disposal system would require observation of at least some contaminants moving through every major component of the system, which could take thousands of years. Instead, the evaluation is based on the study of important features, events, and processes that form the basis of the system model.

The detailed models are evaluated by

- establishing that assumptions are conservative,

- comparing model estimates with observations,

- testing hypotheses, or

- accepting hypotheses on the basis of expert knowledge and judgment.

In addition, the intercomparison of independently developed models (for example, the international INTRAVAL and DECOVALEX test cases discussed elsewhere in these proceedings) are valuable exercises to build confidence in the detailed models.

Tables 1, 2, and 3 give example frameworks for the evaluation of the vault, geosphere, and biosphere models, respectively. These tables are not meant to be definitive or complete, but to illustrate a framework that could be used by analysts and reviewers to evaluate the reliability of the models. The examples are drawn from reports on the development and application of the models [7],[8],[9],[11].

The following section provides more detailed examples of model evaluation in practice.

4 Examples of Model Evaluation

4.1 Radiological Exposure of Humans (Conservative Assumption)

AECB Regulatory Policy Statement R-104 [4] provides guidelines for identifying the individual of concern in long-term safety analyses. It states the following:

> The concept of the critical group is commonly employed when applying individual dose limits to members of the public affected by existing nuclear facilities. This concept involves the identification of a relatively homogeneous group of people that is expected to receive the greatest exposure because of its location, age, habits and diet. Owing to the conservative assumptions usually made in selecting critical groups and in defining their life-styles, the doses actually received by members of the group will in most cases be lower than the estimated mean dose of the critical group. It follows that doses to individuals outside the critical group are even lower.

The critical group, for which the biosphere model estimates radiological impacts, is assumed to draw all of its resources from the most contaminated part of the environment. The assumed location, agricultural practices, life-style and diet of members of the critical group are such as to cause an overestimate of exposures.

4.2 Plutonium Solubility (Estimates vs. Measurements)

The release of fission products and actinides from used natural uranium oxide fuel has been studied at elevated temperatures under a variety of redox conditions and groundwater compositions. The measured steady-state concentrations were compared to solubilities estimated using a thermodynamic equilibrium model that accounts for uncertainty in geochemical conditions [12]. Generally, the measured and calculated concentrations compare favourably.

Figure 1 shows the measured steady-state concentrations and calculated solubilities for plutonium released from CANDU fuels. The concentrations pertain to experiments conducted at 100°C under reducing conditions in several different groundwaters. The calculated solubilities are based on 40 000 simulations conducted with the vault model [7]. Figure 1 indicates that the model overestimates the plutonium solubility relative to the actual

TABLE 1

EXAMPLE FRAMEWORK FOR VAULT MODEL EVALUATION

Feature, Event or Process	Conservative Assumption	Comparison of Prediction with Observation	Test Hypothesis	Accept Hypothesis
Radionuclide inventory	Short interim storage period (10a) is assumed			Law of radioactive decay
Instant release fraction from fuel		Model overestimates observed short-term releases		
Congruent dissolution of fuel matrix		Model overestimates measured plutonium solubilities	Stability of uranium oxide over very long times confirmed at Cigar Lake ore deposit in Saskatchewan	
Initial defects in titanium container				Defect frequency for good quality engineering practice
Crevice corrosion of titanium containers	Propagation of crevices on all containers assumed at highest rates observed experimentally			
Diffusion of contaminants in buffer and backfill		New measurements of diffusion coefficients indicate data inputs are conservative		Radionuclides diffuse according to Fick's Law

TABLE 2

EXAMPLE FRAMEWORK FOR GEOSPHERE MODEL EVALUATION

Feature, Event or Process	Conservative Assumption	Comparison of Prediction with Observation	Test Hypothesis	Accept Hypothesis
Lithostructural model	Assumed hydraulic connection of permeable fracture zone intersecting the horizon of the vault			
Groundwater flow		Good agreement between measured and predicted drawdown of the hydraulic heads during excavation of the URL*		
Location of domestic well	Well assumed to be in the centre of the contaminant plume to maximize the exposure of humans			
Contaminant transport in host rock		In situ migration measurements confirm no retardation of I-129 and strong retardation of Tc-99 under reducing conditions		Contaminants diffuse according to Fick's Law
Fracture propagation			Fracture age and infilling data indicate that a fracture at the URL* has remained dormant for several hundred million years	

* URL: Underground Research Laboratory near Lac du Bonnet, Manitoba, Canada

TABLE 3

EXAMPLE FRAMEWORK FOR BIOSPHERE MODEL EVALUATION

Feature, Event or Process	Conservative Assumption	Comparison of Prediction with Observation	Test Hypothesis	Accept Hypothesis
Contaminant transport in soils		Good agreement between measured and predicted concentration profiles of contaminants in soils under a wide variety of conditions		
Contaminant transport in surface waters		Good agreement between measured and predicted concentration of contaminants in lakes and lake sediments		
Location of domestic well	Well assumed to be in the centre of the contaminant plume to maximize the exposure of humans			
Contaminant transport in the atmosphere		Good agreement between measured and predicted radon emission rates		
Radiological exposure of humans	Assumed location, agricultural practices, lifestyle and diet of members of a critical group such as to overestimate of exposures	Good agreement between measured and predicted ingestion and inhalation rates of selected radionuclides		Linear, no-threshold dose-response relationship is applicable below the stochastic dose limit
Glaciation				Onset of glaciation will not occur in the next 10 000 a
Institutional controls	Long-term monitoring, surveillance or other institutional controls are assumed to be absent			

measurements. This would lead to an overestimate of impacts by the system model.

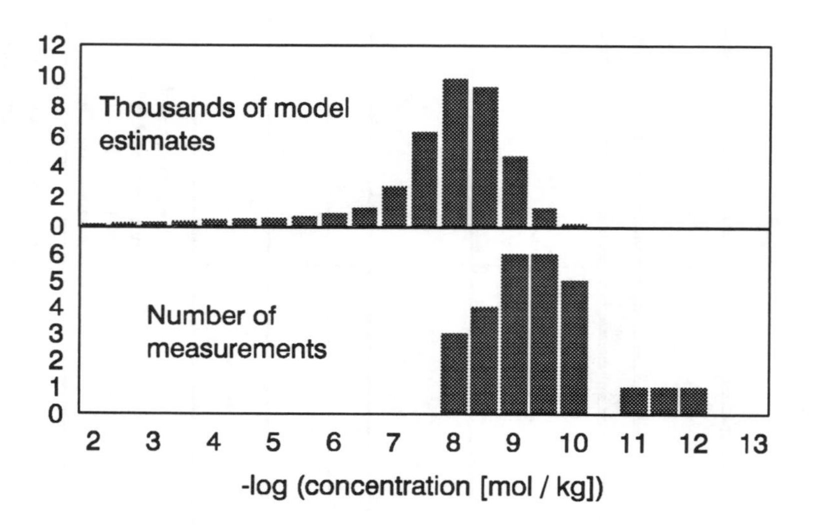

Figure 1: Calculated and measured plutonium concentrations

4.3 Fracture Propagation (Hypothesis Testing)

The postclosure performance assessment case study for the Canadian disposal concept uses geological information from the Whiteshell Research Area in Manitoba. This research area contains AECL's Underground Research Laboratory (URL) and is the most extensively characterized research area in the Canadian program. The geosphere model for the case study assumes that the permeability of the rock surrounding the disposal vault remains constant over the time simulated (10 000 a – 100 000 a). Although sensitivity analysis indicates that the results of the postclosure assessment are not very sensitive to the permeability of the sparsely fractured rock in the vicinity of the vault, the assumption that the fractures that exist in other parts of the rock mass will not propagate significantly so as to increase the permeability has been scrutinized closely.

One test of the assumption was to determine the propagation history of a fracture encountered at the 240-m level of the URL [13]. The initial fracture formation and subsequent episodes of reactivation and propagation leave a record in the form of overlapping mineral infilling and alteration assemblages. The rate of propagation of the fracture was estimated by plotting mineral-assemblage ages against their extent along the fracture, as shown in Figure 2. It is apparent that most of the present-day fracture surface was formed over 2 billion years ago and that the rate of fracture propagation has decreased rapidly with time. The dashed part of the curve indicates some uncertainty in determining age. The exponential form of the curve indicates that the fracture has been essentially dormant for several hundred million years.

This analysis builds confidence in the assumption that propagation of mesoscopic fractures from surface will not significantly affect the permeability of the rock near the disposal vault over the simulated period.

4.4 Effects of Low-Level Radiation (Hypothesis Acceptance)

The main concerns arising from exposures to low doses of radiation are changes in a living cell that produce gross effects such as cancers and genetic damage. The deposition of energy

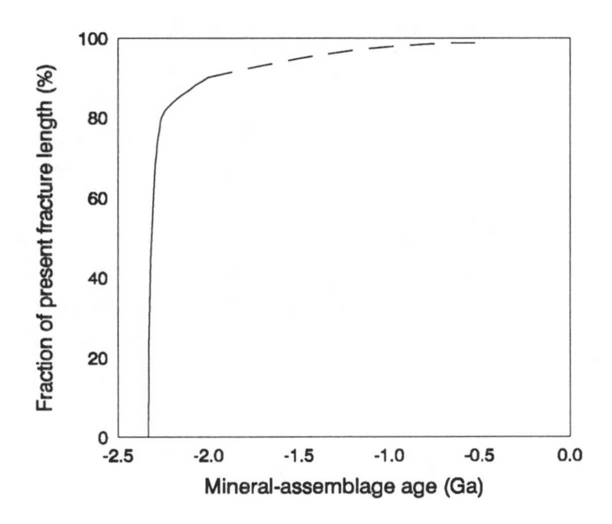

Figure 2: Approximate fracture propagation curve

by ionizing radiation is a stochastic process, so that even at very low average doses sufficient energy can be deposited in critical cellular sites to cause changes in cell function. When unrepaired changes remain in the DNA of even a single cell, gross effects on the whole animal may result. These effects have been called *stochastic*, and it has been assumed that when large populations are exposed to low doses, a few individuals may develop cancers or genetic damage.

Because the frequency of serious health effects caused by low doses near background levels of radiation is so low that it cannot be measured reliably, predictions about the effects of low doses are made by inference from data on effects at high doses. For low doses and low dose rates, effects are estimated by assuming that

- the relationship between low dose and frequency of effect is linear, i.e., halving a given dose halves the frequency of producing an effect; and
- the linear relationship continues down to zero dose, i.e., there is no dose threshold below which effects do not occur.

These dose-response relationships are called the linear, no-threshold hypotheses [14].

The system model used for the postclosure assessment case study calculates the annual dose to members of the critical group as a function of time. Doses are converted to health effects using the linear, no-threshold hypotheses. The probability coefficient for stochastic effects is provided by the AECB [4] on the basis of recommendations of the International Commission on Radiological Protection.

5 Discussion

The evaluation of models for long-term performance assessment plays a vital role in establishing the reliability of the assessment. An example framework for model evaluation has been given along with more specific examples of evaluation tests done in the Canadian postclosure assessment case study. These evaluations do not in themselves allow one to conclude that the use of the model in an assessment is valid. The validity of an assessment must be discussed in the overall context of data used, models used, model results, interpretation of results, and conclusions. The discussion inevitably involves judgment on the part of the safety analysts and the reviewers.

Confidence in the models as used in a specific application will, in general, grow with time as the models are continually calibrated, evaluated, and refined to increase their reliability. This is consistent with the growing need for model reliability with time, because decisions, which will be based in part on the models, become more important to safety as implementation proceeds.

Acknowledgment

This paper describes work in the Canadian Nuclear Fuel Waste Management Program, which is jointly funded by AECL and Ontario Hydro under the auspices of the CANDU Owners Group.

References

[1] D.N.P. Murthy, N.W. Page, and E.Y. Rodin. *Mathematical Modelling.* Pergamon Press, 1990.

[2] K.W. Dormuth. Evaluation of models in performance assessment. In *SAFEWASTE 93, International Conference on Safe Management and Disposal of Nuclear Waste*, Avignon, June 1993.

[3] L.F. Konikow and J.D. Bredehoeft. Groundwater models cannot be validated. *Advances in Water Resources*, 15:75, 1992.

[4] AECB. Regulatory objectives, requirements and guidelines for the disposal of radioactive wastes—long-term aspects. Regulatory Policy Statement R-104, Atomic Energy Control Board, Ottawa, June 1987.

[5] K.W. Dormuth, B.W. Goodwin, and A.G. Wikjord. Long-term safety assessment of the disposal of nuclear fuel waste. In *Proceedings of the 1993 International Nuclear Conference, INC 93*, Toronto, October 1993. Canadian Nuclear Association.

[6] G.R. Simmons and P. Baumgartner. The disposal of Canada's nuclear fuel waste: Engineering for a disposal facility. Technical Report AECL-10715, Atomic Energy of Canada Limited, 1994.

[7] L.H. Johnson, D.M. LeNeveu, D.W. Shoesmith, D.W. Oscarson, M.N. Gray, R.J. Lemire, and N.C. Garisto. The disposal of Canada's nuclear fuel waste: the vault model for postclosure assessment. Technical Report AECL-10714, Atomic Energy of Canada Limited, 1994.

[8] C.C. Davison, T. Chan, A. Brown, M. Gascoyne, D.C. Kamineni, G.S. Lodha, T.W. Melnyk, B.W. Nakka, P.A. O'Connor, D.U. Ophori, N.W. Scheier, N.M. Soonawala, F.E. Stanchell, D.R. Stevenson, G.A. Thorne, S.H. Whitaker, T.T. Vandergraaf, and P. Vilks. The disposal of Canada's nuclear fuel waste: the geosphere model for postclosure assessment. Technical Report AECL-10719, Atomic Energy of Canada Limited, 1994.

[9] P.A. Davis, R. Zach, M.E. Stephens, D.B. Amiro, G.A. Bird, J.A.K. Reid, M.I. Sheppard, S.C. Sheppard, and M. Stephenson. The disposal of Canada's nuclear fuel waste: the biosphere model, BIOTRAC, for postclosure assessment. Technical Report AECL-10720, Atomic Energy of Canada Limited, 1994.

[10] B.W. Goodwin, D.B. McConnell, T.H. Andres, W.C. Hajas, D.M. LeNeveu, T.W. Melnyk, G.R. Sherman, M.E. Stephens, J.G. Szekely, P.C. Bera, C.M. Cosgrove, K.D. Dougan, S.B. Keeling, C.E. Kitson, B.C. Kummen, S.E. Oliver, K. Witzke, L. Wojciechowski, and A.G. Wikjord. The disposal of Canada's nuclear fuel waste: Postclosure assessment of a reference system. Technical Report AECL-10717, Atomic Energy of Canada Limited, 1994.

[11] M. Kumata and T.T. Vandergraaf. Nuclides migration tests under deep geological conditions. In *Proceedings of the Third International Symposium on Advanced Nuclear Energy Research*, Mito City, Japan, 1991.

[12] S. Stroes-Gascoyne. Trends in the short-term release of fission products and actinides to aqueous solution from used CANDU fuels at elevated temperature. *Journal of Nuclear Materials*, 190:87–100, 1992.

[13] R.A. Everitt and A. Brown. An approach for fracture propagation history with application to estimating future occurrence. Technical Report TR-659, Atomic Energy of Canada Limited, (in preparation).

[14] ICRP (International Commission on Radiological Protection). Protection from potential exposure: A conceptual framework (ICRP Publication 64). *Annals of the ICRP*, 23(1), 1993.

Establishing a Technical Basis for Evaluating Performance Assessment Modeling

Eric Smistad
U.S. Department of Energy
Yucca Mountain Site Characterization office
Las Vegas, Nevada

Abraham Van Luik
Civilian Radioactive Waste Management System
Management & Operating Contractor/INTERA, Inc.
Las Vegas, Nevada

(USA)

Abstract

The U.S. Department of Energy's (DOE's) Yucca Mountain Site Characterization Project is determining the suitability of the Yucca Mountain Site in Southwestern Nevada as a potential repository to isolate spent nuclear fuel and high-level radioactive waste. The U.S. Nuclear Regulatory Commission, in its generally applicable regulation governing the disposal of high-level radioactive waste and spent nuclear fuel, makes a statement regarding the necessity of establishing the soundness of models and their applications as part of the Safety Analysis Report that is part of the License Application. The DOE has replanned its efforts from 1995 through the year 2000, resulting in an aggressive, realistic, goal oriented activity schedule. By the end of 1998 the DOE will determine the technical suitability of the site at Yucca Mountain for licensing as a geologic repository, and, if Yucca Mountain is suitable and if the President and the Congress have approved the site, the DOE will submit a License Application for construction of a repository to the U.S. Nuclear Regulatory Commission in 2001. At each step, the decision making process requires that the confidence placed in modeling results is supported through field and laboratory tests and natural analogue studies.

INTRODUCTION

Determining the Suitability of Yucca Mountain as a Site for a Repository

The U.S. Department of Energy's (DOE's) Yucca Mountain Site Characterization Project (Yucca Mountain Project) has been charged by law with determining the suitability of the Yucca Mountain Site in Southwestern Nevada as a potential repository to isolate spent nuclear fuel and high-level radioactive waste (radioactive waste) from the accessible environment.

In the near term, this isolation is to be accomplished primarily through containment of the radioactive waste in engineered waste packages. A recent conceptual design for emplaced waste packages is illustrated in Figure 1. The integrity of these waste packages cannot be assured for all time. Therefore, over longer times, continued isolation is to be achieved through providing a basis for reasonable assurance that the rate at which the radioactive components of the waste are released to the environment is low enough so that no unacceptable risk is presented to the human population in the potentially affected area. The primary basis of that assurance is the long-term performance of the multiple natural barriers of the waste isolation system illustrated in Figure 2 (arid climate, thick unsaturated zone, radionuclide sorption barriers, deep groundwater system).

The long term behavior of the repository system is to be determined through calculations called performance assessments. Results of these calculations must be compared with regulatory performance requirements to determine whether or not compliance with the performance related aspects of the U.S. Nuclear Regulatory Commission regulations is likely to be achieved. The assumptions, models and data used in performance assessments will come under scrutiny in the determination of site suitability, and, if the site is found suitable, in the reviews and hearings that are part of the Licensing process. As part of the License Application, the DOE must provide documentation that shows there is a coherent, transparent, and technically defensible basis for the performance assessments used. The providing of this technical basis is sometimes referred to as "validation."

The Concept of Validation

There is some debate over the meaning of the word "validation." [1] Therefore, it is necessary to describe how the concept of validation is approached in the Yucca Mountain Project performance assessment program. Validation is establishing the soundness of specific computer models and the legitimacy of specific applications being made of those models. This definition is echoed in the Quality Assurance Procedure applicable to the Yucca Mountain Site Characterization Project whereIN validation is defined as: "the process by which the analyst assures himself and others that a model is a representation of the phenomena being modeled that is adequate for the purposes of the study of which it is a part." This definition is not "validation" in the classical sense of matching predictive results with subsequent observations. For the sake of clarity, the word validation is not used beyond the introductory section of this paper. Instead, this paper discusses the activities needed to establish confidence in the modeling done to build the Licensing safety case.

U.S. Regulatory Requirements for Establishing the Technical Basis for Performance Assessment Modeling

The U.S. Nuclear Regulatory Commission (NRC) did not use the word "validation" in its generally applicable regulation governing the disposal of high-level radioactive waste and spent nuclear fuel (10 CFR Part 60). [2] However, the necessity of establishing the soundness of models and their applications is covered in statements specifying the content of the Safety Analysis Report that is part of the License Application. For example: "Analyses and models ... shall be supported using an appropriate combination of such methods as field tests, in situ tests, laboratory tests which are representative of field conditions, monitoring data, and natural analog studies" (60.21(c)(ii)(F)). Similar language occurs in 60.101(a)(2), regarding the "demonstration of compliance."

The U.S. Environmental Protection Agency (EPA) did not mention the word validation in its 1985 regulation on the disposal of spent nuclear fuel, high-level waste and transuranic waste, 40 CFR Part 191. [3] Its paragraph 191.113 (b) states: "Performance assessments need not provide complete assurance that the requirements ... will be met. ... what is required is a reasonable expectation, on the basis of the record before the implementing agency, that compliance ... will be achieved." Thus, both the EPA and the NRC have recognized the difficulty inherent in establishing the soundness of predictions for geologic processes over very long times.

These reasonable expectations reflect awareness of inherent uncertainties in performance assessment models, which derive from uncertainties associated with the conceptual model on which the model is based, measurement errors, sparse data, intrinsic heterogeneities of internal system properties or processes, and uncertainty in the future environmental setting of the repository. These types of uncertainties have been previously identified and discussed by Eisenberg et al. [4]

Finally, the NRC has recognized that all predictive uncertainty need not be minimized prior to the construction of a disposal system. Subpart F of 10 CFR Part 60 provides for a "performance confirmation period," which suggests that the NRC expects the types of activities that can establish the soundness of models to continue from the time of site-characterization to the time of the permanent closure of a repository. This implies that results of confirmatory testing, exploration, and natural analogue work will be part of the documentation accompanying the two scheduled updates of the License Application: the one to allow the receipt and emplacement of waste, and the other to allow permanent closure. Thus, the degree of confidence needed in the modeling done to support the safety argument increases over time as the DOE comes closer to making the decision not to retrieve emplaced waste, and to permanently seal the repository system.

Legal Challenges to Modeling Confidence

In a discussion by Bradley on "The Scientist and Engineer in Court," a number of observations were made that may be generally applicable. [5] These may be summarized as follows: legal decisions "are generally made on the 'weight of evidence,'" and the weight of evidence, when it includes a major modeling effort, consists of two separable aspects of the modeling: 1) the "scientific studies and research" aspect [ie: the model development phase], and 2) the "field justification" aspect [ie: the field calibration and subsequent application phases].

The application phase requires field data for calibration and separate sets of field data for establishing credibility, and is a preferred target for legal challenge. Legal vulnerability may be minimized by 1) assuring that the model user is familiar with the development of the model and the conditions for which it was designed to be useful, 2) assuring the modeler either has involvement in collecting field data or is very familiar with the data used, its nature, limitations, etc., and 3) assuring results are carefully and competently interpreted, and limitations recognized, but not exaggerated.

Challenges to a modeling exercise usually involve a challenge of the field sampling program and its data. Thus, modeling confidence can not be divorced from the adequacy of site characterization as an issue in the courtroom or licensing hearing.

MODEL CONFIDENCE EVALUATION

Hypothesis Testing to Build Confidence in Conceptual Models

Establishing the credibility of the conceptual model is one of the principal aims of the confidence building process. The conceptual model is a set of relationships, selected from among a larger set of possible relationships and conditions, that is sufficient to describe the system for the intended application of the model. A conceptual model is developed on the basis of many sources of data and evidence, and should be consistent with as many of these sources of data and evidence as possible. A conceptual model must address the relationships that describe all important aspects of system performance.

These relationships and their alternatives may be expressed in terms of testable hypotheses. Ideally, a conceptualization of a system may have a number of component hypotheses and counter-hypotheses that can be experimentally evaluated. In such a case, experiments can be constructed that will attempt to falsify a hypothesis, or that will allow discrimination between competing hypotheses: this experimental validation of the component hypotheses of a conceptual model is called "hypothesis testing."

It is expected that knowledge concerning a system under active investigation would increase with time. As new data is evaluated, however, there may be instances where it may be in conflict with aspects of the provisional conceptual model. Therefore, conceptual model development will be an evolving and iterative process of model modification, testing, and refinement, as site characterization, field testing, and laboratory investigations progress.

Natural Analogues and Modeling Confidence

As previously noted, in its 10 CFR Part 60 regulation, the NRC suggests the use of natural analogues as part of the work done to provide a supporting technical basis for models used in system performance assessments, and suggests the need for a "performance confirmation period" in such a way as to suggest that activities to support the technical basis of analyses and models are to continue from the time of site-characterization to the time of permanent closure. Thus, the obligation to continue to perform confirmatory studies well into the next century will provide opportunity to study a few, well selected, additional natural analogue sites in detail.

There are several technically weak areas in current system performance assessments of Yucca Mountain, as has been outlined by Andrews et al. and Wilson et al. [6,7] These weak areas include waste package materials behavior, water availability and likely water contact modes, thermal perturbations and their potentially permanent effects on water flux in the unsaturated zone and on geochemistry, radionuclide dissolution rates, radionuclide solubilities, groundwater flow paths and fluxes toward the accessible environment, geochemistry of flow paths over long times, and radionuclide migration characteristics along flow paths. It should be remembered, however, that current performance assessments are still preliminary: design and materials testing and selection is in progress, site characterization is in progress, thermal tests are planned and in progress, and laboratory testing of materials and processes is ongoing.

It is likely that a number of qualitative and quantitative insights may be obtained through the study of suitable natural analogues. The likely effects of the thermal perturbation of the unsaturated rock, the likely rate of dissolution of the spent fuel waste form, the likely range of groundwater fluxes that may contact the waste packages, likely radionuclide solubilities and retardation behaviors, etc. Examples of preliminary work on gaining insight regarding spent fuel dissolution behavior from natural analogues are that by Curtis et al. [8], and the work of the NRC reported by Pearcy et al. [9].

By observing a number of uranium occurrences in unsaturated volcanic rock under different climatic conditions, if a credible estimate can be made of the geologic history of the analogue site, bounds may be set on the rates at which some of these important processes proceed in comparable natural settings. This is particularly important in addressing issues of scale. There will be no opportunity to observe the flux in unsaturated volcanic rock that is comparable in properties to the Yucca Mountain rock under a range of climatic-change scenarios, there will be no opportunity to observe the dissolution of spent fuel under expected conditions, or to observe the migration of sorbing radionuclides under expected flux conditions over more than a few meters. Yet, predictions of these processes need to be made.

The role of natural analogues is to provide supporting information that allows a determination of the reasonableness of such predictions. Laboratory and field tests that are accelerated to allow meaningful observations in short times may or may not yield realistic results for predicting the progress of very long term processes in natural settings. It is likely that accelerated testing, since it purposely manipulates conditions to enhance rates, would lead to conservative predictions of long term behavior under milder conditions. Natural analogues may provide a particularly suitable check on these types of predictions, although, as noted in Miller et al. [10], it is also possible to misinterpret or overstate the results of natural analogue work.

Miller et al. [10] discuss the status of international natural analogue efforts, and also the status of the use that has been made of natural analogue information in published performance assessments. Their conclusion is that very limited use has been made, to date, of natural analogue information. The difficulties inherent in comparing model results with natural analogue study results are discussed as one reason for this limited use in performance assessment. Three areas are identified, however, where natural analogue information could be most useful in terms of supporting performance assessment. These three areas are: 1) conceptual model construction (qualitative), 2) data acquisition as an adjunct to laboratory testing (quantitative), and 3) model testing (qualitative and quantitative). This last area is billed as the most significant in terms of potential benefit. This last area, for the Yucca Mountain Project, is also the use of natural analogues that is explicitly suggested in the applicable NRC regulations as part of what needs to be done to establish the technical basis of the analyses presented in the Safety Analysis Report of the License Application and its subsequent amendments.

STATUS OF ACTIVITIES FOR ESTABLISHING THE TECHNICAL BASIS OF MODELS

Recent total system performance assessments [6,7] examined, in a limited sense, sensitivities of results to uncertainties in conceptual models and parameter distributions. Results were expressed in terms of both the cumulative release of radionuclides to the accessible environment and the corresponding individual dose associated with that release.

Conceptual Model of Unsaturated Flow

As has also been noted in several previous assessments of the performance of a potential repository at Yucca Mountain, the key modeling issue affecting the ability of the site and engineered barriers to contain and isolate radioactive wastes from the accessible environment is the conceptual model for flow and transport through the fractured-porous media as well as the magnitude of the percolation flux through the unsaturated zone. In the two recent assessments of total system performance, it was assumed, except in the case of the "weeps" model also used by Wilson et al., [7] that transport through the unsaturated zone is dominated by the matrix, either because of the large capillary pressure differences between the fractures and matrix *or* the large matrix diffusion due to the concentration gradient between the fractures and matrix. Most all previous analyses which addressed the hydrologic flow regime in the unsaturated zone at Yucca Mountain (whether under ambient or thermally perturbed conditions) assumed the composite porosity flow model. The validity of this assumption as well as the impact of this assumption on predicted performance should be rigorously evaluated.

As in the previous total system performance assessment iteration, TSPA 1991 [11,12] the ambient percolation flux was considered to be uncertain in TSPA 1993. In contrast to TSPA 1991, the uncertainty has been broken into two distributions, one representing the uncertainty in the current conditions and the other representing the uncertainty in the percolation flux due to future climates. The uncertainty in the expected percolation flux at the repository depth is large, which, in terms of current conceptualizations, implies large uncertainties for total system performance. Percolation flux estimates range from the flux being upward (i.e., corresponding to a current net drying of the unsaturated zone at Yucca Mountain) to essentially zero, to a downward flux of between 0.01 and 1.0 mm/yr. [13-22] Other, ongoing work is contributing to the understanding of unsaturated zone percolation processes through experimentation on blocks and columns of tuff. [23, 24]

Climate Change Effects

The Andrews et al. [6] and Wilson et al. [7] contributions to TSPA 1993 made somewhat different assumptions regarding the percolation flux. In the Andrews et al. case, an exponential distribution was assumed for the current percolation flux with an expected value of 0.5 mm/yr. This distribution implies that 63 percent of the values sampled will be less than 0.5 mm/yr and 37 percent of the values will be greater than 0.5 mm/yr. This distribution is exactly a factor of two less than that used in TSPA 1991. This difference is not so much due to new insights or new information as it is to the fact that climate change is modeled explicitly as a multiplier on the expected ambient flux distribution instead of embedding the climate change implicitly in the distribution as was assumed in TSPA 1991. This distribution was consistent with the range selected by Wilson et al. for their TSPA 1993 contribution, but the latter's implementation of climate change perturbations differed. This implementation difference did not result in an important difference in the outcomes of these two system performance calculations, however.

The representation of the possible increase in flux that may be attributable to future climate changes is uncertain. In the Wilson et al. approach to TSPA 1993, the flux change is described as a linearly increasing and decreasing function from the current value (which is assumed to represent a "dry" climate), based on a proposal put forward by Long and Childs [22]. For the expected value of the flux multiplier of three which applies to the maximum glacial period assumed to be 100,000 years from now, the average percolation flux over 100,000 years would be increased by a factor of two times the ambient flux. Given the large uncertainty in the ambient flux, this factor of two has little added significance. Clearly future iterations of total system performance assessment will continue to investigate alternate conceptualizations of likely climate changes and their impact on the repository level percolation flux.

In the site characterization program, a variety of lines of evidence are being followed to ascertain the regional environmental response to a full glacial and post-glacial climate. Among these lines of evidence are the use of oxygen isotope ratios in modern and fossil mollusc shells [25] to ascertain general differences between past and present climates, and fossil ostracode taxonomy and distributions associated with paleospring and wetland deposits [26] to estimate post-glacial climates. The State of Nevada has also sponsored work on paleospring deposits [27] to establish water table levels for the last full glacial and postglacial climates. This paleospring deposit work, by Quade, suggests that full glacial water table rises in the region are generally under 100 m and localized. Sharpe et al. [25] suggest that past climates included times of greater rainfall and cooler temperatures. Finally, Forester and Smith's [26] preliminary work suggests that during the 14-20 ka late glacial time annual precipitation northwest of Las Vegas may have been four times the current average with a mean annual temperature about 10C cooler than present.

Geosphere Properties

As in TSPA 1991, several geosphere properties were considered to be uncertain in this iteration of TSPA 1993. These include matrix porosity, bulk density, and retardation. In contrast to TSPA 1991, the uncertainty in fracture properties was not included because it was assumed that the matrix controlled the transport of radionuclides in the geosphere. This assumption is based on the idea that since there is generally a higher water potential in the unsaturated matrix than in the flowing fracture, water will move from the fracture into the matrix. It is recognized, however, that fracture properties may be significant for the realizations where the sampled percolation flux exceeds the saturated hydraulic conductivity of the matrix, especially if one assumes that matrix diffusion does not occur. While the matrix properties play a role in retarding the radionuclides from reaching the accessible environment, they are relatively insignificant over the ranges of properties that have been evaluated. The retardation coefficients of some key radionuclides such as ^{237}Np can play a more significant role for releases over the 100,000-year time frame if they are found to be significantly less than the current conservative estimates based on laboratory measurements conducted under varying geochemical and thermal environments. Some preliminary work on evaluating the adequacy of the retardation coefficient approach to modeling radionuclide transport has been reported by Siegel. [28]

Gaseous Radionuclide Transport

More sophisticated analyses of thermally perturbed gaseous flow and ^{14}C transport in the unsaturated zone as well as aqueous flow in the saturated zone have been incorporated in TSPA 1993. Both models rely on the bulk permeability of the fractured-porous media in the vicinity of Yucca Mountain, underscoring the need for a comprehensive description of this parameter along likely paths of radionuclide travel. In general, however, it may be concluded that the gaseous release to the accessible environment is relatively insensitive to the gaseous flow in the unsaturated zone because the ^{14}C travel times are short in comparison to the expected life of the waste packages.

Saturated Zone Flow

Aqueous release to the accessible environment is relatively insensitive to the aqueous flow in the saturated zone because the travel times of radionuclides in this domain is short in comparison to the travel time through the unsaturated zone. However, the dose attributable to aqueous releases at the accessible environment is directly related to the horizontal flux through the saturated zone, because as the flux increases the dilution increases and the dose decreases. Therefore, if a dose-based standard is promulgated, a greater understanding of the flow in the saturated zone will be required.

MODELING IMPROVEMENT NEEDS

In addition to the need for more site data to support the selection of conceptual models and parameter probability distribution functions, there is also a need to improve the modeling capability that is to make use of this new information and data. These modeling improvement needs were cataloged and discussed as part of the results of the TSPA 1993 exercise. [6,7]

These recommended improvements in modeling capability can not, however, be divorced from the need for additional site information and for a more mature design. This underscores the iterative nature of the performance assessment program/site program interactive relationship: site characterization and design need feedback from performance assessment in terms of importance, sensitivity and uncertainty analyses. In turn, the quality of the models used for these analyses is determined, to a significant extent, by the completeness and representativeness of site data and observations. The same is true of the performance assessment/design program interface. To this point, these interfaces have been maintained on a rather informal basis. Beginning in 1995, however, in addition to maintaining these ongoing informal interfaces, there is also to be a formalization of these performance assessment/site/design interfaces through the publication of an annual report from the performance assessment function to the site and design functions. This report is to utilize all analyses performed during the year, and glean from them those data or design features that may be of potential importance to guiding site and design work. Ideally, all these things will already have been made known to site and design personnel through the informal interface, but these documents will assure that nothing is dropped through these interfaces and lost.

Total System Performance Assessment Modeling

While the TSPA 1993 analyses have significantly extended the analyses documented in TSPA 1991 (primarily by the abstraction of detailed process modeling results and the use of recent laboratory understanding of thermally dependent processes and parameters), many uncertainties still remain. These residual uncertainties are, in most instances, to be addressed, and hopefully narrowed, through field and laboratory studies. The performance assessment program, however, needs to have a set of models that can properly address the system and subsystem performance implications of both the data being obtained and the remaining uncertainties. The key to obtaining these models lies in performing systematic sensitivity and uncertainty analyses using the more detailed process models being produced to evaluate design details and new site data.

Waste Package/Engineered Barrier System Modeling

In addition to the uncertainty associated with the expected thermohydrologic environment in the vicinity of the waste packages, considerable uncertainty remains regarding the processes affecting the initiation and rate of aqueous corrosion of the mild steel outer barrier and the Alloy 825 inner barrier. Greater understanding is required of the potential for cathodic protection of the inner barrier, the processes affecting the growth of pits, and even the definition of waste package "failure" in order to provide a more defendable argument for the range of possible waste package lifetimes and early release mechanisms.

Other aspects of the near-field thermohydrology are also uncertain. For example, some of the near-field performance may involve coupled thermo-hydrologic-chemical effects. The dynamics of water removal by the spreading of the boiling isotherm, the potential for condensate flow back onto some of the waste packages, and the dynamics of rewetting as the thermal output wanes are also not well understood. Finally, the uncertainty in engineered system performance is also affected by the uncertainty in the design, the waste stream, and the characteristics of a backfill, if one is to be used. The backfill question has importance to the modeling of waste package temperatures, but it also potentially controls water contact mode and rate, and radionuclide transport mode and rate.

Unsaturated Zone Flow Modeling

The conceptual understanding of how water moves through the unsaturated zone at Yucca Mountain should be improved. In particular, the potential role of continuous localized fracture flow, as considered in an alternate conceptualization used by Wilson et al. [7], or of episodic fracture flow, should be evaluated. Additional testing is required to determine the correct conceptual model or models for unsaturated flow. The ambient unsaturated zone flux remains a very significant parameter in this iteration of total system performance. Any direct or indirect observations to better quantify the expected value and its uncertainty should be used, although it is acknowledged that until underground observations can be made of the undisturbed conditions, the current uncertainty will remain. The likely near-term and longer-term thermal perturbation of the ambient hydrologic system needs to be evaluated, as do the potential changes in flow system characteristics from climate change, seismotectonic and volcanic activity.

Geosphere Transport Modeling

In addition to the conceptual understanding of flow, there is a need to validate the representation of radionuclide transport through the unsaturated zone. The simplified K_d approach has been used to approximate the complex physical-chemical processes affecting transport of dissolved species. This approximation, as well as the parameter ranges used, need to be justified by laboratory and field tests.

Gas Flow Modeling

In terms of gas flow modeling, improvements are needed in both site understanding and modeling approach. Analyses of gaseous and aqueous transport need to be linked through the use of the same data and common geometric descriptions of the system. Gas transport time distributions are needed to match the thermal loading cases being modeled. Further site data on gas permeabilities, preferably as a function of water saturation, are needed.

PLANS FOR EXPANDED SITE CHARACTERIZATION AND DESIGN ACTIVITIES

The DOE has replanned its efforts from 1995 through the year 2000 in the face of diminished funding expectations and rising criticism of the repository program's lack of discernable progress. The result is an aggressive, realistic, goal oriented activity schedule. By the end of 1998 the DOE will determine the technical suitability of the site at Yucca Mountain for licensing as a geologic repository, and develop a draft Environmental Impact Statement for public comment. By the end of 2000, if Yucca Mountain is suitable, the DOE will deliver a final Environmental Impact Statement and a Site Recommendation Report to the President. Finally, in 2001, if the President and the Congress have approved the site, the DOE will submit a License Application for construction of a repository to the U.S.

Nuclear Regulatory Commission, and be ready to support the commission's legal obligation to decide on the repository construction authorization within three to four years of the date of the application. Goals beyond 2001 are shown in Table 1, which shows a timetable for this rescheduled program, referred to currently as the new Program Approach.

Table 1. Timetable for the New Program Approach

Technical Site Suitability Determination	1998
Draft Environmental Impact Statement	1998
Final Environmental Impact Statement	2000
Site Recommendation Report to the President	2000
Record of Decision	2000
License Application to Construct a Repository	2001
Construction Authorization	2004
License Application Update to Receive and Possess Waste	2008
License to Operate a Repository	2010

This new plan has several impacts on the performance assessment program, and therefore on the effort to establish the technical basis of performance assessment models and their applications. First, the priority of the program between the present and 1997 is to expedite underground exploration and provide sufficient site understanding and data to support the suitability finding in 1998. Given reduced funding, compared with previous planning, this reduces resources for performance assessment activities of a confirmatory nature: the activities that establish the technical basis of the models and the confidence that may be had in the results of their applications.

The strategy is to provide realistic, yet conservative where necessary, performance assessments in support of the suitability decision and the draft Environmental Impact Statement. From 1997 through 2000, however, resources will be reallocated to allow additional work to provide support to the performance assessment modeling that will be used in support of the final Environmental Impact Statement and the License Application of 2001. A larger experimental, and modeling effort, using field, laboratory, and natural analogue work, will be funded to provide additional technical support to the assessments that will accompany the updated License Application to receive waste, scheduled for 2008. It is foreseen that this type of confirmatory testing will continue at some level, depending on the results of earlier work as well as monitoring efforts, until the License Application amendment to allow final closure of the repository.

CONCLUSIONS

The DOE has replanned its efforts from 1995 through the year 2000, resulting in an aggressive, realistic, goal oriented activity schedule. If Yucca Mountain is found suitable and if the President and the Congress have approved the site, the DOE will submit a License Application for construction of a repository to the U.S. Nuclear Regulatory Commission in 2001. At this point in time, the confidence that is placed in modeling results must be evaluated, and shown to be firmly supported through field and laboratory tests and natural analogue studies. The DOE is using a wide range of site and laboratory studies, and some natural analogue studies, to address the major uncertainties is system performance assessments. In turn, system performance assessments are being used to evaluate site data and identify areas where reductions in uncertainties may be expected if additional data could be obtained. This iterative process is being formalized by the creation of annual evaluations of the site characterization program's priorities by the performance assessment function, starting in 1995. By the time of the License Application in 2001, this process will have supported the completion of the Technical Site Suitability evaluation, the Draft and Final Environmental Impact Statements, the Site Recommendation Report, and the Safety Analysis Report that is part of the License Application.

REFERENCES

1. C. Pescatore, "Validation: An Overview of Definitions," in this volume.

2. U.S. Nuclear Regulatory Commission, 10 CFR Part 60 Disposal of high-level radioactive wastes in geologic repositories -- technical criteria, U.S. Code of Federal Regulations, U.S. Government Printing Office, Washington, D.C., 1982.

3. U.S. Environmental Protection Agency, "40 CFR Part 191 Environmental standards for the management and disposal of spent nuclear fuel, high-level and transuranic radioactive wastes." U.S. Code of Federal Regulations, U.S. Government Printing Office, Washington, D.C., 1985.

4. N. A. Eisenberg, A. E. Van Luik, L. E. Plansky and R. J. Van Vleet, "A Proposed Validation Strategy for the U.S. DOE Office of Civilian Radioactive Waste Management Geologic Repository Program," in GEOVAL 1987, Symposium In Stockholm April 7-9, 1987, Swedish Nuclear Power Inspectorate (SKI), Stockholm, Sweden, 1988.

5. M. D. Bradley, The Scientist and Engineer in Court. AGU Water Resources Monograph Series # 8, American Geophysical Union, San Francisco, California, 1983.

6. R. Andrews, T. Dale, and J. McNeish, Total system Performance Assessment-1993: An Evaluation of the Potential Yucca Mountain Repository, B00000000-01717-2200-00099-Rev. 01, Civilian Radioactive Waste management System, Management and Operating Contractor, Vienna, Virginia 1994.

7. M. L. Wilson, J. H. Gauthier, R. W. Barnard, G. E. Barr, H. A. Dockery, E. Dunn, R. R. Eaton, D. C. Guerin, N. Lu, M. J. Martinez, R. Nilson, C. A. Rautman, T. H. Robey, B. Ross, E. E. Ryder, A. R. Schenker, S. A. Shannon, L. H. Skinner, W. G. Halsey, J. Gansemer, L. C. Lewis, A. D. Lamont, I. R. Triay, A. Meijer, and D. E. Morris, Total-System Performance Assessment for Yucca Mountain -- SNL Second Iteration (TSPA-1993), SAND93-2675, Sandia National Laboratories, Albuquerque, New Mexico, 1994.

8. D. B. Curtis, J. Fabryka-Martin, P. Dixon, R. Aguilar and J. Cramer, "Radionuclide Release Rates from Natural Analogues of Spent Nuclear Fuel," in High Level Radioactive Waste Management, Proceedings of the Fifth Annual International Conference, Las Vegas, Nevada, May 22-26, 1994, American Nuclear Society, La Grange Park, Illinois, 1994.

9. E. C. Pearcy, J. D. Prikryl, W. M. Murphy and B. W. Leslie, Uranium Mineralogy of the Nopal I Natural Analog Site, Chihuahua, Mexico, CNWRA 93-012, Center for Nuclear Waste Regulatory Analysis, San Antonio, Texas, 1993.

10. W. Miller, R. Alexander, N. Chapman, I. McKinley and J. Smellie, Natural Analogue Studies in the Geological Disposal of Radioactive Wastes, Studies in Environmental Science 57, Elsevier, Amsterdam, The Netherlands, 1994.

11. R. W. Barnard, M. L. Wilson, H. A. Dockery, J. H. Gauthier, P. G. Kaplan, R. R. Eaton, F. W. Bingham, and T. H. Robey, TSPA 1991: An Initial Total-System Performance Assessment for Yucca Mountain, SAND91-2795, Sandia National Laboratories, Albuquerque, New Mexico, 1992.

12. P. W. Eslinger, L. A. Doremus, D. W. Engel, T. B. Miley, M. T. Murphy, W. E. Nichols, M. D. White, D. W. Langford, and S. J. Ouderkirk, Preliminary Total-System Analysis of a Potential High-Level Nuclear Waste Repository at Yucca Mountain, PNL-8444, Pacific Northwest Laboratory, Richland, Washington, 1993.

13. A. L. Flint, L. E. Flint, and J. A. Hevesi, "The Influence of Long Term Climate Change on Net Infiltration at Yucca Mountain, Nevada," in High Level Radioactive Waste Management, Proceedings of the Fourth Annual International Conference, Las Vegas, Nevada, April 26-30, 1993, American Nuclear Society, La Grange Park, Illinois, 1993.

14. L. E. Flint, A. L. Flint and J. A. Hevesi, "Shallow Infiltration Processes in Arid Watersheds at Yucca Mountain, Nevada," in High Level Radioactive Waste Management, Proceedings of the Fifth Annual International Conference, Las Vegas, Nevada, May 22-26, 1994, American Nuclear Society, La Grange Park, Ilinois, 1994.

15. J. A. Hevesi and A. L. Flint, "The Influence of Seasonal Climatic Variability of Shallow Infiltration at Yucca Mountain," in High Level Radioactive Waste Management, Proceedings of the Fourth Annual International Conference, Las Vegas, NV, April 26-30, 1993, American Nuclear Society, La Grange Park, Illinois, 1993.

16. J. A. Hevesi, A. L. Flint and L. E. Flint, "Verification of a 1-Dimensional Model for Predicting Shallow Infiltration at Yucca Mountain," in High Level Radioactive Waste Management, Proceedings of the Fifth Annual International Conference, Las Vegas, Nevada, May 22-26, 1994, American Nuclear Society, La Grange Park, Ilinois, 1994.

17. W. D. Nichols, Geohydrology of the Unsaturated Zone at the Burial Site for Low-Level Radioactive Waste near Beatty, Nye Country, Nevada, USGS-WSP-2312, Water-Supply Paper, U.S. Geological Survey, Washington, D.C., 1987.

18. M. J. Nicholl, R. J. Glass and H. A. Nguyen, "Small-Scale Behavior of Single Gravity-Driven Fingers in an Initially Dry Fracture," in High Level Radioactive Waste Management, Proceedings of the Fourth Annual International Conference, Las Vegas, Nevada, April 26-30, 1993, American Nuclear Society, La Grange Park, Illinois, 1993.

19. M. J. Nicholl, R. J. Glass and H. A. Nguyen, "Wetting Front Instability in an Initially Wet Unsaturated Fracture," in High Level Radioactive Waste Management, Proceedings of the Fourth Annual International Conference, Las Vegas, Nevada, April 26-30, 1993, American Nuclear Society, La Grange Park, Illinois, 1993.

20. R. J. Glass, "Modeling Gravity-Driven Fingering in Rough-Walled Fractures Using Modified Percolation Theory." in High Level Radioactive Waste Management, Proceedings of the Fourth Annual International Conference, Las Vegas, Nevada, April 26-30, 1993, American Nuclear Society, La Grange Park, Illinois, 1993.

21. J. H. Gauthier, "The Most Likely Groundwater Flux Through the Unsaturated Tuff Matrix at USW H-1," in High Level Radioactive Waste Management, Proceedings of the Fourth Annual International Conference, Las Vegas, Nevada, April 26-30, 1993, American Nuclear Society, La Grange Park, Illinois, 1993.

22. A. Long and S. W. Childs, "Rainfall and Net Infiltration Probabilities for Future Climate Conditions at Yucca Mountain," in High Level Radioactive Waste Management, Proceedings of the Fourth Annual International Conference, Las Vegas, Nevada, April 26-30, 1993, American Nuclear Society, La Grange Park, Illinois, 1993.

23. F. Thamir, E. M. Kwicklis and S. Anderton, "Laboratory Study of Water Infiltration Into a Block of Welded Tuff," in High Level Radioactive Waste Management, Proceedings of the Fourth Annual International Conference, Las Vegas, Nevada, April 26-30, 1993, American Nuclear Society, La Grange Park, Illinois, 1993.

24. R. Boehm, M. Krotke, S. Thota and X. Xu, "Experimental Studies to Calibrate Unsaturated Flow Models." in High Level Radioactive Waste Management, Proceedings of the Fourth Annual International Conference, Las Vegas, Nevada, April 26-30, 1993, American Nuclear Society, La Grange Park, Illinois, 1993.

25. S. E. Sharpe, J. F. Whelan, R. M. Forester and T. McConnaughey, "Molluscs as Climate Indicators: Preliminary Stable Isotope and Community Analysis," in High Level Radioactive Waste Management, Proceedings of the Fifth Annual International Conference, Las Vegas, Nevada, May 22-26, 1994, American Nuclear Society, La Grange Park, Illinois, 1994.

26. R. M. Forester and A. J. Smith, "Late Glacial Climate Estimates for Southern Nevada, the Ostracode Fossil Record," in High Level Radioactive Waste Management, Proceedings of the Fifth Annual International Conference, Las Vegas, Nevada, May 22-26, 1994, American Nuclear Society, La Grange Park, Ilinois, 1994.

27. J. Quade, "Spring Deposits and Late Pleistocene Ground-Water Levels in Southern Nevada," in High Level Radioactive Waste Management, Proceedings of the Fifth Annual International Conference, Las Vegas, Nevada, May 22-26, 1994, American Nuclear Society, La Grange Park, Illinois, 1994.

28. M. D. Siegel, "Toward a Realistic Approach to the Validation of Reactive Transport Models for Performance Assessment," in Proceedings, Site Characterization and Model Validation, Focus '93, September 26-29, 1993, Las Vegas, Nevada, American Nuclear Society, La Grange Park, Illinois, 1993.

PANEL SESSION

Chairman:

C. McCombie

Panel Session

Panel Members:

C. McCombie (Chairman)
J. Andersson
J. Bruno
M. Federline
A.J. Hooper
G. de Marsily
S. Neuman
O. Olsson
C. Pescatore
H. Röthemeyer

Chairman

The discussion will be divided into two parts. In the first session we will treat processes that determine the safety of our repositories and make a review of where we are and our understanding of the processes. We will start with four introductory contributions from panel members which will address different types of processes. Questions to be asked are such as:

- How confident are we in our conceptual understanding?
- How defensible are our quantitative models?
- What can we do to improve the "level of validation"?
- When will we have models sound enough to support decision-making?
- Can we expect discrete improvements in any areas?
- Will we have to look for "alternative" approaches in any areas?

We would like to know which are the most open questions today and in which areas corresponding answers may lie.

- Where should we put most effort?
- What, specifically, should we be doing?
- Is there something we can do which can lead to a sudden and dramatic increase of our level of understanding some of the processes?
- When can we expect to have the efficient level of validation we need for taking decisions?
- Is the time scales for reaching adequate levels of validation unrealistic relative to those for the implementation of repository projects?
- Do we have to look at any other kind of approaches?

In the second part we will deal with more generalised issues, such as, the question of long time scales, how we can show to other people our level of understanding without being immodest and also without giving the contrary impression that absolutely nothing is possible in terms of predictability of our models!

The site issue is now very important, in contrast to the situation at the previous GEOVAL symposia. For many years we have complained that site specific data were missing. In many countries we now have sites and we can work on them. Has that given us the extra impetus and the extra level of understanding which we expected?

Perhaps the most important issue of all is how the question of validation will be handled when we come to real decision making processes, that is decisions to implement repositories and to licence them.

Shlomo Neuman

In the context of validation it is important to discuss scale dependence and information, without which, I think, processes cannot be validated. A related topic is models. Two papers, presented earlier this week, dealing with analysis of confidence in site-specific data and with validation strategy are especially interesting in this regard.

The main point is that process phenomenology, including equations and parameters describing the processes, are scale and information dependent. Imagine we have measured hydraulic conductivities on some scale. The scale of measurement, or support, can for instance be represented by the distance between packers in a borehole and we designate this scale ω. The data are represented by the curve in Figure 1. In conformity with suggestions made by Björn Dverstorp, everything that is measured should be consistent with the same scale and everything that is predicted should also hopefully be consistent with ω.

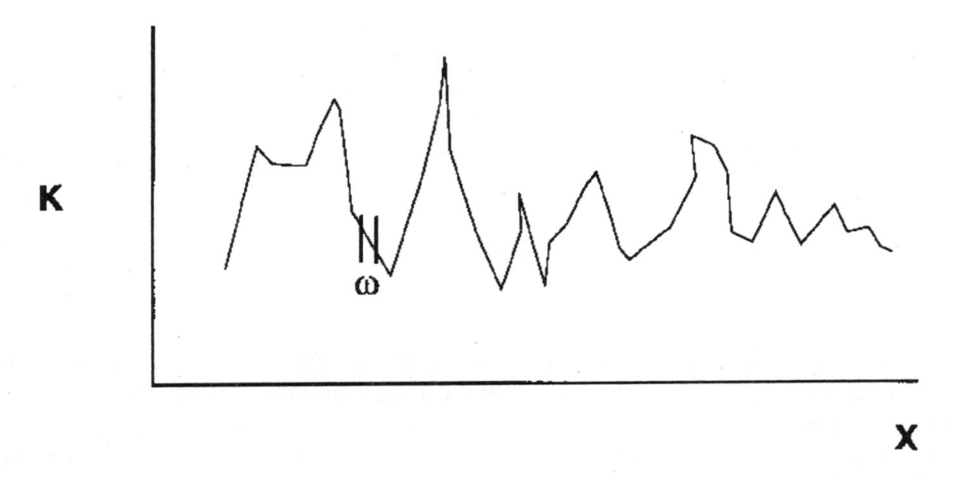

Figure 1. **Permeability K on support scale ω showing random spatial variability.**

As often happens in boreholes in fractured rock, the permeability shows what appears to be a random variation, when it is measured on the scale ω. Groundwater velocity, whatever it means, will also exhibit random fluctuations on the same scale. It does not mean that there are no fluctuations on other scales, but we don't know much about fluctuations on scales smaller than ω.

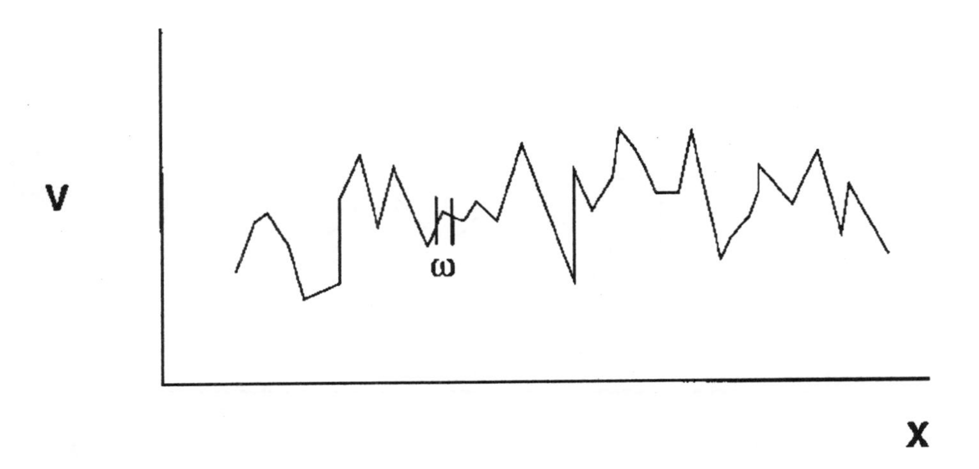

Figure 2. **Velocity on scale ω showing random variability.**

Budhi Sagar and others have suggested that dispersion is a small parameter and can be disregarded. In this example then, we disregard dispersion on the scale of ω and, furthermore, we disregard everything except advective transport. We are also assuming the simplest possible representation of flow, namely steady state, such that the divergence of the velocity is zero.

$$\nabla \cdot \underline{v} = 0$$

Transport is then described by the advection equation.

$$\frac{\partial C}{\partial t} + \underline{v} \cdot \nabla C = 0$$

The velocity in the first equation is defined on the scale ω and the second equation will predict concentrations on the same scale. One way to deal with this equation is to set up a three-dimensional grid of finite elements. To resolve all the fluctuations of the velocity, the grid should be on the scale of ω. With a coarser grid, we would face the problem of smoothing. Such an elaborate grid is computationally demanding and we would like to simplify it. The simplification could go all the way to determinism, but in that case the simplification should be done in a formal way.

Suppose that we could obtain a smooth version of the random velocity field through a process that we call conditioning on the data. If we, for instance, have made some hydraulic measurements, heads or permeabilities, on the scale of ω, maybe we will run a Monte Carlo simulation and average the results to come up with the best estimate of what v is. It would be a smooth estimate, because there is no way to tell exactly what v actually is at every point.

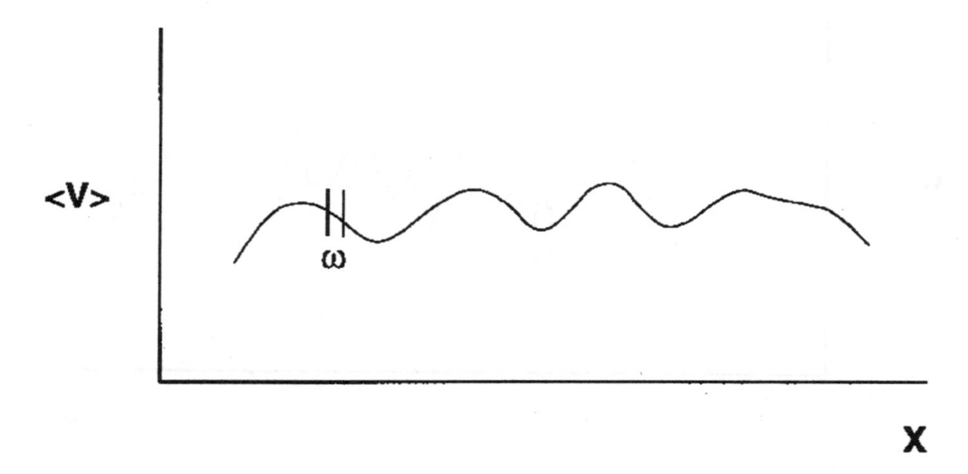

Figure 3. **Conditional mean velocity.**

The figure shows an ensemble mean or conditional mean. Not only is it smooth, it is deterministic and it is information dependent. The reason is that if we collect different pieces of information, we will get different smooth presentations of the actual velocity.

Now, we will associate with this a smooth concentration function, that is the deterministic function we would like to calculate. It still would represent the concentration that we measure over the scale of ω. We refer to it as an optimum predictor.

$<C>$ = conditional mean of C, optimum predictor of C (smooth, deterministic, information dependent)

It can be shown rigorously, mathematically, that whereas v and C satisfy the advection dispersion equation, their smooth counterparts do not.

The deterministic transport equation is:

$$\frac{\partial <C>}{\partial t} + <\underline{v}>\cdot\nabla<C> + \nabla\cdot\underline{Q} = 0$$

The equation, satisfied by the smooth counterparts of the velocity and the concentration, has an advective counterpart. However, there is an additional term, Q, which is the dispersive flux of the smooth concentration. It is given exactly, rigorously, as a space-time integral of some parameter which is a kernel of the convolution integral, which we call α, times the gradient of the concentration. This is shown in the next equation.

$$\underline{Q}(\underline{x},t) = -\int\int \underline{\underline{\alpha}}(\underline{x},t;\underline{\chi},\tau)\nabla<C(\underline{\chi},\tau)>d\underline{\chi}\,d\tau$$

Because the gradient of the concentration sits under the integral, Q at any point in space-time does not depend on the concentration at that same point but on the concentrations throughout the entire field and as such it is non-local. Nor is $\underline{\underline{\alpha}}$ a local property, it depends on two points and on space-time. It is non-Fickian and information dependent because everything, including α, depends on what we know and don't know about the system.

Only in very special cases can we approximate the integral with what appears to be a Fickian expression.

$$\underline{Q}(\underline{x}t) \approx -\underline{\underline{D}}(\underline{x},t)\nabla<C>$$

In this expression we have recovered a data-dependent, time-space-dependent dispersion tensor which did not exist on the local scale. The tensor came from lack of information. If we knew everything about the system, it would never appear.

$\underline{\underline{\alpha}}$ and $\underline{\underline{D}}$ depend on the degree of smoothing (information loss) relative to the correlation scale of $\underline{v}$. Deterministic models which do not resolve the velocity on the scale of ω, implying any smoothed average models, will, if we compare them to experimental data, return a dispersion tensor which is close to $\underline{\underline{D}}$.

We now assume that we have a dispersivity $\tilde{\alpha}$ which relates to D through v according to the equation:

$$\underline{\underline{D}} = \underline{\underline{\tilde{\alpha}}}|v|$$

From stochastic theories we know that if v is statistically homogenous, then the longitudinal component of that apparent dispersivity will first grow at a rate of one-to-one on a logarithmic scale and then reach an asymptote. If the velocity is non-stationary, if its correlation scale never attains a constant asymptote, for example, if it is fractal $\underline{\underline{\tilde{\alpha}}}$ will continue to increase with time.

If we disregard the dispersion coefficient in the smoothed model, we are disregarding loss of information due to smoothing. As long as we do this consciously and can support our decision, this is acceptable. But if we base our decision on the argument that dispersion is small, because we have seen it to be small in a small-scale experiment, it is an incorrect argument. A change in the support scale ω or in information (data) will change $\underline{\underline{\alpha}}$, $\underline{\underline{D}}$ and $<\underline{v}>$ and their behaviour.

Both the stochastic and deterministic transport equations shown are equally valid. So there is no point in arguing which of them are valid or invalid. This leads to a suggestion that if we have difficulties in defining the term "valid", perhaps we should start by trying to define the term "invalid".

The same type of reasoning, as we have just gone through, also holds for steady, transient, unsaturated or multi-phase flow and we find that our smoothed equations usually do not satisfy Darcy's law. That does not mean that Darcy's law is invalid as a physical principle, but when we deal with smoothed functions, they do not as a rule satisfy Darcy's law, except under special conditions.

Jordi Bruno

At this meeting we have not talked much about the geochemistry of our disposal systems. Probably this has to do with the fact that we expect this to be an additional complexity in our way of thinking. There are two main areas to which geochemical modelling can be applied in the safety assessment of radioactive waste. The first is the general area in which we try to predict the most probable evolution in the time scales of waste confinement. This includes understanding of the main processes that control the master variable which are critical for the geochemical stability of the geosphere, primarily Eh and pH but also others.

The second area covers the modelling of the chemical behaviour of the different components of the repository system. This is an art that has a firm scientific rooting in one American and one European school of geochemical modelling based on pioneering work of Garrells and Christ in Chicago and Lars-Gunnar Sillén in Stockholm.

The application of geochemical modelling in the safety assessment of nuclear waste repositories is not yet so well established. The problems are of various kinds, but some are of a general nature. When we do modelling we test our models by doing interpreting modelling in an iterative manner in the analysis of the experiment. When we do this we build our confidence in the model and we include new concepts as we find them. The safety assessments for nuclear repositories will include predictive modelling. In principle also predictive modelling should be done iteratively.

One of the main applications of natural analogue studies has been to test the validity of geochemical models, specifically regarding establishing solubilities and speciation of trace elements in natural systems. Using the so called blind predictive modelling methodology, a kind of site characterisation information is distributed to a group of modellers. They then get the information of the basic composition of the natural waters and have to establish the solubilities, the solubility limiting phase and the speciation of specific trace elements in the particular waters. The results are then compared with the actual information from the sites.

In practice, most of the natural analogue sites have been used for this purpose. The elements modelled are those we think are relevant for the safety assessment of nuclear waste repositories.

For uranium we know the solubilities fairly well. Solubilities are normally overestimated in the models. If a solubility limiting phase is present, that is an advantage. A good example is Cigar Lake with its confined uranium system. In other systems there are associations which we cannot take into account with our thermodynamic data bases.

Thorium is normally easier to deal with. The thorium that we have to consider is colloidal and can be described as amorphous thorium hydroxide.

Nickel poses some problems, as the concentration has been underestimated in many cases because of its associations with iron oxides. Calculations made for hypothetical fluxes of nickel at Poços de Caldas show large differences compared with observations in the field. The conclusion is that there is a lot to gain in better modelling the trace elements in the field.

To summarise, one could say that our capability to predict uranium trace element behaviour is varying due to incomplete geochemical understanding. The mobility of these elements from the source is strongly linked to the dissolution-precipitation of the main components in the ground water. The problem is that our data bases do not consider this process.

I propose two main driving forces for natural water systems: the sulphur–iron redox interface and the calcite–manganese–carbonate precipitation–dissolution. These are the only dynamic systems in the time scales that we are concerned with in terms of equilibrium.

To deal with coprecipitation and solid solutions in a simple but rigorous form, we could use the conditional solubility product. A large amount of experimental laboratory and field data for a range of pH values shows a good match between the individual solubilities of uranium and the actual modelled solubilities using the conditional solubility approach. The concept is not validated but at least it is not falsified.

Another problem is kinetics versus thermodynamics. Because of slow kinetically dependent processes, systems seldom are in thermodynamic equilibrium. When considering the kinetics, we add different degrees of complexity to our models. As a simple approach I have plotted all experimental information from different sites, as Cigar Lake and Poços de Caldas in a feldspar alteration diagram. If I compare the thermodynamic reaction path with the kinetic reaction path and also add the flow component of the system, it is apparent, that even for a site like Cigar Lake with long resident time of water, we must take the kinetic component into account. If we move to younger systems like Poços de Caldas, we have to add the flow component. Such an approach could be a kind of fingerprint of how far we are from thermodynamic equilibrium.

The conclusion is, therefore, that the main shortcomings of applying geochemical modelling to safety assessment, are the quality and quantity of thermodynamic and kinetic data for the system. For near field modelling it is a contradiction that we apply individual solubility for the source terms, when we know that the source terms are controlled by the major components of the groundwater system.

Ghislain de Marsily

I was given the chance to comment on the issue of validation of transport processes in the geosphere. I will take matrix diffusion as an example, and will try to give a brief history of the research in this area, and give my opinion on the present situation.

Although matrix diffusion is treated mostly by those of us working on fractured media, where there is a conducting fracture and matrix diffusion on either side of the fracture, the concept applies to any of the media, where there is a variation of velocities due to heterogeneity. It is certainly possible to have matrix diffusion from a high permeability zone to a low permeability zone.

Kristina Skagius in her thesis in 1986 described laboratory experiments which defined the concept of matrix diffusion. Several others worked on the same issue, Ivars Neretnieks, who probably started it all, Martin Put and others. Their conclusions were that the phenomenon exists and that it can considerably slow down the transport of radionuclides in the geosphere. Some uncertainties remain regarding the phenomenon. One of them concerns the surface diffusion which Ivars Neretnieks introduced some years ago and which should explain why cesium is moving faster than iodine. Principles of anion exclusions were also evidenced, showing that elements would move at different speeds.

Some work on matrix diffusion in France have treated a number of different elements: ytterbium, dysprosium, europium, samarium and lanthanum, all moving and diffusing in a fracture, but the results of these experiments have been difficult to interpret.

Single fracture experiments were performed in a quarry in Cornwall. These experiments revealed the existence of channelling, implying that the whole fracture was not open to sorption and matrix diffusion. This was also observed at Stripa and at Pinawa. Chin-Fu Tsang then developed his concept of the variable aperture model.

Yesterday Jörg Hadermann presented some work in a single fracture at the Grimsel field laboratory. An injection well and a pumping well were used for the experiment. The matrix was not the granite itself but some filling of the fracture on which the diffusion coefficients were estimated. It was shown that the experimental results could only be explained by including matrix diffusion. I would also like to mention the kakirites that the Swiss found. Kakirites are extreme channels, tubes, that somehow connect the repository to the surface. However, it seems as if the discussion on kakirites has stopped.

From these single fracture experiments, it can be concluded that matrix diffusion exists, but its efficiency will depend on the amount of surface in contact with the flowing water.

In the case of multiple fractures I like to refer to an experiment at Fanay-Augères. Tracers were going back and forth into the system from injection wells and transported over distances of tens of meters. Matrix diffusion could not be measured at this scale, but channelling was proven to exist. We were able to show that in this multiple fracture system, the widths of a channel was in the order of 2.5 cm.

In WIPP, testing was done with distances between wells in the order of 10 to 30 metres. These experiments showed clearly the existence of matrix diffusion, provided that certain assumptions were made on other parameters that were not possible to measure directly. In complex experiments *in situ*, uncertainties exist regarding the matrix diffusion coefficient, the surface to volume ratio, the existence of sorption and the geometry of the fractured system.

An early experiment on natural analogues may be of some interest. It was performed in the United Kingdom around 1980. Large blocks of granite were put in a harbour that was built about 400 years ago. Documentation on the blocks and the quarry from which they were brought exists. It was thought that one could get a nice chloride profile of maybe 20 centimetres penetration in them. However, it turned out that the chloride concentration in the investigated block was almost constant over distances of one metre. So the experiment was a major surprise.

Others have continued to work on matrix diffusion and natural analogues and we have now a better understanding of different processes. One important question is if matrix diffusion ends after a certain distance of migration in the rock or if it goes on for unlimited depths. My understanding is that the analogue studies have more or less shown that a limited penetration depth on each side of a fracture is more probable, as the pores get more and more clogged or get less and less connected, as you progress out from a fracture.

The people doing performance assessments need diffusion coefficients, surface to volume ratios and penetration depths for a site specific path. The question is if we are in a situation, after having done all the validation of the concept of matrix diffusion, where we, within two or three years, can give answers on the values of these parameters for the specific site of interest, and develop relevant methods to estimate these parameters at the repository scale, from laboratory and *in situ* measurements. Three years is, by the way, the scheduled time of the licensing application at WIPP.

Claudio Pescatore

Shlomo Neuman has introduced the concept of non-locality. In particular, he has shown that the process of smoothing over spatial scales larger than the observed ones injects non-locality in space and time. Accordingly, the original differential equation for the non-smoothed function turns into an integro-differential equation for the smoothed one. Indeed, we do live in a non-local world: quantum-mechanics tells us that (Bell's theorem) and, in any event, the familiar differential equations of momentum, mass, and energy transport are but approximations of more complex integro-differential ones.

Non-locality considerations arise also from the analysis of the source term in waste disposal. For instance, if "far away" from the source (waste) a process exists that removes radionuclides from solution, the source will respond to such a process. In fact, the existence of a removal process of leached species is a necessary condition for release, hence the importance of coupling the chemical behaviour of the source with species removal processes taking place outside. Yesterday, P. Van Iseghem did link the continued dissolution of the glass matrix to the continuous scavenging of silica in solution by the surrounding clay.

The coupling between the behaviour of the waste matrix and transport phenomena in open systems is an area that deserves further attention. Yesterday, I gave one example of the great difference that exists between open-system and closed-system behaviour. Consider a solubility-limited release process: in an open system, the release can never become zero even if we are at solubility next to the waste form surface. The reason is that a concentration gradient will always exist that will remove leached species through diffusion. Conversely, in closed systems the whole medium will eventually reach saturation and the leaching process will stop.

The above considerations apply both to vitrified waste and to spent nuclear fuel. In the case of the latter, it is often said that natural analogues exist. Is it really the case? I would like to hear more evidence/considerations to that effect. Spent fuel is of variable burnup and composition and, besides the obvious effects of fracturing of the pellets due to swelling and heat, high neutron fluxes may change some of the lattice structures while fissions may produce non-stoichiometric amounts of oxygen. Can we do, defensibly, thermodynamic calculations with this material?

Moving to another subject: matrix diffusion is excellent to have and to rely upon at least in part. Unfortunately, INTRAVAL did not give us the definitive view on this. However, it seems that pore diffusion in most clays is a fact and we have heard here evidence for a limited, but important, extent of matrix diffusion even in advective systems due to the presence of reactive zones in fault gauges. That raises the question of whether we know enough of characterisation of fractures. I detected positive indications in the work of Steven Barnwart, where he found that the fractures are filled with clayey alteration products and that, many times, we may not see the clays because they have been removed by the drilling fluids. A further good finding reported at GEOVAL is that the field tracer tests do not suggest the presence of new, previously unknown processes. Thus we may not have to focus on new questions, but only on exploring deeper some of the present issues.

Finally, Charles, let me respond briefly to one of the questions you asked: When do we expect validation? In a way, "only at the very end", that is when every interested party feels satisfied and a license is close to be granted. A judgement of "validity" will be rendered by the regulator only as a result of an iterative process between proponent and regulator whereby "the proposed" evidence is weighed and counterweighed and new evidence may be elicited. This is an experimental fact, as it has been witnessed in the licensing of other nuclear facilities.

Chairman

These different viewpoints on the various parts of the modelling chain lead to the conclusion that there are still many open questions. I would now like to hear from the audience where you feel that the biggest problems are. Are there still problems that stop us from using performance assessment for decision-making purposes?

Rip Anderson

At this meeting we have spent and are spending a lot of time discussing simplification. I am not sure why we are doing that. Is it because we don't have enough computing power? I don't agree with that position. Is it because we have not settled on a conceptual model that we should use? Is it because we have not been able to address the scaling issue or is it because we use the word validation and don't know how to do validation? In the WIPP program we have lately used a new set of computers similar to about a half a dozen Crays linked together giving us all the computational capability we need without any simplifications at a relatively inexpensive price. So to me it is not a problem of computational power, but a problem of sorting out the other issues identified in the questions above.

Peter Jackson

I certainly feel limited by current computing power. For example, I think that to address realistically very detailed stochastic models in three dimensions with salt transport – the sort of models Shlomo Neuman was alluding to earlier – with Monte-Carlo techniques and current methods would require 10^6 or even 10^9 more computing power than we have at the moment. Even ten Cray supercomputers linked together would not give that sort of computing power.

However, the crucial issue is to convince the public and the regulators that we have a sensible case. That will not be easy if we have to rely on a very complicated model that very few people understand.

Shlomo Neuman

Even if we had all that great computing power, I think that much more has got into these very complex Monte Carlo models than is needed to describe the problems.

Chin-Fu Tsang

In the models, there is some intrinsic need for simplification because of lack of complete deterministic microscopic information. There are also some theoretical problems on scaling, when all details are not known. However, to answer the need of performance assessment, we should not need to know all the microscopic details. A church bell might serve as an illustration. What is important is the sound frequencies that are produced. For that we don't have to study the behaviour of all the molecules in the bell material. We simplify and examine the size and shape of the bell rather than make a study of how all the molecules vibrate.

We have come a long way in the process over the six years of INTRAVAL. A major statement is that no new processes have been identified, which is very good. Concerning channelling and matrix diffusion, we have been studying them for the last six years or more. Even if we cannot say that we have done everything, at least we know where we are. We should also be in a good position to handle new field experiments.

The present emphasis we are following is real field case stochastic modelling including the understanding of some uncertainty due to variability. To me it has been a great advance in recent years to apply stochastic modelling to real field experiments. Earlier, Cacas, de Marsily and others began to look at data against stochastic modelling. Today we try to do that more extensively. However, there is more work to be done. For example, much theoretical work remains: how to get the basic input parameters, how to determine the uncertainty due to variability and how to condition data to obtain the model input parameters. If the data are used for conditioning and the point of interest for prediction is within the data domain, the answer might be very good. However, for neighbouring or further out regions the conditioning may cause bias, which seems to be a big problem.

Chairman

If, Chin-Fu Tsang, you have to do a performance assessment using existing models but were allowed to collect all the data you want to collect could you do it?

Chin-Fu Tsang

I don't know how to answer that question. But we must know the answer to this question when we submit results for license application, which, in the United States, is by definition the year 2001!

Abraham van Luik

What Eric Smistad made clear yesterday was that we will continue to support the technical bases for modelling studies into the year 2008 for the application to receive waste in the repository and for at least 75 years beyond that date when the application to close the repository will be submitted.

Chairman

I will remind you that 1997 is the date for site confirmation.

Allan Duncan

The interface between geochemistry and hydrology is very important, as we finally have to calculate the radionuclide migration back to the biosphere. My impression from this meeting and elsewhere is that geochemical work is much more difficult than hydrologic work, or maybe the level of understanding of geochemistry is less. Don´t you think it would be easier to sustain a safety case if it depended primarily on knowledge of water flow rates, after which any geochemical hold-up would be a bonus?

Chairman

To me geochemistry seems to be easier than hydrology, but I might have listened too much to Shlomo Neuman.

Ghislain de Marsily

I would like to come back to the issue of averaging or model limitations. If we have some kind of virgin material for which we try to determine the spatial variability within a given formation to predict transport, we may use the concept of averaging and make some assumptions of stationarity. In a way, we have then to assume, either that the transport properties are stationary or that they vary in a known way.

However, there are other methods to address this problem, which are essentially developed in the oil industry. The so called "Facies" models will in a stochastic Monte Carlo simulation way represent the heterogeneity of a layer or the variability within a certain formation due to the complexity of the natural processes that led to form these layers.

As Peter Jackson pointed out, if we believe that we have a good representation of the reality, we are faced with a computer problem. Taking the Harwell basin as an example, we have about 10^9 nodes to describe reality in these models. Even if we would like to treat the problem at this scale with many realisations of what it could be in order to represent all the complexities of the transport, diffusion in the matrix and what not, we cannot do that because of limitations in computational power. Instead, we may select 10^6 nodes or perhaps 10^7 if we are very rich and are using ten Crays. But certainly not to 10^9 or 10^{10}. The problem is however at that scale.

So what should we do? Referring to Shlomo Neuman, I propose that we should upgrade and average at an intermediate scale. The alternative to stop at an intermediate scale would be to average up to the major scale, all the way from the repository to the outlet. If we can do the averaging of some of the important factors and stop at an appropriate scale, the problem could be handled by the computers. We could then represent some of the variability that otherwise would be very difficult to represent with averaging up to the upper limit.

Ivars Neretnieks

I would like to respond to those that have the impression that we have stopped discussing and studying phenomena like extreme channelling, for instance kakirites, and surface diffusion. The kakirites are not "dead", we have evidence from Stripa and other places that fracture intersections play an important role in conducting water. That may be very similar to the behaviour of the kakirites and we try to build that into our models.

Nor is surface diffusion by any means "dead". More and more experimental evidence shows that we have some kind of "speeding-up" component. To visualise it, we may think of it as the opposite to ion exclusion. The electrostatic attractive forces will attract cations to the surface. They are not bound as in surface complexation. They are mobile and the higher their concentration, the higher are their transport capacity.

Regarding penetration depths, Lars Birgersson in *in situ* diffusion experiments in rock under virgin stress found that tracers could move at least 30 cm, which was the observation distance. Finnish scientists found clear evidence of salt concentration profiles when sectioning a drillcore adjacent to a flowing fracture. This observation may be matched to the knowledge of the geology after the disappearance of the last ice age which implies that the migration has been taking place during 3000 to 5000 years.

Why don't we have uranium series disequilibration over more than five centimetres? The answer is that 3–5 centimetres in fact is the distance predicted with existing K_d-values and a time span of a quarter of a million to one million years.

Chairman

In the summary by Neil Chapman on INTRAVAL, the conclusions on matrix diffusion were very vague. I hope it is made clear, when INTRAVAL finally is reported, that references only are made to the experiments included in the INTRAVAL study and not to the whole range of experiments that have been made. The comments we have received indicate that there are experiments not within INTRAVAL which provide more evidence on matrix diffusion effects.

Ghislain de Marsily

Regarding especially the matrix diffusion issue, I suggest that NEA collects and summarises what has been done in the last 20 years on the evidence of matrix diffusion for different cases and distances on various rock types. The idea is to collect a data base for granodiorite or granite, a rock type we have in Canada, France, Sweden and Switzerland. For other rock types like dolomite, evidence from WIPP and other sites will help. We like to be able to convince ourselves and others that the phenomena we use in performance assessment and the values of the parameters are not just invented. In fact, we rely more or less on observations that have been performed using a number of different techniques at many places in the world, but a synthesis is lacking at present.

Chairman

Many of you know that the proposal by Ghislain de Marsily could be included in discussions now underway at NEA. The potential project is presently called GEOTRAP. The input from this meeting might be very useful for NEA in its discussions and considerations on that issue.

The next item will be a talk by Helmut Röthemeyer, who will consider the perspective of long time scales.

Helmut Röthemeyer

I would like to start by giving you an example of what a blind predictive model could be. Let us consider an insect with a life span of about a day. If its life experience is gained on a sunny day, its predictive capabilities concerning its environment, if any, will surely be based on this experience without realizing the possibility of rain. We know the interrelationship between rain and sun and know the important cycles of matter influencing our weather and we know why there are principal limitations to our respective weather forecast over a range of days. We know too that these uncertainties do not matter for the understanding of the environment. The far future prediction from the supposed insect's point of view, that would be 10 000 times its life span, would be some ten years. This is mainly dependent on the climate with its long time cycles and not on the weather. The question we have to ask ourselves is: Do our capabilities for far future predictions of the performance of repository systems reflect those of the above insect or those of present day meteorologists?

Before answering that question, let us consider how believable scientific predictions are in general. In the past, science has achieved convincing success in describing and predicting single natural phenomena. Thus, scientists were convinced that the answers to single questions would finally lead to an understanding of Nature as a whole. Today, we know that this approach has failed. The reason is that the experimental method has an influence on the object under investigation and we have talked about that in great detail. Thus, science does not lead to an understanding of Nature itself but to an understanding of our relationship to Nature by means of the respective models of natural processes. So, our science is nothing very spectacular or particular.

How believable are far future predictions for repository systems with that background? Geoscientific models are based on the premise that present cycles and processes are the key to our understanding of the past. Predictive modelling again rests on the assumption that the understanding of the past is a prerequisite for reliable predictions, as long as these predictions do not cover time spans far beyond those of the respective cycles of matter. Predictions about geological, climatological and biospherical cycles thus deal with different types of uncertainties in different dimensions of time. Qualitative assessments indicate the basic possibilities and limitations in far future predictions of repository systems. Predictions mainly influenced by climatic changes may indicate the performance of a repository system for up to a thousand, and, as we have heard yesterday, maybe 10 000 years, whereas predictions of geological conditions deep underground may be done up to millions of years. Further research of the relevant processes and events may give a more quantitative understanding of these limiting time periods and may even reveal, as in meteorology, a chaotic behaviour of some of the processes concerned. Even with these time spans, predicted results cannot be interpreted as reality but only as indicators of reality.

I have explained the reason in basic general terms and I will do it specificly with respect to waste disposal facilities. First, we have the uncertainties in the initial and time dependent boundary conditions from the modelling of the past, which usually has to be bridged by expert judgement, and second, we have and will always have an incomplete data base. A complete data base would practically destroy the natural characteristics of a site. Our goal must therefore be to assure that the processes which have happened in the past have limited consequences if they occur in the future. For this purpose, we can use several indicators which represent the different cycles influencing a repository system, for instance, dose, risk, environmental conditions, flux through barriers and time. I might point out that recently environmental indicators have been introduced for environmental protection purposes and research and this has been done because the systems these people deal with are of a comparable complex nature.

If the assessment of these indicators is based on a careful study of the present mineralogical and chemical composition of the geological repository system and its changes in the geological past, we have a convincing site specific natural analogue. This applies especially if the composition of certain parts of the repository rock has not at all or only to a minor extent been changed since its formation, in spite of the various geological processes over millions of years. This approach leads to believable, far future predictions, if convincing evidence is provided that the analogues will adequately represent the repository system. This approach has proved quite successful with the decision-making process for Gorleben with respect to the vast investments for underground disposal there and also in the licensing process for the Konrad repository. At the moment a draft version of the latter licence has been drawn up.

Nuclear waste cannot be considered exempt even for time spans exceeding our present predictive possibilities. Here, the assessment of consequences, if considered necessary, may be based on toxicity comparisons or on site independent generic assessments. The latter reveal the relatively small consequences of radioactive waste disposed in a geological environment with its special characteristics.

After this overview, we may answer our introductory question. We certainly know more about future predictions than our supposed insect but our understanding of the relevant processes is far from being satisfactory from a scientific point of view.

Future research, as envisaged for instance with GEOTRAP, may further improve the information about the importance of the different uncertainties, the optimisation of site investigations and planning and the confidence building process. Having said that, my conclusion is that the knowledge we have today has proved sufficient for a licensing process and for taking decisive decisions with respect to site investigation processes.

Olle Olsson

I have got the very difficult task to try to answer the question: How good are our site models?. I will do that using a bit of my own perspective that is working with crystalline bedrock.

First of all, we have certain basic requirements on a repository. It should provide a chemically stable environment, hence we need geochemical models. The transport of radionuclides should be limited or slow, hence we need flow and transport models. We need to show that the repository is mechanically stable and that we actually can build a repository at the selected site. We try to describe all these aspects by different process models. However, the structural model is the same for all of them.

A model consists essentially of a set of process parameters with some added geometrical structure. For fractured rock we generally decompose the structural model into fractured zones and rock types.

Then we have the first problem: How do we define the structural features of the models? How do we find these features?

The second problem is: How do we correlate the process parameters to the features that we find?

If we look at the data we actually collect in a site investigation, we can divide them into four different types.

The first type of data give us some material property and process property at the measurement point, normally along a borehole, but sometimes at ground surface. Examples are hydraulic measurements made between packers, electric resistivity and core samples.

The second type of data provides information on material properties and some structural information near the observation point. Such data is for example obtained by televiewers, borehole television, and formation microscanners. The collected information is specific for the observation points and has to be extrapolated to locations outside the borehole. In that process, there is always the question of the representativeness of the data and how far structural information can be extrapolated.

The third type of data provides structural information within a large volume surrounding the observation point and some information on material properties. Methods which provide this type of data are remote sensing methods such as borehole radar and seismics which give us the possibility to see through rock. These methods were developed in the Stripa Project and are today used routinely in many nuclear waste programs.

The fourth type of data only provides information on structure. An example is lineament studies which provide no information on properties, but can be quite useful, especially if they cover large areas.

With the acquired set of basic structural information we must use our basic geological know-how of the site in our extrapolation of the data. In a fractured environment, the fractured zones seem to be essentially planar and can be fairly extensive. Certain rock types may have inclusions, often quite irregular and hence difficult to extrapolate, while intrusive dykes can be fairly regular and easier to extrapolate.

In Stripa we developed an iterative scheme for correlating single hole data to remote sensing data. The scheme resulted in a list of what we could explain with the data we have and what we could not explain.

The amount of data we have about a site is always limited, but the question is how much we really need to know to describe a site. We must look at the reliability of our interpretations. We may have different requirements for different size features. We should be quite certain we know where the large features, regional zones and major zones are. However, we can accept the possibility that there are major zones outside our repository area. Regarding minor zones and individual fractures, we know that we cannot find them all, so we have to describe them in some other way, probably stochastically, if we cannot average their properties within the process model we use.

With the tools we have today, including radar and seismics, that make it possible to see through the bedrock, and the structured approach that has been developed over the years, I believe that we can have good site models for the important features. We should be able to realise models for small scale features, either by averaging them out or by describing them stochastically.

Another problem is how to make the correlation between the structured models and the parameters that go into our process models? Normally plenty of data exist for the hydraulic properties and the procedures are relatively well established. But if we, for instance, are looking at radionuclide transport and matrix diffusion, how do we find the site data for diffusivity, surface to volume ratio, flow wetted surface, the K_d-values, etc? How should the values be correlated to the different types of structures that we identify? We have started some work at Äspö together with NAGRA in order to find out if we can classify features into different sets and give them different values.

Chairman

It is very interesting to note that for years and years I have heard people saying that we know all this process data, but what we need are structural data from a site. Now Olle Olsson just said that we can get all the site structural data, but we need to know more about the processes!

Jörg Hadermann

If we interpret validation as having confidence in, it is a question of having confidence in the processes. The obvious question is then how much our assumed processes relies on fundamental laws. Matrix diffusion is an excellent example as it is based on the fundamental laws of statistical mechanics. As long as the statistical mechanics laws are valid, concentration gradients will be the driving forces for matrix diffusion.

The second question is how much confidence we have in the parameter values describing the process. This is a site specific question and there is no other way than to go to the site and measure those values.

We should distinguish between the processes – and we have not seen any new processes in the last 12 years or so – and the reliability of the parameter values that are involved in the description of processes. What is our confidence in the parameter values?

Shlomo Neuman

What Jörg Hadermann just said exemplifies the tension between physicists and geologists or hydrogeologists. I don´t think that anybody disputes that matrix diffusion takes place. What is unknown is not what the diffusion coefficient might be, but how much area is available for this process to take place. That is a hydrogeological question to which we have difficulties in obtaining site specific answers.

Jörg Hadermann

We have to determine the magnitude of the parameter in our system. For example, if the diffusion coefficient is 10^{-20} m^2/s then we can forget about diffusion, but if the value is 10^{-11} m^2/s then diffusion becomes an important process.

Chairman

This discussion shows that we have to be very careful about semantics. If we discuss whether matrix diffusion occurs or not, people outside this group might think that we are not sure if the process occurs, whereas we actually are discussing to what extent the process occurs.

Tönis Papp

It is important that the assessments we make are not only focused on licensing, but also on the decision-making in the process of building and constructing a good repository. There is a difference because decision-making is usually made in small steps, while licensing represents larger steps. In order to have access to a site, we have to show that it has the potential to be a good site which we really cannot prove until we have investigated the site. We are thus reducing the risk in site selection by taking stepwise decisions even though we believe that we have enough evidence and confidence to take larger steps.

This has a consequence on our use of models in practice. It also has a parallel to Rip Anderson´s question what we need simplified models for? Some of these simplified models could be used to make crude evaluations on questions such as: Do we accept the site? How can we use the site? Where we place the various parts of the repository at the site might depend on our possibility to model the general flow paths rather than the speed of the groundwater. Performance measures like the water transport time to the surface from different positions in the disposal area may be used as a crude approximation or indicator of the span in the transport times of radionuclides.

Not only qualified scientific evaluation of the models are important, but also estimations of the capability of the models in giving limited and bounding answers. Such a validity statement could, for example, include an evaluation of the ability of the model to accurately predict the gradients and general flowpaths, if the structure of the site model is known.

This has bearing on a question Johan Andersson raised, namely how much confidence we really need? If we take small steps and if we also can step backwards, we do not need so much confidence. A more extensive evaluation will be performed at a later stage in connection with the licensing.

An important question for safety assessors concerns the confidence in conservatism. How do we know that a model is conservative and with what confidence? My answer would be that conservatism can partly be connected to the selection of input data and also to the simplification of the models.

Peter Jackson

During this meeting I have seen several plots of confidence as a function of time that imply that the level of confidence increases monotonically with time. I would expect that our level of confidence generally increases as we gain information, but on occasion some new data might lead to a decrease in our level of confidence. What we should be looking for is really assurance that our level of confidence is not going to drop below a certain acceptable level.

Johan Andersson

In answering the question what level of validation is needed and how much we need to know we must not forget performance assessment. This is the only tool we have to put the pieces together. Performance assessment is not necessarily a total system integrated stochastic Monte-Carlo model. It is an iterative process where we have to go back to the information and the data we have in order to rank the problems and the key issues. This is more critical for the parts of the system on which we rely most heavily regarding safety. When going back to the experimental evidence and the scientific knowledge, we might find that some of the initial ideas applied in the assessment were too simple and that our simple models have to be updated. This is particularly true for flow and transport in crystalline rock where we cannot replace our current understanding with simplistic models of average groundwater flow and dispersivity.

Judging from a series of iterations of performance assessment studies and advances from experimental studies my opinion is that the main problem now does not concern validation but rather invalidation. We have found some disturbing evidence in our field experiments. We have seen examples of fast channels in crystalline rock, which make the potential for matrix diffusion doubtful. Earlier this week we heard that experiments and analyses of gas pressure build-up has been included in the WIPP-program. All these concerns are a result of increased understanding of sites and more detailed modelling. We have to do something about these concerns, either to show that they do not exist at our site, change our safety case or change repository design.

Then we still have an almost endless list of assumptions in the performance assessment where we can question the theoretical and empirical support. Major improvements are being made, but there is certainly an end to how much further improvements that can be made during the next ten years or so.

In answering the question on how much validation we need my view is that the only way to find an answer is to try it out for real in the licensing process, and no one can make any promises. Decisions on nuclear waste repositories have to be taken by several groups where maybe the most important ones include implementors, regulators and the political system. The first thing that needs to happen is that the implementor takes a decision to put up an application. In order to do this the implementor has to be convinced that they have a defensible safety case. The regulators then have to decide whether or not they trust the implementor and if they can defend their decision to give a license. When you come to the political system it is even more a question of trust. Do the political decision makers trust the implementors and the regulators? Does the public trust the politicians? In order to obtain the trust we need that one important part is humbleness.

Chairman

The trust will be based on how we all behave in a scientific manner and how believable we are as scientists. That is why I do not like when people make statements which put performance assessors in one box and scientists in another box. Unless we have them in the same box, we will not have this necessary level of trust.

Shlomo Neuman

I have a short comment to Johan Andersson´s discussion regarding confidence. I think that we, as well as others, should have confidence in our calculations. One way to achieve this is to stop pretending that we can do things we in fact cannot do.

Budhi Sagar

I agree with Peter Jackson that the confidence can indeed go down as you proceed in the process. Then you have to have the conviction and strength to walk away, irrespective of how much time and money you have spent. This decision can not be entirely scientific, it could be influenced by political factors.

I am beginning to realise that definition of terms is almost impossible. We have problems defining terms like validity or validation, and I have great trouble defining simplicity. To some people Monte–Carlo simulations would be much simpler than what Shlomo Neuman showed in his equations.

There are certainly discrete decision points in the repository licensing process and there will be different amount of information available and different amount of resources spent at these decision points. When you start the construction you will mainly have access to surface based characterisation data from the site, complemented with some underground data from shafts and drifts. The concepts for the engineered barriers will be more or less fixed at that time and a safety analysis will be made to show that the safety requirements can be met. The safety requirements will be met by using models that are supported by tests, literature and peer review. The assumptions will be conservative and uncertainty bounds will be presented.

The second phase of decision-making will come about ten years later when you have spent a lot of money at the site, which makes it even harder to abandon the site at that time. At this point, the engineering design is final, you have done a detailed analysis regarding hydrology, geochemistry, rock mechanics etc., some long-term testing and updated your models to include site-specific information. At this point your confidence may go down and you may decide this is no good because your models might be incompetent, you might not have enough data or the site might not be good. But you have to decide if you want to proceed or not.

The third and final phase is when you are ready to close the repository. This phase is probably at least 50 years away. Until that time, different types of long-term tests should be done in order to update and test your models. At this point, you have to say that your models are validated, if you decide to close the repository.

A question I want to raise is related to the scientific credibility issue. I feel that there is a time lag between when a concept is proposed by a scientist and when it is actually used. This is because any concept that is proposed has to be repeated, tested and reviewed by other people and finally accepted by the scientific community. I think the time period for this process is 5–10 years. This means that you can not use the state of the art concept-model in a license application. The models that will be presented in a safety case will be at least 5 years old. That doesn´t mean that science should stop. It should carry on, but my question is: Should we wait for the "final" answer before making decisions? I think that is a validation issue, since it involves a judgement on how far you are going to validate your models.

Tom Nicholson

I have prepared a few questions which address the theme of this workshop "validation through model testing using field experiments" which may be answered in part by reviewing certain INTRAVAL results.

The first question is: How critical are repeated experiments for validation strategies? In some cases we only had one experiment to address validation. The Las Cruces studies demonstrated that it was critical that repeated experiments be performed. The testing of models require both characterisation and dynamic observations to allow sufficient quantification to perform rigorous testing.

Next: Should the experimental design require *a priori* modelling to determine information needs and structure to discriminate between models? The Las Cruces data needs and structure were defined using modelling results prior to the initial field studies. These *a priori* modelling efforts assisted in defining the acceptance criteria and performance measures. The modelling also defined the scale of the experiment, the boundary conditions, and water and chemical tracer application schedule. Obviously, the identification of the antecedent moisture and pressure conditions were not known, and therefore limited the *a priori* modelling.

Can adequate experimental modelling be conducted for coupled model testing? For example, thermo-hydro-mechanical models? This question is being answered in part by the DECOVALEX project. For the Apache Leap Tuff site, the experimental design modelling was limited to two-phase isothermal modelling. The introduction of non-isothermal conditions had to be delayed until the processes were better understood for partially-saturated, fractured rock. Similarly, the introduction of such mechanical considerations as deformable fracture apertures, and associated hydrologic responses await further process testing and characterisation.

Finally: In the next ten years what would be the most fruitful research directions to pursue for clarifying validation questions and for reducing uncertainties? As discussed during this workshop, rigorous hypothesis testing using field-scale experiments, in which, *a priori* modelling and definition of discriminatory data needs and structure would be highly advantageous to answering this question. Sufficient time, resources and the use of interdisciplinary teams should be committed to meet this challenge.

Chin-Fu Tsang

We have not discussed sufficiently the quantification of uncertainty. We should not give a single number as the answer in performance assessment because that will have a tremendous uncertainty. This uncertainty will not only be dependent on the parameter uncertainty, but also on uncertainties in the conceptual model and future events. We must help the regulators and the public by telling them our limitation of knowledge, because it is not true that we will be able to know exactly what will happen. We should give performance assessment with a quantified uncertainty and say that this is the limit of our scientific knowledge and then let the public make their decision.

Chairman

My impression is that we have made progress since previous GEOVAL meetings in 86 and 90. There has been a large amount of data collected, a large variety of experiments of different types have been performed and the extent of international cooperation seems to be increasing.

We have been spared hard decisions for a long period of time, partly because time scales have been pushed out for reasons for which we are not responsible. But major decisions, like the site selection in the Swedish program, are coming up. At that time, we have to do an honest evaluation of where we stand and what the remaining uncertainties are. We will not be able to quantify the uncertainties in the way we, at least I, hoped for. I therefore hope that Chin-Fu Tsang is right when he says that the public is much more mature than we sometimes think with respect to making decisions based on information that have large elements of uncertainty.

With that I would like to close the panel discussion.

GEOVAL '94
(11-14 October 1994)

List of Participants

AUSTRALIA

Peter DUERDEN
Counsellor (Nuclear)
Australian Embassy
Mattiellistrasse 2
A-1040 Vienna
Austria

Tel. + 43 (1) 512 8580
Fax. + 43 (1) 504 1178

BELGIUM

Bernard NEERDAEL
SCK/CEN
Boeretang 200
B-2400 Mol

Tel. + 32 (14) 33 37 70
Fax. + 32 (14) 32 12 79

Lorenzo ORTIZ
SCK/CEN
Boeretang 200
B-2400 Mol

Tel. + 32 (14) 33 22 47
Fax. + 32 (14) 32 12 79

Martin PUT
SCK/CEN
Boeretang 200
B-2400 Mol

Tel. + 32 (14) 33 32 21
Fax. + 32 (14) 32 12 79

Pierre VAN ISEGHEM
SCK/CEN
Boeretang 200
B-2400 Mol

Tel. + 32 (14) 33 37 35
Fax. + 32 (14) 32 12 79

Geert VOLCKAERT
SCK/CEN
Boeretang 200
B-2400 Mol

Tel. + 32 (14) 33 32 20
Fax. + 32 (14) 32 12 79

CANADA

Tin CHAN
c/o Ontario Hydro H16 F19
700 University Avenue
Toronto, Ontario
M5G 1X6

Tel. + 1 (416) 592 5296
Fax. + 1 (416) 592 4485

Douglas METCALFE
Assessment Specialist
Wastes and Impacts Division
Atomic Energy Control Board
P.O. Box 1046, Station B
Ottawa K1P 5S9
Tel. + 1 (613) 995 5294
Fax. + 1 (613) 995 5086

Son NGUYEN
Atomic Energy Control Board
P.O. Box 1046, Station B
Ottawa K1P 5S9
Tel. + 1 (613) 996 1436
Fax. + 1 (613) 995 5086

C. ONOFREI
Atomic Energy of Canada Research
Whiteshell Laboratories
Pinawa, Manitoba
R0E 1L0
Tel. + 1 (204) 753 2311
Fax. + 1 (204) 753 8404

A.G. WIKJORD
Head, Environmental and Safety
Assessment Branch
Atomic Energy of Canada Research
Whiteshell Laboratories
Pinawa, Manitoba
R0E 1L0
Tel. + 1 (204) 753 2311
Fax. + 1 (204) 753 8404

FINLAND

Timo ÄIKÄS
Nuclear Waste Management
Teollisuuden Voima Oy (TVO)
Annankatu 42 C
FIN-00100 Helsinki
Tel. + 358 (0) 6180 3730
Fax. + 358 (0) 6180 2570

Jorma AUTIO
Saanio & Riekkola Consulting Engineers
Laulukuja 4
FIN-00420 Helsinki
Tel. + 358 (0) 566 6500
Fax. + 358 (0) 566 3354

Esko ELORANTA
Finnish Centre for Radiation
and Nuclear Safety (STUK)
P.O. Box 14
FIN-00881 Helsinki
Tel. + 358 (0) 759881
Fax. + 358 (0) 75988382

Aimo HAUTOJÄRVI
VTT Energy
P.O. Box 1604
FIN-02044 VTT
Tel. + 358 (0) 456 5052
Fax. + 358 (0) 456 5000

Juhani VIRA
Teollisuuden Voima Oy (TVO)
Annankatu 42 C
FIN-00100 Helsinki
Tel. + 358 (0) 6180 3740
Fax. + 358 (0) 6180 2570

FRANCE

J.-P. BOUCHARD
EDF DER
6, quai Watier
F-78400 Chatou
Tel. + 33 (1) 30 87 78 72
Fax. + 33 (1) 30 87 80 86

Christian BROUSSE
Institut de Protection et de Sûreté Nucléaire
(IPSN) - BP 6
F-92265 Fontenay-aux-Roses Cedex
Tel. + 33 (1) 46 54 79 82
Fax. + 33 (1) 46 54 71 35
+ 33 (1) 45 54 71 22

Christine BRUN-YABA
Institut de Protection et de Sûreté Nucléaire
(IPSN) - BP 6
F-92265 Fontenay-aux-Roses Cedex
Tel. + 33 (1) 46 54 87 33
Fax. + 33 (1) 46 54 71 35

Lionel DEWIERE
Département Etudes Expérimentations Calculs
ANDRA - BP 38
F-92266 Fontenay-aux-Roses Cedex
Tel. + 33 (1) 41 17 80 38
Fax. + 33 (1) 41 17 84 08

Pierre ESCALIER DES ORRES
CEA
IPSN/DES/SESID
Centre d'Etude Nucléaire de Fontenay-aux-Roses
B.P. 6
F-92265 Fontenay-aux-Roses Cedex
Tel. + 33 (1) 46 54 86 20
Fax. + 33 (1) 42 53 91 27

Jean-Marie GRAS
Département Etude des Matériaux
Service Réacteurs Nucléaires et Echangeurs
Direction des Etudes et Recherches
E D F
Les Renardières, Route de Sens, BP 1
F-77250 Moret sur Loing
Tel. + 33 (1) 60 73 68 14
Fax. + 33 (1) 60 73 68 89

Christophe GRENIER
Commissariat à l'Energie Atomique
Centre d'Etudes de Saclay
DRN/DMT/SEMT/LTTMF
F-91191 Gif sur Yvette Cedex
Tel. + 33 (1) 69 08 26 10
Fax. + 33 (1) 69 08 82 29

Jean-Claude GROS
Institut de Protection et
de Sûreté Nucléaire (IPSN)
BP 6
F-92265 Fontenay-aux-Roses Cedex
Tel. + 33 (1) 46 54 90 88
Fax. + 33 (1) 47 35 14 23

Mersh LAVENUE
Ecole Nationale Supérieure des Mines de Paris
Centre d'Informatique Géologique
Laboratoire d'Hydrogéologie Mathématique
35, rue Saint-Honoré
F-77305 Fontainebleau Cedex
Tel. + 33 (1) 64 69 47 02
Fax. + 33 (1) 64 69 47 03

Sophie LEVEEL
Centre d'Etudes de Fontenay-aux-Roses
DESD/SESD - BP 6
F-92265 Fontenay-aux-Roses Cedex
Tel. + 33 (1) 46 54 81 32
Fax. + 33 (1) 46 54 81 29

Pierre-Loment LUCILLE
EDF DER
6, quai Watier
F-78400 Chatou
Tel. + 33 (1) 30 87 78 72
Fax. + 33 (1) 30 87 80 86

Ghislain de MARSILY
Laboratoire de Géologie Appliquée
B 123
Université Pierre et Marie Curie
4, place Jussieu
F-25252 Paris Cedex 05
Tel. + 33 (1) 44 27 51 26
Fax. + 33 (1) 44 27 51 25
E-Mail: gdm@ccr.jussieu.fr

Emmanuel MOUCHE
Commissariat à l'Energie Atomique
Centre d'Etudes de Saclay
DRN/DMT/SEMT/LTTMF
F-91191 Gif-sur-Yvette Cedex
Tel. + 33 (1) 69 08 66 99
Fax. + 33 (1) 69 08 82 29

Denis PERRAUD
ANDRA
BP 38
F-92266 Fontenay-aux-Roses Cedex
Tel. +33 (1) 41 17 81 11
Fax. +33 (1) 41 17 80 13

Jean-Claude PETIT
Commissariat à l'Energie Atomique
DESD - B.P. No. 6
F-92265 Fontenay-aux-Roses Cedex
Tel. + 33 (1) 46 54 81 33
Fax. + 33 (1) 42 53 14 69

Bernard VIALAY
ANDRA
BP 38
F-92266 Fontenay-aux-Roses Cedex
Tel. +33 (1) 41 17 81 57
Fax. +33 (1) 41 17 80 13

Bertrand VIGNAL
ANDRA
Département Etudes Expérimentations Calculs
BP 38
F-92266 Fontenay-aux-Roses Cedex
Tel. + 33 (1) 41 17 81 49
Fax. + 33 (1) 41 17 84 08

A. VINSOT
INTAKTA FRANCE SARL
16 bis, avenue Desgenettes
F-94100 Saint Maur des Fossés
Tel. + 33 (1) 45 11 22 99
Fax. + 33 (1) 45 11 23 03

GERMANY

Horst BESENECKER
Niedersächsisches Umweltministerium
Postfach 41 07
D-30041 Hannover
Tel. + 49 (511) 104 3611
Fax. + 49 (511) 104 3399

Peter BOGORINSKI
Gesellschaft für Anlagen- und Reaktorsicherheit
(GRS) mbH
Schwertnergasse 1
D-50667 Köln
Tel. + 49 (221) 2068 521
Fax. + 49 (221) 2068 442
E-mail bog@mhsgw.grs.de

Dr. BRANDES
Technischer Überwachungs-Verein
Hannover/Sachsen-Anhalt e.V.
Hauptabt. Energietechnik und Anlagensicherheit
Postfach 81 05 51
D-30505 Hannover
Tel.
Fax. + 49 (511) 986 1848

Eckhard FEIN
GSF-Forschungszentrum
Institut für Tieflagerung
Postach 2163
D-38011 Braunschweig
Tel. + 49 (531) 8012 292
Fax. + 49 (531) 8012 200

Dr. RINKLEFF
Technischer Überwachungs-Verein
Hannover/Sachsen-Anhalt e.V.
Hauptabt. Energietechnik und Anlagensicherheit
Postfach 81 05 51
D-30505 Hannover
Tel. + 49 (511) 986 1808
Fax. + 49 (511) 986 1848

Helmut RÖTHEMEYER
Bundesamt für Strahlenschutz
Postfach 100149
D-38201 Salzgitter
Tel. + 49 (531) 592 7600
Fax. + 49 (531) 592 7614

Klaus SCHELKES
Bundesanstalt für Geowissenschaften
und Rohstoffe
Postfach 510153
D-30631 Hannover
Tel. + 49 (511) 643 2820
Fax. + 49 (511) 643 2304

Jürgen WOLLRATH
Bundesamt für Strahlenschutz
ET 2.4: Sicherheitsanalysen
Postfach 100149
D-38201 Salzgitter
Tel. + 49 (531) 592 7704
Fax. + 49 (531) 592 7614

JAPAN

Masakazu CHIJIMATSU
Hazama Corporation
515-1 Nishi-mukai, Karima, Tsukuba-shi
Ibaraki, 305
Tel. + 81 (298) 58 8811
Fax. + 81 (298) 58 8819

Akira KOBAYASHI
Department of Agricultural Engineering
Faculty of Agriculture, Iwate University
3-18-8, Ueda, Morioka
Iwate, 020
Tel. + 81 (196) 23 5171 / 2584
Fax. + 81 (196) 53 8231
E-mail kobadesu@msv.cc.iwate-u.ac.jp

Kunio OTA
Geological Environment Research Section
Tono Geoscience Center
Power Reactor and Nuclear Fuel
Development Corporation (PNC)
959-31 Jorinji, Izumi-cho, Toki
Gifu
Tel. + 81 (572) 53 0211
Fax. + 81 (572) 55 0180

Hiroyuki UMEKI
Isolation Syst. Res. Programme
Radioactive Waste Management Project
P N C, 1-9-13, Akasaka
Minato-Ku, Tokyo 107
Tel. + 81 (3) 3586 3311 x 2592
Fax. + 81 (3) 3586 2786 / 7404

SPAIN

Jordi BRUNO
Intera Information Technologies
Parc Tecnologic del Vallés
E-08290 Cerdanyola
Tel. + 34 (3) 582 4410
Fax. + 34 (3) 582 4411 or 582 4412

Pedro CARBONERAS
ENRESA
Emilio Vargas, 7
E-28043 Madrid
Tel. + 34 (1) 519 5258 / 63
Fax. + 34 (1) 519 5795

J. Jaime GOMEZ-HERNANDEZ
Departemento de Hydraulica
Escuela de Caminos
Universidad Politecnica
E-46071 Valencia

Tel. + 34 (6) 387 7612
+ 34 (6) 385 7104
Fax. + 34 (6) 387 7618

Carlos del OLMO
ENRESA
Calle Emilio Vargas, 7
E-28043 Madrid

Tel. + 34 (1) 519 5206
Fax. + 34 (1) 519 5268

Pedro PRADO
CIEMAT
Instituto de Tecnología Nuclear
Avenida Complutense, 22
E-28040 Madrid

Tel. + 34 (1) 346 6528
Fax. + 34 (1) 346 6005
E-mail: prado@dec.ciemat.es

A. SAHUQUILLO
Departemento de Hidraulica
Escuela de Caminos
Universidad Politecnica
E-46071 Valencia

Tel. + 34 (6) 387 7610
+ 34 (6) 387 7058
Fax. + 34 (6) 387 7618

SWEDEN

Johan ANDERSSON
Swedish Nuclear Power Inspectorate (SKI)
Box 27106
S-102 52 Stockholm

Tel. + 46 (8) 665 4485
Fax. + 46 (8) 661 9086

Peter ANDERSSON
GEOSIGMA
Box 894
S-751 08 Uppsala

Tel. + 46 (18) 650800
Fax. + 46 (18) 121302

Göran BÄCKBLOM
Swedish Nuclear Fuel and Waste
Management Co. (SKB) - Box 5864
S-102 40 Stockholm

Tel. + 46 (8) 665 2831
Fax. + 46 (8) 661 5719

Steven A. BANWART
Royal Institute of Technology
Aquatic Chemistry Group
The Department of Inorganic Chemistry
S-100 44 Stockholm

Tel. + 46 (8) 790 8085
Fax. + 46 (8) 21 26 26

Runo BARRDAHL
Division of Waste Management and
Environmental Protection
Swedish Radiation Protection Institute
S-171 16 Stockholm

Tel. + 46 (8) 729 7258
Fax. + 46 (8) 729 7162

Björn DVERSTORP
Swedish Nuclear Power Inspectorate (SKI)
Box 27106
S-102 52 Stockholm
Tel. + 46 (8) 665 4486
Fax. + 46 (8) 661 9086

Fritz KAUTSKY
Swedish Nuclear Power Inspectorate (SKI)
Box 27106
S-102 52 Stockholm
Tel. + 46 (8) 665 4487
Fax. + 46 (8) 661 9086

Alf LARSSON
Kemakta Konsult AB
Box 12655
S-112 93 Stockholm
Tel. + 46 (8) 654 0680
Fax. + 46 (8) 652 1607

Ivars NERETNIEKS
Royal Institute of Technology
Department of Chemical Engineering
S-100 44 Stockholm
Tel. + 46 (8) 787 8229
Fax. + 46 (8) 10 52 28

Sören NORRBY
Swedish Nuclear Power Inspectorate (SKI)
Box 27106
S-102 52 Stockholm
Tel. + 46 (8) 665 4482
Fax. + 46 (8) 661 9086

Olle OLSSON
Conterra AB
Box 493
S-751 06 Uppsala
Tel. + 46 (18) 12 32 90
Fax. + 46 (8) 730 17 60

Tönis PAPP
Swedish Nuclear Fuel and
Waste Management Co. (SKB)
Box 5864
S-102 40 Stockholm
Tel. + 46 (8) 665 2800
Fax. + 46 (8) 661 5719

Antonio PEREIRA
University of Stockholm
Department of Physics
Vanadisv. 9
S-113 46 Stockholm
Tel. + 46 (8) 164585
Fax. + 46 (8) 347817

Jan-Olof SELROOS
Division of Water Resources Eng.
Royal Institute of Technology (KTH)
S-100 44 Stockholm
Tel. + 46 (8) 790 8693
Fax. + 46 (8) 790 8689
E-mail: josel@wre.kth.se

Kristina SKAGIUS
Kemakta Konsult AB
Box 12655
S-112 93 Stockholm
Tel. + 46 (8) 654 0680
Fax. + 46 (8) 652 1607

John SMELLIE
CONTERRA AB
Box 493
S-751 06 Uppsala
Tel. + 46 (18) 12 32 41
Fax. + 46 (18) 12 32 62

Ove STEPHANSSON
Engineering Geology
Royal Institute of Technology
S-100 44 Stockholm
Tel. + 46 (8) 790 7906
Fax. + 46 (8) 790 6810
E-Mail: ove@ce.kth.se

Anders STRÖM
Swedish Nuclear Fuel and
Waste Management Co. (SKB)
Box 5864
S-102 40 Stockholm
Tel. + 46 (8) 665 2862
Fax. + 46 (8) 661 5719
E-mail: skbas@skb.se

Benny SUNDSTRÖM
Swedish Nuclear Power Inspectorate (SKI)
Box 27106
S-102 52 Stockholm
Tel. + 46 (8) 665 4488
Fax. + 46 (8) 661 9086

Öivind TOVERUD
Swedish Nuclear Power Inspectorate (SKI)
Box 27106
S-102 52 Stockholm
Tel. + 46 (8) 665 4453
Fax. + 46 (8) 661 9086

Anders WINBERG
Conterra AB
Krokslätts Fabriker 30
S-431 37 Mölndal
Tel. + 46 (31) 878302
Fax. + 46 (31) 278705

SWITZERLAND

Urs FRICK
NAGRA
Hardstrasse 73
CH-5430 Wettingen
Tel. + 41 (56) 371318
Fax. + 41 (56) 371214 (290 secretary)

Andreas GAUTSCHI
NAGRA
Hardstrasse 73
CH-5430 Wettingen
Tel. + 41 (56) 371111
Fax. + 41 (56) 371207

Jörg HADERMANN
Paul Scherrer Institute
Waste Management Laboratory
CH-5232 Villigen PSI
Tel. + 41 (56) 99 21 11 x 2415
Fax. + 41 (56) 98 23 27

Charles McCOMBIE
NAGRA
Hardstrasse 73
CH-5430 Wettingen
Tel. + 41 (56) 37 11 11
Fax. + 41 (56) 37 12 07

R. ROMETSCH
Haus "Tiger"
Terrassenweg
CH-3818 Grindelwald
Tel. + 41 (36) 53 21 86
Fax: + 41 (36) 53 16 01

Johannes O. VIGFUSSON
Waste Management Section
Swiss Nuclear Safety Inspectorate
CH-5232 Villigen HSK
Tel. + 41 (56) 99 39 74
Fax. + 41 (56) 99 39 07

Piet ZUIDEMA
NAGRA
Hardstrasse 73
CH-5430 Wettingen
Tel. + 41 (56) 37 11 11
Fax: + 41 (56) 37 12 07

UNITED KINGDOM

Allan ASHWORTH
Pollution Inspector
Business Strategy Division
Her Majesty's Inspectorate of Pollution
Room P3/009
2 Marsham Street
London SW1P 3EB
Tel. + 44 (71) 276 4752
Fax. + 44 (71) 276 6544

Charles BOYLE
AEA Technology
424.4, Harwell
Oxfordshire OX11 ORA
Tel. + 44 (235) 433 849
Fax. + 44 (235) 436 798

Neil CHAPMAN
INTERA Information Techn. Ltd.
Environmental Division
47 Burton Street
Melton Mowbray
Leicestershire LE13 1AF
Tel. + 44 (664) 411 445
Fax. + 44 (664) 411 402

A.G. DUNCAN
Director, Regulatory Systems
H.M. Inspectorate Pollution
Room A508 Romney House
43 Marsham Street
London SW1P 3PY
Tel. + 44 (71) 276 8129
Fax. + 44 (71) 276 8216

M. HARVEY
NRPB
Chilton, Didcot
Oxon

Tel. + 44 (235) 831 600 xt 2
Fax.

C. Peter JACKSON
Assessments Department
AEA Technology
Decommissioning and Waste Management
424.4 Harwell
Oxfordshire OX11 ORA

Tel. + 44 (235) 433 005
Fax. + 44 (235) 436 579

UNITED STATES

D.R. "Rip" ANDERSON
Sandia National Laboratories
P.O. Box 5800, MS1328
Organization 6342
Albuquerque, NM 87185-1328

Tel. + 1 (505) 848 0692
Fax. + 1 (505) 848 0705

Richard L. BEAUHEIM
Sandia National Laboratories
MS 1324
P.O. Box 5800
Albuquerque, NM 87185-1324

Tel. + 1 (505) 848 0675
Fax. + 1 (505) 848 0605
E-Mail: rlbeauh@nwer.sandia.gov

Lokesh CHATURVEDI
Environmental Evaluation Group
7007 Wyoming Boulevard, N.E.
Suite F-2
Albuquerque, NM 87109

Tel. + 1 (505) 828 1003
Fax. + 1 (505) 828 1062

Peter DAVIES
Geohydrology Dept. 6115
Sandia National Laboratories
P.O. Box 5800, MS 1324
Albuquerque, NM 87185

Tel. + 1 (505) 848 0709
Fax. + 1 (505) 848 0605

Linda LEHMAN
L. Lehman & Associates, Inc.
1103 W. Burnsville Parkway
Suite 209
Burnsville, MN 55337

Tel. + 1 (612) 894 0357
Fax. + 1 (612) 894 5028

Schlomo P. NEUMAN
Department of Hydrology and
Water Resources
The University of Arizona
Tucson, AZ 85721

Tel. + 1 (602) 621 7114
Fax. + 1 (602) 621 1422

Thomas J. NICHOLSON
Office of Nuclear Regulatory Research
U.S. Nuclear Regulatory Commission
Mail Stop NL/S-260
Washington, D.C. 20555-0001
Tel. + 1 (301) 415 6268
Fax. + 1 (301) 415 5385
E-mail: tjn@nrc.gov

Budhi SAGAR
Center for Nuclear Waste
Regulatory Analyses (CNWRA)
Southwest Research Institute
Post Office Drawer 28510
6220 Culebra Road
San Antonio, TX 78238
Tel. + 1 (210) 522 5525
Fax. + 1 (210) 522 5155
E-Mail: bsagar@smtp.swri.edu

Eric T. SMISTAD
Yucca Mountain Protection Office
U.S. Department of Energy
P.O. Box 98608
Las Vegas, NV 89193-8608
Tel. + 1 (702) 794 7587
Fax. + 1 (702) 794 7907

John L. SMOOT
Pacific Northwest Laboratory
P.O. Box 999, MS K6-96
Richland, WA 99352
Tel. + 1 (509) 372 6120
Fax. + 1 (509) 372 6328

Chin-Fu TSANG
Earth Sciences Division
Lawrence Berkeley Laboratory (LBL)
1 Cyclotron Road
Berkeley, CA 94720
Tel. + 1 (510) 486 5782
Fax. + 1 (510) 486 5686

Abraham E. VAN LUIK
Intera/M+O Inc.
101 Convention Center Drive
Suite P-110
Las Vegas, NV 89109
Tel. + 1 (702) 794 7441
Fax. + 1 (702) 794 5151

Gordon WITTMEYER
Center for Nuclear Waste
Regulatory Analyses (CNWRA)
Southwest Research Institute
Post Office Drawer 28510
6220 Culebra Road
San Antonio, TX 78238
Tel. + 1 (210) 522 5082
Fax. + 1 (210) 522 5155

INTERNATIONAL ATOMIC ENERGY AGENCY

Shaheed HOSSAIN
Division of Nuclear Fuel Cycle and
Waste Management
International Atomic Energy Agency
P.O. Box 100, A-1400 Vienna
Austria

Tel. + 43 (1) 2360 2668
Fax. + 43 (1) 234564
E-mail: hossain@nepo1.iaea.or.at

OECD NUCLEAR ENERGY AGENCY

Kunihiko UEMATSU
Director General
OECD Nuclear Energy Agency
Le Seine Saint Germain
12, boulevard des Iles
F-92130 Issy-les-Moulineaux

Tel. + 33 (1) 45 24 10 00
Fax. + 33 (1) 45 24 11 10

Jean-Pierre OLIVIER
Head, Radiation Protection and
Waste Management Division
OECD Nuclear Energy Agency
Le Seine Saint Germain
12, boulevard des Iles
F-92130 Issy-les-Moulineaux

Tel. + 33 (1) 45 24 10 40
Fax. + 33 (1) 45 24 11 10

Claudio PESCATORE
Radiation Protection and
Waste Management Division
OECD Nuclear Energy Agency
Le Seine Saint Germain
12, boulevard des Iles
F-92130 Issy-les-Moulineaux

Tel. + 33 (1) 45 24 10 48
Fax. + 33 (1) 45 24 11 10
E-mail: pescatore@nea.fr

Ki-Jung JUNG
Radiation Protection and
Waste Management Division
OECD Nuclear Energy Agency
Le Seine Saint Germain
12, boulevard des Iles
F-92130 Issy-les-Moulineaux

Tel. + 33 (1) 45 24 10 04
Fax. + 33 (1) 45 24 11 10

Bertrand RUEGGER
OECD Nuclear Energy Agency
Le Seine Saint Germain
12, boulevard des Iles
F-92130 Issy-les-Moulineaux

Tel. + 33 (1) 45 24 10 44
Fax. + 33 (1) 45 24 11 10

MAIN SALES OUTLETS OF OECD PUBLICATIONS
PRINCIPAUX POINTS DE VENTE DES PUBLICATIONS DE L'OCDE

ARGENTINA – ARGENTINE
Carlos Hirsch S.R.L.
Galería Güemes, Florida 165, 4° Piso
1333 Buenos Aires Tel. (1) 331.1787 y 331.2391
Telefax: (1) 331.1787

AUSTRALIA – AUSTRALIE
D.A. Information Services
648 Whitehorse Road, P.O.B 163
Mitcham, Victoria 3132 Tel. (03) 873.4411
Telefax: (03) 873.5679

AUSTRIA – AUTRICHE
Gerold & Co.
Graben 31
Wien I Tel. (0222) 533.50.14
Telefax: (0222) 512.47.31.29

BELGIUM – BELGIQUE
Jean De Lannoy
Avenue du Roi 202
B-1060 Bruxelles Tel. (02) 538.51.69/538.08.41
Telefax: (02) 538.08.41

CANADA
Renouf Publishing Company Ltd.
1294 Algoma Road
Ottawa, ON K1B 3W8 Tel. (613) 741.4333
Telefax: (613) 741.5439

Stores:
61 Sparks Street
Ottawa, ON K1P 5R1 Tel. (613) 238.8985
211 Yonge Street
Toronto, ON M5B 1M4 Tel. (416) 363.3171
Telefax: (416)363.59.63

Les Éditions La Liberté Inc.
3020 Chemin Sainte-Foy
Sainte-Foy, PQ G1X 3V6 Tel. (418) 658.3763
Telefax: (418) 658.3763

Federal Publications Inc.
165 University Avenue, Suite 701
Toronto, ON M5H 3B8 Tel. (416) 860.1611
Telefax: (416) 860.1608

Les Publications Fédérales
1185 Université
Montréal, QC H3B 3A7 Tel. (514) 954.1633
Telefax: (514) 954.1635

CHINA – CHINE
China National Publications Import
Export Corporation (CNPIEC)
16 Gongti E. Road, Chaoyang District
P.O. Box 88 or 50
Beijing 100704 PR Tel. (01) 506.6688
Telefax: (01) 506.3101

CHINESE TAIPEI – TAIPEI CHINOIS
Good Faith Worldwide Int'l. Co. Ltd.
9th Floor, No. 118, Sec. 2
Chung Hsiao E. Road
Taipei Tel. (02) 391.7396/391.7397
Telefax: (02) 394.9176

CZECH REPUBLIC – RÉPUBLIQUE TCHEQUE
Artia Pegas Press Ltd.
Narodni Trida 25
POB 825
111 21 Praha 1 Tel. 26.65.68
Telefax: 26.20.81

DENMARK – DANEMARK
Munksgaard Book and Subscription Service
35, Nørre Søgade, P.O. Box 2148
DK-1016 København K Tel. (33) 12.85.70
Telefax: (33) 12.93.87

EGYPT – ÉGYPTE
Middle East Observer
41 Sherif Street
Cairo Tel. 392.6919
Telefax: 360-6804

FINLAND – FINLANDE
Akateeminen Kirjakauppa
Keskuskatu 1, P.O. Box 128
00100 Helsinki

Subscription Services/Agence d'abonnements :
P.O. Box 23
00371 Helsinki Tel. (358 0) 12141
Telefax: (358 0) 121.4450

FRANCE
OECD/OCDE
Mail Orders/Commandes par correspondance:
2, rue André-Pascal
75775 Paris Cedex 16 Tel. (33-1) 45.24.82.00
Telefax: (33-1) 49.10.42.76
Telex: 640048 OCDE

Orders via Minitel, France only/
Commandes par Minitel, France exclusivement :
36 15 OCDE

OECD Bookshop/Librairie de l'OCDE :
33, rue Octave-Feuillet
75016 Paris Tel. (33-1) 45.24.81.81
(33-1) 45.24.81.67

Documentation Française
29, quai Voltaire
75007 Paris Tel. 40.15.70.00

Gibert Jeune (Droit-Économie)
6, place Saint-Michel
75006 Paris Tel. 43.25.91.19

Librairie du Commerce International
10, avenue d'Iéna
75016 Paris Tel. 40.73.34.60

Librairie Dunod
Université Paris-Dauphine
Place du Maréchal de Lattre de Tassigny
75016 Paris Tel. (1) 44.05.40.13

Librairie Lavoisier
11, rue Lavoisier
75008 Paris Tel. 42.65.39.95

Librairie L.G.D.J. - Montchrestien
20, rue Soufflot
75005 Paris Tel. 46.33.89.85

Librairie des Sciences Politiques
30, rue Saint-Guillaume
75007 Paris Tel. 45.48.36.02

P.U.F.
49, boulevard Saint-Michel
75005 Paris Tel. 43.25.83.40

Librairie de l'Université
12a, rue Nazareth
13100 Aix-en-Provence Tel. (16) 42.26.18.08

Documentation Française
165, rue Garibaldi
69003 Lyon Tel. (16) 78.63.32.23

Librairie Decitre
29, place Bellecour
69002 Lyon Tel. (16) 72.40.54.54

Librairie Sauramps
Le Triangle
34967 Montpellier Cedex 2 Tel. (16) 67.58.85.15
Tekefax: (16) 67.58.27.36

GERMANY – ALLEMAGNE
OECD Publications and Information Centre
August-Bebel-Allee 6
D-53175 Bonn Tel. (0228) 959.120
Telefax: (0228) 959.12.17

GREECE – GRÈCE
Librairie Kauffmann
Mavrokordatou 9
106 78 Athens Tel. (01) 32.55.321
Telefax: (01) 32.30.320

HONG-KONG
Swindon Book Co. Ltd.
Astoria Bldg. 3F
34 Ashley Road, Tsimshatsui
Kowloon, Hong Kong Tel. 2376.2062
Telefax: 2376.0685

HUNGARY – HONGRIE
Euro Info Service
Margitsziget, Európa Ház
1138 Budapest Tel. (1) 111.62.16
Telefax: (1) 111.60.61

ICELAND – ISLANDE
Mál Mog Menning
Laugavegi 18, Pósthólf 392
121 Reykjavik Tel. (1) 552.4240
Telefax: (1) 562.3523

INDIA – INDE
Oxford Book and Stationery Co.
Scindia House
New Delhi 110001 Tel. (11) 331.5896/5308
Telefax: (11) 332.5993
17 Park Street
Calcutta 700016 Tel. 240832

INDONESIA – INDONÉSIE
Pdii-Lipi
P.O. Box 4298
Jakarta 12042 Tel. (21) 573.34.67
Telefax: (21) 573.34.67

IRELAND – IRLANDE
Government Supplies Agency
Publications Section
4/5 Harcourt Road
Dublin 2 Tel. 661.31.11
Telefax: 475.27.60

ISRAEL
Praedicta
5 Shatner Street
P.O. Box 34030
Jerusalem 91430 Tel. (2) 52.84.90/1/2
Telefax: (2) 52.84.93

R.O.Y. International
P.O. Box 13056
Tel Aviv 61130 Tel. (3) 49.61.08
Telefax: (3) 544.60.39

Palestinian Authority/Middle East:
INDEX Information Services
P.O.B. 19502
Jerusalem Tel. (2) 27.12.19
Telefax: (2) 27.16.34

ITALY – ITALIE
Libreria Commissionaria Sansoni
Via Duca di Calabria 1/1
50125 Firenze Tel. (055) 64.54.15
Telefax: (055) 64.12.57

Via Bartolini 29
20155 Milano Tel. (02) 36.50.83

Editrice e Libreria Herder
Piazza Montecitorio 120
00186 Roma Tel. 679.46.28
Telefax: 678.47.51

Libreria Hoepli
Via Hoepli 5
20121 Milano Tel. (02) 86.54.46
Telefax: (02) 805.28.86

Libreria Scientifica
Dott. Lucio de Biasio 'Aeiou'
Via Coronelli, 6
20146 Milano Tel. (02) 48.95.45.52
Telefax: (02) 48.95.45.48

JAPAN – JAPON
OECD Publications and Information Centre
Landic Akasaka Building
2-3-4 Akasaka, Minato-ku
Tokyo 107 Tel. (81.3) 3586.2016
Telefax: (81.3) 3584.7929

KOREA – CORÉE
Kyobo Book Centre Co. Ltd.
P.O. Box 1658, Kwang Hwa Moon
Seoul Tel. 730.78.91
Telefax: 735.00.30

MALAYSIA – MALAISIE
University of Malaya Bookshop
University of Malaya
P.O. Box 1127, Jalan Pantai Baru
59700 Kuala Lumpur
Malaysia Tel. 756.5000/756.5425
Telefax: 756.3246

MEXICO – MEXIQUE
Revistas y Periodicos Internacionales S.A. de C.V.
Florencia 57 - 1004
Mexico, D.F. 06600 Tel. 207.81.00
Telefax: 208.39.79

NETHERLANDS – PAYS-BAS
SDU Uitgeverij Plantijnstraat
Externe Fondsen
Postbus 20014
2500 EA's-Gravenhage Tel. (070) 37.89.880
Voor bestellingen: Telefax: (070) 34.75.778

NEW ZEALAND
NOUVELLE-ZÉLANDE
Legislation Services
P.O. Box 12418
Thorndon, Wellington Tel. (04) 496.5652
Telefax: (04) 496.5698

NORWAY – NORVÈGE
Narvesen Info Center – NIC
Bertrand Narvesens vei 2
P.O. Box 6125 Etterstad
0602 Oslo 6 Tel. (022) 57.33.00
Telefax: (022) 68.19.01

PAKISTAN
Mirza Book Agency
65 Shahrah Quaid-E-Azam
Lahore 54000 Tel. (42) 353.601
Telefax: (42) 231.730

PHILIPPINE – PHILIPPINES
International Book Center
5th Floor, Filipinas Life Bldg.
Ayala Avenue
Metro Manila Tel. 81.96.76
Telex 23312 RHP PH

PORTUGAL
Livraria Portugal
Rua do Carmo 70-74
Apart. 2681
1200 Lisboa Tel. (01) 347.49.82/5
Telefax: (01) 347.02.64

SINGAPORE – SINGAPOUR
Gower Asia Pacific Pte Ltd.
Golden Wheel Building
41, Kallang Pudding Road, No. 04-03
Singapore 1334 Tel. 741.5166
Telefax: 742.9356

SPAIN – ESPAGNE
Mundi-Prensa Libros S.A.
Castelló 37, Apartado 1223
Madrid 28001 Tel. (91) 431.33.99
Telefax: (91) 575.39.98

Libreria Internacional AEDOS
Consejo de Ciento 391
08009 – Barcelona Tel. (93) 488.30.09
Telefax: (93) 487.76.59

Llibreria de la Generalitat
Palau Moja
Rambla dels Estudis, 118
08002 – Barcelona
(Subscripcions) Tel. (93) 318.80.12
(Publicacions) Tel. (93) 302.67.23
Telefax: (93) 412.18.54

SRI LANKA
Centre for Policy Research
c/o Colombo Agencies Ltd.
No. 300-304, Galle Road
Colombo 3 Tel. (1) 574240, 573551-2
Telefax: (1) 575394, 510711

SWEDEN – SUÈDE
Fritzes Customer Service
S–106 47 Stockholm Tel. (08) 690.90.90
Telefax: (08) 20.50.21

Subscription Agency/Agence d'abonnements :
Wennergren-Williams Info AB
P.O. Box 1305
171 25 Solna Tel. (08) 705.97.50
Telefax: (08) 27.00.71

SWITZERLAND – SUISSE
Maditec S.A. (Books and Periodicals - Livres et périodiques)
Chemin des Palettes 4
Case postale 266
1020 Renens VD 1 Tel. (021) 635.08.65
Telefax: (021) 635.07.80

Librairie Payot S.A.
4, place Pépinet
CP 3212
1002 Lausanne Tel. (021) 341.33.47
Telefax: (021) 341.33.45

Librairie Unilivres
6, rue de Candolle
1205 Genève Tel. (022) 320.26.23
Telefax: (022) 329.73.18

Subscription Agency/Agence d'abonnements :
Dynapresse Marketing S.A.
38 avenue Vibert
1227 Carouge Tel. (022) 308.07.89
Telefax: (022) 308.07.99

See also – Voir aussi :
OECD Publications and Information Centre
August-Bebel-Allee 6
D-53175 Bonn (Germany) Tel. (0228) 959.120
Telefax: (0228) 959.12.17

THAILAND – THAÏLANDE
Suksit Siam Co. Ltd.
113, 115 Fuang Nakhon Rd.
Opp. Wat Rajbopith
Bangkok 10200 Tel. (662) 225.9531/2
Telefax: (662) 222.5188

TURKEY – TURQUIE
Kültür Yayinlari Is-Türk Ltd. Sti.
Atatürk Bulvari No. 191/Kat 13
Kavaklidere/Ankara Tel. 428.11.40 Ext. 2458
Dolmabahce Cad. No. 29
Besiktas/Istanbul Tel. 260.71.88
Telex: 43482B

UNITED KINGDOM – ROYAUME-UNI
HMSO
Gen. enquiries Tel. (071) 873 0011
Postal orders only:
P.O. Box 276, London SW8 5DT
Personal Callers HMSO Bookshop
49 High Holborn, London WC1V 6HB
Telefax: (071) 873 8200
Branches at: Belfast, Birmingham, Bristol, Edinburgh, Manchester

UNITED STATES – ÉTATS-UNIS
OECD Publications and Information Center
2001 L Street N.W., Suite 650
Washington, D.C. 20036-4910 Tel. (202) 785.6323
Telefax: (202) 785.0350

VENEZUELA
Libreria del Este
Avda F. Miranda 52, Aptdo. 60337
Edificio Galipán
Caracas 106 Tel. 951.1705/951.2307/951.1297
Telegram: Libreste Caracas

Subscription to OECD periodicals may also be placed through main subscription agencies.

Les abonnements aux publications périodiques de l'OCDE peuvent être souscrits auprès des principales agences d'abonnement.

Orders and inquiries from countries where Distributors have not yet been appointed should be sent to: OECD Publications Service, 2 rue André-Pascal, 75775 Paris Cedex 16, France.

Les commandes provenant de pays où l'OCDE n'a pas encore désigné de distributeur peuvent être adressées à : OCDE, Service des Publications, 2, rue André-Pascal, 75775 Paris Cedex 16, France.

5-1995

OECD PUBLICATIONS, 2 rue André-Pascal, 75775 PARIS CEDEX 16
PRINTED IN FRANCE
(66 95 12 1) ISBN 92-64-14467-6 - No. 47967 1995

UNIVERSITY OF STRATHCLYDE
25 AUG 1995
UNIVERSITY LIBRARY